FUNDAMENTALS OF TRANSPORT PHENOMENA

McGraw-Hill Chemical Engineering Series

BUILDING THE LITERATURE OF A PROFESSION

Fifteen prominent chemical engineers first met in New York more than 50 years ago to plan a continuing literature for their rapidly growing profession. From industry came such pioneer practitioners as Leo H. Baekeland, Arthur D. Little, Charles L. Reese, John V. N. Dorr, M. C. Whitaker, and R. S. McBride. From the universities came such eminent educators as William H. Walker, Alfred H. White, D. D. Jackson, J. H. James, Warren K. Lewis, and Harry A. Curtis. H. C. Parmelee, then editor of *Chemical and Metallurgical Engineering,* served as chairman and was joined subsequently by S. D. Kirkpatrick as consulting editor.

After several meetings, this committee submitted its report to the McGraw-Hill Book Company in September 1925. In the report were detailed specifications for a correlated series of more than a dozen texts and reference books which have since become the McGraw-Hill Series in Chemical Engineering and which became the cornerstone of the chemical engineering curriculum.

From this beginning there has evolved a series of texts surpassing by far the scope and longevity envisioned by the founding Editorial Board. The McGraw-Hill Series in Chemical Engineering stands as a unique historical record of the development of chemical engineering education and practice. In the series one finds the milestones of the subject's evolution: industrial chemistry, stoichiometry, unit operations and processes, thermodynamics, kinetics, and transfer operations.

Chemical engineering is a dynamic profession, and its literature continues to evolve. McGraw-Hill and its consulting editors remain committed to a publishing policy that will serve, and indeed lead, the needs of the chemical engineering profession during the years to come.

THE SERIES

Bailey and Ollis: *Biochemical Engineering Fundamentals*
Bennett and Myers: *Momentum, Heat, and Mass Transfer*
Beveridge and Schechter: *Optimization: Theory and Practice*
Carberry: *Chemical and Catalytic Reaction Engineering*
Churchill: *The Interpretation and Use of Rate Data—The Rate Concept*
Clarke and Davidson: *Manual for Process Engineering Calculations*
Coughanowr and Koppel: *Process Systems Analysis and Control*
Fahien: *Fundamentals of Transport Phenomena*
Finlayson: *Nonlinear Analysis in Chemical Engineering*
Gates, Katzer, and Schuit: *Chemistry of Catalytic Processes*
Holland: *Fundamentals of Multicomponent Distillation*
Holland and Liapis: *Computer Methods for Solving Dynamic Separation Problems*
Johnson: *Automatic Process Control*
Johnstone and Thring: *Pilot Plants, Models, and Scale-Up Methods in Chemical Engineering*
Katz, Cornell, Kobayashi, Poettmann, Vary, Elenbaas, and Weinaug: *Handbook of Natural Gas Engineering*
King: *Separation Processes*
Klinzing: *Gas-Solid Transport*
Knudsen and Katz: *Fluid Dynamics and Heat Transfer*
Luyben: *Process Modeling, Simulation, and Control for Chemical Engineers*
McCabe and Smith, J. C.: *Unit Operations of Chemical Engineering*
Mickley, Sherwood, and Reed: *Applied Mathematics in Chemical Engineering*
Nelson: *Petroleum Refinery Engineering*
Perry and Chilton (Editors): *Chemical Engineers' Handbook*
Peters: *Elementary Chemical Engineering*
Peters and Timmerhaus: *Plant Design and Economics for Chemical Engineers*
Probstein and Hicks: *Synthetic Fuels*
Ray: *Advanced Process Control*
Reid, Prausnitz, and Sherwood: *The Properties of Gases and Liquids*
Resnick: *Process Analysis and Design for Chemical Engineers*
Satterfield: *Heterogeneous Catalysis in Practice*
Sherwood, Pigford, and Wilke: *Mass Transfer*
Smith, B. D.: *Design of Equilibrium Stage Processes*
Smith, J. M.: *Chemical Engineering Kinetics*
Smith, J. M., and Van Ness: *Introduction to Chemical Engineering Thermodynamics*
Thompson and Ceckler: *Introduction to Chemical Engineering*
Treybal: *Mass Transfer Operations*
Valle-Riestra: *Project Evolution in the Chemical Process Industries*
Van Ness and Abbott: *Classical Thermodynamics of Nonelectrolyte Solutions: With Applications to Phase Equilibria*
Van Winkle: *Distillation*
Volk: *Applied Statistics for Engineers*
Walas: *Reaction Kinetics for Chemical Engineers*
Wei, Russell, and Swartzlander: *The Structure of the Chemical Processing Industries*
Whitwell and Toner: *Conservation of Mass and Energy*

FUNDAMENTALS OF TRANSPORT PHENOMENA

Ray W. Fahien

Professor of Chemical Engineering
University of Florida, Gainesville

McGraw-Hill Book Company

New York St. Louis San Francisco Auckland Bogotá Hamburg
Johannesburg London Madrid Mexico Montreal New Delhi
Panama Paris São Paulo Singapore Sydney Tokyo Toronto

This book was set in Times Roman by York Graphic Services, Inc.
The editors were Julienne V. Brown, Kiran Verma, and Madelaine Eichberg;
the production supervisor was Leroy A. Young.
The drawings were done by Wellington Studios Ltd.
R. R. Donnelley & Sons Company was printer and binder.

FUNDAMENTALS OF TRANSPORT PHENOMENA

2 3 4 5 6 7 8 9 0 DOCDOC 8 9 8 7 6 5 4 3

ISBN 0-07-019891-8

Library of Congress Cataloging in Publication Data

Fahien, Ray W.
Fundamentals of transport phenomena.

(McGraw-Hill chemical engineering series)
Includes bibliographies and index.
1. Transport theory. 2. Chemical engineering.
I. Title. II. Series.
TP156.T7F33 1983 660.2′8432 82-20807
ISBN 0-07-019891-8

CONTENTS

†Sections so marked may be omitted on first reading.

Part 2 Multidimensional Transport

Tables of Equations

8 Heat and Mass Transport in More than One Direction **317**

*Does not require knowledge of vector notation.

**Combines vector and nonvector notation.

PREFACE

The study of transport phenomena is traditionally approached by taking up momentum transport first, then energy transport, and finally mass transport. For each transport process such topics as molecular transport, shell balances, and multidimensional transport are discussed, so that the similarities and analogies between the transport processes can be inferred. Generalized differential equations of change are then derived, usually expressed in vector-tensor notation. The student is then shown how to simplify these equations for specific physical cases.

An alternative organization, the one used in this book, is to take up similar topics for all three processes simultaneously. This approach has the following advantages: (1) the analogies can be exploited more fully and repetition reduced; (2) the limitations of, and exceptions to, the analogies can be highlighted; (3) the most elementary topics, such as one-dimensional transport, can be taken up first; (4) the physical meaning of such terms as diffusion, convection, generation, and accumulation in the general balance equations can be illustrated initially for each process by simple physical examples, without the complication of generalized equations; (5) the generalized multidimensional equations can then be derived as a logical extension of one-dimensional transport and as an incorporation into a general form of the terms previously illustrated; (6) the simplification of the multidimensional equations can then be carried out for specific cases with a fuller appreciation of their meaning.

In this approach the student not only learns how to use incremental balances to develop differential equations but also is shown how to derive the general equations and then how to simplify them for specific applications. Thus, initially inductive reasoning is used to go from specific physical phenomena to the general equations, and then deductive reasoning is used to go from the general equations to additional specific applications.

As a consequence of the organization used in this book, the difficult topic of laminar momentum transport can be taught after taking up the more familiar (to the student) subject of heat conduction. Thus, one-dimensional molecular momentum transport in Couette flow is shown to be analogous to heat conduction in one dimension. Then the relationship, through Newton's law of motion, between momentum flux and viscous stress is demonstrated, and the physical meaning is discussed.

Throughout Part One (One-Dimensional Transport) mathematical analogies between the differential equations describing the transport processes are used—but not without appropriate words of caution about their limitations; e.g., the physical significance of momentum flux as a second-order tensor and of mass and heat fluxes as vectors is first mentioned in Chap. 1 and later discussed in more detail in Chap. 9, where the physical meaning of the rate-of-deformation tensor is also discussed.

This text is more than a text in transport phenomena; it is also, in part, a text in applied physics, applied mathematics, and mathematical modeling. The physical meaning of mathematical abstractions such as the derivative and the integral is reviewed and emphasized. Likewise, the meaning of the phenomenological laws (Fourier's, Fick's, and Newton's) is carefully presented. In addition, methods of solution by separation of variables and by combination of variables are illustrated.

In Part Two (Multidimensional Transport) the physical meaning of vector and tensor quantities, such as ∇T (the temperature gradient), $\nabla \cdot \mathbf{q}$ (the divergence of the heat-flux vector), or $\underline{\tau}$ (the momentum-flux tensor), is pointed out. The origin of the effective acceleration terms that occur in curvilinear coordinates is discussed and illustrated.

The use of dimensionless variables in the differential balance equations is taken up as early as Chap. 1 and is emphasized throughout—not only as a means of demonstrating the existence of a mathematical analogy but also for the purpose of performing a dimensional analysis on the variables and parameters involved. For situations in which analytical solutions are not available, the use of an experimental program to relate the dimensionless variables to each other is described.

Scaling and ordering of the differential equations of change in order to determine which terms are negligible for various physical conditions is illustrated for boundary-layer flow and for convective heat and mass transport. Applications of the generalized equations are made to typical engineering problems, e.g., entrance-region flow, two-dimensional tubular-reactor modeling, heat-exchanger design, and air-pollutant transport in the atmosphere.

In Chap. 11 the unsteady-state macroscopic balances are derived by integration of the differential balances and applied to engineering problems. In addition, an angular-momentum balance is applied to centrifugal-pump design and a differential macroscopic balance is derived by integration over the cross section and applied to tubular-reactor design, showing the origin of Taylor axial dispersion. Also, a general population balance is derived and applied to an industrial crystallizer.

This text encourages the use of SI units but recognizes that in the nonacademic world—where the vast majority of students will work—English units are still quite common. Therefore, conversion between sets of units is emphasized. Furthermore, since Newton's second law is introduced as a proportionality (rather than as an identity), the gravitational constant g_c is used in the early chapters (after its origin has been carefully explained by an example of a lunar gravitational constant). The use of g_c makes it possible to enhance students' understanding of momentum flux and shear stress through the use of separate symbols for each. The distinction between the direction of transport and the direction of the momentum being transported (or the direction of the shear force) is illustrated, as is the physical meaning of the force and momentum rate balances.

In summary, this book is intended to fill a need for a beginning transport phenomena text that (1) offers a careful explanation of the fundamentals, (2) assumes minimal previous mathematical or scientific background on the part of the student, (3) proceeds gradually from an elementary level into increasingly difficult material, (4) uses a logical step-by-step approach, (5) emphasizes the mathematical similarity of the equations used to describe the processes of energy, mass, and momentum transport while cautioning the student about their dissimilarities, and (6) emphasizes the physical meaning of mathematical quantities and operations.

Historically, courses in chemical engineering unit operations have included not only equipment design but also a modicum of transport phenomena (heat transfer, fluid flow, diffusion). With the publication in 1960 of the pioneering Bird, Stewart, and Lightfoot text, "Transport Phenomena," the question arose whether it should be taught before the unit operations courses or after them; that question has not yet been universally answered. For while it is apparent that some knowledge of transport phenomena is needed for unit operations or equipment design courses, the full three-dimensional vector and tensor treatment is not always justified, especially since, as in the case of turbulent and other complex flows, the equations are not always solvable. In some schools, this has led to the postponement of transport phenomena as a course to the senior-graduate level and concurrently to the inclusion of more transport phenomena in the unit operations texts. Since this text takes up one-dimensional transport first, it is adaptable to a variety of plans; i.e., it can be used (1) before equipment design or unit operations courses, (2) following them, (3) in conjunction with them, or (4) partly before (Part One plus portions of Part Two) and partly after (remainder of Part Two).

The flexibility of the book allows the instructor to choose any one of the following paths through Part Two: (1) a high-level path which, for example, derives the equations of change by using the Gauss divergence and Reynolds transport theorems, (2) an intermediate approach, in which the equations are derived by means of incremental balances in cartesian coordinates, written in Gibbs vector-tensor notation, and transformed into other coordinate systems, and (3) an approach which deemphasizes the manipulation of vectors and tensors but emphasizes the application of the multidimensional equations of change in various coordinate systems (see Introduction to Part Two).

Part One (Chaps. 1 to 7) can be taken up in a 3-quarter-hour or 2-semester-hour beginning undergraduate course. For a 3-semester-hour course at the undergraduate level, the brief treatment of Part Two (described in its Introduction) can be included, with Chap. 8 as an optional supplement. At the graduate level, most, if not all, of the book can be covered in a 3-semester-hour course.

The book has been used in both lecture and nonlecture formats. In the nonlecture format the course is divided into units, each unit covering certain sections in the text, including assigned problems and examples. Usually a unit is assigned for every period; i.e., in a 3-hour course, there would be three units per week. After the material assigned for a unit has been read, the student may question the instructor or teaching assistant. The student then receives personalized instruction showing how, by use of reasoning, the answer could be obtained. When ready, the student is given a short quiz on the unit, which is graded in the presence of the student and supplemented by oral questioning to make sure the principles and concepts involved are understood. A suggested list of sections to be included in each unit and sample quizzes may be obtained from the publisher.

The notes from which the present text emerged have been used with good student response at the undergraduate and graduate levels at the University of Florida since 1965, at the University of Brazil in 1964, at the Universidade de Oriente in Puerto La Cruz, Venezuela, in 1976, and at the University of Minnesota in a graduate course in 1978. The final drafts of the text were used at the University of Florida in graduate and undergraduate courses in 1982.

The author wishes to thank the students in his courses, who not only made helpful comments but urged that the notes be published. The criticisms of Spyros Svoronos, Glenn Fredrickson, Norman Johns, J. V. S. Sharma, and Stewart Daw are especially appreciated. R. J. Gordon contributed to early drafts of certain sections and R. Narayanan contributed to the section on free convection. The effort of Ed Everage in preparing some of the problems is appreciated, as well as the support of the Chemical Engineering Department and the faculty at the University of Florida. The author is aware of his great debt not only to the pioneering Bird, Stewart, and Lightfoot text but also to its authors, who have individually had a very profound personal impact on him through their scholarship and their development of transport-phenomena. In particular his discussion with Professor Bird on such topics as rheology, theology, and the relative advantages of the unit operations and transport phenomena approaches to the design of equipment or to the making of popcorn have been very valuable.

The author especially wishes to thank Carole Yocum for excellent typing and artwork; without her help the book could not have been completed.

Ray W. Fahien

LIST OF SYMBOLS

DIMENSIONS AND UNITS

Symbol	Dimension	Units		Equivalent
		English	SI	
M	Mass	lb	kg	
M_l	Moles	lb mol	kg mol	
L	Length	ft	m	
t	Time	h	s	
T	Temperature	°F	°C, K	
F	Force	lbf	N	kg · m/s^2
E	Energy	Btu	J	kg · m^2/s^2

SYMBOLS

A = magnitude of area $[L^2]$

$\mathbf{A}$ = area vector with direction of velocity $[L^2]$

$\mathcal{A}$ = thermodynamic work function $[ML^2/t^2]$ or $[E]$

a = interfacial area/volume $[L^{-1}]$; area of packing/volume $[L^{-1}]$

a_A = activity of species A, dimensionless

$\mathbf{a}$ = acceleration vector $[L/t^2]$

$B^{(I)}$ = interfacial transport of energy into system $[E/t]$

Bo = Bodenstein number, dimensionless

Br = Brinkmann number, dimensionless

b = breadth $[L]$

C = molar concentration $[M_l/L^3]$; constant

C_A = molar concentration of component A $[M_l/L^3]$
$\hat{C}_p$ = heat capacity at constant pressure (on unit mass basis) $[L/t^2T]$ or $[E/MT]$
$\hat{C}_v$ = heat capacity at constant volume (on unit mass basis) $[L^2/t^2T]$ or $[E/MT]$
D, D_t = tube diameter $[L]$
D_{AB} = mass diffusivity of A in B $[L^2/t]$
Da = Damköhler number, dimensionless
D_{ch} = characteristic length $[L]$
E = activation energy $[ML^2/t^2M_l]$; energy $[ML^2/t^2]$; enhancement factor, dimensionless; effective diffusivity $[L^2/t]$
E_{ax} = axial dispersivity $[L^2/t]$
e_v = viscous loss, dimensionless
$\mathbf{e}$ = total-energy flux vector $[E/tL^2]$
F = magnitude of force $[ML/t^2]$
$\mathbf{F}$ = force acting on a system $[ML/t^2]$ or $[F]$
F_D = drag force $[ML/t^2]$
F_f, F' = force of fluid acting on surroundings $[ML/t^2]$ or $[F]$
$\mathfrak{F}$ = dimensionless heat-flow rate
f = friction factor, dimensionless; function of
$\mathfrak{f}$ = function of
G = Hooke's-law constant $[M/Lt^2]$
G_p = pressure gradient $[F/L^3]$ or $[M/t^2L^2]$
$\mathbf{G}$ = temperature-gradient vector $[T/L]$
$\mathcal{G}$ = free energy $[E]$
Gr = Grashof number for heat, dimensionless
$(\mathrm{Gr})_{AB}$ = Grashof number for mass, dimensionless
g = magnitude of gravity $[L/t^2]$; function of
g_c = conversion factor $[ML/t^2F]$, dimensionless in SI
Ha = Hatta number, dimensionless
h = heat-transfer coefficient $[M/t^2T]$; elevation $[L]$
J_{Ax} = same as j_{Ax} but in molar units instead of mass units $[M_l/L^2t]$
J^*_{Ax} = same as J_{Ax} but with respect to molar average velocity $[M/L^2t]$
$\mathbf{J}_A$ = molar-diffusive-flux vector for species A $[M_l/L^2t]$
$\mathcal{J}$ = dimensionless flux
j_{Ax} = mass diffusive flux of A in x direction with respect to mass-average velocity $[M_l/L^2t]$
j_D = mass-transfer factor, dimensionless
j_H = heat-transfer factor, dimensionless
j_M = momentum-transfer factor, dimensionless
$\mathbf{j}_A$ = mass-diffusive-flux vector for species A $[M_l/L^2t]$
J_0 = Bessel function of zero order, first kind

J_1 = Bessel function of first order, first kind
K = overall mass-transfer coefficient $[L/t]$
k = thermal conductivity $[ML/t^3T]$ or $[E/TLt]$
k_B = Boltzmann constant
k_c = mass-transfer coefficient $[L/t]$
k_{cR} = mass-transfer coefficient with reaction occurring $[L/t]$
k_e = effective thermal conductivity $[ML/t^2T]$
k_R = specific reaction-rate constant $[t^{-1}$ for first order$]$
$\underline{\mathbf{k}}$ = thermal-conductivity tensor $[ML/t^3T]$ or $[E/TLt]$
L = length $[L]$
L_f = film thickness $[L]$
l = mixing length $[L]$
M = molecular weight of mixture $[M/M_l]$
M_A = molecular weight of species A $[M/M_l]$
$\mathfrak{M}$ = dipole moment
m = mass of molecule $[M]$; total mass of system $[M]$; ratio of equilibrium concentrations, dimensionless
$\tilde{N}$ = Avogadro's number, molecules/kg mol
N_{Ax} = molar total flux of A in x direction with respect to fixed axis $[M_l/L^2t]$
N_t = total-molar-flux vector $[M_l/L^2t]$
Nu = Nusselt number for heat transfer, dimensionless
$(\text{Nu})_{AB}$ = Nusselt number for mass transfer, dimensionless
n = molecules/volume; order of reaction
$\mathbf{n}_t$ = total mass flux vector $[M/L^2t]$
n_{Ax} = same as N_{Ax} but in mass units instead of molar $[M/L^2t]$
$\mathbf{P}$ = momentum $[ML/t]$
P = dynamic pressure $[M/Lt^2]$ or $[F/L^2]$
$\dot{P}$ = momentum rate $[ML/t^2]$
Pr = molecular Prandtl number, dimensionless
Q = heat $[E]$
$\dot{Q}$ = rate of heat flow $[ML^2/t^3]$ or $[E/t]$
$\dot{Q}_G$ = rate of gain of heat by system at a surface $[ML^2/t^3]$ or $[E/t]$
$\dot{Q}_L$ = rate of heat loss by system at a surface $[ML^2/t^3]$ or $[E/t]$
$\dot{Q}_t, \dot{Q}_{Gt}$ = total or net rate of heat gain by system from all surfaces $[ML^2/t^3]$ or $[E/t]$
$\dot{Q}_R$ = total rate of heat gain by system due to reaction $[E/t]$
q_n = heat scalar $[E/L^2t]$
q_x = heat flux in $+x$ direction $[M/t^3]$ or $[E/L^2t]$
$\mathbf{q}$ = heat-flux vector $[E/L^2t]$
R = radius $[L]$
R_A = rate of production of A $[M_l/L^3t]$
R_G = gas constant $[ML^2/t^2TM_l]$ or $[E/M_lT]$

R_H = hydraulic radius $[L]$
$\mathcal{R}$ = resistance to transfer $[Tt/E]$ or $[Tt^3/ML^2]$
Ra = Rayleigh number, dimensionless
Re = Reynolds number, dimensionless
r = radial distance in cylindrical coordinates $[L]$; distance from origin in spherical coordinates $[L]$
r_A = mass rate of production of A per unit volume of mixture $[M/L^3t]$
r_c = mass of production of A per unit mass catalyst
S_x = scalar x component of $\mathbf{S}$ $[L^2]$
$\mathbf{S}$ = surface-area vector with direction of unit normal $[L^2]$
$\mathcal{S}$ = entropy
Sc = Schmidt number, dimensionless
Sh = Sherwood number, dimensionless
St = Stanton number, dimensionless
s = element of length $[L]$; dimensionless time
T = absolute temperature $[T]$
T = residence time $[t]$
T_{xy} = total stress on positive surface S_x in y direction $[F/L^2t]$ or $[M/t^2L]$
$\underline{\mathbf{T}}$ = total-stress tensor $[F/L^2t]$ or $[M/t^2L]$
t = time $[t]$
$\mathbf{t}_{(n)}$ = stress vector $[F/L^2t]$
U = internal energy $[E = ML^2/t^2]$
U_i = overall heat-transfer coefficient based on inside area $[M/t^2T]$
U_o = overall heat-transfer area based on outside area $[M/t^2T]$
u = arbitrary coordinate
$\mathbf{u}$ = velocity of moving surface $[L/t]$
V = characteristic velocity $[L/t]$
$\hat{V}$ = volume per unit mass $[L^3/M]$
$\mathcal{V}$ = volume $[L^3]$
$\dot{\mathcal{V}}$ = volume rate of flow $[L^3/t]$
v = magnitude of velocity $[L/t]$
v_a = cross-sectional average velocity $[L/t]$
v_{ch} = characteristic velocity $[L/t]$
v_x = mass-average velocity in x direction $[L/t]$
v_x^* = molar average velocity in x direction $[L/t]$
W = work done on system $[ML^2/t]$ or $[E]$
$\dot{W}_t$ = work done on system per unit time $[E/t]$
$\mathcal{W}_A$ = molal flow rate of species A $[M_l/t]$
w = mass flow rate $[M/t]$
w_A = mass flow rate of species A $[M/t]$
w_G = mass gained by system at a surface $[M/t]$
w_t = total mass flow rate into system from all surfaces $[M/t]$

$\mathbf{w}$ = mass-flow-rate vector in direction of velocity $[M_l/t]$
X = rotated cartesian coordinate $[L]$; fractional conversion, dimensionless
x = cartesian coordinate $[L]$
$\mathbf{z}$ = position vector $[L]$
y = cartesian coordinate $[L]$
y_0 = Bessel function of second kind, zero order
Z = perimeter $[L]$
z = cartesian coordinate $[L]$

Greek

α = thermal diffusivity $[L^2/t]$
β = thermal coefficient of volume expansion $[T^{-1}]$
$\Gamma(x)$ = gamma function
$\mathbf{\Gamma}$ = angular-momentum vector
γ = $\widehat{C}_p/\widehat{C}_v$, dimensionless
$\dot{\gamma}$ = activity coefficient [Eq. (4-80)], dimensionless
γ_{xy} = deformation in x direction per unit length in y direction, dimensionless
$\dot{\gamma}_{xy}$ = deformation rate in x direction per unit length in y direction $[t^{-1}]$
Δ = difference operator
$\underline{\mathbf{\Delta}}$ = rate-of-deformation tensor $[t^{-1}]$
δ = thickness of boundary layer [L]; differential in lagrangian system
$\boldsymbol{\delta}_x$ = unit vector in x direction
$\underline{\boldsymbol{\delta}}$ = unit tensor
ϵ = pipe roughness $[L]$; void fraction, dimensionless
ϵ_{AB} = energy of attraction between molecules A and B $[Mt^2/t^2]$
ζ = coefficient of volume expansion due to concentration difference, dimensionless; dimensionless position
η = dimensionless position; nonnewtonian viscosity $[M/Lt]$
θ = dimensionless temperature; angle [radians]
κ = ratio R_1/R_2, dimensionless; von Kármán constant, dimensionless
Λ = Pr or Sc
λ = heat of reaction $[E/M_l]$; mean free path of molecule $[L]$
$\boldsymbol{\lambda}$ = unit vector in flow direction
μ = viscosity $[M/Lt]$
ν = kinematic viscosity = momentum diffusivity $[L^2/t]$
ξ = dimensionless position
Π = product
π = 3.14159
$\underline{\mathbf{\Pi}}$ = $p\underline{\boldsymbol{\delta}} + \underline{\boldsymbol{\tau}}$ pressure tensor $[M/t^2L]$
ρ = density $[M/L^3]$
ρ_A = mass A per unit total volume $[M/L^3]$

Σ = summation
σ_{AB} = collision diameter $[L]$
σ_{xy} = shear-stress component acting on negative surface S_x in y direction $[F/L^2t]$ or $[M/t^2L]$
$\boldsymbol{\sigma}_{(n)}$ = viscous-stress vector $[F/L^3]$
$\underline{\boldsymbol{\sigma}}$ = viscous-stress tensor $[F/L^3]$ or $[M/t^2L]$
τ_{xy} = flux of y momentum in x direction $[F/L^2t]$ or $[M/L^2t]$
$\boldsymbol{\tau}_{(n)}$ = momentum-flux vector $[M/t^2L]$
$\underline{\boldsymbol{\tau}}$ = momentum-flux tensor $[M/t^2L]$
$\boldsymbol{\Upsilon}$ = torque $[FL]$ or $[ML^3/s^2]$
Φ_H = rate of internal generation of heat per unit volume $[E/L^3t]$
Φ_M = rate of internal generation of momentum per unit volume $[(ML/t)(L^3t)]$
Φ_v = rate of viscous dissipation per unit volume $[E/L^3t]$
ϕ = intermolecular potential-energy function $[ML^2/t^2]$; angle, radians
ϕ_v = viscous-dissipation function $= \Phi_v/\mu$
Ψ = potential energy $[E]$; distribution function $[L^{-4}]$
ψ = stream function for cartesian coordinate $[L^2/t]$
ψ_c = stream function for cylindrical coordinate $[L^2/t]$
ψ_s = stream function for spherical coordinates $[L^3/t]$
$\underline{\boldsymbol{\Omega}}$ = vorticity tensor $[t^{-1}]$
Ω_μ = collision integral for momentum, dimensionless
Ω_k = collision integral for heat, dimensionless
Ω = collision integral for diffusivity, dimensionless
ω_A = mass fraction A, dimensionless
$\boldsymbol{\omega}$ = angular-velocity vector $[t^{-1}]$
ω = angular-velocity magnitude $[t^{-1}]$

Subscripts

$_A$ = component A; phase A
$_a$ = average value
$_B$ = component B; phase B
$_b$ = bulk value
$_{ch}$ = characteristic value
$_{cs}$ = cross section
$_D$ = diffusion, drag
$_f$ = film
$_G$ = gained by system at a surface
$_g$ = gravity, generation
$_H$ = heat
$_I$ = at interface

$_i$ = i component or species
$_j$ = j component or species
$_k$ = k component
$_L$ = lost by system
$_{LM}$ = log mean
$_M$ = momentum
$_n$ = normal
$_O$ = overall
$_o$ = outside
$_P$ = pressure, particle
$_q$ = in direction of heat flux **q**
$_R$ = reaction; reference value; at $r = R$
$_r$ = radial direction
$_s$ = at surface; shaft
$_T$ = tube
$_t$ = total value
$_v$ = viscous; volume
$_{vA}$ = vapor, for species A
$_w$ = at wall
$_x$ = x direction
$_y$ = y direction
$_z$ = z direction
$_{(x)}$ = x face
$_{(n)}$ = face whose unit normal is **n**
$_0$ = at $y = 0$ or $x = 0$, etc.
$_1$ = at location 1
$_2$ = at location 2
$_\infty$ = at infinity

Superscripts

^{A} = phase A
^{B} = phase B
$^{(G)}$ = gas phase
$^{(I)}$ = transfer across interface
$^{(L)}$ = liquid phase
$^{(l)}$ = laminar
$^{(t)}$ = turbulent
$^{(\alpha)}$ = phase α
$^{(\beta)}$ = phase β
$^+$ = dimensionless
^{0} = standard state

$'$ = fluctuating quantity (Chap. 6); force of system on surroundings; derivative with respect to independent variable

$*$ = dimensionless value; with respect to molar average velocity; equilibrium value; tagged molecule

Accents

$\dot{}$ = time derivative

$\bar{}$ = time average (Chap. 6); Laplace transform (Chap. 7); partial molal (Chap. 6, 10)

$\hat{}$ = per unit mass

$\tilde{}$ = per unit mole

Miscellaneous

$\langle\ \rangle$ = average value

$[=]$ = dimension of

$[\![\ ]\!]$ = scalar component of vector

PART ONE

ONE-DIMENSIONAL TRANSPORT

CHAPTER

ONE

BASIC LAWS FOR ONE-DIMENSIONAL TRANSPORT

1-1 INTRODUCTION

What Do We Mean by Transport Phenomena?

Transport phenomena is the collective name given to the systematic and integrated study of three classical areas of engineering science: (1) energy or heat transport, (2) mass transport or diffusion, and (3) momentum transport or fluid dynamics. Of course, heat and mass transport occur frequently in fluids, and for this reason some engineering educators prefer to include these processes in their treatment of fluid mechanics. Since transport phenomena also includes heat conduction and diffusion in solids, however, the subject is actually of wider scope than fluid mechanics. It is also distinguished from fluid mechanics in that the study of transport phenomena makes use of the similarities between the equations used to describe the processes of heat, mass, and momentum transport. These *analogies,* as they are usually called, can often be related to similarities in the physical mechanisms whereby the transport takes place. As a consequence, an understanding of one transport process can readily lead to an understanding of other processes. Moreover, if the differential equations and boundary conditions are the same, a solution need be obtained for only one of the processes since by changing the nomenclature that solution can be used to obtain the solution for any other transport process.

It must be emphasized, however, that while there are similarities between the transport processes, there are also important differences, especially between the transport of momentum (a vector) and that of heat or mass (scalars). Nevertheless, a systematic study of the similarities between the transport processes makes it easier to identify and understand the differences between them.

How We Approach the Subject

In order to demonstrate the analogies between the transport processes, we will study each of the processes in parallel—instead of studying momentum transport first, then energy transport, and finally mass transport. Beside promoting understanding, there is another pedagogical reason for not using the serial approach that is used in other textbooks: of the three processes, the concepts and equations involved in the study of momentum transport are the most difficult for the beginner to understand and to use. Because it is impossible to cover heat and mass transport thoroughly without prior knowledge of momentum transport, one is forced under the serial approach to take up the most difficult subject (momentum transport) first. On the other hand, if the subjects are studied in parallel, momentum transport becomes more understandable by reference to the familiar subject of heat transport. Furthermore, the parallel treatment makes it possible to study the simpler concepts first and proceed later to more difficult and more abstract ideas. Initially we can emphasize the physical processes that are occurring rather than the mathematical procedures and representations. For example, we will study one-dimensional transport phenomena first because it can be treated without requiring vector notation and we can often use ordinary differential equations instead of partial differential equations, which are harder to solve. This procedure is also justified by the fact that many of the practical problems of transport phenomena can be solved by one-dimensional models.

Why Should Engineers Study Transport Phenomena?

Since the discipline of transport phenomena deals with certain laws of nature, some people classify it as a branch of science† instead of as a branch of engineering. For this reason the engineer, who is concerned with the economical design and operation of plants and equipment, quite properly should ask how transport phenomena will be of value in practice. There are two general types of answers to this question. The first requires one to recognize that heat, mass, and momentum transport occur in many kinds of engineering equipment, e.g., heat exchangers, compressors, nuclear and chemical reactors, humidifiers, air coolers, driers, fractionators, and absorbers. These transport processes are also involved in the human body as well as in the complex processes whereby pollutants react and diffuse in the atmosphere. It is important that engineers have an understanding of the physical laws governing these transport processes if they are to understand what is taking place in engineering equipment and to make wise decisions with regard to its economical operation.

The second answer is that engineers need to be able to use their understanding of natural laws to design process equipment in which these processes are occurring. To do so they must be able to predict *rates* of heat, mass, or momentum transport. For example, consider a simple heat exchanger (Fig. 1-1), i.e., a pipe

†Because of their usefulness to the engineer, such subjects are often called *engineering sciences*.

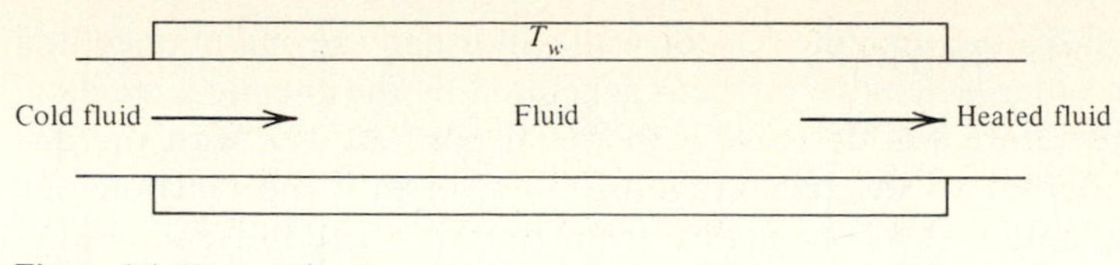

Figure 1-1 Heat exchanger.

used to heat a fluid by maintaining its wall at a higher temperature than that of the fluid flowing through it. The rate at which heat passes from the wall of the pipe to the fluid depends upon a parameter called the heat-transfer coefficient, which in turn depends on pipe size, fluid flow rate, fluid properties, etc. Traditionally heat-transfer coefficients are obtained after expensive and time-consuming laboratory or pilot-plant measurements and are correlated through the use of dimensionless empirical equations. Empirical equations are equations that fit the data over a certain range; they are not based upon theory and cannot be used accurately outside the range for which the data have been taken.

The less expensive and usually more reliable approach used in transport phenomena is to *predict* the heat-transfer coefficient from equations based on the laws of nature. The predicted result would be obtained by a *research engineer* by solving some equations (often on a computer). A *design engineer* would then use the equation for the heat-transfer coefficient obtained by the research engineer.

Keep in mind that the job of designing the heat exchanger would be essentially the same no matter how the heat-transfer coefficients were originally obtained. For this reason, some courses in transport phenomena emphasize only the determination of the heat-transfer coefficient and leave the actual design procedure to a course in unit operations. It is of course a "practical" matter to be able to obtain the parameters, i.e., the heat-transfer coefficients that are used in design, and for that reason a transport-phenomena course can be considered an engineering course as well as one in science.

In fact, there are some cases in which the design engineer might use the methods and equations of transport phenomena *directly* in the design of equipment. An example would be a tubular reactor, which might be illustrated as a pipe, e.g., the heat exchanger described earlier, with a homogeneous chemical reaction occurring in the fluid within. The fluid enters with a certain concentration of reactant and leaves the tube with a decreased concentration of reactant and an increased concentration of product (Fig. 1-2).

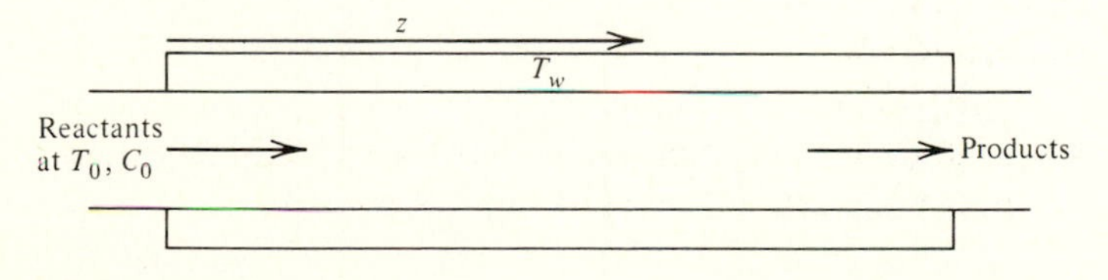

Figure 1-2 Tubular reactor.

If the reaction is exothermal, the reactor wall will usually be maintained at a low temperature in order to remove the heat generated by the chemical reaction. Therefore the temperature will decrease with radial position, i.e., with the distance from the centerline of the pipe (Fig. 1-3). Then, since the reaction rate increases with temperature, it will be higher at the center, where the temperature is high, than at the wall, where the temperature is low. Accordingly, the products of the reaction will tend to accumulate at the centerline while the reactants accumulate near the wall of the reactor. Hence, concentration as well as temperature will vary both with radial position and with length (Figs. 1-4 and 1-5). To design the reactor we would need to know, at any given length, the mean concentration of product. Since this means concentration is obtained from the point values averaged over the cross section, we actually need to obtain the concentration at every point in the reactor, i.e., at every radial position and at every length. But to calculate the concentration at every point we need to know the reaction rate at every point, and to calculate the rate at every point we need to know both the temperature and the concentration at every point! Furthermore, to calculate the temperature we also need to know the rate and the velocity of the fluid at every point. We will not go into the equations involved, but obviously we have a complicated set of partial differential equations that must be solved by sophisticated procedures, usually on a computer. It should be apparent that we could not handle such a problem by the empirical design procedures used in unit operations courses for a heat exchanger. Instead the theory and mathematical procedures of transport phenomena are essential— unless one wishes to go to the expense and take the time to build pilot plants of increasing size and measure the conversion in each. Even then the final scale-up is precarious and uncertain.

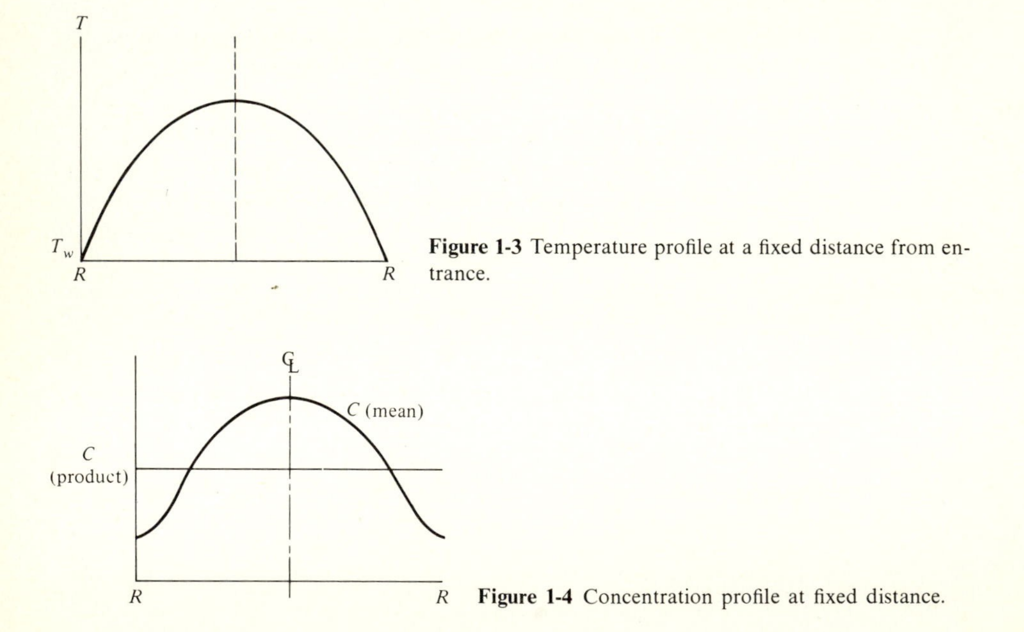

Figure 1-3 Temperature profile at a fixed distance from entrance.

Figure 1-4 Concentration profile at fixed distance.

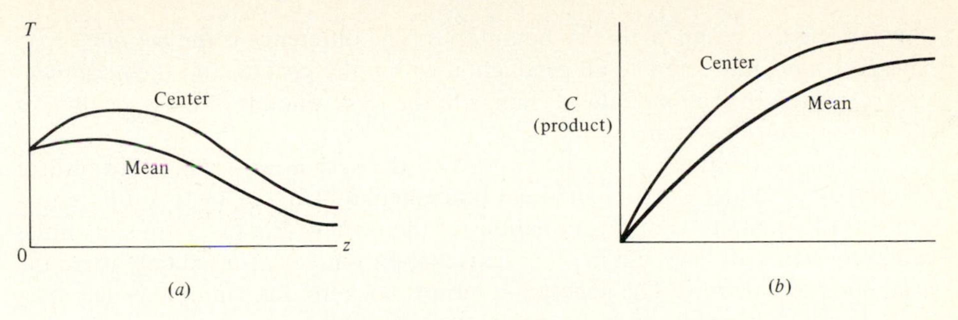

Figure 1-5 (*a*) Mean and center temperature versus axial distance in tubular reactor. (*b*) Mean and center product concentration versus axial distance in tubular reactor.

Of course, not *all* problems today can be solved by the methods of transport phenomena. However, with the development of the computer, more and more problems *are* being solved by these methods. If engineering students are to have an education that is not to become obsolete, they must be prepared, through an understanding of the methods of transport phenomena, to make use of the computations that will be made in the future. Because of its great potential as well as its current usefulness, a course in transport phenomena may ultimately prove to be the most practical and useful course in a student's undergraduate career.

1-2 REVIEW OF BALANCE EQUATIONS

The field of transport phenomena is primarily concerned with the prediction of the temperature, concentration, and velocity variations *within* a medium. In order to obtain these profiles we use two sets of equations: (1) balance (or conservation) equations and (2) rate equations (or "flux laws"). It is assumed in this book that the reader has had some previous experience with balance equations, especially with using the laws of conservation of mass and energy on an overall basis. If not, the following brief review may suffice.† The general balance equation for a system is

$$\text{Input rate} - \text{output rate} + \text{generation rate} = \text{rate of accumulation} \quad (1\text{-}1)$$

The *system* is defined as the portion of the universe under study. The remainder of the universe is called the *surroundings*. The system may be either a specified quantity of matter or a specified volume (in fluid mechanics, the latter is often called a *control volume*). The *input rate* refers to all flow into a system (of the quantity involved) across the system boundaries, and the *output rate* refers to all

†For a more thorough treatment, see various chemical engineering texts on material and energy balances.

flow leaving the system across its boundaries. The difference is the *net input rate*. The *generation rate* refers to all production within the system, and the *accumulation rate* refers to the *time* rate of change in the *total* amount of mass, energy, or momentum in the system.

The balance equation can be applied to the system as a whole (overall or macroscopic balance), to an increment (incremental balance), or to a differential element (differential balance). Equations of the form of Eq. (1-1) are sometimes called conservation laws, but in other texts that terminology is used only when the *generation* term is zero. The generation term is not zero, for example, when mass of a species A is produced by chemical reaction.

We illustrate the meaning of each of the terms in Eq. (1-1) by considering two cases.

Case 1 A tank is being filled with a liquid that is flowing in at a mass rate of flow w_1 kg/s. At the same time liquid is leaving at a mass rate w_2 kg/s. The cross-sectional area of the tank is A, and the height of the liquid in the tank at a given time t is h (see Fig. 1-6). Let us apply the balance equation to the mass of liquid in the tank. The *generation* rate is zero since no mass is produced in the tank, but the accumulation rate is not zero unless $w_1 = w_2$, that is, only if the magnitudes of the rates of flow in and out are equal. In that case we would have a *steady-state* condition because there is no change in the amount of liquid in the tank with time.

However, suppose the mass rate entering w_1 is greater than the mass rate leaving w_2. Then the liquid level in the tank will be changing with time as the tank tends to fill up, and the accumulation rate will be greater than zero. If the total mass of the system is m and the density of the liquid is ρ, the accumulation rate is

$$\frac{dm}{dt} = \frac{d(\rho A h)}{dt} = \rho A \frac{dh}{dt}$$

The overall mass balance is then

$$w_1 - w_2 + 0 = \frac{dm}{dt} \tag{1-1a}$$

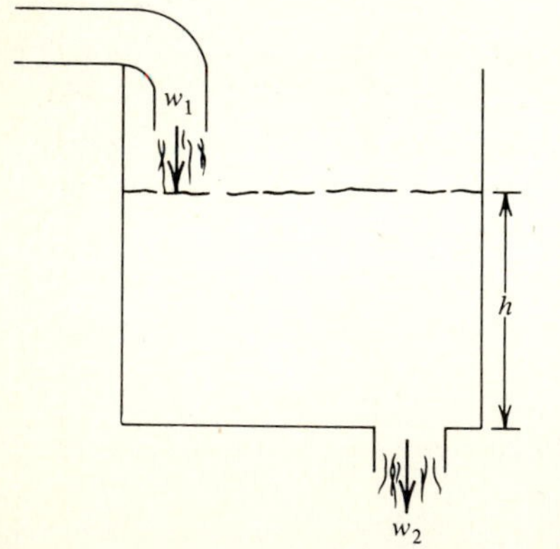

Figure 1-6 Illustration of terms in balance equation.

or if w_t is the net or total rate of *gain* of mass,

$$w_t + 0 = w_1 - w_2 + 0 = \frac{dm}{dt} \tag{1-1b}$$

If w_2 is greater than w_1, the accumulation rate is negative and is called the *depletion*.

Case 2 In this case a chemical reaction is taking place inside the tank. The tank is well stirred, so that the rate of reaction can be considered uniform throughout the tank. Let species A be a product of the reaction and let the rate of production of A per unit volume in the tank be r_A (mass A per unit volume and unit time). Then

$$\text{Generation rate} = r_A \mathcal{V}$$

where $\mathcal{V}$ is the volume of the fluid in the tank. Assume that the entering stream contains a small amount of A such that A enters at a mass rate of flow w_{A1}. Let w_{A2} represent the rate of flow of A leaving. Then the conservation law on species A becomes

$$\underset{\text{Input}}{w_{A1}} - \underset{\text{Output}}{w_{A2}} + \underset{\text{Generation}}{r_A \mathcal{V}} = \underset{\text{Accumulation}}{\frac{dm_A}{dt}} \tag{1-1c}$$

In the example in case 1 there can be no internal generation since total mass is a conserved quantity; but if a chemical reaction is occurring in the liquid and producing species A (as in case 2), the internal generation is not zero and the mass of A is *not* conserved. We use the balance equation on species A as a kind of bookkeeping device. Of course, to be useful each term must have a distinct physical meaning. For example, in case 2 we distinguish between the generation term and the input term by including (as generation) the rate of production of A that occurs at every point *within* the system and (as input) the rate of transfer of A *across the boundary* of the system. The generation rate depends on the *volume* of the system, whereas the input rate depends on the *area* of the boundary over which the transport occurs, e.g., the cross-sectional area of the inlet pipe.

Cases 1 and 2 are examples of the use of *macroscopic* (overall) balances in a physical situation where the temperature, concentration of A, and fluid velocity can be assumed to be uniform throughout the system. As previously mentioned, such problems are *not* the major concern of this text; instead we are primarily interested in problems for which variations (gradients) in the temperature, concentration, and velocity *are* important. Thus, in case 2 the macroscopic balance would include, as the generation rate, $\langle r_A \rangle \mathcal{V}$ where $\langle r_A \rangle$, the average reaction rate, is

$$\langle r_A \rangle \equiv \frac{\iiint r_A \, d\mathcal{V}}{\mathcal{V}}$$

and r_A is the local or point rate. Rigorously r_A is required at every point in the system, since it will depend on the temperature and concentration at the point.

In a well-mixed batch reactor it is convenient to assume that $r_A = \langle r_A \rangle$, but in a tubular reactor (Fig. 1-2) the temperature and concentration vary throughout (Figs. 1-3 to 1-6). This variation of T and C_A with position can theoretically be predicted by means of transport phenomena and such problems *are* the subject of the text.

Energy

In the case of *energy conservation,* Eq. (1-1) rigorously (according to the second law of thermodynamics) must be written as a balance on *all* forms of energy: heat, work, internal energy, kinetic energy, potential energy, etc. Often the work and kinetic-energy terms do not exist or can be neglected and only internal-energy (or enthalpy) and heat terms need appear. If an exothermal chemical reaction takes place, there is no *net* generation of total energy; instead, there is a generation of heat corresponding to the difference between the internal energy of the products and the internal energy of the reactants. This change in internal energy (or enthalpy) can be separated into the change that occurs due to reaction at a fixed T and p and the change required to bring reactants and products to the temperature and pressure of the reaction. Thus the heat of reaction can be treated either as a heat-generation term or as a part of the change in the internal energy of the system.

Likewise, the heat generated by an electric current flowing through a wire is a transformation internally of electric energy into heat. The heat developed in a nuclear reaction can be considered a heat-generation term since it occurs within the medium due to the changes that occur in the individual atoms of the fuel.

Momentum

A similar balance equation of the form of Eq. (1-1) can be written for momentum through the use of Newton's law of motion, which states that the rate of change of the momentum of an object is proportional to the net force acting on it. Thus force and momentum rate are interrelated, and certain forces (such as gravity and pressure) can be thought of as momentum sources. This will be discussed in the text that follows.

A. HEAT TRANSPORT IN ONE DIRECTION

1-3 MODES OF HEAT TRANSPORT

Of the three transport processes we study in this text, heat transport is probably the most familiar since it is a part of our everyday experience. We observe heat transport when we prepare our morning cup of coffee, and we observe it again as the coffee cools. When we go outside on a cold day, our body immediately experiences heat transport and we are constantly reminded in the weather reports of the heat transport that is occurring in the atmosphere. Processes employing heat transport occur frequently in chemical plants and refineries, e.g., heating a crude oil to its boiling point in order to separate it into petroleum fractions in a distillation column or removing the heat of a chemical reaction. In either case, we must first find the *rate* at which heat transport takes place in order to calculate the size of new equipment required or to make improvements in an existing process. This section of the text deals with the methods used to predict these transport rates.

At the outset we should remember that heat is only one of the forms of energy and that it is energy and not heat which, according to the first law of thermodynamics, is conserved. Later we will study the conservation of energy in detail, but in this chapter we are concerned only with heat transport and specifically only with *heat conduction,* i.e., the transport of heat from a high temperature to a lower temperature in a solid or fluid by the motion of molecules or electrons. Hence we exclude heat transport by *radiation,* which is an electromagnetic phenomenon and is illustrated by the transmission of the energy of the sun to the earth through space. We reserve for later treatment the subject of heat transport by *convection,* i.e., heat transport that results from bulk-fluid motion. Thus, heat transport in a solid occurs by conduction alone, whereas in a fluid in motion both convective and conductive heat transport may have to be considered.

1-4 HEAT CONDUCTION IN ONE DIRECTION

In order to define some terms and to illustrate the fundamental laws of heat conduction we will devise a simplified model which approximates physical processes occurring in practice.

Consider a large slab of thin dimension L_x (Fig. 1-7). The other dimensions L_y and L_z are considered large compared with L_x, so that heat will be transported only in the x direction when there is a difference in temperature between the top and bottom faces of the slab. The bottom face of the slab lies in the plane $x = 0$ and the top in the plane $x = L_x$. The slab is initially ($t < 0$) at a uniform temperature T_2. At time $t = 0$ the lower face of the slab is suddenly subjected to a higher uniform temperature T_1, which is maintained while the upper face is maintained at the uniform temperature T_2. We know from experience that heat will begin to flow from the higher temperature T_1 to the lower temperature T_2.

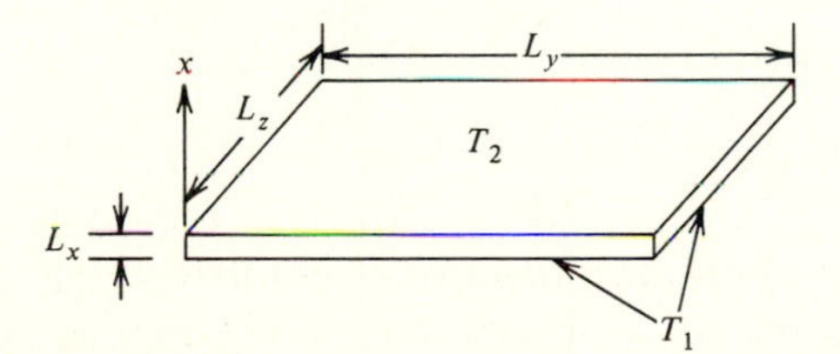

Figure 1-7 Heat conduction in a thin slab.

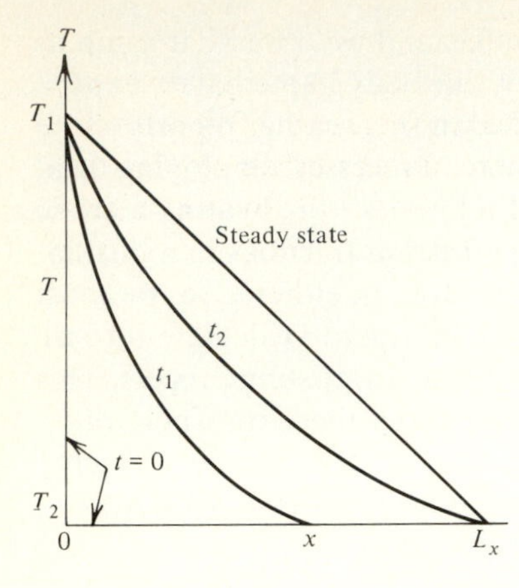

Figure 1-8 Temperature versus position in slab at various times.

The slab might represent the wall of a boiler or furnace that has recently been fired. As engineers, we need to know the answer to the questions: How fast does heat flow through the slab? Suppose the slab were steel or copper, how would the rates of heat transfer compare? How does the temperature vary within the slab?

To help analyze the physical situation we will sketch the variation of temperature T with position x in the slab (Fig. 1-8). Such a plot is called a *temperature profile* or *temperature distribution*. Since the temperature varies with both time t and position x, we will plot T versus x at various times. At $t = 0$ the lower surface ($x = 0$) will be at $T = T_1$, but the rest of the slab ($x > 0$) will be at T_2 throughout. The plot of T versus x will therefore make a right angle; it will exhibit a slope of $-\infty$ at $x = 0$ and a slope of 0 at all other x. At an instant of time later (say $t = t_1$) some heat will have penetrated a certain distance into the slab and will have raised the temperature near the surface ($x = 0$). At a still later time ($t = t_2$) heat has reached the upper surface and is flowing out of the slab. As heat continues to enter the slab, the temperature at each position x keeps increasing with time until eventually these temperatures reach the steady state, a condition in which the temperature at each position no longer varies with time. Note, however, that the temperature will still vary with position; thus we are still not at *equilibrium*. For emphasis, we can formalize this idea as a general definition.

Definition When the physical state of a system† does not change with time, the system is said to be at *steady state*.

Macroscopic Balance

In this case the balance law [Eq. (1-1)] can be written

$$\begin{matrix}\text{Rate of} \\ \text{energy} \\ \text{input} \\ \text{into system}\end{matrix} - \begin{matrix}\text{rate of} \\ \text{energy} \\ \text{output} \\ \text{from system}\end{matrix} + \begin{matrix}\text{rate of} \\ \text{generation} \\ \text{of heat} \\ \text{in system}\end{matrix} = \begin{matrix}\text{rate of} \\ \text{accumulation} \\ \text{of energy} \\ \text{in system}\end{matrix}$$

†System is defined in Sec. 1-1.

We can write a *macroscopic* heat balance over the slab. Since $T_1 > T_2$, the input rate is the rate at which heat enters the lower face of the slab $\dot{Q}_1$ (energy per time), and the output rate is the rate at which it leaves the upper face $\dot{Q}_2$. We note that no heat (energy) is being produced by chemical reaction or nuclear or electrical heating; therefore the generation rate is zero. Also, at steady state, there can be no accumulation or depletion of the energy of the slab. Therefore, the law of conservation of energy tells us that

$$\dot{Q}_1 - \dot{Q}_2 + 0 = 0 \qquad \text{or} \qquad \dot{Q}_1 = \dot{Q}_2$$

The total or net rate of heat flow into the slab is

$$\dot{Q}_t = \dot{Q}_1 - \dot{Q}_2 = 0$$

Heat Flow Rate

Upon what quantities will the rate of heat flow $\dot{Q}_1$ depend?

1. It certainly will be greater, the greater the cross-sectional area of the slab. Since this area $L_y L_z$ lies in a plane normal to the x axis, we will refer to it as A_x to distinguish it from the cross-sectional areas perpendicular to the other coordinate axes, $L_x L_z$ and $L_x L_y$. It seems reasonable that the rate of heat transport in the slab will be directly proportional to the area A_x, that is, if we double the area of the slab, we also double $\dot{Q}_1$.
2. From our intuition (as verified by scientific experiment) we would expect $\dot{Q}_1$ to be an increasing function of the temperature difference $T_1 - T_2$; also, as $T_1 - T_2$ goes to zero, $\dot{Q}_1$ should go to zero. In fact, engineers often say that the temperature difference is the *driving force* for heat transfer (although it is not actually a force as defined by physicists).
3. If $T_1 - T_2$ is maintained at a constant value while the thickness of the sheet L_x is increased, $\dot{Q}_1$ decreases; i.e., the resistance of the slab to heat transport increases because there is a larger distance over which the heat must be transported.
4. $\dot{Q}_1$ is found to depend strongly on the nature of the material. Combining these ideas, it follows that

$$\dot{Q}_1 = f[(T_1 - T_2), A_x, L_x, \text{nature of material}]$$

That is, $\dot{Q}_1$ is a function of temperature difference, normal area, thickness, and type of material. The simplest functional relationship possible would be to assume that $\dot{Q}_1$ varies linearly with the temperature difference and inversely with the thickness L_x. Then we could write

$$\frac{\dot{Q}_1}{A_x} = k\frac{T_1 - T_2}{L_x} \tag{1-2}$$

where k is a proportionately constant characteristic of the material; it is known as the *thermal conductivity*. This expression is verified experimentally for small values of $T_1 - T_2$. For larger values of $T_1 - T_2$ the variation of k with temperature must be taken into account.

Table 1-1 Typical range of values of thermal conductivity
J/s · m · K

Gases	0.001–0.1
Liquids	0.01–1.0
Solids	1.0–100

We could conduct a similar experiment for gases and liquids by enclosing the fluid between two parallel plates at temperatures T_1 and T_2, respectively. In this case we must make sure that $T_1 - T_2$ is sufficiently small or that the top slab is the hotter one to ensure that heat will be transported by conduction only and not by convection currents that arise from large density differences (see Sec. 10-14).

Typical values of k for solids, liquids, and gases are listed in Table 1-1. Note that the SI units of k are J/s · m · K = W/m · K; in the engineering system they are Btu/h · ft · °F. Conversion factors are given in the appendix.

Heat Flux

Since, according to Eq. (1-2), the rate of heat flow $\dot{Q}_1$ is proportional to the area A_x, it is convenient to use a quantity called the *heat flux,* the heat flow rate per unit area, or

$$q_x \Big|_{x=0} = \frac{\dot{Q}_1}{A_x} \tag{1-2a}$$

A more general definition will be discussed later. It is convenient to use the idea of a flux for other quantities, such as mass and momentum. In any case, we can write

$$\text{Flux} = \frac{\text{flow of quantity/time}}{\text{area}} = \frac{\text{flow rate}}{\text{area}} \tag{1-3}$$

where the area is normal to the direction of the flow. Equation (1-3) will be used frequently in this text.

A heat flux may exist not only at a boundary but also at an internal position. For example, at an arbitrary x within the slab, we can write

$$q_x \Big|_{x=x} = \frac{\dot{Q}_x \Big|_x}{A_x} \tag{1-4}$$

where $\dot{Q}_x\big|_x$ is the heat flow rate in the x direction at x. The *sign* of the heat flux depends on the direction of heat flow and the coordinate system used. Here heat is flowing in the positive x direction (since $T_1 > T_2$) and q_x is positive.

The units of q_x are usually given as Btu/h · ft^2 or J/s · m^2 = W/m^2. In specifying the heat flux it is necessary to specify the *orientation* of the area through which the heat is transported. Thus, in this example,† if areas A_y and A_z correspond respectively to areas oriented perpendicular to the y and z axes, the heat fluxes are q_y and q_z.

Fourier's Law

In the previous section we considered the special case of one-dimensional heat conduction at steady state in a rectangular geometry. Equation (1-2) holds *only* for this special case and *cannot* be used for other situations, e.g., a cylindrical geometry or unsteady state. Nor can it be used to predict the variation of temperature with position *within* a medium. Consequently we develop a more general equation that is applicable at any location in any geometry and for either steady- or unsteady-state conditions. For this purpose we return to a plot of the temperature in the slab versus x at some arbitrary time t (see Fig. 1-9). We can relate the rate of heat flow $\dot{Q}_x$ at an arbitrary location x to the heat flux at the same location by using Eq. (1-4); $\dot{Q}_x \equiv q_x A_x$.

In order to get an equation for q_x at any location x we assume that Eq. (1-2) applies over a small increment Δx. Then we let L_x correspond to Δx and let $T_1 - T_2$ correspond to $-\Delta T$. The minus sign is necessary because, according to the calculus definition of the Δ operator [Eq. (A1-1-1)],‡

$$\Delta T = T(x + \Delta x) - T(x)$$

†In this case, the temperature difference occurs between planes perpendicular to the x axis, causing transport in the x direction. The fact that the slab was very thin in the x direction and very wide in the y and z directions meant that there was negligible heat leakage at the edges perpendicular to the y and z axes. Thus q_y and q_z are zero. In general, the rate of heat conduction at any point in a material is characterized by a heat-flux *vector* $\mathbf{q}$, which can be resolved into components along the three coordinate axes. It was possible to ignore the vectorial nature of $\mathbf{q}$ and consider only its scalar *x component* q_x for this simple case of unidirectional heat conduction. The heat-flux vector will be examined in detail when we take up heat transport in more than one direction (Chap. 8).

‡Triple numbers refer to chapter appendixes at the end of the chapter.

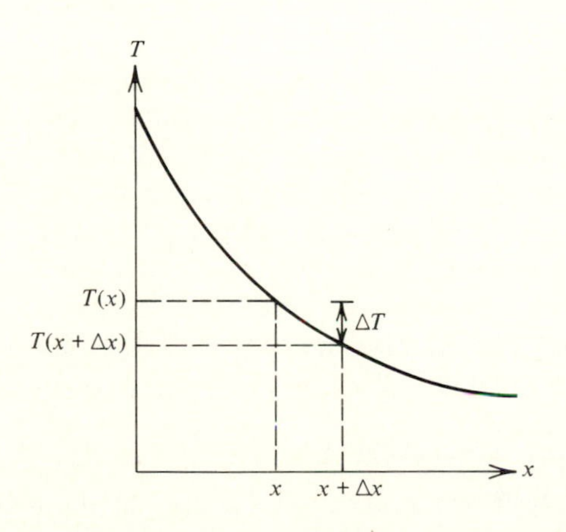

Figure 1-9 Plot of temperature versus position in slab at some arbitrary time t.

Then the average heat flux through a distance Δx is

$$q_x \approx -k\frac{\Delta T}{\Delta x} = -k\frac{T(x + \Delta x, t) - T(x, t)}{\Delta x}$$

From Fig. 1-9 it is clear that $\Delta T/\Delta x$ represents the *average* slope over the region Δx of the T-versus-x curve. We can also see that if we make Δx smaller and smaller, we obtain a better and better approximation to the actual slope at position x. In the limit as $\Delta x \to 0$, we obtain the partial derivative of T with respect to x since, by definition [Eq. (A1-1-3)],

$$\lim_{\Delta x \to 0} \frac{T(x + \Delta x, t) - T(x, t)}{\Delta x} \equiv \frac{\partial T}{\partial x}$$

Hence for *unsteady* state we can write at any location

$$\boxed{q_x = -k\frac{\partial T}{\partial x}} \tag{1-5}$$

which is called *Fourier's law of heat conduction in one dimension*. This equation is taken as the definition of thermal conductivity and as the general cartesian coordinate expression for Fourier's law for one-dimensional transport. We will show that other equations, such as Eq. (1-2), can be derived from it.

Now let us consider the steady-state case, i.e., the case in which T depends only on x. Since there is only one independent variable x instead of two (x and t), the total derivative should be used; it is, by definition [Eq. (A1-1-1)]

$$\frac{dT}{dx} = \lim_{\Delta x \to 0} \frac{T(x + \Delta x) - T(x)}{\Delta x}.$$

Fourier's law of heat conduction for the special case of steady-state transport in one direction is then

$$q_x = -k\frac{dT}{dx} \tag{1-6}$$

which states that the heat flux in the x direction is proportional to the derivative of the temperature with respect to x, the negative sign indicating that heat flows in the direction of decreasing temperatures. Thus if dT/dx were positive, indicating an increase in T as we move in the x direction, Eq. (1-6) indicates that q_x would be negative; i.e., the direction of heat flow is in the opposite ($-x$) direction. The derivative dT/dx is referred to as the *temperature gradient* in the x direction.

In the *general* case where there is heat flow in *all* three coordinate directions, T is a function of more than one independent variable, and these relations are expressed in terms of *partial derivatives*

$$\frac{\partial T}{\partial x} = \lim_{\Delta x \to 0} \frac{T(x + \Delta x, y, z) - T(x, y, z)}{\Delta x}$$

etc. [see Eq. (A1-1-3)]. Then Fourier's law becomes

$$q_x = -k\frac{\partial T}{\partial x} \qquad q_y = -k\frac{\partial T}{\partial y} \qquad q_z = -k\frac{\partial T}{\partial z} \tag{1-7}$$

For the slab in Fig. 1-7, $\partial T/\partial y = 0 = \partial T/\partial z$ and $q_y = 0 = q_z$. Also, the first of Eqs. (1-7) reduces to Eq. (1-6) at steady state. Equation (1-5) and its three-dimensional counterparts (1-7) represent the fundamental law of heat conduction.

1-5 STEADY-STATE INCREMENTAL BALANCE

A typical problem in transport phenomena involves finding the rate of flow of heat energy to or from a system; e.g., an item of equipment or a portion of it. In such problems coordinate systems other than cartesian may be selected, and such effects as heat generation, convection, and accumulation may occur; but even in these more complicated cases the same general procedure is used to obtain the rate of heat flow into and out of the system. This procedure involves using (1) the balance equations or conservation laws and (2) the flux laws. We illustrate this general procedure for the simple example of the slab. The steps are as follows:

1. Use the balance law (of say energy) to derive an incremental balance and a differential equation for the flux as a function of position in the system.
2. Use the flux law, such as Fourier's law, Eq. (1-5), to relate the flux to the temperature. Then, using the boundary conditions, solve the equations in 1 and 2 to obtain the temperature distribution, i.e., the variation of temperature with position.
3. Use the flux law and the equation for the temperature distribution to obtain the flux at each boundary.
4. Obtain the total transport rate or flow at the boundary by multiplying the flux at the boundary by the area perpendicular to the flow.

We now show we would follow these steps to find the steady-state temperature profile in the slab and the heat flow through it.

Step 1: Incremental Balance

The condition of steady state requires that the energy of the slab and also the energy of any small section of the slab we choose to consider not change with time. Consider an extremely thin slice formed by the planes $x = x$ and $x = x + \Delta x$ (Fig. 1-10). The total heat flow into this slice minus the total heat flow out must be zero, or rate in − rate out = 0. From Eq. (1-4) the rate of heat flow into the element at x is equal to the product of the heat flux and the area

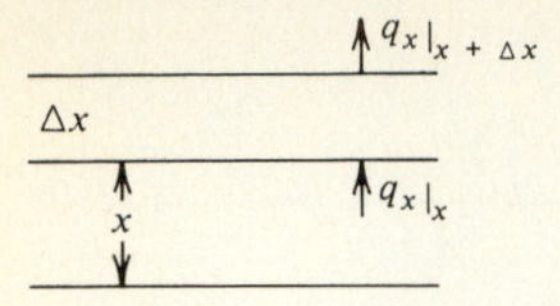

Figure 1-10 Incremental balance over element of slab.

perpendicular to the direction of flow, or

$$\text{Rate in} = q_x A_x \Big|_x$$

and the rate of heat flow *out* of the element is

$$\text{Rate out} = q_x A_x \Big|_{x+\Delta x} \qquad \text{where} \qquad A_x = L_y L_z$$

Then our incremental energy balance becomes

$$q_x A_x \Big|_x - q_x A_x \Big|_{x+\Delta x} = 0 \tag{1-8}$$

Dividing by the volume of the increment $A_x \Delta x$ and changing signs, we obtain

$$\frac{-q_x|_x + q_x|_{x+\Delta x}}{\Delta x} = 0$$

since A_x is constant.

Letting $\Delta x \to 0$ and using the definition of the derivative [Eq. (A1-1-2)] gives the *flux differential equation*

$$\frac{dq_x}{dx} = 0 \tag{1-9}$$

Step 2: Solution for Temperature Distribution

We integrate Eq. (1-9) between an arbitrary x and $x = 0$, obtaining

$$q_x(x) = q_x(0)$$

or the *flux distribution* is

$$q_x = \text{const}$$

We then arbitrarily let $q_x = q_1$ be the heat flux at $x = 0$. Noting that $q_x = -k\, dT/dx$ by Fourier's law, equating to q_1, and rearranging gives

$$dT = -\frac{q_1}{k} dx$$

We now integrate between limits, making use of the known boundary conditions, to get

$$\int_{T_1}^{T_2} dT = -\frac{q_1}{k} \int_0^{L_x} dx$$

or
$$T_2 - T_1 = -\frac{q_1}{k} L_x \tag{1-10a}$$

If we wish to obtain the temperature distribution in terms of T_1, we integrate up to an arbitrary $x = x$

$$\int_{T_1}^{T} dT = -\frac{q_1}{k} \int_0^x dx$$

or
$$T - T_1 = -\frac{q_1}{k} x \tag{1-10b}$$

which gives a linear profile. To eliminate q_1/k we can divide Eq. (1-10*b*) by Eq. (1-10*a*) to obtain

$$\frac{T - T_1}{T_2 - T_1} = \frac{x}{L_x} \tag{1-10c}$$

Rearranging gives the *temperature distribution* or profile

$$\boxed{T = T_1 + \frac{T_2 - T_1}{L_x} x} \tag{1-10d}$$

Step 3: Flux Distribution

To calculate the heat flux q_x we use Fourier's law. First we calculate the gradient by taking the derivative of T in Eq. (1-10*d*)

$$\frac{dT}{dx} = \frac{T_2 - T_1}{L_x}$$

Hence,
$$\boxed{q_x = k \frac{T_1 - T_2}{L_x}} \tag{1-10e}$$

Step 4: Total Rate of Heat Flow

The heat flow is given as the product of the flux at $x = 0$ with the normal area at the same location. Thus from Eqs. (1-2*a*) and (1-10*e*)

$$\dot{Q}_1 = \dot{Q}\Big|_{x=0} = q_x A_x\Big|_{x=0} = kA_x \frac{T_1 - T_2}{L_x} \tag{1-11}$$

which is identical to Eq. (1-2). Since in this case both flux and area are independent of position x, the same result is obtained at $x = L_x$ and at $x =$ arbitrary x.

Example 1-1: Effect of conductivity on temperature profiles Heat is being transported at a steady-state rate of 6000 Btu/min (1.054×10^5 W) through a 5- by 10-ft (1.505- by 3.03-m) steel plate having a thickness of 2 in (5.08 cm). The thermal conductivity is listed as 27.1 Btu/h · ft · °F. The temperature of the cold face of the plate is $T_2 = 20$°C (293 K). (*a*) Calculate (1) the heat flux through the plate in Btu/h · ft² and J/s · m² and (2) the temperature of the warm face in degrees Celsius and kelvins. (*b*) Repeat the calculation if the plate is made of copper and $k = 227$ Btu/h · ft · °F. (*c*) Plot temperature versus position on the same graph for the copper and steel plate in order to illustrate the effect of thermal conductivity on the slope of the temperature curve.

SOLUTION (*a*) (1) We let x be the direction in which the heat is flowing. This is in the thin dimension over which the temperature drop occurs. The area perpendicular to the direction of flow is $A_x = 5(10) = 50\ \text{ft}^2$. Then by Eq. (1-2*a*) the heat flux through the plate is

$$q_x = \frac{\dot{Q}_1}{A_x} = \frac{6000\ \text{Btu/min}}{50\ \text{ft}^2}\frac{60\ \text{min}}{1\ \text{h}} = 7200\ \text{Btu/h}\cdot\text{ft}^2$$

Converting to SI units, we use the table in the appendix to obtain

$$q_x = \frac{7200\ \text{Btu}}{\text{h}\cdot\text{ft}^2}\frac{0.2931\ \text{W}}{\text{Btu/h}}\frac{1\ \text{ft}^2}{(0.305)^2\ \text{m}^2} = 2.27 \times 10^4\ \text{W/m}^2$$

(2) From Eq. (1-10*e*)

$$T_1 - T_2 = \frac{q_x L_x}{k} = \frac{(7200\ \text{Btu/h}\cdot\text{ft}^2)(\frac{2}{12}\ \text{ft})}{27.1\ \text{Btu/h}\cdot\text{ft}\cdot{}^\circ\text{F}}$$

$$= 44.3\ \text{Fahrenheit degrees} = 44.3\ \text{Rankine degrees}$$

$$= 44.3(\tfrac{5}{9}) = 24.6\ \text{K}$$

$$T_1 = 20^\circ\text{C} + 24.6\ \text{K} = 44.6^\circ\text{C}\ (317.6\ \text{K})$$

(*b*) For copper the temperature difference will be inversely proportional to the thermal conductivities

$$T_1 - T_2 = (24.6\ \text{K})\frac{27.1}{227} = 2.94\ \text{K}$$

$$T_1 = 20^\circ\text{C} + 2.9\ \text{K} = 22.9^\circ\text{C}\ (295.9\ \text{K})$$

(*c*) Figure 1-11 shows that the slope is flatter for the better conductor (copper)

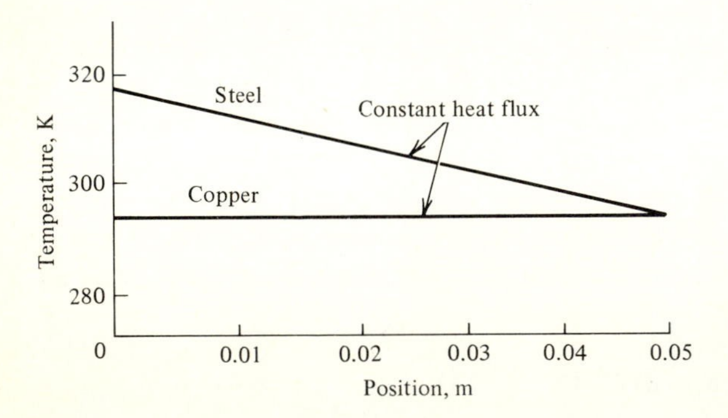

Figure 1-11 Temperature distribution in a metal plate at constant heat flux.

than for the steel. As a rule of thumb we can distinguish a good conductor from an insulator by comparing the slope of the temperature-profile curves; for a perfect conductor the slope would be zero, and for a perfect insulator it would approach infinity. □†

1-6 PHYSICAL MEANING OF FOURIER'S LAW AND SIGN CONVENTION

Let us return to the equation for steady-state heat transport in a thin slab

$$\dot{Q}_1 = kA_x \frac{T_1 - T_2}{L_x} \tag{1-11}$$

We can write this expression as

$$\text{Rate} = \frac{\text{driving force}}{\text{resistance}}$$

where $\dot{Q}_1$ = rate of heat flow (energy per time) in x direction at $x = 0$
$T_1 - T_2$ = driving force
L_x/kA_x = resistance to heat transport

Hence, to obtain the same rate of heat transport in a slab of given dimensions, we find that the temperature difference for a good conductor (high k) would be much less than for a poor conductor (low k) because the resistance of the poor conductor is greater (see Example 1-1).

However, we should again emphasize that expression (1-11) is *not* an adequate statement of Fourier's law since it holds *only* when the temperature gradient is linear, e.g., in the very special case of steady-state heat transport in only one direction in a medium. In the *usual* case, in which the temperature profile is *not* linear, we need to use the derivative of temperature with position (the temperature gradient) instead of the temperature difference per unit length, $(T_1 - T_2)/L_x$. Then, in Fourier's law,

$$q_x = -k\frac{dT}{dx}$$

we can visualize dx as an infinitesimal length corresponding to L_x and $-dT$ as the infinitesimal difference in temperature, corresponding to $T_1 - T_2$, that occurs over the element dx. The driving force is then dT and the resistance is dx/kA_x. Since the temperature gradient is the slope of a plot of T versus x, it will be negative and, according to Fourier's law, Eq. (1-6), the heat flux will be positive (Fig. 1-12*a*).

Now suppose T *increases* with distance x (see Fig. 1-12*b*). Then the gradient dT/dx will be positive and q_x will be negative. We will use the mathematical convention that a positive q_x means that the physical direction of the heat flow is

†The □ symbol indicates the end of the example.

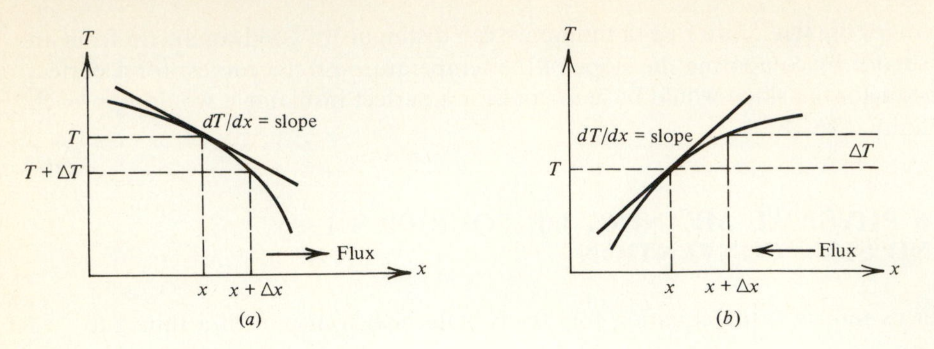

Figure 1-12 (*a*) Positive q_x and (*b*) negative q_x.

in the direction of increasing x and vice versa. *Therefore the physical interpretation of a negative q_x is that heat is actually flowing in the minus x direction.*†

The physical meaning of Fourier's law can be summarized as follows:

1. There is a natural tendency for heat to flow from a high temperature to a low temperature, i.e., "downhill."
2. This tendency depends upon the nature of the material.
3. The larger the magnitude of the temperature gradient or slope, i.e., the greater the driving force per unit distance, the greater the flow of heat will be.
4. The flux and gradient have opposite signs.

We see later that similar gradient laws also apply to momentum and mass transport. If they are understood physically, they can be written down in mathematical form immediately.

Sign Convention for Balances

Problems in transport phenomena frequently deal with cases so complex that the direction of transport at every point in the system cannot be ascertained from the physical situation but must be determined from mathematics. For this and other reasons it is desirable to derive general equations that hold no matter what the physical direction. We must therefore adopt at an early stage an agreed-upon sign convention for our balances. We illustrate the sign convention by returning to heat transport in a slab of which the lower face is held at a temperature T_1 and the upper face is maintained at a lower temperature T_2. As case 1 we take $T_1 > T_2$, and as case 2 we take $T_1 < T_2$. Since, for case 2, dT/dx is positive, q_x must be negative. When writing the energy balance over an increment Δx, we still write

$$\text{Rate in} - \text{rate out} = 0 \tag{1-12}$$

†As another example, consider an ant moving from left to right across the page. If we choose the x direction as increasing from left to right, the velocity of the ant v_x is positive but if the ant is moving from right to left, its velocity v_x is negative

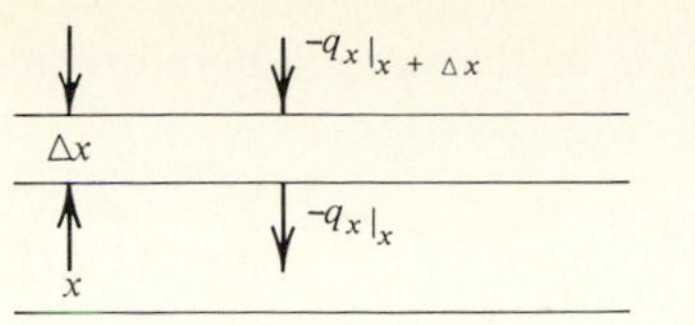

Figure 1-13 Heat flows when $T_1 < T_2$.

but in this case we note (Fig. 1-13) that

$$\text{Rate in} = -q_x A_x \Big|_{x+\Delta x} \qquad \text{and} \qquad \text{Rate out} = -q_x A_x \Big|_x$$

Then

$$-q_x A_x \Big|_{x+\Delta x} - (-q_x A_x)\Big|_x = 0 \tag{1-14}$$

which is algebraically the same equation as Eq. (1-8) (case 1). Thus the flux differential equation will also be the same as Eq. (1-9) and since the boundary conditions are the same, the solution will also be the same as Eqs. (1-10*d*) and (1-10*e*)

$$q_x = k\frac{T_1 - T_2}{L_x} \tag{1-15}$$

This equation automatically tells us that q_x is positive if $T_1 > T_2$ and that it is negative if $T_2 > T_1$. Although the result is obvious here, such will not always be the case.

Since the above results indicate that one obtains the same incremental balance for either positive or negative q_x, we conclude that the *balance can be written on the assumption that q_x is positive*. The actual sign of q_x is then determined after one obtains the final equation and specifies which temperature, T_1 or T_2, is the greater.

Thermodynamic Convention

In thermodynamics if a system is gaining heat, Q is positive; and if it is losing heat to its surroundings, Q is negative. In some books the same convention is now also used for work, but in older books the opposite sign is used. In this book, in order to avoid confusion, we will use $\dot{Q}_G$ for the rate of heat gain by a system at a boundary, $\dot{Q}_L \equiv -\dot{Q}_G$ for a rate of heat loss at a boundary, and $\dot{Q}_{Gt}$ (or simply $\dot{Q}_t$) for the *net* or total rate of heat gain by the system over all boundaries.

Using the above convention, we can write the balance over the element $A_x\,\Delta x$ as

$$\dot{Q}_t \equiv \dot{Q}_G\Big|_x + \dot{Q}_G\Big|_{x+\Delta x} = 0 \tag{1-16}$$

Equation (1-16) shows that the sign of $\dot{Q}_G|_x$ must be opposite that of $\dot{Q}_G|_{x+\Delta x}$; nevertheless, the sign of the heat flow rate $\dot{Q}_x \equiv q_x A_x$ is the same at x as it is at

$x + \Delta x$ since (according to Fourier's law) the sign of q_x is the same. Therefore at $x + \Delta x$ we must use

$$\dot{Q}_G\bigg|_{x+\Delta x} = -q_x A_x\bigg|_{x+\Delta x} \equiv -\dot{Q}_x\bigg|_{x+\Delta x} \tag{1-17}$$

while at x we must use

$$\dot{Q}_G\bigg|_{x} = q_x A_x\bigg|_{x} = \dot{Q}_x\bigg|_{x} \tag{1-18a}$$

These equations result in Eq. (1-10*e*) and would apply to either case 1 or 2, as does Eq. (1-10*e*).

Since the same conventions apply, the macroscopic balance (for case 1 *or* 2) is

$$\dot{Q}_t \equiv \dot{Q}_G\bigg|_{x=0} + \dot{Q}_G\bigg|_{x=L_x} = 0$$

or

$$q_x A_x\bigg|_{x=0} - q_x A_x\bigg|_{x=L_x} = 0 \tag{1-18b}$$

Note that Eqs. (1-17) and (1-18) apply only for surfaces on which the flux q_x does not vary with position. If it does, $\dot{Q}_G$ and A_x must be replaced with $d\dot{Q}_G$ and dA_x as will be discussed later.

1-7 MATHEMATICAL CONVENTION†

The correct signs in Eqs. (1-17) and (1-18) can be automatically obtained from a general expression if we let

$$\dot{Q}_G = -n_x A_x q_x \equiv -n_x \dot{Q}_x \tag{1-19}$$

where n_x is an outwardly directed unit normal to the surface. As its name implies, n_x is a normal of unit magnitude pointing from the interior of the system to the exterior. For example, if the system is the increment $\Delta x\, A_x$ (Fig. 1-14), we note that the unit normal $n_x|_{x+\Delta x}$ points in the *same* direction as that chosen for the x

†This section may be omitted on first reading.

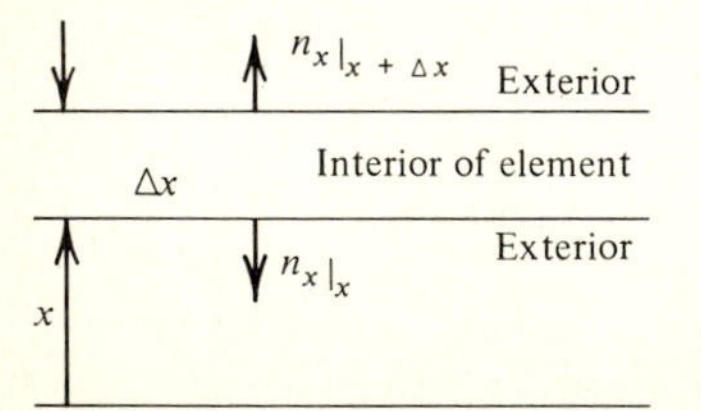

Figure 1-14 Outwardly directed normals.

axis whereas $n_x|_x$ points in the *opposite* direction. Therefore, if we let

$$n_x\Big|_{x+\Delta x} = +1 \qquad \text{and} \qquad n_x\Big|_x = -1$$

the surface at $x + \Delta x$ will be called a *positive* surface and that at x will be called a *negative* surface.

Note that a surface is a vector (it has direction) while an area such as A_x in our usage has magnitude only. As discussed in Chap. 8, we define the scalar x component of the surface vector as $S_x = n_x A_x$. For a positive surface $S_x = A_x$; for a negative surface $S_x = -A_x$. Then using Eq. (1-19) in (1-16) gives

$$-q_x n_x A_x\Big|_x - q_x n_x A_x\Big|_{x+\Delta x} = 0 \tag{1-19a}$$

and inserting the values of n_x gives

$$q_x A_x\Big|_x - q_x A_x\Big|_{x+\Delta x} = 0 \tag{1-19b}$$

which is what we had before in the first of Eqs. (1-7). Note that with this method we do not have to know whether heat is entering or leaving or whether q_x is positive or negative: the signs of n_x and q_x automatically take care of the signs for us.

More General Definition of Heat Flux

In general, q_x may vary from point to point on a surface such as the plane $x = x_1$. In that case, Eq. (1-19) does not apply and the differential rate of heat gain over a differential area dA_x is given by

$$d\dot{Q}_G = -n_x\, dA_x\, q_x = -dS_x\, q_x \tag{1-20}$$

This is a more general definition of heat flux in a one-dimensional system than Eq. (1-4). Integrating Eq. (1-20) over the entire surface $x = x_1$ gives

$$\dot{Q}_G = -\iint (n_x q_x\, dA_x)\Big|_{x=x_1} = -\int_{A_x}\!\!\int (q_x\, dS_x)\Big|_{x=x_1} \tag{1-21a}$$

which reduces to Eq. (1-19) if q_x is constant over A_x.

For two parallel planes each of area A_x, one located at $x = x_1$ and the other at $x = x_2$, the total or net rate of heat gain is

$$Q_{Gt} \equiv \dot{Q}_G\Big|_{x_1} + \dot{Q}_G\Big|_{x_2} = -\iint (q_x\, dS_x)\Big|_{x_1} - \iint (q_x\, dS_x)\Big|_{x_2} \tag{1-21b}$$

Equations (1-21) will be generalized to three dimensions in Chap. 8.

1-8 STEADY-STATE HEAT TRANSPORT IN SYSTEMS WITH CYLINDRICAL GEOMETRY

So far we have been dealing with a plane geometry where the area through which heat is being transported is constant. That gave us a linear temperature profile at steady state. We now consider a cylindrical geometry, in which the area normal to the direction of heat flow is changing, e.g., the conduction of heat through the wall of a metal pipe (Fig. 1-15). Such a geometry is important in the design of heat exchangers, for which the heat transfer through the metal wall of the exchanger must be obtained. In this example, since the boundaries are surfaces of a cylinder, it is convenient to use a polar or cylindrical coordinate system (Table I-1) to describe the process. Let the inner surface of the pipe with radius R_1 be maintained at a temperature T_1, and let the outer surface of the pipe with radius R_2 be maintained at a temperature T_2. Let the length of the pipe be L.

Macroscopic Balance

As in the previous example, we first write an overall (or macroscopic) energy balance on the pipe wall. Since the system is at steady state and there is no internal heat generation, the law of conservation of energy tells us that the rate at which heat enters the wall $\dot{Q}_1$ will equal the rate at which heat leaves the wall $\dot{Q}_2$. From our previous definition for flux, the magnitude of the heat *flux* at R_1 will be given by

$$|q_r|\Big|_{R_1} \equiv \frac{\dot{Q}_1}{A_1} \tag{1-22a}$$

where
$$A_1 \equiv A_r\Big|_{R_1} = 2\pi R_1 L$$

is the area through which the heat is flowing at $r = R_1$. Similarly

$$|q_r|\Big|_{R_2} \equiv \frac{\dot{Q}_2}{A_2} \tag{1-22b}$$

where $A_2 = 2\pi R_2 L$. We note that since $A_2 \neq A_1$ and $\dot{Q}_1 = \dot{Q}_2$, the heat *flux* at R_1 will *not* be equal in magnitude to the heat flux at R_2. Physically this tells us that to maintain the same rate of heat *flow*, the heat *flux* must be smaller when the area

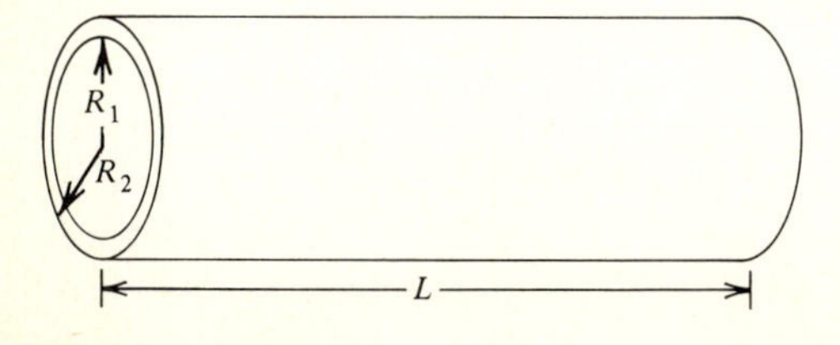

Figure 1-15 Wall of pipe heat exchanger.

becomes larger. We recall that this was not the case with rectangular geometry because the area normal to the direction of heat flow was constant.

Mean Area

Since the area is not constant, an equation like Eq. (1-2) cannot be written for a cylindrical geometry unless it is written in terms of a mean area A_m defined by the analogous equation

$$\dot{Q}_1 \equiv A_m k \frac{T_1 - T_2}{L_r} \tag{1-23}$$

where $L_r = R_2 - R_1$ is the distance over which the heat is transferred. Such an equation would be (and is) convenient to use in heat-exchanger design, but it requires an expression for obtaining the mean area from A_1 and A_2. In order to find this expression we need to derive a rigorous equation for $\dot{Q}_1$ in terms of T_1, T_2, k, A_1, and A_2. Then we can equate the two expressions for $\dot{Q}_1$ and solve for A_m. In other words, A_m is a mean area that gives the right answer when we use the wrong equation!

In order to find $\dot{Q}_1$, we could use either Eq. (1-22*a*) or (1-22*b*) since $\dot{Q}_1 = \dot{Q}_2$. Depending upon which equation we use, we would need to know the heat flux at either R_1 or at R_2. For cartesian coordinates the heat flux is given by Fourier's law, Eq. (1-6). By similar reasoning, Fourier's law for steady-state transport in a radial direction r would be

$$q_r = -k \lim_{\Delta r \to 0} \frac{T(r + \Delta r) - T(r)}{\Delta r} = -k \frac{dT}{dr} \tag{1-24}$$

Thus if we can find an equation for $T(r)$, we can differentiate to find q_r, from which we can obtain $\dot{Q}_1$. How do we find the temperature profile $T(r)$? Recalling our experience with the slab, we follow the same steps:

1. Write an incremental energy balance to find how the heat flux varies with r.
2. Write Fourier's law to relate the heat flux to the temperature gradient.
3. Equate the two expressions for heat flux and then integrate, using the boundary conditions to obtain $T(r)$.

Incremental Energy Balance

Since heat flow is in the r direction only, we consider an arbitrary location $r = r$ and write our energy balance over the incremental volume between r and $r + \Delta r$ (see Fig. 1-16). This volume element is the volume of the outer cylinder minus the volume of the inner cylinder, or $\Delta \mathcal{V} = \pi(r + \Delta r)^2 L - \pi r^2 L = 2\pi r \, \Delta r \, L + \pi \Delta r^2 \, L$. Since Δr is small, the Δr^2 can be neglected and $\Delta \mathcal{V} = 2\pi r \, \Delta r \, L$. Heat will be entering through the cylindrical area $A_r = 2\pi r L$ and leaving through the concentric area $2\pi(r + \Delta r)L$. Since the system is at steady state and there is no

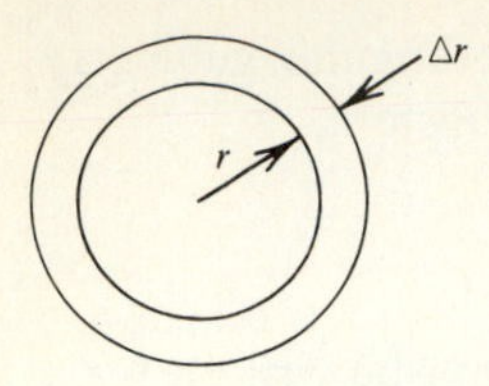

Figure 1-16 Annular increment of pipe wall.

internal heating, the rate of heat flow entering must equal the rate of heat flow leaving, and a physical balance gives†

$$\underset{\substack{\text{Heat rate}\\ \text{in at } r}}{q_r A_r \Big|_r} - \underset{\substack{\text{Heat rate}\\ \text{out at}\\ r+\Delta r}}{q_r A_r \Big|_{r+\Delta r}} = 0$$

Then since $A_r = 2\pi r L$,

$$q_r(2\pi r L)\Big|_r - q_r(2\pi r L)\Big|_{r+\Delta r} = 0$$

Differential Energy Balance

Dividing the last equation by the incremental volume $2\pi r\,\Delta r\,L$ and taking the limit as $\Delta r \to 0$ gives

$$\lim_{\Delta r \to 0} \frac{q_r r|_{r+\Delta r} - q_r r|_r}{r\,\Delta r} = 0 \tag{1-25}$$

Note that since r is a variable, it does not cancel out. Using the definition of the derivative [Eq. (A1-1-2)], with r corresponding to x and rq_r corresponding to f, gives the *flux differential equation*

†We could also write (as in Sec. 1-7)

$$\dot{Q}_G\Big|_r + \dot{Q}_G\Big|_{r+\Delta r} = 0 \qquad \text{where} \qquad \dot{Q}_G = -n_r q_r A_r$$

Since n_r points in the $+r$ direction at $r + \Delta r$ and in the r direction at r,

$$n_r\Big|_r = -1 \qquad \text{and} \qquad n_r\Big|_{r+\Delta r} = +1$$

Then

$$\dot{Q}_G\Big|_r = q_r A_r\Big|_r \qquad \text{and} \qquad \dot{Q}_G\Big|_{r+\Delta r} = q_r A_r\Big|_{r+\Delta r}$$

which gives the same result as the physical balance.

$$\frac{d(q_r r)}{r\,dr} = 0 \tag{1-26}$$

Flux Distribution

Integrating the flux differential equation above gives the flux distribution

$$rq_r = \text{const} \equiv C_1 \tag{1-27}$$

which states that the heat flux varies inversely with radial position r (or with the area A_r, as we observed earlier).

Temperature Distribution

Substituting Fourier's law (1-24) into Eq. (1-27) gives a differential equation for the temperature

$$r\left(-k\frac{dT}{dr}\right) = C_1 \qquad \text{or} \qquad dT = -\frac{C_1}{k}\frac{dr}{r}$$

The boundary conditions are

$$r = \begin{cases} R_1 & \text{at } T = T_1 \\ R_2 & \text{at } T = T_2 \end{cases}$$

If we integrate over T between the limits T_1 and T and over r between R_1 and r, we get

$$T - T_1 = -\frac{C_1}{k}\ln\frac{r}{R_1} \tag{1-28}$$

If we let $r = R_2$ and $T = T_2$, this becomes

$$T_2 - T_1 = -\frac{C_1}{k}\ln\frac{R_2}{R_1} \tag{1-29}$$

Dividing Eq. (1-28) by Eq. (1-29) gives the temperature distribution

$$\frac{T_1 - T}{T_1 - T_2} = \frac{\ln(r/R_1)}{\ln(R_2/R_1)} \tag{1-30}$$

This differs from the case of the slab in that the temperature profile here is not linear, but it is still independent of the thermal conductivity, as it was for the slab. The thermal conductivity k comes into the picture (for this simple example) only when we want to calculate the heat flux.

Heat Transport at the Wall

Solving for T and differentiating Eq. (1-30), we obtain

$$\frac{dT}{dr} = -\frac{T_1 - T_2}{\ln(R_2/R_1)}\frac{R_1}{r}\frac{1}{R_1}$$

From Fourier's law we then get

$$q_r = -k\frac{dT}{dr} = k\frac{T_1 - T_2}{r\ln(R_2/R_1)}$$

Evaluating the rate of heat flow into the pipe at $r = R_1$, we obtain from Eq. (1-22*a*)

$$\dot{Q}_1 = A_r q_r\Big|_{R_1} = 2\pi Lk\frac{T_1 - T_2}{\ln(R_2/R_1)} \tag{1-31}$$

Note that the same result is obtained at $r = R_2$ for $\dot{Q}_2$ and at any r for $\dot{Q}_r = q_r A_r$.

Equating Eqs. (1-23) and (1-31) gives

$$\boxed{A_m = \frac{2\pi L(R_2 - R_1)}{\ln(R_2/R_1)}} \tag{1-31a}$$

This equation fits the definition of a *logarithmic mean,* i.e.,

$$A_{\text{LM}} \equiv \frac{A_2 - A_1}{\ln(A_2/A_1)} = \frac{2\pi R_2 L - 2\pi R_1 L}{\ln(2\pi R_2 L/2\pi R_1 L)} = \frac{2\pi L(R_2 - R_1)}{\ln(R_2/R_1)}$$

A_{LM} is a convenient quantity to use when dealing with heat-conduction problems in cylindrical geometries. In terms of A_{LM}, Eq. (1-31) becomes

$$\boxed{\dot{Q}_1 = kA_{\text{LM}}\frac{T_1 - T_2}{R_2 - R_1}} \tag{1-32}$$

which is equivalent to Eq. (1-23).

1-9 FOURIER'S LAW IN CURVILINEAR COORDINATES

We have now written Fourier's law in cartesian coordinates for transport in the x direction and in cylindrical coordinates for transport in the r direction. Whereas in these cases the coordinate in the direction of transport was a distance, in some cases it is not; e.g., in cylindrical coordinates, one coordinate is the angle $\theta \equiv \arctan(y/x)$, measured counterclockwise from the x axis (see Table I-1). Thus it is incorrect to use $\partial T/\partial\theta$ as the gradient since a gradient, to be consistent with the cartesian coordinate case, must have the dimensions of degrees per unit length. (The dimensions of heat flux and thermal conductivity must be the same regardless of the coordinate system.) Instead, we need to express the gradient in terms of a *distance* ds_θ (see Fig. 1-17) measured along a circular path at constant r

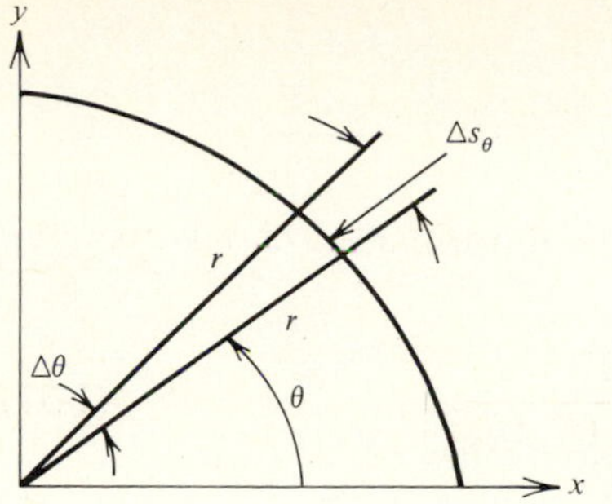

Figure 1-17 Element of distance ds_θ.

and z and varying θ. Thus, if T depends on θ only, Fourier's law is

$$q_\theta = -k \lim_{\Delta s_\theta \to 0} \frac{\Delta T}{\Delta s_\theta} = -k \frac{dT}{ds_\theta}$$

In Example 1-2 we show that the distance $ds_\theta = r\,d\theta$ and therefore $q_\theta = -k(dT/r\,d\theta)$. The distance r by which $d\theta$ must be multiplied to obtain ds_θ is called the *scale factor* h_θ. For an arbitrary coordinate u it is defined by

$$ds_u \equiv h_u\,du \tag{1-33}$$

We can now generalize our expression for Fourier's law to other coordinate systems by letting u be an arbitrary coordinate, say (x, r, θ) and letting an element of distance along the u-coordinate axis be Δs_u. Then for transport in the u direction we can write

$$q_u = -k \lim_{\Delta s_u \to 0} \frac{\Delta T}{\Delta s_u} = -k \frac{\partial T}{\partial s_u} = -k \frac{\partial T}{h_u\,\partial u} \tag{1-34}$$

Example 1-2 Consider a cylindrical coordinate system (r, θ, z), where θ is an angle measured counterclockwise from the x axis, as shown in Fig. 1-17. We are interested in writing the expression for Fourier's law for heat transport in the θ direction, i.e., along the circular path obtained by choosing an arbitrary radial position r and holding r constant.

SOLUTION We let $\Delta\theta$ be a small angle obtained by increasing θ. Now from trigonometry we know that the length of the arc Δs_θ is equal to the product of the radius and the angle, or $\Delta s_\theta = r\,\Delta\theta$. Then

$$\begin{aligned} q_\theta &= -k \lim_{\Delta s_\theta \to 0} \frac{T(r, \theta + \Delta\theta, z) - T(r, \theta, z)}{\Delta s_\theta} \\ &= -k \lim_{\Delta\theta \to 0} \frac{\Delta T}{r\,\Delta\theta} = -k \frac{\partial T}{r\,\partial\theta} \end{aligned} \tag{1-35}$$

Now we consider heat transport in the r direction again. Here $u \equiv r$, $\Delta s_u \equiv \Delta r$, and [see Eq. (1-24)]

$$q_r = -k \lim_{\Delta r \to 0} \frac{T(r + \Delta r, \theta, z) - T(r, \theta, z)}{\Delta r} = -k \frac{\partial T}{\partial r}$$

Similarly, in the z direction, $u \equiv z$, $\Delta s_u \equiv \Delta z$ [see third of Eqs. (1-7)]

$$q_z = -k \lim_{\Delta z \to 0} \frac{T(r, \theta, z + \Delta z) - T(r, \theta, z)}{\Delta z} = -k \frac{\partial T}{\partial z}$$

We can now summarize our expressions for Fourier's law as

$$q_x = -k \frac{\partial T}{\partial x} \qquad \text{cartesian coordinates, } x \text{ direction} \tag{1-6}$$

$$q_u = -k \frac{\partial T}{\partial s_u} \qquad \text{curvilinear coordinates, } u \text{ direction} \tag{1-34}$$

□

B. MOMENTUM TRANSPORT IN ONE DIRECTION

1-10 MOMENTUM TRANSPORT AND FLUID DYNAMICS

Now that we have studied some elementary aspects of heat transport we are in a better position to understand the more complex subject of fluid momentum transport. Since the *momentum* of a body **P** is defined as the product of its *mass and velocity,* we can think of the velocity of a fluid at a given point as its momentum per unit mass. Hence, changes in the velocity of a fluid can result in momentum transport, just as changes in temperature result in heat transport. The mathematical description of this transport forms an important part of the science of fluid mechanics. In this chapter, for pedagogical reasons, we consider momentum transport in those special cases where it *can* be considered analogous to heat or mass transport. Since the concept of momentum transport is usually not emphasized in traditional fluid-flow or physics courses, we start with a review of certain basic definitions.

1-11 NEWTON'S SECOND LAW OF MOTION

Newton's second law of motion states that the force† **F** acting on a body of mass m is proportional to $\dot{\mathbf{P}}$, the time rate of change of its momentum. If the mass is constant, the force is proportional to the product of the mass of the body and its acceleration. Thus,

$$\mathbf{F} = K\dot{\mathbf{P}} = K\frac{d(m\mathbf{v})}{dt} = Km\mathbf{a} \tag{1-36}$$

†Vectors are in boldface in this text.

where $\mathbf{a} = d\mathbf{v}/dt$ is the acceleration of the body and K is a proportionality constant to be determined by the units used. The units of velocity and acceleration are determined when the units of length and time are defined. The units of mass are arbitrary; i.e., mass can be defined relative to a standard piece of platinum-iridium alloy which is assigned a mass of 1 kg. Then the mass of a second body can be determined by comparison. Force can then be defined by Eq. (1-36). It is convenient to devise a system of units such that $K = 1.0$ and is dimensionless. Such an *absolute system* of units has obvious advantages to be described later.

A *gravitational system* of units is one in which the force and mass units are defined such that the body weight at sea level is numerically equal to the body mass. In the commonly used English *gravitational* system of units, the mass unit is taken as the pound mass (lb or lbm) and the force unit, called the pound force (lbf), is defined such that the weight in pounds force of an object at sea level will be *numerically* equal to its mass in pounds mass. Since the acceleration of gravity at sea level is $g = 32.2 \text{ ft/s}^2$, we can find the magnitude of the constant K by letting the weight in pounds force and the mass in pounds mass have the same *numerical* value. When we use $F = W$ (the weight) and $a = g = 32.2 \text{ ft/s}^2$, Eq. (1-36) becomes

$$W = Kmg \qquad \text{or} \qquad W \text{ lbf} = (Km \text{ lbm})(32.2 \text{ ft/s}^2)$$

For $W = m$, this expression implies

$$K = \frac{1 \text{ lbf}}{32.2 \text{ lbm} \cdot \text{ft/s}^2}$$

but we commonly write

$$K = \frac{1}{g_c}$$

where g_c is a *conversion factor*, equal to $(32.2 \text{ lbm} \cdot \text{ft/s}^2)/\text{lbf}$.

Thus, in terms of g_c Newton's second law is

$$\mathbf{F} = \frac{\dot{\mathbf{P}}}{g_c} = \frac{1}{g_c}\frac{d\mathbf{P}}{dt} = \frac{1}{g_c}m\mathbf{a} \tag{1-37}$$

We emphasize that while g_c has the *magnitude* of g at sea level, its units are *not* the same and it is *not* the acceleration due to gravity—or an acceleration of any kind. It is simply a conversion factor required by the selection of units. While g_c is a pure constant, the acceleration of gravity varies with distance from the earth.

An *absolute system* of units is a system in which g_c equals 1.0 and is dimensionless. Examples are the cgs system, the international system (SI), and the English system (pounds mass, poundal, feet, seconds). In these systems, the units in Eq. (1-36) are as shown in Table 1-2. Thus, in the absolute system of units, the force unit is specifically defined in terms of the mass and acceleration units and

$$1 \text{ dyn} = 1 \text{ g} \cdot \text{cm/s}^2 \qquad 1 \text{ N} = 10^5 \text{ dyn} = 1 \text{ kg} \cdot \text{m/s}^2 \qquad 1 \text{ poundal} = 1 \text{ lbm} \cdot \text{ft/s}^2$$

The absolute cgs system of units has been widely used in scientific work, and the English gravitational system unfortunately is still used in the United States for

Table 1-2

System	Force	Mass × acceleration
cgs	dynes (dyn)	$g \cdot cm/s^2$
SI	newtons (N)	$kg \cdot m/s^2$
English	poundals	$lb \cdot ft/s^2$

some engineering work.† However, there recently has been a strong move toward the SI, which will be emphasized in this book.

Beginning students should first write Newton's law in the form of Eq. (1-37). In that way they will remember to include g_c when it is needed but can neglect it when it is not. Let us go through some examples to illustrate the various systems of units.

Case 1 A body is accelerating at 10 ft/s^2. Its mass is 1.0 lbm. What is the force acting on the body in (*a*) poundals, (*b*) pounds force, and (*c*) newtons?

(*a*)
$$F = (m \text{ lbm})(a \text{ ft/s}^2) = 1(10) = 10 \text{ poundals}$$

(*b*) and (*c*)
$$F = \frac{1}{g_c}(m \text{ lbm})(a \text{ ft/s}^2)$$
$$= \frac{1}{32.2 \dfrac{\text{lbm} \cdot \text{ft/s}^2}{\text{lbf}}}(1.0 \text{ lbm})(10 \text{ ft/s}^2)$$
$$= 0.31 \text{ lbf} = 0.31(4.45) = 1.38 \text{ N}$$

Case 2 An international colony has settled on the moon, where the gravitational acceleration is one-sixth that of earth. Some of them want to adopt a lunar gravitational system of units while the others want to use an absolute system. Set up each system of units (using standard earth units for mass).

Gravitational system First we define the new unit "lunar pounds force" lbf$_\ell$). We retain the customary unit of pound mass because it is independent of the gravitational field. Then Eq. (1-36) becomes

$$F \text{ lbf}_\ell = (K_\ell m \text{ lbm})(a \text{ ft/s}^2)$$

For the weight and mass to be numerically equal at moon surface level, we require that $F = W = m$. Since a, the lunar gravitational acceleration, is $32.2/6 = 5.3$ ft/s^2,

$$K_\ell = \frac{1 \text{ lbf}_\ell}{(\text{lbm})(5.3 \text{ ft/s})} \frac{1}{g_{\ell c}}$$

†Some engineers have used an absolute system in which the unit of force is taken as 1 lbf and the unit of mass is defined as 1 lbf/(ft/s^2) and called the *slug*.

where $g_{\ell c}$ is the lunar gravitational constant. Then Newton's second law can be written in the form

$$\mathbf{F} = \frac{1}{g_{\ell c}} m\mathbf{a}$$

where
$$g_{\ell c} = \frac{5.3 \text{ lbm} \cdot \text{ft/s}^2}{1 \text{ lbf}_\ell}$$

The *absolute* system would be the same as that on earth, because its units are independent of the gravitational field. Thus, it is obviously the system of the space age.

1-12 NEWTON'S SECOND LAW APPLIED TO FLUIDS

In elementary physics courses Newton's second law of motion is applied to bodies or particles. For fluids it is necessary to include a term for the flow of momentum into and out of an arbitrary element of volume. The resulting expression is

$$\begin{matrix}\text{Rate of accumulation} \\ \text{of momentum in} \\ \text{arbitrary volume}\end{matrix} = \begin{matrix}\text{net flow of} \\ \text{momentum into} \\ \text{volume}\end{matrix} + \begin{matrix}\text{sum of forces} \\ \text{acting on} \\ \text{volume} \times g_c\end{matrix} \tag{1-38}$$

Note that for a gravitational system of units the forces must be multiplied by g_c in order to convert to momentum units.

1-13 MOMENTUM TRANSPORT BETWEEN PARALLEL PLATES: COUETTE FLOW

Consider a fluid contained between two large parallel plates (Fig. 1-18). The distance between the plates is L_x, which is small compared with the other dimensions, L_y and L_z. At time $t = 0$ the bottom plate is set in motion with a constant velocity $v_{y1} \equiv V$ by applying a force F_y in the y direction while the upper plate is left stationary (or $v_{y2} = 0$). As the lower plate moves, it drags along the layer of fluid immediately adjacent to it, which moves at the velocity of the surface. This so-called *no-slip boundary condition* has strong experimental and theoretical sup-

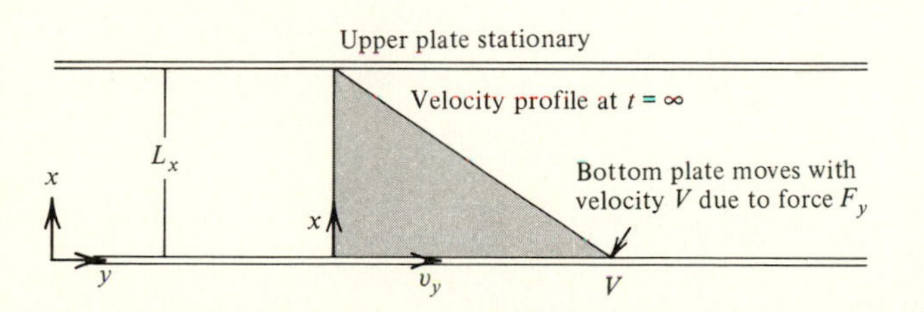

Figure 1-18 Fluid flow between parallel plates.

port. Since the top plate is stationary, the fluid velocity there must be zero. Now the layer of fluid next to the lower plate begins to move relative to the initially stationary layer of fluid immediately above it and in turn starts to drag this layer of fluid along. In this manner the movement of the bottom plate causes a velocity field to be set up in the liquid, with the velocity decreasing continuously from V at the lower plate to zero at the upper plate.

This process is similar to that experienced when the bottom card of a deck of cards is moved with respect to the top card; each card moves because of the movement of the lower card, but its motion is retarded by the friction between it and the upper card. In the case of the fluid, each fluid layer is retarded by the friction between the fluid molecules.

The movement of the lower plate therefore leads to an increase in v_y, the velocity of the fluid in the y direction, from zero to some positive value. Since momentum is proportional to velocity, there is therefore a corresponding increase in the y momentum. In other words, y momentum is being "transported" in the x direction from the plate to the fluid and then from one fluid layer to the next. In Fig. 1-19 the velocity profiles are plotted for various times. For $t = 0$ there is a sharp drop at $x = 0$ from $v_y = V$ to $v_y = 0$. At $t = t_1$ the velocity has increased near the lower plate, but the momentum has not yet penetrated to the fluid near the upper plate. At $t = t_2$ the upper plate begins to sense the motion of the lower plate. Finally, at $t \to \infty$, a steady state is attained,† in which the velocity no longer changes with time. Note that these curves are similar to the plots (Fig. 1-8) of temperature versus position in the heated slab; also the physical description of momentum transport parallels that for the transport of heat. Later we show that in gases the physical mechanism whereby momentum is transported is also similar to that for heat transport, namely, the motion and collision of the molecules within the gas.

The model we have set up is a classical problem in fluid mechanics, referred to as *plane Couette flow*. This flow is approximated in lubrication systems and in Couette viscometers. A Couette viscometer consists of two concentric cylinders

†The concept of infinite time is of course only a mathematical abstraction. For a very viscous fluid it may require only a fraction of a second to achieve 99 percent of the steady-state condition.

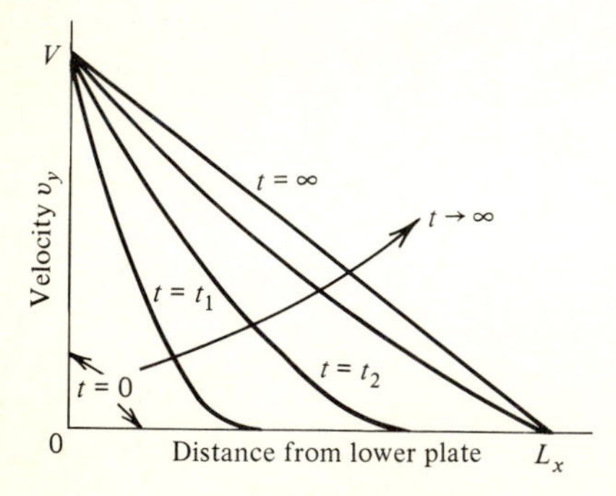

Figure 1-19 Velocity-profile development in flow between parallel plates.

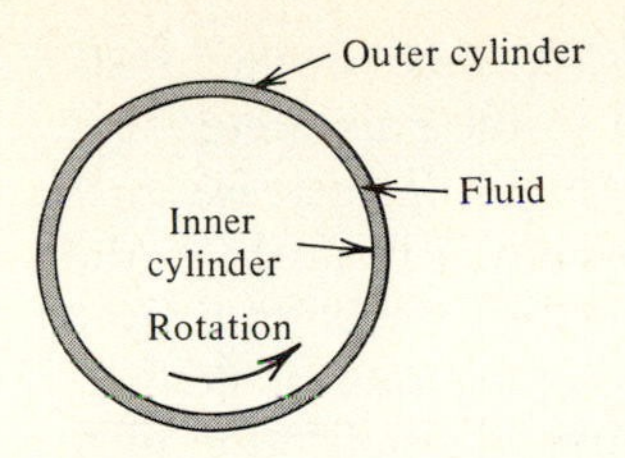

Figure 1-20 Top view of Couette viscometer.

with fluid between them and one (or both) of the cylinders capable of rotation. If the gap between the cylinders is small, the curvature can be neglected (just as we can neglect the curvature of the earth when measuring small relative differences from its surface). Then the fluid behaves as if it were between two infinite flat parallel plates (Fig. 1-20).

1-14 SHEAR STRESS AND MOMENTUM FLUX

In the last section we saw that the application of a force F_y imparted y-directed momentum† to the fluid contained between the plates. F_y is a *shear force*, i.e., a force that acts tangentially to the area over which it is applied.‡ In this case F_y acts tangentially to the area of the plate $A_x = L_y L_z$, where the subscript x indicates that the area is perpendicular to the x axis. If we doubled the area A_x, we would find that a force $2F_y$ is required to maintain the same dynamical state, i.e., the same velocity profiles versus time in the fluid. Hence, F_y is proportional to A_x in this example. Therefore it is convenient to work with the ratio

$$\frac{F_y}{A_x} = \sigma_{xy} \tag{1-39}$$

where σ_{xy} is known as the *shear stress*.§ A stress is any force per unit area. (A normal force per unit area is called a *normal stress*.) Here we are dealing with a *tangential* stress or shear stress.

The notation convention established here will be useful later in the discussion of multidimensional transport phenomena: the first subscript on σ_{xy} refers to the *area* over which it acts (in this case the x area, or the area normal to the x axis), and the second subscript denotes the *direction* in which the shear stress acts.¶ For example, σ_{xz} represents the force per unit area on the x surface acting in the z

†The term "y-directed momentum" refers to the kind of momentum, namely the y component of the momentum vector.

‡Examples of shear forces are the force between the blade of an ice skate and the thin layer of melted ice beneath it, the force exerted by both sides of a knife cutting through a stick of butter, and the force used to break a piece of chalk by twisting it.

§The sign convention will be discussed later; F_y is the scalar y component of the force vector, σ_{xy} is a scalar component of the viscous-stress tensor, and A_x is the magnitude of the area vector.

¶Many texts transpose the subscripts.

direction (this is zero for the present case), σ_{zy} represents the force per unit area in the y direction acting on a surface in the fluid normal to the z axis, etc.

We have observed that the shearing force σ_{xy} results in an increase in the momentum of the fluid and a transport or flow of momentum from the bottom plate to the top plate—just as we observed a flow of heat from the bottom face of the slab (discussed in Part A) to the top face. If we let the rate of flow of y momentum be $\dot{P}_y$, we can define a *flux* of y momentum by application of the general definition

$$\text{Flux} = \text{rate of flow per unit area} \tag{1-3}$$

Since in this case the area through which the momentum is flowing is $A_x = L_y L_z$, the momentum flux at $x = x$ has the magnitude

$$\tau_{xy}\Big|_{x=x} = \frac{\dot{P}_y}{A_x}\Bigg|_{x=x} = \frac{\text{rate of flow of } y \text{ momentum}}{\text{normal area}} \tag{1-40}$$

Again our subscript convention is that the first subscript on τ refers to the *area* through which the momentum is being transported and the second subscript refers to the *direction* with which the momentum is associated.† Note that σ and τ will then always have the same subscripts. The shear stress and momentum flux are clearly related in precisely the same manner as force and momentum rates, namely, through Newton's second law [Eq. (1-37)] (see also Sec. 1-17),

$$\text{Momentum rate} = \text{force} \times g_c \tag{1-41}$$

or

$$\dot{P}_y = F_y g_c \tag{1-42}$$

Dividing by area and using the definitions of momentum flux and shear stress, we obtain

$$\text{Momentum flux} = \text{shear stress} \times g_c \tag{1-43a}$$

or

$$\frac{\dot{P}_y}{A_x} = \frac{F_y g_c}{A_x} \tag{1-43b}$$

or

$$\tau_{xy} = \sigma_{xy} g_c \tag{1-43c}$$

If we use the English gravitational system, the units in Eq. (1-43c) are

$$\tau_{xy} \frac{\text{lbm} \cdot \text{ft/s}}{\text{s} \cdot \text{ft}^2} = \sigma_{xy} \frac{\text{lbf}}{\text{ft}^2} g_c \frac{\text{lbm} \cdot \text{ft}}{\text{lbf} \cdot \text{s}^2}$$

where τ_{xy} is momentum flux (momentum per unit time and unit area) and momentum is mass times velocity.

If SI units are used, g_c is unity and dimensionless; hence

$$\tau_{xy} = \sigma_{xy} \tag{1-44}$$

or

$$\frac{\text{kg} \cdot \text{m/s}}{\text{s} \cdot \text{m}^2} = \frac{\text{N}}{\text{m}^2}$$

†The momentum is of course a vector quantity, and the second subscript refers to a particular component of the momentum.

This relationship indicates that, for an absolute system, shear stress and momentum flux are numerically equivalent quantities; i.e., a force per unit area acting on the x face in the y direction corresponds to the same effect as a flux of y-directed momentum into the x face, and so on for other faces and directions. For this reason most texts use the same symbol for *both* shear stress and momentum flux. This approach has some merit since shear stress and momentum flux both occur together and result from the same condition. However, engineering students must later work in a real world that still uses pounds mass and pounds force, and it is important at this time that they understand the relationship between these units. Otherwise they will constantly be plagued by missing g_c terms in situations far more complex than the above. Furthermore, the interpretation of the shear stress as a momentum flux will be seen to be quite useful in understanding analogies between heat, mass, and momentum transport.

1-15 NEWTON'S LAW OF VISCOSITY

Consider again flow between two parallel plates. After a period of time following the onset of motion of the bottom plate, the velocity profile reaches its final steady-state form (Fig. 1-19). The force F_y and the shear stress σ_{xy} simultaneously attain their steady-state values and remain constant as long as the velocity V is unchanged. These steady-state values of σ_{xy} and F_y clearly depend on the velocity V, the nature of the fluid, and the distance L_x between the plates. This can be expressed mathematically as

$$\sigma_{xy} = f(V, L_x, \text{fluid}) \tag{1-45}$$

where $f(V, L_x, \text{fluid})$ is an unknown function. A simple relationship that holds for many fluids was postulated by Newton for this special case; it is

$$\sigma_{xy} = \frac{\mu}{g_c}\frac{V}{L_x} \tag{1-46}$$

where μ depends on the nature of the fluid and is known as the *viscosity*. However, since this expression is valid *only* for the special case of steady-state Couette flow, we now develop a more general relationship.

Consider the velocity profile at *unsteady state*. Referring to a plot of v_y versus x at constant t (Fig. 1-21), we consider a region of width Δx in which the velocity

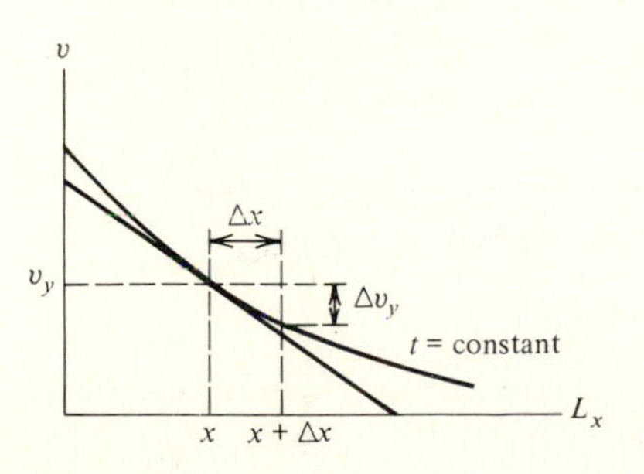

Figure 1-21 Velocity gradient as a slope.

changes by an amount Δv_y, which, using the definition of the difference operator (see Fig. A1-1-1), is

$$\Delta v_y \equiv v_y(x + \Delta x, t) - v_y(x, t) \tag{1-47}$$

An equation consistent with the form of both Eqs. (1-46) and (1-47) is

$$g_c \sigma_{xy} \approx -\mu \frac{\Delta v_y}{\Delta x}$$

where the average slope of the v_y-versus-x curve is $\Delta v_y / \Delta x$. If we take the limit as $\Delta x \to 0$, we approach the true slope at x, which is given by the partial derivative $\partial v_y / \partial x$. The resulting basic equation for *unsteady-state momentum transport in one direction* is

$$\sigma_{xy} g_c = \tau_{xy} = -\mu \frac{\partial v_y}{\partial x} \qquad \text{Newton's law of viscosity in one dimension} \tag{1-48}$$

Fluids which obey Eq. (1-48) are called *newtonian fluids*. We show later that for transport in more than one direction, Newton's law of viscosity is

$$\sigma_{xy} g_c \equiv \tau_{xy} = -\mu \left(\frac{\partial v_y}{\partial x} + \frac{\partial v_x}{\partial y} \right) \tag{1-49}$$

At *steady state* v_y depends on x only, and Eq. (1-48) becomes

$$\sigma_{xy} g_c \equiv \tau_{xy} = -\mu \frac{dv_y}{dx} \tag{1-50}$$

for transport in *one* direction only.

Many industrially and biologically important fluids do not obey Eq. (1-48), e.g., toothpaste, molten plastics, and polymer solutions. Fluids not obeying Eq. (1-48) are known as *nonnewtonian fluids*. We concentrate our attention on newtonian fluids—like water and air—for the time being. For these fluids, the relation between σ_{xy} and $\partial v_y / \partial x$ is totally defined by the specification of the viscosity μ. Examination of Eq. (1-48) shows that μ has dimensions of mass per length and unit time. Previously, viscosity has been expressed in g/cm · s, known as a poise (P), or in units of 0.01 P, known as a centipoise (cP). In SI units the viscosity is

Table 1-3 Viscosity of typical substances

	cP	Pa · s
Water, at 20°C	1.00	0.001
At 100°C	0.28	2.8×10^{-4}
Air, at 20°C	0.018	1.8×10^{-5}
At 100°C	0.022	2.2×10^{-5}

given in pascal-seconds (Pa · s), where 1 Pa · s = 10 P = 10^3 cP = 1 kg/m · s. The viscosities of some common substances are listed in Table 1-3.

1-16 ANALOGY WITH HEAT TRANSPORT IN A SLAB

We can immediately observe certain parallels between the transport of momentum in plane Couette flow and heat transport by conduction in a slab. First, momentum is transported from a high-velocity region ($v_y = V$ at $x = 0$) to a lower-velocity region ($v_y = 0$ at $x = L_x$); second, the flux laws (Newton's law of viscosity, and Fourier's law of heat conduction) are both of the form

$$\text{Flux} = -\text{ transport property} \times \text{potential gradient}$$

where μ and k are the *transport properties* and dv_y/dx and dT/dx are the potential gradients or *driving forces* for momentum and heat transfer, respectively. Furthermore, just as Fourier's law states that heat is transported "downhill," in the direction of decreasing T (corresponding to a negative dT/dx), so also does Newton's law of viscosity tell us that momentum flows downhill, in the direction of decreasing velocity. Thus, in that sense, the two processes are analogous.

We should caution the student that the analogy between heat transfer, momentum transfer, and mass transfer holds only for certain situations; in general, these processes cannot be completely analogous since heat is a scalar and momentum is a *vector*.

1-17 PHYSICAL PICTURE OF MOMENTUM TRANSPORT

The force due to pressure is usually shown in physics or physical chemistry courses to be related to the bombardment of the walls of a container by individual molecules that impart momentum when they hit the wall and rebound. Thus the force due to pressure on a wall A_x is given by

$$F_{px} \equiv \frac{\dot{P}_{px}}{g_c}$$

where $\dot{P}_{px}$, a momentum-transport rate at the wall, is equal to the net change per unit time of the momentum of the molecules hitting the wall and rebounding with the same speed but opposite direction. This rate is equal to the product of the average number of molecules colliding with the wall per unit time and the average momentum transported to the wall during each collision. Note that F_{px} is a *normal* force.

Just as the pressure force is attributable to the momentum transported to a surface by molecular motion, so also is a shear force caused by molecular motion. This momentum transport by molecular motion is related to a shear force or viscous *drag force* F_{vy} by Eq. (1-42)

$$F_{vy} g_c = \dot{P}_{vy}$$

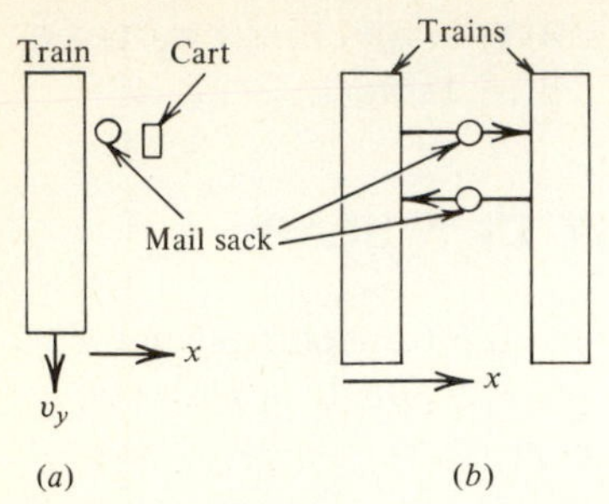

Figure 1-22 Momentum transport.

For example, in Fig. 1-18 at the lower wall ($x = 0$) $\dot{P}_{vy}$ is equal to the time rate of change of the y momentum of the molecules that move laterally (in the x direction) and hit the lower wall, which is moving with velocity v_{y1}. Since these molecules rebound with the y velocity of the wall, which is greater than their previous velocity, y momentum is transferred from the wall to them at a rate proportional to F_{vx}.

At $x = L_x$ molecules moving laterally and having y momentum v_y (on the average) hit the upper wall, which is maintained at $v_{y2} = 0$. Their y momentum is thus transported to the wall, where it is reduced to zero. To hold the wall stationary requires a force which is related to the momentum loss by Eq. (1-42), as indicated above.

A different example might be helpful. Consider a train (there *are* still a few!) moving at high velocity v_y in the y direction through a small town (Fig. 1-22*a*). A mail sack tossed out horizontally (say in the x direction) lands on a small flat cart. We would expect the cart, having very little friction (if unbraked), to start moving in the same direction as the train. The mail sack, which originally had the same y velocity v_y as the train, would transport some of its y momentum to the cart and there would be a flow of y momentum in the x direction. If the cart were held stationary, however, the force required to do so would be proportional to the rate of momentum transport.

Within a fluid at an arbitrary plane $x = x_a$ the situation is more like that in Fig. 1-22*b*. Two trains are traveling in the same direction on parallel tracks with different velocities. Some hobos on each train are throwing mail sacks (or whiskey barrels) back and forth from one train to the other. The faster train in this case tends to be slowed down by the low-momentum (lower v_y) objects coming from the slower train and vice versa; the net effect again is to transport momentum from the higher to the lower velocity.

In Chap. 2 we will see that diffusive momentum transport in gases can be described by a kinetic theory in which gas molecules correspond to the mail sacks and adjacent layers of fluid with different velocities correspond to the trains. At an arbitrary plane $x = x_a$ momentum is therefore transported from the high-velocity fluid below $x = x_a$ to the low-velocity fluid above $x = x_a$ by means of the lateral (x-direction) motion of the molecules. This lateral momentum flow is manifested as a *drag* or frictional flow of the fluid above on the fluid below and an equal but opposite pulling force of the fluid below on the fluid above. The application of this balance between the forces will be discussed next.

1-18 FORCE-MOMENTUM BALANCES

We are now prepared to follow the procedure used in heat transport (which we will use, with minor modifications, in subsequent problems). We first write an incremental momentum balance over an element of volume of width Δx. After taking the limit as $\Delta x \to 0$, we obtain a differential equation for the momentum flux, which is solved to obtain the flux distribution. The flux thereby obtained is equated to the flux as given by Newton's law of viscosity. The resulting equation is integrated, using the boundary conditions, to obtain the velocity distribution. We establish, for the case of parallel plates, that the velocity profile is linear. We are then able to obtain an expression for the total transport of momentum, i.e., the drag due to friction, just as we earlier found an expression for the total transport of heat. The total transport is then related to other variables by means of an overall (or macroscopic) balance.

Consider the volume lying between the planes $x = x$ and $x = x + \Delta x$, and let the area of the planes be A_x (Fig. 1-23). At $x = x$ there is a shear force $|F_y||_x = |\sigma_{xy}A_x||_x$ due to the fluid *below* the element exerting a frictional pull on the fluid *within* the element. As we have seen earlier, this force tends to speed up the fluid in the element and results in a transport of momentum *into* the element at a rate

$$\dot{P}_y\Big|_x = F_y g_c\Big|_x$$

where $F_y|_x$ is the force acting at x *on* the fluid *in* the element.

At $x + \Delta x$ the fluid *in* the element is likewise exerting a force $F'_y|_{x+\Delta x}$ on the fluid *above* (see Fig. 1-23), where the prime indicates that it is the force of the fluid *in* the system *on* the surroundings that we are concerned with. Newton's *third law* says that for every action there is an equal and opposite reaction; therefore, the fluid *above* the element must be exerting a force $-F'_y|_{x+\Delta x} = F_y|_{x+\Delta x}$ on the fluid *in* the element. This force is a frictional *drag* on the fluid in the element that tends

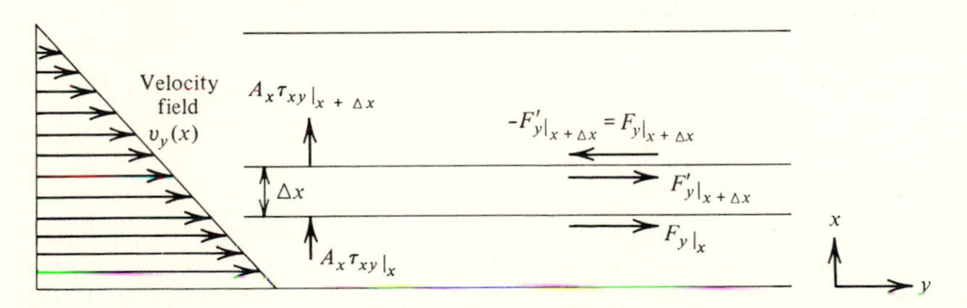

Figure 1-23 Relationship between forces and momentum flows.

to *slow down* the fluid in the element and results in a transport of momentum *out of* the element at a rate $\tau_{xy}A_x|_{x+\Delta x}$. As we have seen above, the action of the shear stresses on the element can be included either as forces in a force balance or as momentum flows† in a momentum balance. Note that there are no other net forces acting in the y direction; e.g., the pressure does not vary with y, and gravity does not act in the y direction.

Force Balance

When Eq. (1-38) is treated as a force balance in the y direction *on* the fluid *in* the element, the momentum-flow term is taken as zero and we have for the total or net force in the y direction

$$F_{ty} \equiv \sum F_y = 0 \tag{1-51a}$$

where the summation is done over all forces acting along the y axis, F_y *being positive if the force physically acts in the* $+y$ *direction and negative if it acts in the* $-y$ *direction.* Then the total force is

$$F_{ty} = F_y\bigg|_x + F_y\bigg|_{x+\Delta x} = 0 \tag{1-51b}$$

Thus

$$F_y\bigg|_x = -F_y\bigg|_{x+\Delta x} \tag{1-51c}$$

i.e., the forces at x and $x + \Delta x$ must have opposite signs, and one of them must be negative. Instead of dealing with negative forces, it is customary in some elementary courses (and in much of engineering practice) to write the force balance in terms of the magnitudes of the forces and then supply the signs on physical grounds. Thus one would write

$$\sum\left|\text{positive forces}\right| - \sum\left|\text{negative forces}\right| = 0 \tag{1-51d}$$

or in our example

$$|F_y|\bigg|_x - |F_y|\bigg|_{x+\Delta x} = 0 \tag{1-51e}$$

Relating the above to the shear stress by means of Eq. (1-39), we obtain

$$\sigma_{xy}A_x\bigg|_x - \sigma_{xy}A_x\bigg|_{x+\Delta x} = 0 \tag{1-52}$$

Momentum Balance

If Eq. (1-38) is written as a momentum balance, we exclude the force terms and write the steady-state momentum rate balance as

†We will see in Chaps. 5 and 9 that (in this example) there is no *net* inflow of bulk momentum in the y direction because v_y does not depend on y.

$$\dot{P}_{yt} = \sum \dot{P}_y = 0 \tag{1-53a}$$

where $\dot{P}_y$ *is positive if y momentum is gained by the system and negative if it is lost.* Then for the increment the total or net rate of gain is

$$\dot{P}_{yt} \equiv \dot{P}_y \Big|_x + \dot{P}_y \Big|_{x+\Delta x} = 0 \tag{1-53b}$$

which shows that if momentum is flowing *into* the element at x (that is, $\dot{P}_y > 0$), it must be flowing *out of* the element at $x + \Delta x$ ($\dot{P}_y < 0$).

However, if we use magnitudes we write

$$|\dot{P}_y| \Big|_x - |\dot{P}_y| \Big|_{x+\Delta x} = 0 \tag{1-53c}$$

Using the definition [Eq. (1-40)] that flux = flow per unit area gives

$$\underset{\text{In}}{\tau_{xy} A_x \Big|_x} - \underset{\text{Out}}{\tau_{xy} A_x \Big|_{x+\Delta x}} = 0 \tag{1-54}$$

Note that all the above momentum-balance equations could have been obtained by multiplying the corresponding force-balance equations by g_c. But in writing the momentum balance one attaches to it the physical significance "in minus out," whereas in writing the force balance one attaches the significance "positive y direction minus negative y direction."

Derivation of Velocity-Profile Equation

Upon division through by A_x, which is constant, Eqs. (1-54) and (1-52) become, respectively,

$$\tau_{xy} \Big|_x - \tau_{xy} \Big|_{x+\Delta x} = 0 \tag{1-55}$$

and

$$\sigma_{xy} \Big|_x - \sigma_{xy} \Big|_{x+\Delta x} = 0 \tag{1-56}$$

Since $g_c \sigma_{xy} = \tau_{xy}$, division of Eq. (1-55) by g_c yields Eq. (1-56) and thus these results are equivalent, as they should be. Dividing Eq. (1-55) by Δx and taking the limit as $\Delta x \to 0$, we obtain

$$\lim_{\Delta x \to 0} \frac{\tau_{xy} \Big|_x - \tau_{xy} \Big|_{x+\Delta x}}{\Delta x} = -\frac{d\tau_{xy}}{dx} = 0 \tag{1-57}$$

and so

$$\tau_{xy} = \text{const} \equiv \tau_1 \qquad \text{say} \tag{1-58}$$

Therefore the flux *distribution* is constant.

From Newton's law of viscosity (for newtonian fluids), Eq. (1-50), the flux law is

$$\tau_{xy} = -\mu \frac{dv_y}{dx}$$

Equating the flux-law and flux-distribution equations and rearranging gives

$$\frac{dv_y}{dx} = -\frac{\tau_1}{\mu}$$

Using the boundary conditions that $v_y = V$ at $x = 0$ and integrating between $x = 0$ and arbitrary x, we obtain

$$\int_V^{v_y} dv_y = -\frac{\tau_1}{\mu} \int_0^x dx$$

or

$$v_y - V = \frac{-\tau_1}{\mu} x \tag{1-59}$$

which shows that the velocity profile is linear. Substituting the boundary condition that $v_y = 0$ at $x = L_x$, we get

$$-V = \frac{-\tau_1}{\mu} L_x \tag{1-60}$$

or

$$\tau_1 = \mu \frac{V}{L_x} \tag{1-61}$$

Since $\tau_1 = \tau_{xy} = \sigma_{xy} g_c$, this is equivalent to Eq. (1-46), obtained earlier by heuristic methods. We can eliminate τ_1 from the velocity profile by dividing Eq. (1-59) by Eq. (1-60). Then

$$\frac{V - v_y}{V} = \frac{x}{L_x} \qquad \text{or} \qquad v_y = V\left(1 - \frac{x}{L_x}\right) \tag{1-62}$$

which is analogous to the linear temperature profile obtained in Part A [Eq. (1-10*d*)].

Overall Balance

For heat transport in a slab we obtained $\dot{Q}_1$ and $\dot{Q}_2$, the total transport of heat entering and leaving the slab, and showed that they were equal at steady state by using an overall balance. In order to obtain the analogous quantities for momentum transport we must multiply the flux at each of the boundaries by the area at that boundary. Then Eqs. (1-40) and (1-61) give

$$\dot{P}_1 \equiv |\dot{\mathbf{P}}_y|\bigg|_{x=0} = |\tau_{xy} A_x|\bigg|_{x=0} = \frac{\mu V}{L_x} A_x \tag{1-63}$$

Just as heat was transported from the high temperature T_1 at the bottom of the slab to the lower temperature T_2 at the top of the slab, so also is y momentum transported from the high velocity V at the lower plate through the fluid to the upper plate, where the velocity is zero. Although it is easy to conceive of heat leaving the slab at the top, it is often difficult for the student to conceive of momentum as leaving the system. But that is precisely what happens as a result of the drag of the upper plate on the fluid.†

The momentum leaving at the upper plate is then [Eqs. (1-40) and (1-61)]

$$|\dot{P}_2| \equiv |\dot{P}_y|\Big|_{x=L_x} = |\tau_{xy}A_x|\Big|_{x=L_x} = \frac{\mu V}{L_x}A_x \tag{1-64}$$

which, as required, is equal to the momentum entering at $x = 0$.

Example 1-3: Numerical calculation In a lubrication system, an oil with a viscosity of 10.0 cP (10^{-2} N · s/m² = 10^{-2} Pa · s) is confined between two metal surfaces, which can be assumed flat and parallel. The lower surface or plate is being pulled at steady state with a force of 0.1 lbf (0.4448 N) and has an area of 50 ft² (4.645 m²). The distance between plates is 0.12 in (3.048×10^{-3} m). Calculate (*a*) the shear stress at the lower surface in N/m² and lbf/ft² and (*b*) the momentum flux entering the lower surface in units involving pound mass, feet, and seconds. Give the actual units. (*c*) Convert the momentum flux in part (*b*) into SI units involving kilograms, meters, and seconds. Give the actual units and compare with part (*a*). (*d*) Calculate the velocity of the lower plate if the upper surface is kept fixed. Use units of meters per second and feet per second. Repeat part (*d*) if the oil is replaced by (*e*) water at 20°C and (*f*) air at 20°C and all other quantities are the same. (*g*) Sketch the velocity-versus-position profiles for parts (*d*) to (*f*).

SOLUTION (*a*)

$$\frac{\text{Force}}{\text{Area}} = \frac{F_x}{A_x} = \sigma_{xy} = \frac{0.4448\ \text{N}}{4.645\ \text{m}^2} = 0.0958\ \text{N/m}^2$$

$$= \frac{0.1\ \text{lbf}}{50\ \text{ft}^2} = 2 \times 10^{-3}\ \text{lbf/ft}^2$$

(*b*)

$$\tau_{xy} = \sigma_{xy} g_c (2 \times 10^{-3}\ \text{lbf/ft}^2)\frac{32.17\ \text{lbm} \cdot \text{ft}}{\text{lbf} \cdot \text{s}^2}$$

$$= 6.434 \times 10^{-2}\ \text{lbm/ft} \cdot \text{s}^2$$

(*c*) Using Table A-1 gives

$$\tau_{xy} = \frac{0.08434\ \text{lbm}}{\text{ft} \cdot \text{s}^2}\frac{0.454\ \text{kg}}{\text{lbm}}\frac{1\ \text{ft}}{0.3048\ \text{m}} = 0.0958\ \text{kg/m} \cdot \text{s}$$

†The fact that the velocity is zero merely means that we have chosen the upper plate as our reference; it does *not* mean that no momentum is being transported, since the rate of momentum transport depends on the velocity *gradient,* not on the velocity.

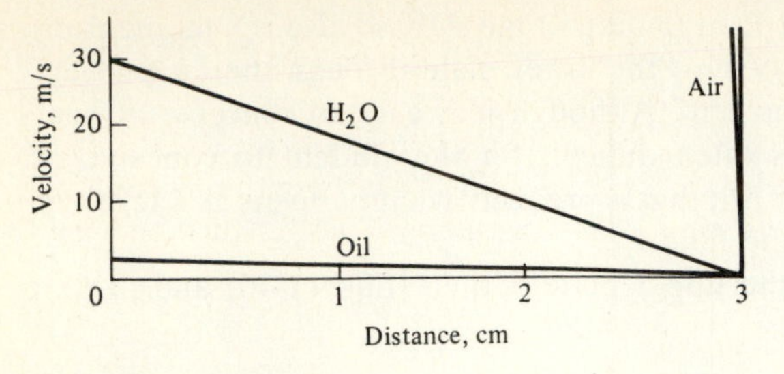

Figure 1-24 Effect of viscosity on velocity profiles in Couette flow.

In part (*a*) we also obtained $\sigma_{xy} = 0.0958\ \text{N/m}^2$, as we should since in the SI $\tau_{xy} = \sigma_{xy}$. Also, since $1\ \text{N} = 1\ \text{kg} \cdot \text{m/s}^2$, the units can be shown to be identical.

(*d*) From Eq. (1-46)

$$g_c\sigma_{xy} = \mu\frac{V}{L_x} \qquad \text{or} \qquad V = \frac{\tau_{xy}L_x}{\mu}$$

$$\mu = 10\ \text{cP} = 0.1\ \text{g/cm} \cdot \text{s} = 10(6.72 \times 10^{-4})$$

$$= 6.72 \times 10^{-3}\ \text{lbm/ft} \cdot \text{s}$$

$$L_x = 0.01\ \text{ft}$$

$$V_{\text{oil}} = \frac{(6.44 \times 10^{-2}\ \text{lbm/ft} \cdot \text{s})(0.01\ \text{ft})(60\ \text{s/min})}{6.72 \times 10^{-3}\ \text{lbm/ft} \cdot \text{s}}$$

$$= 5.75\ \text{ft/min} = \frac{30.48}{60}(5.75) = 2.92\ \text{m/s}$$

(*e*) $\qquad \mu\ (\text{at } 20°\text{C}) = 1\ \text{cP} \qquad$ (Table 1-3)

$$V_{H_2O} = V_{\text{oil}}\frac{\mu_{\text{oil}}}{\mu_{H_2O}} = 5.75(\tfrac{10}{1}) = 57.5\ \text{ft/min} = 29.2\ \text{m/s}$$

(*f*) $\qquad \mu\ (\text{at } 20°\text{C}) = 0.018\ \text{cP} \qquad$ (Table 1-3)

Then
$$V_{\text{air}} = V_{\text{oil}}\frac{\mu_{\text{oil}}}{\mu_{\text{air}}} = 5.75\frac{10.0}{0.018} = 3194\ \text{ft/min} = 1623\ \text{m/s}$$

Note in Fig. 1-24 that, as in the case of heat transport, *the slope is steeper for a poor "conductor" of momentum (air) and flatter for a good conductor of momentum (oil).* □

Sign Convention for Incremental Momentum and Force Balances

In Sec. 1-6 a sign convention for heat flux was adopted in which q_x is taken as positive if heat is physically flowing in the $+x$ direction and as negative if it is flowing in the $-x$ direction. We use a similar convention for momentum transport. Let us again consider the case of Couette flow (Fig. 1-10), in which one of the plates is pulled while the other plate is held stationary. For generality we will let the velocity of the lower plate be v_{y1} and that of the upper plate be v_{y2}. We again consider two cases: (1) $v_{y1} \equiv V_1$ and $v_{y2} = 0$ (pull lower plate) (Fig. 1-23) and (2) $v_{y1} = 0$ and $v_{y2} \equiv V_2$ (pull upper plate) (Fig. 1-25).

In case 1, dv_y/dx is negative and τ_{xy} is positive, whereas in case 2, dv_y/dx is positive and τ_{xy} is negative. However, for case 2 (Fig. 1-25) momentum is entering at $x + \Delta x$ and leaving at x. Thus, since $\tau_{xy} < 0$, our momentum balance becomes

$$\underset{\text{In}}{-\tau_{xy}A_x\Big|_{x+\Delta x}} - \underset{\text{Out}}{(-\tau_{xy}A_x)\Big|_x} = 0 \tag{1-65}$$

which is the same as Eq. (1-37) (obtained for case 1) except that the terms have been rearranged and the in and out terms are at opposite locations. Of course, the flux differential equation would also be the same. Consequently, as was the case for heat transport, the *flux differential equations* will be the same regardless of the sign of the flux and *can be derived by assuming the flux τ_{xy} to be positive.* The actual sign will be determined by the boundary conditions, in this case by whether

$$\underset{\text{Case 1}}{v_{y1} > v_{y2}} \qquad \text{or} \qquad \underset{\text{Case 2}}{v_{y2} > v_{y1}}$$

Similarly, since σ_{xy} is negative for case 2, the force balance would be written

$$\underset{\substack{\text{Force in}\\ +y \text{ direction}}}{-\sigma_{xy}A_x\Big|_{x+\Delta x}} - \underset{\substack{\text{Force in}\\ -y \text{ direction}}}{(-\sigma_{xy}A_x)\Big|_x} = 0 \tag{1-66}$$

which is identical to Eq. (1-52) obtained in case 1. Therefore we again find that *if we write our incremental and differential force balances as though the flux were positive, the resulting equations will apply to negative as well as to positive fluxes.*

We should note at this time, as we did for heat transport, that it would not be necessary to consider the physical direction of the forces or the direction of momentum transport at all if we wrote the balance in the general form of Eq. (1-51*a*)

$$F_y\Big|_x + F_y\Big|_{x+\Delta x} = 0 \tag{1-67}$$

which applies to case 1 or 2. Also, for either case, we would use the equations

$$F_y\Big|_x \equiv \sigma_{xy}A_x\Big|_x \tag{1-68}$$

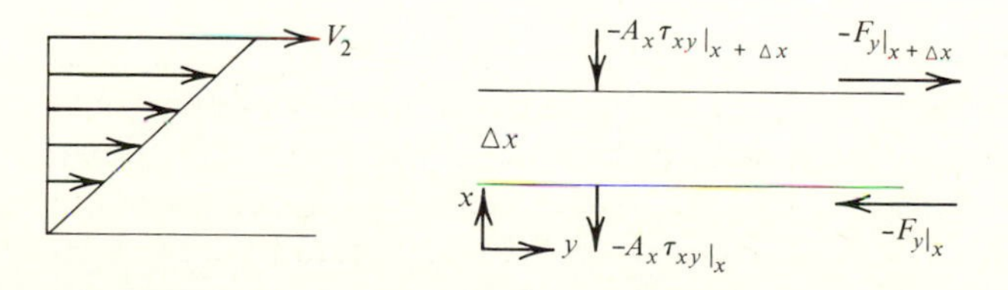

Figure 1-25 Momentum flows and forces for $\tau_{xy} < 0$.

and
$$F_y\Big|_{x+\Delta x} \equiv -\sigma_{xy}A_x\Big|_{x+\Delta x} \tag{1-69}$$

Note that since $F_y|_{x+\Delta x} = -F_y|_x$, it is necessary to use a minus sign at $x + \Delta x$ to maintain the same sign on σ_{xy}.

If we multiply the above equations by g_c, we obtain

$$\dot{P}_y\Big|_x + \dot{P}_y\Big|_{x+\Delta x} = 0 \tag{1-70a}$$

$$\dot{P}_y\Big|_x \equiv \tau_{xy}A_x\Big|_x \tag{1-70b}$$

$$\dot{P}_y\Big|_{x+\Delta x} \equiv -\tau_{xy}A_x\Big|_{x+\Delta x} \tag{1-70c}$$

Stress Definition Used in Fluid Dynamics

Equations (1-68) and (1-69) define σ_{xy} and Eqs. (1-70*b*) and (1-70*c*) define τ_{xy} at x and at $x + \Delta x$ such that $\tau_{xy} = g_c\sigma_{xy}$ will be positive when the velocity gradient is negative, thereby resulting in a negative sign in Newton's law of viscosity [Eq. (1-48)] and making that law analogous to Fourier's law of heat conduction. However, most textbooks in fluid dynamics use a shear stress T_{xy} that is the negative of the σ_{xy} used here. Thus they write Newton's law of viscosity for steady-state flow in one direction [Eq. (1-48)] in the form†

$$-\sigma_{xy} \equiv T_{xy} = \frac{\mu}{g_c}\frac{dv_y}{dx} \tag{1-71a}$$

The sign of T_{xy} would therefore be positive for case 2 and negative for case 1, and Eq. (1-69) becomes

$$F_y\Big|_{x+\Delta x} \equiv T_{xy}A_x\Big|_{x+\Delta x} \tag{1-71b}$$

Total Rate of Transport at a Surface

Since, according to Eq. (1-3), flux is flow rate per unit area, the total rate of flow at a surface is the product of the flux at the surface times the area of the surface. For momentum transport between parallel plates we then obtain

$$\dot{P}_y\Big|_{x=0} = \tau_{xy}A_x\Big|_{x=0} \quad \text{and} \quad \dot{P}_y\Big|_{x=L_x} = -\tau_{xy}A_x\Big|_{x=L_x} \tag{1-72}$$

or, in terms of forces,

$$F_y\Big|_{x=0} = \sigma_{xy}A_x\Big|_{x=0} \quad \text{and} \quad F_y\Big|_{x=L_x} = -\sigma_{xy}A_x\Big|_{x=L_x} \tag{1-73}$$

†Many texts transpose the subscripts.

These equations can be applied to curvilinear coordinates by replacing x and y with the corresponding coordinates. When σ_{xy} and τ_{xy} vary over A_x, we must use dA_x, dF_y, and $d\dot{P}_y$ (see Chap. 9), as discussed in Sec. 1-7 for heat transport.

1-19 MATHEMATICAL EXPRESSION FOR STRESS AND MOMENTUM FLUX†

Equations (1-68) and (1-69) indicate that the expression for F_y at $x + \Delta x$ differs from that at x by a negative sign. Therefore it is again desirable (as in Sec. 1-7) to develop a general expression that will apply at *either* x or $x + \Delta x$. At each surface x and $x + \Delta x$ we again construct an outwardly directed unit normal n_x such that $n_x|_{x+\Delta x}$ points in the *same* direction as that chosen for the x axis, whereas $n_x|_x$ points in the opposite direction. Therefore, if we let

$$n_x \Big|_{x+\Delta x} = +1 \qquad \text{and} \qquad n_x \Big|_x = -1$$

we can write the following general equations that apply at either surface for either positive or negative flux or stress:

$$F_y = -n_x A_x \sigma_{xy} = -S_x \sigma_{xy} \tag{1-74}$$

or

$$\dot{P}_y = -n_x A_x \tau_{xy} = -S_x \tau_{xy} \tag{1-75}$$

Note that σ_{xy} can be *defined* as the force per unit area acting in the $+y$ direction on a *negative* surface of area A_x and the stress T_{xy} can be defined in the same way but for a *positive* surface. The above equations can be used for other coordinates, e.g., the radial coordinate r in a cylindrical coordinate system, by replacing x with r.

1-20 A PREVIEW OF SCALAR, VECTOR, AND TENSOR QUANTITIES

When we specify heat flux, velocity, and momentum, we must use a subscript to indicate direction. This directional quality is one of the properties of a vector. Although the mathematical definition of a vector is more complex, for present purposes we can think of a vector as follows.

Physical definition A vector is a quantity for which not only magnitude but also direction must be specified.

In the case of temperature, only magnitude is involved: temperature does not have direction (although its *gradient* does). We call such quantities *scalars*.

†This section may be omitted on first reading.

Physical definition A scalar is a quantity for which only magnitude must be specified.

Momentum flux is a still different quantity because we must specify not only magnitude but *two* directions. As with heat transport, one direction, say x, gives the direction of transport (which is the same as the area the momentum flows through). But since momentum is a vector, we must also specify the direction of the momentum itself, which, since momentum is mass times velocity, is that of the velocity, say y. The momentum flux is then said to be a second-order tensor. We can state for our purposes the following physical definition.

Physical definition A second-order tensor is a quantity for which not only magnitude but also *two* directions must be specified.

This definition can be generalized to include tensors of higher or lower order; e.g., a vector is a first-order tensor and a scalar is a tensor of zero order. If the order of a tensor is not indicated otherwise, the term "tensor" is taken to mean second-order tensor. The student should *not* conclude that *all* quantities that have magnitude and direction are vectors or that all quantities that have magnitude and two directions are second-order tensors. For a discussion of the definition of a vector, see Aris[1]† or Chap. 8. Rigorously a tensor (or vector) is defined in terms of its mathematical properties (Appendix 8-7).

1-21 CURVILINEAR COORDINATES

It has been pointed out that the analogy between heat and momentum transport cannot be complete or general since momentum is a vector and temperature a scalar. This difference between heat and momentum transport usually shows up in two-dimensional flow—and even in one-dimensional systems when coordinate systems other than cartesian are used. For example, in cylindrical coordinates r, θ, z the momentum flux $\tau_{r\theta}$ for a newtonian fluid is *not* given by an equation of the form of Eq. (1-49) for τ_{xy} but (Chap. 9) by

$$\tau_{r\theta} = -\mu\left[\frac{\partial(v_\theta/r)}{\partial r}r + \frac{\partial v_r}{r\,\partial\theta}\right]$$

Thus, even when $v_r = 0$, the expression for $\tau_{r\theta}$ is of a different form than for cartesian coordinates. On the other hand the component

$$\tau_{rz} = -\mu\left(\frac{\partial v_r}{\partial z} + \frac{\partial v_z}{\partial r}\right)$$

follows the same relationship as the expression for τ_{xy} in cartesian coordinates. Furthermore, if $v_z = v_z(r)$ and $v_r = v_\theta = 0$, the transport of z momentum in the r

†Numbered references appear at the end of the chapter.

direction is described by

$$\tau_{rz} = -\mu \frac{\partial v_z}{\partial r}$$

which is similar to the equation for τ_{xy} for transport of y momentum in the x direction. Therefore an analogy between heat transport in the r direction and z-momentum transport in the r direction is possible.

Expressions for the momentum flux for cartesian, cylindrical, and spherical coordinates are discussed in Chap. 9. Inspection of the relationships for newtonian fluids in Table I-9 indicates that for other cylindrical coordinate components as well as for other coordinate systems (including spherical) an analogy to heat transport is *not* generally valid. Nevertheless, many practical engineering problems can be handled in cartesian coordinates or in special cases in cylindrical coordinates such that an analogy does hold. Since the use of an analogy can result in a greater understanding of the physical processes occurring, this pedagogical device is used in the first part of the text. The student is cautioned, however, against generalizing to other components or other systems and should refer to Table I-9 for the correct form if needed. A case for which the analogy *does* hold is discussed below.

Example 1-4: Momentum transport in an annulus Consider a horizontal metal piston of radius R_1 and length L moving through a lubricating oil contained in the annular† space between the piston and a concentric cylinder of radius R_2 (Fig. 1-25). It is desired to find the force F_z required to pull the piston through the oil with a constant velocity v_1 with no pressure difference or gravity forces. End effects will be neglected in this preliminary analysis.

SOLUTION Let r be the distance from the center of the piston and z the distance along the piston axis. As the piston moves in the z direction, we can observe that the fluid will have a velocity v_z. Now the velocity at the piston surface ($r = R_1$) must equal v_1 because the fluid will cling to the wall. For the same reason, the velocity of the fluid at the wall of the outer cylinder must be $v_z = 0$ since the outer cylinder is stationary. Therefore the velocity of the fluid will vary with radial position r from $v_z = v_1$ to $v_z = 0$. This radial variation means that the piston is imparting z momentum to the fluid and this momentum is being transported radially through successive fluid layers to the outer cylinder. Thus we have a situation similar to the parallel-plate example in Sec. 1-13 except that we must now work in cylindrical coordinates. Actually the problem is directly analogous to the problem of heat transport in a pipe discussed in Sec. 1-8, as we can show by listing the results in parallel columns.

First we write a z-momentum balance over the annular element (Fig. 1-26) within the fluid formed by concentric cylinders of radii r and $r + \Delta r$. Since the

†An annular region consists of the region between two concentric circular cylinders that have a common axis. The inside wall of the outer cylinder and the outside wall of the inner cylinder are the walls of an annulus.

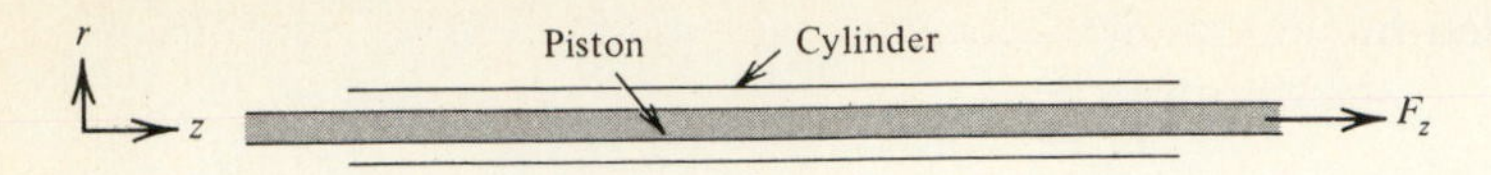

Figure 1-26 Piston and cylinder.

pressure is constant and the piston is horizontal, there are no gravity or pressure forces in the z direction. The steady-state momentum balance can be written (see Fig. 1-27) as

$$\tau_{rz}A_r\Big|_r - \tau_{rz}A_r\Big|_{r+\Delta r} = 0 \tag{1}$$

Rate of momentum flow in at r	rate of momentum flow out at $r + \Delta r$

An alternative is to make a force balance in the z direction. From Eq. (1-39) the magnitude of the force at radial position r is $|F_z||_r = |\sigma_{rz}A_r||_r$ while at $r + \Delta r$ it is $|F_z||_{r+\Delta r} = |\sigma_{rz}A_r||_{r+\Delta r}$, which acts in the z direction. The sum of the forces is zero, or

$$0 = \sigma_{rz}A_r\Big|_r - \sigma_{rz}A_r\Big|_{r+\Delta r} \tag{2}$$

We note that if we multiply Eq. (2) by g_c, we obtain Eq. (1).

If we substitute $A_r = 2\pi rL$ into Eq. (2), divide by $2\pi rL\,\Delta r$, and take the limit as $\Delta r \to 0$, we obtain the flux differential equation

$$\lim_{\Delta r \to 0} \frac{\tau_{rz}r|_r - \tau_{rz}r|_{r+\Delta r}}{r\,\Delta r} = -\frac{d(\tau_{rz}r)}{r\,dr} = 0 \tag{3}$$

Integration gives $\tau_{rz}r = \text{const} \equiv C_1$. Then the flux distribution is

$$\tau_{rz} = \frac{C_1}{r} \tag{4}$$

Newton's law of viscosity is

$$\tau_{rz} = -\mu\frac{dv_z}{dr} \tag{5}$$

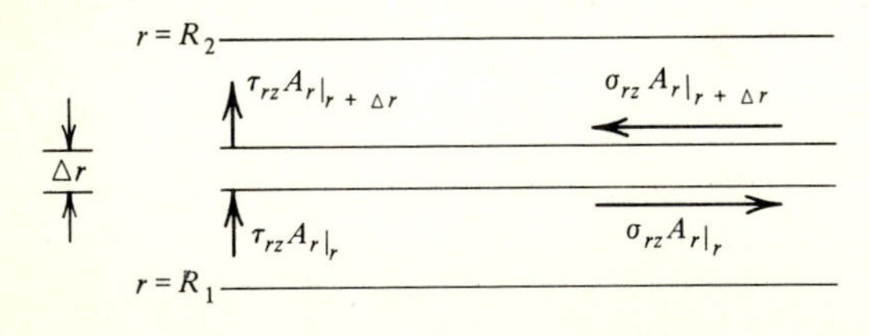

Figure 1-27 Forces and momentum flows in annular increment.

The no-slip assumption states that the velocity of the fluid at each solid surface is equal to the velocity of the surface. Thus

$$v_z = \begin{cases} v_1 & \text{at } r = R_1 \\ 0 & \text{at } r = R_2 \end{cases} \tag{6}$$

To find the velocity distribution we equate the expressions for the flux obtained in (4) and (5), separate variables, and integrate using the boundary conditions. We should note, however, that the differential equation, flux law, and boundary condition are all analogous to the case of heat transport in a metal wall discussed previously if we let $T_2 = 0$. Therefore we can save the effort of integration by making use of this analogy. To do so we replace T with v_z, T_1 with v_1, and T_2 with 0. The equation for the velocity distribution is then given by analogy to Eq. (1-30) as

$$\frac{v_1 - v_z}{v_1} = \frac{\ln (r/R_1)}{\ln (R_2/R_1)}$$

The rate of momentum flow $\dot{P}_1$ is given by the product of the momentum flux and the flow area, both evaluated at $r = R_1$. It is analogous to the rate of heat flow $\dot{Q}_1$ and therefore could be obtained by analogy to Eq. (1-31). However, we will carry out the steps in order to emphasize the procedure.

Since the momentum flow is constant, it can be evaluated at $r = R_1$, $r = R_2$, or arbitrary r. Selecting $r = R_1$ and using Newton's law of viscosity, we have

$$\dot{P}_1 = |\tau_{rz} A_r| \Big|_{r=R_1} = \left(-\mu \frac{dv_z}{dr} \Big|_{r=R_1} \right)(2\pi R_1 L)$$

$$= \mu \left[\frac{v_1}{r \ln (R_2/R_1)} \right] \Bigg|_{r=R_1} (2\pi R_1 L)$$

Then

$$\dot{P}_1 \equiv |\dot{P}_z| \Big|_{r=R_1} = \frac{\mu v_1}{R_1 \ln (R_2/R_1)}(2\pi R_1 L) = \frac{2\pi L v_1 \mu}{\ln (R_2/R_1)}$$

The force with which the piston must be pulled is

$$F_1 = \frac{\dot{P}_1}{g_c} = \frac{2\pi \mu L v_1}{g_c \ln (R_2/R_1)}$$

Note that we can write this equation in the form of Eq. (1-64) by using the log-mean area A_m as defined previously; then

$$g_c F_1 = \mu A_m \frac{v_1}{R_2 - R_1}$$

The various quantities are compared in Prob. 1-8. □

1-22 RELATION BETWEEN NEWTON'S LAW OF VISCOSITY AND HOOKE'S LAW OF ELASTICITY

We have indicated that many fluid dynamicists prefer writing Newton's law of viscosity for one-dimensional flow in terms of the shear stress

$$T_{xy} = +\frac{\mu}{g_c}\frac{dv_y}{dx}$$

We will try to find another meaning for the terms dv_y/dx. To do so we first consider an elastic solid of thickness Δx that is subjected to a shear force $F_y = T_{xy}A_x$ on its *upper* face (Fig. 1-28*a*). As a result, the solid will be deformed a distance of Δs_y (Fig. 1-28*b*). If we define the strain γ_{xy} as the deformation per unit length, it is given by

$$\frac{\Delta s_y}{\Delta x} = \gamma_{xy} \qquad \text{or, in the limit,} \qquad \gamma_{xy} \equiv \frac{\partial s_y}{\partial x}$$

Hooke's law of elasticity states that the shear stress is proportional to the strain or

$$T_{xy} = G\gamma_{xy} \tag{1-76}$$

where G is *Hooke's-law constant.*

A fluid differs from a solid in that it is "deforming" continuously. Therefore the shear stress for a newtonian fluid is postulated as being proportional, not to the strain, but to the *rate* of strain

$$\dot{\gamma}_{xy} = \frac{d\gamma_{xy}}{dt} = \frac{d}{dt}\left(\frac{\partial s_y}{\partial x}\right)$$

If we change the order of differentiation, we get

$$\frac{\partial}{\partial x}\left(\frac{ds_y}{dt}\right) = \frac{\partial v_y}{\partial x} \qquad \text{since} \qquad v_y \equiv \frac{ds_y}{dt}$$

If we now let the proportionality constant be μ/g_c, we have

$$T_{xy} = \frac{\mu}{g_c}\frac{\partial v_y}{\partial x}$$

The term $\partial v_y/\partial x$ is often called the *rate of deformation* or *rate of strain* and given

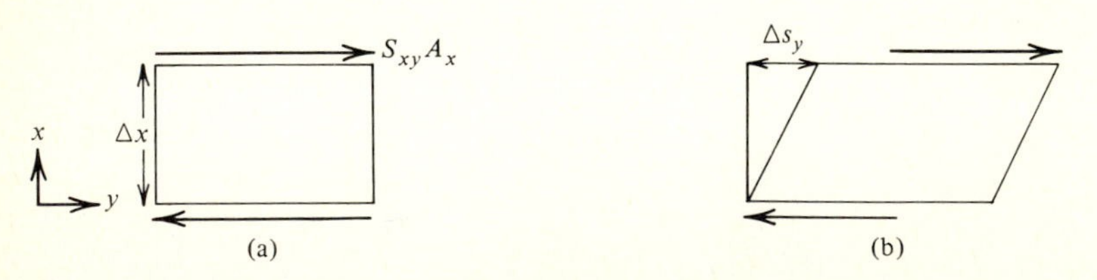

Figure 1-28 Elastic-solid deformation: (*a*) before and (*b*) after.

the symbol $\dot{\gamma}_{xy}$ or Δ_{xy}. It is a component of a tensor since two directions are involved.

We will find (Chap. 9) that for flow in more than one direction

$$\Delta_{xy} = \frac{\partial v_y}{\partial x} + \frac{\partial v_x}{\partial y} \tag{1-77}$$

and

$$T_{xy} g_c = -\tau_{xy} = +\mu \left(\frac{\partial v_y}{\partial x} + \frac{\partial v_x}{\partial y} \right) \tag{1-78}$$

for a newtonian fluid.

As pointed out in Sec. 1-22, the model used in fluid mechanics texts differs from that used in transport phenomena texts (see Fig. 1-18) in that in fluid mechanics the *upper* plate is taken to be moving while the lower plate is held stationary. This results in a positive T_{xy} instead of a negative one. Many texts define a rate-of-strain tensor by

$$\tfrac{1}{2}\Delta_{xy} \equiv d_{xy} \tag{1-79}$$

In summary we can write Newton's law for one-dimensional flow as

$$\boxed{-T_{xy} g_c \equiv \sigma_{xy} g_c = \tau_{xy} = -\mu \frac{\partial v_y}{\partial x} = -\mu \Delta_{xy}} \tag{1-80}$$

Nonnewtonian Fluid

A newtonian fluid can be described as one in which the shear stress is directly proportional to the deformation rate, i.e., one in which the viscosity is constant no matter what the deformation rate or the stress. A plot of σ_{xy} versus Δ_{xy} then gives a straight line for a newtonian fluid but deviates from a straight line for a nonnewtonian fluid (see Fig. 1-29). For these cases the power-law equation is often used

$$\sigma_{xy} g_c \equiv \tau_{xy} = -m \left| \frac{dv_y}{dx} \right|^{n-1} \frac{dv_y}{dx} \tag{1-80a}$$

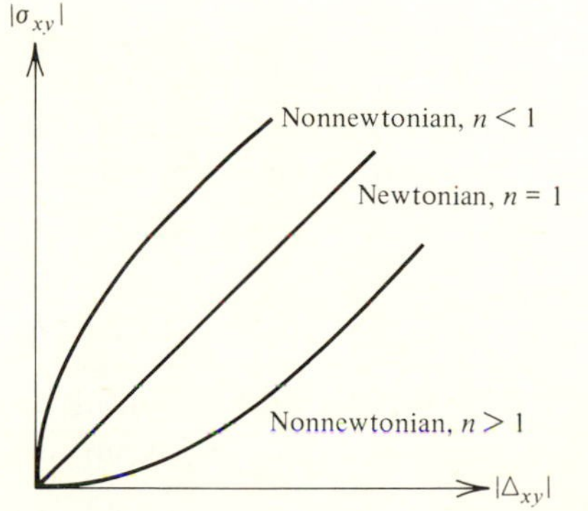

Figure 1-29 Shear stress versus deformation rate for newtonian and nonnewtonian fluids.

where $n = 1$ for a newtonian fluid, $n > 1$ for a dilatant or shear-thickening fluid, $n < 1$ for a pseudo-plastic or shear-thinning fluid, and m is a constant. See Chaps. 2 and 9 for other models.

C. MASS TRANSPORT IN ONE DIRECTION

1-23 TYPES OF MASS TRANSPORT

Examples of mass transport are plentiful: the diffusion of smoke and other pollutants in the atmosphere; the transfer of a solute between phases in a gas absorber, extractor, or cooling tower; the mixing of cream or sugar in a cup of coffee; the drying of clothes (diffusion of water vapor in air); and the O_2-CO_2 exchange in the lungs.

Suppose we drop a few crystals of sugar into a cup of coffee and let them dissolve. The dissolved sugar molecules will slowly diffuse from the high-concentration region at the bottom, and the concentration will tend to become uniform throughout. This form of diffusion is due to the random motion of the molecules and is called *molecular diffusion*. On the other hand, if we release a stream of smoke from the stack of a power plant on a windy day and let the smoke disperse through the atmosphere, the process is due to the fluctuations in wind velocity and is called dispersion or *turbulent diffusion*. In this chapter, we deal only with molecular diffusion. A mechanism of molecular diffusion will be discussed in Chap. 2, and turbulent diffusion will be discussed in Chap. 6.

As in the case of heat transport, mass transport may occur both by *diffusion* and by *convection*, the latter representing the transport of mass that results from a bulk-fluid motion, and the former representing the transport that occurs as a result of a concentration gradient. Again, as in heat transport, convective mass transport consists of two types: *forced* convection, in which the velocity is imparted by an external source, and *free* convection, a buoyancy effect in which the velocity develops naturally as a result of changes in density developing from concentration differences in the medium.

We discuss convective mass transport in more detail in Chap. 5, but the student should be aware of the fact that diffusion and convection generally occur together; i.e., if one species is diffusing through another, the fluid *mixture* will, in general, have an *average* velocity other than zero. However, for dilute solutions this convective effect will be small. We will make that assumption in this chapter.

1-24 DIFFUSIVE AND CONVECTIVE MOLAR FLUXES

In order to define some of the terms used in the study of diffusion, let us consider a simple example, the leaching or extraction of a solute A from a solid surface which is in contact with a solvent B. For example, the solid surface may be the wall of a duct through which water is flowing, and A may be a slightly soluble substance which coats the wall. We will assume that the concentration C_{A1} of A in the solvent at the surface is equal to its solubility in B and that the water contains a small amount of solute A (see Fig. 1-30*a*). Thus there is a concentration difference between the solid and the fluid, and experience tells us that consequently there will be diffusion of A from the wall into the main or bulk stream of liquid. It is postulated that there exists near the surface a thin film of stagnant fluid. The constant and equal to C_{A2}. The thickness of the film is L_x. We will assume a flat concentration of A at the boundary of the film and the main stream is assumed surface and take x to be the distance from the surface.

At time $t = 0$ the surface is suddenly brought into contact with the solvent, which has an initial concentration C_{A2}. A plot of concentration C_A versus distance from the surface x at various constant times is given in Fig. 1-30*b*. The concentration at the edge of the film, $x = L_x$, is assumed to be constant at C_{A2}. At $t = 0$ there is a sharp drop in concentration at the surface from C_{A1} to C_{A2}. At some arbitrary time $t = t_1$ some A has diffused from the surface, and at time $t = t_2$ it has reached the edge of the film. Eventually a steady state is reached at $t \to \infty$, in which the concentration profile no longer changes with time.

If we were involved in the design of an extraction unit, we would want to know how much A would be transported from the surface during a given period of time. To make such a calculation we would need an expression for the rate of diffusion as a function of C_{A1}, C_{A2}, L_x, and any other variables. To obtain this relationship let us now consider the flow of species A across a plane at an arbitrary location x. Using our general definition of a flux, Eq. (1-3), the total molar flux of

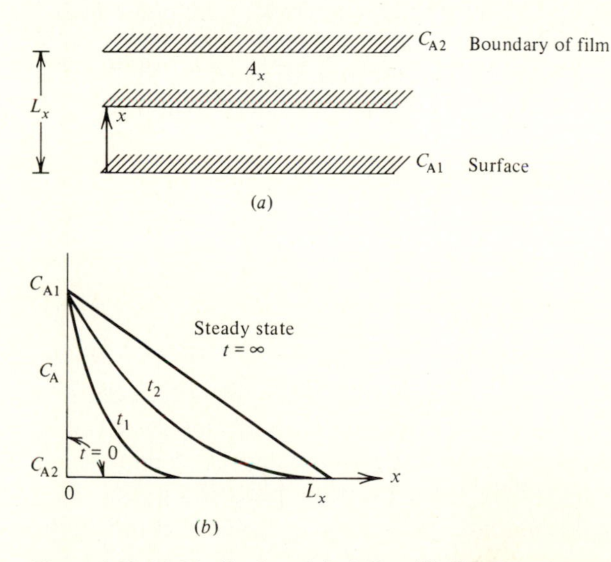

Figure 1-30 (*a*) Idealized model of film. (*b*) Concentration profiles in film.

A across this plane in mol/h · ft² or mol/s · m² is

$$N_{Ax} = \text{total molar flux} = \frac{\text{total flow rate}}{\text{area}} = \frac{\mathcal{W}_{Ax}}{A_x} \tag{1-81}$$

where $\mathcal{W}_{Ax}$ = total rate of flow species A, moles per unit time
A_x = area through which flow occurs

As indicated earlier, *in general,* the transport of A will occur by both diffusion and convection. Hence the total flux is the sum of the diffusive flux and the convective flux. The convective *flow* of A (as will be shown rigorously in Chap. 5) is given by $C_A v_x A_x$, where v_x is the velocity of the stream in the x direction and C_A is the concentration of A in moles per unit volume. The convective *flux* will then be $C_A v_x$. The total flux is then the diffusive flux plus the convective flux or

$$\boxed{N_{Ax} = J_{Ax} + C_A v_x} \tag{1-82}$$

where J_{Ax} is the molar diffusive flux of A (moles of A diffusing per unit time and unit area).†

Since we are assuming dilute solutions, C_A is small and $C_A v_x$ is negligible. Thus, for *dilute solutions*

$$N_{Ax} \approx J_{Ax} \tag{1-83}$$

1-25 FICK'S LAW OF DIFFUSION

In the previous case of energy transport we recognized that the flux law must be written in terms of a temperature gradient (instead of a temperature difference) if it is to apply within an arbitrary medium. For mass transport we can make use of that experience and write an analogous flux law for mass transport in the form of Newton's law of viscosity [Eq. (1-48)] and Fourier's law of heat conduction [Eq. (1-5)]. In general, if

$$\text{Flux} = \text{transport property} \times \text{potential gradient}$$

then for moles of A

$$\boxed{J_{Ax} = -D_{AB}\frac{\partial C_A}{\partial x}} \tag{1-84}$$

where D_{AB}, the transport property, is the binary molecular mass *diffusivity* of species A through B. The above equation, a simplified form of *Fick's first law of diffusion,* holds for ideal dilute binary solutions of constant density. A more rigorous analysis based on the thermodynamics of irreversible processes would show that the correct potential gradient is *not* the concentration gradient but the gradi-

†In Chap. 5 we define v_x and J_{Ax} more rigorously.

Table 1-4 Typical values of binary diffusivity
D_{AB}, cm^2/s

Gas	0.1
Liquid	0.1×10^{-4}
Solid	0.1×10^{-9}

ent of the chemical potential and that, for multicomponent mixtures, the gradients of other species must be included in the equation. But even though Eq. (1-84) holds rigorously only for a binary solution, it is customary in engineering practice to assume, for a multicomponent mixture, that "species B" represents all components other than A and to use Eq. (1-84) as a semiempirical expression.

We see that Fick's law is similar to Fourier's law (and Newton's law of viscosity) and expresses the fact that diffusion results from concentration inequalities (concentration gradients) and occurs in the direction of decreasing concentration (downhill, we might say, just like heat and momentum). Like thermal conductivity and viscosity D_{AB} is a transport property, but it depends upon both the nature of the diffusing substance (A) and the nature of the diffusing medium (B). The dimensions of J_{Ax} are moles per unit area and unit time, and those of C_A are moles per unit volume; hence the *diffusivity* D_{AB} has dimensions of (length)^2 per unit time. D_{AB} is usually reported in cm^2/s; typical values for gases, liquids, and solids are given in Table 1-4.

1-26 BALANCE LAW FOR MASS OF A SPECIES

Just as we applied a conservation law to energy- and momentum-transport problems, so also can we write a similar relationship for mass (or moles) transport of a species

$$\begin{matrix}\text{Net rate of flow} \\ \text{of mass of a} \\ \text{species into an} \\ \text{arbitrary volume}\end{matrix} + \begin{matrix}\text{production of} \\ \text{mass of the} \\ \text{species within} \\ \text{the volume}\end{matrix} = \begin{matrix}\text{accumulation of} \\ \text{mass of the} \\ \text{species within} \\ \text{the volume}\end{matrix} \tag{1-85}$$

The net rate of flow into the volume is the difference between the input rates and the output rates. Production of mass of a species can occur due to chemical reaction although the net mass production of all species must be zero. At steady state the accumulation must be zero.

1-27 OVERALL BALANCE

Applying an overall balance to the entire system at steady state, we find that

$$N_{Ax}A_x\bigg|_{x=0} - N_{Ax}A_x\bigg|_{x=L_x} = 0$$

or, since A_x is constant,

$$N_{Ax}\Big|_{x=0} - N_{Ax}\Big|_{x=L_x} = 0$$

Similarly, if we make our balance between $x = 0$ and an arbitrary x, we obtain

$$N_{Ax}A_x\Big|_{x=0} - N_{Ax}A_x\Big|_{x=x} = 0$$

or, since A_x is constant,

$$N_{Ax}\Big|_{x=0} - N_{Ax}\Big|_{x=x} = 0 \tag{1-86}$$

or N_{Ax} is constant. Then, since we are dealing with a dilute solution,

$$N_{Ax}\Big|_{x=0} \approx J_{Ax}\Big|_{x=0} = J_1 = \text{const} \tag{1-87}$$

1-28 STEADY-STATE INCREMENTAL MOLAR BALANCE

We can now use Fick's law of diffusion to find the concentration profile and the total moles of A transferred. To do so we apply the balance law for species A to an incremental element of volume $A_x\,\Delta x$ located at an arbitrary position x (see Fig. 1-31). The rate of flow of A (moles A per unit time) entering the element at x is equal at steady state to the rate of flow of A leaving the element, or

$$N_{Ax}A_x\Big|_{x} - N_{Ax}A_x\Big|_{x+\Delta x} = 0$$

Dividing by $A_x\,\Delta x$ and taking the limit as $\Delta x \to 0$ gives

$$\lim_{\Delta x\to 0} \frac{N_{Ax}A_x|_{x+\Delta x} - N_{Ax}A_x|_x}{A_x\,\Delta x} = \lim_{\Delta x\to 0} \frac{N_{Ax}|_{x+\Delta x} - N_{Ax}|_x}{\Delta x} = 0$$

or

$$\frac{dN_{Ax}}{dx} = 0 \qquad \text{where} \qquad N_{Ax} = \text{const} \tag{1-88}$$

Since the solution is dilute, the density of the solution is approximately constant and Eq. (1-84) becomes at steady state

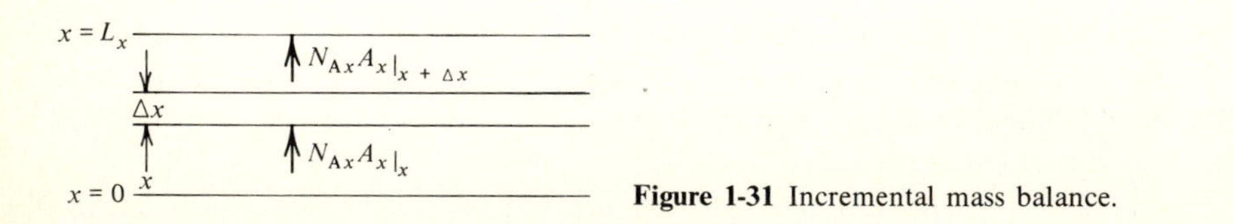

Figure 1-31 Incremental mass balance.

$$J_{Ax} = -D_{AB}\frac{dC_A}{dx} \qquad \text{steady state} \tag{1-89}$$

Using Eq. (1-83) and the above gives

$$N_{Ax} \approx J_{Ax} = -D_{AB}\frac{dC_A}{dx} = J_1 = \text{const}$$

We now integrate this equation subject to the boundary conditions

$$C_A = \begin{cases} C_{A1} = \text{solubility of A in solvent} & \text{at } x = 0 \\ C_{A2} & \text{at } x = L_x \end{cases}$$

Since the differential equation, flux law, and boundary conditions are analogous to Eqs. (1-10c) and (1-62) obtained for heat and momentum transport, we can write down the solution by analogy by replacing T or v_y with C_A. Then we have the linear concentration distribution

$$\frac{C_{A1} - C_A}{C_{A1} - C_{A2}} = \frac{x}{L_x} \tag{1-90}$$

The diffusive flux out is then calculated by differentiating Eq. (1-90)

$$N_{Ax}\Big|_{x=0} \approx J_{Ax}\Big|_{x=0} = -D_{AB}\frac{dC_A}{dx} = D_{AB}\frac{C_{A1} - C_{A2}}{L_x}$$

Then the total molar rate of transport is

$$\mathcal{W}_{A1} = N_{Ax}A_x\Big|_{x=0} = D_{AB}A_x\frac{C_{A1} - C_{A2}}{L_x} \tag{1-91}$$

which is analogous to Eq. (1-2) and (if it is multiplied by g_c) to Eq. (1-46).

1-29 GENERALIZATION OF FLUX LAW

Since concentration, like temperature is a scalar quantity, analogies between heat and mass transfer are more generally valid than between either of these processes and momentum transport. The flux laws for mass transport in more than one direction in cartesian coordinates (unlike those for momentum transport) take the simple forms

$$J_{Ax} = -D_{AB}\frac{\partial C_A}{\partial x} \qquad J_{Ay} = -D_{AB}\frac{\partial C_A}{\partial y} \qquad J_{Az} = -D_{AB}\frac{\partial C_A}{\partial z} \tag{1-91}$$

For curvilinear coordinates expressions can be obtained by using the same procedure outlined for Fourier's law in Sec. 1-9. Thus for an arbitrary coordinate u, if s_u is the distance along the coordinate, the molar diffusive flux in the u direction is

$$J_{Au} = -D_{AB}\frac{\partial C_A}{\partial s_u} \tag{1-92}$$

For transport in one direction only at steady state

$$J_{Au} = -D_{AB}\frac{dC_A}{ds_u} \tag{1-93}$$

For example, in spherical coordinates (see Table I-1)

$$J_{A\theta} = -D_{AB}\frac{\partial C_A}{r\,\partial\theta} \qquad J_{A\phi} = -D_{AB}\frac{\partial C_A}{r\sin\theta\,\partial\phi} \qquad J_{Ar} = -D_{AB}\frac{\partial C_A}{\partial r} \tag{1-94}$$

PART D ANALOGIES BETWEEN ONE-DIMENSIONAL TRANSPORT PHENOMENA

1-30 FLUX LAW

So far we have used the same basic models to develop the following flux laws† for energy, mass, and momentum transport

Energy: $$q_x = -k\frac{\partial T}{\partial x} \quad \text{Fourier's law} \tag{1-5}$$

Mass: $$J_{Ax} = -D_{AB}\frac{\partial C_A}{\partial x} \quad \text{Fick's law} \tag{1-84}$$

Momentum (one-dimensional transport): $$\tau_{xy} = -\mu\frac{\partial v_y}{\partial x} \quad \text{Newton's law of viscosity} \tag{1-48}$$

In each case the equations take the *fundamental form*

$$\text{Flux} \equiv \frac{\text{flow rate}}{\text{area}} = \text{transport property} \times \text{potential gradient} \tag{1-95}$$

where k, D_{AB}, and μ are called the molecular *transport properties* and T, C_A, and v_y are the *potentials*.

While these equations are similar, they are not completely analogous because the transport properties have different units. Noting that the units on mass diffusivity are, say m^2/s, we can define diffusivities for heat and momentum as

$$\boxed{\text{Thermal diffusivity} \equiv \alpha \equiv \frac{k}{\rho\hat{C}_p}} \tag{1-96}$$

where $\hat{C}_p$ = heat capacity at constant pressure

and

$$\boxed{\text{Momentum diffusivity} \equiv \nu \equiv \frac{\mu}{\rho}} \tag{1-97}$$

†The flux laws are also called *constitutive equations* or *phenomenological expressions*.

Table 1-5 Analogous terms in one-dimensional flux laws

	Mass	Energy	Momentum
Flux	J_{Ax}	q_x	τ_{xy}
Transport property	D_{AB}	k	μ
Potential gradient	$\dfrac{\partial C_A}{\partial x}$	$\dfrac{\partial T}{\partial x}$	$\dfrac{\partial v_y}{\partial x}$
Diffusivity	D_{AB}	$\alpha = \dfrac{k}{\rho \widehat{C}_p}$	$\nu \equiv \dfrac{\mu}{\rho}$
Concentration	C_A	$\rho \widehat{C}_p T$	$v_y \rho$
Gradient of concentration	$\dfrac{\partial C_A}{\partial x}$	$\dfrac{\partial(\rho \widehat{C}_p T)}{\partial x}$	$\dfrac{\partial(v_y \rho)}{\partial x}$

where ν is also called the kinematic viscosity. Then, since we are assuming that $\widehat{C}_p$ and ρ are both constant, we can rewrite the flux laws as

Energy:
$$q_x = -\alpha \frac{\partial(\rho \widehat{C}_p T)}{\partial x} \tag{1-98a}$$

Mass:
$$J_{Ax} = -D_{AB} \frac{\partial C_A}{\partial x} \tag{1-98b}$$

Momentum:
$$\tau_{xy} = -\nu \frac{\partial(\rho v_y)}{\partial x} \tag{1-98c}$$

We note that $\rho \widehat{C}_p T$ has the dimensions of (mass per unit volume) times (energy per unit mass), or energy per volume; it can be interpreted as an "energy concentration" by analogy to C_A (moles per volume). Furthermore, ρv_y has dimensions of (mass per volume) times velocity = momentum per volume and can be interpreted as a "momentum concentration." Hence the flux laws can be written in the diffusion form

$$\boxed{\text{Flux} = -\text{diffusivity} \times \text{concentration gradient}} \tag{1-99}$$

These various quantities are summarized in Table 1-5.

1-31 MEANING OF AN ANALOGY

If the physical conditions of a problem lead to mathematical relationships (differential equations, flux laws, and boundary conditions) for heat transport similar to those for momentum transport, we say that there is an analogy between the heat- and momentum-transport problems. By interchanging analogous quantities (such as diffusivities) we can use the known solution of a problem in heat transport to obtain the solution to a problem in momentum transport or vice versa. The same can be done with regard to *heat* and *mass* transport and momentum and mass transport. The use of analogies makes the learning process easier, and because of these similarities we are able to study three subjects (heat and mass transfer and fluid dynamics) as though they were one. In *practice* it makes it possible to take experimental measurements in one system (say heat) in order to obtain information on another (say momentum or mass).

Example of an Analogy

Consider the three processes in transport between parallel planes, discussed previously. In an element of size $\Delta x \, A_x$ we can write the conservation law (at steady state with *no internal generation*)

Input rate	*Output rate*	*Accumulation rate*	*Units*
$q_x A_x \Big\vert_x$	$- q_x A_x \Big\vert_{x+\Delta x}$	$= 0 =$ energy rate	energy/s
$\tau_{xy} A_x \Big\vert_x$	$- \tau_{xy} A_x \Big\vert_{x+\Delta x}$	$= 0 =$ momentum rate	y momentum/s
$J_{Ax} A_x \Big\vert_x$	$- J_{Ax} A_x \Big\vert_{x+\Delta x}$	$= 0 =$ mole rate	mol A/s

If we divide each equation by $A_x \, \Delta x$, let $\Delta x \to 0$, and factor out constant A_x [Eqs. (1-9), (1-57), and (1-88)], we have

$$\lim_{\Delta x \to 0} \frac{q_x A_x|_x - q_x A_x|_{x+\Delta x}}{A_x \, \Delta x} = \frac{-d(q_x A_x)}{A_x \, dx} = \frac{-dq_x}{dx} = 0$$

$$\lim_{\Delta x \to 0} \frac{\tau_{xy} A_x|_x - \tau_{xy} A_x|_{x+\Delta x}}{A_x \, \Delta x} = -\frac{d(\tau_{xy} A_x)}{A_x \, \Delta x} = -\frac{d\tau_{xy}}{dx} = 0$$

$$\lim_{\Delta x \to 0} \frac{J_{Ax} A_x|_x - J_{Ax} A_x|_{x+\Delta x}}{A_x \, \Delta x} = -\frac{d(J_{Ax} A_x)}{A_x \, \Delta x} = -\frac{dJ_{Ax}}{dx} = 0$$

In each case the differential equation for the flux is of the same form. The differential equations are therefore similar as well as the flux laws and boundary conditions (Table 1-6). The solutions [Eqs. (1-10*c*), (1-62), (1-90)] are also similar, as shown below. These equations are written below in a form that is physically more meaningful by subtracting 1 from each side and changing signs on both sides; e.g.,

Table 1-6

	Boundary condition		
	$x = 0$	$x = L_x$	Flux laws
Heat	$T = T_1$	$T = T_2$	$q_x = -k\dfrac{dT}{dx}$
Momentum	$v_y = V$	$v_y = 0$	$\tau_{xy} = -\mu\dfrac{dv_y}{dx}$
Mass, mol A	$C_A = C_{A1}$	$C_A = C_{A2}$	$J_{Ax} = -D_{AB}\dfrac{dC_A}{dx}$

in Eq. (1-10*c*) below

$$-\left(\frac{T - T_1}{T_2 - T_1} - \frac{T_2 - T_1}{T_2 - T_1}\right) = -\left(\frac{x}{L_x} - 1\right)$$

or the fractional temperature rise attained at x is

$$\frac{T - T_2}{T_1 - T_2} = 1 - \frac{x}{L_x}$$

	Original form		*Revised form*	
Heat:	$\dfrac{T - T_1}{T_2 - T_1} = \dfrac{x}{L_x}$	(1-10*c*)	$\dfrac{T - T_2}{T_1 - T_2} = 1 - \dfrac{x}{L_x}$	(1-100*a*)
Mass:	$\dfrac{C_A - C_{A1}}{C_{A2} - C_{A1}} = \dfrac{x}{L_x}$	(1-62)	$\dfrac{C_A - C_{A2}}{C_{A1} - C_{A2}} = 1 - \dfrac{x}{L_x}$	(1-100*b*)
Momentum:	$\dfrac{V - v_y}{V} = \dfrac{x}{L_x}$	(1-90)	$\dfrac{v_y}{V} = 1 - \dfrac{x}{L_x}$	(1-100*c*)

Thus if we knew the solution to the heat-transport problem [Eq. (1-10*c*)], we could readily write down, by analogy, the solution to the mass- (or momentum-) transport problem by replacing temperatures with the corresponding concentrations (or velocities).

1-32 DIMENSIONLESS ANALOGIES

For heat, mass, and momentum transfer, if in Eqs. (1-100) we let

$$\theta_H \equiv \frac{T - T_2}{T_1 - T_2} \qquad \theta_D \equiv \frac{C_A - C_{A2}}{C_{A1} - C_{A2}} \qquad \theta_M \equiv \frac{v_y}{V} \qquad \text{and} \qquad \eta \equiv \frac{x}{L_x} \tag{1-101a}$$

we obtain the solutions

$$\theta_H = 1 - \eta \qquad \theta_D = 1 - \eta \qquad \text{and} \qquad \theta_M = 1 - \eta \tag{1-101b}$$

Thus in dimensionless form, the solutions are identical. This is another characteristic of an analogy. It means that if we have the solution for the heat-transfer case $\theta_H(\eta)$, we can also use it as the solution for the mass- or momentum-transfer cases. Then by examining the boundary conditions (and in some cases the differential equations) we can tell how to define the corresponding dimensionless variables θ_D and θ_M for mass and momentum. This idea will be a big timesaver when the equations are more complicated.

Dimensional Analysis of Differential Equations

Dimensional analysis is also useful when the equations *cannot* be solved mathematically, by letting us know what variables ought to be measured in an *experimental* program. In order to use this procedure, we need to write the differential equation and boundary conditions in dimensionless form. To obtain a differential equation for *temperature* in the example, we substitute the *flux law* (1-6) into the flux differential equation (1-9). If k is constant,

$$\frac{d^2T}{dx^2} = 0 \tag{1-102}$$

We now write this equation in terms of θ. For convenience we drop the subscript H. Solving the first of Eqs. (1-101a) for T gives

$$T = T_2 + (T_1 - T_2)\theta$$

Then $$\frac{dT}{dx} = (T_1 - T_2)\frac{d\theta}{dx} \qquad \text{and} \qquad \frac{d^2T}{dx^2} = (T_1 - T_2)\frac{d^2\theta}{dx^2}$$

Substitution into the first of Eqs. (1-101a) gives

$$\frac{d^2\theta}{dx^2} = 0 \tag{1-103}$$

In order to write this in terms of η we use the chain rule from calculus

$$\frac{d\theta}{dx} = \frac{d\theta}{d\eta}\frac{d\eta}{dx} = \frac{d\theta}{d\eta}L_x^{-1} \qquad \frac{d^2\theta}{dx^2} = \left[\frac{d}{d\eta}\left(\frac{d\theta}{dx}\right)\right]\frac{d\eta}{dx} = \frac{d^2\theta}{d\eta^2}L_x^{-2}$$

Substituting into the first of Eqs. (1-101a), we get

$$\frac{d^2\theta}{d\eta^2} = 0 \tag{1-104}$$

The dimensionless boundary conditions become

$$\begin{aligned} &\text{At } x = 0 \text{ (or } \eta = 0\text{):} \qquad && T = T_1 \qquad \text{or} \qquad && \theta = 1 \\ &\text{At } x = L_x \text{ (or } \eta = 1\text{):} \qquad && T = T_2 \qquad \text{or} \qquad && \theta = 0 \end{aligned} \tag{1-105}$$

We could easily integrate Eq. (1-104) twice, using the above boundary conditions, to obtain θ_M [Eqs. (1-101)]. But suppose we did not know how? In that case we would still know that the solution must contain the terms appearing in the differential equation, i.e., $\theta = \theta(\eta)$ since no *new* terms appear in the boundary conditions.

Thus we could set up an *experiment* to measure θ_H at various η and obtain the dimensionless $\theta_H(\eta)$ plot (it ought to turn out a straight line). This plot would be *universal;* i.e., we could use it for any T_1, T_2, or L_x even though our *data* were taken for only one set of T_1, T_2, L_x. (Actually it is usually a good idea to test the theory with a few extra points at various T_1, T_2, and L_x.) Thus we can greatly reduce our experimental program by using dimensionless variables. Furthermore, the analogy tells us that the *same* plot will hold for the analogous mass and momentum cases as well!

APPENDIX 1-1 DEFINITION OF TOTAL DERIVATIVE

If $y = f(x)$ is any function of x, an *increment* of x, Δx, is any finite increase in the independent variable x and an increment of y, Δy, is defined as

$$\Delta y \equiv f(x + \Delta x) - f(x) \tag{A1-1-1}$$

Note that if y increases when x increases, Δy is positive and if y decreases, Δy is negative; i.e., we do not define Δy as the absolute value of the change in y. Instead it is the *algebraic increase,* a quantity that is negative if y actually decreases. In general, Δ is a mathematical operator which can be thought of as the final value minus the initial value.

The total derivative of y with respect to x is defined by

$$\frac{dy}{dx} = \frac{df}{dx} = \lim_{\Delta x \to 0} \frac{f(x + \Delta x) - f(x)}{\Delta x}$$

$$= \lim_{\Delta x \to 0} \frac{f(x)|_{x+\Delta x} - f(x)|_x}{\Delta x} = \lim_{\Delta x \to 0} \frac{\Delta f}{\Delta x} \tag{A1-1-2}$$

Note (Fig. A1-1-1) that $\Delta f/\Delta x = \tan\theta$ is the slope of the straight line *chord*

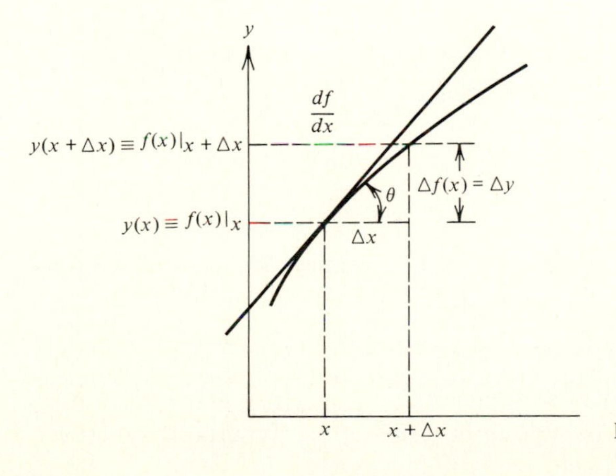

Figure A1-1-1 Definition of derivative.

that passes through the points (y, x) and $(y + \Delta y, x + \Delta x)$. As we make Δx smaller, the slope $\Delta f/\Delta x$ becomes a better approximation of the slope of the straight line tangent to the curve at x. The slope of this tangent line is df/dx.

Definition of Partial Derivative

If $f(x, t)$ is any function of x and t, by definition the partial derivative of f with respect to x (t being understood to be constant) is

$$\frac{\partial f}{\partial x} \equiv \lim_{\Delta x \to 0} \frac{f(x + \Delta x, t) - f(x, t)}{\Delta x}$$

$$\equiv \lim_{\Delta x \to 0} \left(\frac{f|_{x+\Delta x} - f|_x}{\Delta x} \right)_t \equiv \lim_{\Delta x \to 0} \left(\frac{\Delta f}{\Delta x} \right)_t \tag{A1-1-3}$$

We usually do not indicate that t is constant this way because it is assumed that all independent variables other than x are kept constant. The partial derivative occurs when the function f depends upon more than one independent variable (such as t and x).

APPENDIX 1-2 SOLUTION OF A FREQUENTLY OCCURRING ORDINARY DIFFERENTIAL EQUATION

Consider the differential equation

$$\frac{d^2y}{dx^2} - a^2 y = 0 \tag{A1-2-1}$$

which can be written in terms of the operator D as

$$(D^2 - a^2)y = 0$$

where $D \equiv d/dx$. Let us try the solution $y = e^{mx}$. By substitution into Eq. (A1-2-1) we get the auxiliary equation

$$m^2 - a^2 = 0 \qquad \text{with roots} \qquad \left.\begin{matrix} r_1 \\ r_2 \end{matrix}\right\} = \pm a$$

and the solutions are

$$y_1 = e^{ax} \qquad \text{and} \qquad y_2 = e^{-ax}$$

Since any linear combination of y_1 and y_2 is also a solution, the general solution is

$$y = C_1 e^{ax} + C_2 e^{-ax} \tag{A1-2-2}$$

If, as a boundary condition, y or its derivatives are specified at $x = 0$, a more convenient form of solution can be written in terms of the hyperbolic functions, defined by

$$\cosh ax = \frac{e^{ax} + e^{-ax}}{2} \qquad \text{and} \qquad \sinh ax = \frac{e^{ax} - e^{-ax}}{2} \tag{A1-2-3}$$

Then Eq. (A1-2-2) becomes

$$y = A \cosh ax + B \sinh ax \tag{A1-2-4}$$

where $\qquad C_1 = \frac{1}{2}(A + B) \qquad \text{and} \qquad C_2 = \frac{1}{2}(A - B)$

If a is not real, Eq. (A1-2-2) is not convenient and trigonometric functions are usually employed. If we let $a \equiv ib$, we can use the identities

$$e^{ibx} \equiv \cos bx + i \sin bx \qquad e^{-ibx} \equiv \cos bx - i \sin bx$$

in Eq. (A1-2-2) to obtain

$$y = \alpha \cos bx + \beta \sin bx \qquad \text{where} \quad \begin{aligned} \alpha &= C_1 + C_2 \\ \beta &= (C_1 - C_2)i \end{aligned} \tag{A1-2-5}$$

APPENDIX 1-3 DEFINITION OF AN INTEGRAL

Suppose $y = f(x)$ (see Fig. A1-3-1) and we want to find the integral

$$I = \int_{x_1}^{x_2} y(x)\, dx$$

or the area under the curve between x_1 and x_2. First we divide the region between x_2 and x_1 into N increments of width $\Delta x \equiv (x_2 - x_1)/N$ and let y_i be the value of y at arbitrary x_i. We note that the product $y_i\, \Delta x$ is the area of the rectangle of height y_i and width Δx. This area is approximately equal to the area under the curve $y = f(x)$ between x_i and $x_i + \Delta x$, and the summation

$$\sum_{i=1}^{N} y_i\, \Delta x$$

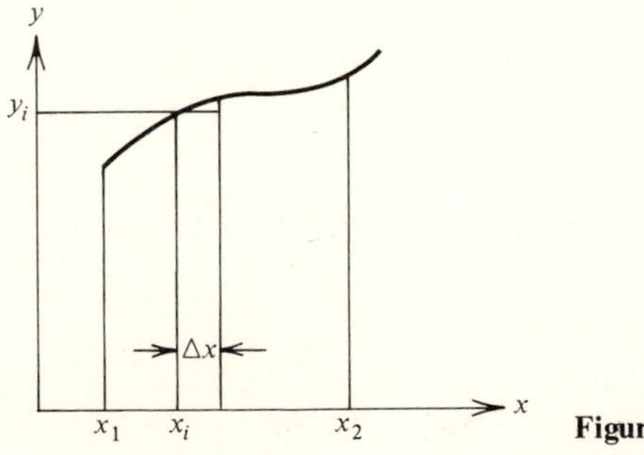

Figure A1-3-1 Definition of integral.

is the total area obtained by adding together all the incremental areas of the strips $y_i\,\Delta x$. This total area is approximately the area I under the curve. Now as $\Delta x \to 0$, the error becomes smaller and smaller and eventually the total area of the strips becomes equal to the area under the curve, i.e., equal to the desired integral of y over x. Thus we can define the integral as

$$I \equiv \int_{x_1}^{x_2} f(x)\,dx = \lim_{\Delta x \to 0} \sum_{i=1}^{N} y_i\,\Delta x \tag{A1-3-1}$$

APPENDIX 1-4 DEFINITION OF A MEAN VALUE

In general, if $y = f(x)$ (see Fig. A1-4-1), the mean or average value of y can be defined as

$$y_a \equiv \frac{\int_{x_1}^{x_2} f(x)\,dx}{x_2 - x_1} \tag{A1-4-1}$$

where the numerator is the area under the curve between x_1 and x_2. Note that $y_a(x_2 - x_1)$ = area under rectangle of height y_a and width $x_2 - x_1$, and

$$\int_{x_1}^{x_2} y\,dx = \text{area under the curve } y = f(x)$$

Then y_a is defined as that value of y for which these areas are equal.

The above definition of a mean applies when x represents a linear distance in cartesian coordinates. For example, the mean temperature in a slab (such as in Fig. 1-1) using Eq. (A1-4-1) is

$$T_a = \frac{\int_0^{L_x} T\,dx}{L_x} \tag{A1-4-2}$$

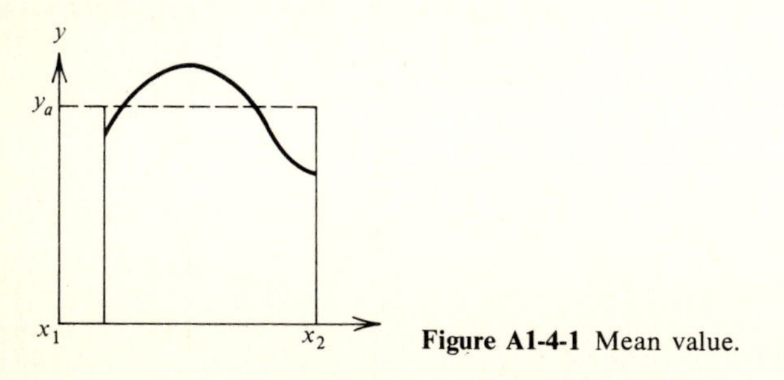

Figure A1-4-1 Mean value.

But more generally the temperature in a rectangular body may vary with *all* the space coordinates x, y, z. Thus the volume-averaged T can be defined as

$$\langle T \rangle = \frac{\int T \, d\mathcal{V}}{\mathcal{V}} \tag{A1-4-3}$$

where in cartesian coordinates $d\mathcal{V} = dx\,dy\,dz$.

We can see that this reduces to the previous equation for the slab by recognizing that, in the slab, T does not depend on y or z and that $d\mathcal{V} = A_x\,dx$, where $A_x = L_y L_z$. Then

$$\langle T \rangle = \frac{\int_0^{L_x} T A_x \, dx}{A_x L_x} = \frac{\int_0^{L_x} T \, dx}{L_x} = T_a$$

since A_x is constant.

For a cylinder, in which T depends only upon r, the distance from the center $d\mathcal{V} = L_z\,dA_z$, where L_z is the length of the cylinder and dA_z is a differential area perpendicular to the z axis (see Fig. A1-4-2), or

$$dA_z = \lim_{\Delta r \to 0} \Delta A_z = \lim_{\Delta r \to 0} [\pi(r + \Delta r)^2 - \pi r^2]$$

$$= \pi \lim_{\Delta r \to 0} [(r^2 + 2r\,\Delta r + \Delta r)^2 - r^2] = 2\pi r\,dr$$

since $\Delta r \to dr$ and $(\Delta r)^2 \to 0$. Then

$$\langle T \rangle = \frac{\int_0^R T L_z (2r\,dr)}{L_z (R^2)} = 2 \frac{\int_0^R Tr\,dr}{R^2} \tag{A1-4-4}$$

Note that one *could* define the mean temperature as

$$T_m = \frac{\int_0^R T\,dr}{R} \tag{A1-4-5}$$

but this would give results different from Eq. (A1-4-4). So which mean is correct? To determine this we ask: What is the physical meaning that the mean value is intended to convey? Usually (for making an energy balance) we want the mean

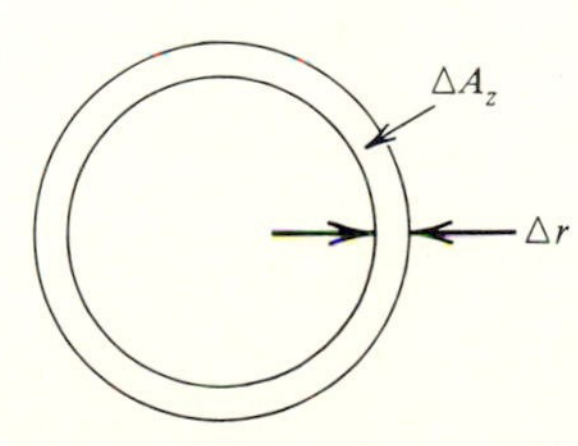

Figure A1-4-2 Element of area.

temperature $\langle T \rangle$ to be such that it can be used to find the internal energy (or enthalpy) of the body above a reference temperature T_R. Thus if U is the internal energy at T and U_R is the value of U at T_R,

$$U - U_R = \hat{C}_v(\langle T \rangle - T_R)\rho \mathcal{V} = \hat{C}_v \langle T \rangle \rho \mathcal{V} - \hat{C}_v T_R \rho \mathcal{V} \qquad \text{(A1-4-6)}$$

where $\hat{C}_v$, the heat capacity at constant volume (energy per mass and kelvin), is assumed constant. However, we can also obtain $U - U_R$ by finding ΔU, the internal energy of an increment of volume $\Delta \mathcal{V}$ with temperature T, summing over all increments, and letting $\Delta \mathcal{V} \to 0$. Then

$$d(U - U_R) = \hat{C}_v(T - T_R)\rho \, d\mathcal{V}$$

and

$$U - U_R = \int \hat{C}_v(T - T_R)\rho \, d\mathcal{V} \qquad \text{(A1-4-7)}$$

Since T_R is constant and if ρ and $\hat{C}_v$ can be assumed constant,

$$U = U_R = \rho \hat{C}_v \int T \, d\mathcal{V} - \rho \hat{C}_v T_R \mathcal{V} \qquad \text{(A1-4-8)}$$

Comparing Eq. (A1-4-8) with (A1-4-6), we see that

$$\langle T \rangle = \frac{\int T \, d\mathcal{V}}{\mathcal{V}}$$

or that the volume average gives the physically meaningful result.

PROBLEMS

1-1 Heat is being transferred through a steel slab 5 ft long, 2 ft wide, and 1 in thick. The temperature of the cooler face of the slab is 25°C, and the temperature of the opposite face is 85°C. The thermal conductivity of steel is 0.112 cal/s · cm · K.

(*a*) Convert all units above to SI units (see Table A-1). Then, using both SI and English units given, obtain (1) the heat flux through the slab in Btu/h · ft² and the heat flow in Btu per hour and (2) the temperature gradient in kelvins per foot.

(*b*) Repeat for copper if $k = 0.92$ cal/s · cm · K.

1-2 A 1-in-thick slab of unknown material is initially at a temperature of 20°C. The top face of the slab is suddenly brought to a temperature of 100°C and maintained at that temperature. The bottom face of the slab is maintained at 20°C. After steady state is obtained, the heat flux is measured and found to be 10,000 Btu/h · ft². The density and heat capacity of the slab are measured to be 300 lbm/ft³ and 3 Btu/lbm · °R, respectively.

(*a*) Express all the above quantities in SI units.

(*b*) Sketch a plot of the heat *flux* versus distance at $t = 0$, $t = t_1$, and $t \to \infty$.

(*c*) Explain graphically why the heat rate entering $|\dot{Q}_1|$ is not always equal to the heat rate leaving $\dot{Q}_2$.

(*d*) Explain why Fourier's law must be written in terms of a temperature derivative instead of the overall temperature difference $T_1 - T_2$, illustrating with a plot.

(*e*) In Chap. 7 it will be shown that the temperature in the slab varies as follows:

$$\frac{T - T_2}{T_1 - T_2} = 1 - \frac{x}{L_x} - \sum_{n=1}^{\infty} \frac{2}{n\pi} \sin \frac{n\pi x}{L_x} \exp\left(\frac{-n^2\pi^2\alpha t}{L_x^2}\right)$$

Derive equations for the heat flow entering and leaving and compare. What is the difference?

(*f*) Neglecting all terms for which $n > 1$ in the infinite series, calculate (1) the time in minutes at which the temperature at the slab center is within 1 percent of the steady-state values and (2) the ratio of the heat flux into the slab at steady state to the heat flux at 0, 10, and ∞ min.

(*g*) Sketch $\partial T/\partial x$ versus x at constant t for $t = 0$, 10, and ∞ min. Explain the physical significance of these curves.

1-3 Compare and contrast the derivation of the temperature distribution and total heat transport in a slab with that in a cylinder by preparing a table showing for each geometry, the incremental balance, flux differential equation, flux distribution, flux law, differential equation for temperature, boundary conditions, temperature profile, total transport rate, and overall balance.

1-4 An aluminum plate with an area of 0.929 m^2 and thickness of 0.0254 m conducts 7000 cal/s at steady state with surface temperatures of $T_1 = 99°C$ and $T_2 = 101°C$. The same plate conducts 9800 cal/s with surface temperature of $T_1 = 299°C$ and $T_2 = 301°C$.

(*a*) Assuming a linear variation of thermal conductivity with the absolute temperature, derive an expression for the thermal conductivity k in units of W/m · K as a function of temperature.

(*b*) If the walls of the aluminum plate described above were maintained at $T_1 = 199°C$ and $T_2 = 201°C$, at what rate in watts would heat be conducted through the plate?

1-5 In a lubrication system, an oil with a viscosity of 10.0 cP is confined between two metal surfaces, which can be assumed flat and parallel. The lower surface or plate is being pulled at steady state with a velocity of 5 ft/s and has an area of 50 ft^2. The distance between the plates is 0.12 in. After converting to SI units:

(*a*) Calculate the shear stress at the lower surface in lbf/ft^2 and SI units.

(*b*) Calculate the momentum flux entering the lower surface in units involving pounds mass, feet, and seconds (give the actual units).

(*c*) Convert the shear stress in part (*a*) and the momentum flux in part (*b*) into SI units involving kilograms, meters, and seconds. Give the actual units and express in terms of newtons,

(*d*) Calculate the total force on the lower plate in newtons,

(*e*) Repeat part (*a*) if the oil is replaced by (1) water at 20°C and (2) air at 20°C all other quantities are the same.

(*f*) Sketch the velocity-versus-position profiles for parts (*d*) to (*e*).

1-6 The fluid between the plates in Prob. 1-5 is initially at rest. Suddenly the top plate is pulled with a constant velocity equal to 5 ft/s. The oil has a specific gravity of 0.9.

(*a*) Sketch plots of velocity as a function of position at various times.

(*b*) Calculate the time required for the velocity at the midpoint to be within 10 percent of the steady-state value.

(*c*) Find the momentum flow into and out of the oil at the condition in part (*b*) in lbm · ft/s and in SI units.

(*d*) Calculate the force in pounds and newtons acting on the fluid at the upper and lower plates at the condition in part (*b*). Assume that (Chap. 7)

$$\frac{v_y}{V} = 1 - \frac{x}{L_x} - \sum_{n=1}^{\infty} \frac{2}{n\pi} \sin \frac{n\pi x}{L_x} \exp\left(\frac{-n^2\pi^2\nu t}{L_x^2}\right)$$

where $\nu \equiv$ kinematic viscosity $\equiv \mu/\rho$.

1-7 A fluid with a viscosity of 7×10^{-4} Pa · s is contained between two flat parallel plates. At steady state the velocity of the lower plate is 0.61 m/s in the positive x direction, and the velocity of the upper plate is 4.57 m/s in the negative x direction. The plate separation is 3.05×10^{-4} m.

(*a*) Calculate the steady-state momentum flux τ_{xy} in lbf/ft^2.

(*b*) Convert the momentum flux calculated in part (*a*) into SI units.

(*c*) If the area of both plates is 2.79 m^2, calculate the steady-state force in newtons exerted on the lower plate.

(*d*) Repeat parts (*a*) and (*c*) if the velocity of the upper plate is 4.57 m/s in the positive x direction.

1-8 Demonstrate the analogy between heat transport in a metal pipe and momentum transport in an annulus by preparing a table listing each of the items in Prob. 1-3.

1-9 Compare the cases of momentum transport in rectangular and cylindrical coordinates by preparing a table of similar quantities to those in Prob. 1-3.

1-10 Referring to the physical situation in Sec. 1-24, show that mass transfer from a solid surface through a liquid film can be treated analogously to Couette-flow momentum transport by preparing a table listing analogous quantities for each transport process, first carrying out the actual derivations and then obtaining the results by analogy.

1-11 A cylindrical solid of pure A is dissolving in a stagnant liquid B that surrounds it completely. The cylinder radius R is very small compared with its length L. Assume that the concentration of the solute A in the liquid at the surface is equal to the solubility, C_{As}, that a thin film of thickness L_f surrounds the solid, that the concentration of A at the film boundary is equal to the bulk-liquid concentration C_{Ab}, that bulk flow is negligible, and that a steady state exists.

(*a*) Derive equations (in terms of C_{As}, C_{Ab}, L_f, L_c, R, and D_{AB}, as needed) for (1) C_A and J_{Ar}, the concentration and flux distributions in the film, and (2) $\mathcal{W}_{A1}$, the molar rate of flow of A through the film.

(*b*) Describe a physical situation (if any exists) for which there is an analogy between the mass-transport case above and a momentum-transport problem.

(*c*) Write the differential equation, flux law, and boundary conditions for the momentum case (if it exists). Define diffusivities and concentrations.

(*d*) Using the analogy (if it exists), write solutions analogous to the results in part (*a*). (Do *not* rederive.)

1-12 Repeat parts (*b*) to (*d*) of Prob. 1-11 for heat transport.

1-13 Repeat Prob. 1-11 for a spherical film.

1-14† Helium is separated from natural gas by a method based on the fact that Pyrex glass is permeable to helium but impermeable to natural gas. In one method a cylindrical Pyrex glass tube is used with the natural gas inside the tube and the helium diffusing radially through the walls of the tube.[4] Let the inside radius of the Pyrex tube be R_1, the outside radius be R_2, and the length be L. Let the concentration of He at the inside be C_{A1} and at the outside be C_{A2}. It is desired to find the total transport of helium in moles per hour through the glass.

(*a*) Make an incremental molar balance over an element of thickness Δr. Draw a diagram showing mass flow into and out of the increment.

(*b*) Obtain a differential equation for the molar flux.

(*c*) Obtain the flux distribution.

(*d*) Write the flux law for *this* case.

(*e*) Write the boundary conditions.

(*f*) Solve for the concentration distribution $C_A(r)$ in terms of R_1 and R_2.

(*g*) Find the total molar transport of He through the wall of the tube.

(*h*) Show that this problem is analogous to that of heat transport in the wall of a metal pipe (discussed previously) by listing your results in parallel columns.

1-15 A spherical shell has internal radius R_1 and an external radius R_2. The interior surface of the shell is maintained at temperature T_1, and the exterior is maintained at T_2.

(*a*) Assuming that the thermal conductivity of the shell varies linearly with temperature and has the value k_1 at $T = T_1$ and k_2 at $T = T_2$, find an expression for the steady-state heat flux q_r in the shell.

(*b*) Rework part (*a*) assuming that the thermal conductivity is constant at its value for $T = (T_1 + T_2)/2$ to show that the heat flux q_r is the same for both cases.

1-16 Derive, if possible, a mass-transfer analogy to Prob. 1-15. Is there a momentum-transport analogy?

1-17 A nuclear reactor is shaped in the form of a sphere of radius R_1. The nuclear reaction occurs *inside* the sphere and generates heat, which is transported through the walls of the reactor. The outer wall of the reactor is a sphere of radius R_2. The temperature at the inside of the wall is T_1, and the

†Adapted by permission from a problem in Ref. 2.

temperature on the outside of the wall is T_2. Assume that there is no heat generation in the wall and that we have steady state. The rate of heat flow from the reactor into the wall is $|\dot{Q}_1|$. Let r be the distance from the center of the reactor and assume that the temperature at any fixed r in the wall of the reactor is constant; that is, $T = T(r)$ only. We define a mean area A_m such that we can use the expression

$$|\dot{Q}_1| \equiv kA_m \frac{T_1 - T_2}{L_r}$$

where $L_r \equiv R_2 - R_1$ is the thickness of the wall.

Find the correct mean area, e.g., arithmetic mean or log mean in terms of the inside area $A_1 = 4\pi R_1^2$ and the outside area $A_2 = 4\pi R_2^2$. In other words, we state that A_m is some function of A_1 and A_2 such that the equation above will hold and you are to find the functional relationship. In your derivation, obtain expressions for the following (showing *all* steps):

(*a*) Overall energy balance
(*b*) Incremental energy balance over element of width Δr
(*c*) Differential energy balance
(*d*) Boundary conditions
(*e*) Flux law (Fourier's law). (*Do not* give equation above.)
(*f*) Temperature distribution
(*g*) Flux distribution
(*h*) Total transport
(*i*) Mean area

Answer: $A_m = \sqrt{A_1 A_2}$ = geometric mean

1-18 Now assume that the temperature in the wall of the reactor in Prob. 1-17 varies with r, θ, and ϕ, where ϕ is the angle of rotation about the z axis [or $\phi = \arctan(y \cdot x)$] and θ is the angle measured from the z axis.

(*a*) Write an expression for the heat flux in the ϕ direction, based on Fourier's law.
(*b*) Explain the equation physically.

1-19 A prolate spherical coordinate system is given by

$$x = a \sinh \eta \sin \theta \cos \psi \qquad y = a \sinh \eta \sin \theta \sin \psi \qquad z = a \cosh \eta \cos \theta$$

where a is a constant. The scale factors in the η, θ, and ψ directions are[3]

$$h_\theta = h_\eta = a(\sinh^2 \eta + \sin^2 \theta)^{1/2} \qquad h_\psi = a \sinh \eta \sin^2 \theta$$

Write Fourier's law and Fick's-law expressions for transport in the η, θ, and ψ directions.

1-20 For steady-state transport of heat, moles A, or z momentum between R_1 and R_2 in an annular geometry with no internal generation, summarize the incremental balances, the differential equations, boundary conditions, and flux laws. State the conditions needed for an analogy to exist and on that basis write the solutions for each.

1-21 Write the differential equations and boundary conditions in Prob. 1-20 in dimensionless form, defining all terms. Then obtain the common solution in terms of dimensionless variables. Describe how one could (theoretically at least) set up an experiment to obtain, say, concentration profiles by measuring temperature or velocity profiles.

PROBLEMS FOR APPENDIXES

A1-1 If $y = x^3$, (*a*) find dy/dx by using the definition of the derivative.

(*b*) Plot y versus x. At $x = 2$, calculate $\Delta y/\Delta x$ for $x = 0.2, 0.1, 0.05, 0.01$ and show the slopes $\Delta y/\Delta x$ for each Δx. Also, calculate dy/dx and show on the graph. Plot $\Delta y/\Delta x$ versus Δx and show dy/dx on the same graph.

A1-2 The temperature in a slab is given by

$$\frac{T - T_2}{T_1 - T_2} = 1 - \frac{x}{L_x} - \frac{2}{\pi}\sin\frac{\pi x}{L_x}\exp\left(\frac{-\pi^2\alpha t}{L_x^2}\right)$$

where L_x is the thickness of the slab, T_1 and T_2 are the temperatues of the faces at $x = 0$, and $x = L_x$. Find $\partial T/\partial x$ and $\partial T/\partial t$.

A1-3 If $y = x^3$, (*a*) find the mean value of y between $x = 1$ and $x = 10$ by integrating analytically.

(*b*) Compare the above with the arithmetic average of the endpoint values and with the value at the average x.

(*c*) Integrate the equation numerically using increments of size $\Delta x = 2, 1, 0.1$. For each case find $\Sigma\, y_i\, \Delta x$. Compare with the analytical result.

REFERENCES

1. Aris, R.: "Vectors, Tensors, and the Basic Equations of Fluid Mechanics," Prentice-Hall, Englewood Cliffs, N.J., 1962.
2. Bird, R. B., W. E. Stewart, and E. N. Lightfoot: "Transport Phenomena," Wiley, New York, 1960.
3. Moon, P., and D. E. Spencer: "Field Theory for Engineers," Van Nostrand, Princeton, N.J., 1961.
4. *Sci. Am.*, July 1958, p. 52.

CHAPTER

TWO

TRANSPORT PROPERTIES†

2-1 TRANSPORT PROPERTIES FROM SIMPLIFIED GAS-KINETIC THEORY

Molecular theories are often useful in strengthening our understanding of various transport processes. Such theories may also serve to predict qualitatively or quantitatively the dependence of the transport coefficients μ, D_{AB}, and k on temperature and pressure. Ideally, a molecular theory should be able to predict μ, k, and D_{AB}, for given substances a priori, so that tedious experimental measurements can be eliminated. Unfortunately, this objective has been realized for the gaseous state only, to which we now turn our attention. We will have more to say about molecular theories of transport properties in liquids and solids later in the chapter.

Mass Transport in Gases at Low Pressure

In order to get a simplified picture of the mechanism of diffusive transport in gases, let us consider a mixture of gas A in gas B at *equilibrium,* i.e., at uniform temperature, pressure, and concentration throughout. We know from kinetic theory that the molecules will be in random motion and will be colliding with each

†This chapter need not be taken up at any particular time. It gives theoretical and empirical equations for predicting molecular mass, thermal, and momentum transport properties (D_{AB}, k, and μ). However, it is desirable that Sec. 2-1 (which shows how the molecular diffusivities of a gas are predicted, by simple kinetic theory, to be proportional to the product of a mean molecular velocity and a mean free path) be taken up before the discussion of the Prandtl mixing-length theory in Chap. 6. The remainder of this chapter can be used as a reference for the prediction of transport properties. Early drafts of this chapter were prepared with the assistance of Professors K. E. Gubbins and R. J. Gordon.

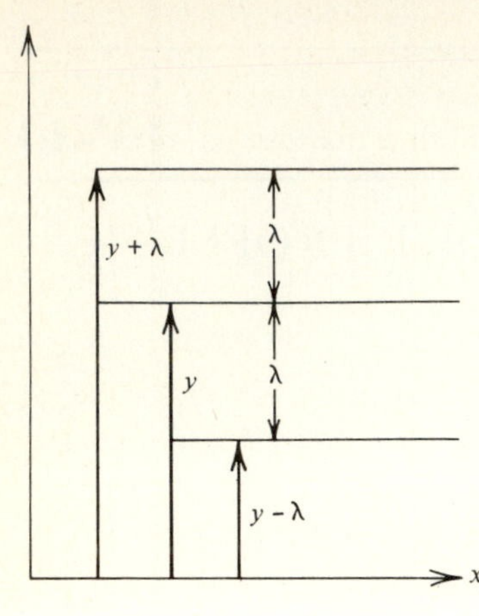

Figure 2-1 Planes at y, $y + \lambda$, and $y - \lambda$.

other at a rate on the order of 10^{21} collisions per second. At a given point and time each molecule will have its own velocity, and will travel a certain distance before it collides with another molecule. Thus there will be a distribution of molecular speeds ranging from 0 to ∞. If we know this distribution, we can calculate a mean speed $\bar{u}$ and a mean distance between collisions λ, called the *mean free path*. Since conditions are uniform within the gas, $\bar{u}$ and λ will not vary with position, and since all directions of molecular motion are equally probable, $\bar{u}$ will be the same for all directions and orientations of the coordinate axes; i.e., it will be a scalar. If we consider an arbitrary plane, say at $y = y$, in the gas (Fig. 2-1), the number of molecules of A in a unit of time that originate from *below* the plane $y = y$ and then cross it must on the average be the same as the number that originate from *above* the plane $y = y$ and then cross it (since the concentration is uniform and all molecules of A have the same average velocity). Thus there is no *net* flow or molecular *diffusion* of molecules of A in the y direction.

Now let x_A be the mole fraction of gas A in B and suppose that there is a gradient of A in the y direction, dC_A/dy or dx_A/dy, but no gradients in the x or z directions.† If the concentration of A is larger at lower values of y (that is, dx_A/dy is negative), there will be more A molecules coming from below and crossing $y = y$ than there are A molecules coming from above and crossing $y = y$, simply because there are more A molecules per unit volume in the region below. Thus there will be a *net* flow of A in the y direction. In order to calculate this net diffusive flow we will assume that the molecules of A coming from below had their last collision at a plane $y - \lambda$ while those coming from above had their last collision at the plane $y + \lambda$, where λ is the mean free path. It is assumed, in this simplified model, that one-third of the molecules are moving along each of the three coordinates. Then along the y coordinate one-sixth of the molecules are moving in the $+y$ direction.

If n is the number of molecules per unit volume, the number of molecules of

†Since $C_A = x_A C$ and C is constant, $dC_A/dy = C\, dx_A/dy$.

A and B passing upward per unit time through a plane of area A_y is $\frac{1}{6}n\bar{u}A_y$ and of these $x_A(\frac{1}{6}n\bar{u}A_y)$ are A molecules. Then if it is assumed that there is no bulk flow of either A or B in the y direction, the net flow of A molecules is the difference between the flow due to A molecules moving upward and that due to those moving downward, or

$$\frac{\text{Molecules A}}{\text{Time}} = A_y(\tfrac{1}{6}n\bar{u}x_A)\bigg|_{y-\lambda} - A_y(\tfrac{1}{6}n\bar{u}x_A)\bigg|_{y+\lambda} \tag{2-1}$$

In order to get the molar diffusive flux of A we must divide by area A_y to get a flux [Eq. (1-3)] and by Avogadro's number $\tilde{N}$ to convert from molecules to moles. Letting $n = C\tilde{N}$, where C is the molar density of the mixture, we get the molar diffusive flux†

$$J_{Ay} = \tfrac{1}{6}\bar{u}Cx_A\bigg|_{y-\lambda} - \tfrac{1}{6}\bar{u}Cx_A\bigg|_{y+\lambda} \tag{2-2}$$

Now if we assume that (at constant T and p) C is constant and that since λ is small, dx_A/dy is constant over 2λ, we get

$$-J_{Ay} = \Delta(\tfrac{1}{6}\bar{u}Cx_A) = \tfrac{1}{6}\bar{u}C\,\Delta x_A = \tfrac{1}{6}\bar{u}C(2\lambda)\frac{dx_A}{dy} = \tfrac{1}{3}\bar{u}\lambda\frac{dC_A}{dy} \tag{2-3}$$

Comparing with Fick's law

$$J_{Ay} = -D_{AB}\frac{dC_A}{dy}$$

gives

$$D_{AB} = \tfrac{1}{3}\lambda\bar{u} \tag{2-4}$$

If we assume that $\bar{u}$ and λ are the same as for *equilibrium* (no concentration gradient), that A and B have roughly the same size and mass, and that they are hard spheres with no attractive forces between them, kinetic theory gives

$$\bar{u} = \left(\frac{8\,k_B T}{\pi m}\right)^{1/2} \tag{2-5}$$

where k_B = Boltzmann's constant = 1.38×10^{-23} J/K = $R_G/\tilde{N}$
m = mass of a molecule = $M/\tilde{N}$
M = molecular weight
R_G = gas constant
$\tilde{N}$ = Avogadro's number

and

$$\lambda = \frac{1}{(2\pi\,d^2 n)^{1/2}} \tag{2-6}$$

†More rigorously we should use J^*_{Ay} (defined in Chap. 5) instead of J_{Ay}, but in this elementary example we are assuming that A and B are of equal or nearly equal molecular weight, for which case ρ and C are both constant and $J_{Ay} \approx J^*_{Ay}$.

where d is the molecular diameter and n the molecules per unit volume. Then Eq. (2-4) becomes

$$D_{\mathrm{AB}} = \frac{2}{3d^2 n}\left(\frac{k_B T}{\pi^3 m}\right)^{1/2} \tag{2-7}$$

If we use $n = p/k_B T$ (from the ideal-gas law), we get

$$D_{\mathrm{AB}} = \frac{2}{3d^2 p}\left(\frac{k_B^3 T^3}{\pi^3 m}\right)^{1/2} \tag{2-8}$$

This equation assumes that A and B have the same size and mass; for example, B is a radioactively tagged molecule of A designated A*. The diffusivity $D_{\mathrm{AA}*}$ is then called the *coefficient of self-diffusion.*

If A and B have different sizes and masses, one can use $d = \frac{1}{2}(d_{\mathrm{A}} + d_{\mathrm{B}})$ and the *reduced* mass

$$\frac{1}{m} = \frac{1}{m_{\mathrm{A}}} + \frac{1}{m_{\mathrm{B}}}$$

This gives

$$D_{\mathrm{AB}} = \frac{2}{3}\left(\frac{k_B^3}{\pi^3}\right)^{1/2}\left(\frac{1}{2m_{\mathrm{A}}} + \frac{1}{2m_{\mathrm{B}}}\right)^{1/2} \frac{T^{3/2}}{p[(d_{\mathrm{A}} + d_{\mathrm{B}})^2/2]} \tag{2-9}$$

Momentum Transport

Now suppose that the fluid is in motion and there is a velocity gradient dv_x/dy while $v_y = 0 = v_z$. Then the molecules that cross $y = y$ originating from below will have a higher velocity than those originating from above. Also, since momentum is mass times velocity, they will have a higher x momentum. Therefore when they collide, the molecules from below with higher momentum will tend to speed up the slower molecules above; and likewise the slower molecules above will tend to slow down the faster ones from below. (See Sec. 1-17 for another physical picture.) Thus there will be an actual net transport of x momentum from a lower y to a higher y (or in the $+y$ direction). Speeding up the fluid above by the fluid below has the effect of a force acting tangentially to the area A_y; this force is the shear force $\sigma_{yx} A_y$. Likewise slowing down the fluid below has the effect of an equal and opposite shear force, often called a *drag force,* i.e., a frictional force. The shear force is related by Newton's law of motion [Eq. (1-42)] to the momentum rate; that is, $F_x g_c = \dot{P}_x$. Thus the result of the random movement of the molecules is both a shear force and a momentum flow.

If we let m be the mass of a molecule, its momentum will be mv_x. Then the rate of flow of x momentum upward from $y - \lambda$ will be

$$\frac{\text{Molecules}}{\text{Time}}\,\frac{\text{momentum}}{\text{molecule}} = (\tfrac{1}{6}A_y n\bar{u})(mv_x)\bigg|_{y-\lambda}$$

and the *net* rate of flow will be

$$\tfrac{1}{6}A_y n\bar{u}mv_x \Big|_{y-\lambda} - \tfrac{1}{6}A_y n\bar{u}mv_x \Big|_{y+\lambda} \tag{2-10}$$

To obtain the flux of x momentum in the y direction we divide by A_y, to get

$$-\tau_{yx} = \tfrac{1}{6}n\bar{u}mv_x \Big|_{y-\lambda} - \tfrac{1}{6}n\bar{u}mv_x \Big|_{y+\lambda} = \tfrac{1}{6}n\bar{u}m\,\Delta v_x \tag{2-11}$$

If we assume a linear velocity gradient over distance 2λ and use $nm = \rho$, we get

$$\tau_{yx} = -\tfrac{1}{3}\rho\bar{u}\lambda\frac{dv_x}{dy} \tag{2-12}$$

Comparing with Newton's law of viscosity [Eq. (1-50)] gives

$$\mu = \tfrac{1}{3}\rho\bar{u}\lambda \tag{2-13}$$

If we note that ρv_x has the units of momentum per unit volume, it can be thought of as a momentum concentration (see Chap. 1, Part D). Then Eq. (2-12) becomes

$$\tau_{yx} = -\tfrac{1}{3}\bar{u}\lambda\frac{d(\rho v_x)}{dy} \tag{2-14}$$

Changing notation in Eq. (1-98*c*) gives

$$\tau_{yx} = -\nu\frac{d(\rho v_x)}{dy} \tag{2-15}$$

where $\nu = \mu/\rho$ is a momentum diffusivity. Comparing Eqs. (2-14) and (2-15) leads to

$$\nu = \tfrac{1}{3}\bar{u}\lambda \tag{2-16}$$

This is the same expression as Eq. (2-4), which we obtained for mass diffusivity. Substituting for λ and $\bar{u}$ from kinetic theory gives the same result for ν as for D_{AB}.

The Schmidt number is defined as the ratio of the momentum to mass diffusivities, or, from Eqs. (2-4) and (2-16),

$$(\mathrm{Sc})_{AB} \equiv \frac{\nu}{D_{AB}} = \frac{\mu}{\rho D_{AB}} \approx 1 \tag{2-17}$$

Substituting Eq. (2-6) into (2-13) gives

$$\mu = \frac{2}{3\pi^{3/2}}(mk_B T)^{1/2} \tag{2-18}$$

Note that μ increases with temperature and is *independent of pressure* or equivalently is *independent of density at constant temperature*. The *rigorous-kinetic-theory* expression for μ for a gas composed of rigid spheres of diameter d is given by

$$\mu = 2.6693 \times 10^{-8}\frac{\sqrt{MT}}{d^2} \tag{2-19}$$

where μ is in pascal-seconds, T is in kelvins, d is in nanometers, and M is the molecular weight of the gas.

We should note in closing that a real gas is, of course, not made of perfectly rigid spheres, and this fact is taken into account in modern kinetic theory. See Ref. 8 for a discussion of theoretical expressions for the prediction of gas-phase viscosity.

Energy Transport

Now suppose that there is a negative temperature gradient dT/dy at the plane $y = y$. If we assume a monatomic gas and neglect rotational and vibrational energy contributions, each molecule will have an internal energy of

$$e = \tfrac{1}{2}m\overline{u^2} = \tfrac{3}{2}k_B T \tag{2-20}$$

As before, the molecules are in random motion, but those originating at $y - \lambda$ will have more energy (higher T) than those originating at $y + \lambda$. Following our previous approach, we divide the net energy flow by the area [Eq. (1-3)] to get the net flux of energy

$$q_y = \tfrac{1}{6}\bar{u}ne\Big|_{y-a} - \tfrac{1}{6}\bar{u}ne\Big|_{y+a} \equiv \tfrac{1}{6}n\bar{u}\,\Delta e = \tfrac{1}{6}n\bar{u}\,\Delta(\tfrac{3}{2}k_B T) \tag{2-21}$$

Assuming that dT/dy is linear and n is constant over 2λ gives

$$q_y = -\tfrac{1}{6}\bar{u}\frac{d(\tfrac{3}{2}nk_B T)}{dy}2\lambda \tag{2-22}$$

Now

$$\tfrac{3}{2}nk_B T = \frac{3}{2}\frac{\rho\tilde{N}}{M}\frac{R_G}{\tilde{N}}T = \rho\hat{C}_v T \tag{2-23}$$

since $\hat{C}_v = \tfrac{3}{2}(R_G/M)$ for a monatomic gas. Then

$$q_y = -\tfrac{1}{3}\bar{u}\lambda\frac{d(\rho\hat{C}_v T)}{dy} \tag{2-24}$$

where $\hat{C}_v$ is the specific heat at constant volume (energy per unit mass and kelvin). The units of $\rho\hat{C}_v T$ are energy per unit volume, which is an "energy concentration," and the units of $\bar{u}\lambda$ are m^2/s, which could be called a thermal diffusivity. However, by convention, the thermal diffusivity is defined by Eq. (1-96) as $\alpha \equiv k/\rho\hat{C}_p$. From Eq. (1-98*a*) we obtain

$$q_y = -\alpha\frac{d(\rho\hat{C}_p T)}{dy} \tag{2-25}$$

Thus Eq. (2-24) can be written

$$q_y = -\tfrac{1}{3}\bar{u}\lambda\frac{\hat{C}_v}{\hat{C}_p}\frac{d(\rho\hat{C}_p T)}{dy} \tag{2-26}$$

which gives

$$\alpha = \frac{1}{3}\frac{\bar{u}\lambda}{\gamma} \tag{2-27}$$

where
$$\gamma \equiv \frac{\widehat{C}_p}{\widehat{C}_v} \tag{2-28}$$

The ratio of the momentum diffusivity to the thermal diffusivity has special meaning in engineering problems in heat transport. It is called the *Prandtl number*

$$\Pr \equiv \frac{\widehat{C}_p \mu}{k} \equiv \frac{\mu/\rho}{k/\widehat{C}_p \rho} \equiv \frac{\nu}{\alpha} \tag{2-29}$$

For a monatomic gas

$$\Pr = \frac{\frac{1}{3}u\lambda}{\frac{1}{6}\mu\lambda/\gamma} = \gamma = \tfrac{5}{3}$$

Actually the Prandtl number is more like $\frac{2}{3}$, or a factor of 2.5 less.

From Eqs. (2-24) to (2-26) and Fourier's law (1-5) we get an equation for the thermal conductivity of a monatomic gas

$$k = \frac{1}{d^2}\left(\frac{k_B^3 T}{\pi^3 m}\right)^{1/2} \tag{2-30}$$

Hence, according to our simplified kinetic model, k should vary roughly as $T^{1/2}$ and should be independent of pressure. These predictions are again borne out qualitatively by rigorous kinetic theory, which gives

$$k = 8.333 \times 10^{-4} \frac{(T/M)^{1/2}}{\sigma^2} \tag{2-31}$$

where k is in W/m · K and σ is in nanometers.

Example 2-1: Generalized treatment Generalize the treatment of kinetic theory given in the previous section by letting $\widetilde{G}$ = "concentration" of transportable quantity (for example $\widetilde{G}_A \equiv n x_A/\widetilde{N}$ = moles A per unit volume, $\widetilde{G}_{xM}$ = x momentum per unit volume, $\widetilde{G}_H = \widetilde{C}_v \rho T$ = heat energy per unit volume, where n = molecules per unit volume and $\widetilde{N}$ = Avogadro's number). Let J_{Gy} be the net flux of $\widetilde{G}$ in the y direction. If the molecular velocity $\bar{u}$ and the number density n are constant, show that the diffusivity in general is proportional to the product of $\bar{u}$ and λ (the mean free path). Write expressions for the diffusivity of each process.

SOLUTION We can generalize the above treatment by letting $G = \widetilde{G}/n$ be the transportable quantity per molecule. For the various processes we can write $\widetilde{G}$ as shown in Table 2-1. If there is a gradient of G in the y direction, the net flux can be written

$$\text{Flux} = \frac{\text{quantity}}{\text{area} \times \text{time}} = \text{net}\left(\frac{\text{molecules}}{\text{area} \times \text{time}}\,\frac{\text{quantity}}{\text{molecule}}\right)$$

or
$$J_{Gy} = \tfrac{1}{6}\bar{u}nG\Big|_{y-\lambda} - \tfrac{1}{6}\bar{u}nG\Big|_{y+\lambda} = -\tfrac{1}{6}\bar{u}\left(\widetilde{G}\Big|_{y-\lambda} - \widetilde{G}\Big|_{y+\lambda}\right)$$

Table 2-1

Process	G	$\tilde{G} = nG$	Units of $\tilde{G}$
Mass	$\dfrac{x_A}{\tilde{N}}$	$C_A = \dfrac{nx_A}{\tilde{N}}$	$\dfrac{\text{Moles A}}{\text{Unit volume}}$
Energy	$\dfrac{m}{2}\overline{u^2}$	$\frac{1}{2}\rho\overline{u^2}$	$\dfrac{\text{Kinetic energy}}{\text{Unit volume}}$
	$e \equiv \frac{3}{2}k_BT$	$\frac{3}{2}k_BTn = \hat{C}_V\rho T$	$\dfrac{\text{Heat energy}}{\text{Unit volume}}$
Momentum	mv_x	ρv_x	$\dfrac{\text{Momentum}}{\text{Unit volume}}$

Also if we again assume a linear gradient over 2λ,

$$J_{Gy} = -\tfrac{1}{6}\bar{u}(2\lambda)\frac{d\tilde{G}}{dy} = -\tfrac{1}{3}\bar{u}\lambda\frac{d\tilde{G}}{dy}$$

This is of the form

$$\text{Flux} = -\text{diffusivity} \times \text{concentration gradient}$$

Then the diffusivities are proportional to the product of a molecular velocity and a mean free path

$$D_{AB} = \frac{\mu}{\rho} = \frac{k}{\rho\hat{C}_V} = \tfrac{1}{3}\bar{u}\lambda = \frac{2}{3d^2n}\left(\frac{k_BT}{m\pi^3}\right)^{1/2} \qquad \square$$

2-2 CHAPMAN-ENSKOG RIGOROUS THEORY FOR DILUTE GASES[9]

The equations above give only a rough estimate of the transport properties. The principal source of error is the assumption that molecules behave like rigid spheres with no interaction. Actually, molecules are compressible, and forces exist between them. The force of interaction between molecules varies with molecular separation r and is related to the potential energy of interaction ϕ by

$$F = -\frac{\partial\phi}{\partial r} \tag{2-32}$$

The form of the potential energy ϕ as a function of separation is shown in Fig. 2-2. At small values of r the molecules repel each other and the potential energy is large and positive. At larger separations molecules attract each other, and at still larger values of r the intermolecular forces tend to zero.

A number of semitheoretical equations have been proposed for $\phi(r)$. One of

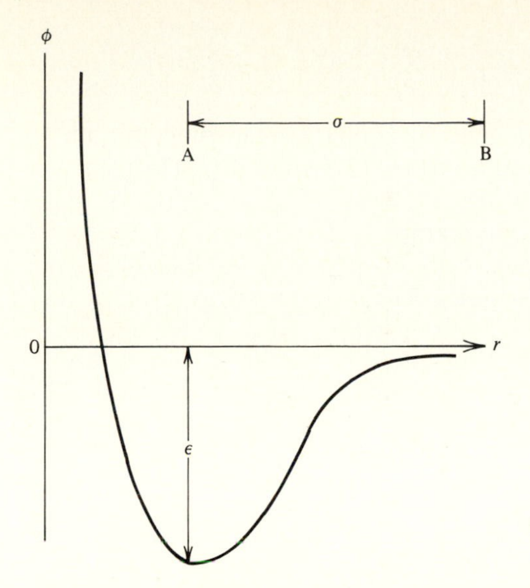

Figure 2-2 Potential energy of interaction between molecules.

the most commonly used is the Lennard-Jones 6-12 potential

$$\phi(r) = 4\epsilon\left[\left(\frac{\sigma}{r}\right)^{12} - \left(\frac{\sigma}{r}\right)^{6}\right] \tag{2-33}$$

σ and ϵ are parameters characteristic of the molecule under study. Values for common molecules are available.[7-9] For molecules for which data are unavailable, values of σ and ϵ can be estimated from thermodynamic properties.[7]

In a mixture of molecules of types A and B there will be interactions between A and B molecules. Equation (2-33) is written as

$$\phi_{AB}(r) = 4\epsilon_{AB}\left[\left(\frac{\sigma_{AB}}{r}\right)^{12} - \left(\frac{\sigma_{AB}}{r}\right)^{6}\right] \tag{2-34}$$

where ϵ_{AB} and σ_{AB} are characteristic of the A–B mixture. These parameters can be estimated from the parameters for pure components by the approximate equations

$$\sigma_{AB} = \tfrac{1}{2}(\sigma_A + \sigma_B) \qquad \text{and} \qquad \epsilon_{AB} = \sqrt{\epsilon_A \epsilon_B} \tag{2-35}$$

Chapman and Enskog have developed a rigorous theory for gases at low pressure assuming that the molecules interact. They obtained the following equations using the Lennard-Jones 6-12 potential, for nonpolar gases.

Viscosity

$$\mu = 2.669 \times 10^{-8} \frac{\sqrt{MT}}{\sigma^2 \Omega_\mu} \tag{2-36}$$

where μ = viscosity, Pa · s
T = absolute temperature, K
M = molecular weight
σ = collision diameter, a parameter in Lennard-Jones function, nm
$\Omega_\mu = \Omega_\mu(T^*)$ = collision integral [see Eqs. (2-37*a*) and (2-37*b*)]; for no interaction between molecules, $\Omega_\mu = 1$
$T^* \equiv k_B T/\epsilon$ = dimensionless temperature

Equation (2-36) holds well for nonpolar gases. The collision integral can be approximated by[4]

$$\Omega_\mu = \frac{1.604}{\sqrt{T^*}} \qquad \text{for } 0.4 < T^* < 1.4 \tag{2-37a}$$

or more exactly by[6]

$$\Omega_\mu = \frac{1.16145}{T^{*0.14874}} + \frac{0.52487}{e^{0.7732T^*}} + \frac{2.16178}{e^{2.43787T^*}} \tag{2-37b}$$

For polar molecules, a different potential should be used. Brokaw[1] recommends correcting Eq. (2-37*b*) for nonpolar molecules by using

$$\Omega_{\mu,\text{polar}} = \Omega_{\mu,\text{nonpolar}} + \frac{0.2\delta^{*}}{T^*} \tag{2-37c}$$

where the dimensionless dipole moment δ is

$$\delta \equiv \frac{\mathfrak{M}^2}{2\epsilon\sigma^3} \tag{2-37d}$$

where $\mathfrak{M}$ is the dipole moment in debyes (1 debye = 3.162×10^{-25} $N^{1/2} \cdot m^2$), ϵ is in newton-meters, σ is in meters. For gas mixtures a semiempirical equation can be used

$$\mu_{\text{mix}} = \sum_{i=1}^{n} \frac{x_i \mu_i}{\sum_{j=1}^{n} x_j \phi_{ij}} \tag{2-38}$$

where $$\phi_{ij} = \frac{1}{\sqrt{8}}\left(1 + \frac{M_i}{M_j}\right)^{-1/2}\left[1 + \left(\frac{\mu_i}{\mu_j}\right)^{1/2}\left(\frac{M_j}{M_i}\right)^{1/4}\right]^2 \tag{2-38a}$$

where n = number of species in mixture
x_i, x_j = mole fraction of i, j
μ_i, μ_j = viscosity of pure i, j at temperature and pressure of mixture
M_i, M_j = molecular weights

Modified forms of Eqs. (2-36) and (2-38) must be used for highly polar gases.

Thermal Conductivity

$$k = 8.322 \times 10^{-4} \frac{\sqrt{T/M}}{\sigma^2 \Omega_k} \qquad \text{for monatomic gases} \qquad (2\text{-}39)$$

where k is the thermal conductivity in W/m · K and $\Omega_k = \Omega_\mu(T^*)$ [see Eqs. (2-37*a*) and (2-37*b*)]. Comparing Eqs. (2-36) and (2-39) gives

$$k = \frac{15}{4}\frac{R_G}{M}\mu = \tfrac{5}{2}\widehat{C}_v \mu = \tfrac{3}{2}\widehat{C}_p \mu \qquad \text{monatomic} \qquad (2\text{-}40)$$

where $\widehat{C}_v$, the heat capacity at constant volume is $\frac{3}{2}(R_G/M)$ for monatomic gases.

The Eucken factor is defined as the group

$$\text{Eu} \equiv \frac{k}{\mu \widehat{C}_v} \equiv \frac{k}{\mu \widehat{C}_p}\frac{\widehat{C}_p}{\widehat{C}_v} = \frac{\gamma}{\text{Pr}} \qquad (2\text{-}40a)$$

Eu is predicted by Eq. (2-40) to be 2.5 for a monatomic gas. Then the Prandtl number is

$$\text{Pr} = \frac{\gamma}{\text{Eu}}$$

For a monatomic gas,

$$\gamma \equiv \frac{\widehat{C}_p}{\widehat{C}_v} \approx \frac{\frac{5}{2}}{\frac{3}{2}} = \frac{5}{3}$$

If Eu = 2.5,

$$\text{Pr} = \frac{\frac{5}{3}}{\frac{5}{2}} = \frac{2}{3}$$

which is close to the experimental value.

Equations (2-39) and (2-40) do not apply to polyatomic gases, which may transfer energy on collision by transfer of vibrational energy. This fact can be approximately taken into account by means of the Eucken correlation

$$\text{Eu} \equiv \frac{k}{\mu C_v} = \tfrac{9}{4}\gamma - \tfrac{5}{4} = 1 + \frac{4.47}{\widehat{C}_v M} \qquad (2\text{-}41)$$

Then

$$\text{Pr} \equiv \frac{\gamma}{\text{Eu}} = \frac{4\gamma}{9\gamma - 5} \qquad (2\text{-}41a)$$

Note that Eq. (2-41*a*) reduces to Eq. (2-40) for a monatomic gas ($\gamma = \frac{5}{3}$). The modified Eucken equation

$$\text{Eu} \equiv \frac{k}{\mu \widehat{C}_v} = 1.32 + \frac{3.52}{\widehat{C}_v M} \qquad (2\text{-}41b)$$

gives larger values of k that are too high, while Eq. (2-41) predicts values that are too low except for polar gases, for which both are too high.[9] The equation for estimating k for mixtures is of the same form as Eq. (2-38) with k replacing μ.

Mass Diffusivity

For the diffusion of gas A in gas B at low densities, assuming the ideal-gas law, $C = p/R_G T$, the diffusivity is

$$D_{\mathrm{AB}} = 8.42 \times 10^{-24} \frac{\left[\frac{T^3}{2}\left(\frac{1}{M_{\mathrm{A}}} + \frac{1}{M_{\mathrm{B}}}\right)\right]^{1/2}}{P\sigma_{\mathrm{AB}}^2 \Omega_D} \tag{2-42}$$

where D_{AB} is in $\mathrm{m^2/s}$, p is in $\mathrm{N/m^2}$, and Ω_D is tabulated [7-9] as a function of $k_B T/\epsilon$ [illegible] is given by

$$\Omega_D = \frac{1.06036}{T^{*0.1561}} + \frac{0.1930}{e^{0.47635}} + \frac{1.03587}{e^{1.52996}} + \frac{1.76474}{e^{3.89411}} \tag{2-43}$$

for nonpolar molecules. For polar molecules, Brokaw[1] recommends

$$\Omega_{D,\mathrm{polar}} = \Omega_{D,\mathrm{nonpolar}} + \frac{0.19\delta_{\mathrm{AB}}^2}{T^*} \tag{2-43a}$$

$$\delta = \frac{1.94 \times 10^3 \mathfrak{M}}{V_b T_b}$$

where V_b is the liquid molar volume at the boiling point in $\mathrm{m^3/g\ mol}$ and T_b is the normal boiling point in kelvins. The parameters are

$$\frac{\epsilon}{k_B} = 1.18(1 + 1.3\delta^*)T_b \qquad \sigma = \left(\frac{1.585 V_b}{1 + 1.3\delta^2}\right)^{1/3} \tag{2-43b}$$

$$\delta_{\mathrm{AB}} = (\delta_{\mathrm{A}}\, \delta_{\mathrm{B}})^{1/2} \qquad \epsilon_{\mathrm{AB}} = \left(\frac{\epsilon_{\mathrm{A}}}{k_{\mathrm{A}}} \frac{\epsilon_{\mathrm{B}}}{k_{\mathrm{B}}}\right)^{1/2} \qquad \sigma_{\mathrm{AB}} = (\sigma_{\mathrm{A}}\sigma_{\mathrm{B}})^{1/2}$$

2-3 RECOMMENDED EQUATIONS FOR PREDICTING TRANSPORT PROPERTIES OF LIQUIDS

Viscosity

The Eyring equation is

$$\mu = \frac{\tilde{N}h}{\tilde{V}} e^{3.8T_b/T} \tag{2-44}$$

where $\tilde{N}$ = Avogadro's number
$\tilde{V}$ = molal volume

h = Planck's constant
T_b = normal boiling temperature

This equation is limited to newtonian liquids and does not hold well for long molecules or near the critical point. It usually predicts values within 40 percent.

Thermal Conductivity

For predicting the thermal conductivity of liquids, Robbins and Kingrea[10] recommend the equation

$$k = (88.0 - 4.94H) \times 10^{-3} \left(\frac{0.55}{T_r}\right)^N \frac{\widehat{C}_p C^{4/3}}{\Delta S^*} \tag{2-45}$$

$$\Delta S^* = \frac{\Delta H_{vb}}{T_b} + R \ln \frac{237}{T_b}$$

where k = thermal conductivity, W/m · K
T_r = reduced temperature = T/T_c
$\widehat{C}_p$ = molal heat capacity, J/g mol · K
C = density, g mol/cm³
ΔH_{vb} = molal heat of vaporization at T_b, J/g mol

The parameters H and N are tabulated in Ref. 9. Sato (see Ref. 9, p. 520) recommends

$$k = \frac{11.05 \times 10^{-3} C_p T_b}{M^{1/2} C_{pb} T} \left(\frac{C}{C_b}\right)^{4/3}$$

where b refers to conditions at the normal boiling point and units are as in Eq. (2-45).

Diffusivity

For the diffusion of large spherical molecules or particles in liquids, e.g., polymer molecules or colloid particles, the Stokes-Einstein equation gives good results

$$\frac{D_{AB}\mu_B}{k_B T} = \frac{1}{6\pi R_A} \tag{2-46}$$

where R is the radius of diffusing molecule A. For the diffusion of normal-sized molecules, no theory exists which gives reasonable agreement with experiment. Empirical relations give the best predictions. The best known of these is the Wilke-Chang equation

$$D_{AB} = 5.89 \times 10^{-15} \frac{(\psi_B M_B)^{1/2} T}{\mu \widetilde{V}_A^{0.6}} \tag{2-47}$$

where D_{AB} = diffusivity, m²/s
μ = viscosity of solution, kg/m · s
$\tilde{V}_A$ = molar volume of liquid A at normal boiling point, m³/m · s
ψ_B = association parameter

$$\psi_B = \begin{cases} 1 & \text{for nonassociated solvents} \\ 1.5 & \text{for ethanol} \\ -2.6 & \text{for water} \end{cases}$$

The Wilke-Chang equation gives estimates that are usually within 10 to 20 percent of experimental values. Equation (2-47) gives values of D_{AB} when A is present only at low concentrations. For most solutions D_{AB} varies with concentration; no satisfactory equations exist for predicting this variation.

Other Transport Properties of Liquids and Solids

For equations for other transport properties of liquids, see Ref. 9. For diffusion in solids, see Ref. 3.

2-4 NONNEWTONIAN FLUIDS†

Fluids that do not obey Newton's law of viscosity are called *nonnewtonian fluids* and include such substances as plastics, polymer solutions, toothpaste, paper-pulp slurries, human blood, and asphalt. In such fluids the momentum flux (or shear stress) is not directly proportional to the velocity gradient (or rate-of-deformation tensor Δ_{xy}) but will depend on the velocity gradient in some complex manner.

This functional relationship between the shear stress and rate of deformation is called a *constitutive equation* referring to the fact that the relation in some sense reflects the "constitution" of the fluid. Many nonnewtonian fluids obey a simple constitutive equation of the form

$$\tau_{yx} = -m \left| \frac{dv_x}{dy} \right|^{n-1} \frac{dv_x}{dy} \tag{2-48}$$

Comparing this equation with Newton's law of viscosity, we see that such a fluid exhibits an apparent viscosity μ_A of the form

$$\mu_A = m \left| \frac{dv_x}{dy} \right|^{n-1} \tag{2-49}$$

where the absolute-value signs assure that μ_A will always be positive. A fluid obeying Eq. (2-49) is called a *power-law fluid.* For most power-law fluids n, *the power-law index,* is less than 1.0, and so μ_A decreases with increasing dv_x/dy. Physically, the decrease in n is a reflection of the alignment of the polymer molecules in the direction of flow, thus reducing their resistance to the sliding of fluid

†Contributed by Professor R. J. Gordon.

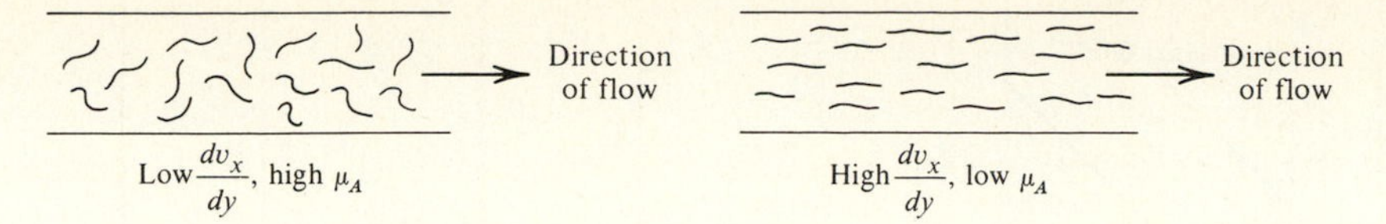

Figure 2-3 Mechanism of viscosity reduction with increasing velocity gradient in high-polymer solutions.

layers past one another (see Fig. 2-3). A fluid for which n is greater than 1.0 is referred to as a shear-thickening fluid. Examples are certain pastes and solid suspensions. Of course, for $n = 1.0$ we have a newtonian fluid.

Although the power-law equation accurately portrays the behavior of a large number of nonnewtonian fluids over a wide range of velocity gradients, some fluids exhibit more complex behavior. In addition, at both very low and very high velocity gradients all fluids appear to exhibit newtonian behavior with viscosities μ_0 and μ_∞, respectively (Fig. 2-4).

To describe the entire viscosity curve, a more complex expression than Eq. (2-49) is necessary. One of the numerous proposed expressions is the *extended Williamson model*

$$\mu_A = \mu_\infty + \frac{\mu_0 - \mu_\infty}{1 + \left(\frac{|dv_x/dy|}{\alpha_1}\right)^{\alpha_2}} \tag{2-50}$$

where α_1 and α_2 are constants. For low and high $|dv_x/dy|$, Eq. (2-50) yields $\mu_A \to \mu_0$ and $\mu_A \to \mu_\infty$, respectively. A summary and experimental test of a variety of such models can be found in Ref. 2.

Nonnewtonian fluids may also exhibit other interesting flow phenomena, such as *yield stresses* and *viscoelasticity*. Fluids with a *yield stress* do not begin to flow until a critical stress τ_0 corresponding to the yield stress is applied. Following this, their behavior is frequently similar to that of newtonian fluids (Fig. 2-5). This behavior can be expressed mathematically as

$$\frac{dv_x}{dy} = \begin{cases} 0 & \tau_{yx} < \tau_0 \\ -\frac{1}{\mu_0}(\tau_{yx} - \tau_0) & \tau_{yx} \geq \tau_0 \end{cases} \tag{2-51}$$

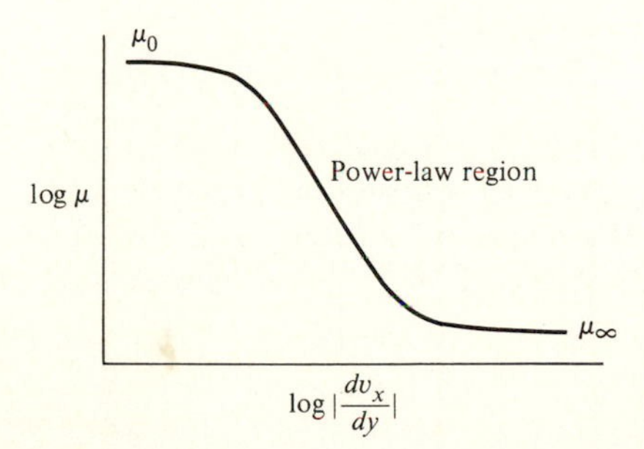

Figure 2-4 Typical viscosity behavior of nonnewtonian shear-thinning fluid, illustrating upper and lower limiting viscosities.

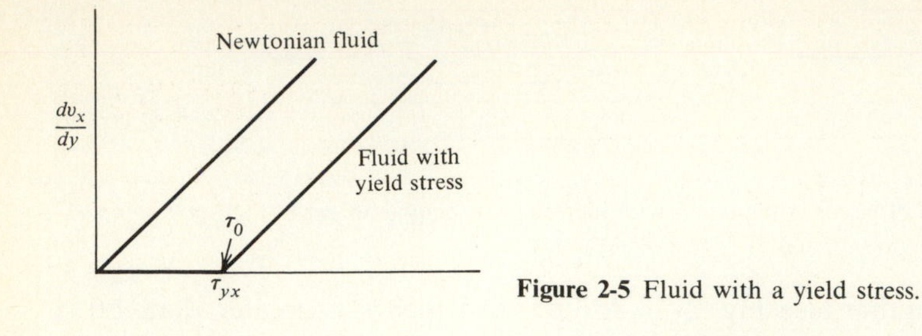

Figure 2-5 Fluid with a yield stress.

where μ_0 and τ_0 are parameters. A fluid obeying Eq. (2-51) is referred to as a *Bingham plastic.*

Common materials with yield stresses include catsup, toothpaste, and mayonnaise. If toothpaste did not have a yield stress, it would flow out whenever the cap was off the tube. Many solid suspensions also exhibit yield stresses; in addition, human blood has a yield stress, which plays an important role in a number of circulatory disorders (see the excellent article Rheology of Blood[5]).

PROBLEMS

2-1 Using Chapman-Enskog theory, predict the viscosity of oxygen at 200°C and atmospheric pressure. Compare the results with an experimental value found in the literature.

2-2 Using Chapman-Enskog theory, predict the thermal conductivity of argon at 120 K and 70 atm pressure. Compare the results with an experimental value found in the literature.

2-3 Using the Eucken equation, predict the thermal conductivity of methane at 122°F and 1 atm if $\mu = 1.116 \times 10^{-4}$ g/cm · s and $\widetilde{C}_p = 8.55$ cal/g mol · K.

2-4 Using Chapman-Enskog theory, predict the diffusivity for the methane-oxygen system at 500°C and 1 atm and compare your result with the experimental value.

REFERENCES

1. Brokaw, R. S.: *Ind. Eng. Chem. Proc. Des. Dev.*, **8**:240 (1964).
2. Cramer, S. D., and J. M. Marchello: *AIChE J.*, **14**:980 (1968).
3. Geiger, G. H., and D. R. Poirer: "Transport Phenomena in Metallurgy," Addison-Wesley, Reading, Mass., 1973.
4. Kim, S. K., and J. Ross: *J. Chem. Phys.*, **46**:818 (1967).
5. Merrill, E. W.: *Physiol. Rev.*, **49**:863 (1969).
6. Newfeld, P. D., A. R. Janzen, and R. A. Azig: *J. Chem. Phys.*, **57**:1100 (1972).
7. Perry, J. H., and C. H. Chilton: "Chemical Engineers' Handbook," 5th ed., McGraw-Hill, New York, 1973.
8. Reed, T. M., and K. E. Gubbins: "Applied Statistical Mechanics," McGraw-Hill, New York, 1973.
9. Reid, R. C., J. M. Prausnitz, and T. K. Sherwood: "Properties of Gases and Liquids," 3d ed., McGraw-Hill, New York, 1977.
10. Robbins, L. A., and C. L. Kingrea: *Hydrocarbon Proc. Petrol. Refiner*, **41**:5 (1962).

CHAPTER

THREE

TRANSPORT WITH INTERNAL GENERATION

3-1 BALANCE EQUATION

In Chap. 1 we used the flux laws and balance equations for one-directional energy, momentum, and mass transport without internal generation in order to obtain expressions for the temperature, velocity, or concentration distribution and to predict the total rate of transport of heat, mass, or momentum in various geometries. In this chapter we are interested in deriving the same expression for systems in which the generation term (also called the source or production term) is *not* zero. In energy transport the source term may arise as a result of chemical or nuclear reaction or electrical heating. In mass transport, the production is zero for the entire system, but the production of an individual species may take place if chemical reaction occurs. For momentum transport we will see later that the presence of pressure or gravity forces can be interpreted as a source of momentum.

3-2 ENERGY TRANSPORT WITH INTERNAL GENERATION

Consider a cylindrical fuel element in a nuclear reactor. This element generates heat due to the nuclear reaction at a rate Φ_H Btu/ft^3 · h or J/m^3 · s. Although in practice Φ_H varies with position, we assume it to be the same at all locations in the rod. This heat must be removed by surrounding the fuel element with a cooling medium, which maintains the surface temperature at some constant value T_R. We wish to calculate the rate of heat flow to the coolant per unit length of rod and the maximum temperature in the rod.

At steady state, the energy-balance equation (1-1) reduces to

$$\begin{matrix}\text{Rate of} \\ \text{heat flow in}\end{matrix} - \begin{matrix}\text{Rate of} \\ \text{heat flow out}\end{matrix} + \begin{matrix}\text{Rate of} \\ \text{energy production}\end{matrix} = 0$$

Let us apply this expression to the annular volume element illustrated in Fig. 3-1. The length of the element is L, and its thickness is Δr. As discussed in Sec. 1-8, the volume of the element is $2\pi r\,\Delta r\,L$. The various terms in the energy balance (1-1) are then found to be

$$\text{Rate of heat flow} = \begin{cases} A_r q_r \Big|_r = 2\pi r L q_r \Big|_r & \text{in} \\ A_r q_r \Big|_{r+\Delta r} = 2\pi r L q_r \Big|_{r+\Delta r} & \text{out} \end{cases}$$

$$\text{Rate of energy production} = \text{volume} \times \frac{\text{rate of production}}{\text{unit volume}}$$

$$= (2\pi r\,\Delta r\,L)\Phi_H$$

The energy balance is then

$$2\pi r L q_r \Big|_r - 2\pi r L q_r \Big|_{r+\Delta r} + 2\pi r L\,\Delta r\,\Phi_H = 0 \tag{3-1}$$

Dividing by $2\pi r L\,\Delta r$ and taking the limit as $\Delta r \to 0$, we have

$$\lim_{\Delta r \to 0} \frac{q_r(2\pi r L)|_r - q_r(r\pi r L)|_{r+\Delta r}}{2\pi r\,\Delta r\,L} + \Phi_H = 0$$

Since $2\pi L$ is a constant, the *flux differential equation* becomes

$$\boxed{-\frac{d(q_r r)}{r\,dr} + \Phi_H = 0} \tag{3-2}$$

In this problem we can solve directly for the flux distribution by integrating

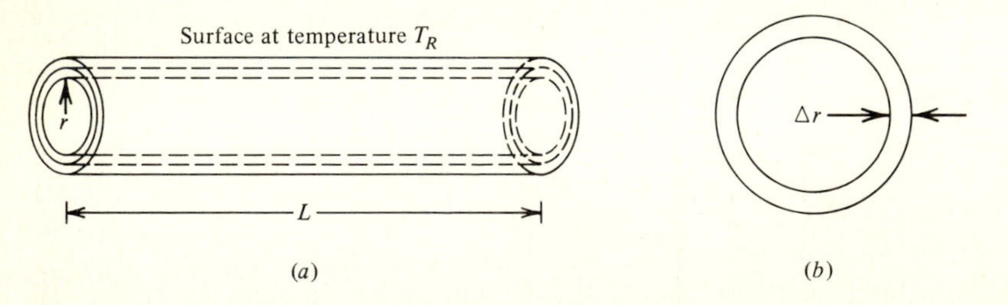

Figure 3-1 (*a*) Nuclear-reactor fuel element. (*b*) Incremental element of area.

the above expression between the limits $r = 0$ and $r = r$. Since we can expect the temperature to reach a maximum at $r = 0$, the gradient dT/dr, and hence the flux, ought to be zero at the center and $q_r|_{r=0} = 0$ could be used as the boundary condition. However, since Eq. (3-2) contains $d(q_r r)$ as the differential rather than dq_r, we can use the less restrictive condition that $q_r r|_{r=0} = 0$ or that $q_r \neq \infty$. Then

$$\int_0^{q_r r} d(q_r r) = \int_0^r \Phi_H r\, dr \qquad \text{or} \qquad rq_r = \frac{\Phi_H r^2}{2}$$

and the *flux distribution* is

$$\boxed{q_r = \frac{\Phi_H r}{2}} \tag{3-3}$$

indicating that q_r is indeed zero at $r = 0$. Equating the flux given by the flux law (1-5) to that given by (3-3), we obtain

$$-k\frac{dT}{dr} = \frac{\Phi_H r}{2}$$

Integrating this expression using the boundary condition that $T = T_R$ at $r = R$ gives

$$-\int_T^{T_R} dT = \frac{\Phi_H}{2k}\int_r^R r\, dr$$

Then our final expression for the *temperature distribution* in the rod becomes

$$\boxed{T - T_R = \frac{\Phi_H}{4k}R^2\left(1 - \frac{r^2}{R^2}\right)} \tag{3-4}$$

The temperature profile is therefore parabolic, with the maximum temperature occurring at $r = 0$, at the center of the rod. The maximum temperature rise is equal to $\Phi_H R_2/4k$, and the rise is seen to increase with the *square* of the rod radius.

Total Heat Flow Rate at Wall

It is useful to calculate the *total heat flow rate* through the surface of the rod. Its magnitude is given by Eq. (1-3) as the product of the heat flux at the surface and the area at $r = R$. Since, by thermodynamic convention, $\dot{Q}_G$ is positive if heat is gained by the system, $\dot{Q}_G$ will be negative in this case because the rod is losing heat to the coolant (see Sec. 1-7). However, q_r is positive according to Eq. (1-5) since dT/dr is negative; i.e., the heat is actually flowing in the $+r$ direction.

Therefore the rate of heat *loss* is†

$$\dot{Q}_L = -\dot{Q}_G = q_r A_r \bigg|_{r=R} = q_r \bigg|_{r=R} (2\pi RL)$$

If we obtain the flux from Eq. (3-3) as

$$q_r \bigg|_R = \frac{\Phi_H R}{2}$$

substituting gives

$$-\dot{Q}_G = \frac{\Phi_H R}{2}(2\pi RL) = \Phi_H(\pi R^2 L) \tag{3-5}$$

Physically, Eq. (3-5) can be interpreted as

$$\text{Heat loss at wall} = \frac{\text{Production}}{\text{Volume}} \times \text{Volume} = \text{Energy production inside}$$

Thus we could have written down the result in Eq. (3-5) immediately by taking an overall or *macroscopic balance* over the entire rod. Since that is not always easy to do, students will find it worthwhile to learn to use the method above. As an alternative we could have evaluated q_r at the surface by applying Fourier's law [Eq. (1-5)], i.e., by differentiating the temperature profile [Eq. (3-4)] and then letting $r = R$.

3-3 MOMENTUM TRANSPORT WITH INTERNAL GENERATION

For momentum transport, the general conservation law for one-directional transport takes the form [Eq. (1-38)]

$$\begin{matrix}\text{Rate of accumulation} \\ \text{of momentum in} \\ \text{arbitrary volume}\end{matrix} = \begin{matrix}\text{Net rate of flow} \\ \text{of momentum} \\ \text{into volume}\end{matrix} + \begin{matrix}\text{Sum of forces} \\ \text{acting on the} \\ \text{volume} \times g_c\end{matrix} \tag{3-6}$$

This expression is similar to Eq. (1-1) except that in place of a production or generation term we have a force term. As will become clearer later, this means that certain forces can be interpreted as generation terms, i.e., as production sources for momentum. To illustrate this idea we consider fluid flow in a pipe under a pressure drop.

A newtonian fluid is flowing upward at steady state through a long cylindrical pipe or tube of length L. Consider a section of the tube which is far removed from the entrance or exit (Fig. 3-2). We let z be the distance upward and let r be distance from the center. The velocity in the $+z$ direction is v_z, which depends on radial position r. At $z = 0$ the pressure acting uniformly over the cross section of area πR^2 is p_0; at $z = L$ it is p_L.

†The same result could be obtained using the mathematical convention [Eq. (1-19)] $\dot{Q}_G = -n_r A_r q_r|_{r=R}$. Since at $r = R$, a normal pointing *outward* from the surface of the rod is pointing in the $+r$ direction (away from the center), $n_r = +1$ and $\dot{Q}_G = -A_r q_r|_{r=R}$.

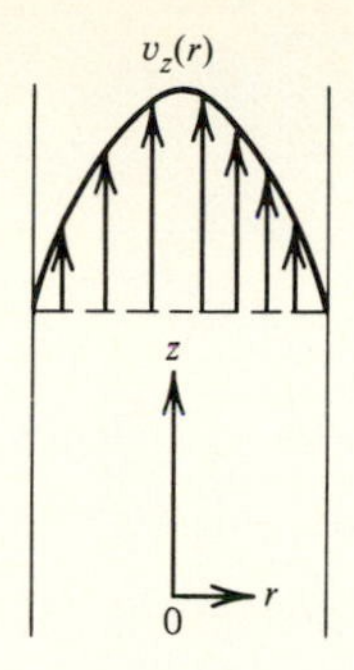

Figure 3-2. Velocity profile for laminar flow in a pipe.

The density of the fluid ρ is assumed to be constant. (We call such a fluid an *incompressible fluid.*) Under these conditions the flow is said to be *fully developed,* meaning that the velocity v_z does not change with distance z but varies only with distance r from the center. We further assume that the fluid is in *laminar flow*. This means that a tracer particle placed at a given radial and angular position stays at the same radial and angular position as it moves along with the fluid in the axial direction. For a newtonian fluid (such as water or low-viscosity oils) laminar flow exists for values of a dimensionless group called the Reynolds number (Re) of less than 2100. The Reynolds number is given by

$$\mathrm{Re} \equiv \frac{2 R v_a \rho}{\mu} \tag{3-7}$$

where R = tube radius
μ = fluid viscosity
v_a = average velocity

Design Problem

An engineer may wish to obtain the *pressure difference* $p_0 - p_L$ required to pump a fluid with a specified viscosity μ through a pipe of radius R and length L at an average velocity v_a. The result can be used to find the size of pump needed to develop the pressure difference $p_0 - p_L$.

Average Velocity†

If ρ is constant, the average velocity $v_a \equiv \langle v_z \rangle$ is defined in terms of the volumetric flow rate $\dot{\mathcal{V}}$ m^3/s by

$$\dot{\mathcal{V}} \equiv A_{cs} v_a \tag{3-8}$$

where A_{cs} is the cross-sectional area (πR^2 in the case of the tube). To find v_a from the distribution of the point velocity $v_z(r)$ we must first develop an equation for $\dot{\mathcal{V}}$ in terms of v_z. To do so, consider an element of area $\Delta A_z = 2\pi r\, \Delta r$, as in Fig. 3-3.

†See Appendix 1-4 for a discussion of averages.

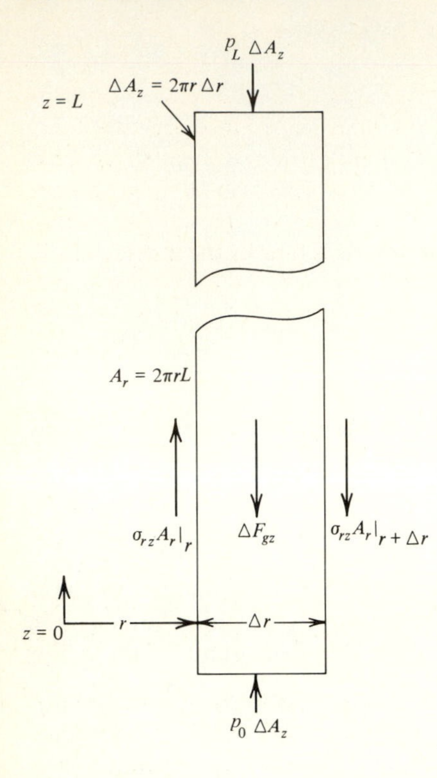

Figure 3-3. Forces acting on incremental element of volume for flow in a pipe.

The volumetric flow rate through this element at an arbitrary z is

$$\Delta\dot{\mathcal{V}} = v_z(r)2\pi r\,\Delta r$$

As $\Delta r \to 0$, this becomes

$$d\dot{\mathcal{V}} = v_z 2\pi r\,dr$$

Then to find the volumetric rate of flow for the entire tube we must integrate over the cross section

$$\dot{\mathcal{V}} = \int_{A_{cs}} d\dot{\mathcal{V}} = \int_0^R v_z(r)2\pi r\,dr$$

Comparing this expression with Eq. (3-8), we have

$$v_a \equiv \frac{1}{A_{cs}}\int_0^R v_z(r)2\pi r\,dr \tag{3-9}$$

More generally for a tube of arbitrary cross section A_{cs}

$$v_a = \frac{1}{A_{cs}}\int v_z\,dA_{cs} \tag{3-10}$$

where dA_{cs} is a differential element of area normal to the direction of flow. [In the above example $dA_{cs} = d(\pi r^2) = 2\pi r\,dr$.] From Eq. (3-9) it is clear that calculation of v_a requires knowledge of $v_z(r)$ at every radial position. Our next objective is to calculate $v_z(r)$.

Incremental Force-Momentum Balance in Tube Flow

In Chap. 1 we saw that we can interpret the balance law [Eq. (1-38)] as a momentum balance, as a force balance, or as a combination of both. Since most students are more familiar with the use of force balances, we consider that approach first. We will write a force balance in the z direction (the direction of flow) on an incremental element of volume $2\pi r\,\Delta r\,L$ (Fig. 3-3). Since we are at steady state, there is no net acceleration and the sum of the forces acting in the z direction is zero. These forces can be classified as (1) surface forces and (2) body forces (such as gravity, ΔF_{gz}) that act throughout the fluid. The surface forces consist of both forces due to pressure ΔF_{pz} and viscous forces ΔF_{vz} due to fluid friction. Thus the force balance in the z direction can be written

$$\Delta F_{vz} + \Delta F_{pz} + \Delta F_{gz} = 0$$

In general the viscous forces can act either perpendicular to a fluid surface (normal forces) or tangential (shear forces) to it. However, for a newtonian incompressible fluid there is no normal viscous stress σ_{zz} since it would depend on $\partial v_z/\partial z$, which is zero in fully developed flow. Note that the pressure force ΔF_{pz} is a normal force that acts on an area $\Delta A_z = 2\pi r\,\Delta r$ (Fig. 3-3). The body force ΔF_{gz} includes in general the force due to gravity as well as electric or magnetic forces, but in this example only gravity is involved.

We note that ΔF_{gz} is *negative* since we have taken the z direction as being upward whereas the gravitational force actually acts downward. (Since force has direction, it is a vector and we must follow the definition that a component of a vector F_z is positive if it points in the direction of the $+z$ axis and negative if it points in the direction of the $-z$ axis.) Since we *define* g_z as the z component of the acceleration due to gravity, it is also negative. (In *this* example, $g_z = -g = -9.80\ \text{m/s}^2 = -32.2\ \text{ft/s}^2$; but if the z axis were taken as pointing downward, g_z would be $+9.80\ \text{m/s}^2$ and ΔF_{gz} and g_z would both be positive.) If Δm is the mass of the fluid, $2\pi r\,\Delta r\,L$ its volume, and ρ its density, then, by Newton's law of motion [Eq. (1-37)],

$$\Delta F_{gz} = \frac{\Delta m\, a_z}{g_c} = \frac{\Delta m\, g_z}{g_c} = -\rho\frac{(2\pi r\,\Delta r\,L)g}{g_c} \tag{3-11}$$

The other forces acting in the z direction are then

$$\text{Tangential viscous forces at } r \text{ (up, or } +z) = \sigma_{rz}A_r\Big|_r$$

$$\text{Tangential viscous forces at } r+\Delta r \text{ (down, or } -z) = -\sigma_{rz}A_r\Big|_{r+\Delta r}$$

$$\text{Net tangential viscous forces} = \sigma_{rz}A_r\Big|_r - \sigma_{rz}A_r\Big|_{r+\Delta r} \equiv \Delta F_{vz}$$

$$\text{Pressure force} = \begin{cases} p_0\,\Delta A_z & \text{up} \\ -p_L\,\Delta A_z & \text{down} \end{cases}$$

$$\text{Net pressure force} = (p_0 - p_L)\,\Delta A_z \equiv \Delta F_{gz}$$

Since $A_r = 2\pi rL$ and $\Delta A_z = 2\pi r\,\Delta r$, the net force acting in the $+z$ direction is

$$\sigma_{rz}(2\pi rL)\Big|_r - \sigma_{rz}(2\pi rL)\Big|_{r+\Delta r} + (p_0 - p_L)(2\pi r\,\Delta r) + \frac{\rho g_z(2\pi r\,\Delta r\,L)}{g_c} = 0 \qquad (3\text{-}12)$$

Let us now write this equation as a momentum balance by multiplying through by g_c and making use of the relation $\sigma_{rz} g_c = \tau_{rz}$. Then

$$\tau_{rz}(2\pi rL)\Big|_r - \tau_{rz}(2\pi rL)\Big|_{r+\Delta r} + g_c(p_0 - p_L)(2\pi r\,\Delta r) + g_z(2\pi r\,\Delta r\,L)\rho = 0 \qquad (3\text{-}13)$$

Note that the same equation is obtained regardless of whether the flow is up or down but the sign of individual terms (g_z, for example) may change. We can compare this equation with Eq. (3-6) and with our general balance law

$$\text{Input rate} - \text{output rate} + \text{generation rate} = \text{accumulation rate} \qquad (1\text{-}1)$$

where the first two terms represent, respectively, the flow of z momentum into and out of the increment (see Fig. 3-4).

We now show that the terms involving pressure and gravity forces correspond to momentum-generation terms just as the term involving Φ_H in the incremental energy balance in the nuclear reactor [Eq. (3-1)] corresponded to a heat-production term. Dividing by $2\pi r\,\Delta r\,L$ and letting $\Delta r \to 0$, we have

$$\lim_{\Delta r \to 0} \frac{\tau_{rz} r|_r - \tau_{rz} r|_{r+\Delta r}}{r\,\Delta r} + \frac{g_c(p_0 - p_L)}{L} + \rho g_z = 0 \qquad (3\text{-}14)$$

and the *flux differential equation* becomes

$$\boxed{-\frac{d(\tau_{rz} r)}{r\,dr} + \frac{g_c(p_0 - p_L) + \rho g_z L}{L} = 0} \qquad (3\text{-}15)$$

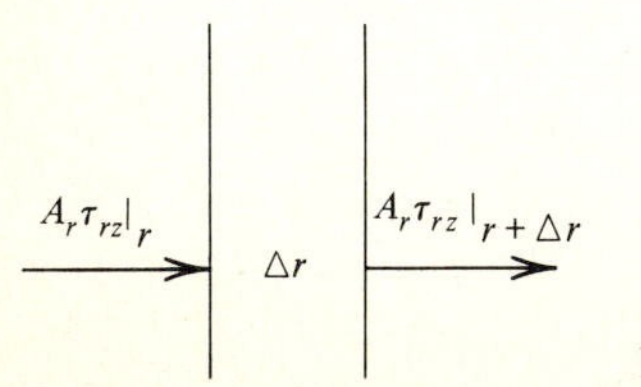

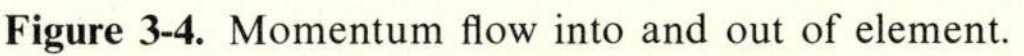
Figure 3-4. Momentum flow into and out of element.

We refer now to the flux differential equation derived for heat transport

$$\frac{d(rq_r)}{r\,dr} = \Phi_H \tag{3-2}$$

the physical meaning of which is

$$\begin{matrix}\text{Net rate of diffusion} \\ \text{of heat per unit} \\ \text{volume at arbitrary location } r\end{matrix} = \begin{matrix}\text{Rate of heat production} \\ \text{per unit volume at} \\ \text{arbitrary location } r\end{matrix}$$

If we define a quantity analogous to Φ_H as

$$\Phi_M \equiv \frac{(p_0 - p_L)g_c + \rho g_z L}{L} \tag{3-16}$$

we note that it has the dimensions of momentum per unit time and unit volume (the same as force per unit volume times g_c); thus it can be regarded as a source term for momentum. Then Eq. (3-15) becomes

$$\frac{d(r\tau_{rz})}{r\,dr} = \Phi_M \tag{3-17}$$

and has the physical meaning

$$\begin{matrix}\text{Net rate of diffusion of} \\ \text{momentum per unit volume at } r\end{matrix} = \begin{matrix}\text{Rate of production of momentum} \\ \text{per unit volume at } r\end{matrix}$$

The flux laws for the two examples are also analogous, since

$$\boxed{q_r = -k\frac{dT}{dr} \qquad \tau_{rz} = -\mu\frac{dv_z}{dr}}$$

We indicated in Chap. 1 that in order for an analogy to exist the flux differential equations, the flux laws, and the boundary conditions must all be similar. Let us now compare the boundary conditions for the present example with that for the nuclear reactor:

	Reactor	*Pipe flow*	
At $r = 0$:	$rq_r = 0$	$r\tau_{rz} = 0$	(3-18)
At $r = T$:	$T = T_R$	$v_z = 0$	

As in Chap. 1, we have used the no-slip boundary condition that $v_z = 0$ at the wall.† Since T_R is a constant, the above boundary conditions are analogous if we use $T - T_R$ as an energy potential analogous to v_z. The boundary condition at the center specifies that the flux is finite, a condition required to integrate the flux

†This condition appears to hold for all fluid systems except gases at extremely low pressures and certain polymer melts during extrusion through a die.

differential equation between the limits $r = 0$ and some arbitrary r; that is,

$$\int_0^{r\tau_{rz}} d(r\tau_{rz}) = \Phi_M \int_0^r r\,dr \qquad r\tau_{rz} = \frac{\phi_M r^2}{2}$$

and
$$\tau_{rz} = \frac{\Phi_M r}{2} = \frac{\tau_R r}{R} \tag{3-19}$$

where τ_R is the momentum flux at $r = R$.

Equation (3-19), called the *law of linear shear-stress distribution,* indicates that the flux distribution is linear. Note that we could also write (3-19) as

$$\sigma_{rz} = \frac{\sigma_R r}{R} \tag{3-20}$$

where σ_{rz} is τ_{rz}/g_c and σ_R is the shear stress at the wall. Equating the flux given by the flux law, Eq. (1-50), with that given by Eq. (3-19), we obtain

$$-\mu \frac{dv_z}{dr} = \frac{\Phi_M r}{2}$$

After separating variables, this equation can be integrated between the limits of an arbitrary r to $r = R$ to obtain the equation for the velocity profile

$$v_z = \frac{\Phi_M R^2}{4\mu}\left[1 - \left(\frac{r}{R}\right)^2\right] \tag{3-21}$$

Note the similarity between the results for energy and momentum transport. The flux distributions and the temperature and velocity profiles are illustrated in Fig. 3-5*a, b* for the two examples. In each case, we have a *linear* flux distribution and a parabolic potential profile. The fluxes are maximum at the wall and zero at the center. This is consistent with the fact that the profiles are maximum at the

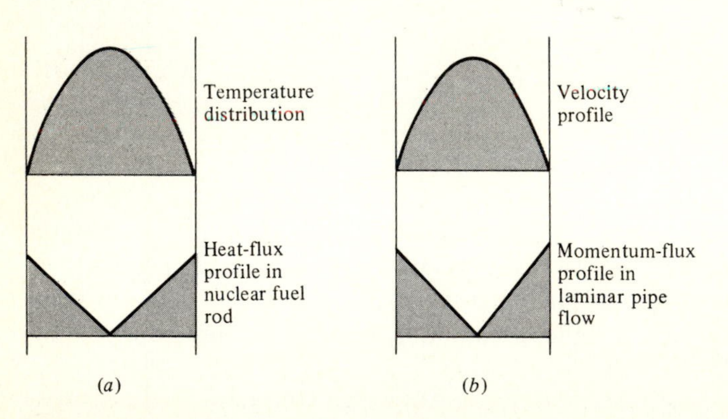

Figure 3-5 (*a*) Temperature and heat flux in a nuclear fuel reactor. (*b*) Velocity and momentum flux in laminar pipe flow.

centerline, so that

$$\left.\frac{dT}{dr}\right|_{r=0} \quad \text{and} \quad \left.\frac{dv_z}{dr}\right|_{r=0} = 0$$

Since the two problems are analogous, it was unnecessary to carry out the mathematical steps for momentum transport. Having observed that the differential equations, flux laws, and boundary conditions were similar in each case, we could have found the momentum flux and velocity distributions by analogy to the heat-flux and temperature distributions. Thus in Eqs. (3-3) and (3-4) we would merely need to replace q_r with τ_{rz}, Φ_H with Φ_M, and $T - T_R$ with v_z in order to obtain Eqs. (3-19) and (3-21).

In general, if the solution to one transport problem is available, say in the literature, the solution to an analogous problem can be written down by the above procedure. And even in the more complex problems in which the analytical solution *cannot* be obtained it is often possible to make use of the analogy by experimentally measuring one quantity and using these results to obtain another. For example, if one wished to estimate the maximum temperature in a nuclear reactor more complicated than the one described here, it might be easier (and safer) to measure the maximum velocity in an analogous fluid-flow experiment and then use the analogy to calculate the temperature in the reactor before trying to measure temperatures in the reactor directly. This procedure is illustrated in Chap. 7.

Total Transport Rate and Macroscopic Balance

In the case of heat transport we found the total transport of heat at the wall by evaluating the heat flux at the wall and multiplying by the wall area [see Eq. (3-5)]. Similarly, the magnitude of the total momentum-transport rate at the wall in the pipe-flow problem can be obtained by evaluating the momentum flux at the wall from Eq. (3-19) and multiplying by the wall area. Since momentum is lost, $\dot{P}_z$ will be negative, just as $\dot{Q}_G$ was negative in the energy case. Thus, since the velocity gradient at the wall is negative, τ_{rz} will be positive, and

$$-\left.\dot{P}_z\right|_{r=R} = \left.\tau_{rz}A_r\right|_{r=R} = \Phi_M(\pi R^2 L) \tag{3-22}$$

The *physical interpretation* is that $-\dot{P}_z|_R$, the rate of momentum loss at the wall, must equal the rate of momentum generation inside, which is given by the product of the volume and the momentum-generation rate per unit volume. It is apparent that this equation is the same as that obtained by means of an *overall* momentum balance on the *entire* system since there is no momentum *input* at the boundaries and the momentum output is that lost at the wall $-\dot{P}_z|_R$, which is exactly balanced by the internal generation $\Phi_M(\pi R^2 L)$.

Overall Force Balance

It is instructive to interpret Eq. (3-22) as an overall force balance. To do so we divide by $-g_c$ and substitute for Φ_M its definition [Eq. (3-16)]. This gives the viscous force at the wall

$$F_{vz}\Big|_R = \frac{1}{g_c}\dot{P}_z\Big|_R = -\frac{[(p_0 - p_L)g_c + \rho g_z L]}{L g_c}(\pi R^2 L)$$

$$= -(p_0 - p_L)(\pi R^2) - \frac{\rho g_z(\pi R^2 L)}{g_c} \tag{3-23}$$

The force $F_{vz}|_R$ is the frictional viscous force of the wall on the fluid. Since the direction of F_{vz} will be opposite the direction of flow, it is actually negative and it is customary to use its magnitude F_D, called the drag force. Equation (3-23) is an application of the statement that at steady state the sum of all forces acting on the entire fluid is zero; i.e.,

$$\underset{\substack{\text{Viscous}\\\text{force}\\\text{at wall}}}{F_{vz}\Big|_R} + \underset{\substack{\text{Net pressure}\\\text{force on}\\\text{cross section}}}{F_{pz}} + \underset{\substack{\text{Net gravity}\\\text{force on total}\\\text{volume}\\\text{of fluid}}}{F_{gz}} = 0 \tag{3-24}$$

The viscous force is negative in our example, since it acts in the negative z direction (down), while the net pressure force is positive and the gravity force is negative since we have assumed upward flow in the $+z$ direction (Fig. 3-6). In order for flow to occur in the $+z$ direction, that is, for v_z to be positive, p_0 must be sufficiently greater than p_L to overcome the effects of gravity and the friction of the wall on the fluid. If we observe our sign conventions, Eqs. (3-23) and (3-24) will hold regardless of the direction of flow; but if we use the magnitudes of the

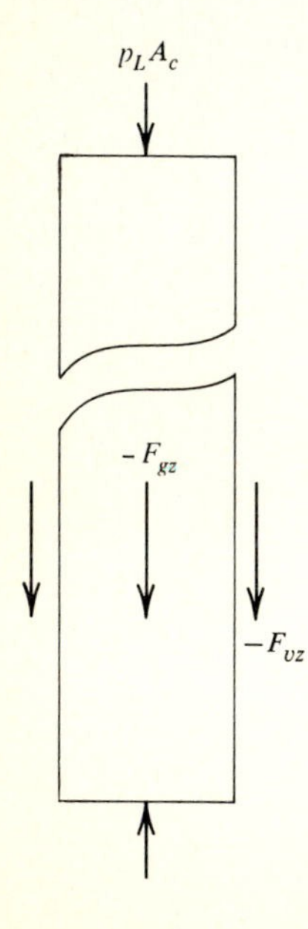

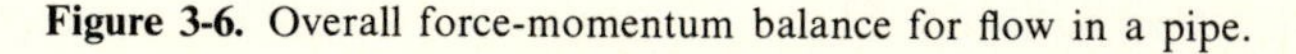
Figure 3-6. Overall force-momentum balance for flow in a pipe.

forces, we would write Eq. (3-24) for *upward* flow as

$$F_D = F_p - F_g$$

where $F_g \equiv |F_{gz}| = \dfrac{\rho(\pi R^2 L)g}{g_c}$

$$F_D = |F_{vz}|\Big|_R$$

Hagen-Poiseuille Equation

Returning now to our earlier objective of obtaining an expression for the pressure drop in the tube in terms of the average velocity, we first integrate Eq. (3-21) to obtain the average velocity, using Eq. (3-9)

$$v_a = \frac{\int_0^R v_z(2\pi r\,dr)}{\int_0^R 2\pi r\,dr} = \frac{2}{R^2}\int_0^R v_z r\,dr \tag{3-25}$$

It is convenient to write v_z in terms of the maximum velocity v_M

$$v_z = v_M\left[1 - \left(\frac{r}{R}\right)^2\right] \tag{3-26}$$

where
$$v_M \equiv \frac{\Phi_M R^2}{4\mu} \tag{3-27}$$

Then
$$v_z = \frac{2v_M}{R^2}\int_0^R \left[1 - \left(\frac{r}{R}\right)^2\right] r\,dr = \frac{2v_M}{R^2}\left(\frac{r^2}{2} - \frac{r^4}{4R^2}\right)\Bigg|_0^R$$

or
$$v_a = \frac{v_M}{2} \tag{3-28}$$

The maximum velocity is therefore twice the average. Substituting Eq. (3-27) for v_M in terms of Φ_M and using the definition of Φ_M in Eq. (3-16), from Eq. (3-28) we get

$$v_a = \frac{\Phi_M R^2}{8\mu} = \frac{g_c(p_0 - p_L) + \rho g_z L}{8\mu L} R^2 \tag{3-29}$$

We define the *dynamic* or *modified pressure* by

$$\mathcal{P} \equiv p - \frac{\rho g_z z}{g_c} \tag{3-30}$$

Thus $\mathcal{P}_0 = p_0$, $\mathcal{P}_L = p_L - \rho g_z L/g_c$, and

$$p_0 - p_L + \frac{\rho g_z L}{g_c} \equiv \mathcal{P}_0 - \mathcal{P}_L \tag{3-31}$$

Solving Eq. (3-29) for $\mathcal{P}_0 - \mathcal{P}_L$ gives the *Hagen-Poiseuille equation*

$$\mathcal{P}_0 - \mathcal{P}_L = \frac{8\mu L v_a}{R^2 g_c} \tag{3-32}$$

The quantity $\mathcal{P}_0 - \mathcal{P}_L$, termed the *dynamic pressure difference*, is the actual driving force causing flow.† As we have seen, the term

$$\Phi_M \equiv g_c \frac{\mathcal{P}_0 - \mathcal{P}_L}{L} = \frac{(p_0 - p_L)g_c + \rho g_z L}{L} \tag{3-16}$$

represents the momentum-generation rate per unit volume. We note that it is positive, negative, or zero depending upon the sign of $p_0 - p_L$ and of g_z. If the momentum-generation term is negative, the physical interpretation is that flow is in the $-z$ direction and negative z momentum is being generated, i.e., that $v_z < 0$. If Φ_M is zero, we have the condition of no flow.

It is often convenient to express Eq. (3-29) in terms of the volumetric rate of flow $\dot{\mathcal{V}}$ is multiplying v_a by the cross-sectional area $A_c = \pi R^2$. Then

$$\mathcal{P}_0 - \mathcal{P}_L = \frac{8\mu L \dot{\mathcal{V}}}{\pi R^4 g_c} \tag{3-33}$$

In terms of the mass rate of flow w, Eq. (3-32) becomes

$$\mathcal{P}_0 - \mathcal{P}_L = \frac{8\mu L w}{\pi R^4 \rho g_c} \tag{3-34}$$

Equation (3-32) was first obtained experimentally by Jean Louis Poiseuille in 1826, during a study of the flow of water through capillary tubes, and independently in 1839 by G. Hagen. Today, Eqs. (3-29) and (3-32) are known as the Hagen-Poiseuille equation and are perhaps the best-known relations in all of fluid mechanics.

†In order to see this, consider the case in which there is *no* flow; i.e., the fluid is at rest in the tube. Then $v_a = 0$, and, from Eq. (3-32), $\mathcal{P}_0 - \mathcal{P}_L$ must be zero and

$$p_0 + \frac{\rho g_z L}{g_c} = p_L$$

If we multiply by the cross-sectional area A_c, we obtain the force balance

$$p_0 A_c + \frac{\rho g_z L A_c}{g_c} = p_L A_c$$

This equation properly states that the pressure force acting downward at $z = L$ is greater than the pressure force acting upward at $z = 0$ by the force due to gravity $\rho g_z A_c L/g_c$ when the fluid is at rest. If we divide the gravitational force by ρA_c, we obtain a quantity $g_z L/g_c$, called the *hydrostatic head*, which has the English gravitational units of ft · lbf/lbm. But since the ratio g_z/g_c is numerically equal to unity, the hydrostatic head is frequently measured in units such as feet of water. If the fluid were not at rest, the term $(\mathcal{P}_0 - \mathcal{P}_L)/\rho$ would be the total difference in head. Some engineers define the dynamic pressure as $\mathcal{P} \equiv p + \rho g h/g_c$, where h is the distance *above* a reference point and g is *always* positive. In the above example $h = L$ at $z = L$ which agrees with Eq. (3-30) since $g_z = -g$.

3-4 TRANSPORT IN AN ANNULUS WITH INTERNAL GENERATION

Now let us consider two similar transport problems involving an annular geometry. In the first instance, we study the laminar flow of a newtonian fluid in the annulus, while in the second case we deal with heat transfer from an annular-shaped nuclear fuel element. These two situations are illustrated in Fig. 3-7. The inner and outer radii of the annular regions are denoted by R_1 and R_2, respectively, and $\kappa \equiv R_1/R_2$ is the radius ratio. The inner and outer walls of the fuel element are maintained at T_0, and the fluid velocity at R_1 and R_2 is zero (no-slip boundary condition). The pressure drop in the laminar-flow problem can be treated as a source term for momentum, just as in the previous pipe-flow problem. Hence, we anticipate that these annular problems will be exactly analogous. Let us prove it by a careful analysis of the systems.

Heat Transport

In this problem we are interested in the temperature profile and in the total heat loss $-\dot{Q}_t$ from *both* surfaces. Again we consider an incremental annular element of area $2\pi r\,\Delta r$ and length L, lying between R_1 and R_2 at arbitrary r. The energy

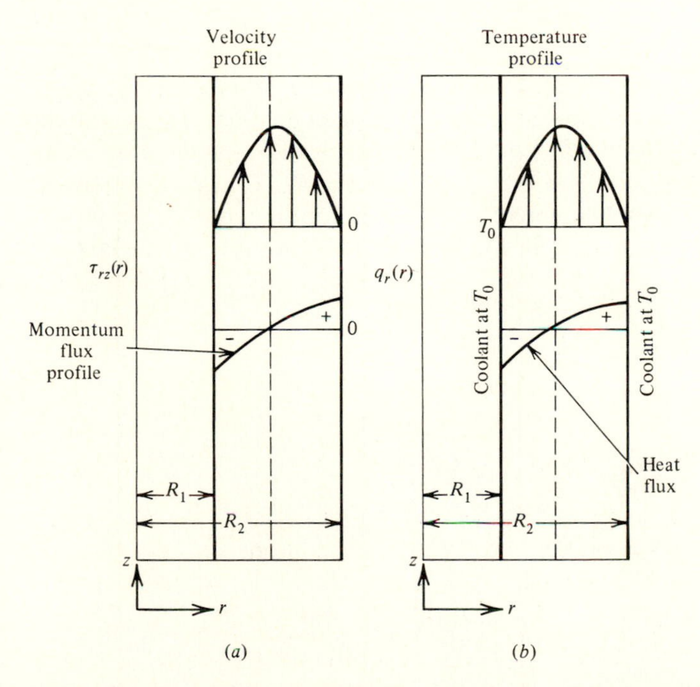

Figure 3-7 (*a*) Laminar flow in an annulus (momentum transport). (*b*) Heat generation in an annular fuel element (heat transport).

balance Eq. (1-1) can be written, at steady state, as

$$\underset{\text{Output rate}}{2\pi r L q_r\Big|_{r+\Delta r}} - \underset{\text{Input rate}}{2\pi r L q_r\Big|_{r}} = \underset{\text{Generation rate}}{2\pi r\,\Delta r\, L\Phi_H} \tag{3-35}$$

where Φ_H is the heat generated by the fuel rod per unit volume and $2\pi r\,\Delta r\,L$ is the volume of the differential element. Dividing Eq. (3-35) by $2\pi L\,\Delta r$ and taking the limit as $\Delta r \to 0$ gives

$$\frac{d(rq_r)}{dr} = r\Phi_H \tag{3-36}$$

Note that this equation is identical with Eq. (3-2), obtained earlier for the cylindrical rod. This is true because each equation is based on the same incremental balance and these incremental (or differential) changes are independent of the boundary conditions since they apply at an arbitrary internal point. Integrating Eq. (3-36), we obtain

$$rq_r = \frac{\Phi_H r^2}{2} + C_1$$

and

$$q_r = \frac{\Phi_H r}{2} + \frac{C_1}{r} \tag{3-37}$$

In this case, we cannot eliminate C_1 by arguing, as we did in Eq. (3-2), that q_r is not infinite at $r = 0$, because the equations developed hold only in the annular region $R_1 \leq r \leq R_2$ and for $r < R_1$ Eq. (3-35) is inapplicable. On the other hand, we have a new boundary condition that $T = T_0$ at $r = R_1$. To use this boundary condition, however, we must first derive an equation for the temperature profiles by using a flux law to eliminate q_r in Eq. (3-37). Substituting Fourier's law into Eq. (3-37) gives

$$-k\frac{dT}{dr} = \frac{\Phi_H r}{2} + \frac{C_1}{r} \qquad \text{and} \qquad \frac{dT}{dr} = -\frac{\Phi_H r}{2k} - \frac{C_1}{kr}$$

Integrating, we get

$$T = -\frac{\Phi_H}{4k}r^2 - \frac{C_1}{k}\ln r + C_2 \tag{3-38}$$

To eliminate C_1 and C_2 we use the two boundary conditions

$$T_0 = \begin{cases} -\dfrac{\Phi_H R_1^2}{4k} - \dfrac{C_1}{k}\ln R_1 + C_2 & \text{at } r = R_1 \\[2ex] -\dfrac{\Phi_H R_2^2}{4k} - \dfrac{C_1}{k}\ln R_2 + C_2 & \text{at } r = R_2 \end{cases}$$

Solving for C_1 and C_2 gives

$$C_1 = -\frac{\Phi_H(R_2^2 - R_1^2)}{4\ln(R_2/R_1)} \tag{3-39}$$

$$C_2 = T_0 + \frac{\Phi_H R_1^2}{4k} + \frac{\Phi_H \ln R_1}{4k}\frac{R_1^2 - R_2^2}{\ln(R_2/R_1)} \tag{3-40}$$

Substituting these results into Eq. (3-38), we obtain the temperature profile

$$\boxed{T - T_0 = \frac{\Phi_H R_1^2}{4k}\left(1 - \frac{r^2}{R_1^2}\right) + \frac{\Phi_H(R_2^2 - R_1^2)}{4k\ln(R_2/R_1)}\ln\frac{r}{R_1}} \tag{3-41}$$

We can now find the flux distribution with the use of Fourier's law by differentiating Eq. (3-41)

$$q_r = -k\frac{dT}{dr} = \frac{\Phi_H r}{2} - \frac{\Phi_H}{4r}\frac{R_2^2 - R_1^2}{\ln(R_2/R_1)} \tag{3-42}$$

These profiles are illustrated in Fig. 3-7*b*.

Net Rate of Heat Loss

To calculate the *net rate of heat loss* $-\dot{Q}_t$ we can use an overall energy balance. Since there is no input, the *overall balance* is

$$\underset{\substack{\text{Rate}\\\text{of}\\\text{out-}\\\text{flow}}}{-\dot{Q}_t} = \underset{\text{Generation}}{\Phi_H(\pi R_2^2 - \pi R_1^2)L} \tag{3-43}$$

where $(\pi R_2^2 - \pi R_1^2)L$ is the total volume of the fuel element. We could also calculate $\dot{Q}_t$ from the fluxes at R_1 and R_2. We recall that the magnitude of the heat flow rate at a surface is the product of the flux q_r and the area $2\pi rL$, both evaluated at the same location. Now at $r = R_2$, q_r is positive because dT/dr is negative; hence $-q_rA_r$ will be negative, meaning that there is a heat *loss*. But at $r = R_1$, q_r is negative because dT/dr is positive; hence we must use $-q_rA_r|_{R_1}$ to obtain the contribution of the inner wall to the heat loss. Since by convention $\dot{Q}_t$ is positive if heat is *gained*, the net rate of heat loss is†

$$-\dot{Q}_t = q_rA_r\Big|_{r=R_2} - q_rA_r\Big|_{r=R_1}$$

Then

$$-\dot{Q}_t = 2\pi R_2L(q_r)\Big|_{r=R_2} - 2\pi R_1L(q_r)\Big|_{r=R_1}$$

†The mathematical notation gives $\dot{Q}_t = -n_rA_rq_r|_{R_2} - n_rA_rq_r|_{R_1}$ where $n_r|_{R_1} = -1$ and $n_r|_{R_2} = +1$ [see Eq. (1-19)].

Using Eq. (3-42), we have

$$q_r\Big|_{r=R_2} = \frac{\Phi_H R_2}{2} - \frac{\Phi_H}{4R_2}\frac{R_2^2 - R_1^2}{\ln(R_2/R_1)}$$

and

$$q_r\Big|_{r=R_1} = \frac{\Phi_H R_1}{2} - \frac{\Phi_H}{4R_1}\frac{R_2^2 - R_1^2}{\ln(R_2/R_1)}$$

Substituting the above expression for the fluxes and simplifying gives

$$-\dot{Q}_t = (\pi R_2^2 - \pi R_1^2)L\Phi_H$$

which is identical to Eq. (3-43).

Momentum Transport

In this case we are interested in obtaining expressions for the velocity profile and drag force on the wall as well as a relationship between the flow rate and pressure drop. We could proceed as before, by making a force-momentum balance on a differential element $2\pi r\,\Delta r\,L$; but this is unnecessary since the forces acting on the element are the same as for the pipe because the boundary conditions are not involved, and therefore the same differential equation will be obtained, namely,

$$\frac{d(r\tau_{rz})}{r\,dr} = \frac{g_c(p_0 - p_L)}{L} + \rho g_z \tag{3-15}$$

or

$$\frac{d(r\tau_{rz})}{r\,dr} = \Phi_M \tag{3-44}$$

where

$$\Phi_M \equiv g_c\frac{p_0 - p_L}{L} + \rho g_z \tag{3-45}$$

The boundary conditions are $v_z = 0$ both at $r = R_1$ and at $r = R_2$. We note that the flux differential equations and flux laws for the heat and momentum cases differ only in their notation and that the boundary conditions will be similar if we replace $T - T_0$ with v_z. Therefore the problems *are* analogous, and we can simply use the heat-transfer result, with Φ_H replaced by Φ_M and k replaced by μ. The velocity profile is thus

$$\boxed{v_z = \frac{\Phi_M R_1^2}{4\mu}\left(\frac{1 - r^2}{R_1^2}\right) + \frac{\Phi_M(R_2^2 - R_1^2)}{4\mu\ln(R_2/R_1)}\ln\frac{r}{R_1}} \tag{3-46}$$

Similarly, the momentum flux is obtained by analogy to Eq. (3-42) as

$$\tau_{rz} = \frac{\Phi_M r}{2} - \frac{\Phi_M}{4r}\frac{(R_2^2 - R_1^2)}{\ln(R_2/R_1)} \tag{3-47}$$

Equation (3-46) is plotted in Fig. 3-7*a* along with Eq. (3-47).

The total or net momentum loss at the walls is given by

$$-\dot{P}_{zt} = \tau_{rz}\bigg|_{r=R_2}(2\pi R_2 L) - \tau_{rz}\bigg|_{r=R_1}(2\pi R_1 L) \tag{3-48}$$

where again a minus sign is required for the term at $r = R_1$. At this point we can turn to our heat-transfer result Eq. (3-43) and write by analogy

$$-\dot{P}_{zt} = (\pi R_2^2 - \pi R_1^2)L\Phi_M \tag{3-49}$$

Dividing by $-g_c$ and using the definition of Φ_M [Eq. (3-45)], we find the net viscous force at the walls to be

$$F_{vzt} = \frac{\dot{P}_{zt}}{g_c} = -\frac{(\pi R_2^2 - \pi R_1^2)L}{g_c}\left[\frac{g_c(p_0 - p_L)}{L} + \rho g_z\right]$$

Then

$$\boxed{F_{vzt} = -(\pi R_2^2 - \pi R_1^2)(p_0 - p_L) + (\pi R_2^2 - \pi R_1^2)L\frac{g_z}{g_c}\rho} \tag{3-50}$$

or

$$-F_{vz} = F_{pz} + F_{gz} = 0 \tag{3-51}$$

where F_{pz} = pressure force
F_{gz} = weight of fluid
$-F_{vzt}$ = total drag force

Hence, Eq. (3-50) states that the drag force exerted by the walls exactly balances the forces due to gravity and pressure.

Pressure Drop in Annulus

The most useful relation is that between the average velocity v_a and the dynamic-pressure difference

$$\mathcal{P}_0 - \mathcal{P}_L \equiv p_0 - p_L + \frac{\rho g_z L}{g_c}$$

This can easily be obtained by integrating Eq. (3-46) across the cross section of the annulus using the general equation (3-10) for average velocity

$$v_a = \frac{1}{\pi R_2^2 - \pi R_1^2}\int_{R_1}^{R_2} v_z 2\pi r\, dr$$

We find that

$$v_a = \frac{(\mathcal{P}_0 - \mathcal{P}_L)g_c R_2^2}{8\mu L}\phi(\kappa) \tag{3-52}$$

where

$$\phi(\kappa) \equiv \frac{1-\kappa^4}{1-\kappa^2} - \frac{1-k^2}{\ln(1/\kappa)} \qquad \kappa \equiv \frac{R_1}{R_2} \tag{3-53}$$

3-5 MASS TRANSPORT WITH INTERNAL GENERATION[1,2]

The general balance equation (1-1) can be written in the units of mass per unit time or moles per unit time for an individual species or for the total of all species. Of course, on a total-mass basis, the net production of all species must be zero.

As illustrated in Chap. 1, input and output terms are obtained by multiplying the total flux (convective plus diffusive) at the boundary of the system by the area. The production term represents the net rate of generation of a species due to all simultaneous chemical reactions. If the species is a reactant, this term is negative. Just as we previously found it convenient to define the term Φ_H as the rate of heat generation per unit volume and the term Φ_M as the rate of momentum generation per unit volume, we can also define a term R_A as the rate of production of moles of species A per unit volume. For example, for an nth-order irreversible reaction, if A is the only reactant,

$$R_A = -k_R C_A^n \tag{3-54}$$

where k_R is the specific reaction-rate constant and n is the order of the reaction. For a zero-order reaction ($n = 0$), R_A would be constant; this could result in a physical situation analogous to the problems in heat and momentum transport discussed previously. However, since very few reactions are actually zero order, if the mass-transport problem is to have the same solution as an analogous transport problem, it would be necessary for Φ_H to vary with temperature in the same manner that R_A varies with concentration in the mass-transport problem.

In summary, while the process of mass transport with internal generation obeys the same general conservation law as the other transport processes, the functional relationship for the generation terms is not likely to be the same. This means that we can make use of the analogy in writing incremental differential or overall balances but not in obtaining final solutions.

Example 3-1 A gas stream composed of a reactant A and product B is flowing past a spherical catalyst pellet in a reactor. The concentration C_{As} of A at the outer surface of the pellet is assumed constant. The reactant A diffuses from the surface through the pores of the catalyst and reacts in the pellet. Product B is formed and diffuses back to the outer surfaces and into the gas stream. Find the concentration distribution in the pellet and the total transport of A at the surface.

SOLUTION We write an incremental molar balance on A over a spherical shell of thickness Δr. Since the surface area (perpendicular to the r axis) of the shell is $A_r = 4\pi r^2$ and the volume is $4\pi r^2\,\Delta r$, the incremental molar balance becomes

$$\underbrace{N_{Ar}A_r\Big|_r}_{\text{Input}} - \underbrace{N_{Ar}A_r\Big|_{r+\Delta r}}_{\text{Output}} + \underbrace{R_A(4\pi r^2\,\Delta r)}_{\text{Generation}} = \underbrace{0}_{\text{Accumulation}} \tag{3-55}$$

where R_A is the molar rate of production of species A per unit volume of catalyst. Since A is a reactant, R_A is negative.

Substituting for A_r, dividing by $4\pi r^2 \Delta r$, and letting Δr approach zero gives

$$\lim_{\Delta r \to 0} \frac{N_{Ar}(4\pi r^2)\big|_r - N_A(4\pi r^2)\big|_{r+\Delta r}}{4\pi r^2 \Delta r} + R_A = 0$$

or

$$\frac{-d(N_{Ar}r^2)}{r^2\,dr} + R_A = 0 \tag{3-56}$$

Neglecting bulk flow and using Fick's law of diffusion, we have

$$N_{Ar} = -D_{Ac}\frac{dC_A}{dr}$$

where D_{Ac} is the diffusivity of A through the catalyst pores. Assuming a first-order reaction leads to

$$R_A = -k_R C_A$$

where k_R is called the specific rate constant. Equation (3-56) becomes

$$D_{Ac}\frac{d\left(r^2\dfrac{dC_A}{dr}\right)}{r^2\,dr} - k_R C_A = 0 \tag{3-57}$$

If we let

$$C_A(r) \equiv \phi\frac{(r)}{r} \tag{3-58}$$

we have

$$\frac{dC_A}{dr} = -\frac{\phi}{r^2} + \frac{1}{r}\frac{d\phi}{dr}$$

and

$$\frac{d}{dr}\left(r^2\frac{dC_A}{dr}\right) = \frac{d}{dr}\left(-\phi + r\frac{d\phi}{dr}\right) = r\frac{d^2\phi}{dr^2}$$

Then Eq. (3-57) becomes

$$\frac{D_{Ac}}{r}\frac{d^2\phi}{dr^2} - k_R\frac{\phi}{r} = 0$$

or

$$D_{Ac}\frac{d^2\phi}{dr^2} - k_R\phi = 0 \tag{3-59}$$

The solution to this second-order equation (see Appendix 1-2) is

$$rC_A \equiv \phi = C_1\cosh\sqrt{\frac{k_R}{D_{Ac}}}r + C_2\sinh\sqrt{\frac{k_R}{D_{Ac}}}r \tag{3-60}$$

where $\quad \cosh u \equiv \frac{1}{2}(e^u + e^{-u}) \qquad \sinh u \equiv \frac{1}{2}(e^u - e^{-u})$

The boundary conditions are

$$\phi = \begin{cases} 0 & C_{\mathrm{A}} \text{ finite} \quad \text{at } r = 0 & \text{(3-61}a\text{)} \\ r_s C_{\mathrm{A}s} & C_{\mathrm{A}} = C_{\mathrm{A}s} \quad \text{at } r = r_s & \text{(3-61}b\text{)} \end{cases}$$

where r_s is the radius of the spherical pellet. The first boundary condition requires that $C_1 = 0$ since cosh 0 = 1. The second boundary condition gives

$$C_2 = \frac{r_s C_{\mathrm{A}s}}{\sinh (k_R / D_{\mathrm{A}c} r_s)^{1/2}}$$

Then

$$C_{\mathrm{A}} = \frac{C_{\mathrm{A}s} r_s \sinh (k_R / D_{\mathrm{A}c} r)^{1/2}}{r \sinh (k_R / D_{\mathrm{A}c} r_s)^{1/2}} \tag{3-62}$$

The total transport at the surface is

$$\mathcal{W}_{\mathrm{A}s} = N_{\mathrm{A}r} A_r \Big|_{r=r_s}' = -4\pi r_s^2 D_{\mathrm{A}c} \frac{dC_{\mathrm{A}}}{dr} \Big|_{r=R}$$

Differentiating Eq. (3-62) and substituting into the above gives

$$\mathcal{W}_{\mathrm{A}s} = 4\pi r_s D_{\mathrm{A}c} C_{\mathrm{A}s} \left\{ 1 - \left(\frac{k_R}{D_{\mathrm{A}c}} \right)^{1/2} r_s \coth \left[\left(\frac{k_R}{D_{\mathrm{A}c}} \right)^{1/2} r_s \right] \right\} \tag{3-63}$$

We now postulate a perfectly effective catalyst pellet, in which all interior pores are at C_s. The total transport at the surface would then be equal to the total production inside, or

$$\mathcal{W}_{\mathrm{A}p} = (\tfrac{4}{3}\pi r_s^3)(k_R C_{\mathrm{A}s}) \tag{3-64}$$

The ratio between the transport rates for an actual catalyst and the perfectly effective catalyst is called the *effectiveness factor*. For a sphere it is given by

$$\eta \equiv \frac{\mathcal{W}_{\mathrm{A}s}}{\mathcal{W}_{\mathrm{A}p}} = \frac{3}{\alpha^2}(\alpha \coth \alpha - 1) \tag{3-65}$$

where $\alpha = \sqrt{k_R / D_{\mathrm{A}c}}\, r_s$ is a dimensionless group called the *Thiele modulus*. For nonspherical particles, one can use an equivalent radius defined by

$$(r_s)_{\mathrm{eq}} = \frac{3V_p}{A_p}$$

where V_p and A_p are the volume and surface areas of the pellet. Thus, for a sphere $(r_s)_{\mathrm{eq}} = r_s$. □

PROBLEMS

3-1 Calculate the value of $p_0 - p_L$ for the upward flow of pure water through a 0.3-m-long tube of 0.05 m diameter if the Reynolds number is 2000.

3-2 Derive an expression analogous to the Hagen-Poiseuille equation (1-80*a*) for the laminar flow of a power-law fluid in a long cylindrical tube. Check to see that the expression reduces to the Hagen-Poiseuille equation for $n = 1.0$.

Answer:
$$\mathcal{P}_0 - \mathcal{P}_L = \frac{2(3n+1)^n}{n^n} \frac{mL\dot{\mathcal{V}}_v^n}{\pi^n R^{3n+1} g_c}$$

3-3 Rework Prob. 3-1 for a dilute aqueous polymer solution. Assume the density to be that of water and the viscosity of the polymer solution to be adequately described by $m = 0.03$ and $n = 0.4$ if τ_{yx} is in dyn/cm^2 and dv_y/dx has units of s^{-1}. Assume the mass rate of flow to be the same as in Prob. 3-1. Sketch the velocity profiles for water and the polymer solution.

3-4 Consider the steady-state fully developed flow of an incompressible fluid through an annulus and its heat-transport analogy: uniform heat generation in an annular nuclear fuel element with walls at constant temperature T_0. In each case, obtain the incremental balance, differential equation, flux distribution, flux law, boundary conditions, potential distribution (velocity and temperature), average potential, maximum potential, and total transport rate. Prepare a table listing these quantities in parallel for heat and momentum transport.

3-5 A fluid is flowing downward in a thin film on the outside of a vertical flat plate. Neglect end effects; i.e., assume fully developed flow. The film thickness is L_f.

(*a*) Write an incremental momentum balance over an element of the *film* (outside the plate).

(*b*) Write an incremental force balance over the element of film.

(*c*) Write the boundary conditions. *Note:* You must assume that the air outside the film exerts negligible drag on the film, i.e., that the momentum flux at the air interface is zero.

(*d*) Derive an equation for the velocity distribution in the film.

(*e*) Derive an equation for the volume rate of flow in the film.

(*f*) Devise a heat-transfer analogy, giving the physical situation involved, the incremental balance, the differential equation and boundary conditions, and the equations analogous to parts (*c*) and (*d*).

3-6 Work Prob. 3-5 for flow down the outside of a vertical circular tube.

3-7 An incompressible newtonian fluid is flowing under a pressure difference $p_0 - p_L$ in fully developed steady-state flow *upward* through a thin rectangular duct of thin dimension $2B$ and wide dimension W. Assume that $W/B \to \infty$; hence the flow reduces to that between a slit formed by two vertical parallel plates of essentially infinite width. Under these conditions we can assume that the velocity in the z direction (where z is measured as distance *upward*) depends only upon the distance from the midplane between the two large vertical walls. Take that distance as x so that the walls are located at $x = \pm B$. The length of the plates in the z direction is L. (Assume symmetry of velocity about the midplane; see Fig. P3-7.)

(*a*) Write incremental force and momentum balances, giving the physical meaning of all terms and including terms for the bulk flow of momentum into and out of the increment.

(*b*) Derive an equation for the momentum-flux distribution and draw a sketch showing how it varies across the slit.

(*c*) Derive an equation for the velocity profile.

(*d*) Derive an equation for the average velocity.

Answer:
$$v_a = \frac{(\mathcal{P}_0 - \mathcal{P}_L)g_c B^2}{3\mu L}$$

(*e*) Derive an equation for the volumetric rate of flow.

3-8 Consider a system of two parallel plates (assumed infinite). The upper plate is moving with velocity V in the positive z direction, and the lower plate is stationary. Let y = distance up from the bottom plate. The applied pressure gradient dp/dz is constant.

(*a*) Set up a momentum balance over a shell of thickness Δy and length L and derive the differential equation for the momentum flux.

(*b*) Assuming the fluid to be newtonian, determine the velocity profile $v_z(y)$ between the two plates.

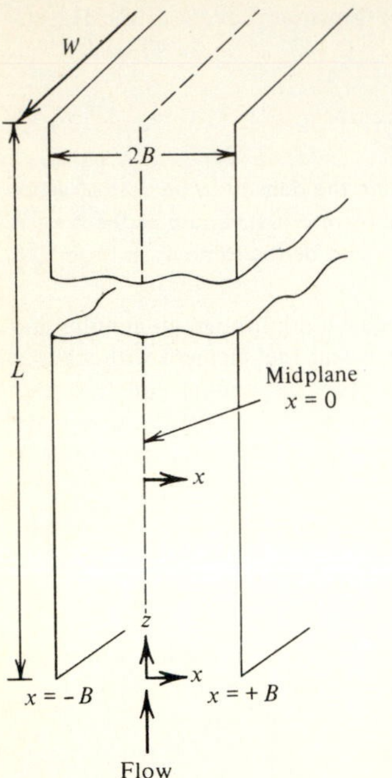

Figure P3-7 Problem 3-7.

(*c*) Derive an expression for the volumetric flow rate per unit width.

(*d*) Determine the value of dp/dz required to reduce the volumetric flow rate to zero. Sketch the velocity profile at the zero-flow condition.

(*e*) Devise a heat-transfer analogy.

(*f*) For the heat-transfer analogy determine the temperature distribution.

3-9 The power-law model for nonnewtonian fluids [Eq. (1-80*a*)] is a two-parameter equation that reduces to Newton's law for $n = 1$ and $m = \mu$. Show that the velocity profile for a power-law fluid between two infinite horizontal flat plates where the applied pressure gradient dp/dz is constant is

$$v_z = \frac{n}{n+1}\left(\frac{dp/dz}{m}\right)^{1/n}\left[(x-h)^{(n+1)/n} - h^{(n+1)/n}\right]$$

where h is one-half the plate separation and x is the distance from the lower plate.

3-10 An incompressible newtonian fluid is flowing in the annular region between two vertical coaxial cylinders of radius κ and R. The inner cylinder is moving in the $+z$ direction at a constant velocity V. The gravitational force is in the $-z$ direction.

(*a*) Derive the differential equation describing the momentum-flux and velocity distribution.

(*b*) Solve the equations to determine the velocity distribution and the momentum-flux distribution.

(*c*) Find the average velocity $\langle v_{-z} \rangle$ in the $-z$ direction.

(*d*) Find the volume rate of flow $\dot{\mathcal{V}}$ in the $-z$ direction.

(*e*) What value of V would reduce the volume flow rate to zero?

3-11 A fluid is being pumped through a well-insulated cylindrical copper pipe into the shell of a heat exchanger. In a deliberate attempt to heat the fluid, a young engineer proposes that an electric current be allowed to flow through the walls of the pipe. Since the transmission of electric current is an irreversible process and some electric energy is converted into heat, the current would result in a uniform heat-production rate in the wall of the pipe. Assuming that the insulation allows zero heat loss from the outer wall of the pipe, that the fluid is nonconducting, and that the inner pipe wall is kept at temperature T_1, derive and solve the equation describing the temperature distribution in the pipe wall. Let Φ_e Btu per unit volume be the heat-production rate in the pipe wall of outside diameter R_2 and inside diameter R_1.

3-12 In circular tubes the flow remains laminar up to a Reynolds number of 2100. Above this value turbulent flow results except in very special situations.

(*a*) What is the smallest-diameter pipe that can carry 80 gal/min of water at 70°F in laminar flow?

(*b*) What is the maximum flow rate in gallons per minute of 70°F water that can occur in a 3-in pipe under laminar conditions?

(*c*) If the Reynolds number is 4000 for the flow of 70°F water in a 2-in-diameter pipe, to what diameter should the pipe be increased to result in laminar flow if the volumetric flow rate remains constant?

REFERENCES

1. Aris, R., *Chem. Eng. Sci.,* **6:** 265 (1957).
2. Thiele, E. W., *Ind. Eng. Chem.,* **31:** 916 (1957).

CHAPTER
FOUR

TRANSFER COEFFICIENTS AND MULTIPHASE SYSTEMS

4-1 NATURE OF INTERFACIAL TRANSPORT

In the previous chapters we were primarily concerned with the transport that takes place *within* a phase. However, for application in engineering design, it is usually necessary to relate this information to the transfer† of heat, mass, or momentum from one phase to another, i.e., across or at an interface.‡ For example, in a simple double-pipe heat exchanger (Fig. 1-1) in which the inner fluid is being heated, the following processes occur:

1. Heat transport within the inner fluid toward the inner wall
2. Heat transfer through the liquid-solid interface to the wall
3. Heat transport through the wall
4. Heat transfer through the solid-fluid interface to the outer fluid
5. Heat transport within the outer fluid
6. Heat transfer through the liquid-solid interface to the outer wall
7. Heat transport through the outer wall
8. Heat transfer at the interface between wall and insulation
9. Heat transport through the insulation
10. Heat transfer from the insulation to the outside air
11. Heat transport through the outside air

†In this text we will use the term *transport* to refer primarily to *intra*phase processes and the term *transfer* to refer to *inter*phase movement.

‡An interface is the boundary between two phases; it may be solid-solid, solid-liquid, liquid-liquid, or liquid-gas, etc.

In addition to these heat-transport processes, the following momentum-transport processes occur:

1. Momentum transport within the inner fluid toward the inner wall
2. Momentum transfer to the wall
3. Momentum transport through the outer fluid toward both walls of the annulus
4. Momentum transfer to the walls of the annulus

Similarly, mass transfer at interfaces occurs in absorption columns (gas-liquid), extractors (liquid-liquid), leaching equipment (solid-liquid), and humidifiers and water coolers (gas-liquid).

The first part of this chapter is concerned with transfer between phases. In particular we are interested in using the methods and results of the previous chapters to determine the transfer coefficients: the heat-transfer coefficient h, the mass-transfer coefficient k_c, and the friction factor or drag coefficient f. These coefficients are commonly used in industrial practice for engineering design. Although they are frequently obtained empirically, it is a goal of transport phenomena to use theoretical methods to assist in their prediction.

In the latter part of this chapter, we will consider the overall transport through composite systems—systems in which transport is occurring through two or more phases, e.g., the heat exchanger discussed previously. We will show how to add the various individual resistances to heat or mass transfer in order to find an overall resistance.

A. TRANSFER COEFFICIENTS

4-2 MOMENTUM TRANSFER: FRICTION FACTOR

Consider the flow of a fluid through a pipeline, as discussed in Sec. 3-3. In the case of fully developed *laminar* flow we can use the Hagen-Poiseuille equation (3-32) to calculate the pressure drop at a certain flow rate. Similarly, for laminar flow in an annulus we can use Eq. (3-52) for the same purpose. But we derived both equations using Newton's law of viscosity, which, as indicated in Chap. 3, applies only in laminar flow, i.e., when the Reynolds number, $\mathrm{Re} \equiv 2Rv_a\rho/\mu$, is less than 2100. For turbulent flow, which occurs at larger Re and which is more prevalent in practice, there has been no generally accepted theory that could be used to predict the turbulent momentum flux (although various attempts have been made, as discussed in Chap. 6). Therefore in the past, engineers have had to rely on empirical, i.e., nontheoretical, correlations of experimental data. Since to carry out an elaborate experimental program is usually more expensive than using a theoretical result, it is desirable to reduce the amount of data taking required as much as possible. We show how this can be done through the use of *dimensionless variables*.

We recall (Sec. 1-31) that the solution to the problem of heat transport in a slab can be expressed in terms of two dimensionless variables instead of *five dimensional* quantities. Let us apply this idea to the pipe.

Dimensional Analysis

For flow in a pipe (Fig. 3-2), the dynamic-pressure drop $\mathcal{P}_0 - \mathcal{P}_L$ would be expected to depend upon the length of the pipe L, the pipe diameter D, the fluid viscosity μ, the average velocity v_a, and the fluid density ρ. Since for fully developed flow the pressure drop ought to be directly proportional to the length L, we can simplify matters by using the length-independent quantity $(\mathcal{P}_0 - \mathcal{P}_L)/L$. We postulate that

$$\frac{(\mathcal{P}_0 - \mathcal{P}_L)g_c}{L} \equiv -\frac{\Delta\mathcal{P}\, g_c}{L} = \phi(D, \mu, v_a, \rho) \tag{4-1}$$

where g_c has to be added in a gravitational system to provide the conversion from force to mass units.

In order to determine the nature of this relationship *without* using dimensional analysis, we might decide to measure the pressure drop as a function of average velocity v_a. Then we might do similar runs with pipes of different diameters and lengths. Following that we might carry out runs with fluids of different viscosities and densities, again for different diameters and lengths. We can see that the total number of runs would be quite large because of the large number of variables and parameters involved and that it would be difficult to correlate the results and to use them for conditions different from those for which the data were taken. Consequently it is highly desirable to write Eq. (4-1) in terms of dimensionless groups.

Using the procedure given in Appendix 4-1, we determine that only two dimensionless groups are needed: a dimensionless dynamic-pressure drop and a dimensionless velocity. As indicated earlier, it has been determined experimentally that the transition from laminar to turbulent flow occurs when the dimensionless Reynolds number $\mathrm{Re} \equiv 2Rv_a\rho/\mu$ is less than 2100. Therefore it would be convenient to use the Reynolds number as our dimensionless velocity. A dimensionless dynamic-pressure drop can be obtained by using the formal technique of dimensional analysis in Appendix 4-1 or as follows. Since the kinetic energy of a unit volume of the fluid is $\frac{1}{2}\rho v_a^2$, and since this has the units of g_c times pressure, a dimensionless dynamic pressure is $g_c(\mathcal{P}_0 - \mathcal{P}_L)/\frac{1}{2}\rho v_a^2$. Also, since R is a characteristic length, a dimensionless dynamic-pressure drop per unit length can be *defined for a pipe* as

$$\boxed{f \equiv \frac{g_c(\mathcal{P}_0 - \mathcal{P}_L)}{\rho v_a^2}\frac{R}{L}} \tag{4-2}$$

Then the desired dimensionless form of Eq. (4-1) is

$$f = f(\mathrm{Re}) \tag{4-3}$$

where the quantity f is called the *friction factor.*

The use of the dimensionless variables f and Re means that we need to obtain far fewer data than if we were to try to vary independently the eight quantities mentioned earlier. Therefore, in turbulent flow we would presumably merely need to vary the average velocity v_a and measure the pressure drop for the flow of *one* fluid, e.g., water, through *one* pipe (one R and one L). In practice *empirical* relationships are usually tested with more than one fluid to be sure that we have ascertained the proper dimensionless groups, but even so, the amount of experimentation is enormously reduced by the dimensionless approach.

Laminar Flow

For laminar flow through a pipe the Hagen-Poiseuille equation (3-32)

$$\frac{\mathcal{P}_0 - \mathcal{P}_L}{L} = \frac{8\mu v_a}{R^2 g_c} \tag{4-4}$$

was obtained analytically in Chap. 3. Substituting the above equation into (4-2), the definition of the friction factor, we obtain

$$f \equiv \frac{(\mathcal{P}_0 - \mathcal{P}_L) g_c R}{\rho v_a^2 L} = \frac{16}{2R v_a \rho / \mu} = \frac{16}{\mathrm{Re}} \tag{4-5}$$

General Definition of Friction Factor

It should be emphasized that Eq. (4-5) holds *only* for fully developed laminar flow of a newtonian fluid through a pipe. In order to extend this idea to other cases, a more *general* definition of friction factor is needed. It is desirable that this general definition also include fluid flows other than flow in a pipe, e.g., flow past an object (such as flow past a catalyst pellet or past the wing of an airplane). In these cases, the term *drag coefficient* (rather than friction factor) is more commonly used, and the frictional force or *drag* of the solid acting on the fluid is the desired variable instead of the pressure drop per unit length. The *general* definition of friction factor is therefore expressed in terms of a dimensionless *drag force* per unit area instead of a dimensionless pressure drop per unit length:

$$\boxed{f \equiv \frac{g_c \times \text{drag force/characteristic area}}{\text{kinetic energy/volume}}} \tag{4-6}$$

where the kinetic energy per unit volume is $\frac{1}{2}\rho v_{\mathrm{ch}}^2$ and v_{ch} is a characteristic velocity. For flow past submerged objects the characteristic area is the *projected* area, and the characteristic velocity is the approach velocity.

We restrict ourselves in this chapter to flow through *conduits,* for which the characteristic area is the wall area A_w and the characteristic velocity is the mean velocity of the fluid v_a. The drag force is then

$$F_D = \iint_{A_w} \sigma_w \, dA_w = \iint_{A_w} \frac{\tau_w}{g_c} \, dA_w \tag{4-7}$$

where A_w is the total area of the conduit wall ($2\pi RL$ for a circular pipe) and σ_w and τ_w are the magnitudes of the shear stress and momentum flux evaluated at the wall. For fully developed flow in a conduit of constant cross section, τ_w will not depend upon axial distance z or length L. Also for a pipe of circular cross section τ_w will not be dependent upon position along the perimeter of the pipe and will be a constant. Hence†

$$F_D g_c = \langle \tau_w \rangle A_w \tag{4-8}$$

Substituting this into the general definition (4-6) of friction factor, we have the *definition of f for a conduit*

$$\boxed{f \equiv \frac{g_c F_D / A_w}{\frac{1}{2}\rho v_a^2} = \frac{\langle \tau_w \rangle}{\frac{1}{2}\rho v_a^2}} \tag{4-9}$$

In order to obtain the previous expression (4-2) for friction factor we need to relate F_D or τ_w to pressure drop. In Chap. 3 a macroscopic force balance gave [Eq. (3-23)]

$$\underset{\text{Drag force}}{F_D} = \Bigg| \underset{\text{Net pressure force}}{(p_0 - p_L)(\pi R^2)} + \underset{\text{Gravity force}}{\frac{\rho g_z}{g_c}(\pi r^2 L)} \Bigg| = \Big| \mathcal{P}_0 - \mathcal{P}_L \Big| (\pi R^2) \tag{4-10}$$

The definition of the friction factor for a pipe can therefore be obtained by substituting Eq. (4-10) into Eq. (4-9)

$$f \equiv \frac{g_c(\mathcal{P}_0 - \mathcal{P}_L)}{\frac{1}{2}\rho v_a^2} \frac{R}{2L}$$

which agrees with Eq. (4-2). This shows that our *general* definition (4-6) is consistent with the special definition used previously.

Substituting for $(\mathcal{P}_0 - \mathcal{P}_L)/L$ from the Hagen-Poiseuille equation (3-32) also results in the relationship obtained previously

$$f = \frac{16}{\text{Re}} \tag{4-5}$$

†For a duct of noncircular cross section, e.g., rectangular, τ_w will depend upon location along the perimeter of the cross section. For an annulus, each wall will have a different value of τ_w. Therefore in general we use an *average* wall flux $\langle \tau_w \rangle$.

This equation tells us that, for laminar flow in a pipeline, if we know the Reynolds number, we can calculate the friction factor. Then Eq. (4-2) tells us that if we know the friction factor, we can calculate the pressure drop. An alert student will ask: Why do we need to go through the intermediate calculation of the friction factor; why don't we just use the Hagen-Poiseuille equation directly? The main reason is that it is traditional among practicing engineers to use a friction-factor–Reynolds-number relationship, especially in *turbulent* flow, where we are unable to obtain an *exact* analytical solution similar to the Hagen-Poiseuille equation. Therefore we must either use approximate equations, usually for certain ranges of Re, or, more commonly, we must prepare a graph of f versus Re based on actual experimental data. In laminar flow we can easily plot friction factor versus Reynolds number on a log-log plot as a straight line with slope of -1, but it would take numerous separate plots to show the same information with the dimensional variables.

While a friction-factor–Reynolds-number plot is actually not necessary for laminar flow (an equation being available), it is convenient to include it along with the corresponding curve obtained empirically for turbulent flow. In this manner, the design engineer can use the same plot whether the Reynolds number turns out to be greater or less than 2100. Such a chart is given in Chap. 6.

Application to Arbitrary Cross Section

Consider flow through a conduit of length L and of arbitrary cross-sectional area A_{cs}, as shown in Fig. 4-1. Let the area of the wall be

$$A_w = ZL \tag{4-11}$$

where Z is the *wetted perimeter*, i.e., the total perimeter of all wall surfaces in contact with the fluid. For example, for the annulus in Fig. 4-2, $Z = 2\pi(R_1 + R_2)$.

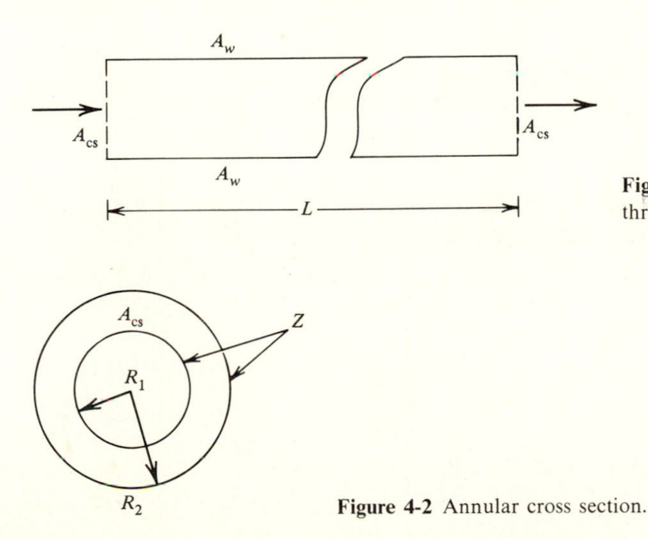

Figure 4-1 Area terms for flow through a conduit.

Figure 4-2 Annular cross section.

In order to obtain a defining expression for the friction factor that is applicable to an arbitrary cross section, it is most convenient to evaluate the drag force F_D and substitute this expression into the general definition (4-9). An overall force balance gives

$$F_D = \left| (p_0 - p_L)A_{cs} + \frac{\rho g_z}{g_c} A_{cs} L \right| \tag{4-12}$$

If we equate this to F_D, as obtained from the definition of the friction factor (4-9), we get

$$\begin{aligned} F_D g_c \equiv \left(\frac{f}{2}\rho v_a^2\right) A_w &= \left| [(p_0 - p_L)g_c + \rho g_z L] A_{cs} \right| \\ &= \left| (\mathcal{P}_0 - \mathcal{P}_L) g_c A_{cs} \right| \end{aligned} \tag{4-13}$$

Then solving for f gives

$$f \equiv \frac{(\mathcal{P}_0 - \mathcal{P}_L) g_c A_{cs}}{(\frac{1}{2}\rho v_a^2) A_w} = \frac{(\mathcal{P}_0 - \mathcal{P}_L) g_c}{(\frac{1}{2}\rho v_a^2) L} \frac{A_{cs}}{Z} = \frac{(\mathcal{P}_0 - \mathcal{P}_L) g_c}{\frac{1}{2}\rho v_a^2} \frac{R_H}{L} \tag{4-14}$$

where

$$\boxed{R_H \equiv \frac{A_{cs}}{Z} = \text{hydraulic radius} = \frac{\text{cross-sectional area}}{\text{wetted perimeter}}} \tag{4-15}$$

For a pipe we see that $R_H = \pi R^2 / 2\pi R = R/2 = D_t/4$, where D_t is the tube diameter. Hence we frequently define an *equivalent diameter*

$$\boxed{D_{eq} \equiv 4 R_H} \tag{4-16}$$

A different friction factor f' defined in some texts as

$$f' \equiv 4f \equiv \frac{(\mathcal{P}_0 - \mathcal{P}_L) g_c}{\frac{1}{2}\rho v_a^2} \frac{D_{eq}}{L} \tag{4-17}$$

is called the *Blasius friction factor,* whereas f is called the *Fanning friction factor.*

Example 4-1: Friction factor for an annulus Define friction factor for an annulus and relate it to the Reynolds number in the case of laminar flow.

SOLUTION For the annulus, the definition of friction factor and the application of the overall force or momentum balance again gives

$$f = \frac{[(\mathcal{P}_0 - \mathcal{P}_L)/L] g_c R_H}{\frac{1}{2}\rho v_a^2} \tag{4-18}$$

where $$R_H \equiv \frac{A_{cs}}{Z} = \frac{\pi(R_2^2 - R_1^2)}{2\pi(R_1 + R_2)} = \frac{R_2 - R_1}{2}$$

Again the relation between the average velocity and $\mathcal{P}_0 - \mathcal{P}_L$ would be obtained from the conservation law, the flux law, and the boundary conditions. This relation is [Eq. (3-52)]

$$v_a = \frac{(\mathcal{P}_0 - \mathcal{P}_L) g_c R_2^2}{8\mu L} \phi(\kappa) \tag{4-19}$$

where $\kappa \equiv \dfrac{R_1}{R_2}$ and $\phi(\kappa) \equiv \dfrac{1 - \kappa^4}{1 - \kappa^2} - \dfrac{1 - \kappa^2}{\ln(1/\kappa)}$

If we substitute this expression into the equation for f, Eq. (4-14), we get

$$f = \frac{8\mu v_a}{\frac{1}{2}\rho v_a^2} \frac{R_H}{R_2^2 \phi(\kappa)} = \frac{16}{4R_H \rho v_a/\mu} \frac{4R_H^2}{R_2^2 \phi(\kappa)}$$

$$= \frac{16}{(\mathrm{Re})_{\mathrm{eq}}} \frac{(1-\kappa)^2}{\phi(\kappa)} \tag{4-20}$$

where
$$(\mathrm{Re})_{\mathrm{eq}} = \frac{D_{\mathrm{eq}} \rho v_a}{\mu} = \frac{4R_H \rho v_a}{\mu} \tag{4-21}$$

We recall that for laminar flow in a tube $f = 16/\mathrm{Re}$. Hence the relationship between f and Re for an annulus is not the same as for a tube even if the equation for the latter is modified by the use of an equivalent diameter equal to $4R_H$. For an annulus this empiricism can lead to errors as large as 50 percent. However, this empirical rule is more accurate for other cross sections and for turbulent flow. In the absence of other information, it can be used as an approximation (see Problem 4.1).

Note that using a modified Reynolds number based on equivalent diameter is actually an arbitrary choice: we could have used any characteristic lateral dimension such as R_1, R_2, $2R_1$, or $4R_2$. For example, had we selected R_2 the equation would be

$$f = \frac{8\mu v_a}{\frac{1}{2}\rho v_a^2} \frac{R_H}{R_2^2 \phi} = \frac{16}{R_2 \rho v_a/\mu} \frac{R_H}{R_2 \phi} = \frac{8(1-\kappa)}{(\mathrm{Re})_2 \phi} \tag{4-22}$$

where
$$(\mathrm{Re})_2 \equiv \frac{R_2 \rho v_a}{\mu}$$

So we note that, no matter which choice we make for Re, some additional function of κ enters into the relation. Hence we can say that, for an annulus, $f = f(\mathrm{Re}, \kappa)$; that is, there are *three* dimensionless groups involved. We should observe, for future reference, that the added group κ enters into the relationship as a result of the *boundary conditions*. Later, when we deal with equations that cannot be solved analytically, it will be important to know how to identify the dimensionless groups involved to minimize any experimental work required. The procedure for doing so is discussed in Appendix 4-2. □

4-3 HEAT-TRANSFER COEFFICIENT

In engineering design problems we frequently need to be able to calculate the transport of heat, mass, or momentum to or from a surface or interface, i.e., a boundary between phases. Consider, for example, a simple heat exchanger (Fig. 4-3*a*) consisting of a fluid flowing through a pipe whose inside wall is at a temperature T_w. If the fluid enters at a uniform temperature T_0, and if $T_0 > T_w$, the fluid will be cooled immediately by transport at the pipe wall and this transport toward the center will continue as it flows through the pipe. At an arbitrary distance $z = z_1$ the temperature of the fluid will increase with distance from the wall and will be at a maximum at the center. The bulk mean temperature T_b (defined in Appendix 4-3) will vary with z (see Fig. 4-3*b*).

A typical engineering *design* problem is for a given mass rate of flow w (1) to find the length of pipe z_1 needed to reduce the bulk mean temperature of the fluid to some value T_b and (2) to find $\dot{Q}_w = -\dot{Q}_G|_{r=R}$, the total heat flow rate (energy per unit time) at the wall of the pipe. Using an overall energy balance, Eq. (1-1), over the pipe between $z = 0$ and $z = z_1$, assuming constant specific heat $\widehat{C}_p$, and an incompressible fluid gives

$$\underset{\text{Input}}{w\widehat{C}_p T_0} - \underset{\text{Output}}{w\widehat{C}_p T_{b1}} + \dot{Q}_w + \underset{\text{Generation}}{0} = \underset{\text{Accumulation}}{0}$$

or

$$-\dot{Q}_w = w\widehat{C}_p(T_0 - T_{b1}) \tag{4-23}$$

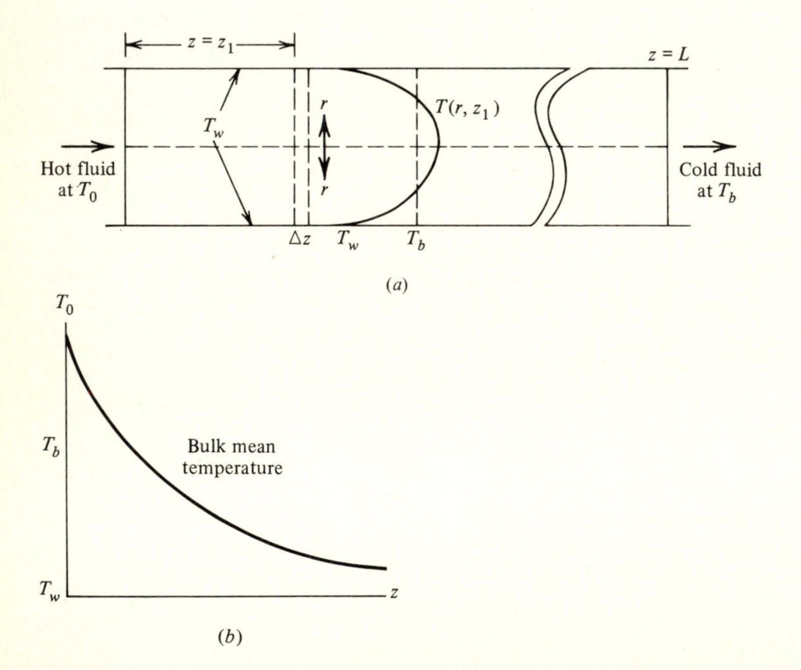

Figure 4-3 (*a*) Notation for heat exchanger. (*b*) Variation of bulk mean temperature with axial distance.

To relate the length z_1 to T_{b1} we proceed as follows. Consider an element of pipe of length Δz located at an arbitrary distance z from the entrance. Let $\Delta z \to dz$. The rate of heat removal over dz will be given by Eq. (1-4) as

$$d\dot{Q}_w = q_w \, dA_w = q_w Z \, dz \tag{4-23a}$$

where $Z = 2\pi R$
dA_w = area of wall through which heat flows
q_w = heat flux at wall

Writing an energy balance over a differential length dz gives

$$-d\dot{Q}_w = w\widehat{C}_p \, dT_b \tag{4-24}$$

Equating to (4-23*a*) gives

$$-w\widehat{C}_p \, dT_b = q_w Z \, dz \tag{4-25}$$

which can be integrated from $z = 0$ to $z = z_1$ to find a relationship between T_{b1} and z_1 if q_w is known. Our design problem therefore reduces to one of finding q_w. In order to find the heat flux q_w, engineers have historically used Newton's law of cooling,

$$|q_w| = h \left| T_b - T_w \right| \tag{4-26}$$

where T_b represents the bulk mean or average temperature of the fluid and h is called the *heat-transfer coefficient*.† At an *interface* between, say, a gas and a liquid or between two immiscible liquids we let

$$|q_I| \equiv h|T_I - T_b| \tag{4-26a}$$

where q_I = heat flux at interface
T_I = temperature at interface
T_b = bulk mean temperature of fluid

Although Eq. (4-26) is empirical rather than theoretical, it has a certain logic which, if understood, makes it possible to write it down without resorting to memory. For it says that the driving force for heat transfer to the wall is the difference between the *average* temperature of the fluid T_b and the temperature of the wall; i.e., the larger the value of the driving force $|T_b - T_w|$ the larger the heat transfer at the wall; furthermore, if the wall and the fluid are at the same temperature, there will be no heat flow.

Design of a Simple Heat Exchanger

Substituting Eq. (4-26) into Eq. (4-25) and introducing a minus sign gives

$$-w\widehat{C}_p \, dT_b = h(T_b - T_w) Z \, dz \tag{4-26b}$$

†We actually deal with the magnitude of q and we take h as positive regardless of the direction of heat transport.

Although in general h will vary with z (and so may T_w), we will assume in this simple case that T_w is constant. Since $dT_b = d(T_b - T_w)$, upon rearranging Eq. (4-26b) becomes

$$-\frac{d(T_b - T_w)}{T_b - T_w} = \frac{hZ}{w\hat{C}_p}dz \tag{4-26c}$$

Integrating from $z = 0$ (where $T_b = T_0$), to $z = z_1$ gives

$$-\ln\frac{T_{b1} - T_w}{T_0 - T_w} = \frac{Z}{w\hat{C}_p}\int_0^{z_1} h\,dz = \frac{Z\langle h\rangle z_1}{w\hat{C}_p} \tag{4-26d}$$

where

$$\langle h\rangle \equiv \frac{1}{z_1}\int_0^{z_1} h\,dz \tag{4-26e}$$

is the *mean* heat-transfer coefficient and $h(z)$ is the *local* value. Solving for z_1 gives

$$z_1 = -\frac{w\hat{C}_p}{\langle h\rangle Z}\ln\frac{T_{b1} - T_w}{T_0 - T_w} \tag{4-26f}$$

which is the desired expression for z_1 as a function of T_{b1}. Once T_{b1} is known, $\dot{Q}_w$ can be obtained from Eq. (4-23). But instead of using Eq. (4-26f) it is more conventional practice to let

$$\dot{Q}_w \equiv \langle h\rangle Z z_1 \langle T_b - T_w\rangle \tag{4-26g}$$

where $\langle T_b - T_w\rangle$ is a mean temperature difference or driving force. To find an equation for calculating it, we use (4-26g), (4-23) and (4-26d) to obtain

$$\langle T_b - T_w\rangle \equiv \frac{\dot{Q}_w}{\langle h\rangle Z z_1} = \frac{w\hat{C}_p(T_{b1} - T_0)}{w\hat{C}_p \ln[(T_{b1} - T_w)/(T_0 - T_w)]}$$

$$= \frac{(T_{b1} - T_w) - (T_0 - T_w)}{\ln[(T_{b1} - T_w)/(T_0 - T_w)]} = \langle T_b - T_0\rangle_{\mathrm{LM}} \tag{4-27}$$

where $\langle T_b - T_0\rangle_{\mathrm{LM}}$ is the log-mean temperature difference. (The log mean is defined in Chap. 1.) Thus if one specifies a desired T_{b1} and knows $\langle h\rangle$ [or $h(z)$], the required length z_1 can be calculated by equating (4-26) and (4-23) and using (4-27).

It should be noted that if the temperature and velocity are known as functions of radial position, $T_b(z_1)$ can be obtained directly (without having to know h) by integration of Tv_z over the cross section (see Appendix 4-3). Thus a heat-transfer coefficient is not needed to find T_{b1} at z_1. However, it is widely used in engineering practice, and consequently its prediction is an important application of transport phenomena.

In Chap. 10 we use a two-dimensional energy balance to obtain an analytical expression for $T(r, z)$ for laminar flow. But since analytical solutions cannot be obtained in all cases, especially for turbulent flows, simplified models are often used. One such model is the film theory, discussed briefly in this chapter and more thoroughly in Chap. 6.

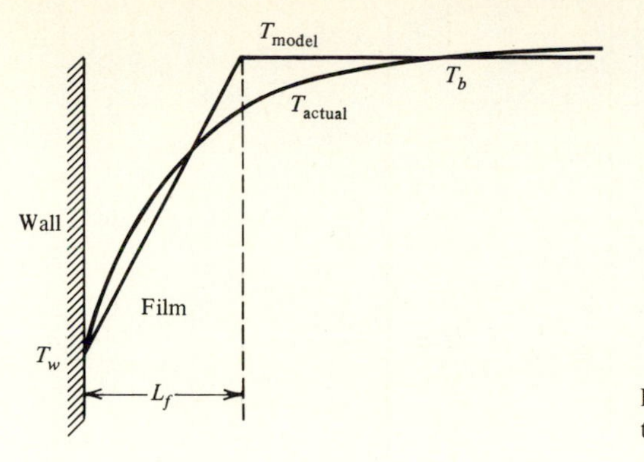

Figure 4-4 Film-theory model for heat transfer.

Film Theory

It has been found experimentally that the temperature profile in turbulent flow increases rapidly near the wall and is very flat away from the wall (see Fig. 4-4). We observed qualitatively in Example 1-1 that a steep temperature gradient corresponds to a small resistance to heat transport. Therefore, since the major resistance is near the wall, it is customary in engineering practice to postulate that *all* the resistance to heat transport occurs in the thin film of fluid near the wall. Outside this film the temperature is assumed to be the bulk temperature. Within the film, laminar flow is assumed, even though the flow outside the film may be turbulent. Figure 4-4 shows the actual case and the model. The distance L_f is called the *effective film thickness,* and the heat-transfer coefficient is often called a *film coefficient*.

4-4 MASS-TRANSFER COEFFICIENT

Now consider the case of mass transfer. If a fluid, e.g., water, were flowing through a pipe whose wall was made of a soluble material, the solid would dissolve and diffuse radially toward the center of the tube and the concentration C_A of the soluble material would decrease with distance from the wall (Fig. 4-5*a,b*). At an arbitrary downstream location there would be an average concentration, which we will call C_{Ab}. We define the mass-transfer coefficient k_c (similarly to the heat-transfer coefficient) by the equation

$$N_{Aw} = k_c(C_{Aw} - C_{Ab}) \tag{4-28}$$

where N_{Aw} is the molar flux of A at the wall.

More generally we could write

$$N_{AI} = k_c(C_{AI} - C_{Ab}) \tag{4-29}$$

where N_{AI} is the molar flux at the interface and C_{AI} is the concentration of A at

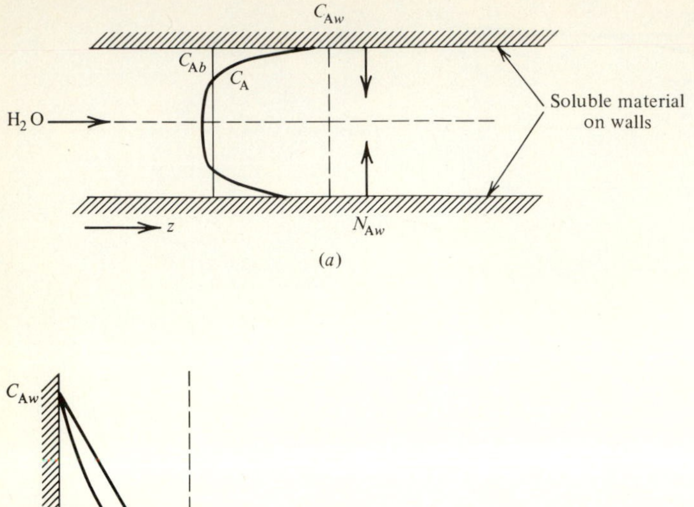

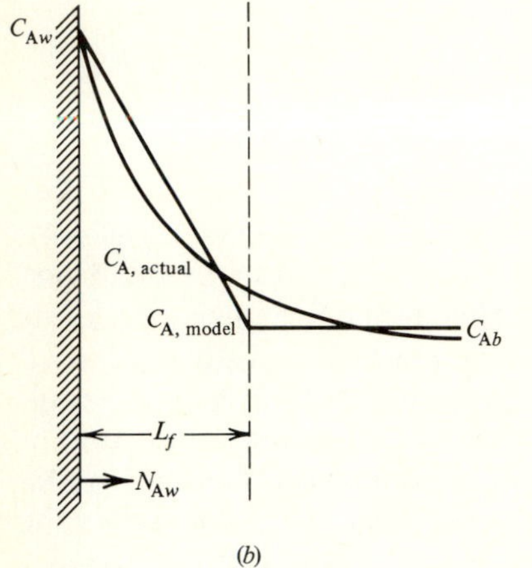

Figure 4-5 (*a*) Mass transport from soluble wall. (*b*) Film-theory model for mass transfer.

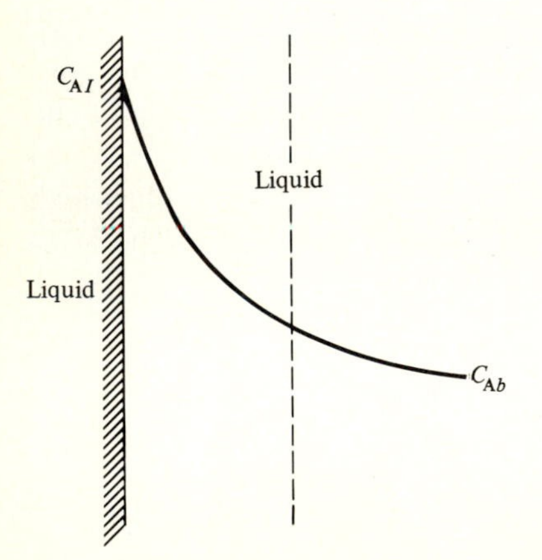

Figure 4-6 Mass transfer from liquid-liquid interface.

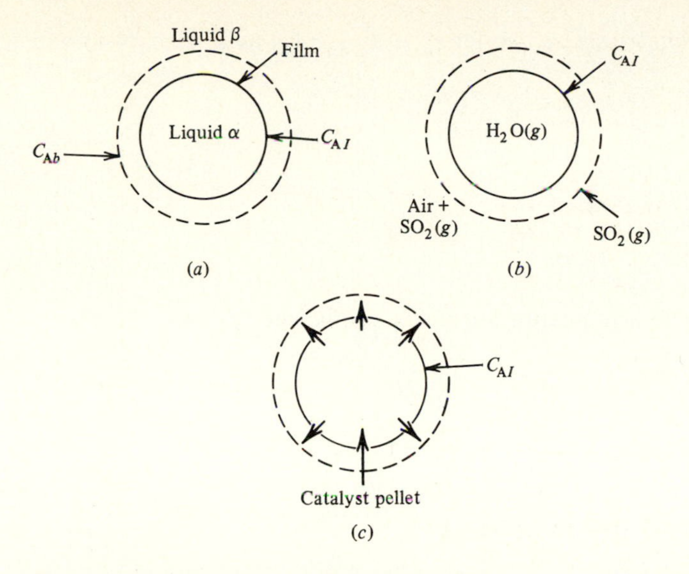

Figure 4-7 Examples of mass transfer: (*a*) liquid drop α immersed in liquid β, (*b*) absorption of SO_2 in H_2O drop, and (*c*) transport from catalyst pellet.

the interface. This equation would be especially useful if we were dealing with the design of liquid-liquid extraction equipment (Fig. 4-6). Here the picture would consist of a droplet of one liquid phase α dispersed in another liquid phase β, solute A being transferred from one phase to the other (Fig. 4-7*a*). We could write an equation like (4-29) for each phase in terms of the concentration at the interface and the mean concentration within the phase. As other examples we might consider the absorption of the SO_2 in an airstream by a water stream (Fig. 4-7*b*), the leaching of a substance from a solid, and the transport of mass from a catalyst pellet in a reactor (Fig. 4-7*c*).

4-5 MOMENTUM-TRANSFER COEFFICIENT

For *momentum transfer* we generally use the *friction factor* f (instead of a momentum-transfer coefficient). It is defined by Eq. (4-6)

$$f \equiv \frac{F_D g_c / A_w}{\frac{1}{2}\rho v_{\text{ch}}^2}$$

where F_D = drag at wall
A_w = area of wall
v_{ch} = characteristic velocity

For flow parallel to a *single surface,* we let $\tau_w = F_D g_c / A_w$, as in Eqs. (4-8) and (4-9),

$$f = \frac{\tau_w}{\frac{1}{2}\rho v_b^2}$$

where f = friction factor
τ_w = momentum flux at surface
v_b = bulk velocity of stream

Note that we *could* define a momentum-transfer coefficient $\frac{1}{2}fv_b$ by

$$\tau_w = \tfrac{1}{2}fv_b(\rho v_{zb} - \rho v_{zw}) = \tfrac{1}{2}fv_b(\rho v_b - 0) = \frac{f}{2}\rho v_b^2$$

but in practice the friction factor is used instead.

4-6 NUSSELT NUMBERS FOR HEAT AND MASS TRANSFER

Although the friction factor is dimensionless, the heat- and mass-transfer coefficients are not. Since it is desirable to use dimensionless variables (as we have seen in the case of momentum transport), it is customary to make the heat- and mass-transfer coefficients dimensionless as follows:

$$\text{Nu} \equiv \frac{hD_{\text{ch}}}{k} \qquad \text{heat-transfer Nusselt number} \tag{4-30}$$

$$\text{Sh} \equiv (\text{Nu})_{\text{AB}} = \frac{k_c D_{\text{ch}}}{D_{\text{AB}}} \qquad \text{mass-transfer Nusselt number or Sherwood number} \tag{4-31}$$

where D_{ch} = characteristic distance
k = thermal conductivity
D_{AB} = diffusivity of A through B

For flow in a pipe the characteristic length is the tube diameter. For flow in other conduits it is usually taken to be the equivalent diameter (4 times the hydraulic radius) or the hydraulic radius itself. For flow past spheres the characteristic length is the diameter of the sphere; for nonspherical objects the equivalent diameter is used.

Example 4-2: Nusselt numbers from film theory Consider the leaching of a solute A from a surface and its subsequent diffusion through a stagnant film. The total transport of A from the surface is given by Eq. (1-91)

$$\mathcal{W}_{\text{A}}\Big|_{x=0} = N_{\text{A}}A_x\Big|_{x=0} = \frac{D_{\text{AB}}A_x(C_{\text{A}I} - C_{\text{A}b})}{L_f} \tag{1}$$

(*a*) Using the definition of a mass-transfer coefficient, express k_c in terms of film thickness L_f. (*b*) Find the Sherwood number. (*c*) Consider heat transport through a film of thickness L_f. The temperature at the interface is T_I, and that at the film boundary is T_b. By analogy to Eq. (1), write an expression for the overall transport of heat from the interface; then, using the definition of heat-transfer coefficient, express h in terms of film thickness; and finally, obtain the heat-transfer Nusselt number, Eq. (4-30).

SOLUTION (*a*) By definition, $N_{AI} \equiv k_c(C_{AI} - C_{Ab})$. By Eq. (1-91), since $A_x = A_w = \text{const}$,

$$N_{AI} = \frac{D_{AB}(C_{AI} - C_{Ab})}{L_f} \tag{4-31a}$$

Using these equations, we get

$$k_c = \frac{D_{AB}}{L_f} \tag{4-31b}$$

(*b*)

$$(\text{Sh})_{AB} = \frac{k_c D_{ch}}{D_{AB}} = \frac{D_{ch}}{L_f} \tag{4-31c}$$

(*c*) By analogy

$$q_w = \frac{k(T_I - T_b)}{L_f}$$

and since, by definition, $q_w \equiv h(T_w - T_b)$,

$$h = \frac{k}{L_f} \quad \text{and} \quad \text{Nu} \equiv \frac{hD_{ch}}{k} = \frac{D_{ch}}{L_f} \tag{4-31d}$$

□

Example 4-3: Nusselt numbers for a sphere In absorbers, extractors, pebble heat exchangers, catalytic reactors, and other equipment, it is frequently necessary to find the total transport of mass or heat from the surface of a sphere. This transport rate is usually expressed in terms of a heat-transfer coefficient or a mass-transfer coefficient k_c. The transfer coefficients are in turn related to Nusselt numbers for heat or mass transfer [Eqs. (4-30) and (4-31)].

For a sphere it is customary to let $D_{ch} = 2R$, where R is the radius of the sphere. In general the Nusselt numbers will depend upon the rate of flow past the sphere. However, at low velocities and for small potential differences, we can assume that the sphere is surrounded by a stagnant region of infinite extent. Our purpose is to find an expression for both the heat- and mass-transfer Nusselt numbers. But for simplicity we will first carry out the derivations for heat transfer and then, by analogy, obtain the results for mass transfer (see Table 4-1).

In keeping with the concept of an infinite stagnant region, we will assume that $T \to T_b$ (a constant bulk temperature) as $r \to \infty$, where r is the distance from the center of the sphere. We will let the temperature at the surface of the sphere be

Table 4-1

	Heat	Mass
Flux differential equation	$\frac{d(r^2 q_r)}{r^2\,dr} = 0$	$\frac{d(r^2 J_{Ar})}{r^2\,dr} = 0$
Flux law	$q_r = -k\frac{dT}{dr}$	$J_{Ar} = -D_{AB}\frac{dC_A}{dr}$
Boundary conditions	$T = \begin{cases} T_R & r = R \\ T_b & r \to \infty \end{cases}$	$C_A = \begin{cases} C_{AR} & r = R \\ C_{Ab} & r \to \infty \end{cases}$
Potential distribution	$T - T_b = (T_R - T_b)\frac{R}{r}$	$C_A - C_{Ab} = (C_{AR} - C_{Ab})\frac{R}{r}$
Flux distribution	$q_r = k(T_R - T_b)\frac{R}{r^2}$	$J_{Ar} = D_{AB}(C_{AR} - C_{Ab})\frac{R}{r^2}$
Total transport	$\dot{Q}\Big\vert_R = 4\pi Rk(T_R - T_b)$	$\mathcal{W}_A\Big\vert_R \equiv 4\pi RD_{AB}(C_{AR} - C_{Ab})$
Transfer coefficient	$h = \frac{q_r\vert_R}{T_R - T_b} = \frac{k}{R}$	$k_c = \frac{J_{Ar}\vert_R}{C_{AR} - C_{Ab}} = \frac{D_{AB}}{R}$
Nusselt number	$\frac{hD_{ch}}{k} = 2$	$(\mathrm{Nu})_{AB} = \frac{k_c D_{ch}}{D_{AB}} = 2$

T_R. (*a*) Find the total transport at the surface of the sphere. (*b*) Show that the heat-transfer Nusselt number is Nu = 2.

SOLUTION (*a*) Using a shell balance over an increment Δr (exercise for the student), we obtain an expression for the total heat transport at the surface

$$\dot{Q}\Big\vert_{r=R} \equiv \dot{Q}_R = 4\pi Rk(T_R - T_b) \tag{4-32}$$

The intermediate equations are given in Table 4-1.

(*b*) Applying the definition of h [Eq. (4-26)] gives

$$h \equiv \frac{q_r\Big\vert_R}{T_R - T_b} = \frac{\dot{Q}_R}{A_r\Big\vert_R (T_R - T_b)} \tag{4-32a}$$

Using the expressions for $\dot{Q}_R$ and A_r, we get

$$h = \frac{4\pi Rk(T_R - T_b)}{4\pi R^2(T_R - T_b)} = \frac{k}{R} \tag{4-32b}$$

If $D_c \equiv 2R$,

$$\mathrm{Nu} \equiv \frac{hD_c}{k} = \frac{k(2R)}{Rk} = 2 \tag{4-32c}$$

Similarly it is found that $\mathrm{Sh} \equiv (\mathrm{Nu})_{\mathrm{AB}} = 2$. These results are quite useful in estimating heat- and mass-transfer rates from a sphere surrounded by a stagnant fluid. If the fluid has an appreciable velocity, an additional term must be added (see Chap. 6). □

Interpretation of Film Theory

At this point, the pragmatic student might well ask: What good is film theory if it merely permits us to replace an empirical parameter h with another empirical parameter L_f, the film thickness? One answer is that the film thicknesses are more easily visualized physically and this qualitative picture can be used to help understand h. Thus the results for a flat surface [Eq. (4-31d)]

$$h = \frac{k}{L_f} \qquad \text{and} \qquad \mathrm{Nu} = \frac{D_{\mathrm{ch}}}{L_f}$$

indicate that in order to increase h we need to reduce the film thickness. This can be done, for example, by increasing the velocity of the fluid past the surface. We also note that h will increase if the thermal conductivity of the fluid increases. However, we must be cautious about assuming that L_f is independent of k. As we will see in Chap. 6, L_f depends on the nature of the fluid (and hence on k), and in turbulent flow it varies roughly with k raised to the $-\frac{1}{3}$ power.

The Nusselt number, as given by film theory in the second of Eqs. (4-31d), is also capable of physical interpretation as the reciprocal of the dimensionless film thickness. Thus, a high Nusselt number corresponds to a film thickness that is thin relative to the characteristic size of the system D_{ch}, and vice versa. So when we encounter a Nusselt number for a flat surface, we can take its reciprocal and physically think of the result as a relative effective film thickness.

For a curved surface such as a sphere, the above interpretation requires modification since for an infinite film we have obtained Nu = 2 instead of Nu = 0. In order to understand this result let us consider a sphere surrounded by a *finite* film of effective thickness $L_f \equiv R_f - R$. Using the same procedure as in Examples 4-2 and 4-3 (the details are left as a student exercise), we obtain

$$h = \frac{k}{L_f} + \frac{k}{R} = h_f + h_g \tag{4-32d}$$

where h_f is the heat-transfer coefficient due to the resistance of the film and h_g is the contribution due to the curved geometry, i.e., due to the increasing area surrounding the sphere as r increases.

The Nusselt number is then

$$\mathrm{Nu} \equiv \frac{2Rh}{k} = \frac{2R}{L_f} + 2 = (\mathrm{Nu})_f + (\mathrm{Nu})_g \tag{4-32e}$$

The first term $(\mathrm{Nu})_f$ is the contribution to the Nusselt number due to the film thickness and is a reciprocal dimensionless film thickness. The second term $(\mathrm{Nu})_g$ is the contribution to the Nusselt number due to the increased area; i.e., since the area increases, the resistance to heat transport decreases and the Nusselt number increases. Thus, for a nonflat surface, we must generalize our physical picture of Nu still further to include the geometrical effect. In that case the Nusselt number can be thought of as a reciprocal dimensionless resistance which is the sum of the dimensionless resistances due to the film and to the increase in transfer area. For solids of shapes other than spherical one might expect a similar limiting $(\mathrm{Nu})_g$ to appear.

Use of Transfer Coefficient in Calculating Transport from a Sphere

Note that while the use of a heat- or mass-transfer coefficient in design is widespread, it is not actually necessary. If, for example, one wishes to find the total rate of heat flow to or from a sphere in a stagnant fluid, it can be obtained directly from an incremental balance (as in Example 4-3). The result is

$$\dot{Q}_R = 4\pi Rk(T_b - T_R) \tag{4-32f}$$

In practice however, it is customary to use a transfer coefficient h together with the rate equation [Eq. (4-32*a*)], giving

$$\dot{Q}_R = hA_R(T_R - T_b)$$

where $A_R = 4\pi R^2$. The value of h will in general depend greatly on the flow conditions around the sphere: if the flow past the sphere becomes turbulent, h will be much larger than if the flow is laminar (turbulent flow will be discussed in Chaps. 6 and 10). In the limit of no flow (stagnant fluid), we have found that

$$\mathrm{Nu} \equiv \frac{2Rh}{k} = 2 \tag{4-32g}$$

For a finite stagnant film transport phenomena would yield (see the Problems)

$$\dot{Q}\Big|_R = 4\pi k\left(\frac{R}{L_f} + 1\right)R(T_R - T_b) \tag{4-32h}$$

but to use this equation L_f must be obtained from theory or estimated from physical considerations.

Estimate of Transfer Coefficient in a Pipe Heat Exchanger

Consider the laminar flow of a fluid through a pipe heated at the wall. Assume that the laminar region extends all the way to the center and that the variation of the heat flux with respect to radial and axial position can be neglected. Then the heat flux at the wall is

$$q_w = -k\frac{\partial T}{\partial r}\bigg|_R = k\frac{dT}{dy}\bigg|_w$$

where y is the distance from the wall. If T_c is the temperature at the center, $y = R$, and if we *assume* that the temperature profile is linear over the cross section, then

$$q_w = k\frac{T_c - T_w}{R}$$

Substituting into the definition Eq. (4-26) gives

$$h \equiv \frac{q_w}{T_b - T_w} = \frac{k}{R}\frac{T_c - T_w}{T_b - T_w} \tag{4-32i}$$

For constant flux Fourier's law can be integrated to give

$$T - T_w = \frac{q_w}{k}y \tag{4-32j}$$

If we neglect curvature and assume a flat velocity profile, Eq. (A4-3-3) gives

$$T_b - T_w = \frac{\int_0^R (T - T_w)\,dy}{R} = \frac{q_w}{Rk}\int_0^R y\,dy = \frac{q_w R}{2k} \tag{4-32k}$$

Then

$$\frac{T_b - T_w}{T_c - T_w} = \frac{1}{2} \tag{4-32l}$$

and

$$h = \frac{2k}{R} \quad \text{and} \quad \mathrm{Nu} = \frac{2Rh}{k} = 4 \tag{4-32m}$$

This is very close to the theoretical value of 3.67, obtained using a two-dimensional model in which velocity, temperature, and heat flux vary with radial position and the latter two vary with axial position (see Chap. 10).

If we do not neglect curvature but do neglect the variation of the velocity with position, we get

$$T_b - T_w = T_a - T_w = \frac{\int_0^R (T - T_w)(R - y)\,dy}{\int_0^R (R - y)\,dy}$$

$$= \frac{2}{R^2}\int_0^R \frac{q_w y}{k}(R - y)\,dy = \frac{q_w R}{3k} \tag{4-32n}$$

Then

$$h \equiv \frac{q_w}{T_b - T_w} = \frac{k(T_c - T_w)}{R(T_b - T_w)} = \frac{3k}{R} \quad \text{and} \quad \mathrm{Nu} = \frac{2Rh}{k} = 6 \tag{4-32o}$$

which compares with a range of 5 to 7 for low-velocity flow of liquid metals found in the literature. Because of the high thermal conductivity of liquid metals, molecular transport occurs over the entire pipe. For other cases see the Problems.

For *turbulent* flow the temperature profile is very steep near the wall (where the flow is laminar and the greatest resistance occurs) and nearly flat near the center (where the flow is turbulent and there is little resistance). (The difference in resistance is due to the fact that turbulent diffusion is several orders of magnitude larger than molecular diffusion, which occurs in laminar flow.) Therefore the film thickness is much smaller and the Nusselt number may be several orders of magnitude larger at large values of Re and Pr ($\equiv \nu/\alpha$) than the values of 2 to 6 we have estimated.

Also, the Nusselt number is much larger near the entrance to the heated or cooled section of a pipe, especially for laminar flow, because the effective film thickness is smaller due to the fact that heat transfer first occurs only near the wall and does not occur at the center until a certain *thermal entrance length* has been reached. This region is discussed in Chap. 10, but the reader should recognize that the Nusselt numbers of 2 to 6 estimated here do not apply to the thermal entrance region or to turbulent flow.

4-7 DIFFUSION WITH INTERNAL GENERATION (CHEMICAL REACTION)

Although the film theory in its application seems limited to systems in which the film thickness is known, it is quite useful in predicting the effect of chemical reaction on the mass-transfer coefficient. The need for the effective film thickness is often handled by measuring the mass-transfer rate for nonreacting conditions and calculating the film thickness from it. The film thickness is then assumed to be the same for reacting conditions. This use of the film theory at unsteady state is discussed in Chap. 7. In this chapter we discuss only steady-state conditions.

Consider a liquid film of thickness L_f through which a dissolved gas A is diffusing. The gas is dissolved at the gas-liquid interface $x = 0$, where the equilibrium concentration is C_{A0}. Assume that the concentration in the bulk stream at the edge of the film is C_{AL}. Previously we have considered the case (Example 4-2) of no reaction, but now we assume that as A diffuses it is also reacting with the liquid. If A is a reactant, the internal generation of A is negative and R_A (which is defined as the rate of generation of A per unit volume) is negative. Then a molar balance over a volume element $A_x\,\Delta x$ gives

$$N_{Ax}A_x\Big|_x - N_{Ax}A_x\Big|_{x+\Delta x} + R_A\,\Delta x A_x = 0$$

which results in

$$\frac{-dN_{Ax}}{dx} + R_A = 0$$

Neglecting bulk flow and using Fick's law gives

$$D_{AB}\frac{d^2C_A}{dx^2} + R_A = 0$$

If the reaction is first order and irreversible,

$$R_A \equiv -k_R C_A \qquad \text{and} \qquad \frac{d^2C_A}{dx^2} - \frac{k_R}{D_{AB}}C_A = 0$$

We now define the dimensionless groups

$$\eta \equiv \frac{x}{L_f} \qquad C_A^* \equiv \frac{C_A}{C_{A0}} \qquad \text{and} \qquad \frac{k_R L_f^2}{D_{AB}} \equiv \beta^2$$

Then

$$\frac{d^2C^*}{d\eta^2} - \beta^2 C^* = 0$$

The boundary conditions are

At $\eta = 0$: $\quad C^* = 1 \quad$ boundary condition 1

At $\eta = 1$: $\quad C_L^* \equiv \dfrac{C_{AL}}{C_{A0}} \quad$ boundary condition 2

Then from Appendix 1-2 the solution is

$$C^* = \text{A} \sinh \beta\eta + \text{B} \cosh \beta\eta$$

From the first boundary condition 1 = B, and from the second

$$C_L^* = \text{A} \sinh \beta + \cosh \beta$$

Solving for A and substituting for A and B gives

$$C^* = \frac{C_L^* \sinh \beta\eta + \sinh \beta(1-\eta)}{\sinh \beta}$$

where we have used the identity

$$\cosh v \sinh u - \cosh u \sinh v = \sinh (u - v)$$

with

$$v \equiv \beta\eta \qquad u \equiv \beta$$

Hatta[2] assumed that $C_L^* \approx 0$, which applies for $\beta > 0.3$. Then

$$C^* = \frac{\sinh \beta(1-\eta)}{\sinh \beta}$$

The molar flux at the interface is

$$N_{A0} = -D_{AB}\frac{\partial C_A}{\partial y}\bigg|_{y=0}$$

$$= \frac{D_{AB}C_{A0}}{L_f}\frac{\beta \cosh \beta}{\sinh \beta} = \frac{D_{AB}C_{A0}}{L_f}\beta \coth \beta$$

and for the reacting case the mass-transfer coefficient is

$$k_{cR} \equiv \frac{N_{A0}}{C_{A0}} = \frac{D_{AB}}{L_f} \beta \coth \beta$$

For the case of no reaction we found [(Eq. 4-31*b*)]

$$k_c = \frac{D_{AB}}{L_f}$$

The enhancement factor E is defined[1] as the factor by which chemical reaction increases the rate of absorption compared with pure physical absorption with $C^*_{A,L} = 0$, or

$$E \equiv \frac{(N_{A0})_R}{N_{A0}} = \frac{k_{cR} C_{A0}}{k_c C_{A0}} = \frac{k_{cR}}{k_c} = \beta \coth \beta$$

which gives k_{cR} if D_{AB}, k_R, and k_c are known.

For a fast reaction with a small diffusivity, β is large and $\coth \beta \rightarrow 1$. Then $E = \beta$, and

$$N_{A0} = \frac{D_{AB} C_{A0} \beta}{L_f} = D_{AB} C_{A0} \sqrt{\frac{k_R}{D_{AB}}} = C_{A0} \sqrt{\frac{k_R}{D_{AB}}}$$

which becomes independent of L_f since the molecules react before reaching L_f. The mass-transfer coefficient is then

$$k_{cR} \equiv \frac{N_{A0}}{C_{A0}} = \sqrt{k_R D_{AB}}$$

The dimensionless group β is often called the *Hatta number*.

B. MULTIPHASE SYSTEMS

4-8 GENERAL BOUNDARY CONDITIONS AT AN INTERFACE

We have indicated that in many engineering applications we must deal with composite systems consisting of two or more phases. In solving problems of this type the general procedure is as follows:

1. Derive differential equations for the flux in each phase
2. Write flux laws for each phase
3. Solve the above using boundary conditions at the interface and at the system boundaries

In general there may be a resistance to transport at an interface, and in such cases equations may have to be written to describe transport through the interface. In this part of the text, however, we neglect interfacial resistances.† *Our boundary conditions then will correspond to statements concerning* (1) *the equality of potentials and* (2) *the equality of fluxes.* We will let superscript (α) refer to phase α and superscript (β) refer to phase β. Then at the interface (represented by subscript I), the boundary conditions for transport in the x direction are

Energy:
$$T^{(\alpha)}\Big|_I = T^{(\beta)}\Big|_I \tag{4-33a}$$

$$q_x^{(\alpha)}\Big|_I = q_x^{(\beta)}\Big|_I \tag{4-33b}$$

y momentum:‡
$$v_y^{(\alpha)}\Big|_I = v_y^{(\beta)}\Big|_I \tag{4-34a}$$

$$\tau_{xy}^{(\alpha)}\Big|_I = \tau_{xy}^{(\beta)}\Big|_I \tag{4-34b}$$

Mass of A:
$$C_A^{(\alpha)}\Big|_I = mC_A^{(\beta)}\Big|_I \tag{4-35a}$$

$$N_{Ax}^{(\alpha)}\Big|_I = N_A^{(\beta)}\Big|_I \tag{4-35b}$$

where m is obtained from the equilibrium curve. Although m will be a function of concentration, it is often assumed constant over small ranges of concentration.

From thermodynamics[6] we know that the chemical *potentials* at the interface are equal, or

$$\bar{G}_A^{(\alpha)}\Big|_I = \bar{G}_A^{(\beta)}\Big|_I \tag{4-36}$$

where $\bar{G}_A^{(\alpha)}\big|_I$ = partial molar free energy or chemical potential of species A in phase α at interface

$\bar{G}_A^{(\beta)}\big|_I$ = chemical potential of species A in phase β at interface

Also from thermodynamics $\bar{G}_A^{(\alpha)}$, the chemical potential of a species A in phase α at a concentration $C_A^{(\alpha)}$, can be related to $\bar{G}_A^{\circ(\alpha)}$, the chemical potential at a standard-state concentration, by the relationship

$$\bar{G}_A^{(\alpha)} = \bar{G}_A^{\circ(\alpha)} + RT \ln a_A^{(\alpha)} \tag{4-37a}$$

†Surface tension or interfacial tension variations can result in surface flows (often called the *Marangoni effect*).

‡The second equation neglects the effects of interfacial tension. If there is a surface-tension gradient along the interface, the momentum fluxes will not be equal.

in which $$a_A^{(\alpha)} = \text{activity of A in phase } \alpha = \gamma_A^{(\alpha)} C_A^{(\alpha)} \tag{4-37b}$$

where $\gamma_A^{(\alpha)}$ is the activity coefficient of A in α. Writing expressions similar to Eqs. (4-37*a*) and (4-37*b*) for phase β and using Eq. (4-36) gives Eq. (4-35*a*), where

$$m = \frac{\gamma_A^{(\beta)}}{\gamma_A^{(\alpha)}} \exp \frac{\bar{G}_A^{\circ(\beta)} - \bar{G}_A^{\circ(\alpha)}}{RT} \tag{4-38}$$

If we let $\bar{G}_A^{\circ(\beta)} = \bar{G}_A^{\circ(\alpha)}$,

$$a_A^{(\beta)}\Big|_I = a_A^{(\alpha)}\Big|_I \qquad \text{and} \qquad m = \frac{\gamma_A^{(\beta)}}{\gamma_A^{(\alpha)}} \tag{4-39a}$$

Other Boundary Conditions at an Interface†

For gas-liquid interfaces, one may assume Raoult's law (for ideal solutions)

$$x_A^{(L)} p_{vA} = x_A^{(G)} p \tag{4-39b}$$

where $x_A^{(L)}$ = mole fraction of A in liquid phase
$x_A^{(G)}$ = mole fraction of A in gas phase
p_{vA} = vapor pressure of A
p = total pressure in gas phase

For dilute gases, Henry's law may apply

$$x_A^{(G)} p = H_A x_A^{(L)} \tag{4-39c}$$

where H_A is Henry's-law constant.

For *impermeable* phases, the flux is zero at the interface; i.e., for

1. *Heat transfer* at an insulated surface

$$q_x\Big|_I = 0 \qquad \text{or} \qquad \frac{dT}{dx}\Big|_I = 0$$

2. *Mass* transfer at an impermeable solid

$$J_{Ax}\Big|_I = 0 \qquad \text{or} \qquad \frac{dC_A}{dx}\Big|_I = 0$$

3. *Momentum* transfer at a gas-liquid interface

$$\tau_{xy}\Big|_I \approx 0 \qquad \text{or} \qquad \frac{dv_y}{dx}\Big|_I \approx 0$$

since $\mu^{(G)}/\mu^{(L)} \approx 0$. For turbulent flow in the gas phase, this assumption may not be valid.

†For an excellent treatment see the review article by Rosner.[5]

For a *heterogeneous chemical reaction* occurring at the interface, the surface rate in mol/m^2 · s can be specified, e.g.,

$$r'_A = k'_R C_A^n \bigg|_I \tag{4-39d}$$

where k' is the specific rate constant and n the order.

4-9 TRANSPORT IN COMPOSITE SYSTEMS WITHOUT INTERNAL GENERATION

Heat Transfer through Composite Solids with Rectangular Coordinates

As an example we first consider heat transport through two solid slabs A and B (Fig. 4-8). The length L of each slab is so large compared with its other dimensions that heat transport need be considered only in the x direction (perpendicular to the interface). The thickness of slab A is X_A, and that of slab B is X_B. The temperatures at each surface are assumed constant and are shown in the figure. Note that we have already used the first boundary condition in Eq. (4-33*a*) that the temperatures of each phase at the interface are equal; i.e.,

$$T_A \bigg|_{x=X_A} = T_B \bigg|_{x=X_A} \equiv T_2 \tag{4-40}$$

For a positive heat flux to occur, we must have $T_1 > T_2 > T_3$.

Since we are at steady state and there is no internal generation of heat, we can make an energy balance over an increment Δx in *each* phase and obtain the differential equations

$$\frac{dq_{xA}}{dx} = 0 \qquad \frac{dq_{xB}}{dx} = 0 \tag{4-41}$$

Then q_{xA} = constant and q_{xB} = const. But at the interface, by the boundary con-

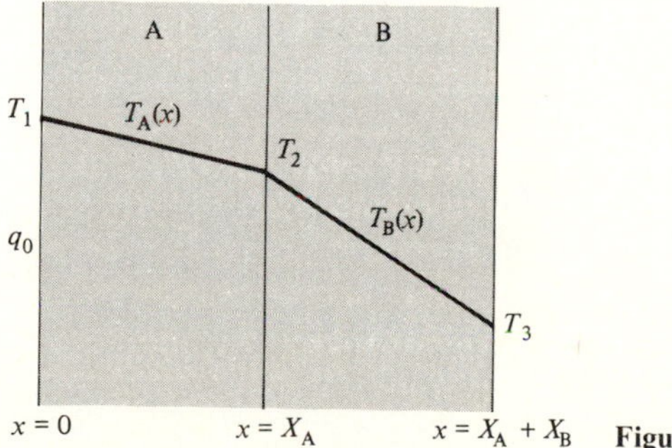

Figure 4-8 Composite solid.

dition [Eq. (4-33*b*)],

$$q_{xA}\Big|_{x=X_A} = q_{xB}\Big|_{x=X_A} = q_I = \text{const} \tag{4-42}$$

Writing Fourier's law of heat conduction for each phase gives

$$q_{xA} = -k_A \frac{dT_A}{dx} \quad \text{and} \quad q_{xB} = -k_B \frac{dT_B}{dx} \tag{4-43}$$

Since the fluxes are constant, we can write the gradients in terms of the total temperature differences and total lengths

$$-\frac{dT_A}{dz} = \frac{T_1 - T_2}{X_A} = \frac{-\Delta T_A}{X_A} \quad \text{and} \quad -\frac{dT_B}{dx} = \frac{T_2 - T_3}{X_B} = \frac{-\Delta T_B}{X_B} \tag{4-44}$$

where we write $-\Delta$ because there is actually a temperature drop. Then

$$\frac{\dot{Q}_I}{A} = q_I = \text{const} = k_A \frac{-\Delta T_A}{X_A} = k_B \frac{-\Delta T_B}{X_B} \tag{4-45}$$

where $\dot{Q}_I$ is the interface value of the heat-flow rate.

We noted in Sec. 1-6 that this equation is of the form

$$\text{Rate} = \frac{\text{driving force (or potential)}}{\text{resistance}}$$

or

$$\dot{Q}_I = \frac{-\Delta T_A}{\mathcal{R}_A} \quad \text{and} \quad \dot{Q}_I = \frac{-\Delta T_B}{\mathcal{R}_B} \tag{4-46}$$

where

$$\mathcal{R}_A = \frac{X_A}{k_A A} \quad \text{and} \quad \mathcal{R}_B = \frac{X_B}{k_B A} \tag{4-47}$$

The total or overall potential is $-\Delta T_t = T_1 - T_3$. This can be written as the sum of the individual potentials and Eq. (4-46) can be substituted

$$-\Delta T_t = -\Delta T_A - \Delta T_B = \mathcal{R}_A \dot{Q}_I + \mathcal{R}_B \dot{Q}_I = (\mathcal{R}_A + \mathcal{R}_B)\dot{Q}_I \tag{4-48}$$

Then

$$\dot{Q}_I = \frac{-\Delta T_t}{\mathcal{R}_A + \mathcal{R}_B} = \frac{-\Delta T_t}{\mathcal{R}_t} \tag{4-49}$$

where the overall resistance is

$$\mathcal{R}_t = \mathcal{R}_A + \mathcal{R}_B \tag{4-50}$$

Then from Eqs. (4-47) and (4-49)

$$\frac{Q_I}{A} = q_I = \frac{T_1 - T_3}{(X_A/k_A) + (X_B/k_B)} \tag{4-51}$$

This can be extended to n phases i by

$$\frac{\dot{Q}_I}{A} = q_I = \frac{-\Delta T_t}{\sum_{i=1}^{n} \frac{X_i}{k_i}} \tag{4-52}$$

where
$$\Delta T_t = \sum_{i=1}^{n} \Delta T_i \tag{4-53}$$

Note the analogy to electric resistances in series given by Ohm's law:

$$I = \frac{\mathcal{E}_t}{\sum_{i=1}^{n} \mathcal{R}_i}$$

where $\mathcal{E}_t$ is the overall potential drop and I is the current.

Example 4-4: Conduction through cylindrical wall An insulated metal steam pipe (Fig. 4-9) carries steam at a bulk temperature T_i. The air surrounding the insulation is at a bulk temperature T_o. The thermal conductivity of the metal wall of the pipe is k_A, and that of the insulation is k_B. The temperature of the inside metal surface of the pipe is T_1. The inside and outside heat-transfer coefficients are h_i and h_o respectively. The temperature at the interface between the metal wall and the insulation is T_2, and the temperature at the outside surface of the insulation is T_3. The inside and outside radii of the metal wall are respectively r_1 and r_2. The outside radius of the insulation is r_3. The temperature variation with distance from the center is shown in Fig. 4-10. For convenience we assume that none of the temperatures vary along the length of the steam pipe. (*a*) Derive an equation for the total rate of heat loss in terms of the total temperature drop $-\Delta T_t \equiv T_i - T_o$, the thermal conductivities k_A and k_B, and the heat-transfer coefficients h_i and h_o for the inside and outside films. (*b*) The *overall heat-transfer coefficients* U_i and U_o are defined by

$$\dot{Q}_f \equiv U_i A_i(-\Delta T_t) \equiv U_o A_o(-\Delta T_t) \tag{4-54}$$

where A_i and A_o are, respectively, the inside area of the pipe and the outside area of the insulation (that is, $A_i = 2\pi r_1 L$ and $A_o = 2\pi r_3 L$) and where U_i and U_o are the overall heat-transfer coefficients based on the inside and outside areas, respectively. Derive equations relating U_o and U_i to k_A, k_B, h_i, and h_o.

SOLUTION (*a*) First we consider the solid phases A and B. The heat-flow rate $\dot{Q}_f$ must be the same at all radial positions since we are at steady state; however, the heat *fluxes* will vary with r due to the varying areas. As we have seen in Chap. 1,

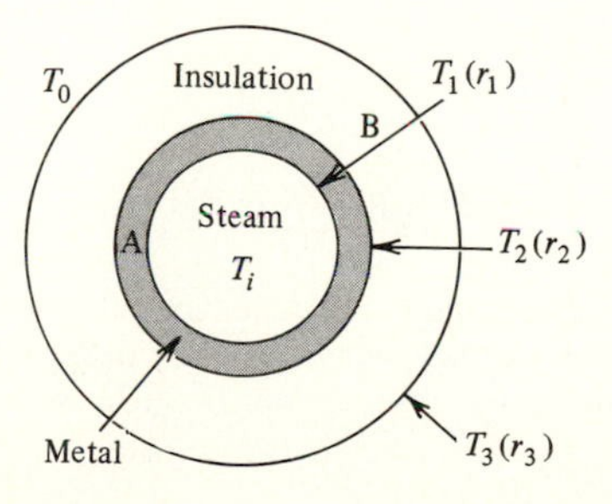

Figure 4-9 Insulated steam-pipe cross section.

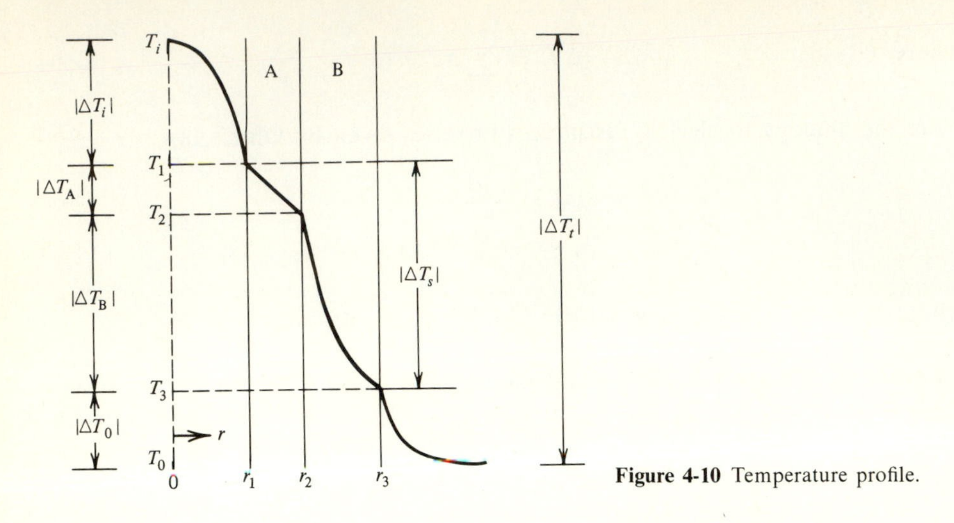

Figure 4-10 Temperature profile.

we must therefore use a mean area (which in cylindrical coordinates is the log mean) to relate the heat flow to the temperature drop as follows [see Eq. (1-32)].† In the *metal*

$$\dot{Q}_f = k_A A_{MA} \frac{T_1 - T_2}{r_2 - r_1} \tag{4-55}$$

where

$$A_{MA} = \frac{A_2 - A_1}{\ln (A_2/A_1)} \tag{4-56}$$

In the *insulation*

$$\dot{Q}_f = k_B A_{MB} \frac{T_2 - T_3}{r_3 - r_2} \tag{4-57}$$

where

$$A_{MB} = \frac{A_3 - A_2}{\ln (A_3/A_2)} \tag{4-58}$$

If we let

$$\dot{Q}_f = \frac{-\Delta T_A}{\mathcal{R}_A} = \frac{-\Delta T_B}{\mathcal{R}_B} \tag{4-59}$$

then

$$\mathcal{R}_A \equiv \frac{X_A}{k_A A_{MA}} \qquad \mathcal{R}_B \equiv \frac{X_B}{k_A A_{MB}} \tag{4-60}$$

and

$$X_A = r_2 - r_1 \qquad X_B = r_3 - r_2 \tag{4-61}$$

Now the resistances in the steam and air must be added. The flows are given by

†Had the temperatures been varying with length (as in a heat exchanger), the equations above would apply over a differential length dz. In that case the areas and heat flow would also be differentials, but the final result [Eqs. (4-69) and (4-72)] would be the same since only area ratios are involved.

$$\dot{Q}_f = \begin{cases} h_i A_i (T_i - T_1) = \dfrac{-\Delta T_i}{\mathcal{R}_i} & \text{inside} \quad (4\text{-}62) \\ h_o A_o (T_3 - T_1) = \dfrac{-\Delta T_o}{\mathcal{R}_o} & \text{outside} \quad (4\text{-}63) \end{cases}$$

where $$\mathcal{R}_i = (h_i A_i)^{-1} \quad \text{and} \quad \mathcal{R}_o = (h_o A_o)^{-1} \tag{4-64}$$

Then $$-\Delta T_t \equiv T_i - T_o = -\Delta T_i - \Delta T_A - \Delta T_B - \Delta T_o \tag{4-65}$$

Using Eqs. (4-59), (4-62), and (4-63), we can write each temperature difference above in terms of a resistance or

$$\dot{Q}_f \mathcal{R}_t = \dot{Q}_f \mathcal{R}_i + \dot{Q}_f \mathcal{R}_A + \dot{Q}_f \mathcal{R}_B + \dot{Q}_f \mathcal{R}_o$$

or $$\mathcal{R}_t = \mathcal{R}_i + \mathcal{R}_A + \mathcal{R}_B + \mathcal{R}_o \tag{4-66}$$

Then $$\dot{Q}_f = \frac{-\Delta T_t}{\mathcal{R}_t} = \frac{T_i - T_o}{\dfrac{1}{h_i A_i} + \dfrac{X_A}{k_A A_{MA}} + \dfrac{X_B}{k_B A_{MB}} + \dfrac{1}{h_o A_o}} \tag{4-67}$$

(*b*) Rearranging the first of Eq. (4-54) gives

$$\frac{1}{U_i} = \frac{A_i (T_i - T_o)}{\dot{Q}_f} \tag{4-68}$$

or $$\boxed{\frac{1}{U_i} = \frac{1}{h_i} + \frac{X_A A_i}{k_A A_{MA}} + \frac{X_B A_i}{k_B A_{MB}} + \frac{A_i}{h_o A_o}} \tag{4-69}$$

Noting that

$$\frac{X_A A_i}{A_{MA}} = \frac{(r_2 - r_1)(2\pi r_1 L)}{2\pi L (r_2 - r_1) \ln (A_2/A_1)} = r_1 \ln \frac{r_2}{r_1} \tag{4-70}$$

and using a similar equation for phase B, we get

$$\frac{1}{U_i} = \frac{1}{h_i} + \frac{r_1 \ln (r_2/r_1)}{k_A} + \frac{r_1 \ln (r_3/r_2)}{k_B} + \frac{r_1}{h_o r_3} \tag{4-71}$$

If we wanted to find the overall heat-transfer coefficient based on the *outside* area A_o, we would get

$$\boxed{\frac{1}{U_o} \equiv \frac{A_o (T_i - T_o)}{\dot{Q}_f} = \frac{A_o}{h_i A_i} + \frac{X_A A_o}{k_A A_{MA}} + \frac{X_B A_o}{k_B A_{MB}} + \frac{1}{h_o}} \tag{4-72}$$

Note that

$$U_o A_o = U_i A_i = \frac{\dot{Q}_f}{-\Delta T_t} = \mathcal{R}_t^{-1} \tag{4-73}$$

□

Design of a Double-Pipe Heat Exchanger

The design of a double-pipe countercurrent (or cocurrent) heat exchanger often involves finding the length $z = L$ required to cool (say) the inner fluid to some bulk mean temperature $T_i|_L$. (In this example we will drop the subscript b for convenience.) We assume that the inner fluid enters at $z = 0$ with temperature $T_i|_0$ while the outer fluid enters at $z = 0$ with temperature $T_o|_0$ and leaves at $z = L$ at temperature $T_o|_L$. If the outer pipe is perfectly insulated, a differential energy balance over an element of length dz gives

$$-(w\widehat{C}_p)_i\, dT_i = -d\dot{Q}_{Gi} = (w\widehat{C}_p)_o\, dT_o = d\dot{Q}_{Go} = d\dot{Q}_f \tag{4-73a}$$

Using an overall heat-transfer coefficient based on the inside area as given by Eq. (4-69) gives

$$d\dot{Q}_f = U_i\, dA_i(T_i - T_o) = U_i Z_i\, dz\,(-\Delta T_t) \tag{4-73b}$$

Since our dependent variable is $\Delta T_t \equiv T_o - T_i$, we obtain

$$d(\Delta T_t) \equiv dT_o - dT_i$$

Substituting for dT_i and dT_o from Eqs. (4-73a) gives

$$d(\Delta T_t) = -\frac{d\dot{Q}_{Gi}}{(w\widehat{C}_p)_o} + \frac{d\dot{Q}_{Gi}}{(w\widehat{C}_p)_i} \equiv d\dot{Q}_f\, \mathfrak{F} \tag{4-73c}$$

where
$$\mathfrak{F} \equiv -(w\widehat{C}_p)_o^{-1} + (w\widehat{C}_p)_i^{-1} \tag{4-73d}$$

Substituting Eq. (4-73b) into Eq. (4-73c) gives

$$d(\Delta T_t) = U_i Z_i \mathfrak{F}\, \Delta T_t\, dz \tag{4-73e}$$

Separating variables and integrating from $z = 0$ to $z = L$ gives

$$\int_{\Delta T_t|_0}^{\Delta T_t|_L} \frac{d(\Delta T_t)}{\Delta T_t} = \int_0^L \frac{U_i Z_i}{\mathfrak{F}}\, dz = \frac{\langle U_i\rangle Z_i L}{\mathfrak{F}} \qquad \langle U_i\rangle \equiv \frac{1}{L}\int_0^L U_i\, dz$$

Integrating the left-hand side and solving for L gives

$$L = \frac{\mathfrak{F}}{\langle U_i\rangle Z_i} \ln \frac{\Delta T_t\big|_L}{\Delta T_t\big|_0} \tag{4-73f}$$

In practice engineers use the rate expression

$$\dot{Q}_f = \langle U_i\rangle Z_i L \langle -\Delta T_t\rangle_{\text{LM}} \tag{4-73g}$$

where the LM refers to the log mean of $\Delta T_t|_0$ and $\Delta T_t|_L$, together with the overall energy balances

$$\dot{Q}_f = (w\widehat{C}_p)_i\left(T_i\Big|_0 - T_i\Big|_L\right) = (w\widehat{C}_p)_o\left(T_o\Big|_L - T_o\Big|_0\right) \tag{4-73h}$$

From Eqs. (4-73*h*) and (4-73*d*) we get

$$\mathfrak{F} = \frac{T_o\Big|_0 - T_o\Big|_L}{\dot{Q}_f} - \frac{T_i\Big|_0 - T_i\Big|_L}{\dot{Q}_f}$$

$$= \frac{\left(T_o\Big|_0 - T_i\Big|_0\right) - \left(T_o\Big|_L - T_i\Big|_L\right)}{\dot{Q}_f} = \frac{\Delta T_t\Big|_0 - \Delta T_t\Big|_L}{\dot{Q}_f}$$

Substituting into Eq. (4-73*f*) results in Eq. (4-73*g*).

Some heat exchangers contain multiple passes, in which the inside tube loops back and forth in the outside shell. In these cases the log mean ΔT_t is still used, but it is multiplied by a correction factor which depends on the number of passes. For the design of such exchangers see Ref. 4.

Two-Film Theory in Mass Transfer

Gas absorbers consist of columns, i.e., cylindrical tubes, packed with small solid particles of various shapes (spherical, cylindrical, hyperbolic, paraboloids, etc.) (Fig. 4-11*a*). A gas stream containing a substance A that is to be removed, say SO_2, is passed upward through the column, and a liquid stream in which the substance A is soluble is passed downward. The liquid coats the packing (Fig. 4-11*b*), which facilitates contact with the gas. The gas-phase concentrations of A entering at the bottom and leaving at the top are respectively $C_{A1}^{(G)}$ and $C_{A2}^{(G)}$. The liquid-phase concentrations of A entering at the top and leaving at the bottom

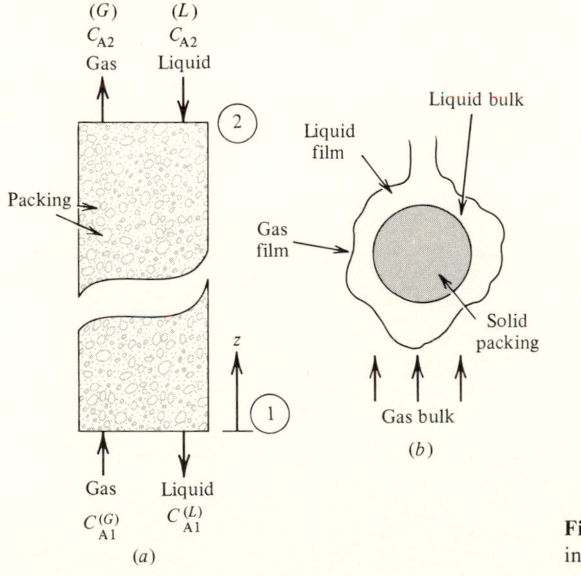

Figure 4-11 (*a*) Gas absorber; (*b*) packing.

are, respectively, $C_{A2}^{(L)}$ and $C_{A1}^{(L)}$. Since at an arbitrary distance z from the bottom the concentrations of A in the liquid and gas streams will vary from point to point over the cross section, we will at first consider the bulk average values $C_{Ab}^{(G)}$ and $C_{Ab}^{(L)}$, respectively. Since solute A is being absorbed by the liquid, $C_{Ab}^{(L)}$ will decrease as we move up the column and $C_{Ab}^{(G)}$ will also decrease. This variation is shown in Fig. 4-12*a*.

A plot of $C_{Ab}^{(G)}$ at a given z versus $C_{Ab}^{(L)}$ at the same z (Fig. 4-12*b*) is called an *operating line;* its equation can be derived as follows. The overall rate of transport $\mathcal{W}_{AI}$ of A over $z = 0$ to $z = z$ is given by a molar balance on each phase

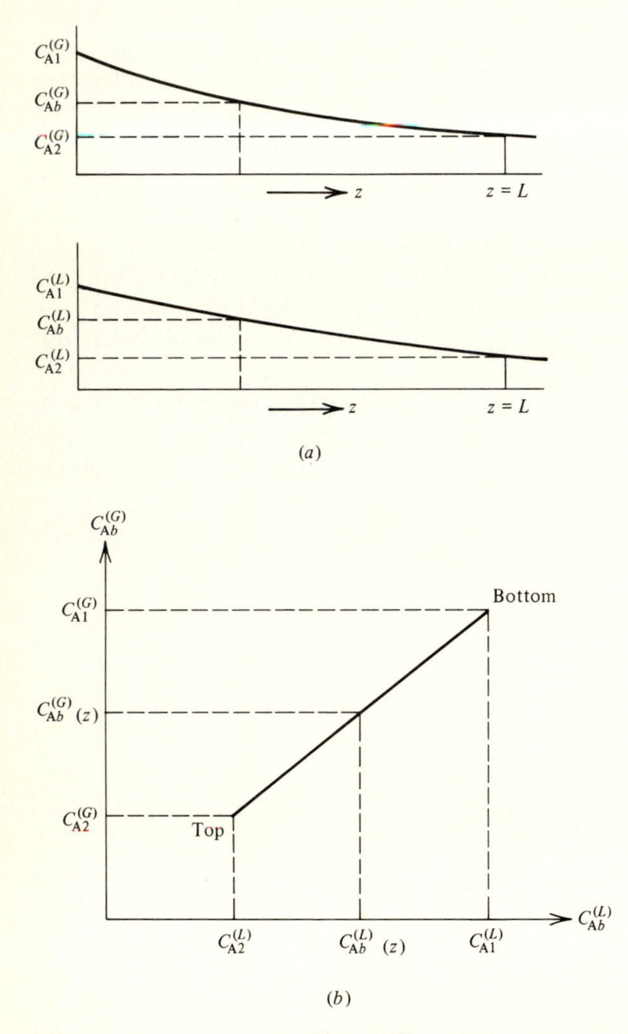

Figure 4-12 (*a*) Variation of $C_{Ab}^{(G)}$ and $C_{Ab}^{(L)}$ with distance from bottom of absorber. (*b*) Variation of $C_{Ab}^{(G)}$ with $C_{Ab}^{(L)}$.

$$\mathcal{W}_{AI} = \dot{\mathcal{V}}_1^{(G)} C_{A1}^{(G)} - \dot{\mathcal{V}}^{(G)} C_{Ab}^{(G)}\Big|_z$$

$$= \dot{\mathcal{V}}^{(L)} C_{Ab}^{(L)}\Big|_z - \dot{\mathcal{V}}_1^{(L)} C_{A1}^{(L)} \tag{4-74}$$

where $\dot{\mathcal{V}}$ is the volumetric rate of flow of a stream. If the volumetric flows $\dot{\mathcal{V}}$ in each phase are nearly constant,

$$C_{Ab}^{(G)}\Big|_z = C_{A1}^{(G)} - \frac{\dot{\mathcal{V}}^{(L)}}{\dot{\mathcal{V}}^{(G)}}\left(C_{Ab}^{(L)}\Big|_z - C_{Ab}^{(L)}\right) \tag{4-75}$$

and the operating line is a straight line.

The design of the column might typically entail the prediction of the height of column needed if all entering and leaving flow rates and concentrations were known. To carry out the design, we need to have a relationship between $\mathcal{W}_{AI}$ and z (or L). Proceeding as we did for the heat exchanger, we divide the region from $z = 0$ to $z = L$ into increments and focus our attention on an element Δz at an arbitrary z. We recall that for the heat exchanger the transfer was at the wall and we needed to predict q_w and multiply it by the wall area. However, in the absorber the transfer occurs at the *interface* between gas and liquid and we need to predict $N_{AI}\,\Delta A_I$, the flux at the interface, multiplied by the interfacial area. If we consider a typical element in the column of volume $\Delta\mathcal{V} = A_c\,\Delta z$, where A_c is the cross-sectional area, and let the interfacial area per unit volume be a, the interfacial area in $\Delta\mathcal{V}$ is

$$\Delta A_I = a\,\Delta\mathcal{V} = aA_c\,\Delta z$$

Then for the increment Δz, the molar rate of transport of A through the interface is

$$\Delta\mathcal{W}_{AI} = N_{AI}\,\Delta A_I = N_{AI}aA_c\,\Delta z$$

Letting $\Delta z \to 0$ gives

$$d\mathcal{W}_{AI} = N_{AI}aA_c\,dz \tag{4-76a}$$

which can be integrated to obtain L if N_{AI} is known.

Thus our problem now becomes one of predicting N_{AI}, just as in heat transport it was one of predicting q_w. To do so in this case requires that we examine the conditions surrounding an individual particle (Fig. 4-13). The solute A with concentration $C_{Ab}^{(G)}$ diffuses from the bulk gas stream through a gas film, through the gas-liquid interface, and then through a liquid film and into the liquid bulk stream. This model of the process is known as the *two-film model*. For simplicity, it is assumed that the packing surface is flat and therefore cartesian coordinates can be used. Figure 4-14 shows how the concentrations change between gas and liquid bulk streams.

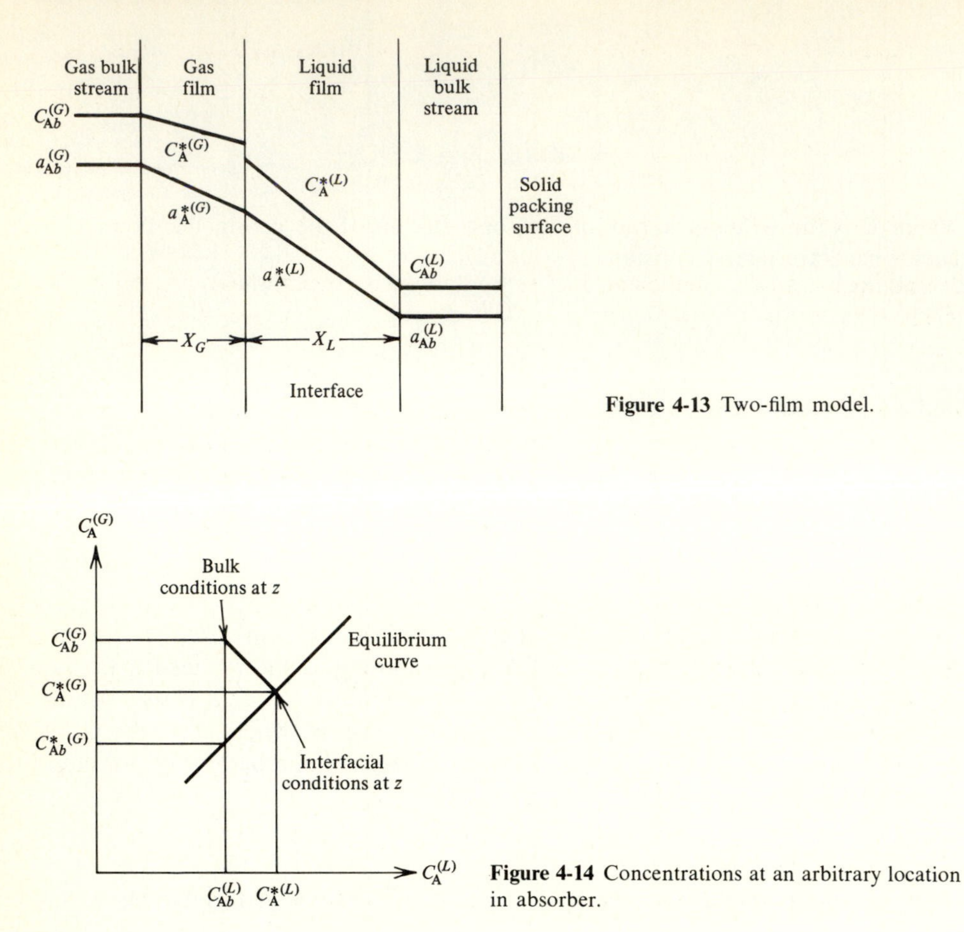

Figure 4-13 Two-film model.

Figure 4-14 Concentrations at an arbitrary location in absorber.

Boundary condition at gas-liquid interface As indicated by Eq. (4-35*b*), the fluxes of mass in each phase can be taken as equal at the interface, or

$$N_A^{(G)}\bigg|_I = N_A^{(L)}\bigg|_I \qquad \text{at interface} \tag{4-76b}$$

If we assume thermodynamic equilibrium at the interface, the chemical potentials (*not* the concentrations) of A in each phase will be equal and there will be an equilibrium relationship between the concentration of A in the gas phase and that in the liquid phase; or, from Eq. (4-35*a*),

$$C_A^{*(G)} = mC_A^{*(L)} \tag{4-77}$$

where the asterisk refers to equilibrium conditions. The equilibrium "constant" m is actually not constant but can be assumed so over small ranges of concentration.

Using Eq. (4-29), we define a mass-transfer coefficient for each phase

$$N_A\Big|_I \equiv \begin{cases} k_G(C_{Ab}^{(G)} - C_A^{*(G)}) \equiv k_G|\Delta C_A^{(G)}| & \text{gas} \qquad (4\text{-}78) \\ k_L(C_A^{*(L)} - C_{Ab}^{(L)}) \equiv k_L|\Delta C_A^{(L)}| & \text{liquid} \qquad (4\text{-}79) \end{cases}$$

The subscript b refers to the concentration in the bulk stream.

Overall mass-transfer coefficient Just as we defined an overall heat-transfer coefficient U in terms of a total temperature drop $-\Delta T_t$, so also can we define an *overall or total mass-transfer coefficient* K_t in terms of an overall driving force ΔC_{At}. However, we cannot use $C_{Ab}^{(G)} - C_{Ab}^{(L)}$ as the driving force because it is not zero at equilibrium; that is, $C_A^{*(G)} \neq C_A^{*(L)}$. As discussed earlier, the correct driving force is the difference in chemical potentials (or activities if the standard states are the same) [the first of Eqs. (4-39*a*)]. Therefore we *could* use chemical potentials or activities instead of concentrations and define an *overall* mass-transfer coefficient by

$$N_A\Big|_I = K_t'(C_{Ab}^{(G)}\gamma_A^{(G)} - C_{Ab}^{(L)}\gamma_A^{(L)}) \qquad (4\text{-}80)$$

where $\gamma_A^{(G)}$ and $\gamma_A^{(L)}$ are activity coefficients for the gas and liquid, respectively. Then at the interface

$$a_A^{(G)}\Big|_I \equiv C_A^{(G)}\gamma_A^{(G)}\Big|_I = C_A^{(L)}\gamma_A^{(L)}\Big|_I \equiv a_A^{(L)}\Big|_I \qquad (4\text{-}81)$$

Equations (4-78) and (4-79) would then be written

$$N_A\Big|_I \equiv k_G'(C_{Ab}^{(G)}\gamma_A^{(G)} - C_A^{*(G)}\gamma_A^{(G)}) \qquad k_G' \equiv \frac{k_G}{\gamma_A^{(G)}}$$

and

$$N_A\Big|_I \equiv k_L'(C_A^{*(L)}\gamma_A^{(L)} - C_{Ab}^{(L)}\gamma_A^{(L)}) \qquad k_L' \equiv \frac{k_L}{\gamma_A^{(L)}}$$

The γ's are assumed independent of concentration over the range involved. The effect of multiplying the concentrations by the γ's is to move the gas and liquid curves in Fig. 4-12*a* so that they meet at the interface where Eq. (4-81) holds and can be written

$$a_A^{*(G)} \equiv C_A^{*(G)}\gamma_A^{(G)} = C_A^{*(L)}\gamma_A^{(L)} = a_A^{*(L)}$$

But instead of multiplying each concentration by an activity coefficient it is usually more convenient to multiply all liquid-phase concentrations by m, that is, to use Eq. (4-77), where

$$m \equiv \frac{\gamma_A^{(L)}}{\gamma_A^{(G)}}$$

is an equilibrium "constant" (or *partition coefficient* in the case of two immiscible liquids) as defined in Eq. (4-38). Then Eq. (4-78) need not be changed, but in Eq. (4-79) one replaces $C_{Ab}^{(L)}$ with the gas-phase concentration $C_{Ab}^{*(G)} \equiv mC_{Ab}^{(L)}$ that would be in equilibrium with $C_{Ab}^{(L)}$. Thus instead of shifting *both* curves we shift only the liquid-concentration curve. (Figure 4-14 shows these concentrations at a given point in the absorber.) Then we can define an overall mass-transfer coefficient K_t

$$N_A \equiv K_t(C_{Ab}^{(G)} - mC_{Ab}^{(L)}) \equiv K_t|\Delta C_{At}| \tag{4-82}$$

Relation between K_t, k_G, and k_L The total driving force $|\Delta C_{At}|$ can be expressed in terms of the individual driving forces $|\Delta C_A^{(G)}|$ and $|\Delta C_A^{(L)}|$. Thus

$$|\Delta C_{At}| \equiv C_{Ab}^{(G)} - mC_{Ab}^{(L)} \tag{4-83}$$

Because of Eq. (4-77) we can subtract $C_A^{*(G)}$ if we add $mC_A^{*(L)}$ to the right-hand side, or

$$\begin{aligned} |\Delta C_{At}| &= C_{Ab}^{(G)} - C_A^{*(G)} + mC_A^{*(L)} - mC_{Ab}^{(L)} \\ &= |\Delta C_A^{(G)}| + m|\Delta C_A^{(L)}| \end{aligned} \tag{4-84}$$

At steady state for a rectangular geometry, N_A will be constant. Then from Eqs. (4-78) and (4-79)

$$\frac{1}{K_t} = \frac{1}{k_G} + \frac{m}{k_L} \tag{4-85}$$

Design of Gas Absorber

Now that we have an expression for N_{AI} [Eq. (4-82)], we can substitute it into Eq. (4-76*a*) to obtain

$$d\mathcal{W}_{AI} = K_t\left(C_{Ab}^{(G)} - mC_{Ab}^{(L)}\right)aA_c\,dz \tag{4-86a}$$

We now note that there is an analogy between the gas absorber and the double-pipe heat exchanger. This means that since the differential and macroscopic balances are of the same form, the mathematics will be the same in each case. Making use of this idea permits us to write equations for the absorber that are identical to those for the heat exchanger. The overall rate equation is then (see the Problems)

$$\mathcal{W}_{AI} = \langle K_t \rangle aA_c\langle -\Delta C_{At}\rangle_{LM} \tag{4-86b}$$

where

$$\langle C_{At}\rangle_{LM} \equiv \frac{\Delta C_{At}|_0 - \Delta C_{At}|_L}{\ln(\Delta C_{At}|_0/\Delta C_{At}|_L)} \qquad \langle K_t\rangle \equiv \frac{1}{L}\int_0^L K_t\,dz \qquad \Delta C_{At} = C_{Ab}^{(G)} - mC_{Ab}^{(L)}$$

and the overall mass balance is given by Eq. (4-74) with $z = L$.

To find the height of the absorber, one needs $\langle K_t \rangle$, which is given by Eq. (4-85) in terms of the individual mass-transfer coefficients for each phase. These coefficients must either be obtained experimentally or calculated from an empirical or theoretical equation. The development of methods for predicting mass-transfer coefficients is an important goal of transport phenomena.

4-10 COMPOSITE SYSTEMS WITH INTERNAL GENERATION

We now consider two analogous problems, one in energy transport and the other in momentum transport, in a two-phase system with internal generation (Fig. 4-15). The momentum-transport problem is the flow of two immiscible liquids A and B between infinite parallel plates under a common pressure drop $p_0 - p_L$. It is assumed that each fluid is in laminar flow and that the flow fields are fully developed. (Thus, entrance and exit effects are neglected.) The viscosity μ^A of fluid A is less than μ^B, and fluid A is in the top layer. This requires that $\rho_A < \rho_B$. Problems of this nature are important in the design of large-scale fluid-transport systems, such as oil transport, where a thin annular ring of water can be used to surround the oil phase, greatly reducing viscous drag.

Let $x = 0$ correspond to a plane interface between the two fluids located so that $x = h_A$ and $x = -h_B$ refer to the top and bottom plate, respectively. A momentum balance on an element in phase A leads to

$$\frac{d\tau_{xz}^A}{dx} = \frac{p_0 - p_L}{L} g_c \equiv \Phi_M \tag{4-87}$$

and similarly for phase B

$$\frac{d\tau_{xz}^B}{dx} = \frac{p_0 - p_L}{L} g_c \equiv \Phi_M \tag{4-88}$$

Here Φ_M is the momentum generation, with dimensions of momentum per unit time and unit volume. Integrating

$$\int_{\tau_0}^{\tau_{xz}^A} d\tau_{xz}^A = \int_{x=0}^{x=x} \Phi_M \, dx \qquad \text{and} \qquad \boxed{\int_{\tau_{xz}^B}^{\tau_0} d\tau_{xz}^B = \int_{x=0}^{x=x} \Phi_M \, dx}$$

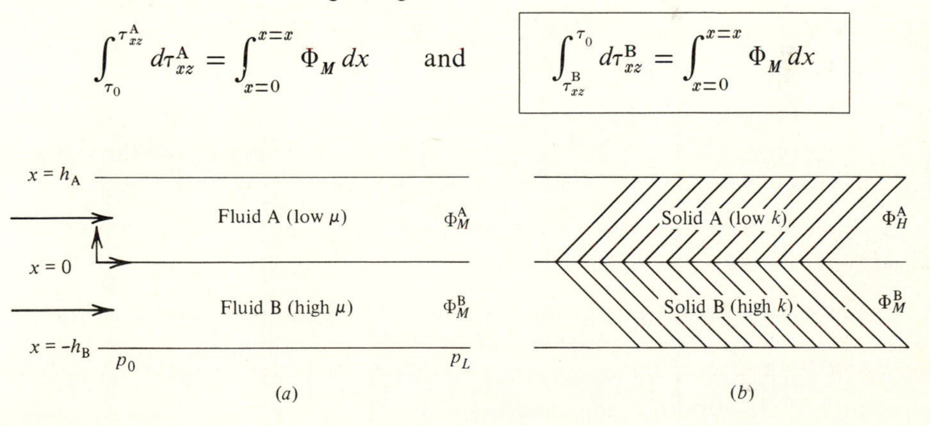

Figure 4-15 (*a*) Momentum transport and (*b*) energy transport with internal generation.

gives

$$\tau_{xz}^{A} = \tau_0 = \Phi_M x \tag{4-89}$$

and

$$\tau_0 - \tau_{xz}^{B} = -\Phi_M x \tag{4-90}$$

where τ_0 is the momentum flux at the centerline. Note that τ_0 is the same for both phases since the momentum flux is not discontinuous; i.e., it does not jump from one value to another at a given point in the fluid. (We assume that there is no interfacial tension gradient in the y direction.) When Newton's law of viscosity is introduced, Eqs. (4-98) and (4-90) yield

$$-\mu^{A}\frac{dv_z^{A}}{dx} = \tau_0 + \Phi_M x \qquad \text{and} \qquad -\mu^{B}\frac{dv_z^{B}}{dx} = \tau_0 + \Phi_M x \tag{4-91}$$

Integrating

$$\int_{v_z^{A}}^{0} dv_z^{A} = -\frac{1}{\mu^{A}}\int_{x}^{h_A}(\tau_0 + \Phi_M x)\,dx \qquad \boxed{\int_{0}^{v_z^{B}} dv_z^{B} = -\frac{1}{\mu^{B}}\int_{-h_B}^{x}(\tau_0 + \Phi_M x)\,dx}$$

gives

$$v_z^{A} = \frac{\tau_0}{\mu^{A}}(h_A - x) + \frac{\Phi_M h_A^2}{2\mu^{A}}\left(1 - \frac{x^2}{h_A^2}\right) \tag{4-92}$$

and

$$v_z^{B} = -\frac{\tau_0}{\mu^{B}}(x + h_B) + \frac{\Phi_M h_B^2}{2\mu^{B}}\left(1 - \frac{x^2}{h_B^2}\right) \tag{4-93}$$

To eliminate the unknown quantity τ_0 we use the condition that $v_z^{A} = v_z^{B}$ at $x = 0$. If these two velocities were not equal, this would imply that the fluids could slip past each other with zero resistance, since in this case the velocity gradient would be infinite. Using this condition leads to

$$\frac{\tau_0}{\mu^{A}}h_A + \frac{\Phi_M h_A^2}{2\mu^{A}} = -\frac{\tau_0}{\mu^{B}}h_A + \frac{\Phi_M h_A^2}{2\mu^{B}}$$

and

$$\tau_0 = \frac{(\Phi_M/2)[(h_B^2/\mu^{B}) - h_A^2/\mu^{A}]}{(h_A/\mu^{A}) + h_B/\mu^{B}} \tag{4-94}$$

Using this result for τ_0, we can now calculate final expressions for τ_{xz}, v_z^{A}, and v_z^{B} in terms of μ^{A}, μ^{B}, Φ_M, h_A, and h_B; they are given in Table 4-2, and the velocity and momentum profiles are sketched in Fig. 4-16.

Energy Transport

Let us now consider the energy analog of the above problem, illustrated in Fig. 4-16*b*. Here we have two contiguous solids of thermal conductivities k^{A} and k^{B}, where $k^{A} < k^{B}$. Each solid is generating heat at Φ_H (J/s · m^3). A good example of a problem of this sort arises in connection with the calculation of the temperature distribution in the human body, where solid B might represent the fatty tissues in the arm, for example, and solid A would correspond to the skin.[3]

Table 4-2 Composite systems with internal generation

Flux differential equation	$\frac{d\tau_{xz}^{A}}{dx} = \frac{p_0 - p_L}{L} g_c \equiv \Phi_M$	$\frac{d\tau_{xz}^{B}}{dx} = \frac{p_0 - p_L}{L} g_c \equiv \Phi_M$
Flux distribution	$\tau_{xz}^{A} = \tau_{0A} + \Phi_M x$	$\tau_{xz}^{B} = \tau_{0B} + \Phi_M x$
Flux law	$\tau_{xz}^{A} = -\mu^{A} \frac{dv_z}{dx}$	$\tau_{xz}^{B} = -\mu^{B} \frac{dv_z}{dx}$
Boundary conditions	$v_z = 0 \quad \text{at } x = h_A \text{ and } x = h_B$	
Continuity of potential	$v_z^{A} = v_z^{B} \quad \text{at } x = 0$	
Continuity of flux	$\tau_0^{A} = \tau_0^{B} \equiv \tau_0 = \frac{\Phi_M h_A \gamma}{2} \quad \text{at } x = 0 \quad \text{where} \quad \gamma \equiv \frac{h_B^2 \mu^{A} - h_A^2 \mu^{A}}{h_A(h_B \mu^{A} + h_A \mu^{B})}$	
Flux distribution	$\tau_{xz} = \Phi_M h_A \left(\frac{x}{h_A} + \frac{1}{2}\gamma \right)$	
Potential distribution	$v_z^{A} = \frac{\Phi_M h_A^2}{2\mu^{A}} \left[1 - \gamma\left(\frac{x}{h_A} - 1 \right) - \frac{x^2}{h_1^2} \right]$	$v_z^{B} = \frac{\Phi_M h_B^2}{2\mu^{B}} \left[1 - \gamma\left(\frac{x}{h_B} + 1 \right) - \frac{x^2}{h_B^2} \right]$

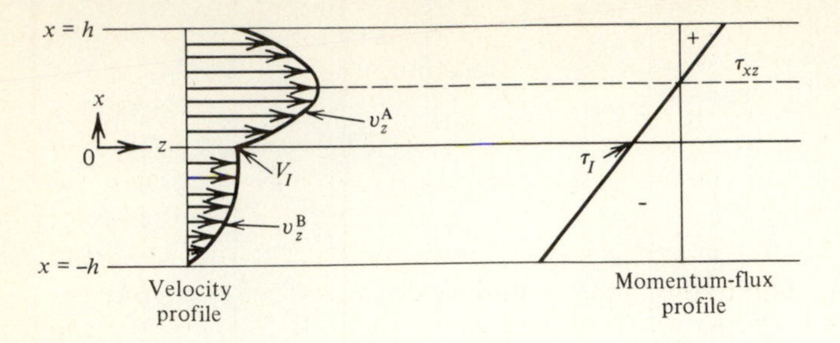

Figure 4-16 Velocity- and momentum-flux profiles.

The flux distributions in phases A and B are found from energy balances

$$\frac{dq_x^A}{dx} = \Phi_H \qquad \frac{dq_x^B}{dx} = \Phi_H$$

These equations are analogous to Eqs. (4-87) and (4-88). To complete the analogy between the two examples we need only show that the boundary conditions are also analogous. In the present problem the boundary conditions are

At $x = 0$:	$q_x^A = q_x^B$	continuity of heat flux
At $x = h_A$ and $x = -h_B$:	$T = T_o$	fixed surface temperature
At $x = 0$:	$T^A = T^B$	continuity of temperature

which correspond to the boundary conditions for the momentum-transfer problem. Hence, the problems are exactly analogous. By analogy the equations for the heat flux and temperature profile can be obtained by replacing τ_{xz} with q_x, Φ_M with Φ_H, μ^A with k^A, and μ^B with k^B.

Interpretation

The physical picture of what is happening inside each medium is as follows. Internal generation of heat (or momentum) is taking place uniformly in both phases. This tends to produce a high temperature (or velocity). Since the boundaries at $x = h_A$ and at $x = -h_B$ are at a lower temperature (or velocity) than inside the medium, there is a tendency for heat (or momentum) to be transported toward *each* boundary. Near the upper boundary $x = h_A$ the temperature (or velocity) must be decreasing with increasing x, and dT/dx (or dv_z/dx) is negative; therefore, by the flux law, the flux is positive. However, near the lower boundary ($x = -h_B$) the temperature (or velocity) is decreasing with *decreasing* x, so the gradient is positive and the flux must be negative. A negative flux then represents transport to the lower boundary while a positive flux represents transport to the upper boundary. A zero flux means that the tendencies for transport toward each boundary are balanced. As we observed in Chap. 1, the resistance to transport between two points is inversely proportional to the thermal conductivity (or viscosity) and increases with the distance between the points. If we start near the lower boundary, we can see that the initial gradient is flatter in phase B than it is near the upper boundary in phase A. This is because phase B has a higher thermal conductivity (or viscosity) and provides an intrinsically low resistance path for transport to the nearest boundary. However, if we move upward, the resistance of the path to the lower boundary increases due to the greater distance while the resistance of an alternate path to the upper boundary decreases due to the shorter distance. Eventually we reach the point $x = x_0$, where the resistance of a path to the lower boundary is equal to that of a path to the upper boundary. At this point the flux must be zero.

At $x > x_0$ transport is to the upper boundary because the shortened distance overcomes the effect of the relatively smaller thermal conductivity.

APPENDIX 4-1 DIMENSIONAL ANALYSIS AND EMPIRICAL CORRELATIONS

The traditional procedure for carrying out a dimensional analysis can be illustrated for flow (laminar or turbulent) in a pipe (Sec. 3-3). The steps are as follows.

Step 1 Postulate that the dependent variable is a function of certain independent variables and parameters; e.g., assume that, for fully developed flow,

$$\frac{\mathcal{P}_0 - \mathcal{P}_L}{L} g_c \equiv \Phi_M = \Phi_M(v_a;\, D, \rho, \mu) \tag{A4-1-1}$$

where v_a is the independent variable and D, ρ, μ are the parameters.†

Step 2 Assume that a typical term in the correlation has the form

$$\Phi_M = C v_a^{\alpha} D^{\beta} \rho^{\alpha} \mu^{\delta}$$

where C is a dimensionless constant.

Step 3 Insert the dimensions of each variable and parameter, letting M be mass, L be length, and t be time

$$\Phi_M [=] \frac{ML/t}{L^3 t} [=] C \left(\frac{L}{t}\right)^{\alpha} \left(L\right)^{\beta} \left(\frac{M}{L^3}\right)^{\gamma} \left(\frac{M}{Lt}\right)^{\delta}$$

where $[=]$ means "has dimensions of."

†A parameter is a quantity that is constant during an experimental run; a universal parameter is constant under all conditions.

Step 4 Since C is dimensionless, and since the dimensions of each side must be the same, the following equations must hold:

$$M: \quad 1 = \gamma + \delta$$

$$L: \quad -2 = \alpha + \beta - 3\gamma - \delta$$

$$t: \quad -2 \equiv -\alpha - \delta = -\alpha + \gamma - 1$$

Since we have three equations and four unknowns, all but one exponent can be eliminated, resulting in one dimensionless group involving v_a, which we will call N_1 and which will be raised to a power, and another dimensionless group N_d involving Φ_M, the dependent variable. In other words, whereas here

$$N_d = CN_1^{\alpha_1} \tag{A4-1-2}$$

in general, we would have

$$N_d = C \prod_{i=1}^{n-1} N_i^{\alpha_i} \equiv CN_1^{\alpha_1} N_2^{\alpha_2} \cdots N_{n-1}^{\alpha_{n-1}} \tag{A4-1-3}$$

Also, in general *the number of dimensionless groups is equal to the number of variables* (here five, counting Φ_M) *minus the number of dimensions* (here three). This relation is called the *Buckingham pi theorem* after the Greek letter Π (pi). A formal proof of the theorem is given in books on matrix methods.

Arbitrarily choosing α as the exponent in terms of which the others are to be expressed gives

$$\gamma = \alpha - 1 \qquad \beta = \alpha - 3 \qquad \delta = 1 - \gamma = 2 - \alpha$$

Then

$$\Phi_M = Cv_a^{\alpha} D^{\alpha-3} \rho^{\alpha-1} \mu^{2-\alpha}$$

Isolating the Reynolds number $Dv_a \rho/\mu$ gives

$$\Phi_M = C \frac{v_a^{\alpha-2} D^{\alpha-2} \rho^{\alpha-2}}{\mu^{\alpha-2}} \frac{v_a^2 \rho}{D}$$

or

$$\frac{\Phi_M D}{v_a^2 \rho} = C \left(\frac{Dv_a \rho}{\mu} \right)^{\alpha-2} \tag{A4-1-4}$$

where the group N_1 is $Dv_a \rho/\mu$, the Reynolds number, and the group

$$N_d \equiv \frac{\Phi_M D}{v_a^2 \rho} \equiv \frac{(\mathcal{P}_0 - \mathcal{P}_L) g_c D}{L v_a^2 \rho} \equiv 2f \tag{A4-1-5}$$

where f is the friction factor as defined herein. Thus (A4-1-4) gives $f = f\,(\text{Re})$.

Multiplying both sides of (A4-1-5) by Re gives an alternate group N_d' on the left-hand side†

†Any product of groups raised to a power can be used as long as the number of groups does not change.

$$N'_d \equiv 2f\,\mathrm{Re} = \frac{\Phi_M D}{v_a^2 \rho} \frac{D v_a \rho}{\mu} = C\left(\frac{D v_a \rho}{\mu}\right)^{\alpha-1}$$

$$= \frac{\Phi_M D^2}{v_a \mu} = C\left(\mathrm{Re}\right)^{\gamma} \qquad \text{(A4-1-6)}$$

where C and $\gamma \equiv \alpha - 1$ must be obtained by experiment or from theory. If we use theory, the Hagen-Poiseuille equation gives, for laminar flow, Eq. (3-32),

$$\Phi_M = \frac{32\,\mu v_a}{D^2} \qquad \text{or} \qquad \frac{\Phi_M D^2}{v_a \mu} = 32$$

Thus for laminar pipe flow, the theory indicates that $\gamma = 0$, $\alpha = 1$, and $C = 32$ in Eq. (A4-1-6).

In the above example the assumption was made that the dynamic-pressure drop was proportional to the length L due to the existence of fully developed flow. Had that assumption not been made, we would have had to introduce L into the correlating equation (A4-1-1), resulting in six quantities, three dimensions, and three dimensionless groups; the additional group being $N_3 = L/D$. But if we were to take data at various L/D ratios, we would find that when we plotted f versus Re, the data for all L/D ratios would fall on the same curve at fully developed flow. This means that if we *do* happen to throw in an unnecessary parameter, the experiment will tell us that we have. Of course, we would have to go to the trouble and expense of taking the added data, in this case that of varying L/D.

Actually, in the dimensional analysis illustrated here, if the flow is laminar, we have already included an unnecessary variable, the density. As reference to the Hagen-Poiseuille equation (3-32) shows, the density does not appear and need not have been included except as part of the definition of the dynamic-pressure drop [Eq. (3-30)]. (Had we been correlating *actual* pressure drop, it *would* be involved.) Without ρ there would be four quantities and three dimensions or only one group, which would be either

$$N'_d \equiv \frac{\Phi_M D^2}{v_a \mu}$$

or its reciprocal or some constant a times N'^b_d, where b is any constant. This result tells us that $N'_d = C$, a constant, for *all* v_a, μ, or D. Dimensional analysis will not give us C, but *one* data point would be enough to find it. If we felt we needed to test our assumption, we might take data at other values of v_a, μ, D and even ρ and L. These results would give us a good average value of C which could be compared with the theoretical value of 32 obtained in Eq. (3-32).

Note that the result $N'_d = C$ is equivalent to $f\,\mathrm{Re} = C'$ where f is the friction factor and Re the Reynolds number. From theory, $C' = 16$. Thus for laminar pipe flow we only need *one* group, $f\,\mathrm{Re} \equiv N'_d$ or C.

In *turbulent* flow, the Hagen-Poiseuille equation does not hold. Thus in Eq. (A4-1-4) $\alpha \neq 1$ and in Eq. (A4-1-6) $\gamma \neq 0$. Instead the experimental data indicate that $\alpha \approx 0.8$ and $\gamma \approx -0.2$, but both α and γ vary slightly with Re. This means that the pressure drop varies with the average velocity raised to a power of about 1.8 rather than 1.0 in laminar flow. From the standpoint of dimensional analysis, it means that we can no longer write $f\,\text{Re} = C$.

This type of dimensional analysis requires that we specify which variables are involved. It will then tell us what dimensionless variables are possible but will not give the functional relationship. For example, Eq. (A4-1-4) *will* tell us that $f = f(\text{Re})$ but it will not tell us whether α in Eq. (A4-1-4) is constant or varies with Re. Only theory or experiment can provide that information.

APPENDIX 4-2 DIMENSIONAL ANALYSIS OF DIFFERENTIAL EQUATIONS

In Sec. 1-32 for heat transport in a slab we showed how to obtain the dimensionless groups by means of a dimensional analysis of the dimensionless differential equation without having to solve the differential equation itself. Thus it was not necessary to assume that certain variables were involved, as we did in Eq. (A4-1-1), or to know the form of the functional relationship. We now outline this procedure and illustrate it for laminar flow in a pipe.

Step 1 Derive the flux differential equation using an incremental momentum balance. In Chap. 3 this resulted in Eq. (3-17)

$$\frac{d(r\tau_{rz})}{r\,dr} = \Phi_M \equiv \frac{(\mathcal{P}_0 - \mathcal{P}_L)g_c}{L} \tag{1}$$

Step 2 Substitute the flux law, Eq. (1-50),

$$\frac{-d(r\mu\,dv_z/dr)}{r\,dr} = \Phi_M \tag{2}$$

Step 3 Write the boundary conditions [Eqs. (3-18)]

At $r = 0$:

$$r\frac{dv_z}{dr} = 0 \tag{3}$$

At $r = R$:

$$v_z = 0 \tag{4}$$

Step 4 Define v_a [Eq. (3-10)]

$$v_a = \frac{\int_0^R v_z\,dA_c}{\int dA_c} = \frac{\int_0^R v_z r\,dr}{\int_0^R r\,dr} \tag{5}$$

Step 5 Use the above results to find the naturally occurring variables or parameters. Since our goal is to predict $v_a(\Phi_M)$ (or the inverse) in dimensionless form, v_a and Φ_M are of course involved. Also Eq. (2) gives μ and the boundary condition gives R (or D); but since Eq. (5) repeats v_a and R we have Φ_M, v_a, R, and μ.

Step 6 Form natural groups such as geometric ratios or ratios of variables to parameters (or combinations of parameters); a logical group is $r^* \equiv r/R$. We are tempted to use v_z/v_a, but since we eventually want to relate v_a to Φ_M, let us temporarily let v_{ch} be an unspecified characteristic velocity, so that

$$v^* \equiv \frac{v_z}{v_{\text{ch}}} \tag{6}$$

Step 7 Solve for the dimensional variables and substitute into the differential equation and boundary conditions

$$\frac{d[r^* R\mu \, d(v^* v_{\text{ch}})/d(r^* R)]}{r^* R \, d(r^* R)} = \Phi_M \tag{7}$$

Since R, v_{ch}, and μ are constants, we can divide both sides by $\mu v_{\text{ch}}/R^2$

$$\frac{d(r^* \, dv^*/dr^*)}{r^* \, dr^*} = \frac{\Phi_M R^2}{\mu v_{\text{ch}}} = \frac{\Phi_M D^2}{4\mu v_{\text{ch}}} = 1 \tag{8}$$

where for convenience we define v_{ch} as

$$v_{\text{ch}} \equiv \frac{\Phi_M D^2}{4\mu} \tag{9}$$

so that the right-hand side becomes unity.

In this case we *could* easily integrate twice to get $v^*(r^*)$; but suppose this were a problem in which the solution could *not* be obtained. Examination of (7) tells us that

$$v^* = v^*(r^*; \text{boundary conditions}) \tag{10}$$

where the boundary conditions are:

At $r^* = 0$: $$r^* \frac{dv^*}{dr^*} = 0 \tag{11}$$

At $r^* = 1$: $$v^* = 0 \tag{12}$$

Here the boundary conditions provide no new information, but in some cases they might give geometric ratios. Writing (5) in dimensionless form gives

$$\frac{v_a}{v_{\text{ch}}} \equiv v_a^* = \frac{\int_0^1 v^* r^* \, dr^*}{\int_0^1 r^* \, dr^*} \tag{13}$$

Step 8 Ascertain the dimensionless groups that would be involved if the dimensionless differential equation were solved analytically using the dimensionless boundary conditions.

We mentally note that Eqs. (10) to (12) indicate that the solution to Eq. (10) would contain *no* new groups and that substitution into (13) and integrating would give a function of the limits, which in *this* case are pure numbers. Therefore, the right-hand side of (13) gives a pure number C''

$$v_a^* \equiv \frac{v_a}{v_{\text{ch}}} \equiv \frac{4 v_a \mu}{\Phi_M D^2} = C'' \tag{13a}$$

By comparison with our previous result (A4-1-6)

$$C'' = \frac{4}{C} \tag{14}$$

This example further demonstrates that nondimensionalizing the differential equation is a much better way to obtain the dimensionless groups than the traditional method described previously.

Noncircular Cross Sections

For an annular cross section with radii R_1 and R_2 the above dimensional analysis would be the same as far as the differential equation and flux laws are concerned, but the dimensionless boundary conditions would be different and we would need to specify a characteristic length scale to replace D. As discussed in Example 4-2, we could use the equivalent diameter, or R_1 or R_2, as length scales. If we were to choose R_2, the dimensionless boundary conditions would be

$$v^* = 0 \qquad \text{at } r^* = \frac{R_1}{R_2} \equiv \kappa \text{ and at } r^* = 1$$

Also in expression (10) for v_a^* we would need to integrate between $r^* = \kappa$ and $r^* = 1$ instead of between 0 and 1. Thus if we were unable to solve the differential equation, dimensional analysis would tell us that

$$v_a^* = \frac{R_1}{R_2} \tag{15}$$

Thus we would have to construct ducts of various R_1/R_2 ratios in order to plot Eq. (15) graphically. But it would *not* tell us the nature of the functional relationship, which we found from transport theory [Eq. (3-52) or (4-19)] to be

$$v_a^* \equiv \frac{v_a}{\Phi_M R_2^2} = \frac{8}{\phi(\kappa)} \tag{A4-2-1}$$

where

$$\phi(\kappa) \equiv \frac{1-\kappa^4}{1-\kappa^2} = \frac{1-\kappa^2}{\ln(1/\kappa)} \tag{A4-2-2}$$

Obviously it is unlikely that Eq. (4-2-2) would be discovered empirically.

For certain cross sections, e.g., a rectangle of sides A and B, the flow becomes two-dimensional; that is, v_z is a function of two variables (say x and y). For two-dimensional laminar flow, the flux differential equation and flux laws are more complex and will be taken up in Chaps. 9 and 10. Dimensional analysis of these results in a Hagen-Poiseuille type of equation similar to Eq. (3-52) or (4-2-1)

$$v_a^* \equiv \frac{v_a \mu}{\Phi_M R_{\mathrm{ch}}^2} = v_a^*(\text{geometric ratio})$$

where for a rectangle $R_{\mathrm{ch}} \equiv B$ and the geometric ratio is B/A.

For cross sections with one characteristic dimension, e.g., an equilateral triangle or a square cross section, one obtains $v_a^* = C''$. This result leads to $f\,\mathrm{Re} = C'$, where C'' and C' are different numbers for each cross section.

APPENDIX 4-3 DETERMINATION OF BULK MEAN TEMPERATURE

In Chap. 3 we defined the average velocity for flow through a pipe of cross-sectional area $A_{\mathrm{cs}} = \pi R^2$ as that velocity v_a which permitted us to calculate the volumetric rate of flow by multiplying it by A_{cs} [Eq. (3-8)]; that is, $\dot{\mathcal{V}}_v \equiv v_a A_{\mathrm{cs}}$. Since v_z varies over the cross section, the volumetric rate of flow was found by considering the flow through an element of area $dA_{\mathrm{cs}} = 2\pi r\,dr$ (Fig. 3-3) and integrating over the cross section, or

$$\dot{\mathcal{V}}_v = \int d\dot{\mathcal{V}}_v = \int_{A_{\mathrm{cs}}} v_z\,dA_{\mathrm{cs}}$$

Equating the above to (3-8) gave (3-9)

$$v_a = \frac{\int v_z\,dA_{\mathrm{cs}}}{A_{\mathrm{cs}}} = \frac{\int_0^R v_z(2\pi r\,dr)}{\int_0^R 2\pi r\,dr} \tag{A4-3-1}$$

Now suppose that the temperature varies with radial position but the velocity does not. We *could* define a volume average temperature as

$$T_a \equiv \frac{\int_0^R T(r) r\,dr}{\int_0^R r\,dr} \tag{A4-3-2}$$

But if the velocity were also varying over the cross section, it would have more physical meaning if we defined a bulk mean temperature T_b as that temperature which would permit us to calculate $\dot{Q}_{\mathcal{H}}$, the flow of enthalpy (heat content) through the cross section, from the average velocity and the total area. For an incompressible fluid the enthalpy per unit mass is given by

$$\hat{\mathcal{H}} - \hat{\mathcal{H}}_R = \hat{C}_p(T - T_R)$$

where T_R is a reference temperature and $\mathcal{H}_R$ is the enthalpy at T_R. Since the mass of fluid flowing through dA_{cs} is

$$\rho\, d\dot{\mathcal{V}}_v = \rho v_z\, dA_c = \rho v_z(2\pi r\, dr)$$

the flow of enthalpy through dA_{cs} is

$$d\dot{Q}_{\mathcal{H}} = (\hat{\mathcal{H}} - \hat{\mathcal{H}}_R)\rho\, d\dot{\mathcal{V}}_v = \rho\hat{C}_p(T - T_R)v_z\, dA_{cs}$$

and the total flow of enthalpy over the cross section is

$$\dot{Q}_{\mathcal{H}} = \int_0^R \rho\hat{C}_p(T - T_R)v_z\, dA_{cs}$$

This must be equal to

$$\dot{Q}_{\mathcal{H}} = \rho\hat{C}_p(T_b - T_R)v_z\, dA_{cs}$$

Equating these two expressions gives

$$T_b - T_R = \frac{\int \rho\hat{C}_p(T - T_R)v_z\, dA_{cs}}{\rho\hat{C}_p(T_b - T_R)v_a A_{cs}} = \frac{\int_0^R (T - T_R)v_z r\, dr}{\int_0^R v_z r\, dr}$$

for flow through a pipe of a fluid with constant $\rho\hat{C}_p$. Since T_R is constant, this gives

$$T_b = \frac{\int_0^R (Tv_z)(2\pi r\, dr)}{\int_0^R v_z(2\pi r\, dr)} = \frac{\int_0^R Tv_z r\, dr}{\int_0^R v_z r\, dr} \tag{A4-3-3}$$

which differs from the *average* temperature T_a by the inclusion of the velocity as a weighting factor. Note that $v_a \equiv v_b$ but $T_a \neq T_b$ unless the velocity profile is flat.

The bulk mean and average molar concentrations C_{Ab} and C_{Aa} are similarly defined by replacing T with C_A.

PROBLEMS

4-1 (*a*) Use the results obtained for the average velocity in an annulus to obtain a relation between friction factor and Reynolds number modified for the case of an annulus by either (1) the use of the outer radius R_2 as a characteristic length or (2) the use of an equivalent diameter equal to 4 times the hydraulic radius.

(*b*) It has been common practice to assume that the *same* relationship between friction factor and Reynolds number obtained in Eq. (4-5) for pipe flow ($f = 16/\mathrm{Re}$) would also hold for annular flow if one used an equivalent diameter ($D_e \equiv 4R_H \equiv 4A_c/z$) in the Reynolds number. Show that this empirical rule does not give the same results as those obtained rigorously in (2) of part (*a*). Then, taking the case where $R_1/R_2 = \frac{1}{2}$, find the percentage error in $\mathcal{P}_0 - \mathcal{P}_L$ resulting from using the old rule for the same fluid, average flow velocity, and length.

4-2 Using the results of Prob. 3-2, derive an expression for friction factor versus Reynolds number for a power-law fluid. Define the modified Reynolds number Re′ such that the same relation between f and Re′ that holds in Eq. (4-5) would also hold in this case.

4-3 Rework Prob. 3-3 using the friction-factor–Reynolds-number equation derived in Prob. 4-2.

4-4 Consider the laminar fully developed steady-state flow of a newtonian incompressible fluid between infinite parallel plates separated by a distance $2B$ where x is distance from the midplane and z is distance in direction of flow (see Fig. 3-8). If the average velocity is given by

$$v_a = \frac{(\mathcal{P}_0 - \mathcal{P}_L)g_c B^2}{3\mu L_z}$$

derive an equation for the friction factor as a function of the Reynolds number.

Answer:
$$f = \frac{24}{\text{Re}}$$

4-5 An incompressible newtonian fluid is flowing at steady state in laminar fully developed flow in the z direction through a vertical rectangular duct of length L_z and cross-sectional dimensions $2B$ and $2D$. The average velocity in the duct has been found (by methods studied later) to be

$$\langle v_z \rangle \equiv V = \frac{(\mathcal{P}_0 - \mathcal{P}_L)g_c R_H 2}{\mu L}\phi\left(\frac{B}{D}\right)$$

where $\phi(B/D)$ is a dimensionless function that depends only on the ratio B/D.

(*a*) Apply the definition of the friction factor in general to the duct by specifying the characteristic terms.

(*b*) Write an overall or macroscopic force balance over the duct, giving the physical meaning of each term and stating how it can be calculated.

(*c*) Write an overall momentum balance, giving the meaning of each term and defining by equation ϕ_M, the momentum generation rate per unit volume.

(*d*) Define hydraulic radius in general, defining each term used and giving equations for these for the duct.

(*e*) Define the equivalent diameter.

(*f*) Use the results above to derive an equation for the friction factor for the duct in terms of the Reynolds number (based on D_e) and $\phi(B/D)$ only.

4-6 A bus bar, i.e., a solid-metal conductor of electricity, generates heat at a uniform rate of Φ_e J/s · m^3. The long dimension of the bus bar is L_z, and the cross-sectional dimensions are $2B$ and $2D$. The surfaces of the bus bar are maintained at a constant temperature T_1.

(*a*) Using an analogy to Prob. 4-5 write an equation for the average temperature over the cross section of the bus bar.

(*b*) Write a *macroscopic* energy balance for the bus bar and *interpret it* in terms of physical processes.

4-7 A catalytic converter consists of an array of small ducts through which the exhaust gases from an automobile engine flow in parallel while being oxidized. Various cross sections have been suggested for the individual ducts, including triangular, elliptical, and hexagonal. Important design considerations include adequate heat transport (because of the high heat of reaction) and minimization of pressure drop (due to the small cross section and the existence of laminar flow).

For a fully developed steady laminar flow in a noncircular cross section of an incompressible newtonian fluid the average velocity can in general be related to the dynamic pressure drop $\mathcal{P}_0 - \mathcal{P}_L$ by

$$\langle v_z \rangle \equiv V = \frac{(\mathcal{P}_0 - \mathcal{P}_L)D_e^2}{C_1 \mu L_z}$$

where D_e is the equivalent diameter and C_1 is a parameter which depends on the shape of the cross section; e.g., for a circular cross section $C_1 = 32$. The friction factor is defined as

$$f \equiv \frac{F_D g_c / A_{ch}}{\frac{1}{2}\rho V^2}$$

where A_{ch} is the characteristic area. Relate the friction factor to the Reynolds number (based on equivalent diameter) and C_1 *only*.

4-8 An incompressible newtonian fluid is flowing in laminar fully developed steady-state flow in the z direction (up) in a duct whose length is L_z and whose cross section is an equilateral triangle with side B. The cross-sectional area is $\sqrt{3}\,B^2/4$. The average velocity is found (from a momentum balance) to be

$$v_a \equiv \langle v_z \rangle = \frac{(\mathcal{P}_0 - \mathcal{P}_L) g_c B^2}{80 \mu L_z}$$

(*a*) Define the hydraulic radius and apply to the duct.

(*b*) Write an overall or macroscopic force balance over the duct (in detail, interpreting each term).

(*c*) Define friction factor in general and apply to the duct.

(*d*) Relate friction factor to a Reynolds number defined on the basis of $D_e = B$.

(*e*) An empirical rule of thumb is that the expression for friction factor in a circular pipe ($f = 16/\text{Re}$) can be used if the equivalent diameter D_e is used to replace the pipe diameter. Determine whether this is correct and find the percentage error made in calculating $\mathcal{P}_0 - \mathcal{P}_L$ if incorrect.

4-9 Devise a heat-transfer analogy to Prob. 4-8, stating the physical conditions involved. Then by analogy write (*a*) an equation for the average temperature and (*b*) a macroscopic balance, interpreting each term.

4-10 In a fermenter O_2 diffuses from a gas bubble to the liquid and from the liquid to a microorganism. The diameter of the microorganism is about 1 μm and the diffusivity of O_2 is 10^{-9} m^2/s. Find the mass-transfer coefficient for transfer through the liquid film to an individual microorganism. Assume that the diameter is so small that the effect of agitation is negligible, i.e., that the film can be assumed to be infinite and stagnant.

4-11 Assuming that the temperature profile is linear and that the velocity profile is parabolic in a laminar-flow pipe heat exchanger with constant wall temperature, find the Nusselt number.

4-12 Repeat Prob. 4-11 if both the temperature and velocity profiles are parabolic. Let

$$\frac{T - T_w}{T_c - T_w} = 1 - \left(\frac{r}{R}\right)^2$$

4-13 Water is flowing through a pipe heat exchanger 9.144 m long at a Reynolds number of 1000. If $\text{Pr} = 3.00$, $D = 0.0254$ m, $T_0 = 50°\text{F}$, $T_w = 100°\text{F}$, and $k = 0.363$ Btu/h · ft^2 · °F, calculate $\dot{Q}_w$ and T_b. Use the estimated value of $\langle \text{Nu} \rangle = 4.0$.

4-14 If $\dot{Q}_w$ in Prob. 4-13 is to be increased 10 percent, find the increased length required if $\text{Nu} = 4$ and other parameters are the same as above.

4-15 The bus bar in Prob. 4-5 is insulated with a material of thickness L_I and thermal conductivity k_I. The temperature at the surface of the insulation is T_2. Outside the insulation is an air film that has a heat-transfer coefficient h_o. The bulk temperature of the air is T_a. The overall heat-transfer coefficient between the air and the interface between the insulation and the bus bar is U_o.

Derive an equation for U_o in terms of L_I, k_I, and h_o, assuming all area ratios are unity. (First show that the total resistance R_T is the sum of the individual resistances and then obtain expressions for each based on the equations defining U_o and h_o.)

4-16 Suppose the fluid in Prob. 4-5 is being cooled by a fluid surrounding it with bulk temperature T_{b0}. The inside fluid bulk temperature is T_{b1}, the inside wall temperature is T_{wi}, and the outside wall temperature is T_{wo}. The inside film coefficient is h_i, and the outside film coefficient is h_o.

(*a*) Write equations *defining* the individual film coefficients and defining the overall coefficient U.

(*b*) Derive an equation relating U to h_i, h_o, the metal wall thickness L_w, and the metal thermal conductivity k.

4-17 A nuclear reactor is shaped in the form of a sphere of radius R_1. The nuclear reaction occurs inside, generating heat in the reactor at ϕ_N W/m^3. The metal wall of the reactor has a uniform thickness $X_A = R_2 - R_1$ and is surrounded by a coolant. Assume that the reaction occurs only inside the reactor. The temperature at the inner surface of the reactor is T_1, that between the metal and the shielding is T_2, and that between the shielding and the coolant is T_3. The bulk temperature of the coolant is T_c. All the above temperatures are assumed constant, and the reactor operates at steady state. The thermal conductivity of the metal is k_A, and that of the shielding is k_B. The heat-transfer coefficient for the coolant is h_c.

(*a*) Write the complete boundary conditions at the interface between the metal wall and the insulation.

(*b*) Write the complete boundary condition at the interface between the insulation and the coolant.

(*c*) Considering the metal only and using an analogy to the results of Prob. 4-10, derive an expression for A_{MA}, the geometric-mean area $\sqrt{A_1 A_2}$ as defined by

$$\dot{Q} = k_A A_{MA} \frac{-\Delta T_A}{X_A}$$

(*d*) Derive an expression for the overall heat-transfer coefficient U_c in terms of k_A, k_B, and h_c and the various radii. Note that U_c is defined by $\dot{Q} = U_c A_c(-\Delta T_t)$, where A_c is the outside area of the shielding of the reactor and $-\Delta T_t \equiv T_1 - T_c$.

4-18 Derive the equations for gas-absorber design showing that the log-mean driving force is valid.

4-19 Using parallel columns for heat and for mass, compare the derivation of the design equations for a double-pipe countercurrent heat exchanger with those for a gas absorber. Show how the final result for a gas absorber is obtained by analogy to heat transport. In your table include notation for rate of transport ($\dot{Q}_f$ and $\mathcal{W}_A$), potentials, driving forces, and transfer coefficients. Also give differential and overall balances for each case.

4-20 Derive equations for the design of a cocurrent double-pipe heat exchanger. Show that the log-mean driving force still applies and define the driving forces.

4-21 A furnace wall is constructed of 0.127 m of heat-resistant brick, 0.1524 m of insulating brick, and 6.25 m of steel plate for mechanical protection. As a result of thermal-expansion effects, it is suspected that layers of air are present between the layers of brick and steel. An accurate heat balance over the furnace indicated the heat loss from the wall to be 12.62 W/s while the interior wall of the furnace was at 1630 K and the outside surface of the steel was at 306.5 K.

(*a*) What thickness of insulating brick is thermally equivalent to the air layers present?

(*b*) If the heat is transferred by conduction only in the air layers, what is the total thickness of air layers present?

Material	Thermal conductivity, Btu/h · ft · °F
Heat-resistant brick	3.6
Insulating brick	1.8
Steel	26.1
Air	0.073

4-22 In a pilot-plant operation 18.14 kg/h of feed to a reactor must be preheated from 972 to 1098 K in a shell-and-tube heat exchanger. The tube, which is copper, is 25.4 mm outside diameter with 2.032-mm walls. The inside wall of the tube is maintained at 1215 K by the condensation of steam on the shell side. If $\langle h \rangle = 40$ Btu/h · ft^2 · °F and $\hat{C}_p = 0.49$ Btu/lbm · °F for the feed, what length of heat exchanger is required?

4-23 As part of a coating operation, a plastic coating of thickness δ is flowing downward on the inner surface of a vertical cylindrical tube of radius R. The plastic is being washed by an upward countercurrent flow of water. The pressure at a bottom cross section ($z = 0$) is p_0 (in both water and plastic). At an upper cross section ($z = L$) the pressure is p_L, where z is longitudinal distance measured upward. Radial distance from the tube center toward the walls is r. At the interface between water and plastic, $r = r_I$. Entrance and exit effects are to be neglected. It is desired to determine the momentum flux and velocity profiles in the water and plastic phases, assuming complete immiscibility, laminar flow, and no rippling of the film.

Water phase

(*a*) Using a momentum balance over an increment of volume, obtain a differential equation for the momentum flux.

(*b*) Integrate this equation employing the proper boundary condition at $r = 0$.

(*c*) Obtain the velocity of the water as a function of r and in terms of the velocity v_I at $r = r_I$.

(*d*) Obtain the average velocity and volumetric flow rate of the water.

Plastic phase In this part use the superscript P to designate the plastic phase and the superscript W to designate the water phase.

(*e*) Repeat parts (*a*) to (*d*) as necessary to obtain a differential expression for the momentum flux τ_{rz}^P in the plastic.

(*f*) Write an integrated expression for the momentum flux in the plastic in terms of the momentum flux at the interface τ_I^P.

(*g*) Write expressions for all the boundary conditions in the water-plastic problem.

(*h*) If the plastic is newtonian, derive the equations for the velocity and flux distributions. Prepare sketches of τ_{rz}^W and τ_{rz}^P versus r and for v_z^P and v_z^W versus r.

4-24 If the plastic in Prob. 4-23 is a Bingham plastic, draw sketches of flux and velocity distributions to illustrate how part (*h*) might be modified if $\tau_0^P > \tau_I^P$.

4-25 (*a*) Devisė a heat-transport analogy to Prob. 4-23 (for a newtonian fluid) by stating the physical situation that would be required and the boundary conditions and giving the equation for the temperature profile. Sketch flux and temperature profiles showing positive and negative regions.

(*b*) State the physical situation analogous to the flow of a Bingham plastic fluid (Prob. 4-24).

4-26 A copper wire with radius R_1 and length L is insulated with plastic to an outer radius R_2. An internal generation of heat due to an electric current takes place in the wire at a uniform rate ϕ_e Btu/h · ft^3, but although radial heat conduction occurs, no heat generation takes place in the plastic. Assume that the outside surface of the plastic is maintained at a constant temperature T_0. Use superscript P for the plastic and W for the wire.

(*a*) Obtain flux differential equations for the wire and for the plastic.

(*b*) Obtain the temperature and flux distribution for the wire and plastic *and* sketch them.

4-27 Develop a momentum-transport analogy to Prob. 4-26 as follows:

(*a*) State *clearly* what physical conditions would have to exist in a two-phase fluid-flow problem for the differential equations in each phase and the boundary conditions to be analogous. Use superscript A for the inside fluid and B for the outside fluid. Write the analogous differential equations for momentum flux in each phase.

(*b*) Draw a sketch of velocity- and momentum-flux distributions and describe physically what is occurring in each phase.

4-28 Modify Prob. 4-26 by postulating that there is a stagnant air film outside the plastic. The heat-transfer coefficient for the air-plastic film is h_o and the *overall* heat-transfer coefficient for the solid plastic and the air film is U_o, based on the outside area of the plastic. Let

T_1 = temperature of the wire-plastic interface
T_2 = temperature of the outside plastic surface, i.e., air-plastic interface
T_0 = temperature of air
k = thermal conductivity of plastic

A_{lM} = mean area of plastic
A_o = outside area of plastic

(*a*) Write expressions for the heat *flow* through the plastic and through the air film (in terms of the appropriate temperature differences).

(*b*) Write expressions for resistance of the plastic R_p, resistance of the air film R_a, and overall resistance R_T.

(*c*) Derive an equation for U_o as a function of h_o, k, and any other variables needed and defined in the statements of Prob. 4-26.

4-29 Carry out the derivations in Table 4-2 for heat transport. Then, by analogy, write the corresponding equations for momentum. Explain the presence of a maximum in the velocity curve (Fig. 4-16). Explain why there is a maximum in medium A rather than in medium B. Can there be more than one maximum?

REFERENCES

1. Danckwerts, P. V.: "Gas-Liquid Relations," McGraw-Hill, New York, 1970.
2. Hatta, S.: *Tohoku Imp. Univ. Tech. Rep.*, **8**:1 (1928); **10**:119 (1932).
3. Huckaba, C. E., L. W. Hansen, J. A. Downey, and R. C. Darling: Calculation of the Temperature Distribution in the Human Body, *AIChE J.*, **19**:527 (1973).
4. Perry, R. H., and C. H. Chilton: "Chemical Engineers' Handbook," 5th ed., McGraw-Hill, New York, 1973.
5. Rosner, D. E.: *Chem. Eng. Educ.*, **14**:193 (1980).
6. Smith, J. M., and H. Van Ness: "Chemical Engineering Thermodynamics," McGraw-Hill, New York, 1975.

CHAPTER
FIVE

CONVECTIVE TRANSPORT

5-1 CONVECTIVE MASS TRANSPORT

Not only can heat, mass, and momentum be transported due to gradients, but they may also be transported by bulk-fluid motion, or *bulk flow*. In fact, in most mass-transport processes bulk transport is concomitant with diffusive transport.

To illustrate this interaction between diffusion and bulk flow, we consider a simple example. Suppose a small quantity of liquid A, say bromine, is placed in the bottom of a test tube (Fig. 5-1). The bromine, being very volatile, will rapidly evaporate and diffuse through the air (B) above. Note that the process is initially *not* at steady state because at $t < 0$ there is no vapor A in the air (B) above the liquid. At $t = 0$ the vapor (A) starts moving up from the surface. As it does, B molecules *also* must move upward so that C, the total molar density, will remain

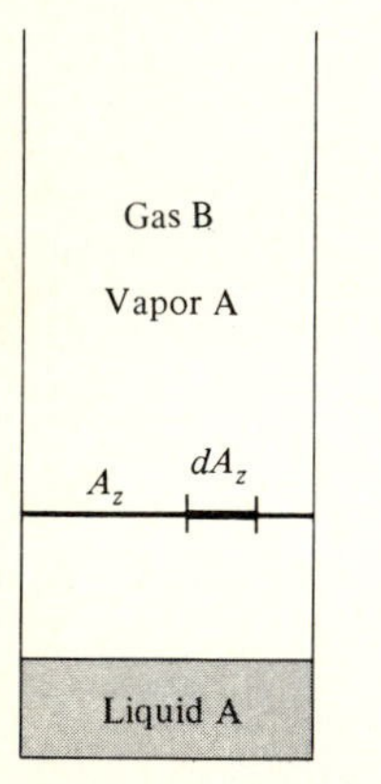

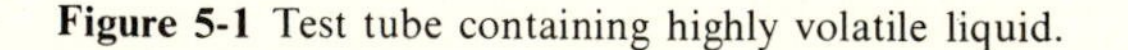
Figure 5-1 Test tube containing highly volatile liquid.

constant. (By the ideal-gas law, $C = p/R_G T$, where R_G is the gas constant and p and T are constant.) We assume that B is not soluble in liquid A so it cannot move down into the liquid. Eventually the bromine will reach the top of the tube, where it will be dispersed into the air. We will let the air be denoted as a single species B and measure z upward from the level of the liquid bromine.

Mass Average Velocity of Species A

Each individual molecule will have its own velocity but there will be an *average* velocity v_{Az} for all of the molecules of A and another average velocity v_{Bz} for the molecules of B. There will also be an average velocity for the mixture of A and B. Since the concentrations of A and of B are in general not equal, a simple average cannot be used. Our goal is to show how we can relate the average velocity to the species velocities by either a mass-fraction weighted average or a mole-fraction weighted average. Let ρ_A be the mass concentration of A in kg/m³ or lb/ft³ in the vapor phase. Consider an element of area dA_z in a plane of cross-sectional area† A_z at some arbitrary value of z and time t. The total mass rate of flow of species A through dA_z is denoted by dw_A kg/s or lb/h; it is equal to $\rho_A\, d\dot{\mathcal{V}}_A$, where $d\dot{\mathcal{V}}_A$, the volumetric rate of flow of A, is equal to the product $v_{Az}\, dA_z$. Thus

$$dw_A = \rho_A v_{Az}\, dA_z \tag{5-1}$$

This is a defining equation for v_{Az}. Dividing both sides by dA_z and recalling the general definition of a flux [Eq. (1-3)], we can define the total mass flux of A, n_{Az} kg/m² · s, as

$$\frac{dw_A}{dA_z} = \rho_A v_{Az} \equiv n_{Az} \tag{5-2}$$

Initially the air molecules will also have an average velocity since they are pushed up and out of the tube as the bromine evaporates. The mass flux of the B molecules can then be expressed as

$$n_{Bz} = \rho_B v_{Bz} \tag{5-3}$$

where v_{Bz} is the mass-average velocity of species B. Note that the species velocities v_{Az} and v_{Bz} are *both* positive (Fig. 5-2*a*) and that the mass-*average* velocity v_z must be between v_{Az} and v_{Bz}. The total flux of A and B will then be (Fig. 5-2*b*)

$$n_{tz} = n_{Az} + n_{Bz} = \rho_A v_{Az} + \rho_B v_{Bz} \tag{5-4}$$

By an argument similar to that used for Eqs. (5-2) and (5-3) we can now define a total mixture mass-average velocity or simply a mass-average velocity v_z by

$$n_{tz} = \frac{dw}{dA_z} \equiv \rho v_z \tag{5-5}$$

†We use a differential area dA_z instead of A_z to allow for variations of velocity over the cross section (see Fig. 5-1).

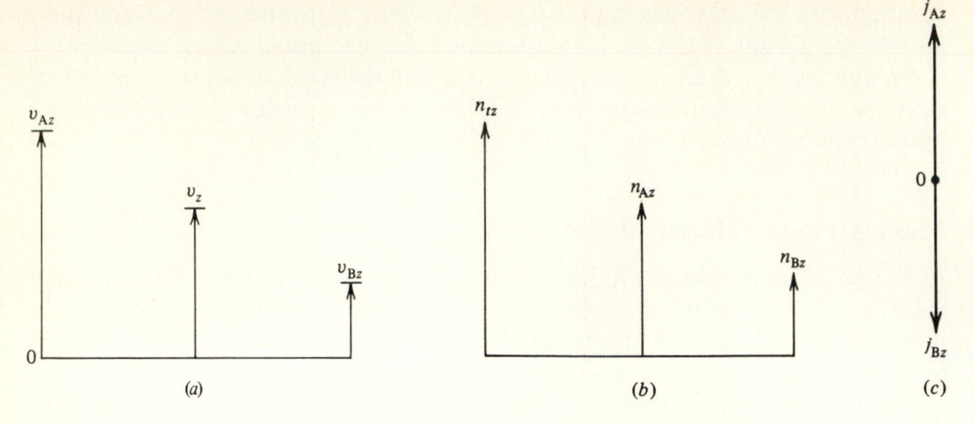

Figure 5-2 (*a*) Velocity vectors; (*b*) total-flux vector; (*c*) diffusive-flux vector.

Equating (5-4) and (5-5), we obtain

$$v_z = \frac{\rho_A v_{Az} + \rho_B v_{Bz}}{\rho} = \omega_A v_{Az} + \omega_B v_{Bz} \tag{5-6}$$

where $\rho = \rho_A + \rho_B$ is the density of the vapor and $\omega_A \equiv \rho_A/\rho$ and $\omega_B \equiv \rho_B/\rho$ are respectively the mass fractions of A and of B. Note that the mass-average velocity is simply a weighted average based on the relative *amounts* of A and B in the mixture. Equation (5-6) is defined specifically for a binary mixture. For a mixture of s different species, the mass-average velocity can be defined by

$$v_z = \frac{\sum_{i=1}^{s} \rho_i v_{iz}}{\sum_{i=1}^{s} \rho_i} = \sum_{i=1}^{s} \omega_i v_{iz} \tag{5-7}$$

Diffusive flux It is essential that the student understand precisely what is meant by diffusion, or diffusive flux. When we speak of the diffusion of a species A (say), we mean its flow *relative to the average velocity of the stream,* not its total flow. For example, consider the general case of a binary mixture moving through a tube at some average velocity v_z. If the concentration of A and B were the same throughout, there would be no diffusive flow—even though there would be a flow of each species—because the flow of either A or B *relative to the mass-average velocity of the stream* would be zero.

However, in Fig. 5-1 the mass concentration of A, ρ_A, is not constant; it is highest at the liquid-vapor interface and decreases with distance z whereas the mass concentration of B increases with z. Thus, species A will have a larger velocity v_{Az} than species B, and v_{Az} will also be larger than the mass-average velocity v_z (Fig. 5-2*a*).

The *mass diffusive flux* of A is thus defined by

$$j_{Az} \equiv \rho_A(v_{Az} - v_z) \tag{5-8}$$

and similarly

$$j_{Bz} \equiv \rho_B(v_{Bz} - v_z) \tag{5-9}$$

Adding Eqs. (5-8) and (5-9) gives

$$j_{Az} + j_{Bz} = \rho_A v_{Az} + \rho_B v_{Bz} - (\rho_A + \rho_B)v_z = 0$$

from Eq. (5-6). Thus the sum of the diffusive fluxes must be zero; therefore (see Fig. 5-2*c*) j_{Az} and j_{Bz} have *opposite* signs; j_{Az} is positive and j_{Bz} is negative; that is, A is diffusing in the $+z$ direction and B is diffusing in the $-z$ direction. However v_{Bz} is *positive* and the *total* flux of B, as given by Eq. (5-3), is also positive.

We now use the definitions of n_{Az} and j_{Az} to relate these quantities to each other. Adding and subtracting $\rho_A v_z$ to the right-hand side of Eq. (5-2) gives

$$n_{Az} \equiv \rho_A v_{Az} \equiv \rho_A(v_z + v_{Az} - v_z) \equiv \rho_A v_z + \rho_A(v_{Az} - v_z)$$

or, using Eq. (5-8),

$$\boxed{n_{Az} = \rho_A v_z + j_{Az}} \tag{5-10}$$

The first term, $\rho_A v_z$, is referred to as the *bulk flux* or *convective flux* of species A. Hence Eq. (5-10) states that

Total mass flux of A (with respect to fixed axis)	=	Bulk mass flux of A (with respect to fixed axis)	+	Diffusive mass flux of A (with respect to mass-average velocity)

Similarly, for species B we can write

$$n_{Bz} = \rho_B v_z + j_{Bz} \tag{5-11}$$

For a binary mixture, Fick's law of diffusion is expressed in the form

$$j_{Az} = -\rho D_{AB} \frac{\partial \omega_A}{\partial z} \qquad \text{where} \qquad \omega_A = \frac{\rho_A}{\rho} \tag{5-12}$$

ω_A being the mass fraction of species A. Note that the potential gradient is rigorously written as $\partial\omega_A/\partial z$ rather than $\partial\rho_A/\partial z$. For constant density, Eq. (5-12) becomes

$$j_{Az} = -D_{AB} \frac{\partial \rho_A}{\partial z} \tag{5-13}$$

Molar units It is often convenient to use molar units instead of mass units. If we divide Eq. (5-10) by M_A, the molecular weight of A, we have

$$N_{Az} = C_A v_z + J_{Az} \tag{5-14}$$

where

$$N_{Az} = \frac{n_{Az}}{M_A} \text{ mol/s} \cdot \text{m}^2 \qquad C_A = \frac{\rho_A}{M_A} \text{ mol/m}^3 \qquad J_{Az} = \frac{j_{Az}}{M_A} \text{ mol/s} \cdot \text{m}^2 \tag{5-15}$$

J_{Az} is the molar diffusive flux of species A with respect to the mass-average velocity. In terms of J_{Az}, Fick's law for constant ρ is

$$\boxed{J_{Az} = -D_{AB}\frac{\partial C_A}{\partial z} \qquad \rho = \text{const}} \tag{5-16}$$

Molar-Average Velocity

It is sometimes convenient in diffusion problems (although perhaps confusing to the beginning student) to introduce other types of average velocities, such as a volume-average velocity or a molar-average velocity. The molar-average velocity turns out to be particularly useful for diffusion in gases, where C is nearly constant, whereas the mass-average velocity is more useful for liquids, where ρ is nearly constant.

Consider again a binary mixture of A and B. The *total molar flux* N_{tz} is given by

$$N_{tz} = N_{Az} + N_{Bz} \tag{5-17a}$$

where $$N_{Az} \equiv \frac{n_{Az}}{M_A} = C_A v_{Az} \qquad \text{and} \qquad N_{Bz} \equiv \frac{n_{Bz}}{M_B} = C_B v_{Bz} \tag{5-17b}$$

Thus, from Eq. (5-17*a*)

$$N_{tz} = C_A v_{Az} + C_B v_{Bz} \tag{5-18}$$

The *molar-average velocity* v_z^* is then defined (in a similar fashion to v_z) so that if $d\mathcal{W}$ is the molar rate of flow through an area dA_z,

$$\frac{d\mathcal{W}}{dA_z} \equiv N_{tz} = Cv_z^* \tag{5-19}$$

Using Eq. (5-18) gives

$$\boxed{v_z^* = \frac{C_A v_{Az} + C_B v_{Bz}}{C} = x_A v_{Az} + x_B v_{Bz}} \tag{5-20}$$

where $$C = C_A + C_B \tag{5-21}$$

is the total moles per volume and $x_A \equiv C_A/C$ and $x_B \equiv C_B/C$ are, respectively, the mole fractions of A and B. Thus v_z^* is a mole-fraction-weighted average velocity. Note the similarity of Eqs. (5-6) and (5-20). For a multicomponent mixture v_z^*

is given by

$$v_z = \frac{\sum_{i=1}^{s} C_i v_{iz}}{\sum_{i=1}^{s} C_i} = \sum_{i=1}^{s} x_i v_{iz} \tag{5-22}$$

The molar flux N_{Az} of species A can be expressed in terms of v_z^* by the identities

$$N_{Az} \equiv C_A v_{Az} \equiv C_A[v_z^* + (v_{Az} - v_z^*)] \equiv C_A v_z^* + C_A(v_{Az} - v_z^*) \equiv C_A v_z^* + J_{Az}^* \tag{5-23}$$

where

$$J_{Az}^* \equiv C_A(v_{Az} - v_z^*) \equiv \text{molar diffusive flux of species A with respect to molar-average velocity} \tag{5-24}$$

$$C_A v_z^* \equiv \text{molar bulk flux of species A} \tag{5-25}$$

In terms of J_{Az}^*, Fick's law of diffusion (see Prob. 5-1) is

$$J_{Az}^* = -D_{AB} C \frac{\partial x_A}{\partial z} \tag{5-26}$$

For constant molar density C

$$J_{Az}^* = -D_{AB} \frac{\partial C_A}{\partial z} \qquad C = \text{const} \tag{5-27}$$

Combining Eqs. (5-6), (5-10), and (5-12) gives

$$n_{Az} = \omega_A(n_{Az} + n_{Bz}) - \rho D_{AB} \frac{\partial \omega_A}{\partial z} \tag{5-28}$$

Combining Eqs. (5-23), (5-20), (5-17*b*), (5-26), and (5-24) gives

$$N_{Az} = x_A(N_{Az} + N_{Bz}) - C D_{AB} \frac{\partial x_A}{\partial z} \tag{5-29a}$$

It can be shown (student exercise) that for a binary mixture

$$J_{Az}^* = -J_{Bz}^* \qquad \text{and} \qquad D_{AB} = D_{BA} \tag{5-29b}$$

Special Cases

There are a number of instances where Eqs. (5-28) and (5-29*a*) simplify considerably. For example, in *equimolar counterdiffusion* equal molar fluxes of A and B are moving in opposite directions. An example of this occurs in distillation. In this case, $N_{Az} = -N_{Bz}$, and Eq. (5-29*a*) becomes

$$N_{Az} = -CD_{AB}\frac{\partial x_A}{\partial z} \qquad \text{equimolar counterdiffusion} \tag{5-30}$$

Diffusion through a stagnant gas A very important special case is the diffusion of a gas A through a stagnant gas B. Consider the situation described earlier (Fig. 5-1) but at steady state. Let us write a molar balance on species B over an element of width Δz (Fig. 5-3). Then

$$\underset{\text{Input}}{N_{Bz}A_z\Big|_z} - \underset{\text{Output}}{N_{Bz}A_z\Big|_{z+\Delta z}} + \underset{\text{Generation}}{0} = \underset{\text{Accumulation}}{0}$$

Dividing by $A_z\,\Delta z$ and letting $\Delta z \to 0$ gives

$$\frac{dN_{Bz}}{dz} = 0$$

Thus N_{Bz} = const. Since no B enters or leaves the liquid at $z = 0$,

$$N_{Bz}\Big|_{z=0} = 0$$

and $N_{Bz} = 0$ at all z. Thus $v_{Bz} = 0$, and B is called a *stagnant* gas. Equation (5-29*a*) then becomes

$$N_{Az} = x_A(N_{Az} + \cancel{N_{Bz}}) - CD_{AB}\frac{\partial x_A}{\partial z}$$

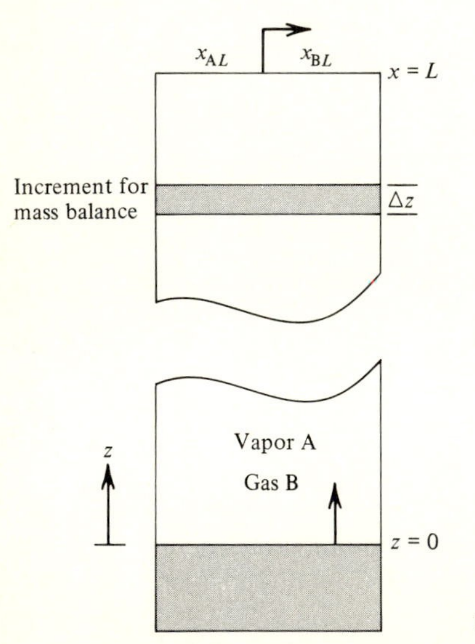

Figure 5-3 Increment used for mass balance.

or
$$N_{Az} = \frac{-CD_{AB}}{1 - x_A}\frac{dx_A}{dz} \tag{5-31}$$

Next we make a molar balance around a small increment Δz and obtain

$$\frac{dN_{Az}}{dz} = 0$$

$$N_{Az} = \text{const} \equiv N_{A0} \equiv N_{Az}\Big|_{z=0}$$

and, from Eq. (5-31),

$$\frac{-CD_{AB}}{1 - x_A}\frac{dx_A}{dz} = N_{A0} \tag{5-32}$$

Now C, the total molar density, is constant since the temperature and pressure are constant. Thus, (5-32) can be integrated directly

$$N_{A0}\int_{z=0}^{z} dz = -CD_{AB}\int_{x_{A0}}^{x_A}\frac{dx_A}{1 - x_A} = +CD_{AB}\int_{1-x_{A0}}^{1-x_A}\frac{d(1 - x_A)}{1 - x_A}$$

giving
$$N_{A0}z = CD_{AB}\ln\frac{1 - x_A}{1 - x_{A0}} \tag{5-33}$$

where x_{A0} is the concentration of species A at the gas-liquid interface. This is assumed to be equal to the equilibrium composition of A at the temperature T and pressure p of the system. Thus, if p_{vA} is the vapor pressure of A at T and p, then $x_{A0} = p_{vA}/p$ by Raoult's law [Eq. (4-39*b*)]. Integrating Eq. (5-32) from $x = 0$ to L gives

$$N_{A0}L = CD_{AB}\ln\frac{1 - x_{AL}}{1 - x_{A0}} \tag{5-34}$$

Combining this with Eq. (5-33), we have

$$\frac{z}{L} = \frac{\ln[(1 - x_A)/(1 - x_{A0})]}{\ln[(1 - x_{AL})/(1 - x_{A0})]} \tag{5-35}$$

which gives the profile $x_A(z)$. Since $x_A + x_B = 1.0$, Eq. (5-35) can be written alternatively as

$$\frac{z}{L} = \frac{\ln(x_B/x_{B0})}{\ln(x_{BL}/x_{B0})} \tag{5-36}$$

The profiles for x_A and x_B are plotted in Fig. 5-4. The total molar flux of A is given from Eq. (5-34) as

$$N_{A0} = \frac{CD_{AB}}{L}\ln\frac{1 - x_{AL}}{1 - x_{A0}}$$

and the total molar flow rate $\mathcal{W}_A^*$ is

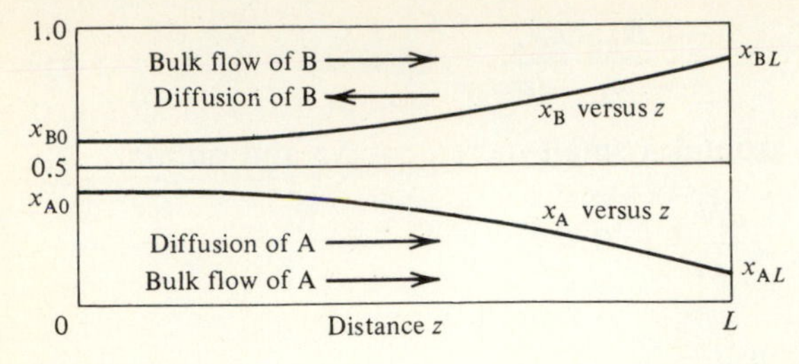

Figure 5-4 Concentration profiles for diffusion of A through stagnant B.

$$\mathcal{W}_A^* = N_{A0}A_z = \frac{CD_{AB}A_z}{L}\ln\frac{1-x_{AL}}{1-x_{A0}} \tag{5-37}$$

If the weight of A evaporated during a given time is measured, $\mathcal{W}_A^*$ can be calculated.

Discussion Consider the expression for the diffusive molar flux of species B with respect to the molar-average velocity

$$J_{Bz}^* = -D_{AB}C\frac{dx_B}{dz}$$

Calculating the gradient dx_B/dz from Eq. (5-36), we find that J_{Bz}^* is not zero. In other words, even though the *net* flow of B is zero, there is diffusion arising from the mass-fraction gradient (Fig. 5-4). To explain how this can occur without altering the concentration profile we turn to Eq. (5-29*a*)

$$N_{Bz} = 0 = \underset{\text{Bulk flux}}{x_B(N_{Az}+0)} - \underset{\text{Diffusive flux}}{CD_{BA}\frac{dx_B}{dz}} \tag{5-38}$$

This expression indicates that the diffusive flux is exactly balanced by the bulk flux $x_B N_{Az}$ (Fig. 5-4).

Effect of Bulk Flow on Mass-Transfer Coefficient

Now let us reconsider the film-theory model discussed in Sec. 4-6 (Figs. 4-5 and 4-6) for diffusion of A through *stagnant* B when bulk flow *cannot* be neglected. To avoid confusion with the mass fraction x_A we will use z for the distance from the interface instead of x, which was used in Chap. 4. An incremental balance over an element Δz and the use of Eq. (5-31) gives Eqs. (5-32) to (5-34). Then, using the definition of k_c as given by Eq. (4-28) and replacing x_{AL} in Eq. (5-34) with x_{Ab}, we have

$$N_{A0} \equiv k_c(C_{A0} - C_{Ab}) = \frac{CD_{AB}}{L}\ln\frac{1-x_{A0}}{1-x_{Ab}} = \frac{CD_{AB}}{L}\ln\frac{x_{B0}}{x_{Bb}} \tag{5-39}$$

Thus
$$k_c = \frac{D_{AB}}{L} \frac{\ln (x_{B0}/x_{Bb})}{x_{B0} - x_{Bb}} = \frac{D_{AB}}{L\langle x_B \rangle_{LM}} \tag{5-40}$$

where $\langle x_B \rangle_{LM}$ is the log-mean mole fraction of B. Also, from Eq. (4-31), the definition of the mass-transfer Nusselt number,

$$(\text{Nu})_{AB} \equiv \frac{k_c D_{ch}}{D_{AB}} = \frac{D_{ch}}{L\langle x_B \rangle_{LM}} \tag{5-41}$$

5-2 CONVECTIVE FLOW OF ENERGY AND MOMENTUM

We have seen that for mass transport the total flux is given by

$$n_{Az} = j_{Az} + \rho_A v_z \qquad \text{or} \qquad N_{Az} = J_{Az} + C_A v_z$$

$$\text{Total flux} = \text{diffusive flux} + \text{bulk flux}$$

Just as the bulk mass (or molar) flux of species A can be expressed as the product of a concentration of A, ρ_A (or C_A), and the velocity v_z, so also can the bulk flux of energy be expressed as the product of an "energy concentration" and the velocity v_z. Since the mass (or molar) concentration has the units of mass A (or moles A) per unit volume, the energy concentration should have the units of energy per unit volume. As indicated in Chap. 1, Part D, a thermal-energy concentration can be taken as $\rho \widehat{C}_p T$ and the bulk flow of thermal energy would be $(\rho \widehat{C}_p T) v_z A_z$. Then the bulk flux would be $(\rho \widehat{C}_p T) v_z$:

$$\underset{\substack{\text{Total} \\ \text{energy flux}}}{e_z} = \underset{\substack{\text{Diffusive} \\ \text{flux}}}{q_z} + \underset{\substack{\text{Bulk} \\ \text{flux}}}{(\rho \widehat{C}_p T) v_z} \tag{5-42}$$

Similarly, the bulk flux of z momentum in the z direction would be given by the product of the z-momentum "concentration" and the mass-average velocity in the z direction. Since momentum is mass times velocity, the z-momentum concentration would be ρv_z, which has the units of z momentum per unit volume. Thus the total z-momentum flux in the z direction is given by

$$\pi_{zz} = \tau_{zz} + (\rho v_z) v_z \tag{5-43}$$

Since momentum is a vector (whereas temperature and concentration are scalars), it is also possible to have a bulk flux of x momentum in the z direction, which would be given by the product of the x-momentum concentration ρv_x and the average velocity in the z direction, or $(\rho v_x) v_z = \rho v_x v_z = \rho v_z v_x$. A similar expression for π_{zy} can be written. The product $v_z v_x$, called a *dyadic* product, will be considered further in Chap. 9.

As an illustration of bulk flow of momentum, consider the case of fully developed laminar flow in a pipe treated in Sec. 3-3. When we made the force-momentum balances, Eqs. (3-13) and (3-14), we did not include bulk flow because the bulk flows of z momentum in at $z = 0$ and out at $z = L$ were equal; i.e., the rate of bulk flow of z momentum through an element of area $\Delta A_z = 2\pi r\,\Delta r$ is

$$(\rho v_z v_z \,\Delta A_z)\Big|_{z=0} = (\rho v_z v_z \,\Delta A_z)\Big|_{z=L} = (\rho v_z v_z \,\Delta A_z)\Big|_{z=z}$$

because for fully developed flow v_z is not a function of z and for an incompressible fluid ρ is constant. Note that since $v_r = 0$, there is no bulk flux of r momentum in the z direction or of z momentum in the r direction. The same would be true of the terms involving v_θ, which is also zero.

5-3 BULK FLOW AND INTERNAL GENERATION

The general conservation equation can be written for an arbitrary volume of fluid as follows:

$$\begin{matrix}\text{Net rate of input} \\ \text{due to diffusive} \\ \text{transport}\end{matrix} + \begin{matrix}\text{Net rate of input} \\ \text{due to bulk} \\ \text{transport}\end{matrix} + \begin{matrix}\text{Rate of} \\ \text{generation}\end{matrix} = \begin{matrix}\text{Rate of} \\ \text{accumulation}\end{matrix} \tag{5-44}$$

where the net input is the difference between the input and output flows of the system boundaries. We will illustrate the use of Eq. (5-44) with an example.

Tubular Reactor

A packed tubular reactor (Fig. 5-5) consists of a pipe of length L packed with catalyst particles. The reactants enter at the bottom, and chemical reaction takes

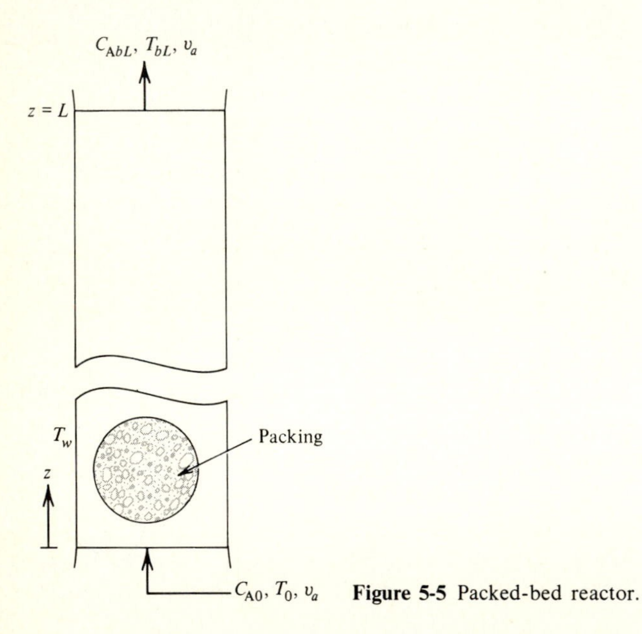

Figure 5-5 Packed-bed reactor.

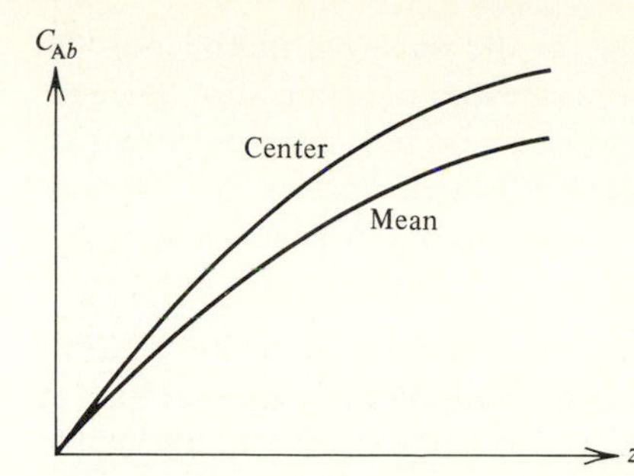

Figure 5-6 Concentration versus length in tubular reactor.

place on the catalyst surfaces as they pass through the reactor. Hence the concentration of a product is increasing with axial distance z, and the concentration of a reactant is decreasing (Fig. 5-6). In reactor design, we are interested in predicting the concentration of product in the exit stream at $z = L$. The general procedure is to derive and solve differential equations for the temperature and concentration in the reactor.

Rigorously the concentration (and temperature) of the fluid and that inside the porous catalyst pellets will be different, requiring us to write a separate mass and energy balance for each phase. Also, as indicated in the introduction to Chap. 1, there may be large radial as well as axial variations in temperature and concentration, making the problem multidimensional and beyond the scope of this chapter. (We discuss these more complicated models in Chap. 10.) At this time we will reduce the problem to one of one-dimensional transport by making an assumption that is widely used in practice, namely, that over any cross section at any z we can replace the fluid-solid medium with an equivalent homogeneous medium with an average or bulk mean temperature T_b and with a bulk mean concentration C_{Ab}, where the averages are taken over both fluid and solid phases.

First we postulate that diffusion in the axial direction can be described by a Fick's-law type of expression

$$J_{Az} = -E_{ax}\frac{dC_{Aa}}{dz} \tag{5-45}$$

where E_{ax} is an effective axial diffusivity that takes into account not only the properties of the packing and the fluid but also the effect of neglecting radial diffusion and velocity variation.† Similarly, we can define an effective axial thermal conductivity k_{ax} by

$$q_z = -k_{ax}\frac{dT_a}{dz} \tag{5-46}$$

where k_{ax} includes transport to and through solid and fluid phases.

†E_{ax} is often called an *axial dispersivity* to distinguish it from the molecular diffusivities defined in Chap. 1. Taylor has developed a method for predicting E_{ax} from the velocity profile and from the diffusivity; hence the dispersion process is often called *Taylor diffusion* (see Chap. 11).

As indicated previously, the reaction occurs on the surfaces of the porous catalyst, and the rate will depend upon the temperature and concentration at each point. However, we again assume that we can use an average for the entire cross section and let R_A represent the net rate of generation of species A per unit volume and unit time. If A is a product, R_A is positive; and if A is a reactant, R_A is negative. For a single irreversible reaction, if A is a product,

$$R_A = k_f C_A^n \tag{5-47}$$

where n is the order of the reaction. According to the Arrhenius equation, the specific rate constant k_f varies with temperature as

$$k_f = Ae^{-E/R_G T} \tag{5-48}$$

where E is the activation energy, R_G is the gas constant, and A is often called the *frequency factor*.

The heat generated by a chemical reaction is

$$\Phi_H = R_A(-\Delta H_R) \tag{5-49}$$

where ΔH_R is the enthalpy change for the reaction. For an exothermal reaction ΔH_R is negative.

Example 5-1: Axial dispersion in packed-bed reactors Consider a packed tubular reactor operating adiabatically (no heat transfer at the wall) (Fig. 5-7). (*a*) Derive the differential equations for energy and mass. (*b*) Assuming k_f is constant, solve the mass equation for a first-order reaction ($n = 1$), using appropriate boundary conditions.

SOLUTION (*a*) ***Molar balance on*** A Consider an element of volume $A_z\,\Delta z$ located at some arbitrary distance z. If we make a molar balance on A, we get

$$\underset{\text{Input rate}}{N_{Az}A_z\Big|_z} - \underset{\text{Output rate}}{N_{Az}A_z\Big|_{z+\Delta z}} + \underset{\text{Generation rate}}{R_A A_z\,\Delta z} = \underset{\text{Accumulation rate}}{0} \tag{5-50}$$

or

$$-\Delta(N_{Az}A_z) + R_A A_z\,\Delta z = 0$$

Note that, by definition, R_A represents molar generation but it will be a negative number if A is a reactant. Also note that by using N_{Az} we automatically include

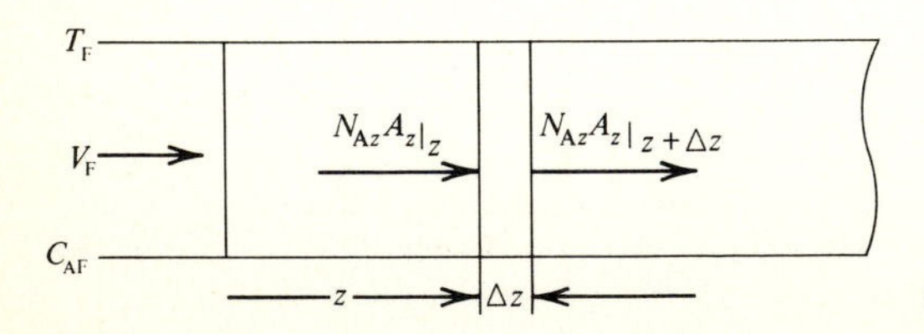

Figure 5-7 Incremental balance on reactor.

both the convective and diffusive fluxes, since $N_{Az} = J_{Az} + C_A v_z$. Consistent with our other assumptions concerning temperature and concentration, we assume that the velocity v_z can be replaced by v_a, the average velocity over the cross section. Then, substituting for N_{Az} and using the difference operator Δ gives†

$$\underset{\text{Net diffusive flow}}{-\Delta(J_{Az}A_z)} - \underset{\text{Net bulk flow}}{\Delta(C_A v_a A_z)} + \underset{\text{Net generation}}{R_A A_z \, \Delta z} = 0$$

Again the accumulation is zero at steady state. Dividing by the volume $A_z \, \Delta z$ and letting $\Delta z \to 0$ gives

$$\lim_{\Delta z \to 0} \left[\frac{-\Delta(J_{Az}A_z)}{A_z \, \Delta z} - \frac{\Delta(C_A v_a A_z)}{A_z \, \Delta z} \right] + R_A = 0$$

or, since A_z is constant,

$$\frac{-dJ_{Az}}{dz} - \frac{d(C_A v_a)}{dz} + R_A = 0$$

Since v_a is constant, if E_{ax} is constant, use of (5-46) gives

$$E_{ax} \frac{d^2 C_A}{dz^2} - v_a \frac{dC_A}{dz} + R_A = 0 \tag{5-51}$$

Energy balance over Δz

$$\begin{matrix}\text{Energy flow} \\ \text{rate in}\end{matrix} - \begin{matrix}\text{Energy flow} \\ \text{rate out}\end{matrix} + \begin{matrix}\text{Energy} \\ \text{generated}\end{matrix} = \begin{matrix}\text{Energy} \\ \text{accumulated}\end{matrix}$$

We define a total energy flux at z as the sum of the diffusive and bulk fluxes [see Eq. (5-42)]

$$e_z = q_z + \rho \widehat{C}_p v_z (T - T_R)$$

where T_R is a reference temperature. Since flow equals flux times area,

$$\text{Net flow} = e_z A_z \Big|_z - e_z A_z \Big|_{z+\Delta z} = -\Delta(e_z A_z)$$

$$\underset{\text{Diffusive}}{-\Delta(q_z A_z)} - \underset{\text{Bulk}}{\Delta[\rho \widehat{C}_p v_a (T - T_R) A_z]} + \underset{\text{Reaction}}{\Phi_H (A_z \, \Delta z)} = 0$$

Dividing by $A_z \, \Delta z$ and letting $\Delta z \to 0$ gives

$$\lim_{\Delta z \to 0} - \frac{\Delta(q_z \cancel{A_z})}{\cancel{A_z} \, \Delta z} - \frac{\Delta[\rho \widehat{C}_p v_a (T - T_R) \cancel{A_z}]}{\cancel{A_z} \, \Delta z} + \Phi_H = 0$$

†For convenience the subscript a will be dropped, but it is to be understood that C_A refers to the average concentration C_{Aa}.

Since A_z is constant,

$$\frac{-dq_z}{dz} - \frac{d(\rho \widehat{C}_p v_a T)}{dz} + \Phi_H = 0$$

where

$$\Phi_H = -\Delta H_R \, R_A$$

If k_{ax}, ~~k_{ax}~~, ρ, $\widehat{C}_p$, and v_a are constant, use of Eq. (5-46) gives

$$+k_{\text{ax}} \frac{d^2T}{dz^2} - \rho \widehat{C}_p v_a \frac{dT}{dz} + R_A(-\Delta H_R) = 0 \tag{5-52}$$

(*b*) ***Solution to mass equation*** As indicated previously, the design of a tubular reactor requires the calculation of the concentration $C_{\text{A}a}$ at a specified distance from the entrance z. However, in order to solve the mass equation (5-51) it is necessary to know the rate R_{A}, which, according to Eqs. (5-47) and (5-49) depends upon temperature. Therefore one would presumably have to solve the differential equation (5-52) for the mean temperature; but this equation also includes the rate as a function of concentration (as well as temperature). Therefore, we need to solve the two equations simultaneously, either by numerical methods or by making some assumption that uncouples the equations. The solution of these equations is discussed in texts on chemical reactor design. To illustrate bulk transport we consider here the solution to a simplified case, namely, one in which the heat of reaction is small so that the reactor is nearly isothermal and k_f can be taken as constant in the mass equation.

Boundary condition It has been found that as a result of axial dispersion the concentration at the entrance to a reactor ($z = 0$) depends upon the entrance conditions preceding the reactor and that the concentration at $z = L$ depends upon the exit conditions following the reactor. In 1956 Wehner and Wilhelm[2] investigated this case by using a model that included nonreacting entrance and exit zones in which $N_{\text{A}z}$ is constant. At $z = -\infty$ the concentration of the feed is $C_{\text{A}F}$ and the concentration gradient is zero. The boundary condition at $z = 0$, based on the equality of the total flux, can be shown to be†

$$C_{\text{A}F} v_F = -E_{\text{ax}} \frac{dC_{\text{A}}}{dz} + C_{\text{A}} v_a \tag{5-53}$$

At $z = \infty$, the concentration gradient is also zero and, since $N_{\text{A}z}|_{\infty} = N_{\text{A}z}|_{L}$,

$$\frac{dC_{\text{A}}}{dz} = 0 \qquad \text{at} \qquad z = L \tag{5-54}$$

First-order irreversible reaction If A is a reactant and the reaction is first order and irreversible.

$$R_{\text{A}} = -k_f C_{\text{A}} \tag{5-55}$$

†This is often called the *Danckwerts' boundary condition.*

Then from Eq. (5-51) we get the second-order linear differential equation

$$\frac{d^2C_A}{dz^2} - \frac{v_a}{E_{ax}}\frac{dC_A}{dz} - \frac{k_f}{E_{ax}}C_A = 0 \tag{5-56}$$

Solution of mass equation in terms of dimensionless variables As observed in Chaps. 1 and 2, it is convenient to express the solution to a differential equation in terms of dimensionless variables. We now show that it can also be advantageous to write the differential equation and the boundary conditions in terms of dimensionless variables and that in so doing we can identify important dimensionless parameters in the system.

In order to make Eq. (5-56) dimensionless we first must identify a characteristic length and a characteristic concentration for the system. The usual choices are L and C_{AF}, which lead to the dimensionless variables

$$C^* \equiv \frac{C_A}{C_{AF}} \qquad \text{and} \qquad z^* \equiv \frac{z}{L}$$

Then, since C_{AF} and L are constants,

$$\frac{d^2C_A}{dz^2} = \frac{d}{dz}\left(\frac{dC_A}{dz}\right) = \left[\frac{d}{dz^*}\left(\frac{dC_A}{dz}\right)\right]\frac{dz^*}{dz} = \frac{C_{AF}}{L^2}\frac{d^2C^*}{dz^{*2}}$$

Substituting this and a similar expression for dC_A/dz into Eq. (5-56) gives

$$\frac{C_{AF}}{L^2}\frac{d^2C^*}{dz^{*2}} - \frac{v_a C_{AF}}{E_{ax}L}\frac{dC^*}{dz^*} - \frac{k_f C_{AF}C^*}{E_{ax}} = 0$$

or, multiplying by L^2/C_{AF},

$$\frac{d^2C^*}{dz^*} - \text{Pe}\frac{dC^*}{dz^*} - \text{Da Pe } C^* = 0 \tag{5-57}$$

where $\text{Pe} \equiv Lv_a/E_{ax}$ is called the *length Peclet number* for mass transfer (or the *Bodenstein number*) and $\text{Da} \equiv k_f L/v_a$ is a dimensionless reaction-rate group often called a *Damköhler number.*

Assuming that $v_F = v_a$, we can write the boundary conditions (5-53) and (5-54) in dimensionless form, or

$$-E_{ax}C_{AF}\frac{dC^*}{L\,dz^*} + v_a C_{AF}C^* = v_a C_{AF} \qquad \text{at } z^* = 0$$

or, after dividing by $v_a C_{AF}$,

$$\left(-\frac{1}{\text{Pe}}\frac{dC^*}{dz^*} + C^*\right)_{z^*=0} = 1 \tag{5-58}$$

At $z = L$ or $z^* = 1$

$$\left.\frac{dC^*}{dz^*}\right|_{z^*=1} = 0 \tag{5-59}$$

Equation (5-57) is a second-order linear differential equation whose general solution (see Appendix A1-2) is

$$C^* = A_1 e^{r_1 z^*} + A_2 e^{r_2 z^*}$$

where r_1 and r_2 are roots of the auxiliary equation

$$m^2 - \text{Pe}\, m - \text{Da Pe} = 0$$

Solving for r_1 and r_2 gives

$$\left.\begin{matrix} r_1 \\ r_2 \end{matrix}\right\} = \tfrac{1}{2}\text{Pe} \pm \tfrac{1}{2}\sqrt{\text{Pe}^2 + 4\,\text{Da Pe}} \tag{5-60a}$$

$$= \tfrac{1}{2}\text{Pe}\left(1 \pm \sqrt{1 + 4\frac{\text{Da}}{\text{Pe}}}\right) \tag{5-60b}$$

Using the boundary conditions to evaluate A_1 and A_2 gives the equation[2] for $C^*(z^*)$

$$C^* = B\left(\exp\frac{\text{Pe}\, z^*}{2}\right)\left\{(1 + a)\exp\left[\frac{a\,\text{Pe}\,(1 - z^*)}{2}\right] - (1 - a)\exp\frac{a\,\text{Pe}\,(z^* - 1)}{2}\right\} \qquad 0 < z^* < 1.0 \tag{5-61}$$

where

$$B \equiv 2\left[(1 + a)^2 \exp\left(\frac{a\,\text{Pe}}{2}\right) - (1 - a)^2 \exp\left(\frac{-a\,\text{Pe}}{2}\right)\right]^{-1}$$
$$a \equiv \sqrt{1 + \frac{4\,\text{Da}}{\text{Pe}}} \tag{5-62}$$

The desired concentration at $z = L$ is

$$C^*(1) \equiv C_L^* = 2aBe^{\text{Pe}/2} \tag{5-63}$$

We can see that the concentration depends upon the dimensionless group $v_a L/E_{ax} \equiv \text{Pe}$, the length Peclet number for mass transport. The Peclet number frequently enters into heat- and mass-transport problems. In general a Peclet number is defined by

$$\text{Pe} \equiv \frac{\text{characteristic velocity} \times \text{characteristic length}}{\text{characteristic diffusivity}} \tag{5-64}$$

In problems involving radial transport in a conduit the characteristic length is usually taken as the pellet diameter or the tube diameter, and the characteristic velocity is the average velocity. For laminar-flow mass transport the molecular mass diffusivity would be used, and for heat transport the molecular thermal diffusivity α would be used as a characteristic diffusivity. Levenspiel[1] calls the group $E_{ax}/v_a L \equiv (\text{Pe})^{-1}$ the *vessel dispersion number.*

Frequently in reactor design the exit concentration is expressed in terms of the fractional conversion, which is related to C^* as follows:

$$X_L = \frac{C_{AF} - C_{AL}}{C_{AF}} = 1 - C_L^*$$

From Eqs. (5-62) and (5-63) we can see that

$$\left.\begin{matrix} X_L^* \\ C_L^* \end{matrix}\right\} = f(\text{Pe}, \text{Da}) \qquad \text{only}$$

Therefore a design chart can be prepared giving X_L as a function of Pe† for various constant values of the parameter Da. Using this single chart, we can predict the exit conversion for any reactor for which L, v_a, k_f, E_{ax}, and C_{AF} are specified. Thus, instead of the relationship

$$C_A = C_A(C_{AF}, L, v_a, k_f, E_{ax})$$

involving six dimensional variables, the problem is reduced to a relationship involving only three dimensionless variables.

It should be emphasized that the value of using dimensionless variables is not limited to cases for which an analytical solution is available. Once the differential equation and boundary conditions have been written in dimensionless form, the nature of the dimensionless variables and the relationship between them can be established. In the present example the differential equation and the boundary conditions indicate that

$$C^* = C^*(z^*, \text{Pe}, \text{Da})$$

Then
$$C_L^* = C_L^*(1, \text{Pe}, \text{Da})$$

which means that one could prepare a design chart by taking experimental data for various values of Pe and Da and use it to predict the conversion.† □

PROBLEMS

5-1 An apparatus of the type shown in Fig. 5-3 is used for the experimental determination of the diffusivity of a vapor through a gas. For the air-water system, calculate the value of $\mathfrak{D}_{AB}$ from the following information. The entire apparatus is enclosed in a constant-temperature environment at 130°F and 1 atm, and the air is circulated over a desiccant so that the concentration of water vapor in the air is zero. If 29 h is required for the level to fall from 5 to 6 in below the top of the tube, what is the diffusivity in cm²/s in the air-water system at 130°F? List any assumptions.

5-2 (*a*) If the bulk flow is neglected in the diffusion of A through stagnant B, the expression for the flux reduces to

$$N_{Az} = \mathfrak{D}_{AB} C \frac{x_{A0} - x_{AL}}{L}$$

Comparing this with the exact expression for N_{Az}, discuss under what conditions the neglect of bulk flow will lead to considerable error. (*b*) Repeat Prob. 5-1 neglecting bulk flow.

5-3 Water is being removed from cylindrical wood chips suspended in a steam of gas B. The chip radius is R_1, and there is a cylindrical stagnant gas film surrounding it of radius R_2. The mole fraction of water in the gas phase is x_{A1} at R_1 and x_{A2} at R_2.

†Since both Pe and Da contain L, the group Da/Pe $\equiv k_f E_{ax}/v_a^2$ would be more convenient if L is desired.

(*a*) Derive an equation for the molar total flux of water in terms of x_{B1}, x_{B2}, and $\mathfrak{D}_{H_2O,B}$.

(*b*) Find the mass-transfer coefficient k_c.

5-4 A vapor A is diffusing from a tube containing a stagnant gas, as shown in Fig. 5-3. The distance between the liquid A level and the top of the tube is 20.0 cm. The total pressure on the system is 760 mmHg, and the temperature is 0°C. The vapor pressure of A at that temperature is 304 mmHg. The cross-sectional area of the diffusion tube is 0.80 cm². The diffusivity D_{AB} is 0.07 cm²/s. The molecular weight of A is 150, and that of B is 32. Assuming that the concentration of A at the top of the tube is negligible and considering bulk flow:

(*a*) At the endpoints and at the midpoint, that is, at $z = 0$, $z = \frac{1}{2}Z$, and $z = Z$, calculate N_{Az}, N_{Bz}, J_{Az}, J_{Bz}, J^*_{Az}, J^*_{Bz}, n_{Az}, n_{Bz}, j_{Az}, j_{Bz}, x_A, and x_B.

(*b*) Sketch the above versus z.

(*c*) At $z = 0$ draw vectors showing magnitudes of v_z, v^*_z, v_{Az}, v_{Bz}, J_{Az}, J_{Bz}, J^*_{Az}, and J^*_{Bz}.

(*d*) If bulk flow is neglected, plot x_A and x_B versus z and calculate the percentage error made at the midpoint.

5-5 A catalyst pellet is shaped like a cylinder of length L and radius R, where L/R is very large. The product A formed inside the porous pellet diffuses through the pores of the pellet to the surface and from the surface through a film of gas B. Assume that B is stagnant, that at the surface of the pellet the concentration of A is constant and equal to C_{A1}, that there is no reaction in the film, that the film has a thickness $R_2 - R_1 = L_r$, and that the concentration at the boundary of the film is constant and equal to C_{A2}. Bulk flow is *not* negligible.

(*a*) Derive a differential equation for N_{Ar}, the total molar flux of A.

(*b*) Relate N_{Ar} to J_{Ar}, the molar diffusive flux of A with respect to the molar-average velocity.

(*c*) Solve for the total molar flux at the surface $N_{Ar}|_{R1}$, assuming constant molar volume C.

(*d*) Derive an equation for the mass-transfer coefficient k_c using the result in part (*c*).

(*e*) Derive an equation for the mass-transfer Nusselt number $(\mathrm{Nu})_{AB}$ using the result in part (*d*). Let $D_c = 2R_1$.

5-6 A droplet of liquid A is suspended in a stream of gas B. The droplet radius is R_1. A spherical *stagnant* film of radius R_2 surrounds the drop. The concentration of A in the gas phase at R_1 is C_{A1} and at R_2 is C_{A2}.

(*a*) Derive a differential equation for the total molar flux N_{Ar}.

(*b*) Relate N_{Ar} to the molar diffusive flux J^*_{Ar}. Do *not* neglect bulk.

(*c*) Define the mass-transfer coefficient k_c and obtain an equation for it.

(*d*) Define $(\mathrm{Nu})_{AB}$ and obtain an equation for it.

5-7 A liquid A is evaporating into a gas B as indicated in Fig. 5-1. The liquid level is maintained at $Z = Z_1$, and the gas-phase concentration of A at the liquid-gas interface is constant at x_{A1} (mole fraction). The mole fraction of A is maintained constant at the top of the column at a value x_{A2}. In general, the evaporation of liquid A results in a temperature gradient through the column. Assuming that the temperature profile in the column is linear and that the temperature dependence of the diffusivity is given by $D_{AB} = aT^b$, where a and b are constant, derive an expression for the mole fraction of B in the column at steady state.

5-8 In a packed tubular reactor a fluid enters with concentration C_{A0}, temperature T_0, and average velocity v_a.

(*a*) Assuming that k_f is roughly constant, solve the mass equation for a first-order reaction $(n = 1)$ if $dC_A/dz \to 0$ at $z = \infty$, $C_A = C_{A0}$ at $z = 0$.

(*b*) Prepare a design chart using dimensionless variables, $v_a z/E_{ax}$ and kE_{ax}/v_a^2.

REFERENCES

1. Levenspiel, O.: "The Chemical Reactor Omnibook," OSU Book Stores, Inc., Corvallis, Oreg., 1979.
2. Wehner, J. F., and R. H. Wilhelm: *Chem. Eng. Sci.*, **6**:89 (1956).

CHAPTER

SIX

TURBULENT TRANSPORT

In previous chapters we have dealt exclusively with laminar-flow conditions in order to illustrate certain basic concepts and procedures. In practice, however, most processes of industrial importance take place under turbulent- rather than laminar-flow conditions, and it is important that the engineer know how to handle such problems. Unfortunately, turbulent flow is extremely complex, and at present *no* theory exists that will permit the prediction of turbulent flows with the same degree of certainty as can be done for laminar flows. For this reason, this chapter does not emphasize theory but instead tries to show the reader how experiment and theory can be used to develop simple models for turbulent-flow situations. In Chap. 10 turbulence theory is discussed in greater detail.

6-1 OBJECTIVE OF A TURBULENCE MODEL

An ultimate objective of a model is to predict the turbulent velocity, temperature, and/or concentration distributions in a system. From them one can obtain the transfer coefficients as a function of other variables (such as average velocity, fluid properties, and characteristic length, e.g., tube diameter). Once these transfer coefficients are known, one can calculate an overall rate of transport of heat, mass, or momentum and obtain the size of the equipment required in the process.

In the above procedure it is important to distinguish between the *design* process, which is carried out industrially by a design engineer, and the prediction, data taking, and correlation processes, which are carried out by a research engineer in industry, government laboratories, or academic life. When engineers *design* a heat exchanger, they use a heat-transfer coefficient to predict, say, the length or area required to transfer a given amount of heat (as outlined briefly in Chap. 4). The design procedure is usually the *same* whether the heat-transfer coefficient is obtained from a theoretical equation or from an empirical equation that fits the data.

In the past, it was not possible to predict, say, the friction factor from turbulent-flow theory as accurately as for fully developed laminar flow (as in Chap. 4). Therefore, engineers to a large extent have traditionally used empirical correlations. However, the resort to purely empirical correlations has its limitations. The first is that building equipment and taking data are very time-consuming and expensive tasks as compared with deriving some theoretical equations. But although theory cannot always eliminate the need for *any* experimentation, it can often reduce the amount needed and consequently the cost of data taking. Such is the case because if we have a theoretical relationship between two variables, it usually requires fewer data points to find the parameters (i.e., the constants) in the equation than if a strictly empirical equation is used. Secondly, many empirical equations cannot be extrapolated to regions or systems other than those for which data have been taken. Thus, theory can be useful in determining a functional relationship or in predicting what might happen in a *new* situation that differs moderately from a situation in which theory and experiment have been found to be in good agreement.

The following sections illustrate this interplay between theory and experiment in the prediction of the rates of transport and the transfer coefficients for momentum, heat, and mass in fully developed turbulent flow in a pipe. With the advent of large computers, turbulence models have been used for more complicated cases such as the transport of pollutants in the atmosphere and in bodies of water, recirculating flows (e.g., after a sudden enlargement), combustion in furnaces, spread of smoke from a fire, transport in stirred tanks and reactors, boundary layers on air foils, and jets. This type of modeling is beyond the scope of this text.

6-2 THE NATURE OF TURBULENT FLOW

Consider a fluid flowing upward through a long pipe or tube under the influence of a pressure drop. If the Reynolds number (as defined in Chap. 3) is under 2100,† the flow will be laminar (Fig. 6-1*a*). Under these conditions a dye injected into the fluid at any point will move with the fluid velocity at that point along a rectilinear path. If the Reynolds number is greater than 2100, a dye injected at the same point will move with the eddies and will swirl and zigzag as it moves up the tube (Fig. 6-1*b*). The flow is then said to be turbulent.

Consider an unsteady-state flow starting at $t = 0$ with the velocity equal to zero. In laminar flow (Fig. 6-2*a*) the velocity at an arbitrary point increases steadily and approaches the steady-state value, but in turbulent flow (Fig. 6-2*b*), the

†The exact figure depends upon the entrance conditions and roughness.

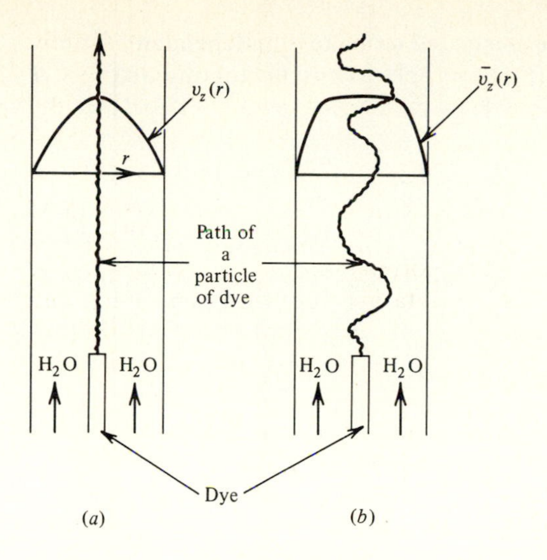

Figure 6-1 Comparison between (*a*) laminar and (*b*) turbulent flow.

velocity fluctuates about the time-averaged value $\overline{v}_z$ during both steady- and unsteady-state flow. At unsteady state this mean or time-averaged value is defined by

$$\overline{v}_z = \frac{\int_t^{t+\Delta t} v_z \, dt}{\Delta t} \tag{6-1}$$

where in practice Δt must be taken large enough to smooth out the small fluctuations but small enough to include any large-scale unsteady-state variations. The selection of Δt is thus somewhat arbitrary, in fact for atmospheric flow its value is usually specified. Having defined Δt, we can express the instantaneous velocity v_z as the sum of a fluctuating velocity v_z' and a time-averaged velocity $\overline{v}_z$, or (see Fig. 6-2*c*)

$$v_z = \overline{v}_z + v_z' \tag{6-2}$$

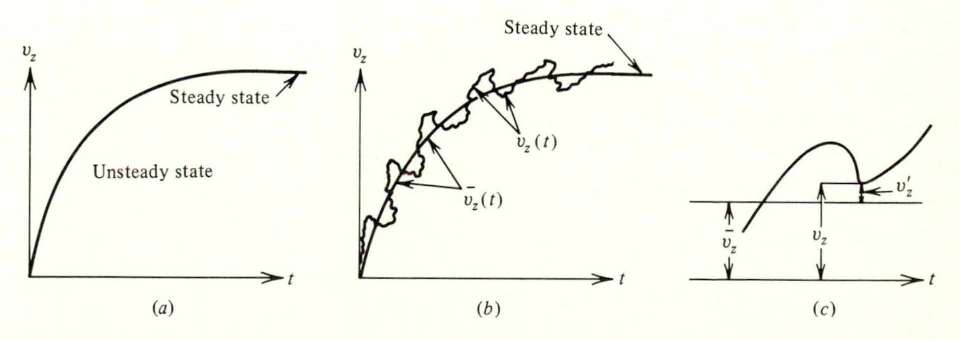

Figure 6-2 Velocity as a function of time for (*a*) laminar flow and (*b*) turbulent flow; (*c*) mean, fluctuating, and instantaneous velocities.

At steady state Δt is taken as large as needed to make $\overline{v}_z$ independent of time. For that case, we can generalize Eq. (6-1) to apply to any fluctuating quantity A by using the definition (see Appendix 6-1)

$$\overline{A} \equiv \lim_{\Delta t \to \infty} \frac{\int_t^{t+\Delta t} A \, dt}{\Delta t} \tag{6-3}$$

If we substitute $\overline{A}$ itself into Eq. (6-3), we obtain the time average of the time average

$$\overline{\overline{A}} \equiv \lim_{\Delta t \to \infty} \frac{\int_t^{t+\Delta t} \overline{A} \, dt}{\Delta t} = \overline{A} \tag{6-3a}$$

If we let

$$A(t) = \overline{A} + A'(t) \tag{6-3b}$$

and substitute into Eq. (6-3), we obtain

$$\overline{A} = \lim_{\Delta t \to \infty} \frac{\int_t^{t+\Delta t} (\overline{A} + A') \, dt}{\Delta t} \equiv \overline{\overline{A} + A'} = \overline{\overline{A}} + \overline{A'} = \overline{A} + \overline{A'}$$

or

$$\overline{A'} = 0 \tag{6-3c}$$

Turbulent Intensities

Although $\overline{v}_r = 0 = \overline{v}_\theta$ for fully developed flow, v_r' and v_θ' are not zero; nor are the mean-square values $\overline{v_r'^2}$ and $\overline{v_\theta'^2}$ [as obtained by the time-averaging procedure defined by Eq. (6-3)]. Experimentally they, along with $\overline{v_z'^2}$, are found to vary with radial position (Fig. 6-3). The sum

$$\overline{v_r'^2} + \overline{v_\theta'^2} + \overline{v_z'^2} = 2\widehat{K}^{(t)}$$

is equal to twice the turbulent kinetic energy per unit mass, and the ratio $\sqrt{\frac{1}{3}\widehat{K}^{(t)}}/\langle \overline{v}_z \rangle$, where $\langle \overline{v}_z \rangle$ is the average of $\overline{v}_z$ over the cross section, is called the *intensity* of turbulence.

Temperature and Concentration Fluctuations

If we were to use probes to measure the instantaneous temperature and concentration in a turbulent flow, we would find these quantities also fluctuating about a mean and similarly we could write

$$T = \overline{T} + T' \tag{6-4}$$

and

$$C_A = \overline{C}_A + C_A' \tag{6-5}$$

where $\overline{T}$ and $\overline{C}_A$ are time-averaged quantities defined by an equation analogous to Eq. (6-3) and T' and C_A' represent fluctuations about the mean.

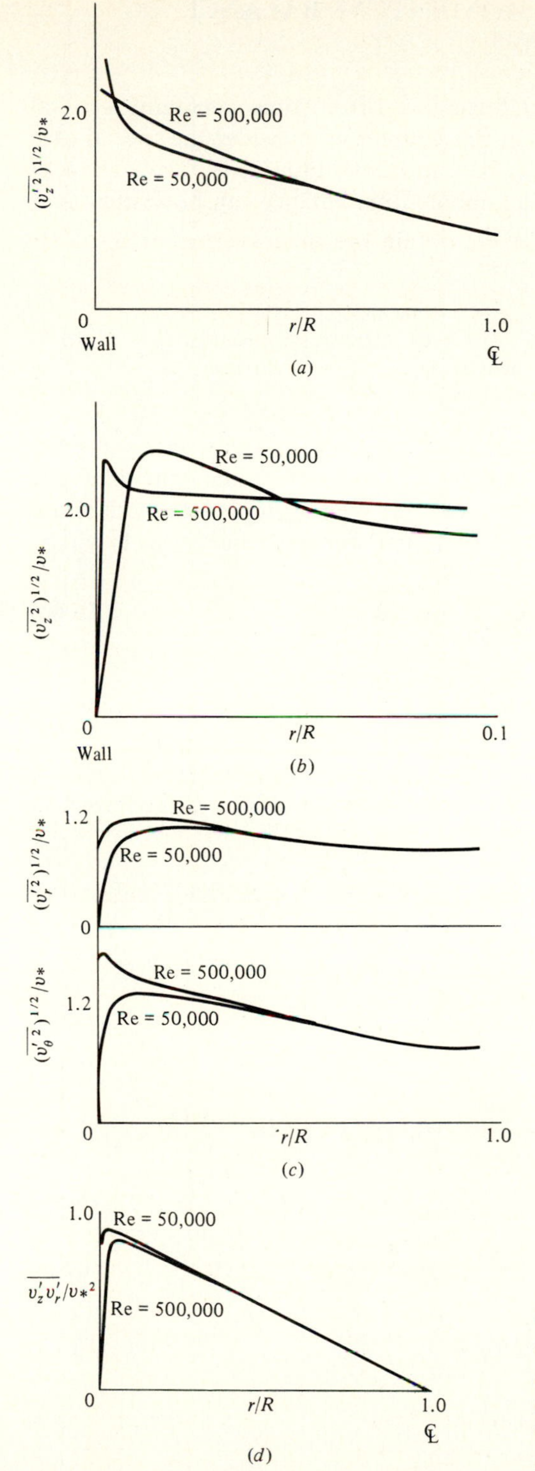

Figure 6-3 Variation with position in a pipe (*a*) of $(\overline{v_z'^2})^{1/2}/v_*$, (*b*) of the same near wall, (*c*) of $(\overline{v_r'^2})^{1/2}/v_*$ and $(\overline{v_\theta'^2})^{1/2}/v_*$ and (*d*) $\overline{v_z'v_r'}/v_*$ (note that $v_* \equiv \sqrt{\tau_0/\rho}$, where τ_0 = flux at wall). (*From J. Laufer, NACA Rep. 1174, 1954.*)

6-3 TIME AVERAGING THE MOMENTUM BALANCE FOR TURBULENT PIPE FLOW†

In Chap. 10 we show how turbulent fluxes arise from time averaging general multidimensional balance equations. In this chapter we consider the special case of fully developed flow in a pipe under the same conditions as in Sec. 3-3 except that the flow is turbulent. As before, a z-momentum balance can be written over an element of volume $\Delta A_z L = 2\pi r \Delta r L$.

The momentum balance at $r = r$ is

$$\underset{(1)}{\begin{array}{c}\text{Net rate of } z\text{-momentum} \\ \text{flow in by} \\ \text{molecular diffusion}\end{array}} + \underset{(2)}{\begin{array}{c}\text{Net rate of } z\text{-momentum} \\ \text{flow in by} \\ \text{bulk transport}\end{array}} + \underset{(3)}{\begin{array}{c}\text{Rate of} \\ z\text{-momentum} \\ \text{generation}\end{array}} = 0 \tag{6-6}$$

We first write this equation in terms of instantaneous conditions; then we time-average it term by term to obtain an equation in terms of the time-averaged values (which are more commonly measured). If we let $\tau_{rz}^{(l)}$ represent the laminar or molecular flux at $r = r$, the net molecular momentum transport rate (1) will be

$$\tau_{rz}^{(l)} A_r \Big|_r - \tau_{rz}^{(l)} A_r \Big|_{r+\Delta r} \equiv -\Delta(\tau_{rz}^{(l)} A_r) \tag{6-6a}$$

The rate of generation of momentum (Chap. 3) will be

$$\Phi_M (2\pi r \Delta r L) \equiv [(p_0 - p_L) g_c + \rho g_z L] \Delta A_z \tag{6-6b}$$

In Chap. 3 we did not include bulk-flow terms because they cancelled out, but we need to reconsider them for turbulent flow. As indicated in Chap. 5, the bulk flow of z momentum in the z direction is

$$\frac{\text{Volumetric flow in } z \text{ direction}}{\text{Time}} \times \frac{z \text{ momentum}}{\text{volume}} = \frac{z \text{ momentum}}{\text{time}}$$

The net bulk flow of z momentum in the z direction (2) is the bulk flow in at $z = 0$ less the bulk flow out at $z = L$, or

$$[(v_z \Delta A_z)(\rho v_z)] \Big|_0 - [(v_z \Delta A_z)(\rho v_z)] \Big|_L \tag{6-6c}$$

The net bulk flow of z momentum in the r direction will be

$$[(v_r A_r)(\rho v_z)] \Big|_r - [(v_r A_r)(\rho v_z)] \Big|_{r+\Delta r} \tag{6-6d}$$

Since the flow is turbulent, v_z and v_r are the *instantaneous* velocities, v_z being given by Eq. (6-2) and v_r by

$$v_r = \bar{v}_r + v_r' \tag{6-6e}$$

†This section may be omitted on first reading.

Since we intend to write our momentum balance in terms of the time-*averaged* values, we will substitute Eq. (6-2) for v_z and Eq. (6-6*e*) for v_r into the bulk-flow terms and take the time average. Rigorously we should also use $\rho = \overline{\rho} + \rho'$, but we will assume that $\overline{\rho} = \rho$ since the fluid is incompressible. Then the time-averaged bulk flux of z momentum in the r direction at r is

$$\overline{\rho(\overline{v}_r + v_r')(\overline{v}_z + v_z')} = \underset{(1)}{\overline{\rho\overline{v}_r\overline{v}_z}} + \underset{(2)}{\overline{\rho v_r'\overline{v}_z}} + \underset{(3)}{\overline{\rho\overline{v}_r v_z'}} + \underset{(4)}{\overline{\rho v_r' v_z'}} \tag{6-6f}$$

Term (1) is, according to our averaging definition [Eqs. (6-3) and (6-3*a*)],

$$\overline{\rho\overline{v}_r\overline{v}_z} = \rho\overline{v}_r\overline{v}_z \tag{6-6g}$$

which is the bulk flux due to the time-averaged velocities and is zero. Term (2) is

$$\overline{v_r'}\,\rho\overline{\overline{v}}_z = 0 \tag{6-6h}$$

due to Eq. (6-3*c*); term (3) is similarly zero. Term (4), $\rho\overline{v_r'v_z'}$, can be measured experimentally; it is not zero except at the wall, where $v_z' = 0 = v_r'$, and at the center. Experimentally it is found to vary with radial position, reaching a maximum near the wall (Fig. 6-3*d*). It has the units of a momentum flux and physically represents the transport of z momentum in the r direction due to the correlation of v_r' with v_z'. Therefore it is called the *turbulent momentum flux* and is defined as†

$$\tau_{rz}^{(t)} \equiv \rho\overline{v_r'v_z'} \tag{6-6i}$$

The mathematical and physical meaning of the correlation $\overline{v_r'v_z'}$ can be understood as follows (Fig. 6-4*a*). Suppose at A a velocity fluctuation occurs toward the wall; that is, $v_r' > 0$. Since $\overline{v}_z$ decreases as we move toward the wall, fluid with a mean velocity $\overline{v}_z$ will be moving into a stream at B with lower $\overline{v}_z$; thus at the new position the fluctuation $v_z'|_B = v_z|_B - \overline{v}_z|_B$ will tend to be positive. On the other hand, if the flow is away from the wall ($v_r' < 0$) (Fig. 6-4*b*), fluid with a lower mean velocity $\overline{v}_z$ will be moving into a stream of higher velocity, a process which results in a negative fluctuation v_z'. Thus, in either case v_r' and v_z' will be of the same sign and if this tendency persists over a sufficiently long averaging time Δt, they will be said to be positively correlated and the correlation $\overline{v_r'v_z'}$ given by Eq. (6-3) will be positive.

The physical meaning of $\rho\overline{v_r'v_z'}$ is then as follows. A lateral fluctuation toward the wall results in fluid of high z momentum per unit volume $\rho\overline{v}_z$ being transported toward the wall (like the mail sacks in Sec. 1-17) by a radial flow $v_r'A_r$. Consequently if v_z' and v_r' are correlated over Δt, there is a transport of z momentum in the r direction which can be expressed as $\rho\overline{v_r'v_z'}\,A_r$. The flux is then $\rho\overline{v_r'v_z'} \equiv \tau_{rz}^{(t)}$.

According to Newton's second law, $\tau_{rz}^{(t)}$ is also manifested as a stress $\sigma_{r\theta}^{(t)}$ and $\rho\overline{v_z'v_\theta'}/g_c \equiv \sigma_{z\theta}^{(t)}$. These, together with $\rho\overline{v_r'^2}/g_c \equiv \sigma_{rr}^{(t)}$, $\rho\overline{v_\theta'^2}/g_c \equiv \sigma_{\theta\theta}^{(t)}$, and $\rho\overline{v_z'^2}/g_c \equiv \sigma_{zz}^{(t)}$, are called the *Reynolds stresses*.

†We use $\tau^{(t)}$ instead of $\overline{\tau^{(t)}}$ for convenience.

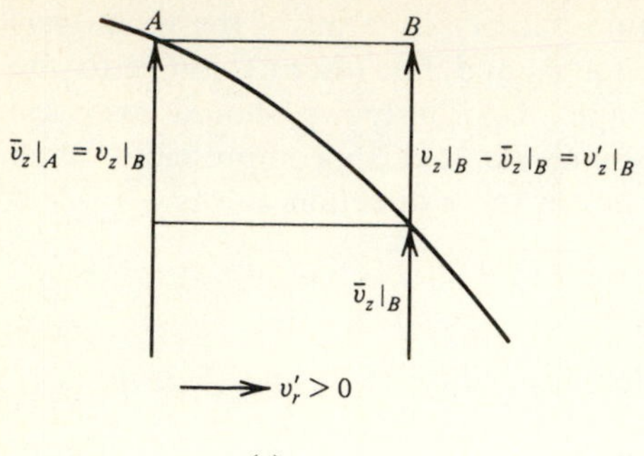

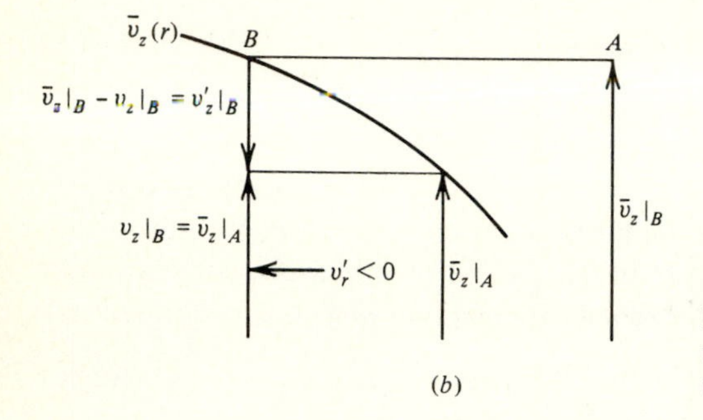

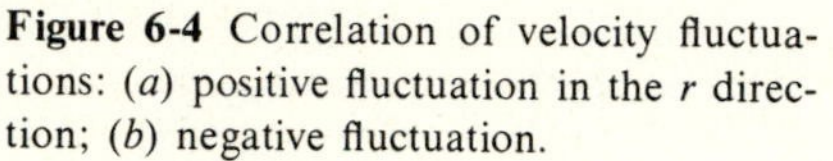
Figure 6-4 Correlation of velocity fluctuations: (*a*) positive fluctuation in the *r* direction; (*b*) negative fluctuation.

A similar procedure for the bulk flux in the *z* direction gives

$$\rho\overline{v_z v_z} = \rho\bar{v}_z\bar{v}_z + \rho\overline{v_z'^2} \tag{6-6j}$$

where the first term is the bulk flux due to the time-averaged velocities and the second is the turbulent flux of *z* momentum in the *z* direction

$$\tau_{zz}^{(t)} \equiv \rho\overline{v_z'^2} \tag{6-6k}$$

In fully developed pipe flow, neither of the terms in Eq. (6-6*j*) is zero, but since neither of them changes between $z = 0$ and $z = L$, they drop out.

If we time-average the laminar transport flux in Eq. (6-6*a*), assuming a newtonian fluid, we get

$$\overline{\tau_{rz}^{(l)}} = -\mu\frac{\overline{dv_z}}{dr} = -\mu\frac{d\bar{v}_z}{dr} \tag{6-6l}$$

where we assume that the processes of time averaging and differentiation can be interchanged. Therefore $\overline{\tau_{rz}^{(l)}}$ is obtained from τ_{rz} by replacing v_z with $\bar{v}_z$.

If we time-average the momentum-generation term in Eq. (6-6*b*), we get

$$\overline{\Phi_M} = \frac{(\bar{p}_0 - \bar{p}_L)g_c + \rho g_z L}{L}$$

Substituting these results into Eq. (6-6) gives

$$-\Delta(\overline{\tau_{rz}^{(l)}A_r}) - \Delta(\tau_{rz}^{(t)}A_r) + \overline{\Phi}_M \Delta A_z L = 0 \tag{6-7}$$

We note, however, that the first two terms can be combined if we define a total flux as

$$\overline{\tau}_{rz} = \overline{\tau_{rz}^{(l)}} + \tau_{rz}^{(t)} \tag{6-8}$$

Then our balance, Eq. (6-7), becomes

$$-\Delta(\overline{\tau}_{rz}A_r) + \overline{\Phi}_M \Delta A_z L = 0 \tag{6-9}$$

which is identical to Eq. (3-13) except that we use the total flux $\overline{\tau}_{rz}$ instead of the laminar flux τ_{rz}.

This result means that in the future it is not necessary to carry out the time-averaging procedure. Instead, one need merely replace τ_{rz} in the balance equation with the total flux $\overline{\tau}_{rz}$ and use time-averaged values of other quantities instead of instantaneous values. This general rule also holds for turbulent heat and mass transport, as discussed later.

6-4 TIME-AVERAGED FLUXES FOR TURBULENT TRANSPORT

In a fluid in turbulent flow the turbulent momentum flux of z momentum in the r direction is given by

$$\tau_{rz}^{(t)} \equiv \rho\overline{v_r' v_z'} \tag{6-10}$$

Similarly in Chap. 10 we demonstrate the existence of a turbulent molar flux of species A in the r direction, defined by

$$J_{\mathrm{A}r}^{(t)} \equiv \overline{C_{\mathrm{A}}' v_r'} \tag{6-11}$$

and of a turbulent energy flux, defined by

$$q_r^{(t)} \equiv \rho\widehat{C}_p \overline{T' v_r'} \tag{6-12}$$

In each case, the turbulent fluxes will appear in the balance equation in such a way that we could write for the total fluxes

$$\overline{J}_{\mathrm{A}r} = \overline{J_{\mathrm{A}r}^{(l)}} + J_{\mathrm{A}r}^{(t)} \tag{6-13}$$

and

$$\overline{q}_r = \overline{q_r^{(l)}} + q_r^{(t)} \tag{6-14}$$

These definitions then permit us to write differential balance equations for heat and mass in the same form as in laminar flow except that instead of using the laminar flux we use the total flux.

To summarize, *when balance equations are written for turbulent-flow conditions, we use* (1) the time-averaged total flux instead of the laminar flux and (2) the time-averaged potentials $\overline{v}_z$, $\overline{C}_{\mathrm{A}}$, and $\overline{T}$, instead of v_z, C_{A}, and T.

6-5 PREDICTION OF TURBULENT FLUXES

In order to obtain the turbulent velocity profile $\overline{v}_z(r)$ in a pipe we need to relate $\tau_{rz}^{(t)}$ in Eq. (6-10) to $\overline{v}_z$. Since we need another equation to have a closed set of equations, this is called the problem of *closure*. It is the major unsolved problem of turbulence. The most common approach is to write an equation that relates $\tau_{rz}^{(t)}$ to the velocity gradient $d\overline{v}_z/dr$ through an *eddy viscosity* $\mu^{(t)}$ in a manner analogous to Newton's law for laminar flux

$$\tau_{rz}^{(t)} = -\mu^{(t)} \frac{d\overline{v}_z}{dr} \tag{6-15}$$

This expression is called the *Boussinesq hypothesis* or the *mean-velocity field closure*.

One problem with this approach is that, unlike the molecular viscosity μ, the turbulent eddy viscosity $\mu^{(t)}$ depends upon the local flow conditions and is not a property of the fluid. It varies with distance from the wall, with Reynolds number, and other flow conditions, as well as with the fluid molecular viscosity. In order to use it, either an empirical equation for it must be known or the parameters in a theoretical equation must be known. For this reason some authors[7] prefer to use other approaches, such as deriving differential equations for the turbulent kinetic energy or for $\tau_{rz}^{(t)}$ itself. Others[16] use dimensional analysis or *scaling*. Nevertheless, for pipe flow the above model gives good results, and we illustrate its use because of its simplicity and widespread adoption.

Since in a turbulent flow there may be laminar as well as turbulent contributions to the total transport, we must add the fluxes contributed by each process to get a total flux. Thus, for transport in the y direction only,

$$\overline{\tau}_{yz} = \overline{\tau_{yz}^{(l)}} + \tau_{yz}^{(t)} = -(\mu + \mu^{(t)}) \frac{\partial \overline{v}_z}{\partial y} \tag{6-16}$$

Similar expressions would apply for heat and mass transport. These flux laws are summarized below:

Laminar	*Turbulent*	*Total*	
$\tau_{yz}^{(l)} = -\mu \frac{\partial v_z}{\partial y}$	$\tau_{yz}^{(t)} = -\mu^{(t)} \frac{\partial \overline{v}_z}{\partial y}$	$\overline{\tau}_{yz} = -(\mu + \mu^{(t)}) \frac{\partial \overline{v}_z}{\partial y}$	(6-17)
$J_{Ay}^{(l)} = -D_{AB} \frac{\partial C_A}{\partial y}$	$J_{Ay}^{(t)} = -D_{AB}^{(t)} \frac{\partial \overline{C}_A}{\partial y}$	$\overline{J}_{Ay} = -(D_{AB} + D_{AB}^{(t)}) \frac{\partial \overline{C}_A}{\partial y}$	(6-18)
$q_y^{(l)} = -k \frac{\partial T}{\partial y}$	$q_y^{(t)} = -k^{(t)} \frac{\partial \overline{T}}{\partial y}$	$\overline{q}_y = -(k + k^{(t)}) \frac{\partial \overline{T}}{\partial y}$	(6-19)

6-6 APPLICATION TO PIPE FLOW

We now illustrate the application of the above rules to fully developed steady-state turbulent flow. As indicated for laminar flow, one goal of the engineer is to predict the pressure drop in the pipe as a function of average velocity or (in dimensionless variables) to relate the friction factor for the pipe to the Reynolds number.

Example 6-1: Turbulent pipe flow For fully developed turbulent flow in a pipe (*a*) using the rule given in Sec. 6-4, write a time-averaged incremental momentum balance and use it to derive an equation for the variation of momentum flux with radial position; (*b*) write the total-flux law; (*c*) derive an equation which, if integrated, would give the mean velocity distribution $\overline{v}_z(r)$ in the pipe; (*d*) state what information is needed before the equation in part (*c*) can be integrated to obtain $\overline{v}_z(r)$; (*e*) describe how the pressure drop and the friction factor can be obtained from the velocity profile; (*f*) carry out a dimensional analysis of the equation derived in part (*c*).

SOLUTION (*a*) Using the rule in Sec. 6-4, we write an incremental rate balance in terms of $\overline{\tau}_{rz}$, $\overline{v}_z$, etc.,

$$\underbrace{\overline{\tau}_{rz}A_r\Big|_r}_{\text{Rate in by diffusion}} - \underbrace{\overline{\tau}_{rz}A_r\Big|_{r+\Delta r}}_{\text{Rate out by diffusion}} + \underbrace{(\rho\overline{v}_z\overline{v}_z)(\Delta A_z)\Big|_{z=0}}_{\text{Rate in by bulk flow}} - \underbrace{(\rho\overline{v}_z\overline{v}_z\,\Delta A_z)\Big|_{z=L}}_{\text{Rate out by bulk flow}} + \underbrace{\overline{\Phi}_M\,\Delta\mathcal{V}}_{\text{Generation rate}} = 0$$

where
$$\overline{\Phi}_M = \frac{(\overline{\mathcal{P}}_0 - \overline{\mathcal{P}}_L)g_c}{L} \equiv \frac{(\overline{p}_0 - \overline{p}_L)g_c + \overline{\rho}g_zL}{L}$$

Note that since $\overline{v}_z$ does not depend on z, the bulk-flow terms cancel each other out. Then, proceeding as in Chap. 3, we obtain equations similar to (3-15), (3-17), and (3-19)

$$\frac{-d(\overline{\tau}_{rz}r)}{r\,dr} = \overline{\Phi}_M \tag{6-20}$$

$$\overline{\tau}_{rz} = \frac{\overline{\Phi}_M r}{2} \tag{6-21}$$

Equation (6-21) is the same as Eq. (3-19), derived for laminar flow. Written in terms of $\overline{\sigma}_{rz}$ [Eq. (3-20)], it is called *the law of linear shear-stress distribution.* The momentum flux at the wall is

$$\overline{\tau}_{rz}\Big|_R \equiv \tau_w = \frac{\Phi_M R}{2} = \frac{\mathcal{P}_0 - \mathcal{P}_L}{L}g_c\frac{R}{2} \tag{6-21a}$$

(*b*) We write the total flux as the sum of laminar and turbulent fluxes, as in Eqs. (6-8) and (6-16),

$$\overline{\tau}_{rz} = \overline{\tau_{rz}^{(l)}} + \tau_{rz}^{(t)} = -\mu\frac{d\overline{v}_z}{dr} - \mu^{(t)}\frac{d\overline{v}_z}{dr}$$

$$= -(\mu + \mu^{(t)})\frac{d\overline{v}_v}{dr} \tag{6-22}$$

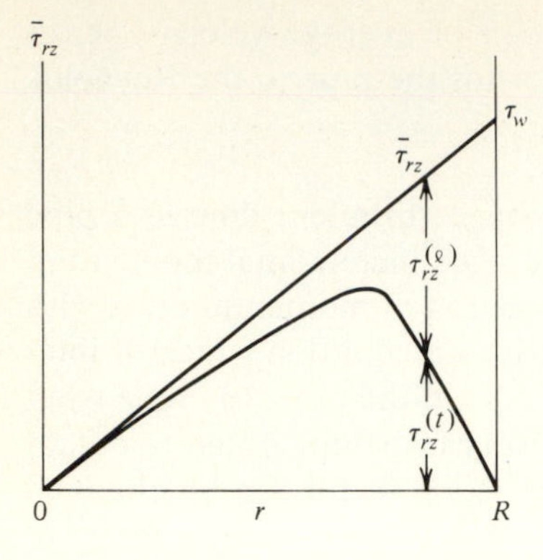

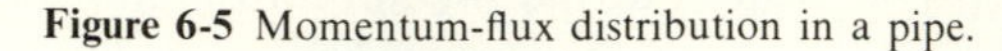

Figure 6-5 Momentum-flux distribution in a pipe.

The various fluxes are plotted in Fig. 6-5. Note that the laminar flux is directly proportional to the velocity gradient since μ is constant. It is zero at the center and reaches its largest value at the wall, where there is little turbulence. For the same reason, the turbulent flux will be zero at the wall. Since it will also be zero at the center, where the velocity gradient is zero, it must reach a maximum. This maximum occurs near the wall.

(*c*) Equating (6-21) to (6-22) and solving for $\bar{v}_z$ gives

$$d\bar{v}_z = \frac{\Phi_M r\,dr}{2(\mu + \mu^{(t)})} \tag{6-23}$$

where $\mu^{(t)}$ is a function of r (Fig. 6-6).

(*d*) To integrate Eq. (6-23) we need an equation describing how $\mu^{(t)}$ varies with r. Such expressions are complicated, and it is customary to divide the radial

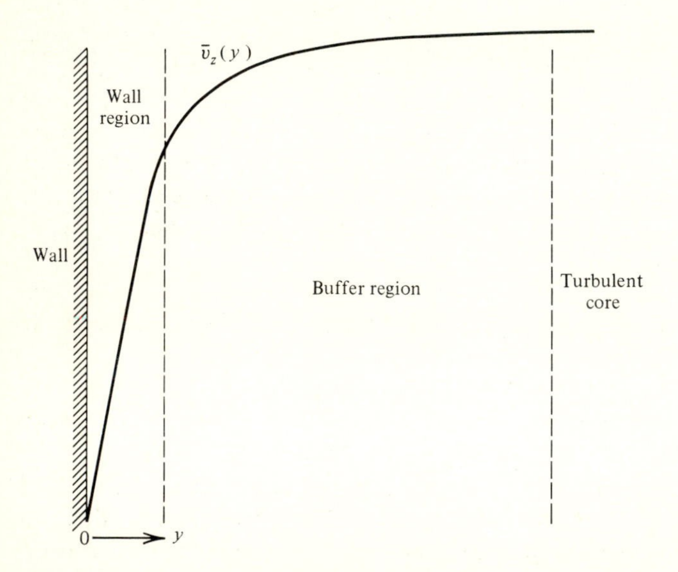

Figure 6-6 Regions used in analysis of turbulent transport.

distance from $r = 0$ to $r = R$ into separate regions and to consider the variation separately in each region. The region near the wall contains the most resistance to momentum transport (even though it is the smallest in thickness) because the eddy viscosity is nearly zero there.

(*e*) As in the case of laminar flow [see Eq. (3-9)], the average velocity $\overline{v}_a$ can be obtained as a function of Φ_M by integration of $\overline{v}_z$ over the cross section

$$\overline{v}_a(\Phi_M) = \frac{\int_0^R \overline{v}_z(r) r \, dr}{\int_0^R r \, dr} \tag{6-24}$$

Since $\overline{v}_z(r)$ depends upon $\overline{\Phi}_M$, which is proportional to the pressure drop, $\overline{v}_a$ can theoretically be obtained as a function of the pressure drop if $\overline{v}_z(r)$ is known; conversely, one can obtain the pressure drop if the flow rate is known. The pressure drop can then be used for calculating the size of pump needed to pump a fluid through a given length of pipe at a specified flow rate.

Usually it is desirable to generalize the results to all fluids, pipe diameters, and flow rates by using the dimensionless friction factor defined by Eqs. (4-9) and (4-2).

If the expression for v_a obtained by integration of Eq. (6-24) is substituted into the definition of friction factor [Eq. (4-2)], an expression for the friction factor as a function of Reynolds number can be obtained, as in laminar flow. We again emphasize one important difference: in turbulent flow, $\mu^{(t)}$ is *not* a constant but depends upon position and Reynolds number. Therefore we do *not* get a parabolic velocity profile, as in laminar flow [Eq. (3-26)]. Furthermore the *shape* of the relative velocity profile $\overline{v}_z(r)/\overline{v}_a$ (Fig. 6-7*a*) changes with Re, whereas it is a single function of Re (Fig. 6-7*b*) in laminar flow. As can be seen, the experimental velocity profile becomes flatter at the center as Re increases and the slope at the wall is steeper as Re increases. Recall from Example 1-1 that a steep profile implies a high resistance and a flat profile implies a low resistance to transport. Since $\mu + \mu^{(t)}$ is a reciprocal resistance, the steepness at the wall can be attributed to the fact that there is very little turbulence near the wall ($\mu^{(t)} \approx 0$) and the transport is largely molecular, a high-resistance process. At the center, however, $\mu^{(t)}$ is several orders of magnitude larger than μ, and the resistance to transport is low, resulting in a flat velocity profile. A sketch of $\mu^{(t)}/\mu$ versus r is shown in Fig. 6-7*c* for a typical Re. For a larger Re, $\mu^{(t)}$ would increase but μ would remain the same.

Because of the large variation of $\mu^{(t)}$ with position, a single function that applies at all radial positions is complicated; integration of Eq. (6-23) is therefore carried out after dividing the pipe into regions, as discussed in the next section.

(*f*) For pipe flow, expressions for $\mu^{(t)}(r)$ are now available, and their use will be discussed later. But for flow situations in which they are not available, an experimental program to correlate the variables must be carried out. As discussed in Appendix 4-2, the use of dimensionless variables greatly reduces the amount of

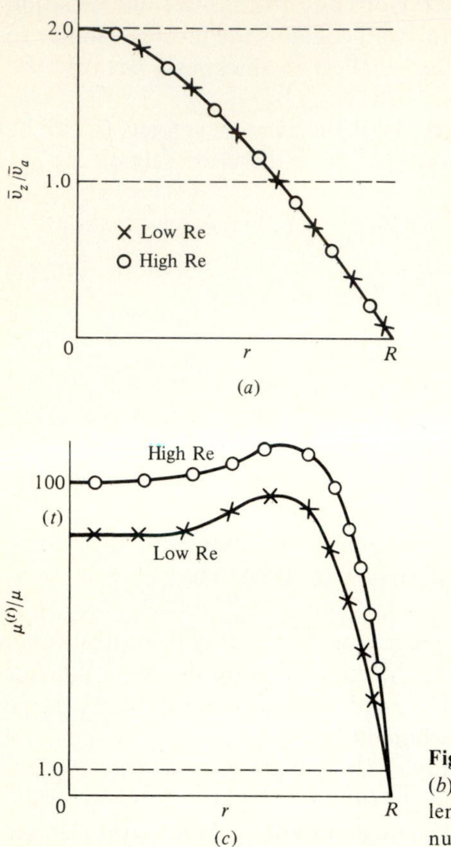

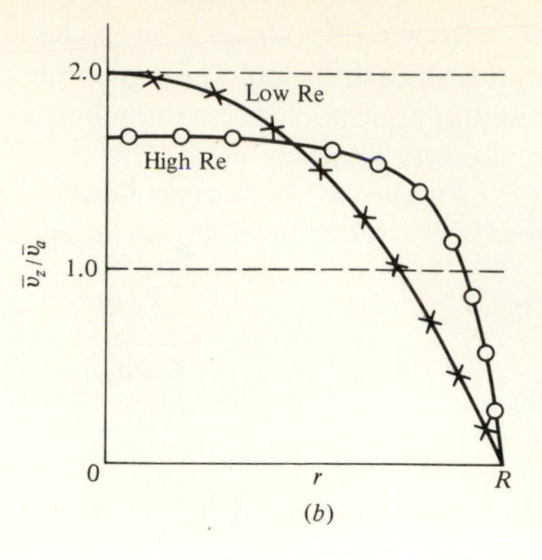

Figure 6-7 (*a*) Laminar-flow velocity profile; (*b*) turbulent-flow velocity profile; (*c*) ratio of turbulent to molecular viscosities at low and high Reynolds numbers.

experimental data that need be taken, but first one must know which dimensionless groups are involved. The procedure for finding them by analysis of the differential equations is illustrated in Appendix 4-2 for laminar pipe flow. We now illustrate the procedure for turbulent pipe flow. We let $\mu^* \equiv \mu^{(t)}/\mu$, and, as in the laminar-flow case, we let $r^* \equiv r/R$ and $v^* \equiv \bar{v}_z/v$. Introducing these into Eq. (6-23), integrating, and letting $v^* = 0$ at $r^* = 1$, we obtain

$$\frac{\bar{v}_z}{v_{\text{ch}}} \equiv v^* \equiv \frac{\bar{v}_z \mu}{\Phi_M R^2} = \int_{r^*}^{1} \frac{r^*\, dr^*}{1 + \mu^*} \tag{6-25}$$

where, for convenience, we have let

$$v_{\text{ch}} \equiv \frac{\Phi_M R^2}{\mu} \tag{6-26}$$

Since in turbulent flow μ^* depends on the Reynolds number (as well as r^*), and since the lower limit is r^*, Eq. (6-25) is of the form

$$v^* \equiv v^*(r^*; \text{Re})$$

whereas in laminar flow we found $v^* = v^*(r^*)$ only. Integrating over the cross section gives

$$v_a^* = 2\int_0^1 v^*(r^*, \mathrm{Re})r^*\,dr^* = v_a^*(\mathrm{Re}) \tag{6-27}$$

which is not constant (as it is in laminar flow).

Thus for turbulent pipe flow one could correlate v_a^* with Re, but engineers prefer to correlate with Re the traditional friction factor, as defined by Eq. (4-2)

$$f \equiv \frac{\Phi_M R}{\rho v_a^2}$$

Then

$$f\,\mathrm{Re} \equiv \frac{\Phi_M R}{\rho v_a^2}\frac{2Rv_a\rho}{\mu} \equiv \frac{2v_{\mathrm{ch}}}{v_a} = \frac{2}{v_a^*} \tag{6-27a}$$

Using Eq. (6-27), we get an equation of the form

$$f = f(\mathrm{Re})$$

In laminar flow, $v_a^* = 8$, and Eq. (6-27a) reduces to $f = 16/\mathrm{Re}$. □

In the above analysis we used R as our characteristic length; when Eq. (6-23) was made dimensionless, this resulted in a characteristic velocity [Eq. (6-26)]. This approach was intended to show that v^* depends on r^* and Re and that v_a^* depends on Re; but it does not provide insight into how the mean velocity varies with position and Reynolds number. Since this variation depends on local turbulence conditions, which vary as we move from the wall toward the center, it is more meaningful to define characteristic scales indicative of the changing turbulence conditions from the wall to the center.

If we let v_* (not to be confused with v^*) represent a characteristic velocity scale to be determined, we can define a local *position Reynolds number* by

$$y^+ \equiv \frac{yv_*}{\nu} \tag{6-28}$$

where $y = R - r$ is the distance from the wall and we have used ν/v_* as our characteristic length. (We follow the conventional practice in fluid mechanics of using a superscript plus to designate a dimensionless variable.) Then

$$v^+ \equiv \frac{\overline{v_z}}{v_*} \tag{6-28a}$$

First we write Eq. (6-21) in terms of y as

$$\bar{\tau}_{rz} = \frac{\Phi_M(R-y)}{2} = \frac{\Phi_M R}{2}\left(1 - \frac{y}{R}\right)$$

At the wall,

$$|\tau_w| \equiv \tau_0 \equiv |-\tau_{yz}|\Big|_{y=0} = |\tau_{rz}|\Big|_{r=R} = \frac{\Phi_M R}{2}$$

Using this to eliminate Φ_M gives

$$\bar{\tau}_{rz} = \tau_0 \left(1 - \frac{y}{R}\right)$$

Using Eq. (6-16) and the laminar flux law gives

$$\mu \frac{d\bar{v}_z}{dy} + \tau_{rz}^{(t)} = \tau_0 \left(1 - \frac{y}{R}\right) \tag{6-28b}$$

Although $\tau_{rz}^{(t)} \equiv \overline{\rho v_z' v_r'}$ varies with radial position (Fig. 6-3*d*), it is never larger than the total flux, which is never larger than the flux at the wall τ_0. Therefore τ_0 is a convenient characteristic flux and we can make Eq. (6-28*b*) dimensionless by dividing by τ_0. Making $\bar{v}_z$ and y dimensionless at the same time gives

$$\frac{\mu v_*^2}{\tau_0(\mu/\rho)} \frac{dv^+}{dy^+} + \frac{\tau_{rz}^{(t)}}{\tau_0} = 1 - \frac{y^+}{R^+} \qquad R^+ \equiv \frac{y v_*}{\nu} \tag{6-28c}$$

Then if we let $\tau_0 \equiv \rho v_*^2$, we can define a characteristic *friction velocity* by

$$v_* \equiv \sqrt{\frac{\tau_0}{\rho}} \tag{6-29}$$

Equation (6-28*c*) then becomes

$$\frac{dv^+}{dy^+} + \frac{\tau_{rz}^{(t)}}{\rho v_*^2} = 1 - \frac{y^+}{R^+} \tag{6-30}$$

which can be integrated if $\tau_{rz}^{(t)}$ is known as a function of position (see Fig. 6-3*d*). In the next sections we will show how to obtain $\tau_{rz}^{(t)}$ and how to apply Eq. (6-30) in the various regions.

One method for doing this is to use the eddy viscosity. Substituting the flux law [Eq. (6-15)] into Eq. (6-30) and letting $\mu^* \equiv \mu^{(t)}/\mu$ gives

$$(1 + \mu^*)\frac{dv^+}{dy^+} = \left(1 - \frac{y^+}{R^+}\right) \approx 1 \tag{6-31}$$

This means that if μ^* depends only on y^+, the solution will be of the form

$$v^+ = v^+(y^+) \qquad \text{only}$$

and we will have combined the effects of position y (or r) with that of Re. Such a relationship is called a *universal velocity profile* since data for all Re should fall on the same curve. Next we discuss how to obtain such a profile from Prandtl mixing-length theory.

6-7 PRANDTL MIXING-LENGTH THEORY†

Let us again consider fully developed steady-state turbulent flow in a pipe at a distance $y \equiv R - r$ from the wall. Since the time-averaged velocity $\bar{v}_z$ increases

†This section may be omitted on first reading. However, the results will be used later.

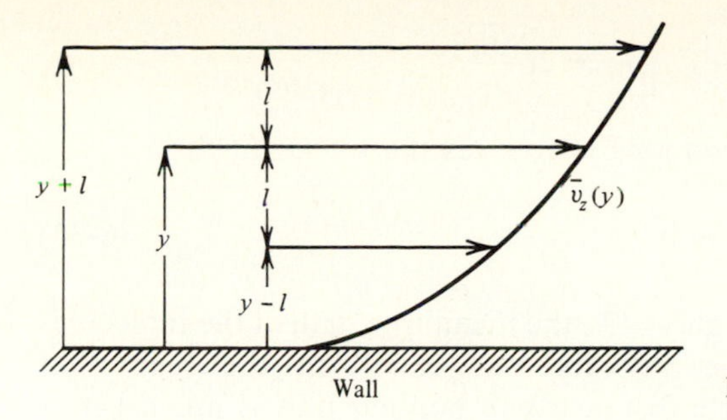

Figure 6-8 Prandtl mixing-length theory.

with distance from the wall, the gradient $d\bar{v}_z/dy = -d\bar{v}_z/dr$ will be positive and $\tau_{yz}^{(t)} = -\tau_{rz}^{(t)}$ will be negative, meaning that there is z-momentum transport toward the wall. Proceeding as we did in Chap. 2 for molecular transport (except that we are no longer dealing with molecules), we consider an eddy originally at a distance $y + l$ from the wall, where l is the mixing length (Fig. 6-8). Prandtl assumed that this eddy will move a distance l in the $-y$ direction, i.e., toward the wall, with a lateral velocity v_y' and will "deposit" its momentum; i.e., it will merge with the fluid at that point. Since the mean velocity at $y + l$ is higher than at y, this will tend to increase the z momentum at y and there will be a subsequent transport of z momentum per unit volume that is proportional to

$$\rho \, \Delta\bar{v}_z \Big| = \rho\left(\bar{v}_z\Big|_{y+l} - \bar{v}_y\Big|_y\right)$$

At the same time, an eddy originally at $y - l$ will move a distance l in the $+y$ direction with a velocity v_y'. Since the mean velocity at $y - l$ is lower than at y, this will tend to decrease the z momentum at y and there will be a transport of z momentum per unit volume that is proportional to

$$-\rho \, \Delta\bar{v}_z \Big| = -\rho\left(\bar{v}_z\Big|_y - \bar{v}_z\Big|_{y-l}\right)$$

The net average momentum change per unit volume will be one-half the algebraic sum of the above contributions or

$$\tfrac{1}{2}\rho\left(\bar{v}_z\Big|_{y+l} - \bar{v}_z\Big|_{y-l}\right)$$

The net average rate of turbulent transport of z momentum in the y direction will then be the product of this momentum change per unit volume and the average lateral volumetric flow rate $|v_y'|A_y$, or

$$\tau_{yz}^{(t)} A_y = -(\overline{|v_y'|}A_y)\left[\tfrac{1}{2}\left(\rho\bar{v}_z\Big|_{y+l} - \rho\bar{v}_z\Big|_{y-l}\right)\right] \tag{6-32}$$

where the minus sign is used to make $\tau_{yz}^{(t)}$ negative when $\Delta\bar{v}_z$ is positive and $\overline{|v_y'|}$ is used to designate an average magnitude.

If the velocity gradient is assumed linear over distance l,

$$\bar{v}_z \bigg|_{y+l} - \bar{v}_z \bigg|_{y-l} = 2l \frac{d\bar{v}_z}{dy}$$

Substituting into Eq. (6-32) and dividing by A_y gives the turbulent flux

$$\tau_{yz}^{(t)} = \overline{|v'|} \rho l \frac{d\bar{v}_z}{dy} \tag{6-33}$$

The mixing length l corresponds roughly to λ, the mean free path of the molecules in molecular transport. However, since λ is very small ($\approx 10^{-7}$ m), the assumption of a constant velocity gradient over a distance λ in laminar flow is much better than the assumption of a constant gradient over a distance l in turbulent flow since l may approach the dimensions of the duct. Similarly, the lateral velocity of the eddy corresponds in molecular transport to u, the mean velocity of the molecules. The mean lateral eddy velocity can be taken as the lateral rms fluctuating velocity, or $\overline{|v'|} = (\overline{v_r'^2})^{1/2}$, which is a routinely measured quantity.

By equating (6-33) to the turbulent flux law [Eq. (6-17)] one obtains the expression for the turbulent momentum diffusivity

$$\frac{\mu^{(t)}}{\rho} = l\overline{|v'|} \tag{6-34}$$

which compares with Eq. (2-4) except that any proportionality constants are absorbed into the mixing length l. (This is permissible because the mixing length of a fluid element is not as well defined as the mean free path of a molecule.)

Prandtl next assumed that the mean lateral fluctuating velocity was equal to the mean fluctuating velocity in the direction of flow

$$\overline{|v'|} \equiv (\overline{v_r'^2})^{1/2} = (\overline{v_z'^2})^{1/2}$$

and that the velocity fluctuation in the flow direction was roughly equal to the change in the mean velocity over the distance l or

$$(\overline{v_z'^2})^{1/2} = l \left| \frac{d\bar{v}_z}{dr} \right| = \overline{|v'|} \tag{6-35}$$

This assumption can be explained as follows. Consider an element of fluid at $y + l$ moving a distance l in the $-y$ direction (toward the wall) (Fig. 6-8) with a lateral fluctuating velocity $v_y' = -v_r' < 0$. Since the mean velocity $\bar{v}_z$ decreases as we approach the wall, the fluid element is moving into a stream with lower $\bar{v}_z$ and this creates a positive fluctuation due to the increase in $\bar{v}_z$

$$v_z' = \bar{v}_z \bigg|_{y+l} - \bar{v}_z \bigg|_{y} = l \frac{d\bar{v}_z}{dy} > 0$$

Similarly an element of fluid at $y - l$ moving a distance l in the $+y$ direction with a lateral velocity $v_y' > 0$ will cause a negative fluctuation

$$v_z' = \bar{v}_z \bigg|_{y-l} - \bar{v}_z \bigg|_{y} = -l \frac{d\bar{v}_z}{dy}$$

So in either case, Eq. (6-35) holds.

Substituting Eq. (6-35) into Eq. (6-34) gives

$$\mu^{(t)} = \rho l^2 \left| \frac{d\bar{v}_z}{dr} \right| \tag{6-36}$$

Note that the Prandtl theory consists of two parts: the first part given by Eq. (6-34) and the second by Eq. (6-35). Equation (6-34) is often used [even in cases where Eq. (6-35) does not apply] by incorporating other hypotheses into the theory. Launder and Spalding[7] discuss these methods.

The mixing length cannot yet be accurately predicted from a general theory that would apply to flows other than in a pipe. Of course, we would not expect the mixing length to be larger than the diameter of the conduit, since it is a measure of the size of an eddy. Also, since the eddies are largest in the core region and smallest near the wall, Prandtl assumed that l is proportional to the distance from the wall; i.e.,

$$l = \kappa y \tag{6-37}$$

The constant κ is called the *von Kármán constant;* its experimental value is about 0.4. Substituting into Eq. (6-33) gives

$$\mu^{(t)} = \rho \kappa^2 y^2 \left| \frac{d\bar{v}_z}{dr} \right| \tag{6-38}$$

Equation (6-15) makes it possible to relate the Reynolds stress or turbulent flux to the velocity by

$$\tau_{rz}^{(t)} = -\rho \kappa^2 y^2 \left| \frac{d\bar{v}_z}{dr} \right| \frac{d\bar{v}_z}{dr} \tag{6-39}$$

Combining with Eqs. (6-20) and (6-22) gives

$$\bar{\tau}_{rz} = \overline{\tau_{rz}^{(l)}} + \tau_{rz}^{(t)} = \frac{\Phi_M r}{2} \tag{6-40}$$

and

$$\bar{\tau}_{rz} = \mu \frac{d\bar{v}_z}{dy} - \rho(\kappa y)^2 \left(\frac{d\bar{v}_z}{dy} \right)^2 = \frac{\Phi_M (R - y)}{2} \tag{6-41}$$

This quadratic equation can be solved for $d\bar{v}_z/dy$, which can in principle be integrated to get $\bar{v}_z(r)$ (see the Problems). As we will see next, the equation is usually simplified for the various regions first.

6-8 THREE-REGION MODEL FOR MOMENTUM TRANSPORT IN A PIPE

It is customary in the engineering analysis of turbulent pipe flow to divide the flow field into three distinct regions, each having certain specific properties. These regions (see Fig. 6-6) are defined as follows:

1. *Wall region or viscous sublayer.* This is the portion of the flow field immedi-

ately adjacent to the pipe wall. Although turbulent fluctuations still exist, the turbulent shear stress is small. Thus, $\tau_{rz}^{(t)}$ or $\mu^{(t)} \approx 0$ and can be neglected.

2. *Turbulent core.* This is the region in the central portion of the tube. It usually occupies by far the greatest fraction of the tube cross section. Here turbulent motion is intense, and $\mu^{(t)}$ is very large. Thus, $\mu/\mu^{(t)} \to 0$, and the laminar momentum flux can be neglected.
3. *Buffer region.* This is the region between the viscous sublayer and the core, where μ and $\mu^{(t)}$ are of comparable magnitude; i.e., both turbulent and viscous momentum flux are important.

Viscous Sublayer

In the viscous sublayer turbulent momentum flux is unimportant, and

$$1 - \frac{y}{R} \approx 1$$

Equation (6-30) then becomes

$$\frac{dy^+}{dy^+} = 1$$

or

$$\boxed{v^+ = y^+} \tag{6-42}$$

which implies that in the viscous sublayer the same velocity profile $v^+ = y^+$ exists for *all* newtonian fluids, for *all* tube diameters, and at *all* flow rates (as long as the main flow is turbulent). Experimentally, Eq. (6-42) is found to hold up to $y^+ \approx 5.0$.

Turbulent Core

In the buffer region and turbulent core we need to know $\mu^{(t)}$. To find it we make use of Prandtl's mixing-length theory. In applying mixing-length theory to the core, Prandtl assumed that $1 - y/R \approx 1$; that is, he neglected the variation of $\bar{\tau}_{rz}$ with position even though it is known to vary linearly with distance from the wall. This assumption is considered justified since (see Prob. 6-5) the resulting equation fits experimental data even near the center. Using it and neglecting molecular transport, Eq. (6-39) gives

$$\tau_0 = \rho \kappa^2 y^2 \left| \frac{d\bar{v}_z}{dy} \right| \frac{d\bar{v}_z}{dy} \tag{6-43}$$

Taking the square root of both sides, using the sign that gives the proper physical meaning, and rearranging gives

$$\frac{d\bar{v}_z}{dy} = \frac{\sqrt{\tau_0/\rho}}{\kappa y} = \frac{v_*}{\kappa y} \tag{6-44}$$

and then

$$\frac{d\bar{v}_z}{v_*} = \frac{1}{\kappa}\frac{dy}{y} \tag{6-45}$$

or in dimensionless form

$$dv^+ = \frac{1}{\kappa}\frac{dy^+}{y^+} \tag{6-46}$$

where $\kappa \approx 0.4$ is called the *von Kármán constant*. Integrating and letting v_1 represent the velocity at a position y_1 at the edge of the core region gives

$$v^+ - v_1^+ = \frac{1}{\kappa}\ln\frac{y^+}{y_1^+} \qquad \text{logarithmic velocity profile} \tag{6-47}$$

It has been found to describe the turbulent velocity profile in the *core* region accurately even though it does not give a zero velocity gradient at the centerline as it should. However, this is not very important because we are usually interested in the velocity rather than the gradient. When y_1^+ and v_1^+ are evaluated experimentally, the final working equation is the Nikuradse equation

$$v^+ = 2.5 \ln y^+ + 5.5 \qquad y^+ > 26 \text{ to } 30 \tag{6-48}$$

This is an example of how experiment and theory interact. In a more general case, if v_1^+ is not known experimentally, it must be obtained theoretically from an equation for the buffer region, as indicated below.

Buffer or Transition Region

The buffer region or buffer zone lies between the points where the viscous sublayer profile $v^+ = y^+$ ceases to apply and the logarithmic profile starts to apply (see Fig. 6-9). This corresponds approximately to $5 < y^+ < 26$ to 30. Physically, the buffer region is that zone where both viscous and turbulent contributions to $\bar{\tau}_{rz}$ are important. Since both μ and $\mu^{(t)}$ are involved, it is difficult to develop an equation that fits this region and is continuous with equations for the laminar and core regions. For this reason, the buffer region is sometimes handled approximately by extrapolating an expression for the turbulent core region into the buffer region. However, to illustrate the principle involved we will discuss one of the many approaches developed to handle this region.

In any case, both μ and $\mu^{(t)}$ are needed, and, as in the laminar region, $1 - y/R$ is assumed to be unity. Then, Eqs. (6-15), (6-28*b*), and (6-29) give

$$\rho v_*^2 \equiv \tau_0 = (\mu + \mu^{(t)})\frac{d\bar{v}_z}{dy} \tag{6-49}$$

Making this dimensionless gives [see also Eq. (6-31)]

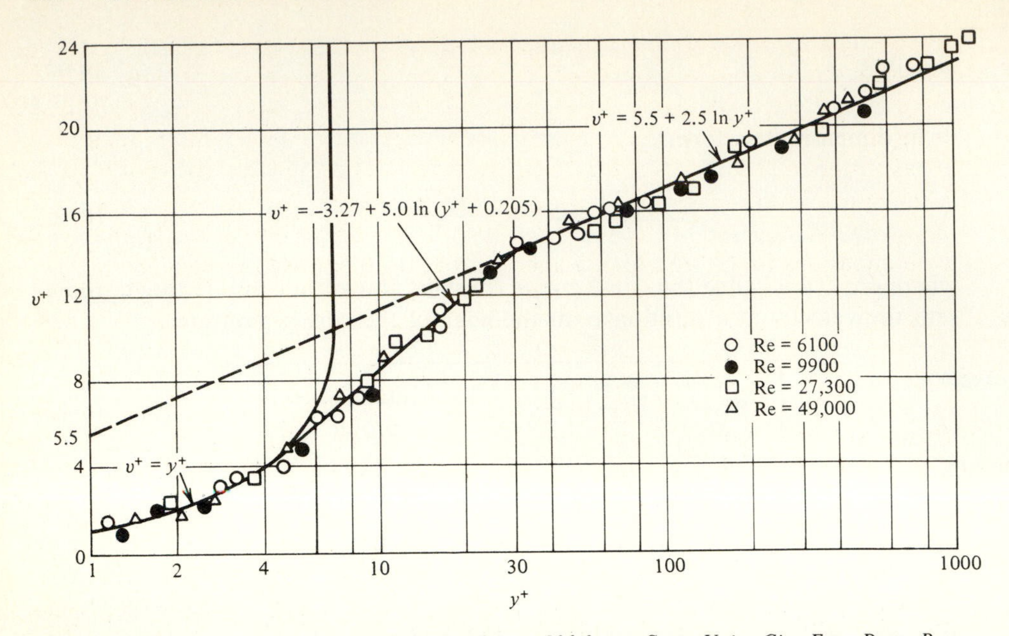

Figure 6-9 Velocity profiles. (*From E. R. Lindgren, Oklahoma State Univ. Civ. Eng. Dept. Rep. IAD621072, 1965.*)

$$dv^+ = \frac{dy^+}{1 + \mu^{(t)}/\mu} \tag{6-50}$$

Integration of this equation requires knowledge of how $\mu^{(t)}/\mu$ varies with y^+ in this region. Several semiempirical equations for this ratio have been proposed, including modification of the Prandtl mixing-length theory—they will be discussed later. For the present, we illustrate the procedure by using the equation recommended by Lin, Moulton, and Putnam[9]

$$\frac{\mu^{(t)}}{\mu} = \frac{y^+}{5} - 0.959 \tag{6-51}$$

for the region $5 < y^+ < 30$. This equation can readily be integrated, when substituted into Eq. (6-50), to give

$$v^+ = -3.27 + 5.0 \ln (y^+ + 0.205) \qquad 5 < y^+ < 30 \tag{6-52}$$

This equation and those obtained for the other regions are plotted in Fig. 6-9. More widely used than (6-52) is the *empirical* equation obtained by Nikuradse

$$v^+ = -3.05 + 5 \ln y^+ \qquad 0 < y^+ < 26 \tag{6-53}$$

Universal Velocity Profile

The purpose of using dimensionless variables such as v^+ and y^+ is to attempt to correlate the effects of tube diameter, average velocity $\langle v_z \rangle \equiv v_a$, and fluid properties (such as ρ and μ) so that equations of the form $v^+(y^+)$ will hold for any newtonian fluid and for any Reynolds number $D\rho v_a/\mu$ in the turbulent regime. Actually there is some variation with Re, but the plots for various Re are usually brought much closer together through the use of the variables v^+ and y^+. A plot of v^+ as a function of y^+ that holds for all Re is said to be a universal velocity profile although different equations are generally used for each of the three regions. However, an equation has been developed[2] that interpolates between the wall-core equations (6-42) and (6-53) and approaches them asymptotically

$$(u^+)^{-2} = (y^+)^{-2} + \frac{0.16}{\ln^2 9y^+} \tag{6-53a}$$

Summary of Velocity Profiles

$$v^+ = \begin{cases} y^+ & 0 < y^+ < 5 & \text{viscous sublayer} & (6\text{-}42) \\ -3.05 + 5\ln y^+ & 0 < y^+ < 26 \text{ to } 30 & \text{viscous sublayer and buffer zone} & (6\text{-}53) \\ 2.5\ln y^+ + 5.5 & y^+ > 26 & \text{core region} & (6\text{-}48) \end{cases}$$

These velocity profiles, often called the *law of the wall,* have been verified experimentally for numerous newtonian fluids such as water, air, and typical organic solvents. Recent studies have also been performed on dilute polymer solutions, and it has been found that for certain of them the viscous sublayer is greatly thickened, so that the expression $v^+ = y^+$ may hold out to $y^+ = 25$ or greater. The result is that the velocity gradient and thus the momentum flux at the wall are reduced, and the pressure drop required for flow is lower relative to that of the solvent. The effect is called *drag reduction.*

Interpretation of Total Resistance

A graphical integration of Eq. (6-50) can be carried out by plotting $(1 + \mu^{(t)}/\mu)^{-1}$ versus y^+ and finding the area under the curve as a function of y^+. Such a plot permits us to examine the physical meaning of Eq. (6-50) and how it influences the shape of the velocity profile upon integration. Referring back to Eq. (6-49), we see that we have an equation of the form (discussed in Sec. 1-6)

$$\text{Flux} = \frac{\text{driving force}}{\text{resistance}}$$

where $-d\bar{v}_x/dy$ is the driving force and $(\mu + \mu^{(t)})^{-1}$ is the resistance, the only difference being that here the resistance varies with y (because $\mu^{(t)}$ does). We recall from Example 1-6 that, for a constant flux, if the resistance is large, the magnitude of the gradient (slope) is large and vice versa. This means that near the wall, where $\mu^{(t)} \approx 0$, the resistance will be large and the slope $(d\bar{v}_x/dz)$ will be large in magnitude, whereas near the center (where $\mu^{(t)} \gg \mu$) the resistance will be small and the slope will be small (see Fig. 6-5).

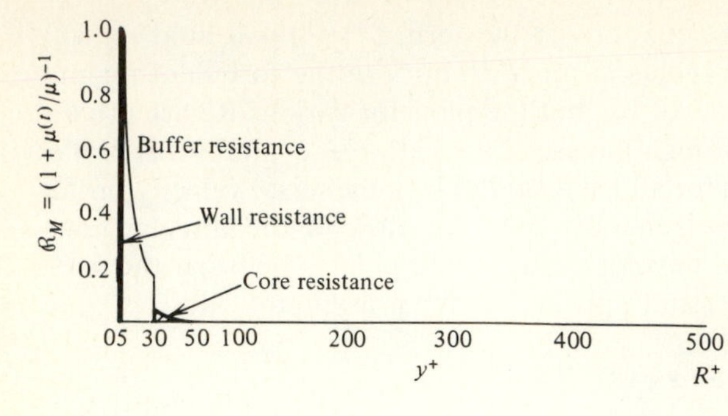

Figure 6-10 Ratio of total resistance to laminar wall resistance in turbulent pipe flow

We make Eq. (6-49) dimensionless by dividing by $\tau_0 \equiv \rho v_*^2$ and using the definitions of v^+ and y^+. This gives us a rearranged form of Eq. (6-50)

$$1 = \frac{1 + \mu^{(t)}}{\mu} \frac{dv^+}{dy^+} = \frac{dv^+/dy^+}{\mathcal{R}_M} \tag{6-54}$$

where

$$\left(1 + \frac{\mu^{(t)}}{\mu}\right)^{-1} \equiv \frac{(\mu + \mu^{(t)})^{-1}}{\mu^{-1}} \equiv \mathcal{R}_M \tag{6-55}$$

represents a *relative* resistance, i.e., the ratio of the total (turbulent plus laminar) resistance to the laminar resistance alone. Thus, $\mathcal{R}_M$ will be unity at the wall, where $\mu^{(t)} = 0$, but will be small ($\ll 1$) near the center, where $\mu^{(t)} \gg \mu$ and $\mathcal{R}_M \approx \mu/\mu^{(t)}$. Note also from Eq. (6-54) that $\mathcal{R}_M$ is equal to the dimensionless velocity gradient dv^+/dy^+ for unit dimensionless flux.

Now in order to obtain v^+ we must evaluate the integral

$$v^+ = \int_0^{y^+} \mathcal{R}_M \, dy^+$$

We can visualize this graphically if we plot $\mathcal{R}_M$ versus y^+ and examine the area under the curve as a function of y^+, as illustrated in Fig. 6-10 for $\mathcal{R}_M^+ = 500$.† Inspection of the figure reveals, as expected, that the largest area is near the wall, where the resistance is largest, and that the smallest area is near the center, where the resistance is small. This means that we do not need to know the velocity as accurately near the center, a result we use later. Note that for $y^+ > 30$, by Eq. (6-46), $\mathcal{R}_M = 0.4/y^+$. For $y^+ > 70$, $\mathcal{R}_M$ is too small to plot. This further justifies the assumption of constant momentum flux even at the center (see Prob. 6-5).

Velocity-Defect Equations

If we integrate Eq. (6-44) using an arbitrary lower limit and an upper limit at the center (where $y^+ = R^+$ and $v^+ = v^+_{\max}$), we get

†An $\mathcal{R}_M^+$ of 500 corresponds to an Re of about 16,000, which is not a very large value.

$$v^+_{max} - v^+ = \frac{1}{\kappa_1}\ln\frac{R^+}{y^+} = \frac{-1}{\kappa_1}\ln\frac{y}{R} = \frac{-1}{\kappa_1}\ln\left(1 - \frac{r}{R}\right) \tag{6-56}$$

These are forms of the *velocity-defect equation,* which has been found to fit experimental data for both rough and smooth pipes quite accurately (except near the wall), the constant $1/\kappa_1$ being taken as 2.5. The parameter v^+_{max} depends on the roughness, as discussed below.

Rough Surfaces

For flow past a rough surface the surface is presumed to have protuberances or bumps of height ϵ, and the velocity is assumed to be zero at some value $y = y_0$. Letting $v_1^+ = 0$ at $y_1^+ = y_0^+$ in Eq. (6-47) gives

$$v^+ = 2.5(\ln y^+ - \ln y_0^+) \qquad y_0^+ \equiv \frac{y_0 v_*}{\nu} \tag{6-57}$$

If the pipe is completely rough, the roughness elements develop large eddies, which mask the viscous forces. An experiment by Nikuradse on artificially roughened pipe resulted in

$$\ln y_0^+ = \ln \epsilon^+ - 3.4 \tag{6-58}$$

The velocity distribution is

$$v^+ = 2.5 \ln\frac{y}{\epsilon} + 8.5 \tag{6-59}$$

The maximum velocity is obtained by letting $y = R$

$$v^+_{max} = 2.5 \ln\frac{R}{\epsilon} + 8.5 \tag{6-60}$$

Comment on Velocity-Profile Equations

The previous treatment follows the classical approach to the prediction of velocity profiles in conduits. A more sophisticated theoretical treatment (Ref. 16, p. 147) involves the use of dimensional analysis. Interestingly enough, the result, i.e., the log velocity profile, is the same.

6-9 COMPARISON BETWEEN LAMINAR AND TURBULENT DIFFUSIVITIES

For laminar flow of gases the transport of mass, heat, or momentum occurs as a result of the motion of individual molecules (Chap. 2). In turbulent diffusion, however, the transport occurs due to the motion of "globs" or *elements* of fluid containing many molecules (see Fig. 6-1*b*). Like the molecules in laminar flow, these elements also have lateral velocities, but they are much larger than molecular mean velocities and are in fact of the order of magnitude of the lateral velocity fluctuations v_r'.

Also in turbulent flow, the mixing length, the average distance traveled by an eddy before losing its identity, may be a million times greater than the mean free path of the molecules and will vary from zero to roughly the size of the container (say the pipe diameter). As discussed in Chap. 2, a simple model for transport in gases indicates that the molecular diffusivity is roughly proportional to the product of the mean molecular velocity and the mean free path of the molecules. According to the Prandtl mixing-length theory (Sec. 6-7), the turbulent (or eddy) diffusivity is similarly roughly proportional to the product of the rms velocity fluctuation $(\overline{v_r'^2})^{1/2}$ and the mixing length. Since both are usually larger in turbulent flow than the corresponding molecular quantities in laminar flow, the eddy transport properties will usually also be several orders of magnitude larger than the corresponding molecular property; hence turbulent diffusion is a much more effective mixing process.

Since the molecular-transport properties do not vary much with position, they are never actually zero; but they are negligible near the center of a pipe compared with the turbulent properties. Furthermore, the turbulent transport properties vary with the Reynolds number (or mean velocity) since they depend on $(\overline{v'^2})^{1/2}$, which varies almost linearly with the mean velocity. Finally, it must be recognized that *during* the motion of a turbulent eddy, it is "leaking" momentum, mass, or energy to the surrounding fluid due to molecular transport. Thus the eddy viscosity $\mu^{(t)}$ depends on μ, $D_{AB}^{(t)}$ depends on D_{AB}, and $k^{(t)}$ depends on k.

These qualitative considerations can be summarized as follows. The eddy or turbulent transport properties, viscosity $\mu^{(t)}$, diffusivity $D_{AB}^{(t)}$, and the thermal conductivity $k^{(t)}$, are *not* constant, but depend on:

1. Distance from solid surfaces
2. Time-averaged velocity or Reynolds number
3. Molecular properties of the fluid (such as μ, ρ, k, D_{AB}, etc.)
4. The flow; i.e., whether the flow is through a pipe or over a flat plate, or in a jet, etc.

6-10 EXPRESSIONS FOR EDDY TRANSPORT PROPERTIES†

There is a plethora of expressions available for the variation of the eddy transport properties with position—just as there are many medications for treating the common cold. And just as there is no cure yet for the cold, there is no theory yet of turbulent diffusion that will predict eddy transport properties for all flow conditions and geometries. The expressions we use are generally empirical or semiem-

†This section is primarily for reference purposes and may be omitted on first reading.

pirical and therefore not always valid under conditions other than those for which the data are correlated. Students should know how to *use* them by substituting into the proper flux expression, but they need not memorize them or try to derive them and should not despair if they do not always understand them. Since the fundamental form of the variation near the wall is different from that near the center of a pipe, separate equations are generally used in the two regions.

Diffusivities near Wall

Lin, Moulton, and Putnam[9] have suggested that the eddy viscosity is not actually zero in the viscous sublayer but follows the equation

$$\frac{\mu^{(t)}}{\mu} = \left(\frac{y^+}{14.5}\right)^3 \qquad y^+ < 5 \tag{6-61}$$

It is commonly assumed that the eddy diffusivities for momentum, heat, and mass are equal. The Lin Moulton Putnam equations then become

$$\frac{\nu^{(t)}}{\nu} = \frac{D_{AB}^{(t)}}{\nu} = \frac{\alpha^{(t)}}{\nu} = \left(\frac{y^+}{14.5}\right)^3 \qquad 0 < y^+ < 5 \tag{6-61a}$$

and

$$\frac{\mu^{(t)}}{\mu} = \frac{y^+}{5} - 0.959 \qquad 0 \leq y^+ \leq 30 \tag{6-61b}$$

where the last equation was used to obtain Eq. (6-52) for the velocity profile in the buffer region. As another example, Notter and Sleicher[12] have used the empirical equation

$$\frac{\mu^{(t)}}{\mu} = 0.0083(y^+)^2 \qquad y^+ < 45 \tag{6-61c}$$

For heat or mass transfer they recommend

$$\frac{\alpha^{(t)}}{\nu} = \frac{9.0 \times 10^{-4}(y^+)^3}{[1 + 6.7 \times 10^{-3}(y^+)^2]^{1/2}} \qquad 0 < y^+ < 45 \tag{6-61d}$$

The van Driest equation has been used successfully by Launder and Spalding[7] and by others. Van Driest[17] modified the Prandtl mixing-length theory by multiplying the mixing length by a factor F to make the value near the wall lower than that obtained by the linear equation. Then Eq. (6-37) becomes

$$l' = \kappa y F \qquad \text{where} \qquad F = 1 - e^{-y^+/A} \tag{6-61e}$$

in which $\kappa = 0.4$ = von Kármán's constant and $A = 26$, a universal constant. From Eqs. (6-34) and (6-35) an expression for the eddy momentum diffusivity can be obtained.

Deissler[3] similarly modified the Prandtl theory, obtaining

$$\frac{\mu^{(t)}}{\mu} = n^2 v^+ y^+ [1 - \exp(-n^2 v^+ y^+)] \tag{6-61f}$$

where the empirical constant n was found from experimental data to be 0.124. The Deissler equation gives good results for the velocity profile, but its use does not result in the most accurate expressions for the Nusselt numbers as a function of Prandtl or Schmidt numbers. Von Kármán's *similarity hypothesis* gives

$$\nu^{(t)} = \kappa \left| \frac{(d\bar{v}_z/dy)^3}{(d^2\bar{v}_z/dy^2)^2} \right| \tag{6-61g}$$

where $\kappa = 0.4$. It also results in a log-velocity profile.

Eddy Diffusivities in the Turbulent Core

In the turbulent core of a pipe of radius R, an equation similar to one proposed by Reichardt[14] has been found by Pinho and Fahien[13] to fit data for momentum, heat, and mass eddy diffusivities for Re $> 20{,}000$. If a cross-sectional average thermal diffusivity is defined as

$$\langle \alpha^{(t)} \rangle = \frac{\int_0^1 \alpha^{(t)} r^* \, dr^*}{\int_0^1 r^* \, dr^*}$$

the dimensionless ratio of point diffusivity to the average in the core region ($r^* < 0.9$) is found to be

$$\alpha_c^* \equiv \frac{\alpha_c^{(t)}}{\langle \alpha^{(t)} \rangle} = 1.123(1 + 2.345 r^{*2})(1 - r^{*2}) \tag{6-61h}$$

The same equation also holds for mass or momentum diffusivities, and the average diffusivity is given by

$$\langle \alpha^{(t)} \rangle = 0.069 v_* R \tag{6-61i}$$

For ducts R can be replaced by twice the hydraulic radius. In the wall region ($y^+ < 25$) Hughmark[6] recommends

$$\frac{\alpha_w^{(t)}}{\nu} = 0.00096 y^{+3} \tag{6-61j}$$

which can also be written

$$\frac{\alpha_w^{(t)}}{\langle \alpha^{(t)} \rangle} \equiv \alpha_w^* = 0.000137(\mathrm{Re})^{1.75}(1 - r^*)^3 \tag{6-61k}$$

In the buffer region, Pinho and Fahien[13] found that the diffusivity can be correlated by

$$\frac{1}{\alpha^*} = \frac{1}{\alpha_w^*} + \frac{1}{\alpha_c^*} \tag{6-61l}$$

or

$$\alpha^* = \{[1.123(1 + 2.345r^{*2})(1 - r^{*2})]^{-1} + [0.00137(\mathrm{Re})^{1.75}(1 - r^*)^3]^{-1}\}^{-1} \tag{6-61m}$$

6-11 CORRELATIONS FOR FRICTION FACTOR

As stated previously, our objective is to correlate the transfer coefficients with other variables. We first illustrate this procedure with the case of momentum transport. The friction factor for a pipe is defined as

$$f = \frac{\tau_0}{\frac{1}{2}\rho \overline{v}_b^2} \tag{6-62}$$

where for turbulent flow in a pipe Eq. (3-2) is

$$\overline{v}_b = \langle \overline{v}_z \rangle \equiv \frac{\int_0^R \overline{v}_z r\, dr}{\int_0^R r\, dr} \equiv \overline{v}_a \tag{6-63}$$

the bulk mean or average velocity. If we recall, from Eq. (6-40), that $v_* \equiv \sqrt{\tau_0/\rho}$ and substitute it into Eq. (6-62), we get

$$f = 2\left(\frac{v_*}{v_b}\right)^2 = \frac{2}{(v_b^+)^2} \tag{6-64a}$$

where
$$v_b^+ \equiv \frac{\overline{v}_b}{v_*} \tag{6-64b}$$

Then to obtain v_b^+ we make Eq. (6-63) dimensionless as follows:

$$v_b^+ \equiv \langle v_z^+ \rangle \equiv \frac{\int_0^{R^+} v^+ r^+\, dr^+}{\int_0^{R^+} r^+\, dr^+} = \frac{\int_{R^+}^0 v^+(R^+ - y^+)(-dy^+)}{(R^+)^2/2} \tag{6-65}$$

where
$$R^+ \equiv \frac{R v_* \rho}{\mu}$$

So if we have a relationship for the velocity profile $v^+(y^+)$, we can integrate the above equation to find $v_b^+(R^+)$ and then use Eq. (6-64) to find the friction factor as a function of R^+. In order to express the friction factor in terms of the Reynolds number we rearrange the above definition and use Eq. (6-64*a,b*) as follows:

$$R^+ \equiv \frac{R v^* \rho}{\mu} \equiv \frac{2 R v_b \rho}{\mu}\frac{1}{2}\frac{v_*}{v_b} \equiv \frac{\mathrm{Re}}{2 v_b^+} = \frac{\mathrm{Re}}{2}\sqrt{\frac{f}{2}} \tag{6-66}$$

The result will be of the form $f = g(\text{Re}\sqrt{f})$, which can sometimes be written explicitly as $f \equiv f(\text{Re})$.

Example 6-2: Friction factor from power law Derive an expression for $f(\text{Re})$ for turbulent velocity profiles at low Re using the *Blasius* $(1/n)$*th power law,* which says that the velocity varies as the $(1/n)$th power of the distance from the wall, where $n \approx f^{-1/2}$. At $\text{Re} \approx 10^5$, $n \approx 7$.

Solution In dimensionless form the velocity profile for $n = 7$ is

$$v^+ = 8.56(y^+)^{1/7} \tag{6-67}$$

Use of Eqs. (6-65) and (6-66) results (see the Problems) in

$$v_b^+ = 8.56(0.817)(R^+)^{1/7} \tag{6-68}$$

Substituting into Eq. (6-64*a*), using Eq. (6-66), and rearranging gives the widely used *Blasius law of friction,* which holds for $4000 < \text{Re} < 10^5$

$$f = 0.0791(\text{Re})^{-0.25} \tag{6-69}$$

Frequently a power of 0.2 is used instead of 0.25 and the equation is written

$$\boxed{f = 0.046(\text{Re})^{-0.2} \qquad \text{for } 10^4 < \text{Re} < 10^6} \tag{6-70}$$

The actual functional relationship between f and Re will of course depend upon the form of the velocity-profile equation (which depends upon the expression used for the eddy viscosity). But since the velocity profile is integrated over the entire cross section, the velocity need not be known accurately near the wall because it is small there and does not contribute greatly to the overall integral. Therefore in practice it is possible to use expressions derived for the turbulent core or buffer regions [such as the logarithmic profile, Eq. (6-48)] and assume that they apply approximately over the entire cross section. □

Example 6-3: Friction factor from velocity profile (*a*) Assuming that the logarithmic velocity profile [Eq. (6-48)]

$$v^+ = 2.5 \ln y^+ + 5.5$$

holds for the entire radius of a pipe (see Prob. 6-1), derive an equation for the friction factor as a function of Reynolds number. (*b*) Justify the assumption made in part (*a*).

Solution (*a*) Substituting Eq. (6-48) into Eq. (6-65) gives

$$v_b^+ = \frac{2}{(R^+)^2} \int_0^{R^+} (2.5 \ln y^+ + 5.5)(R^+ - y^+)\, dy^+ \tag{6-71}$$

$$v_b^+ = 1.75 + 2.5 \ln R^+ \tag{6-72}$$

Using Eq. (6-64*a*), we obtain

$$\frac{1}{\sqrt{f}} = \frac{v_b^+}{\sqrt{2}} = \frac{1}{\sqrt{2}}(1.75 + 2.5 \ln R^+) \tag{6-73}$$

From Eq. (6-66)

$$\boxed{\frac{1}{\sqrt{f}} = 4.07 \log (\text{Re} \sqrt{f}) - 0.60} \tag{6-74}$$

which is Prandtl's universal law of friction for smooth pipes.

(*b*) Equation (6-65) can be written

$$v_b^+ = \frac{2}{(R^+)^2} I_t(R^+)$$

where I_t is the total integral or area under the curve (see Fig. 6-11). Thus

$$I_t(R^+) \equiv \int_0^{R^+} v^+(R^+ - y^+)\, dy^+ = I_l + I_b + I_c \tag{6-75}$$

where I_l, I_b, and I_c are, respectively, the integrals over laminar, buffer, and core regions.

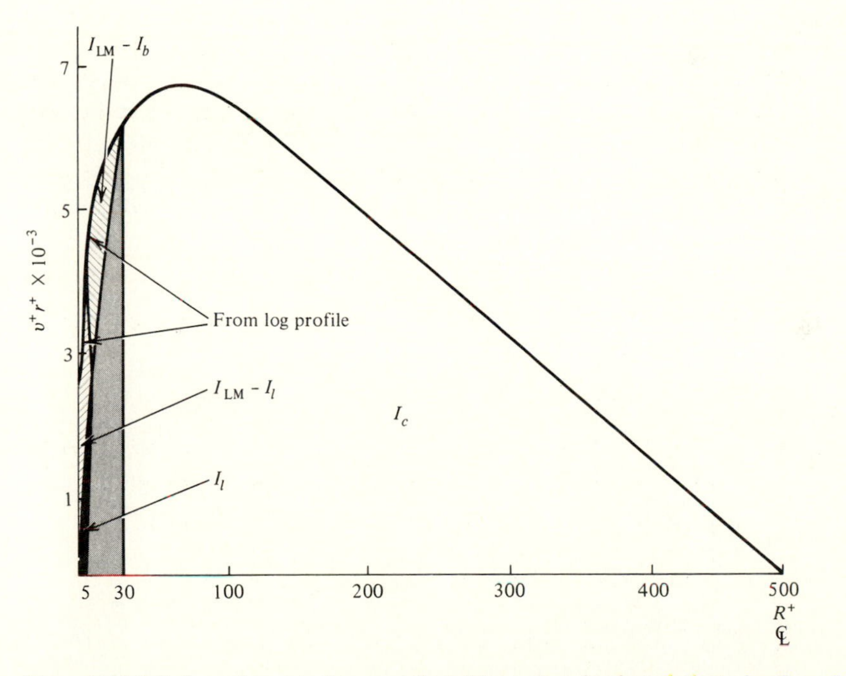

Figure 6-11 Relative volumetric flow contributed by each region in turbulent pipe flow; I_l = laminar sublayer, I_b = buffer region, I_c = core region, I_{LM} = area under logarithmic velocity profile.

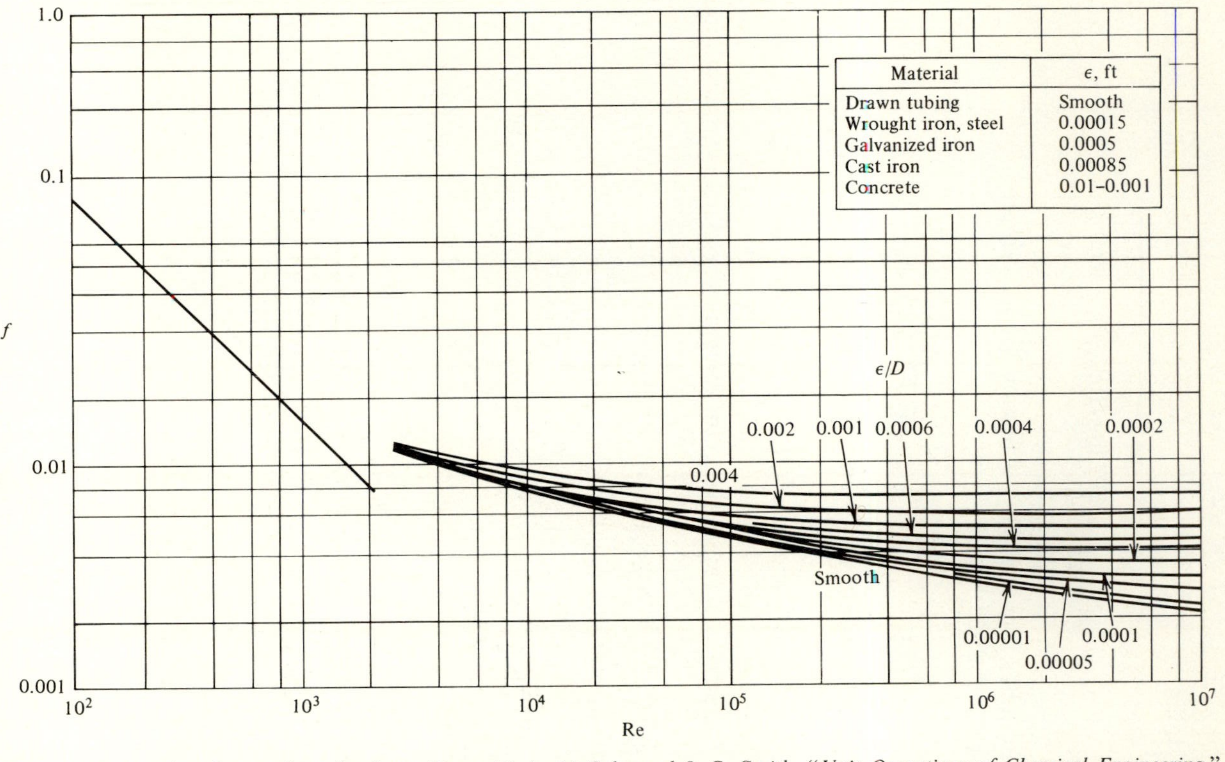

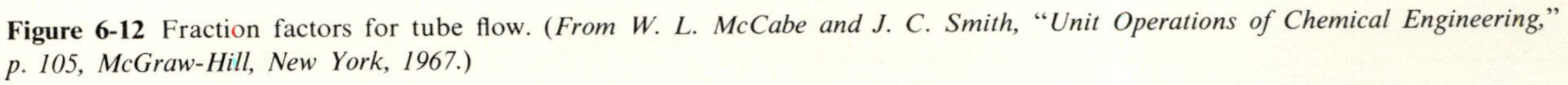

Figure 6-12 Fraction factors for tube flow. (*From W. L. McCabe and J. C. Smith, "Unit Operations of Chemical Engineering," p. 105, McGraw-Hill, New York, 1967.*)

For $R^+ = 500$, the integrals I_l, I_b, and I_c are shown in Fig. 6-11 as areas under the curve $v^+(R^+ - y^+)$ versus y^+. The total area, $I_t \equiv I_l + I_b + I_c$, can then be compared with the approximate area I_{LM}, the area under the logarithmic-velocity-profile curve extrapolated back to $y^+ = 0$.

The percentage error involved in using I_{LM} instead of I_t is $(I_{\text{LM}} - I_t)(100)/I_t$, where

$$I_{\text{LM}} - I_t = \int_0^{30} v_c^+(R^+ - y^+)\,dy^+ - I_l - I_b$$

By substituting Eqs. (6-42), (6-53), and (6-48) for the velocity distribution in the laminar v_l^+, buffer v_b^+, and core v_c^+ regions one can obtain the desired integrals. For $R^+ = 500$ the error is found to be only 1.4 percent. Thus in order to find v_b^+ from v^+ it is not necessary to know the velocity profile very accurately near the wall; however, if one did not know v_1^+, the velocity at the edge of the turbulent core, or the constant 5.5 in Eq. (6-48), it would be necessary to find it by integration of Eq. (6-50) as described previously. □

Rough Walls

The above analysis applies only to *smooth* pipes. For rough pipes a roughness ratio ϵ/D must be estimated. Empirical curves such as Fig. 6-12 are used frequently in practice. The empirical Colebrook equation

$$\frac{1}{\sqrt{f}} = -4.0 \log\left(\frac{\epsilon}{D} + \frac{4.67}{\text{Re}\sqrt{f}}\right) + 2.28 \tag{6-75a}$$

represents the data well for rough pipes. In the limit of large Re this becomes independent of Re, or

$$\frac{1}{\sqrt{f}} = -4.0 \log\frac{\epsilon}{D} + 2.28 \tag{6-75b}$$

By combining equations for rough pipe and for laminar and turbulent flow regimes Churchill[1] has developed a single equation for the friction factor in laminar, transition, and turbulent flow and for smooth or rough pipes.

6-12 TURBULENT TRANSPORT TO A SINGLE SURFACE

Turbulent Energy Transport

Consider a heat exchanger consisting of a pipe through which a fluid is flowing in turbulent flow (see Sec. 4-2). If the temperature of the wall of the heat exchanger $\overline{T}_0$ is less than that of the fluid, the fluid will be cooled as it passes through the pipe. To design the heat exchanger, e.g., to find the length of pipe required to cool the inside fluid to a certain temperature $\overline{T}_b$, we must be able to predict the rate at which heat is being transported from the fluid to the wall of the pipe. This heat flux is usually expressed in terms of a heat-transfer coefficient, as in Eq. (4-26) (with notation changed as appropriate)

$$|q_w| = q_0 = h|\overline{T}_b - \overline{T}_0| \tag{6-76}$$

where $\overline{T}_b$ is the bulk temperature of the fluid. Our goal at present is to use our knowledge of transport phenomena to develop generalized methods and equations for predicting heat-transfer coefficients.

In an actual heat exchanger q_0, $\overline{T}_b$, $\overline{T}_0$ (and possibly h) vary with the axial distance z (measured along the length of the pipe), but in this chapter we consider a simplified model in which we analyze conditions at a given z and neglect bulk flow in the z direction in comparison with the transport from the fluid to the wall. This assumption makes an incremental energy balance unnecessary. In Chap. 10 we consider the case where bulk transport in the z direction is included.

Turbulent Mass Transport

Now consider an interface through which a solute A is being transported from one phase to another, e.g., the gas-liquid interface in a gas absorber, the liquid-liquid interface in an extractor, or the liquid-solid interface in a pipe whose wall is coated with a soluble material (Sec. 4-3). If C_{A0} is the concentration of C_A at the wall and C_{Ab} the bulk mean concentration, the definition of the mass-transfer coefficient k_c is given by Eq. (4-28) by changing notation. Then the molar flux at the interface is

$$N_{A0} \approx \overline{J}_{A0} = k_c|\overline{C}_{A0} - \overline{C}_{Ab}| \tag{6-77}$$

where bulk flow is neglected.

Film-Theory Model

The model that historically has been used is the film-theory model (Chap. 4), in which the resistance to transport is assumed to occur entirely in a thin film adjacent to the surface (Fig. 6-13). Since the flow is primarily laminar in the film, the temperature, concentration, and velocity change rapidly until the turbulent region is reached. There the temperature, concentration, and velocity are nearly equal to the bulk mean values. Usually the curvature of the surface is neglected, so that the model can be used for pipes, flat walls, packing surfaces, etc.

The actual variation of $\overline{T}$ is illustrated in Fig. 6-14. The film-theory model is also plotted in Fig. 6-14, where $\overline{T}_0$ is the potential at the wall and $\overline{T}_b$ and $\overline{v}_b = v_a = \langle \overline{v}_z \rangle$ are the potentials in the bulk stream. The effective film thickness L_H is *not* the thickness of the laminar sublayer but that of a fictitious film which would be required to account for the entire resistance if *only* molecular transport were involved. To illustrate this idea, consider the case of heat transfer: if the flux at $y = 0$ is $\overline{q}_0$ and turbulent transport is neglected, we obtain, by integration of the flux law (assuming constant q)

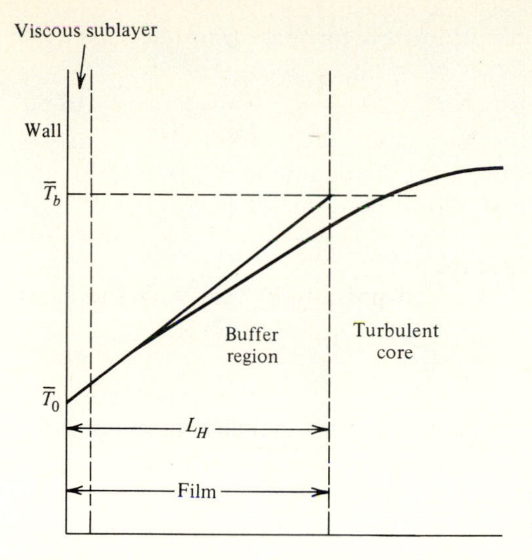

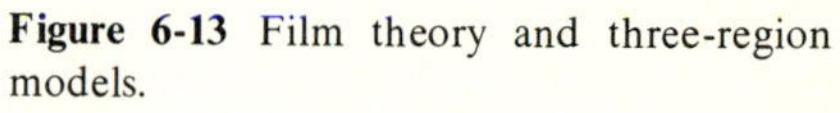
Figure 6-13 Film theory and three-region models.

$$q_0 = \frac{k}{L_H}(\overline{T}_b - \overline{T}_0) \tag{6-78}$$

By definition [Eqs. (4-26) and (4-30)] the heat-transfer coefficient h and Nusselt number are

$$q_0 \equiv h(\overline{T}_b - \overline{T}_0) \qquad \text{and} \qquad \text{Nu} \equiv \frac{hD}{k}$$

Hence

$$h = \frac{k}{L_H} \qquad \text{and} \qquad \text{Nu} \equiv \frac{D}{L_H} \tag{6-79}$$

where the first of Eqs. (6-79) can be taken as the definition of L_H.

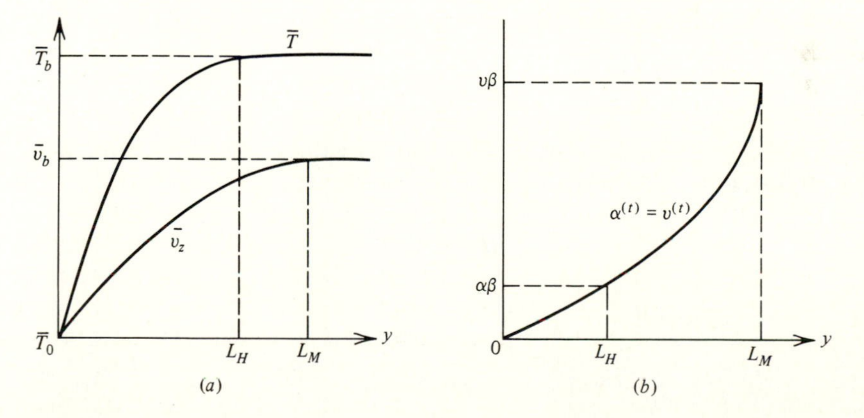

Figure 6-14 (*a*) Temperature and velocity profiles. (*b*) Diffusivities versus distance from wall.

For mass transport, if L_D is the film thickness for mass transfer,

$$k_c = \frac{D_{AB}}{L_D} \qquad \text{and} \qquad (\text{Nu})_{AB} \equiv \frac{k_c D}{D_{AB}} \equiv \frac{D}{L_D} \tag{6-80}$$

For momentum transport, if we equate τ_0 from the definition of f [Eq. (4-8)] and assume that the velocity profile is linear within the film, we get

$$\frac{1}{2} f \rho \bar{v}_b^2 \equiv \tau_0 \approx \mu \frac{v_b}{L_M}$$

or

$$\frac{f}{2} = \frac{\mu}{L_M \rho v_b} = \frac{\mu}{D \rho \bar{v}_b} \frac{D}{L_M} = \frac{D/L_M}{\text{Re}} \tag{6-81}$$

Thus $f\,\text{Re}$ is a reciprocal dimensionless film thickness for momentum just as Nu is for heat.

To obtain Nu from the second of Eqs. (6-79) requires L_H, and therefore we need to relate L_H to L_M. To do so, we assume that the momentum thickness L_M is the distance at which the turbulent momentum diffusivity $\nu^{(t)}$ is a factor β times the molecular momentum diffusivity ν and that L_H is the distance at which the turbulent thermal diffusivity $\alpha^{(t)}$ is the same factor β of the molecular thermal diffusivity α, or

$$\nu^{(t)} = \beta \nu \qquad \text{and} \qquad \alpha^{(t)} = \beta \alpha \tag{6-82}$$

where β is on the order of 10^2 or 10^3.

We then make use of the empirical observation [Fig. (6-14) and Eq. (6-61*a*)] that the *eddy* diffusivities for both heat and momentum are roughly equal and both vary with the third power of the distance from the wall; i.e., if

$$\nu^{(t)} = \alpha^{(t)} = C y^3 \tag{6-83}$$

where $C = \text{const}$, then from Eqs. (6-82)

$$\beta \nu = C L_M^3 \qquad \text{and} \qquad \beta \alpha = C L_H^3 \tag{6-84}$$

Taking the cube root of the ratio gives

$$\frac{L_M}{L_H} = \left(\frac{\nu}{\alpha}\right)^{1/3} = (\text{Pr})^{1/3} \tag{6-85}$$

where $\text{Pr} \equiv \nu/\alpha$ is the Prandtl number. For typical values of Pr, see Table 6-1. Then from Eqs. (6-79) and (6-81)

$$\text{Nu} = \frac{f}{2} \text{Re}\,(\text{Pr})^{1/3} \tag{6-86}$$

Similarly for mass transfer

$$(\text{Nu})_{AB} = \frac{f}{2} \text{Re}\,(\text{Sc})^{1/3} \tag{6-87}$$

Table 6-1 Typical values of the Prandtl number[19]

	Pr	T(K)
Liquid sodium	0.0102	394
Air	0.708	300
Steam	1.008	450
Water	5.88	300
Kerosine	109	300
Hydraulic fluid (MIL-M5606)	1,024	256
Glycerin	83,300	272

where $\text{Sc} \equiv \nu/(D)_{\text{AB}}$ is called the *Schmidt number*. For diffusion in gases $\text{Sc} \approx 1$, but for diffusion in liquids $\text{Sc} \approx 2000$.

The j factor is defined for heat transfer by

$$\frac{\text{Nu}}{\text{Re}\,(\text{Pr})^{1/3}} \equiv j_H \tag{6-88a}$$

for mass transfer by

$$\frac{(\text{Nu})_{\text{AB}}}{\text{Re}\,(\text{Sc})^{1/3}} \equiv j_D \tag{6-88b}$$

and for momentum transfer by

$$j_M \equiv \frac{f}{2} \tag{6-88c}$$

Equations (6-82) to (6-84) then become

$$j_D = j_H = j_M = \frac{f}{2} \tag{6-89}$$

Since the Blasius equation (6-69) and (6-70) gives

$$j_M = \frac{f}{2} = \begin{cases} 0.023(\text{Re})^{-0.2} & 4000 < \text{Re} < 10^6 \quad (6\text{-}90a) \\ 0.079(\text{Re})^{-0.25} & \text{Re} > 10^5 \quad (6\text{-}90b) \end{cases}$$

we can obtain heat-, mass-, and momentum-transfer coefficients from this set of equations. For heat transfer in fluids whose viscosity varies greatly with temperature it is advisable to multiply the j factor as defined by Eq. (6-88a) by the *Seider-Tate correction* $(\mu_b/\mu_0)^{-0.14}$, where μ_b is the bulk-fluid viscosity and μ_0 is the viscosity at the temperature of the wall. For mass transfer the left-hand side of Eq. (6-87) should be multiplied by $p/(p_{\text{B}})_{\text{LM}}$ or $C/(C_{\text{B}})_{\text{LM}}$ to account for bulk flow, as discussed in Chap. 5.

The above relationship, called the *Chilton-Colburn analogy,* has been tested experimentally on a large number of fluids (with widely varying Sc and Pr num-

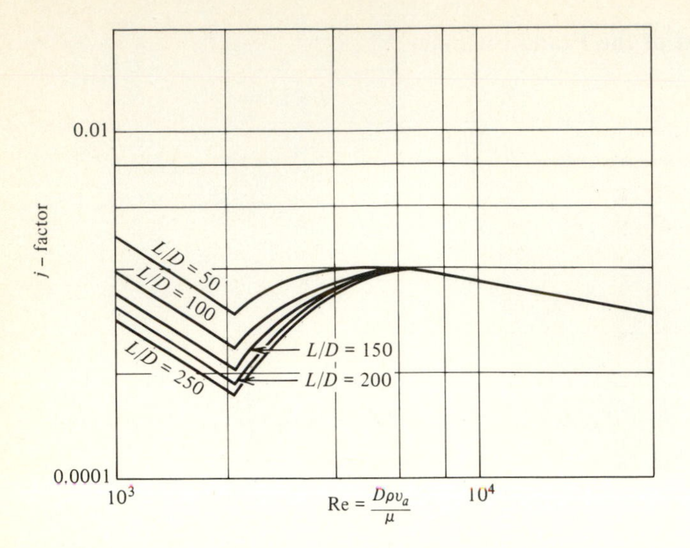

Figure 6-15 Heat transfer in transition range. (*From W. H. McAdams "Heat Transmission," 3d ed., McGraw-Hill, New York, 1954.*)

bers) and has been found to be quite accurate. It is widely used in industry and has the advantage that one need merely plot one function $j(\text{Re})$ on a log-log plot to obtain a correlation for heat, mass, or momentum transport (see Fig. 6-15).

Expressed in terms of the Nusselt number, Eq. (6-90*a*) becomes

$$\text{Nu} = 0.23(\text{Re})^{0.8}(\text{Pr})^{1/3} \tag{6-90c}$$

which agrees favorably with the widely used Dittus-Boelter equation

$$\text{Nu} = 0.023(\text{Re})^{0.8}(\text{Pr})^{b} \qquad b = \begin{cases} 0.4 & \text{for heating} \\ 0.3 & \text{for cooling} \end{cases} \tag{6-90d}$$

The difference in the exponent b arises from the variation of viscosity with temperature and its effect on the viscous sublayer. For mass transfer, the same equation is often used with $b = 0.44$ and Pr replaced by Sc.

6-13 THREE-REGION MODEL FOR HEAT AND MASS

We have been able to show how elementary turbulent transport theory can be used to obtain results in agreement with the empirical correlations for the transfer coefficients that appear in the literature [Eqs. (6-70) and (6-88*b*)]. In the three-region model, the procedure followed for heat and mass transport is the same as that used previously for momentum transport in a pipe, and the same three regions are used. Bulk flow is neglected, and steady state is assumed. We illustrate the development for energy transport; the case of mass transport is left to the reader.

Viscous Sublayer

If we neglect $k^{(t)}$ and let the heat flux $= q_0 =$ const, we get

$$\int_{T_0}^{\bar{T}} d\bar{T} = \bar{T} - \bar{T}_0 = \int_0^y \frac{q_0\,dy}{k} = \frac{q_0 y}{k} \tag{6-91}$$

As before, we express y in terms of a dimensionless distance $y^+ \equiv yv_*/\nu$. To obtain a dimensionless temperature we postulate that $\bar{T}_0$, the wall temperature, can be taken as a reference temperature and $\rho\hat{C}_p(\bar{T} - \bar{T}_0)$ can be interpreted as an enthalpy, which has heat or energy units. Dividing this by q_0 will cancel the heat unit and will tend to put situations with different heat fluxes on the same physical basis. The time and length units this introduces can be canceled by using $v_* \equiv \sqrt{\tau_0/\rho}$. Thus our dimensionless temperature becomes

$$T^+ \equiv \frac{\rho\hat{C}_p v_*(\bar{T} - \bar{T}_0)}{q_0} \tag{6-92}$$

If we multiply both sides of Eq. (6-91) by $\rho\hat{C}_p v_*/q_0$, we obtain

$$T^+ = \frac{\hat{C}_p\mu}{k}\frac{yv_*}{\nu} = \text{Pr}\,y^+ \tag{6-93}$$

where

$$\text{Pr} \equiv \frac{\hat{C}_p\mu}{k} = \frac{\nu}{\alpha} \tag{6-94}$$

is the molecular Prandtl number. It is the ratio of the momentum to the thermal molecular diffusivities. Equation (6-93) indicates that the slope of a plot of $\bar{T}^+$ versus y^+ is the Prandtl number and that the dimensionless temperature increases slowly with dimensionless distance for fluids that are good conductors, e.g., liquid metals.

For *mass* transfer, similar reasoning leads to the definition

$$C_A^+ \equiv \frac{v_*(\bar{C}_A - \bar{C}_{A0})}{J_{A0}} \tag{6-95}$$

Then if the molecular Schmidt number is defined as

$$\text{Sc} \equiv \frac{\mu}{\rho D_{AB}} = \frac{\nu}{D_{AB}} = \frac{\text{molecular momentum diffusivity}}{\text{molecular mass diffusivity}} \tag{6-96}$$

the analogous equation for mass transport is

$$C_A^+ = \text{Sc}\,y^+ \tag{6-97}$$

Buffer Region

Here we must include both turbulent and molecular transport properties. For heat transport

$$q_0 = (k + k^{(t)}) \left| \frac{d\bar{T}}{dy} \right|$$

or

$$\bar{T} - \bar{T}_0 = \int_{\bar{T}_0}^{\bar{T}} d\bar{T} = q_0 \int_0^y \frac{dy}{k + k^{(t)}} \tag{6-98a}$$

It is useful to write this equation in dimensionless form. Thus to obtain T^+ we multiply $\bar{T} - \bar{T}_0$ by $(\rho \hat{C}_p v_* / q_0)(\nu/\nu)$, or

$$T^+ = \int_0^y \frac{\rho \hat{C}_p \nu}{k + k^{(t)}} \frac{v_* \, dy}{\nu} = \int_0^{y^+} \frac{dy^+}{\alpha/\nu + \alpha^{(t)}/\nu} \qquad \alpha^{(t)} \equiv \frac{k^{(t)}}{\rho \hat{C}_p} \tag{6-98b}$$

If we define a *turbulent* Prandtl number as

$$\mathrm{Pr}^{(t)} \equiv \frac{\nu^{(t)}}{\alpha^{(t)}} = \frac{\mu^{(t)} \hat{C}_p}{k^{(t)}}$$

we can write Eq. (6-98*b*) as

$$T^+ = \int_0^{y^+} \frac{dy^+}{\dfrac{\alpha}{\nu} + \dfrac{\alpha^{(t)}}{\nu^{(t)}} \dfrac{\nu^{(t)}}{\nu}} = \int_0^{y^+} \frac{dy^+}{\dfrac{1}{\mathrm{Pr}} + \dfrac{1}{\mathrm{Pr}^{(t)}} \dfrac{\mu^{(t)}}{\mu}} \tag{6-99}$$

If we have $\alpha^{(t)}$ as a function of y^+, we can integrate Eq. (6-98*b*); alternatively we can use Eq. (6-99) if we have the turbulent Prandtl number $\mathrm{Pr}^{(t)}$ as a function of y^+ and $\mu^{(t)}$ as a function of y^+. The latter can be obtained from the equations for $\mu^{(t)}/\mu$ discussed previously.

For mass transport the corresponding equations are

$$C^+ = \int_0^{y^+} \frac{\nu \, dy^+}{D_{AB} + D_{AB}^{(t)}} \tag{6-100}$$

and

$$C^+ = \int_0^{y^+} \frac{dy^+}{\dfrac{1}{\mathrm{Sc}} + \dfrac{1}{\mathrm{Sc}^{(t)}} \dfrac{\mu^{(t)}}{\mu}} \tag{6-101}$$

where

$$\mathrm{Sc}^{(t)} \equiv \frac{\nu^{(t)}}{D_{AB}^{(t)}} = \frac{\text{turbulent momentum diffusivity}}{\text{turbulent mass diffusivity}}$$

$$= \text{turbulent Schmidt number} \tag{6-102}$$

Unfortunately very few data are available on the variation of $\mathrm{Pr}^{(t)}$ or $\mathrm{Sc}^{(t)}$ with position; but since the mechanisms of turbulent momentum, mass, and energy transport are essentially the same, it is common practice to assume that

$$\mathrm{Pr}^{(t)} = \mathrm{Sc}^{(t)} = 1.0 \qquad \text{or} \qquad \alpha^{(t)} = D_{AB}^{(t)} = \nu^{(t)}$$

If this assumption is made, only an expression for $\mu^{(t)}(y^+)$ is needed (Sec. 6-10). This can be obtained by differentiating an expression for $v^+(y^+)$ and using Eq. (6-50).

Turbulent Core

Here we neglect k with respect to $k^{(t)}$. It is convenient to relate the dimensionless temperature profile to the dimensionless velocity profile. We write the flux laws

$$\tau_0 = \mu^{(t)} \frac{d\bar{v}_z}{dy} = \nu^{(t)} \frac{d(\rho \bar{v}_z)}{dy} \tag{6-103}$$

and

$$q_0 = k^{(t)} \frac{d\bar{T}}{dy} = \alpha^{(t)} \frac{d(\rho \widehat{C}_p \bar{T})}{dy} \tag{6-104}$$

Solving for dy in each case and equating gives

$$dy = \frac{\nu^{(t)}}{\tau_0} d(\rho \bar{v}_z) = \frac{\alpha^{(t)} d(\rho \widehat{C}_p \bar{T})}{q_0} \tag{6-105}$$

If we multiply the two expressions on the right by $v_*/\nu^{(t)}$, we have

$$\frac{v_* d(\rho \bar{v}_z)}{\tau_0} = \frac{\alpha^{(t)} v_* d(\rho \widehat{C}_p \bar{T})}{\nu^{(t)} q_0} \quad \text{or} \quad dv^+ = \frac{dT^+}{\text{Pr}^{(t)}} \tag{6-106}$$

If we let $\langle \text{Pr}^{(t)} \rangle$ be the average value of $\text{Pr}^{(t)}$ and integrate from v_1^+ and T_1^+, we get

$$v^+ - v_1^+ = \frac{1}{\langle \text{Pr}^{(t)} \rangle} (T^+ - T_1^+) \tag{6-107}$$

Previously we found that we can write [Eq. (6-47)]

$$v^+ - v_1^+ = \frac{1}{\kappa} \ln \frac{y^+}{y_1^+} \tag{6-108}$$

If $\text{Pr}^{(t)} = 1.0$, the same equation holds for heat and momentum; or

$$T^+ - T_1^+ = \frac{1}{\kappa} \ln \frac{y^+}{y_1^+} \tag{6-109}$$

Limited experimental data indicate that although $\text{Pr}^{(t)}$ in the core is approximately unity, it actually varies with position and the nature of the fluid. In the absence of sufficient experimental evidence to the contrary, it is usually assumed to be 1. Launder and Spalding[7] recommend a value of 0.9.

Equation (6-109) will give the temperature profile provided T_1^+ is known; it must be obtained by letting $y^+ = y_1^+$ in an equation for $T^+(y^+)$ derived for the buffer region, such as Eq. (6-98*b*) or (6-99). It is apparent that T_1^+ will depend upon the molecular Prandtl number Pr and upon the expression for eddy diffusivity used.

Similarly for mass transport we obtain

$$v^+ - v_1^+ = \frac{1}{\langle \text{Sc}^{(t)} \rangle} (C_A^+ - C_{A1}^+) = \frac{1}{\kappa} \ln \frac{y^+}{y_1^+} \tag{6-110}$$

If $\langle \mathrm{Sc}^{(t)} \rangle = 1.0$, $C_\mathrm{A}^+ - C_{\mathrm{A}1}^+ = \kappa^{-1} \ln (y^+/y_1^+)$. Again $C_{\mathrm{A}1}^+$ must be obtained by letting $y^+ = y_1^+$ in an equation for $C_\mathrm{A}^+(y^+)$ derived for the buffer region, such as Eqs. (6-100) and (6-101). Note that $C_{\mathrm{A}1}^+$ will depend upon the molecular Schmidt number and the equation used for the eddy diffusivity.

Reynolds Analogy

Rearranging Eq. (6-105) gives

$$\frac{q_0}{\tau_0} = \frac{\alpha^{(t)}}{\nu^{(t)}} \frac{d(\rho \widehat{C}_p \overline{T})}{d(\rho \overline{v}_z)}$$

Using Eqs. (6-76) and (6-62) on the left and integrating on the right (assuming that $\langle \mathrm{Pr}^{(t)} \rangle = 1.0$ and that the equations for the turbulent core can be extrapolated to the wall) gives

$$\frac{h(T_b - T_0)}{\frac{1}{2} f \rho \overline{v}_b^2} = \frac{\int_{T_0}^{\overline{T}_b} \rho \widehat{C}_p \, d\overline{T}}{\int_0^{v_b} \rho \overline{v}_z} = \frac{\widehat{C}_p(\overline{T}_b - \overline{T}_0)}{\overline{v}_b} \tag{6-111}$$

Then rearranging gives

$$\frac{h}{\rho \widehat{C}_p \overline{v}_b} = \tfrac{1}{2} f \tag{6-112}$$

where

$$\mathrm{St} \equiv \frac{h}{\rho \widehat{C}_p \overline{v}_b} \tag{6-113}$$

is the *Stanton* number. Rewriting in terms of the Nusselt number gives

$$\mathrm{Nu} \equiv \frac{hD}{k} \equiv \frac{h}{\rho \widehat{C}_p v_p} \frac{D \rho \overline{v}_b}{\mu} \frac{\rho \widehat{C}_p}{h} \frac{\mu}{\rho} \equiv \mathrm{St}\,\mathrm{Re}\,\mathrm{Pr} = \frac{f}{2} \mathrm{Re}\,\mathrm{Pr} \tag{6-114}$$

Equations (6-112) and (6-114) are called the *Reynolds analogy*. Since we have neglected the wall region, it holds only when $\mathrm{Pr} \approx 1.0$.

The Stanton number for mass transfer is

$$(\mathrm{St})_{\mathrm{AB}} \equiv \frac{k_c}{v_b}$$

and the Reynolds analogy is

$$(\mathrm{St})_{\mathrm{AB}} = \frac{f}{2} = \mathrm{St} \tag{6-114a}$$

Just as the heat-transfer case is restricted to $\mathrm{Pr} = 1$, the mass-transfer analogy is limited to $\mathrm{Sc} = 1$.

6-14 NUSSELT NUMBERS FOR HEAT AND MASS

Since the Reynolds analogy holds only for Pr = 1, we now illustrate a more general way to predict the local heat-transfer coefficient for turbulent flow through a pipe. By definition [Eq. (4-26) or (6-76)]

$$h \equiv \frac{q_0}{\overline{T}_b - \overline{T}_0}$$

where $\overline{T}_b$ is the bulk mean temperature of the fluid. Thus, if we can find $\overline{T}_b$, we can use Eq. (6-76) to obtain h. But $\overline{T}_b$ is defined (Appendix 4-2) as

$$\overline{T}_b = \frac{\int_0^R (\overline{T}\overline{v}_z)(2\pi r\, dr)}{\int_0^R \overline{v}_z(2\pi r\, dr)} = \frac{\int_0^R \overline{T}\overline{v}_z r\, dr}{\int_0^R \overline{v}_z r\, dr} \tag{6-115}$$

Therefore if we have velocity and temperature profiles for each region, we can use the above equations to get h. However, it is desirable to generalize the above procedure by using dimensionless variables. Combining the definitions of the Nusselt number, $\text{Nu} \equiv 2Rh/k$, and that of h, we obtain

$$\text{Nu} = \frac{2Rq_0}{k(\overline{T}_b - \overline{T}_0)} \tag{6-116}$$

Then using the definition of T^+ [Eq. (6-92)] and y^+ [Eq. (6-42*a*)], we obtain

$$T_b^+ \equiv \frac{(\overline{T}_b - \overline{T}_0)\rho \widehat{C}_p v_*}{q_0} \tag{6-117}$$

and

$$R^+ \equiv \frac{Rv_*}{\nu} \tag{6-118}$$

Solving for $\overline{T}_b - \overline{T}_0$ in Eq. (6-117) and for R in Eq. (6-118) and then substituting into Eq. (6-116) gives

$$\text{Nu} = 2\,\text{Pr}\frac{R^+}{T_b^+} \tag{6-119}$$

If $\overline{T}_b$ is put in dimensionless form, we get

$$T_b^+ = \frac{\int_0^{R^+} T^+ v^+ r^+\, dr^+}{\int_0^{R^+} v^+ r^+\, dr^+} \tag{6-120}$$

Combining Eqs. (6-119) and (6-120) gives

$$\text{Nu} = \frac{2\,\text{Pr}\,R^+ \int_0^{R^+} v^+ r^+\,dr^+}{\int_0^{R^+} T^+ v^+ r^+\,dr^+} = \frac{\text{Pr Re}\sqrt{\dfrac{f}{2}}\,I_M}{I_H} \tag{6-121}$$

where $R^+ \equiv \frac{1}{2}\text{Re}\,\sqrt{f/2}$ by Eq. (6-66),

$$I_H \equiv \int_0^{R^+} T^+ v^+ r^+\,dr^+ = \int_0^{R^+} \left[\int_0^{y^+} \frac{dy^+}{\dfrac{1}{\text{Pr}} + \dfrac{\alpha^{(t)}(y^+)}{\nu}}\right] [v^+(y^+)](R^+ - y^+)\,dy^+ \tag{6-122}$$

and

$$I_M \equiv \int_0^{R^+} v^+ r^+\,dr^+ = \int_0^{R^+} \left[\int_0^{y^+} \frac{dy^+}{1 + \mu^*(y^+)}\right] (R^+ - y^+)\,dy^+ \tag{6-123}$$

where

$$\mu^* \equiv \frac{\mu^{(t)}}{\mu} = \frac{\nu^{(t)}}{\nu}$$

Thus if one has expressions for $\alpha^{(t)}$ and $\nu^{(t)}$ as functions of y^+, the integrals above can be evaluated and Nu can be obtained as a function of Pr, Re, and f. By methods discussed in Sec. 6-11 f and Re can be related, resulting in Nu = Nu(Pr, Re), as desired. A widely used assumption is to let $\alpha^{(t)} = \nu^{(t)}$, in which case only $\nu^{(t)}$ as a function of y^+ is needed.

The integrations above can be carried out region by region, as described previously. Note that two integrations are required to get I_H; the first is a running integration up to an arbitrary y^+ to obtain $T^+(y^+)$, and the second is over the entire cross section. Also, if $v^+(y^+)$ is not known, it must be obtained by integration, and then another integration is needed to get I_M. This method will be illustrated next, and approximate methods will be discussed later.

Determination of Local Nusselt Number by Integration of Temperature Profiles†

To illustrate graphically the procedure used for obtaining Nu [or $(\text{Nu})_{AB}$] from Eq. (6-121) we note first that the right-hand side of Eq. (6-99) [or (6-101)] is the same as Eq. (6-50) if $\text{Pr}^{(t)} = 1.0$ and Pr = 1.0. As we have seen earlier, the turbulent Prandtl number is on the order of magnitude of 1, (and similarly $\text{Sc}^{(t)}$ is nearly 1). In dilute gases the molecular Prandtl number Pr is on the order of 1, but for liquids it can vary from <0.01 for liquid metals to 7 for water and 83,000 for glycerin (Table 6-1). [For mass transfer $\text{Sc}^{(t)} = 1$, but since molecular diffusion in liquids is very slow ($D_{AB} \approx 10^{-5}$ m^2/s), molecular Schmidt numbers around 1000 and 2000 are not uncommon.] Therefore dimensionless temperature [Eq. (6-99)]

†This subsection illustrates the evaluation of the integrals in Eq. (6-121) and may be omitted on first reading.

and concentration [Eq. (6-101)] profiles will be the same as dimensionless velocity profiles [Eq. (6-50)] only if both molecular *and* turbulent Prandtl or Schmidt numbers are 1.

To illustrate the effect of the molecular Prandtl number, let us try to interpret Eq. (6-99) as we did Eq. (6-50). For simplicity we let $\Pr^{(t)} = 1.0$ or $\nu^{(t)} = \alpha^{(t)}$. Just as we interpreted $\mathcal{R}_M \equiv (1 + \mu^{(t)}/\mu)^{-1}$ [Eq. (6-55)] as the ratio of the total (laminar plus turbulent) resistance to the laminar resistance to momentum transport, so also can we interpret the group

$$\mathcal{R}_H \equiv \left(\frac{1}{\Pr} + \frac{\mu^{(t)}}{\mu}\right)^{-1} \equiv \left(\frac{\alpha}{\nu} + \frac{\nu^{(t)}}{\nu}\right)^{-1} = \left(\frac{\alpha}{\nu} + \frac{\alpha^{(t)}}{\nu}\right)^{-1} = \frac{(\alpha + \alpha^{(t)})^{-1}}{\nu^{-1}} \tag{6-124}$$

as the ratio of the total (laminar plus turbulent) heat-transport resistance to the molecular *momentum*-transport resistance.† Note that

$$\frac{dT^+}{dy^+} = \mathcal{R}_H \tag{6-125}$$

which states that since the flux is constant, the driving force equals the resistance (both dimensionless). Integrating, we obtain

$$T^+(y^+) \equiv \int_0^{y^+} \mathcal{R}_H \, dy^+ \tag{6-126}$$

If we plot $\mathcal{R}_H$ versus y^+, the above equation states that the area under the curve from 0 to y^+ gives T^+ at that y^+. Figure 6-16 is such a plot for $R^+ = 500$ and $\Pr \gtrsim 10$.

Let us now observe the relative magnitude of the resistance throughout the cross section. We note that the major resistance is near the wall, where $\mathcal{R}_H = dT^+/dy^+ = \Pr$. Since $\Pr > 1$, the slope of the temperature profile is steeper

†The reason why the denominator of $\mathcal{R}_H$ is not molecular heat transport goes back to the definition of y^+ in terms of ν instead of α, a choice that was made when we were discussing momentum transport. To compensate, we could multiply both sides of Eq. (6-124) by α/ν to obtain α^{-1} in the denominator and use $\mathcal{R}'_H = \mathcal{R}_H\alpha/\nu$. However, since both ν and α are assumed constant, either of them can serve as a basis for nondimensionalizing the total resistance.

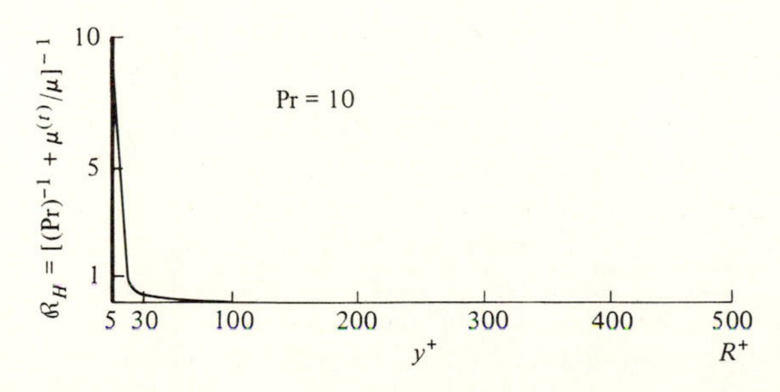

Figure 6-16 Relative heat-transfer resistance in turbulent pipe flow.

than that of the velocity profile (which is 1); but near the center $dT^+/dy^+ \approx \nu/\alpha^{(t)}$, which is small. Therefore the resistance near the center contributes relatively little to the area under the curve and consequently to T^+.

We can obtain T_1^+, the temperature at the beginning of the core region, by letting $y^+ = y_1^+$ in Eq. (6-126). Usually we take $y_1^+ = 30$, but it is actually determined by the region of applicability of the equations used for the eddy transport properties.

To find Nu we need to use Eq. (6-119), which requires T_b^+, given by Eq. (6-120). Proceeding as in Example 6-2 and using Eq. (6-65) gives

$$T_b^+ = \frac{2}{(R^+)^2 v_b^+} \int_0^{R^+} T^+ v^+ (R^+ - y^+)\, dy^+ \tag{6-127}$$

As before, the integral can be separated into laminar, buffer, and core contributions, or

$$I_{tH} \equiv \int_0^{R^+} T^+ v^+ (R^+ - y^+)\, dy^+ = I_{lH} + I_{bH} + I_{cH}$$

In the laminar layer Eq. (6-91), $T^+ = \Pr y^+$, and Eq. (6-42*c*), $v^+ = y^+$, apply, and

$$I_{lH} = \int_0^5 \Pr (y^+)^2 (R^+ - y^+)\, dy^+$$

$$= \Pr \left[\frac{(R^+)(y^+)^3}{3} - \frac{(y^+)^4}{4} \right] \Bigg|_0^5 = \Pr \left(\tfrac{125}{3} R^+ - \tfrac{625}{4} \right)$$

If $R^+ = 500$ (say), this equals 20,600 Pr. Notice that I_{lH} is proportional to Pr, which means that a fluid with a high Pr (low k) will make a larger contribution to T_b^+ than one with a low Pr.

For the buffer region an equation for T^+ and v^+, as obtained by integration of Eqs. (6-99) and (6-50) (using an expression for the eddy diffusivity), could in principle be substituted into Eq. (6-119) and integrated up to $y = y_1^+$ (say) to obtain I_{bH}; but an analytical integration is usually not possible, and numerical or graphical integration is used. The latter involves plotting $T^+ v^+ (R^+ - y^+)$ versus y^+ and finding the area under the curve from $y^+ = 5$ to 30, as illustrated in Fig. 6-17. A similar procedure can be followed in the turbulent core to find I_{cH}. Then the integrals can be added to give I_H.

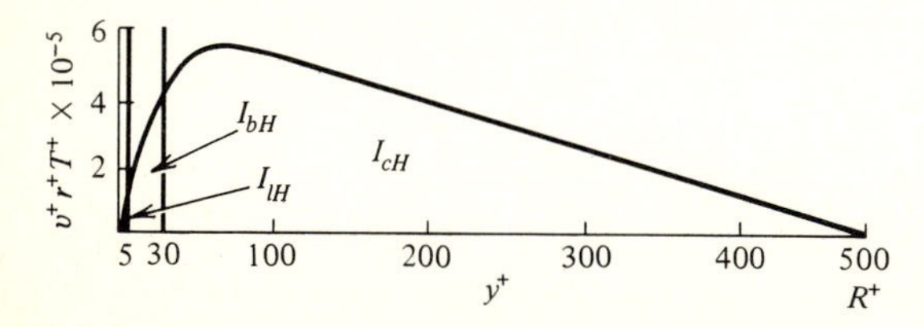

Figure 6-17 Areas show amount contributed by each region to the enthalpy of the fluid stream.

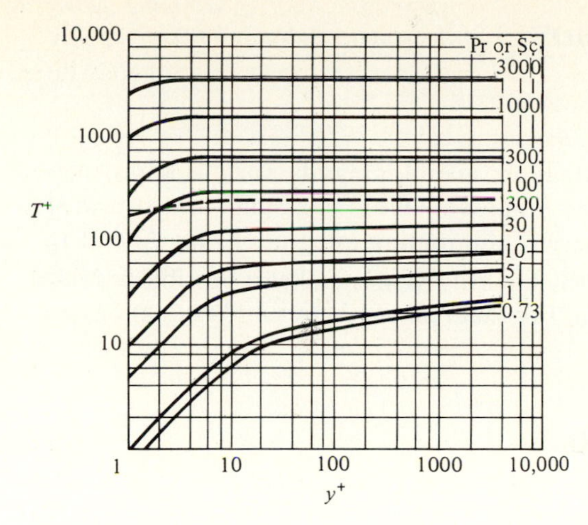

Figure 6-18 Universal temperature profiles: solid curve = present analysis; dashed curve = $\mu^{(t)} = n^2\bar{v}_2 y$. *(From R. G. Deissler, NACA Rep. 1210, 1955.)*

After the integral I_H for a given R^+ and Pr has been evaluated, one would calculate T_b^+ from Eq. (6-120) using the v_b^+ calculated as discussed previously in Eq. (6-65). Then one uses Eq. (6-119) to find Nu for the given R^+. This is repeated at various R^+. The R^+ values are then related to Re by Eq. (6-66), using previously determined expressions for relating f and Re (see Example 6-3). The process can then be repeated for each Pr (or Sc).

Deissler[3] has carried out calculations of Nu like the above for Pr = 0.73 (air) to Pr = 3000, using the expression [Eq. (6-61f)] he developed for $\mu^{(t)}/\mu$ and assuming $\text{Pr}^{(t)} = 1$. His results are shown graphically in Figs. 6-18 and 6-19.

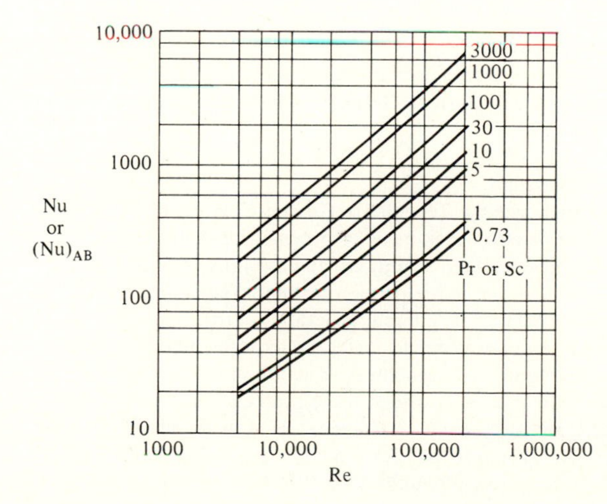

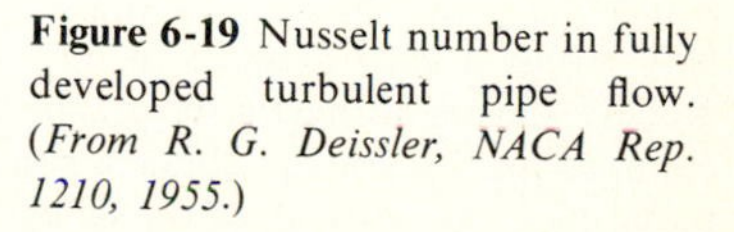
Figure 6-19 Nusselt number in fully developed turbulent pipe flow. *(From R. G. Deissler, NACA Rep. 1210, 1955.)*

6-15 APPROXIMATE METHODS†

Analogies

Because of the complications of evaluating several integrals, various approximations for finding T_b^+ and v_b^+ have been proposed. One is to assume that at $y = y_1^+$, $T^+|_{y_1^+} \equiv T_1^+ = T_b^+$ and $v_b^+ = v_1^+$, thereby eliminating integrations for I_H and I_M. Then the assumed equality of the turbulent diffusivities makes it possible to relate Nusselt numbers to the friction factor. For that reason, the resulting equations are called *analogies*.‡

Generalized von Kármán Analogy

Previously we derived the Reynolds analogy, which holds for $\Pr \approx 1$. We now illustrate how to extend this to other Pr by deriving a generalized von Kármán analogy that reduces to the Reynolds analogy for $\Pr = 1$ [Eq. (6-112)]. Von Kármán used a three-region model in which he obtained the eddy diffusivities in the buffer region by differentiating empirical equations for the velocity distribution $v^+(y^+)$. In the turbulent core he made use of the similarity of the logarithmic temperature and velocity profiles [Eq. (6-107)]. As we have seen, prediction of the temperature profile in the turbulent core can be carried out by neglecting the molecular transport and assuming that $\Pr^{(t)} = 1$. Then, from Eqs. (6-107) to (6-109) and (6-50),

$$T^+ - T_1^+ = v^+ - v_1^+ = \int_{y_1^+}^{y^+} \frac{dy^+}{\nu^{(t)}/\nu} = \int_{y_1^+}^{y^+} \mathcal{R}_{Mc}\, dy^+ \tag{6-128}$$

where $\mathcal{R}_{Mc}$ is the resistance in the core [see Eq. (6-55)].

To find v_1^+ we could integrate Eq. (6-50) between $y^+ = 0$ and $y^+ = y_1^+$. Then

$$v_1^+ = \int_0^{y_1^+} \frac{dy^+}{1 + \nu^{(t)}/\nu} = \int_0^{y_1^+} \mathcal{R}_{Mw}\, dy^+ \tag{6-129}$$

where $\mathcal{R}_{Mw}$ is the resistance in the wall region, which consists of the laminar

†This section may be omitted on first reading.

‡The word "analogy" as used in connection with the Reynolds or Chilton-Colburn analogies has a meaning different from that in Chap. 1. There we were speaking of a mathematical analogy in the sense that the differential equations, flux laws, and boundary conditions differed only in their notation. However, for turbulent momentum and heat transport in a pipe, the differential equations are *not* the same because (1) the momentum equation contains an internal-generation term whereas the heat equation does not and (2) the heat equation rigorously should include a term for bulk flow in the z (axial) direction (since $\overline{T}$ is varying with z) whereas the momentum equation should not because the flow is fully developed and $\overline{v}_z$ is not varying with z. We avoided these problems in the three-region model by assuming that the fluxes were constant and equal to the values at the wall, thereby eliminating the need for balance equations. We then wrote the flux laws for momentum and heat transport and found that *these* were all of the same form, as were the boundary conditions. In this sense, then, an analogy existed.

sublayer and buffer regions. Similarly, Eq. (6-99) gives

$$T_1^+ = \int_0^{y_1^+} \frac{dy^+}{(1/\text{Pr}) + \alpha^{(t)}/\nu} = \int_0^{y_1^+} \mathcal{R}_{Hw}\, dy^+ \tag{6-130}$$

Substituting these equations for T_1^+ and v_1^+ into the left-hand side of Eqs. (6-128) gives

$$T^+ = v^+ + \int_0^{y_1^+} (\mathcal{R}_{Hw} - \mathcal{R}_{Mw})\, dy^+ \tag{6-131}$$

If $v^+(y^+)$ is known, it can be differentiated and substituted into a rearranged Eq. (6-50) to obtain

$$\frac{\nu^{(t)}}{\nu} = \left(\frac{dy^+}{dy^+}\right)^{-1} - 1 \tag{6-131a}$$

Then $\alpha^{(t)}$ can be found by letting $\alpha^{(t)} = \nu^{(t)}$, with which one can evaluate the integrand in Eq. (6-131),

$$\mathcal{R}_{Hw} - \mathcal{R}_{Mw} \equiv \frac{1}{(1/\text{Pr}) + \mu^*} - \frac{1}{1 + \mu^*} = \frac{\text{Pr} - 1}{(1 + \mu^*\, \text{Pr})(1 + \mu^*)} \tag{6-131b}$$

as a function of y^+. Substituting the result into Eq. (6-131) and integrating gives $T^+(y^+)$, which can then be used to find T_b^+.

A great simplification was achieved by von Kármán by assuming that at $y^+ = y_1^+$, $T^+ = T_1^+ = T_b^+$ and $v^+ = v_1^+ = v_b^+$ (see Fig. 6-14). This assumption avoids having to integrate over the cross section to get I_H. With this assumption Eqs. (6-129) and (6-130) can be combined to yield

$$T_b^+ = v_b^+ + \int_0^{y_1^+} (\mathcal{R}_{Hw} - \mathcal{R}_{Mw})\, dy^+$$

In order to compare our result with the Reynolds analogy, we will find the reciprocal Stanton number instead of the Nusselt number. From Eqs. (6-113), (6-76), (6-92), and (6-64b)

$$\frac{1}{\text{St}} \equiv \frac{\rho \widehat{C}_p \bar{v}_b}{h} \equiv \frac{\rho \widehat{C}_p(\overline{T}_b - T_0) v_*}{q_0} \frac{\bar{v}_b}{v_*} \equiv T_b^+ v_b^+$$

Letting $T_b^+ = T_1^+$ and using Eq. (6-130) gives

$$\frac{1}{\text{St}} = v_b^+ \int_0^{y_1^+} \mathcal{R}_{Hw}\, dy^+ = v_b^+ \int_0^{y_1^+} [\mathcal{R}_{Mw} + (\mathcal{R}_{Hw} - \mathcal{R}_{Mw})]\, dy^+$$

$$= v_b^+ \left[v_1^+ + \int_0^{y_1^+} (\mathcal{R}_{Hw} - \mathcal{R}_{Mw})\, dy^+\right]$$

Letting $v_1^+ = v_b^+$ gives

$$\frac{1}{\text{St}} = (v_b^+)^2 + v_b^+ \int_0^{y_1^+} (\mathcal{R}_{Hw} - \mathcal{R}_{Mw})\, dy^+$$

Using $v_b^+ \equiv \sqrt{2/f}$ [Eqs. (6-64*a*), (6-124), and (6-130)] gives

$$(\mathrm{St})^{-1} = \frac{2}{f} + \sqrt{\frac{2}{f}}(T_1^+ - v_1^+) = \frac{2}{f} + \sqrt{\frac{2}{f}}g(\mathrm{Pr}) \tag{6-132}$$

where

$$g(\mathrm{Pr}) \equiv \int_0^{y_1^+} (\mathcal{R}_{Hw} - \mathcal{R}_{Hw})\, dy^+ \tag{6-133}$$

is a function of the Prandtl number and the velocity profile. In the von Kármán analogy Eqs. (6-42), (6-53), and (6-48) were used for $v^+(y^+)$ in Eq. (6-131*a*). When the result was substituted into Eqs. (6-131*b*) and (6-133), we obtained

$$g(\mathrm{Pr}) = 5(\mathrm{Pr} - 1) + \ln \frac{1 + 5\,\mathrm{Pr}}{6}$$

In the Prandtl-Taylor analogy, which was based on a two-region model, the last term is omitted, and in the Reynolds analogy [Eq. (6-112)] $g(\mathrm{Pr}) = 0$. Note that Eq. (6-132) separates the total resistance into the part $2/f$, due to the core (Reynolds analogy), and the remaining part, due to the wall and buffer regions. For mass transfer one would use $g(\mathrm{Sc})$ instead of $g(\mathrm{Pr})$.

Approximate Solution for Large Pr or Sc

For fluids with large molecular Prandtl numbers (or large Schmidt numbers in the case of mass transport), the low thermal conductivity (or diffusivity) will result in a steep temperature gradient at the wall and a relatively flat temperature (or concentration) profile near the center. This means, for example, that the bulk mean temperature of the fluid will be approximately equal to the local temperature in the core. Thus, instead of integrating Eq. (6-120) to obtain T_b^+ one can assume that $T^+ \to T_b^+$ as $y^+ \to \infty$.

For $\mathrm{Pr} > 200$ (or $\mathrm{Sc} > 200$) Deissler obtained

$$\mathrm{Nu} = \frac{0.248}{\pi} \mathrm{Re} \sqrt{\frac{f}{2}} (\mathrm{Pr})^{1/4} \tag{6-134}$$

where Sc replaces Pr for mass transfer. This equation gives results that underestimate the effect of Pr on Nu at large Pr.

In a review article Sherwood[15] indicates that the most important unsolved problem is how to handle systems with large Schmidt numbers such as are encountered in diffusion in liquids, where Sc may be as large as 2000. For such systems the film thickness for mass transfer will be only a few micrometers. Since $L_D^+ \approx 2$, the assumption of no turbulent transport in the sublayer is not valid in this case and eddy diffusion cannot be neglected. Therefore an equation such as (6-61) must be used near the wall. Unfortunately, however, it is not possible to check predicted velocity or concentration profiles with experimental data near the wall because the difficulty of measurement means that very few data are available in that region.

If Eq. (6-61) is assumed to hold throughout the pipe, we can integrate Eq.

(6-99) analytically to obtain an expression for the temperature profile. Then

$$T^+ = \int_0^{y^+} \frac{dy^+}{(1/\text{Pr}) + (y^+/14.5)^3} \tag{6-135}$$

If we then let $y^+ \to \infty$, so as to approximate T_b^+, and if we substitute into Eq. (6-119), we obtain[8]

$$(\text{Nu})_\infty = \frac{9}{14.5(2)(\sqrt{3}\pi)} \text{Re} \sqrt{\frac{f}{2}} (\text{Pr})^{1/3} \tag{6-136}$$

which agrees well with the Chilton-Colburn analogy [Eq. (6-89)] and the data, except that the data indicate that Nu depends on Re $(f/2)$ rather than Re $\sqrt{f/2}$. This discrepancy has been overcome[18] by replacing y^+ with the variable

$$y^{++} \equiv y^+ \sqrt{\frac{f}{2}} \tag{6-137a}$$

and using

$$\frac{\mu^{(t)}}{\mu} = 1.77(y^{++})^3 \tag{6-137b}$$

to obtain

$$(\text{Nu})_\infty = \frac{3\sqrt{3}}{2\pi}(1.77)^3 \left(\text{Re}\frac{f}{2}\right)(\text{Pr})^{1/3} \tag{6-137c}$$

which agrees very well with the data (Fig. 6-20).

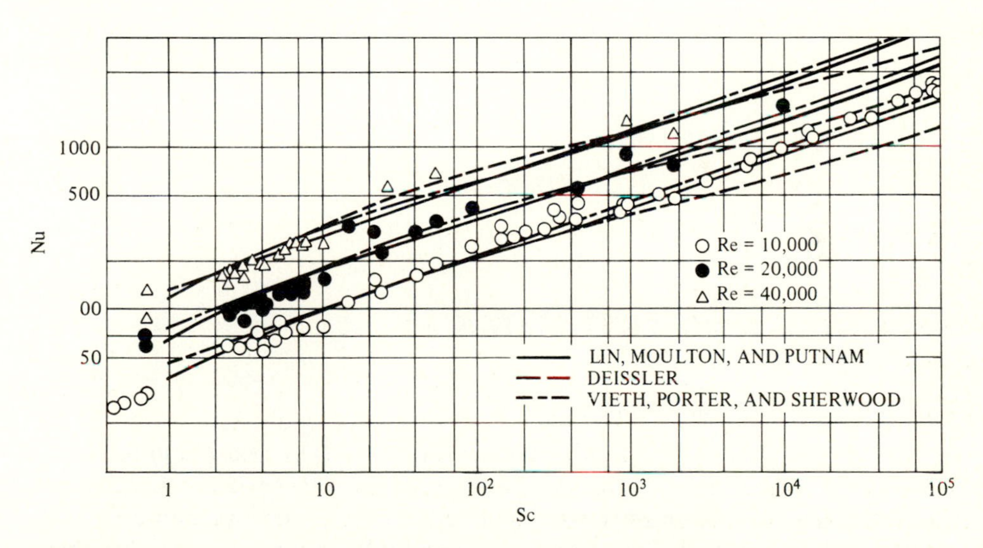

Figure 6-20 Comparison of various correlations with experimental data. Solid curve = Lin, Moulton, and Putnam;[9] dashed curve = Deissler;[3] dotted curve = Vieth, Porter, and Sherwood,[18] [*From D. W. Hubbard and E. N. Lightfoot, Ind. Eng. Chem. Fund.,* **5**:*370 (1966).*]

Low Prandtl Number (Liquid Metals)

For liquid metals, $\Pr \approx 0.01$ (due to the high thermal conductivity), and the assumptions made in obtaining Eqs. (6-134) and (6-136) are no longer valid. The mechanism is primarily one of conduction, and axial transport is important. The heat flux is no longer constant, and the problem becomes two-dimensional. Such problems will be discussed in Chap. 10. Theoretical studies of this case have been made by Martinelli,[11] who extended the von Kármán analogy, but his results required the use of a complex equation and two sets of tables. Lyon[10] has found that the results of Martinelli fit the empirical equation

$$\text{Nu} = (\text{Nu})_0 + 0.025(\text{Pe})^{0.8} \tag{6-138}$$

for $L/D > 60$ and $\text{Pe} \equiv \text{Re}\,\Pr \equiv Dv_b/\alpha > 100$. (As defined more generally in Chap. 5, Pe is a Peclet number.) The expression for Nu for liquid metals contains the constant-intercept term $(\text{Nu})_0$, which Nu approaches as $\text{Pe} \rightarrow 0$. $(\text{Nu})_0$ was found to be 5.0 when the tube wall temperature was held constant and 7.0 when the wall heat flux was constant. This term can be shown to arise as a result of the high thermal conductivity of the metal, which extends the region of molecular conduction to the center of the pipe; thus eddy transport is of less importance, and Eq. (6-99) can be approximated (Chap. 4) as

$$T^+ = \Pr y^+$$

Then if we assume slug flow, i.e., a nearly flat velocity profile in Eq. (6-120), we obtain

$$T_b^+ \approx \frac{\displaystyle\int_0^{R^+} \Pr y^+ r^+ \, dr^+}{\displaystyle\int_0^{R^+} r^+ \, dr^+} = \frac{2R^+ \Pr}{6}$$

Substitution into Eq. (6-119) gives

$$(\text{Nu})_0 \approx 6 \tag{6-139}$$

which compares with values of 5 to 7 given by Lyon.

6-16 OTHER TURBULENT FLOWS

Ducts of Noncircular Cross Section

In the foregoing development of equations for $\bar{v}_z(y)$, $\bar{T}(y)$, and $\bar{C}_A(y)$ for transport to or from the wall of a pipe the curvature at the wall was actually neglected and the fluxes were assumed constant and equal to the values at the wall: τ_0, q_0, and C_{A0}. The shape of the cross section appeared only in the determination of the average velocity, temperature, or concentration, where (in the case of the pipe) integration over the cross section $dA_c = 2\pi r\,dr = -2\pi(R - y)\,dy$ was involved.

Table 6-2 Examples of turbulent flow†

	Example	Problem
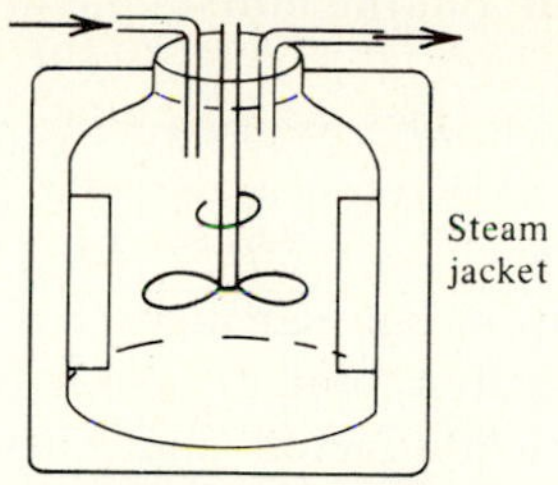	Fluid to be heated in a tank stirred by a paddle wheel	Calculate distribution of heat transfer at wall by predicting flow pattern, distribution of turbulence energy, and length
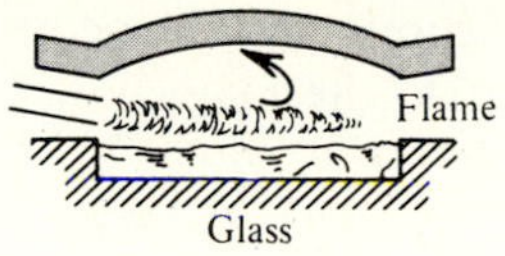	Turbulent oil flame in glass furnace	Predict flame shape, temperature distribution, concentrations of oil droplets and smoke particles, and radiant-heat fluxes to roof and glass (growth and decay of turbulence are controlling processes)
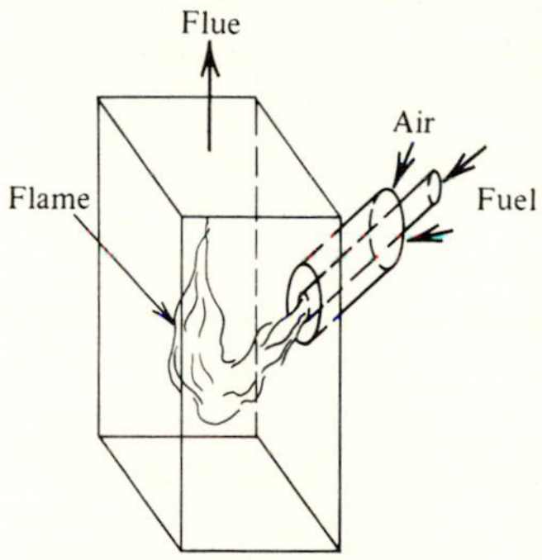	Power-station furnace	Predict the distribution of velocity, temperature, turbulence energy, etc., and hence the heat fluxes to all parts of the wall
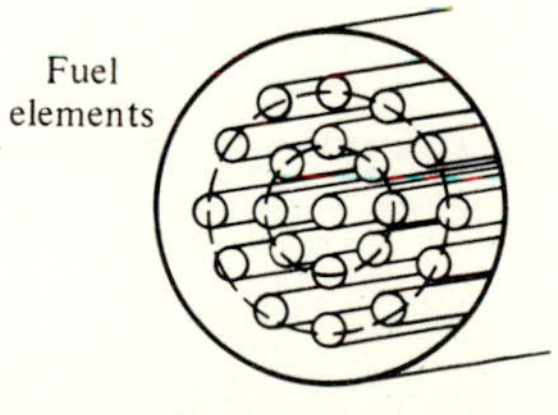	Advanced gas-cooled nuclear reactor	Predict the temperature distribution over the fuel-rod surfaces by calculating the distributions of velocity, turbulence energy, length, etc., in the intervening spaces
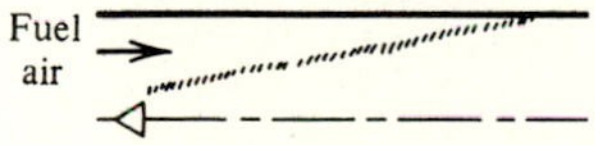	Spread of flame through a combustible mixture flowing along a duct containing a flameholder	Predict the flame angle, distributions of velocity, temperature, etc.
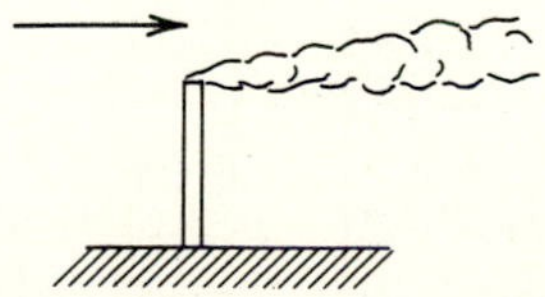	Smoke from a chimney	Predict how wind velocity, effluent temperature, etc., influence the smoke dispersal

†Adapted by permission from B. E. Launder and D. B. Spalding, "Mathematical Models of Turbulence," Copyright 1972, Academic Press Inc. (London) Ltd.

This meant integration from $y = 0$ to $y^+ = R^+$. For a noncircular cross section the same equations for $v^+(y^+)$, etc., can be assumed to hold, but the differential area dA_c and the limit y^+ will differ.

Since the major part of the resistance is near the wall, where curvature is not important, the friction factors for circular and noncircular ducts are similar for similarly sized systems. Thus the use of the equivalent diameter $D_e = 4R_H$, as defined in Chap. 4, gives more reasonable results than in laminar flow.

Deissler and Taylor[4] have used Eq. (6-61f) to calculate v^+ and T^+ in equilateral triangles and square ducts and have predicted friction factors and Nusselt numbers for these geometries as well as for flow between rods in triangular and square arrays. A rigorous analysis of noncircular cross sections would be complicated by the existence of secondary flows, perpendicular to the main flow.

Complex Turbulent Flows

Examples of other turbulent flows are depicted in Table 6-2. In many of them a turbulent-flux-law expression is not valid because the mean velocity gradient cannot be assumed constant over distances of the order of magnitude of the length scale or because there is more than one turbulent length scale and more than one velocity scale. This means that the turbulent flux must be predicted in some other way. An equation can be derived to relate it to other quantities which are also unknown but not to known or calculable quantities. For a discussion of such methods, see Hill.[5] For complex geometries and flow patterns additional equations are needed to describe the flow. For example, a differential equation for the kinetic-energy distribution can be derived and can be used instead of Prandtl's assumption [Eq. (7-32)] to predict the mean velocity fluctuation.[7] In some cases as many as 20 to 28 differential equations have been solved.

APPENDIX 6-1 TIME AVERAGING

Suppose for a fluctuating property A we write

$$A = \bar{A} + A' \tag{1}$$

where the time average is

$$\bar{A} \equiv \lim_{\Delta t \to \infty} \frac{1}{\Delta t} \int_t^{t+\Delta t} A \, dt \tag{2}$$

Then by substituting (1) into (2) we can verify that

$$\bar{A'} = 0 \tag{3}$$

and that

$$\bar{\bar{A}} = \bar{A} \tag{4}$$

If we represent the fluctuation of another quantity B as B', we can use (2) and (3) to show that

$$\overline{\bar{A}B'} = \bar{A}\bar{B}' = 0 \tag{5}$$

In general, however,

$$Q_{AB} \equiv \overline{A'B'} \neq 0 \tag{A6-1-1}$$

The averaged product $\overline{A'B'}$ is called the *correlation* between A and B. If we divide by the rms averages of A' and B', the ratio

$$R_{AB} = \frac{\overline{A'B'}}{(\overline{A'^2})^{1/2}(\overline{B'^2})^{1/2}} \tag{A6-1-2}$$

is obtained. It is called the *correlation coefficient*.

Suppose $A' = kB'$, where k is a constant (see Fig. A6-1-1*a*). This relationship means that a fluctuation in A *always* causes a proportional fluctuation in B. Then

$$R_{AB} = \frac{k\overline{B'^2}}{(\overline{k^2B'^2})^{1/2}(\overline{B'^2})^{1/2}} = \pm 1.0$$

Thus A and B are said to be perfectly correlated; if k is positive, the correlation is positive; and if k is negative, the correlation is negative. In the latter case, an

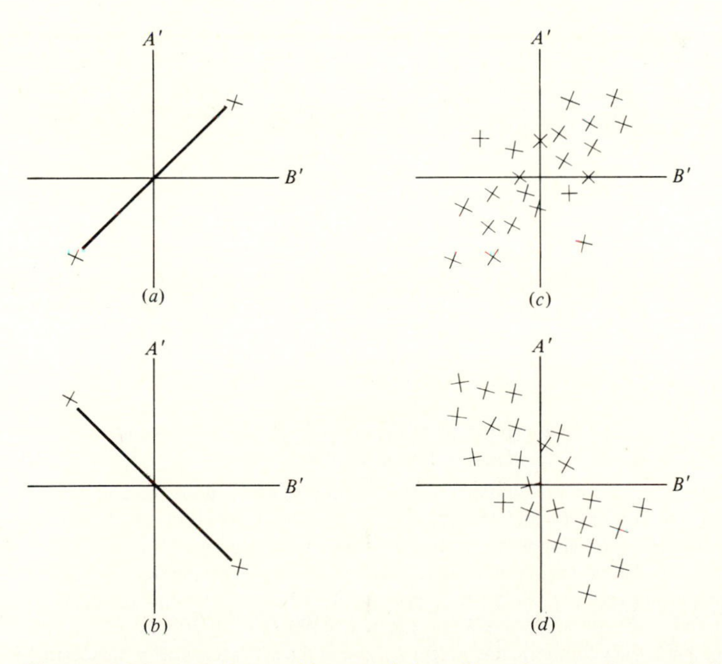

Figure A6-1-1 (*a*) Perfect positive correlation, (*b*) perfect negative correlation, (*c*) imperfect positive correlation, and (*d*) imperfect negative correlation.

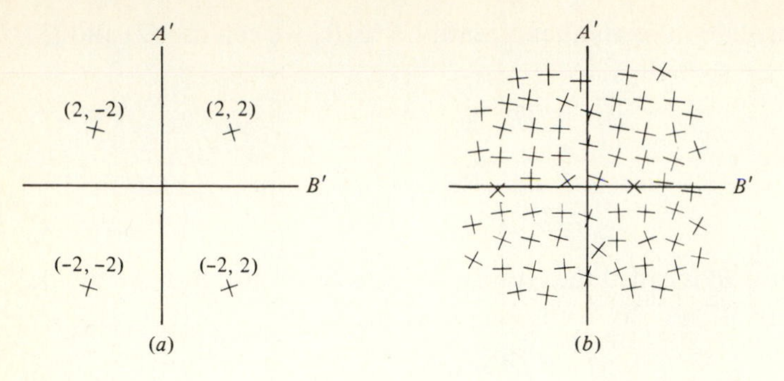

Figure A6-1-2 (*a*) Zero correlation; (*b*) approximate zero correlation.

increase in A' always causes a proportional decrease in B'. These cases are illustrated in Fig. A6-1-1 with two data points shown.

Now suppose we have taken the four data points illustrated in Fig. A6-1-2*a*. As in the previous case, the mean of A' and the mean of B' are both zero. However, the rms value of A' is

$$(\overline{A'^2})^{1/2} = \{\tfrac{1}{4}[(-2)^2 + (2)^2 + (-2)^2 + (2)^2]\}^{1/2} = 2$$

and the same result is obtained for B'. The correlation is

$$Q_{AB} = \{\tfrac{1}{4}[2(2) + (-2)(2) + (-2)(-2) + (-2)(2)]\}^{1/2} = 0$$

and the correlation coefficient is $R_{AB} = 0$. In general if $R_{AB} = 0$, the variables A' and B' are said to be uncorrelated.

In practice the data will not be perfectly correlated or uncorrelated and $-1 \leq R_{AB} \leq +1$. Figures A6-1-1*c* and *d* and A6-1-2 show typical positive, negative, and near zero correlations.

PROBLEMS

6-1 For Reynolds numbers in the range 10^4 to 10^5, experimental data on velocity profiles are roughly correlated by the Blasius equation

$$\bar{v}_z \approx \bar{v}_{\text{max}}\left(1 - \frac{r}{R}\right)^{1/7}$$

Assuming this relation to hold throughout the pipe, show that $\langle \bar{v}_z \rangle / \bar{v}_{\text{max}} \approx 0.81$ in turbulent flow. Compare this with the value of $\langle \bar{v}_z \rangle / \bar{v}_{\text{max}}$ for laminar flow in a tube.

6-2 Water is flowing at 20°C through a horizontal 3-in-ID pipeline under a pressure gradient of 1.242 lb/in^2 per kilometer. It is desired to find the velocity profile and the volumetric flow rate.

(*a*) Find the shear stress and momentum flux at the wall in engineering and SI units.

(*b*) Plot $\bar{v}_z$ in meters per second versus actual distance from the wall y in millimeters, obtaining points at least at the following locations: $y^+ = 5$ [from Eq. (6-42)]; $y^+ = 17.5$ [from Eq. (6-52) or (6-53)]; $y^+ = 30$, $y^+ = 50$, $(30 + R^+)/2$, and $y^+ = R^+$, using Eq. (6-48) at $y^+ \geq 30$.

(*c*) Plot $\bar{v}_z r$ versus y and find the volumetric flow rate in m^3/s and the average velocity. Determine what fraction of the total volumetric flow (area under the curve) occurs in the wall, buffer, and core regions. Illustrate the area under the curve for each region.

(*d*) Calculate the friction factor, using the definition in Chap. 4.

(*e*) Calculate the Reynolds number to check whether the flow is turbulent.

6-3 Using the van Driest equation (6-61*d*) and (6-61*e*), show that

$$v^+ = \int_0^{y^+} \frac{dy^+}{1 + \dfrac{-1 + \{1 + 4(\kappa y^+)^2 [1 - \exp(-y^+/A)^2]\}^{1/2}}{2}}$$

Describe a procedure for finding $v^+(y^+)$ if $A = 26$ and $\kappa = 0.4$.

6-4 For the pipe in Prob. 6-2 use an equation in Sec. 6-10 to calculate $\mu^{(t)}/\mu$ and dv^+/dy^+ at $y^+ = 5$, 17.5, 30, 50, and R^+. Compare your results with values calculated from the equations used in Prob. 6-2. Plot $(1 + \mu^{(t)}/\mu)^{-1}$ versus y^+ and show the areas for each region. Then use a rough numerical or graphical integration to find $v^+(y^+)$.

6-5 For $R^+ = 500$ estimate the error at $y^+ = 5$ and 30 in assuming in Eq. (6-50) that the momentum flux is constant and equal to the value at the wall.

6-6 The Blasius equation for the velocity profile says that the velocity varies as the one-seventh power of the distance from the wall. In dimensionless variables, this relation is

$$v^+ = 8.56(y^+)^{1/7}$$

Show that this equation results in (*a*) $v_b^+ = 8.56(0.817)(R^+)^{1/7}$ and (*b*) $f = 0.0791(\text{Re})^{-0.25}$. *Hint:* Find f^8 and take the eighth *root*.

6-7 Assuming that the velocity profile in each of the three regions can be approximated by a straight line, find an equation relating f and Re only for the conditions of Prob. 6-2 and the following values of y^+ at the boundaries of each region (Fig. 6-9):

v^+	0	5	14	v^+_{max}
y^+	0	5	30	R^+

6-8 Consider steady fully developed turbulent flow, under a pressure difference $p_0 - p_L$, of an incompressible newtonian fluid between two horizontal parallel plates separated by a distance $2B$ such that $B/W \to 0$, where W is the width of each plate, and $B/L \to 0$, where L is the length of each plate.

(*a*) Starting with an incremental momentum balance, derive an equation for $\bar{\tau}_{yz}(y)$, where y is distance from the wall of the lower plate and z is distance along the length of the plate.

(*b*) Assuming that $v^+(y^+)$ is the same as for a pipe [Eq. (6-48)], derive an equation relating f and $\text{Re} = D_{eq}\bar{v}_b/\nu$, where $D_{eq} = 4R_H$.

(*c*) Compare the results above with that obtained in Eq. (6-75) for a pipe if the diameter of the pipe is replaced by $4R_H$ for the slit. Calculate the percentage error at various f.

6-9 Determine whether the von Kármán equation (6-61g) is satisfied by a logarithmic velocity profile.

6-10 The wall of a pipe contains a soluble material A which dissolves in a stream of water flowing through the pipe. The concentration of A in the water at the wall is C_{A0}, and the bulk mean concentration is C_{Ab}.

(*a*) Derive equations for C_A^+ versus y^+ for each of three regions by using the assumptions and definitions used to derive analogous equations for heat transfer. Assume that $J_{A0} \approx N_{A0}$ = total molar flux of A at the wall (neglect bulk flow and assume constant ρ).

(*b*) Derive integral equations for C_{Ab}^+ and $(\text{Nu})_{AB}$.

(*c*) Describe exactly how one could obtain the mass-transfer Nusselt number (1) if equations for the eddy mass and momentum diffusivities as a function of position are known and (2) if an equation only for eddy viscosity is known.

6-11 Consider a pipe like that in Probs. 6-2 and 6-10 coated with benzoic acid, which has a solubility of 0.025 kg mol/m³ and a molecular diffusivity in water of 1.2×10^{-5} cm²/s. The dimensions and other conditions are as in Prob. 6-2.

(*a*) Calculate and plot the dimensionless concentration profiles for the y^+ points used in Prob. 6-2, using the theoretical equations developed in Prob. 6-10 and making any *necessary* assumptions. For comparison, plot $v^+(y^+)$ on the same graph.

(*b*) Plot $C_A^+ r^+$ versus y^+ and show areas under the curve for each region. Then integrate to get C_{Ab}^+.

(*c*) Calculate the Sherwood number, that is, $(\mathrm{Nu})_{AB}$, for the above case and compare with that obtained from the Chilton-Colburn analogy.

(*d*) Calculate the wall flux J_{A0} kg mol/s · m² and compare with that obtained from the Chilton-Colburn analogy.

6-12 Consider heat transport in a pipe heat exchanger at a constant wall temperature of 330 K. Assume that the fluid is water and the conditions are as in Prob. 6-2.

(*a*) Calculate and plot the dimensionless temperature at the same y^+ values as in Prob. 6-2, using the theoretical equations developed in the text and making any *necessary* assumptions. Show $v^+(y^+)$ on the same graph.

(*b*) Plot $T^+ v^+ r^+$ versus y^+ and show areas under the curve for each region. Then integrate to get T_b^+.

(*c*) Calculate the heat-transfer Nusselt number using the theoretical equations developed in the text and compare with that obtained from the Chilton-Colburn analogy.

(*d*) For a section of pipe 10 m long calculate the wall heat flux and compare with the values obtained from the Chilton-Colburn and Reynolds analogies. Assume that the bulk-fluid temperature is 293 K and that $C_{Ab} \approx 0$ (due to the low solubility of A).

6-13 Derive (*a*) Eq. (6-134), (*b*) Eq. (6-136), and (*c*) Eq. (6-139).

REFERENCES

1. Churchill, S. W.: *Chem. Eng.*, Nov. 7, 1977, p. 91.
2. Churchill, S. W., and B. Choi: *AIChE J.*, **19**:196 (1973).
3. Deissler, R. G.: *NACA Rep.* 1210, 1955.
4. Deissler, R. G., and M. F. Taylor: *NASA Rep.* R-31, 1959.
5. Hill, J. C.: *Chem. Eng. Educ.*, **13**:34 (1979).
6. Hughmark, G. A.: *AIChE J.*, **17**:902 (1971).
7. Launder, B. E., and D. B. Spalding: "Mathematical Models of Turbulence," Academic Press, London, 1972.
8. Lightfoot, L. N.: *AIChE Cont. Educ. Ser.* 4, 1969.
9. Lin, C. S., R. W. Moulton, and G. R. Putnam: *Ind. Eng. Chem.*, **45**:636 (1953).
10. Lyon, R. N.: *Chem. Eng. Prog.*, **47**:75 (1951).
11. Martinelli, R. C.: *Trans ASME*, **49**:947 (1947).
12. Notter, R. H., and C. A. Sleicher: *Chem. Eng. Sci.*, **26**:161 (1971).
13. Pinho, M., and R. W. Fahien: *AIChE J.*, **27**:170 (1981).
14. Reichardt, H.: *Angew. Math.*, **31**:208 (1951).
15. Sherwood, T. K.: *Chem. Eng. Educ.*, **8**:204 (1974).
16. Tennekes, H., and J. L. Lumley: "A First Course in Turbulence," M.I.T. Press, Cambridge, Mass., 1972.
17. Van Driest, E. R.: *J. Aero. Sci.*, **23**:1007 (1952).
18. Vieth, W. R., J. H. Porter, and T. K. Sherwood: *Ind. Eng. Chem.*, **2**:1 (1963).
19. Welty, J. R.: "Engineering Heat Transfer/SI Version," Wiley, New York, 1978.

CHAPTER

SEVEN

UNSTEADY-STATE TRANSPORT

In this chapter we return to the basic models discussed in Chap. 1, namely, heat transport in a slab heated at one side, momentum transport in Couette flow, and mass transport with no bulk flow, but this time we derive and solve the partial differential equations for the unsteady state.

7-1 UNSTEADY-STATE HEAT TRANSPORT

Again consider heat transport in a metal slab of thickness L_x. At time $t = 0$ the slab is at uniform temperature T_2. At $t \geq 0$ the lower surface at $x = 0$ is maintained at a constant temperature T_1 while the upper surface is maintained at a constant temperature T_2. At any position x the heat flux and the temperature within the slab will depend upon time. We must therefore make our energy balance not only over an increment of volume $A_x \, \Delta x$ but also over an increment of time Δt.

We are especially interested in the accumulation term, which is not zero at unsteady state. It is the total rate of accumulation of energy in the element over a time increment Δt. According to thermodynamics, $\hat{U}$, the internal energy per unit mass of a pure substance, can be expressed as a function of temperature and volume. Since for a solid the effect of volume can be neglected, the net change in internal energy per unit mass for a given temperature change ΔT is

$$\Delta \hat{U} = \int_T^{T+\Delta T} \hat{C}_v \, dT \approx \hat{C}_v \, \Delta T$$

where $\widehat{C}_v$, the specific heat at constant volume, is taken as constant (or at its average value over the temperature range involved). Furthermore, for a solid, it is convenient to use the approximation $\widehat{C}_v \approx \widehat{C}_p$, since values of the latter are more readily available.

For the increment of slab of volume $A_x \Delta x$ the energy accumulation is $\rho A_x \Delta x \widehat{C}_p \Delta T$, and the accumulation *rate* is this quantity divided by Δt. Then, since there is no generation, our incremental balance is

$$\underset{\text{Rate in}}{q_x A_x \Big|_x} - \underset{\text{Rate out}}{q_x A_x \Big|_{x+\Delta x}} \equiv \underset{\text{Net rate in}}{-\Delta(q_x A_x)} = \underset{\text{Accumulation rate}}{\frac{\rho A_x \Delta x \widehat{C}_p \Delta T}{\Delta t}} \tag{7-1}$$

where the net rate has been written in terms of the difference operator Δ. Dividing by $A_x \Delta x$, we have

$$-\frac{\Delta q_x}{\Delta x} = \rho \widehat{C}_p \frac{\Delta T}{\Delta t} \tag{7-2}$$

We now take the limit of both sides of this equation as $\Delta x \to 0$, holding t constant.

$$-\lim_{\Delta x \to 0} \frac{\Delta q_x}{\Delta x} = \lim_{\Delta x \to 0} \rho \widehat{C}_p \frac{\Delta T}{\Delta t}$$

or

$$-\frac{\partial q_x}{\partial x} = \rho \widehat{C}_p \frac{\Delta T}{\Delta t} \tag{7-3}$$

since $\Delta T/\Delta t$ does not change as Δx becomes increasingly smaller.

Repeating the limiting process for $\Delta t \to 0$, holding x constant, gives

$$\lim_{\Delta t \to 0} -\frac{\partial q_x}{\partial x} = \lim_{\Delta t \to 0} \rho \widehat{C}_p \frac{\Delta T}{\Delta t}$$

$$\boxed{-\frac{\partial q_x}{\partial x} = \rho \widehat{C}_p \frac{\partial T}{\partial t}} \tag{7-4}$$

Note that we obtain partial derivatives since there are two independent variables.

We can now substitute the flux law into Eq. (7-4). If k is constant, we obtain

$$k \frac{\partial^2 T}{\partial x^2} = \rho \widehat{C}_p \frac{\partial T}{\partial t}$$

or

$$\boxed{\alpha \frac{\partial^2 T}{\partial x^2} = \frac{\partial T}{\partial t}} \tag{7-5}$$

Equation (7-5) is frequently referred to as the *one-dimensional diffusion equation.*

Solution of Partial-Differential Equation for Unsteady-State Heat Transport in a Slab

We now discuss the techniques involved in obtaining a solution to Eq. (7-5). The actual details of the solution are worked out in Appendix 7-1.

In solving partial differential equations it is often convenient to use dimensionless variables defined so that they are equal to unity or to zero at the boundaries of the system (or at the initial conditions). For example, if we define a dimensionless temperature θ

$$\theta \equiv \frac{T - T_2}{T_1 - T_2} \tag{7-6}$$

the value of θ will be unity at $x = 0$ and zero at $x = L_x$; it will also be zero at $t = 0$. We could accomplish the desired result if we defined a variable $\phi \equiv 1 - \theta$, but the mathematical solution turns out to be easier if we use θ because the problem then falls into a class for which there are general proofs and solution procedures.

Our first step is to transform from the variable T in the differential equation to the variable θ. Since $T = T_2 + \theta(T_1 - T_2)$, differentiation gives

$$\frac{\partial T}{\partial t} = 0 + \frac{\partial \theta}{\partial t}(T_1 - T_2) \qquad \frac{\partial T}{\partial x} = 0 + \frac{\partial \theta}{\partial x}(T_1 - T_2) \qquad \frac{\partial^2 T}{\partial x^2} = \frac{\partial^2 \theta}{\partial x^2}(T_1 - T_2)$$

Then from Eq. (7-5)

$$\alpha(T_1 - T_2)\frac{\partial^2 \theta}{\partial x^2} = (T_1 - T_2)\frac{\partial \theta}{\partial t}$$

or

$$\alpha \frac{\partial^2 \theta}{\partial x^2} = \frac{\partial \theta}{\partial t} \tag{7-7}$$

It is also possible, and in fact convenient, to transform the variables x and t into dimensionless variables. This will be done later. At this time we use x and t to help us see the physical picture and find the appropriate dimensionless variables to use.

Our next step is to recall that this problem has a steady-state solution, Eq. (1-10). Let us denote the dimensionless steady-state solution by $\theta_\infty(x)$ and the dimensional form by $T_\infty(x)$. In terms of θ, the steady-state solution obtained from Eq. (1-10*c*) is

$$\theta_\infty = 1 - \frac{x}{L_x} \tag{7-8}$$

The general solution for $\theta(x,t)$ can then be written in the form

$$\theta(x,t) = \theta_\infty(x) - \theta_t(x,t) \tag{7-9}$$

where $\theta_t(x,t)$ is the *transient* contribution to θ; it must go to zero as $t \to \infty$, since

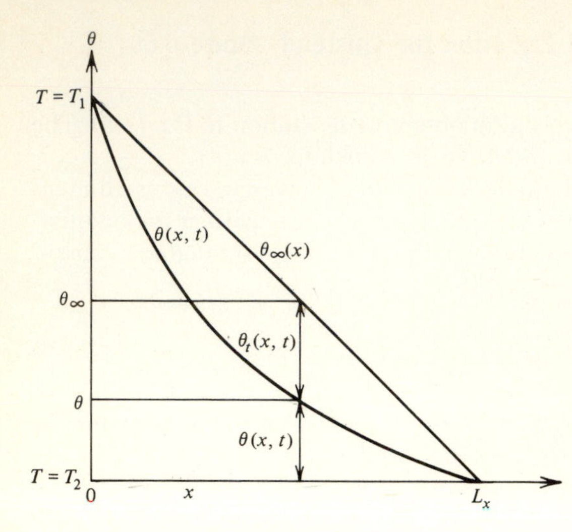

Figure 7-1 Dimensionless temperature versus distance in slab.

then $\theta(x,t) = \theta_\infty(x)$. Rewriting Eq. (7-9) in the form

$$\theta_t(x,t) = \theta_\infty(x) - \theta(x,t) \tag{7-10}$$

we see that $\theta_t(x,t)$ is a measure of the unaccomplished temperature change. If we draw sketches of θ as a function of x at an arbitrary time and of θ_∞ as a function of x, we see that θ_t is the difference between curves θ_∞ and θ, as illustrated in Fig. 7-1.

Let us now prepare a list of boundary and initial conditions (Table 7-1) on all the independent and dependent variables. In preparing this table we need conditions for each of the independent variables t and x *at constant values of the other variable*. The initial condition is that at $t = 0$ the temperature is T_2 at *all* values of x except at $x = 0$, where it is T_1. Then, from its definition, θ is zero for $x \geq 0$. The steady-state solution does not depend upon time, so we merely write it as $\theta_\infty(x)$. The transient function $\theta_t(x, t)$, by its definition [Eq. (7-9)], must therefore be equal to θ_∞ at $t = 0$.

The first boundary condition tells us that at $x = 0$ the temperature is equal to T_1 at *all* values of $t \geq 0$. Then by its definition θ is equal to unity, as is θ_∞; hence θ_t must be $1 - 1 = 0$. The second boundary condition similarly says that at $x = L_x$, $T = T_2$ for *all* values of $t \geq 0$. Then by definition $\theta = 0$, and since θ_∞ is also zero, θ_t must be zero.

Table 7-1 Table of conditions

Condition	t	x	T	θ	θ_∞	θ_t
Initial	0	>0	T_2	0	θ_∞	θ_∞
Boundary 1	≥ 0	0	T_1	1	1	0
Boundary 2	≥ 0	L_x	T_2	0	0	0

We now derive a partial differential equation for θ_t by substituting Eq. (7-9) into Eq. (7-7)

$$\alpha\left(\frac{d^2\theta_\infty}{dx^2} - \frac{\partial^2\theta_t}{\partial x^2}\right) = 0 - \frac{\partial\theta_t}{\partial t} \tag{7-11}$$

But at steady state, Eq. (7-7) becomes

$$\frac{d^2\theta_\infty}{dx^2} = 0 \tag{7-12}$$

whose solution is given by Eq. (7-8). Subtracting Eq. (7-11) from Eq. (7-12) gives

$$\alpha\frac{\partial^2\theta_t}{\partial x^2} = \frac{\partial\theta_t}{\partial t} \tag{7-13}$$

which is the desired equation for θ_t. This equation is solved in Appendix 7-1, and the resulting solution is

$$\theta_t = \sum_{n=1}^{\infty}\frac{2}{n\pi}\left(\sin\frac{n\pi x}{L_x}\right)\exp\left(-\frac{n^2\pi^2\alpha t}{L_x^2}\right) \tag{7-14}$$

Thus, from Eqs. (7-8) and (7-9),

$$\boxed{\theta \equiv \frac{T - T_2}{T_1 - T_2} = 1 - \frac{x}{L_x} - \sum_{n=1}^{\infty}\frac{2}{n\pi}\left(\sin\frac{n\pi x}{L_x}\right)\exp\left(-\frac{n^2\pi^2\alpha t}{L_x^2}\right)} \tag{7-15}$$

7-2 DIMENSIONLESS VARIABLES IN UNSTEADY-STATE HEAT TRANSPORT

Dimensionless-Variables Solution

The solution (7-15) to the partial differential equation for heat transport in a slab can be placed in dimensionless form by letting

$$\eta \equiv \frac{x}{L_x} \qquad \text{and} \qquad s \equiv \frac{\alpha t}{L_x^2} \tag{7-16}$$

where s is a dimensionless time called the *Fourier number:*

$$\theta(\eta, x) = 1 - \eta - \sum_{n=1}^{\infty}\frac{2}{n\pi}(\sin n\pi\eta)e^{-n^2\pi^2 s} \tag{7-17}$$

Hence we see that certain dimensionless groups arise quite naturally. In this case we have reduced the number of variables and parameters involved to three (θ, η, s) from an original total of seven $(T, x, t;\ T_1, T_2, \alpha, L_x)$. Hence a single plot (Fig. 7-1) of dimensionless temperature versus dimensionless position at constant values of dimensionless time will describe the relationship, whereas it could have taken several plots to do so in terms of dimensional variables.†

In cases where we cannot solve the partial differential equations and must obtain a correlation from experimental data we need fewer data to obtain a correlation between three variables than if we had to correlate six or seven dimensional variables. Furthermore, we can take our measurements using an inexpensive material at easily achieved conditions and use them to obtain results for more expensive materials or at hazardous conditions. In addition, we will see that dimensionless solutions to heat-transport problems will also apply to analogous mass- or momentum-transport problems.

Because of the many advantages of using dimensionless variables, it is common practice in the literature to write the original partial differential equation and its boundary conditions in dimensionless form and to obtain the solution *directly* in terms of dimensionless variables (see Appendix 4-2). Of course, the question arises as to how one can determine *a priori* which dimensionless variables to use without the benefit of the solution, but in practice the variables are usually suggested by the differential equation and the boundary conditions. For example, if we want the boundary conditions on temperature to involve numbers such as 0 and 1, our choice might be to let $\theta \equiv (T - T_2)/(T_1 - T_2)$. However, we also *could* use

$$\phi \equiv 1 - \theta = \frac{(T_1 - T_2) - (T - T_2)}{T_1 - T_2} = \frac{T_1 - T}{T_1 - T_2} \tag{7-20}$$

a choice which would eliminate 1 in the solution and change some signs.

It is quite natural to select as a dimensionless distance $\eta = x/L_x$ since the obvious characteristic length is L_x, and this choice gives $\eta = 1$ at $x = L_x$. The choice of a dimensionless time seems less obvious, but we might arrive at it by recognizing that the thermal diffusivity should be involved and that it has the units of (length)^2 per unit time. Hence L_x^2/α would represent a characteristic time. Otherwise we could write the differential equation first in terms of a dimension*al* time but a dimension*less* length and temperature and determine what would be needed to make time dimensionless. This procedure is illustrated below.

†Actually the number of variables and parameters can be reduced to five by letting $T' \equiv T - T_2$ and $T_1' \equiv T_1 - T_2$. Then the solution in dimensional variables would be of the form

$$T'(x, t; T_1', \alpha, L_x) \tag{7-18}$$

This is permitted since the result is independent of the temperature *level*. However, if the thermal conductivity were permitted to vary with T (as it does, but mildly), the level *would* be involved. Then the dimensionless groups might be taken as

$$\frac{T}{T_1} = \frac{T}{T_1}\left(\frac{x}{L_x}; \frac{\alpha t}{L_x^2}; \frac{T_2}{T_1}\right) \tag{7-19}$$

and three groups would be needed.

Diffusion Equation in Dimensionless Form

To illustrate the principles involved consider the equation developed earlier [Eq. (7-7)]

$$\alpha \frac{\partial^2 \theta}{\partial x^2} = \frac{\partial \theta}{\partial t}$$

Now let

$$\frac{\partial^2 \theta}{\partial x^2} = \frac{\partial}{\partial x}\left(\frac{\partial \theta}{\partial x}\right)$$

Then

$$\frac{\partial \theta}{\partial x} = \frac{\partial \theta}{\partial \eta}\frac{d\eta}{dx} = \frac{\partial \theta}{\partial \eta}\frac{1}{L_x}$$

since $\eta = x/L_x$. Also

$$\frac{\partial^2 \theta}{\partial x^2} = \left[\frac{\partial}{\partial \eta}\left(\frac{\partial \theta}{\partial x}\right)\right]\frac{d\eta}{dx} = \frac{\partial^2 \theta}{\partial \eta^2}\frac{1}{L_x^2}$$

With this substitution the equation becomes

$$\frac{\alpha}{L_x^2}\frac{\partial^2 \theta}{\partial \eta^2} = \frac{\partial \theta}{\partial t}$$

We can see that a logical choice for dimensionless time would be $s = \alpha t/L_x^2$, as before. Then,

$$\frac{\partial \theta}{\partial t} = \frac{\partial \theta}{\partial s}\frac{ds}{dt} = \frac{\partial \theta}{\partial s}\frac{\alpha}{L_x^2}$$

Substituting, we get

$$\frac{\partial^2 \theta}{\partial \eta^2} = \frac{\partial \theta}{\partial s} \tag{7-21}$$

Actually, since the parameters α, L_x, T_1, and T_2 are all constants, we can write an equation in dimensionless form more readily by merely multiplying the differential equation by the necessary terms and consolidating those terms which make up a dimensionless group. For example, if we start with Eq. (7-7) in terms of θ, we would need to multiply it by L_x^2/α to obtain

$$\frac{\partial^2 \theta}{\partial (x/L_x)^2} = \frac{\partial \theta}{\partial (t\alpha/L_x^2)} \quad \text{or} \quad \frac{\partial^2 \theta}{\partial \eta^2} = \frac{\partial \theta}{\partial s} \tag{7-22}$$

This procedure would have been used to go from T to θ, but we would first need to subtract T_2 and recognize that its derivatives are zero. We would then divide $T - T_2$ by $T_1 - T_2$. The boundary and initial conditions can be written in tabular form (Table 7-2). The equation could be solved by the same procedures used previously, but the solution would be in terms of $\theta(\eta, s) \equiv \theta_\infty(\eta) - \theta_t(\eta, s)$.

Table 7-2 Table of conditions

Condition	t	x	T	s	η	θ	θ_∞	θ_t
Initial	0	>0	T_2	0	>0	0	θ_∞	θ_∞
Boundary 1	≥ 0	0	T_1	≥ 0	0	1	1	0
Boundary 2	≥ 0	L_x	T_2	≥ 0	1	0	0	0

7-3 UNSTEADY-STATE MASS OR MOMENTUM TRANSPORT

For the analogous mass-transport problem, the accumulation term for mass (moles of A) is an element of volume $A_x \, \Delta x$ would be $\Delta C_A \, A_x \, \Delta x$. The resulting differential equations in terms of flux and concentration, *neglecting bulk flow,* are

$$\frac{-\partial J_{Ax}}{\partial x} = \frac{\partial C_A}{\partial t} \qquad \text{and} \qquad D_{AB} \frac{\partial^2 C_A}{\partial x^2} = \frac{\partial C_A}{\partial t} \tag{7-23}$$

For the *momentum*-transport case, the accumulation term for y momentum is $\rho \, \Delta v_y \, A_x \, \Delta x$, and the appropriate differential equations are

$$\frac{-\partial \tau_{xy}}{\partial x} = \rho \frac{\partial v_y}{\partial t} \qquad \text{and} \qquad \nu \frac{\partial^2 v_y}{\partial x^2} = \frac{\partial v_y}{\partial t} \tag{7-24}$$

We will distinguish between the various transport processes by using the dimensionless variables

$$\theta_H = \frac{T - T_2}{T_1 - T_2} \qquad \theta_D = \frac{C_A - C_{A2}}{C_{A1} - C_{A2}} \qquad \theta_M = \frac{v_y}{V} \tag{7-25}$$

$$s_H = \frac{\alpha t}{L_x^2} \qquad s_D = \frac{D_{AB} t}{L_x^2} \qquad s_M = \frac{\nu t}{L_x^2} \tag{7-26}$$

The general solution, applicable to all three processes, is then

$$\theta(\eta, s) = 1 - \eta - \sum_{n=1}^{\infty} \frac{2}{n\pi} (\sin n\pi\eta) \, e^{-n^2\pi^2 s} \tag{7-17}$$

For heat transport we let $\theta = \theta_H$, $s = s_H$, etc.

The student should summarize the derivation of these equations for each case (see the Problems).

7-4 MACROSCOPIC BALANCE OVER SLAB

If we multiply the flux differential equation (7-4) by the volume $A_x \, dx$ and integrate over all x from $x = 0$ to L_x, we obtain an overall energy rate balance

$$-\int_0^{L_x} \frac{\partial q_x}{\partial x} A_x\, dx = \int_0^{L_x} \rho \widehat{C}_p \frac{\partial T}{\partial t} A_x\, dx = \rho \widehat{C}_p A_x \int_0^{L_x} \frac{\partial T}{\partial t} dx$$

$$= A_x \rho \widehat{C}_p \frac{d}{dt} \int_0^{L_x} T\, dx$$

since the limits of integration are constant (see Appendix 7-4). Integrating the first term and using an average temperature gives

$$A_x q_x \Big|_{x=0} - A_x q_x \Big|_{L_x} = \rho \widehat{C}_p A_x L_x \frac{d}{dt} \langle T \rangle \qquad \langle T \rangle \equiv \frac{1}{L_x} \int_0^{L_x} T\, dx \tag{7-27}$$

Heat flow in at $x = 0$	Heat flow out at $x = L_x$	Total rate of accumulation of energy in slab

where $\langle T \rangle$ is the average temperature in the slab (see Appendix 1-4). If we define a dimensionless heat flux as

$$\mathcal{J}_H \equiv \frac{q_x}{q_x|_{t=\infty}} \equiv \frac{q_x L_x}{(T_1 - T_2)k} \tag{7-28}$$

the macroscopic balance can be written

$$\boxed{\mathcal{J}_H \Big|_{\eta=0} - \mathcal{J}_H \Big|_{\eta=1} = \frac{d\langle \theta_H \rangle}{ds_H}} \tag{7-29}$$

Similarly for mass and momentum transfer we could write

$$\mathcal{J}_D \equiv \frac{N_{Ax} L_x}{(C_{A1} - C_{A2}) D_{AB}} \qquad \text{and} \qquad \mathcal{J}_M \equiv \frac{\tau_{xy} L_x}{V \mu} \tag{7-30}$$

and then merely replace $\mathcal{J}_H$, θ_θ, and s_H with their corresponding values. By the same procedure, a dimensionless *flux* law can be written

$$\mathcal{J} = -\frac{\partial \theta}{\partial \eta} \tag{7-31}$$

7-5 USE OF ANALOGY

Example 7-1 When an analytical solution is not available, it is often possible to take data, say, for the heat-transport case, and use them to predict the results for momentum or mass transport. Suppose we were unable to obtain an analytical solution to the unsteady-state heat-diffusion equation (7-7) for the slab held at T_1 at $x = 0$ and at T_2 at $x = L_x$. But suppose we did have a Couette viscometer (with narrow gap so that it approximated two parallel plates; see Sec. 1-13) and we *were* able to measure, as a function of time, the force F_1 required to move the lower plate with a steady velocity v_1 while the upper plate is stationary. Show how to use the data above to predict, as a function of time, the rate of heat flow into the slab $\dot{Q}_1$ if T_1 and T_2 are constant.

SOLUTION We first recognize that $F_1 g_c \equiv \dot{P}_1$, the rate of flow of momentum into the fluid, is analogous to $\dot{Q}_1$, the rate of heat flow into the slab. Also, v_1 is analogous to $T_1 - T_2$, and $T - T_2$ is analogous to v_y. We note that the flux differential equations (7-4), the first of (7-24), and the flux laws are analogous as are the initial conditions. Therefore the problems are analogous and in terms of dimensionless variables the solutions will be identical. In general we would next nondimensionalize the differential equations and boundary conditions to determine whether any new dimensionless parameters appear. In this case, however, we have already done so, with the result that $\theta = \theta(\eta, s)$, where $\eta \equiv x/L_x$ and $s = \alpha t/L_x^2$. Then the heat flow rate into the slab is

$$\dot{Q}_1 = A_x q_x \Big|_{x=0} \qquad \text{or} \qquad \dot{Q}_1 = -kA_x \frac{\partial T}{\partial x}\Big|_{x=0} = -kA_x \frac{T_1 - T_2}{L_x} \frac{\partial \theta_H}{\partial \eta}\Big|_{\eta=0}$$

In dimensionless form this becomes

$$\frac{\dot{Q}_1 L_x}{kA_x(T_1 - T_2)} \equiv \mathfrak{F}_H = -\frac{\partial \theta_H}{\partial \eta}\Big|_0 \tag{7-32}$$

where $\mathfrak{F}_H$ is a dimensionless heat flow rate. Likewise

$$\dot{P}_1 \equiv \tau_{xy} A_x \Big|_{x=0} = -\mu A_x \frac{\partial v_y}{\partial x}\Big|_{x=0} = -\mu A_x \frac{v_1}{L_x} \frac{\partial \theta_M}{\partial \eta}\Big|_{\eta=0} \tag{7-33}$$

or in dimensionless form

$$\frac{\dot{P}_1 L_x}{\mu A_x v_1} \equiv \mathfrak{F}_M = -\frac{\partial \theta_M}{\partial \eta}\Big|_0 \tag{7-34}$$

Now since $\theta_M(\eta, s_M) \equiv \theta_H(\eta, s_H)$, the gradients at $\eta = 0$ must also be equal. Thus

$$\mathfrak{F}_M(s_M) = \mathfrak{F}_H(s_H)$$

Then if we measure $F_1(t)$, we can convert it into $\mathfrak{F}_M(s_M)$ by using

$$\mathfrak{F}_M = \frac{F_1 g_c L_x}{\mu A_x v_1} \qquad \text{and} \qquad s_M \equiv \frac{\nu t}{L_x^2}$$

Now $\mathfrak{F}_M$ must equal $\mathfrak{F}_H$, from which we get

$$\dot{Q}(t) = \mathfrak{F}_H(s_H) \frac{kA_x(T_1 - T_2)}{L_x} \qquad \text{where } t = s_H \frac{L_x^2}{\alpha}$$

In some cases, the macroscopic balance can be used to relate macroscopic quantities so that we need not measure the same quantity for each process; e.g., if we wished to know the average velocity as a function of time in the Couette-flow case, we could *measure* the heat in and heat out in a slab, from which the dimensionless fluxes at $\eta = 0$ and $\eta = 1$ could be calculated from Eq. (7-28). Then the dimensionless macroscopic balance [Eq. (7-29)] could be used to obtain a plot of $d\langle\theta_H\rangle/ds_H$. This plot could be integrated numerically to obtain $\langle\theta_H\rangle$ versus s_H, which is equivalent to $\langle\theta_M\rangle$ versus s_M. From this $\langle v_y\rangle$ versus t could be calculated. □

7-6 UNSTEADY-STATE HEAT TRANSPORT IN A SLAB COOLED ON BOTH SIDES

Example 7-2 A long, thin slab of metal initially at temperature T_0 is suddenly subjected on two opposite sides to a colder temperature T_1 such that heat is conducted in an x direction from the interior of the slab to the sides, which remain at $T = T_1$ at all times. By symmetry we let the location of these sides be $x = +a$ and $x = -a$, respectively, so that $x = 0$ is the centerline between them. Let L be the length of the slab in the z direction and W be the width in the y direction. (*a*) Sketch and label curves of T versus x for $t = 0$, $t = t_1$, $t = t_2$ ($>t_1$), and $t = \infty$. (*b*) Using an increment of width Δx, write an energy balance. (*c*) Derive the partial differential equation for the flux. (*d*) Write the flux law and use it to obtain the partial differential equation for temperature. (*e*) Write the above equation, the boundary conditions, and the initial conditions in dimensionless form. (*f*) Obtain the solution.

SOLUTION (*a*) See Fig. 7-2. At $t = 0$, we have $T = T_0$ at all x except $x = \pm a$, where $T = T_0$. This sharp (∞) gradient results in rapid heat transfer, which de-

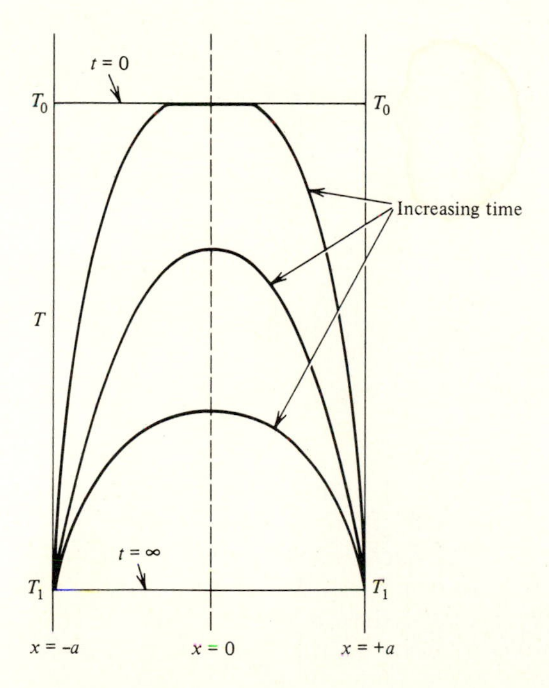

Figure 7-2 Temperature in a slab.

creases T near the surface ($t = t_1$). At $t = t_2$ the center has also cooled. At $t = \infty$ the temperature is T_1 everywhere.

(*b*) to (*d*) The same procedure is used as in Sec. 7-1, and the same equation results

$$\alpha \frac{\partial T}{\partial x^2} = \frac{\partial T}{\partial t} \tag{7-5}$$

(*e*) We use the definitions

$$\eta \equiv \frac{x}{a} \qquad s \equiv \frac{\alpha t}{a^2} \qquad \theta \equiv \frac{T - T_1}{T_0 - T_1} \tag{7-35}$$

where θ is the fraction unaccomplished temperature change. This definition of θ results in the steady-state solution θ_∞. Then to make Eq. (7-5) dimensionless we must multiply both sides by $a^2/\alpha(T_0 - T_1)$, resulting in

$$\frac{\partial^2 \theta}{\partial \eta^2} = \frac{\partial \theta}{\partial s}$$

The initial condition and the boundary conditions are

Condition	t	x	T	s	η	θ
Initial	0	$\neq \pm a$	T_0	0	$\neq \pm 1$	1
Boundary 1	$t \geq 0$	$-a$	T_1	$t \geq 0$	-1	0
Boundary 2	$t \geq 0$	$+a$	T_1	$t \geq 0$	$+1$	0

(*f*) See Appendix 7-2 for method used to obtain the solution

$$\theta = \sum_{n=0}^{\infty} A_n \exp(-\gamma_n^2 s) \cos \gamma_n \eta \tag{7-36}$$

where $$A_n = \frac{2(-1)^n}{(n + \frac{1}{2})\pi} \qquad \text{and} \qquad \gamma_n = (n + \tfrac{1}{2})\pi \tag{7-37}$$

□

7-7 UNSTEADY-STATE TRANSPORT WITH EXTERNAL RESISTANCE

For the problem of unsteady-state heat transfer in a slab cooled from both sides (Example 7-2) we used the boundary condition that the temperature at each boundary ($x = \pm a$) was constant at $T = T_1$. That boundary condition, however, applies only in the special case where there is no resistance at the surface, i.e., when the temperature of the surface T_1 is equal to the temperature of the ambient medium T_a. In practice, however, this is not always the case, and the resistance of the film surrounding the slab must be considered. Customary practice does this by using the *radiation boundary condition,* based on Eq. (4-26), or

$$q_s = h(T_1 - T_a) \tag{7-38}$$

where q_s is the heat flux at the surface and h is a heat-transfer coefficient whose reciprocal is a measure of the resistance to heat transfer in the external film. Thus, if $h \to \infty$, $T_1 - T_a \to 0$ or $T_1 \to T_a$, as in Example 7-1. But if $h \to 0$, $q_s \to 0$ and we have the case of a perfect insulator.

The analytical solution of the heat-conduction equation (7-5) with the above boundary condition can be carried out after using the flux law and writing Eq. (7-38) in the form

$$-k\frac{\partial T}{\partial x}\bigg|_{x=a} = h(T_1 - T_a) \tag{7-39}$$

But before writing this equation in dimensionless form, we must redefine θ, since T_1 is not constant, by replacing T_1 with T_a. Then

$$\theta \equiv \frac{T - T_a}{T_0 - T_a} \tag{7-40}$$

and Eq. (7-39) becomes

$$-k\frac{T_0 - T_a}{a}\frac{\partial\theta}{\partial\eta}\bigg|_{\eta=1} = h(T_1 - T_a) \tag{7-41}$$

or

$$\frac{-\partial\theta}{\partial\eta}\bigg|_{\eta=1} = \mathrm{Bi}\,\theta_1$$

where

$$\mathrm{Bi} \equiv \frac{ha}{k} \tag{7-42}$$

is the *Biot number*. It is a measure of the resistance to conduction in the solid (a/k) to the resistance to convection in the external film. Note that the Biot number looks like a Nusselt number but differs in that the Nusselt number is based on the thermal conductivity of the fluid in the *external* film and not that of the solid. Although the partial differential equation in this case will be the same as in Example 7-1, its solution with the radiation boundary condition will be of the form

$$\theta = \theta(\eta, s; \mathrm{Bi})$$

This solution can be obtained analytically by the method discussed in Appendix 7-2,† but it is usually more convenient to use a plot of the solution prepared by Gurney and Lurie[6] (Fig. 7-3), who also plotted solutions to similar problems for

†The analytical solution of the unsteady-state heat-conduction equation has been carried out for numerous boundary conditions and geometries. An important source for solutions is Carslaw and Jaeger [3].

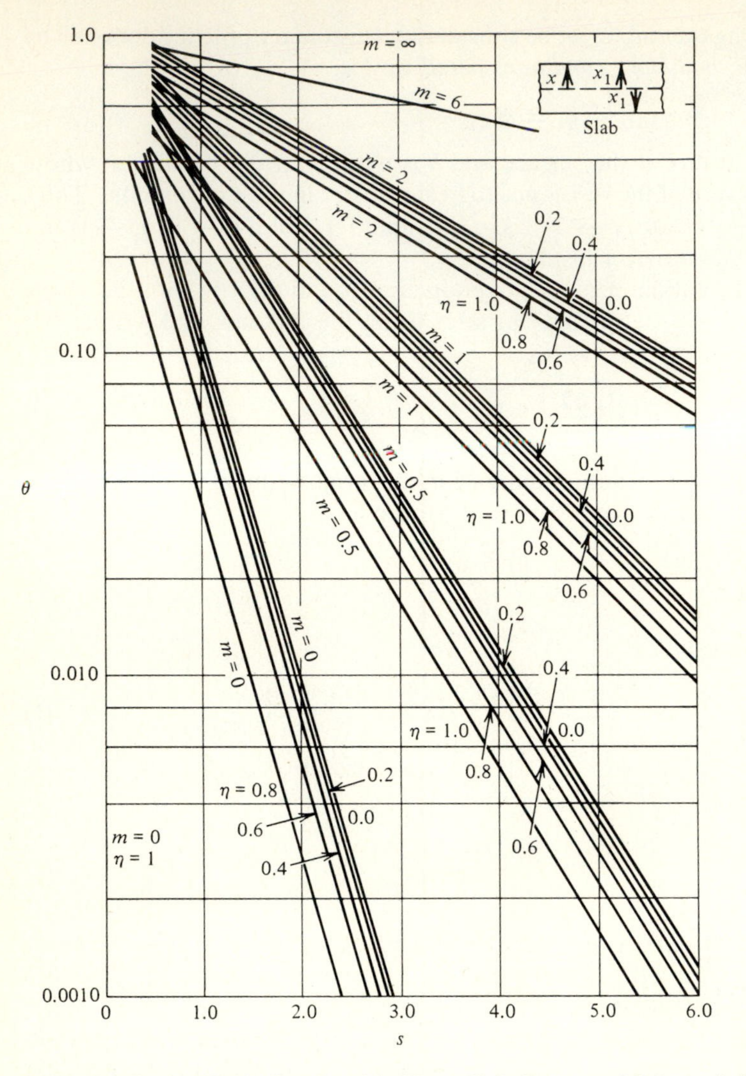

Figure 7-3 Gurney-Lurie chart for slab. [*From H. P. Gurney and J. Lurie, Ind. Eng. Chem.,* **15:** *1170 (1923).*]

the infinite cylinder and sphere (Figs. 7-4 and 7-5). In these charts x_1 is the distance from the center (as defined in the figures) and $\eta = x/x_1$ is a dimensionless distance. The parameter m is the reciprocal of the Biot number Bi, sometimes called the *internal-external conductance ratio*.

Limiting Equation at Large Times

Note that, as plotted, the lines of $\log \theta$ versus s in Figs. 7-3 to 7-5 are straight lines except when s (or t) is small. This effect can be understood by reference to the

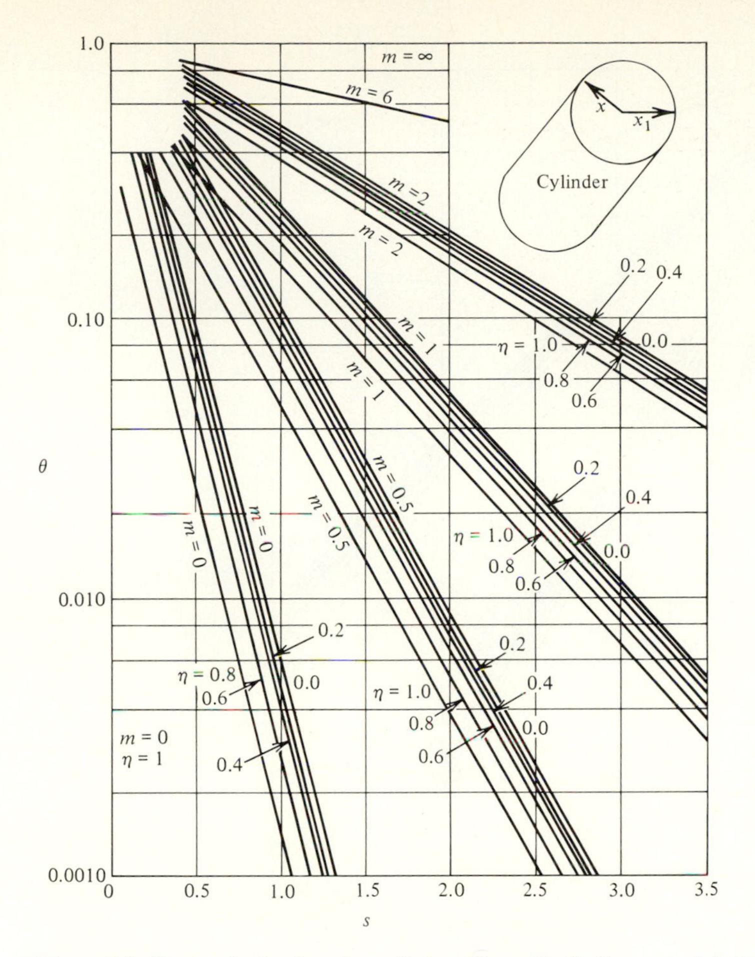

Figure 7-4 Gurney-Lurie chart for cylinder. [*From H. P. Gurney and J. Lurie, Ind. Eng. Chem.*, **15:** *1170* (*1923*).]

analytical solution [Eq. (7-36)] for the slab. If we examine the exponential term, we note that the ratio of the magnitude of the $n + 1$ term to that of the n term at $x = 0$ is

$$\left|\frac{A_n + 1}{A_n}\right| \exp\left[-(\gamma_{n+1}^2 - \gamma_n^2)s\right] = \frac{n + \frac{1}{2}}{n + \frac{3}{2}} \exp\left\{-\pi^2 s[(n + \tfrac{3}{2})^2 - (n + \tfrac{1}{2})^2]\right\}$$

$$= \frac{2n + 1}{2n + 3} \exp\left[-\pi^2 s(\tfrac{1}{2}n + 2)\right]$$

which approaches zero when $s \to \infty$. Therefore for sufficiently large s, the infinite series reduces to one term, and

$$\theta = \frac{4}{\pi} \cos \frac{\pi x}{2a} e^{-\pi^2 \alpha t/4a^2} = B e^{-\pi^2 s/4}$$

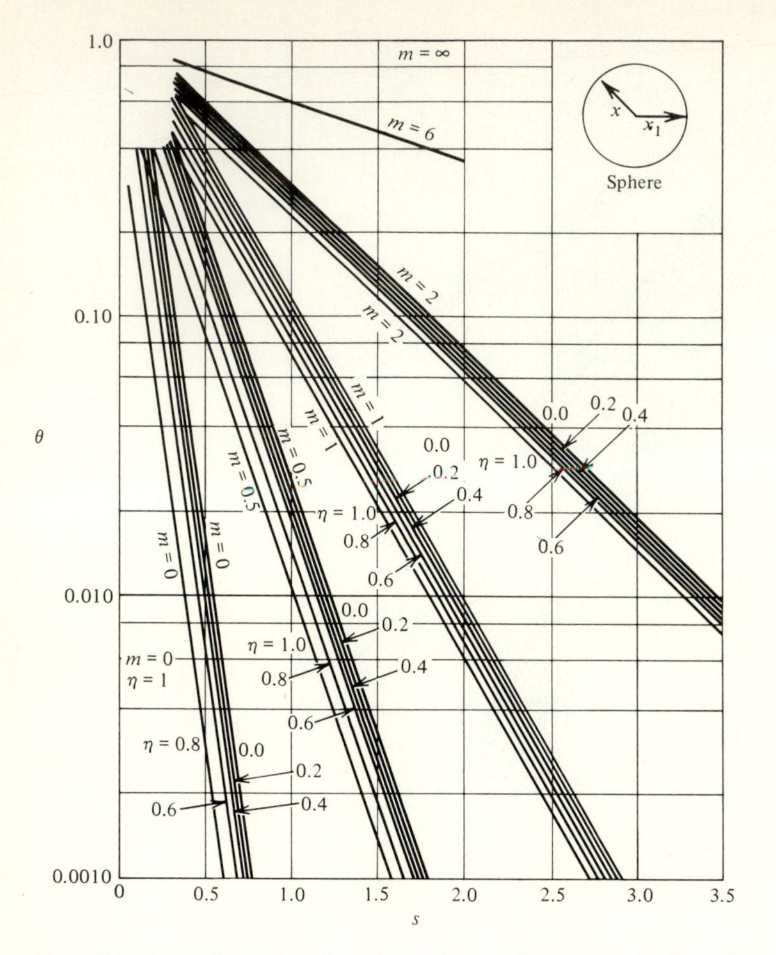

Figure 7-5 Gurney-Lurie chart for sphere. [*From H. P. Gurney and J. Lurie, Ind. Eng. Chem.,* **15:** *1170 (1923).*]

Taking logarithms gives

$$\ln\theta = \ln B - \frac{\pi^2 s}{4}$$

So a plot of θ versus s will give a straight line of slope $-\pi^2/4$ and an intercept B which will depend on η.

Low Internal and High External Resistance

Originally we discussed the case where $m \approx 0$ and $T_1 \approx T_a$, for which the external resistance h^{-1} is low relative to the internal resistance x_1/k. An example is a large nonmetallic object in a well-stirred bath of liquid, for which the temperature gradients in the medium must be considered, as indicated above. The opposite extreme is a small metal object in a bath of still air, where k is large and x_1 is small, so that the internal resistance is low but the external resistance is high and m is large since h is small. Referring to Figs. 7-3 to 7-5, we note that for $m = 6$ or $\text{Bi} = \frac{1}{6}$ only one curve is given for all values of η. This means that internal gradients can be neglected and the solid can be assumed to be at a uniform temperature T_s. If the solid is being cooled, $T_s > T_a$ and a macroscopic balance gives

$$-\rho \widehat{C}_p \mathcal{V} \frac{dT_s}{dt} = hA_s(T_s - T_a)$$

Rate of depletion of enthalpy — Rate of heat loss at surface

where A_s is the heat surface area and $\mathcal{V}$ is the volume of the solid. Rearranging gives

$$-\frac{d(T_s - T_a)}{T_s - T_a} = \frac{hA_s}{\rho \widehat{C}_p \mathcal{V}} dt$$

Integrating, using the lower limit $T = T_0$ at $t = 0$, leads to

$$\theta_s \equiv \frac{T_s - T_a}{T_0 - T_a} = \exp\left(-\frac{hA_s t}{\mathcal{V} \rho \widehat{C}_p}\right) = e^{-t/t_c} \tag{7-43}$$

where the group $\mathcal{V}\rho\widehat{C}_p/hA_S$ is a time constant t_c. It can be defined as the time required for θ_s to reach $e^{-1} = 0.368$ or for $1 - \theta_s$ to reach 0.632, where

$$1 - \theta_s \equiv \frac{(T_0 - T_a) - (T_s - T_a)}{T_0 - T_a} = \frac{T_0 - T_s}{T_0 - T_a}$$

Then t_c is the time required for 63.2 percent of the total temperature change to take place (Fig. 7-6).

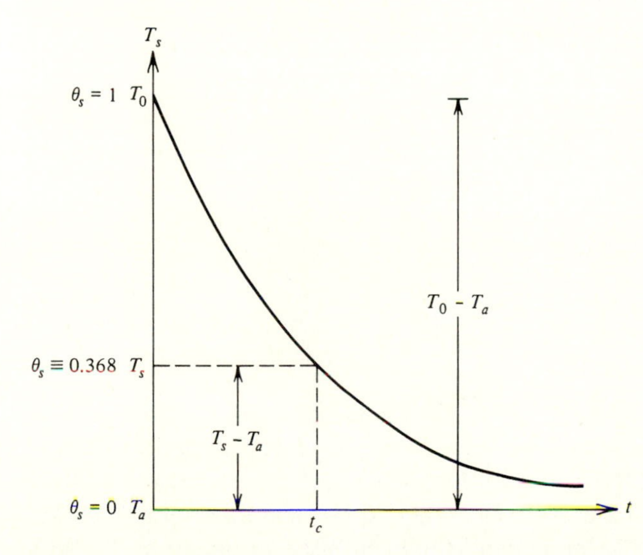

Figure 7-6 Variation of temperature with time for small Bi.

The volume-to-area ratio is a characteristic length $L_{ch} \equiv \mathcal{V}/A$. If we define the Biot and Fourier numbers in terms of it as

$$\text{Bi} \equiv \frac{hL_{ch}}{k} \quad \text{and} \quad s \equiv \frac{\alpha t}{L_{ch}^2} \tag{7-44}$$

we obtain

$$\frac{hA_s t}{\mathcal{V}\rho\hat{C}_p} \equiv \frac{hL_{ch}}{k}\frac{kt}{\rho\hat{C}_p L_{ch}^2} \equiv \text{Bi}\, s$$

Then Eq. (7-43) becomes

$$\theta_s = e^{-\text{Bi}\, s} \tag{7-45}$$

A plot of $\ln \theta_s$ versus s would then be a straight line with slope $-\text{Bi}$.

The above is an example of a *lumped*-system analysis using a one-lump system. In this example the temperature variation with position was neglected and only variation with time was considered. The results depend only on the volume-to-area ratio.

We can express L_c in terms of x_1, the length of the heat-flow path, for various shapes. For a sphere, the path x_1 is the radius R, and

$$\frac{\mathcal{V}}{A} = \frac{4}{3}\frac{\pi R^3}{4\pi R^2} = \frac{R}{3} = \frac{x_1}{3}$$

For an infinite cylinder, $x_1 = R_c$ and $\mathcal{V}/A = R_c/2 = x_1/2$. For a slab, $x_1 = a$ and $\mathcal{V}/A = 2aA_x/2A_x = a = x_1$.

Use of Gurney-Lurie Charts

Example 7-3 A steel ball 0.1524 m in diameter is being quenched in a large water bath maintained at 20°C. The initial temperature of the ball is 600°C. Calculate the temperature at the center of the ball at 2.5, 5, and 7.5 min. The physical properties are:

T, °C	ρ, lb/ft^3	$\hat{C}_p$, Btu/lb · °F	k, Btu/h · ft · °F	α, ft^2/h
20	450	0.12	25	0.46
600	440	0.18	22	0.25

The water bath is well stirred, and h is estimated to be approximately 2000 Btu/h · ft^2 · °F.

Solution In this problem α varies considerably. Since the Gurney-Lurie charts are based on a constant value of α, they cannot be used to obtain an "exact" solution to the problem. However, an approximate solution can be obtained by using an average value of $\alpha = 0.35\ \text{ft}^2/\text{h}$. Then $x_1 = \frac{1}{2}(0.1524\ \text{m}) = 0.25$ ft and

$$s = \frac{\alpha t}{x_1^2} = \frac{0.35}{(0.25)^2} t = 5.6t$$

where t is in hours. Then

$$t = 0.18\,s \qquad \eta = \frac{x}{x_1} = 0$$

$$m = \frac{k}{hx_1} = \frac{23.5}{2000(0.25)} = 0.047$$

at the center. Since m is so small, we assume it to be zero. Then $T_a \approx T_1$. From the definition of θ

$$T = T_1 + \theta(T_0 - T_1) = 20 + 580\theta$$

The values of s and θ at the three times of interest are found from Fig. 7-5 and T is then calculated:

Time, min	s	θ	T, °C
2.5	0.23	0.22	148
5	0.47	0.017	31
7.5	0.70	0.0017	21

We see that the ball cools rather rapidly. Note that the solution is also rather insensitive to h in the region of small m. □

7-8 MASS-TRANSFER ANALOGY

The solutions obtained for unsteady-state heat transport, as well as the Gurney-Lurie charts, will apply to mass transfer without bulk flow by change of notation. The mass-transfer Fourier number is $s_D \equiv D_{AB}t/x_1^2$, and the mass-transfer Biot number is

$$(\text{Bi})_D = \frac{k_c x_1}{D_{AB}}$$

where D_{AB} is the diffusivity of A in the medium B and k_c is an external mass-transfer coefficient. An important application is in drying, where the transport of water vapor through a porous solid can often be described as a diffusion process. Another application is in the diffusion of a reacting chemical species to and from a catalyst pellet.[2]

7-9 TRANSPORT INTO A SEMI-INFINITE MEDIUM

Heat Transport

Consider the problem of unsteady-state heat transport in the slab heated from one side (as discussed in Sec. 1-1 and 7-1). At small values of time for all practical purposes, the heat entering the lower surface of the slab at $x = 0$ has not yet reached the upper surface at $x = L_x$ (see Figs. 1-2 and 7-7). Thus, the slab can be considered to be of infinite extent in the x direction, and we can use the condition

$$T \to T_2 \qquad \text{as } x \to \infty \tag{7-46}$$

instead of $T = T_2$ at $x = L_x$. We retain the initial condition that at $t = 0$

$$T = \begin{cases} T_2 & x > 0 \quad (7\text{-}47a) \\ T_1 & x = 0 \quad (7\text{-}47b) \end{cases}$$

and the boundary condition that at $x = 0$

$$T = T_1 \qquad t > 0 \tag{7-48}$$

Examination of Fig. 7-7 reveals that the plots of θ versus x at t_1 and t_2 are similar in shape but differ in that at t_2 heat has penetrated further into the slab than at t_1. Thus it seems that each curve can be characterized by a different *penetration thickness* $\delta(t)$ and we ask whether there exists a variable, defined as

$$\xi \equiv \frac{x}{\delta(t)} \tag{7-49}$$

that will bring curves for all times together. For example, consider the time $\theta = \theta_p$, as shown in Fig. 7-7. At $x = x_1$ this value of θ_p occurs at time t_1 whereas at $x = x_2$ it occurs at t_2. So if we can define $\delta(t)$ such that $x_1/\delta_1(t_1) = x_2/\delta_2(t_2) = \xi_p$,

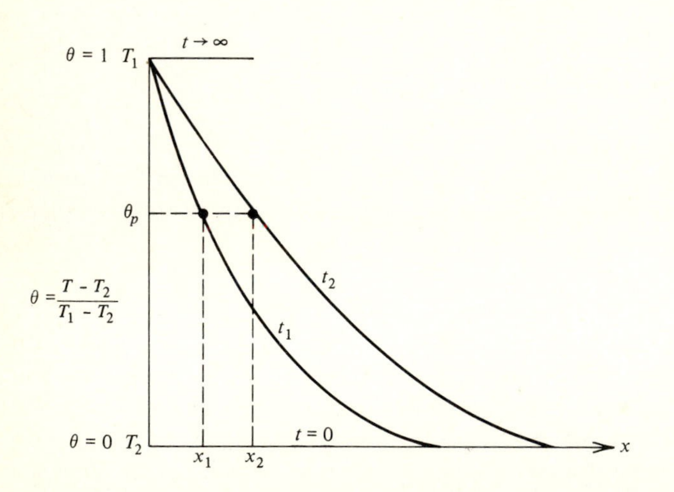

Figure 7-7 Temperature variation at two small times.

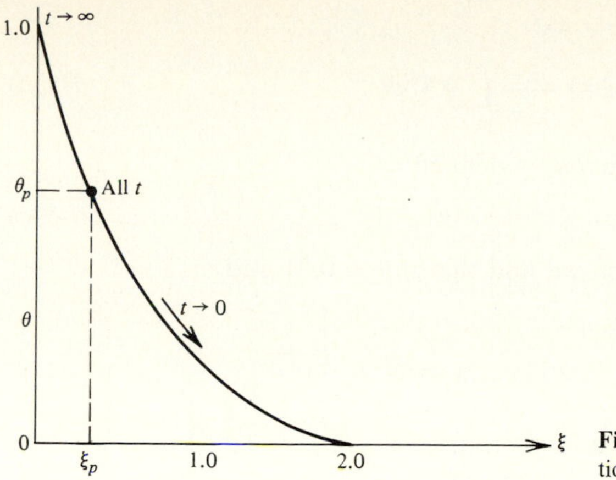

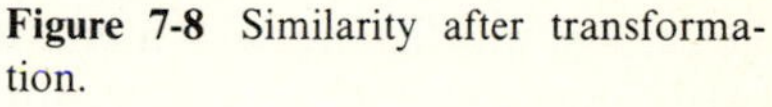
Figure 7-8 Similarity after transformation.

both points will transform to a single point $\theta_p(\xi_p)$, as shown in Fig. 7-8. If such *is* the case, we can eliminate x and t from the partial differential equation

$$\alpha \frac{\partial^2 \theta}{\partial x^2} = \frac{\partial \theta}{\partial t} \tag{7-7}$$

and reduce it to an ordinary differential equation of the form $\theta(\xi)$, where

$$\theta \equiv \frac{T - T_2}{T_1 - T_2}$$

is the dimensionless temperature. This kind of transformation is often called a *similarity transformation.*

It turns out (see Appendix 7-3) that if we let

$$\delta(t) \equiv \sqrt{4\alpha t} \tag{7-50}$$

we obtain the ordinary differential equation

$$\frac{d^2\theta}{d\xi^2} + 2\xi \frac{d\theta}{d\xi} = 0 \tag{7-51}$$

with the boundary conditions

$$\theta = \begin{cases} 0 & \text{at} \quad \xi = \infty \qquad (7\text{-}52a) \\ 1 & \text{at} \quad \xi = 0 \qquad (7\text{-}52b) \end{cases}$$

Note that the horizontal line in Fig. 7-7 for $t = \infty$ is now the point ($\theta = 1, \xi = 0$). As we move down the curve, we go backward in time or forward in distance.

This method of solution, called the *combination-of-variables method,* is also useful in boundary-layer theory in fluid mechanics and in mass-transfer problems. The solution is given in Appendix 7-3 as

$$\theta = 1 - \operatorname{erf} \xi \equiv \operatorname{erfc} \xi \tag{7-53a}$$

where the *error function* is defined as

$$\operatorname{erf} \xi \equiv \frac{2}{\sqrt{\pi}} \int_0^{\xi} e^{-\xi^2} \, d\xi \tag{7-53b}$$

and the *complementary error function* is defined as

$$\operatorname{erfc} \xi \equiv 1 - \operatorname{erf} \xi \tag{7-53c}$$

From tables[4] of the error function we find that if $\theta = 0.01$ and $\operatorname{erf} \xi = 0.99$ (see Fig. 7-8),

$$\xi = \frac{x}{\delta} = 2 \qquad \text{and} \qquad x = 2\delta = 4\sqrt{\alpha t}$$

Thus 2δ is the distance at which the temperature change $T - T_2$ is 1 percent of the total change $T_1 - T_2$. This quantity is often called the *penetration depth*. Note that we cannot define the penetration depth as the distance at which $T = T_2$ since that distance, according to our model, is infinite.

It is of practical importance to calculate the rate of transport into the medium. Using Fourier's law, we have

$$\begin{aligned} q_x \Big|_{x=0} &= -k \frac{\partial T}{\partial x} \Big|_{x=0} = -k(T_1 - T_2) \frac{d\theta}{d\xi} \Big|_{\xi=0} \left(\frac{\partial \xi}{\partial x} \right) \Big|_{x=0} \\ &= \frac{-k(T_1 - T_2)}{\sqrt{4\alpha t}} \frac{d\theta}{d\xi} \Big|_{\xi=0} \end{aligned}$$

From Eqs. (7-53*a*) and (A7-4-1) for differentiating an integral

$$\frac{d\theta}{d\xi} \Big|_{\xi=0} = \frac{-2e^{-\xi^2}}{\sqrt{\pi}} \Big|_{\xi=0} = -\frac{2}{\sqrt{\pi}}$$

and thus

$$q_x \Big|_{x=0} = \frac{k(T_1 - T_2)}{\sqrt{\pi \alpha t}} \tag{7-54}$$

The quantity $\sqrt{\pi \alpha t}$ is often taken as the penetration depth [instead of $\sqrt{4\alpha t}$ as in Eq. (7-50)] since by substitution into Eq. (7-53*a*) one finds that it is the distance at which the temperature difference has been decreased to 20 percent of its total value $T_1 - T_2$.

Heat Transport into a Semi-Infinite Fluid

Suppose that instead of finding the heat transport into a solid medium (as above), we are interested in finding the heat transport into a fluid bounded at $x = 0$ by a solid surface maintained at a uniform temperature T_1 at all $t > 0$. If the fluid is stagnant and is initially at T_2 throughout, we have the same differential equation and boundary conditions as for the slab and therefore the same solution. Referring to the definition of the heat-transfer coefficient (4-26) (letting T_w be T_1 and T_b

be T_2), we obtain

$$q_x \bigg|_{x=0} = h(T_1 - T_2)$$

Then
$$h = \frac{k}{\sqrt{\pi \alpha t}} = \sqrt{\frac{k\rho \widehat{C}_p}{\pi t}} = \sqrt{\frac{\alpha}{\pi t}}\rho \widehat{C}_p \tag{7-55}$$

Hence the heat-transfer coefficient is proportional to the square root of k or α.

Momentum Transport into a Semi-Infinite Medium

In the analogous momentum-transport problem, a large horizontal plate is suddenly pulled at $t = 0$ with velocity V, causing the fluid above it to move in the y direction with velocity $v_y(x)$, where x is the distance from the moving plate. At small values of time the y momentum imparted by the lower plate has penetrated only a distance $\delta(t)$ (as shown in Fig. 1-19). If $\delta < L_x$, the effect of an upper plate (if there is one) is negligible. The momentum balance gives, from the second of Eqs. (7-24),

$$\frac{\partial v_y}{\partial t} = \nu \frac{\partial^2 v_y}{\partial x^2}$$

with the boundary condition

$$v_y \to 0 \qquad \text{as } x \to \infty$$

replacing the condition that $v_y = 0$ at $x = L_x$. We can readily see that if we let $\theta_M \equiv v_y / V$,

$$\xi_M \equiv \frac{x}{\sqrt{4\nu t}} = \frac{x}{\delta_M(t)} \tag{7-56}$$

This problem is directly analogous to the heat-transport case, and the solution is therefore identical to Eq. (7-53*a*).

Note that the boundary-layer thickness for momentum transfer δ_M is not in general the same as that for heat transfer δ_H and that their ratio is

$$\frac{\delta_M}{\delta_H} = \left(\frac{\nu}{\alpha}\right)^{1/2} = (\text{Pr})^{1/2} \tag{7-57}$$

Mass Transfer into a Semi-Infinite Fluid

Consider unsteady-state mass transfer from an interface, such as the evaporation of a liquid A into a gas mixture initially of concentration C_{A2} (as described in Chap. 5). If bulk flow can be neglected, the differential equation is, from the second of Eqs. (7-23),

$$\frac{\partial C_A}{\partial t} = D_{AB} \frac{\partial^2 C_A}{\partial x^2}$$

If the interface equilibrium concentration in the vapor phase is C_{A1}, and if the bulk-stream concentration is C_{A2}, we can let

$$\theta_D \equiv \frac{C_A - C_{A2}}{C_{A1} - C_{A2}} \qquad \text{and} \qquad \xi_D \equiv \frac{x}{\delta_D(t)} = \frac{x}{\sqrt{4D_{AB}t}} \tag{7-58}$$

Then if we assume that $C_A \to C_{A2}$ as $x \to \infty$, the solution is identical to Eq. (7-53*a*).

The rate of mass transfer at the interface is found [either by analogy to Eq. (7-54) for heat transfer or by differentiation] to be

$$J_{A0} \equiv J_{Ax}\bigg|_{x=0} = -D_{AB}\frac{\partial C_A}{\partial x}\bigg|_{x=0} = \frac{D_{AB}(C_{A1} - C_{A2})}{\sqrt{\pi D_{AB}t}} \tag{7-59}$$

Thus the penetration depth for mass transfer is $\sqrt{\pi D_{AB}t}$, compared with $\sqrt{\pi \nu t}$ for momentum transfer. Also the ratio δ_M/δ_D varies as $\sqrt{\mathrm{Sc}}$, where $\mathrm{Sc} \equiv \nu/D_{AB}$ is the Schmidt number. Since D_{AB} in liquid systems is small, the penetration depth is also quite small.

Penetration Theory

The above results are the basis for the penetration theory proposed by Higbie[7] in 1935 for predicting mass transfer between phases in absorbers (see Fig. 4-11), extractors, etc. He argued that steady state seldom is obtained in practice, although it is usually assumed to occur in the stagnant-film model (Chap. 4). For example, a gas bubble of diameter D_p rising through the liquid (Fig. 7-9) flowing countercurrently in an absorption column will transfer its solute to the liquid moving past the bubble. As the bubble rises with velocity V, the interface is renewed with fresh liquid during the time t_D it takes the bubble to move one diameter or $t_D \approx D_p/V$. (We can neglect the curvature of the bubble since the penetration distance is small; see Fig. 7-10.) Thus if the concentration at the interface is C_{A1} and the bulk concentration of the liquid is C_{A2}, Eq. (7-59) will describe the mass-transfer rate that occurs at time t. If bulk flow is neglected, and if the mass-transfer coefficient k_c is defined by Eq. (4-28), the flux is

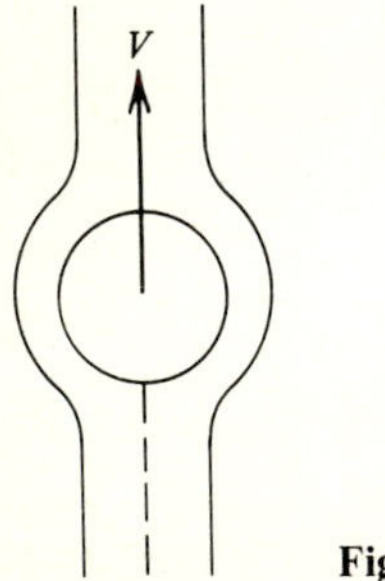

Figure 7-9 Gas bubble rising in liquid.

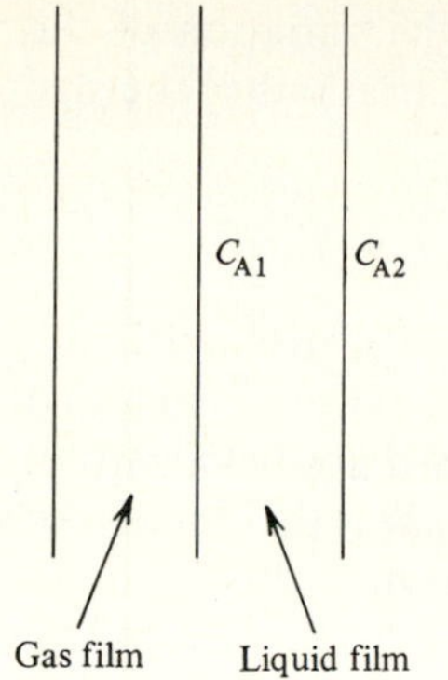

Figure 7-10 Two-film model.

$$J_{A0} \equiv J_{Ax}\Big|_{x=0} \equiv k_c(C_{A1} - C_{A2}) \tag{7-60}$$

If we equate the two expressions (7-59) and (7-60) for the flux, we can obtain an instantaneous mass-transfer coefficient

$$k_c = \left(\frac{D_{AB}}{\pi t_D}\right)^{1/2} \tag{7-61}$$

The average flux over time t_D is given by integrating Eq. (7-59) from $t = 0$ to $t = t_D$ and dividing by t_D

$$\tilde{J}_{A0} \equiv \tilde{J}_{Ax}\Big|_{x=0} = 2(C_{A1} - C_{A2})\frac{(D_{AB})^{1/2}}{(\pi t_D)^{1/2}} \tag{7-62}$$

The average mass-transfer coefficient, defined in terms of $\tilde{J}_{Ax}\Big|_{x=0}$, is then

$$\tilde{k}_c = 2\left(\frac{D_{AB}}{\pi t_D}\right)^{1/2} \tag{7-63}$$

Since $t_D = D_p/V$,

$$\tilde{k}_c = 2\left(\frac{D_{AB}V}{D_p}\right)^{1/2} \tag{7-64}$$

This equation applies to pure water for $0.3 \text{ cm} < D_p < 0.5 \text{ cm}$. Equations (7-61) and (7-63) indicate that k_c and $\tilde{k}_c$ vary with the square root of the diffusivity. By comparison, the stagnant-film model [Eq. (6-80)] gave $k_c = D_{AB}/L_D$, which is a linear dependence on D_{AB} (if L_D is assumed independent of diffusivity). However, if L_D varies as $\sqrt{D_{AB}}$, the results are the same.

The *j*-factor plots, based on the Chilton-Colburn analogy and considerable experimental data (see Chap. 6), predict that k_c varies as $D_{AB}{}^{2/3}$. Actually, powers on D_{AB} from 0 to 0.75 have been reported in the literature, but the most generally accepted value is $\frac{2}{3}$. More rigorous analyses would include the variation of the velocity with position within *both* the liquid and the gas and variations of the size and shape of the surface, e.g., as in drop formation.

Surface-Renewal Model

Instead of using the same contact time for each element, Danckwerts[5] extended the penetration theory to provide for a distribution of contact times from zero to infinity. He let ϕ be an age-distribution function such that $\phi(t)\,dt$ is the fraction of the surface that has ages between t and $t + dt$. He then assumed that

$$\phi(t) = se^{-st} \tag{7-65}$$

where s, the fractional rate of replacement of the elements belonging to any age group, was taken to be constant. The rate of absorption for all elements is then obtained by multiplying Eq. (7-59) by ϕ and integrating over all ages. Defining this average as

$$\langle J_{A0} \rangle \equiv \int_0^\infty se^{-st} J_{A0}\, dt \tag{7-66}$$

gives

$$\langle J_{A0} \rangle = \int_0^\infty \frac{se^{-st} D_{AB}(C_{A1} - C_{A2})}{\sqrt{\pi D_{AB} t}}\, dt \equiv (C_{A1} - C_{A2}) \sqrt{D_{AB} s} \tag{7-67}$$

Letting $\langle J_A \rangle \equiv \langle k_c \rangle (C_{A1} - C_{A2})$ gives

$$\langle k_c \rangle = \sqrt{D_{AB} s} \tag{7-68}$$

Although this equation again predicts a power of $\frac{1}{2}$ on the diffusivity, it is limited otherwise by the need to obtain s, just as the film-theory model is limited by the need to obtain the film thickness and the penetration theory is limited by the need to know t_D. In the case of diffusion and chemical reaction (to be discussed later), this need is often obviated by measuring the mass-transfer rate (or mass-transfer coefficient) without reaction and calculating the needed parameter from it. This parameter is then assumed to be the same in the reacting case. The ratio of the mass-transfer coefficients for the reaction and nonreacting cases, called the *enhancement factor,* is then found. (See Sec. 7-10.)

Film-Penetration Model

Toor and Marchello[10] have developed a model that combines the features of the film and penetration models by solving the unsteady-state diffusion equation [the second of Eqs. (7-23)] using a finite constant film thickness L_D. They postulated that "young" fluid elements obey the penetration model and "old" elements obey the film model. Thus the exponent on Sc in the equation for $(\mathrm{Nu})_{AB}$ varies from 0.5 to 1.0, or an average of about $\frac{2}{3}$, as given by the Chilton-Colburn analogy. The model reduces to the penetration model when $D_{AB}/L_D^2 s$ is small and to the film model when it is large.

Stagnation-Flow Model

In a review series, Scriven[8] discussed the *stagnation-flow model,* which takes into consideration bulk flow normal to the interface and describes other models that deal with the stretching or contraction of portions of the fluid interface.

7-10 UNSTEADY-STATE TRANSPORT WITH INTERNAL GENERATION

Unsteady-State Mass Transport with Chemical Reaction†

Consider a gas A being absorbed at a gas-liquid interface as in Secs. 4-6 and 7-3 but also undergoing chemical reaction while diffusing in the x direction from the interface. If A is a reactant, we have *negative* internal generation (consumption) of A and R_A, the rate of generation of A, is negative. Then, by means of an incremental balance, we obtain

$$D_{AB}\frac{\partial^2 C_A}{\partial x^2} + R_A = \frac{\partial C_A}{\partial t} \tag{7-69}$$

If the reaction is first order and irreversible,

$$R_A = -k_R C_A \tag{7-70}$$

and

$$D_{AB}\frac{\partial^2 C_A}{\partial x^2} - k_R C_A = \frac{\partial C_A}{\partial t} \tag{7-71}$$

The initial condition is

$$C_A = C_{A2} = 0 \text{ (negligible)} \qquad \text{at } t = 0, \text{ all } x \tag{7-72}$$

and the boundary conditions are

$$C_A = \begin{cases} C_{A1} = \text{solubility of A} & \text{at } x = 0,\ t > 0 \quad (7\text{-}73) \\ C_{A2} = 0 & \text{at } x \to \infty \quad (7\text{-}74) \end{cases}$$

Using surface-renewal theory, we assume that the surface age distribution is given by Eq. (7-65) and that the flux is given by Eq. (7-67), which happens also to be s multiplied by the Laplace transform‡ of $J_{A0}(t)$, as given by Eq. (7-59). This trans-

†This section may be omitted on first reading.

‡The Laplace transform is defined as

$$\mathcal{L}\{F\} \equiv \bar{F}(s) \equiv \int_0^\infty e^{-st} F\,dt$$

formed flux can be obtained in the reacting case from the gradient at $x = 0$ of the transformed concentration $\bar{C}_A$, which in turn can be obtained from the solution of the transformed diffusion equation. Taking the Laplace transform of Eq. (7-71) gives

$$D_{AB} \frac{d^2\bar{C}_A}{dx^2} - k_R \bar{C}_A = s\bar{C} \tag{7-75}$$

The Laplace transforms of the boundary conditions are

$$\bar{C}_A = \begin{cases} \dfrac{C_{A1}}{s} & \text{at } x = 0 \\ 0 & \text{at } x = \infty \end{cases}$$

The solution is[5]

$$\bar{C}_A = \frac{C_{A1}}{s} \exp\left[-\left(\frac{k_R + s}{D_{AB}} x\right)^{1/2}\right] \tag{7-76}$$

Then using the flux law and Eq. (7-66), we obtain

$$\begin{aligned} \langle J_{A0} \rangle &= -\int_0^\infty s e^{-st} D_{AB} \frac{\partial C_A}{\partial x}\bigg|_0 dt \\ &= -sD_{AB} \frac{d}{dx} \int_0^\infty (e^{-st}\, dt)\bigg|_{x=0} = -sD_{AB} \frac{\partial \bar{C}}{\partial x}\bigg|_{x=0} \end{aligned}$$

and
$$\langle J_{A0} \rangle = C_{A1} \sqrt{D_{AB}(k_R + s)} \tag{7-77}$$

Comparing with Eqs. (7-67) and (7-68) and defining an *enhancement factor* (or reaction factor)† by

$$E \equiv \frac{k_{cR}}{k_c} \tag{7-78}$$

where k_{cR} is the mass-transfer coefficient for the reaction case, we obtain

$$E = \left(1 + \frac{k_R}{s}\right)^{1/2} \tag{7-79}$$

If we use Eq. (7-68) to obtain $s = k_c^2/D_{AB}$,

$$E = \left(1 + \frac{k_R D_{AB}}{k_c^2}\right)^{1/2} \tag{7-80}$$

This result can be compared with that obtained in Sec. 4-7 from film theory

$$E = \left(\frac{k_R}{D_{AB}}\right)^{1/2} L_x^2 \coth\left(\frac{k_R}{D_{AB}}\right)^{1/2} L_x$$

Figure 7-11 compares results for various theories.

†The reaction factor is defined similarly but with $C_{A2} \neq 0$.

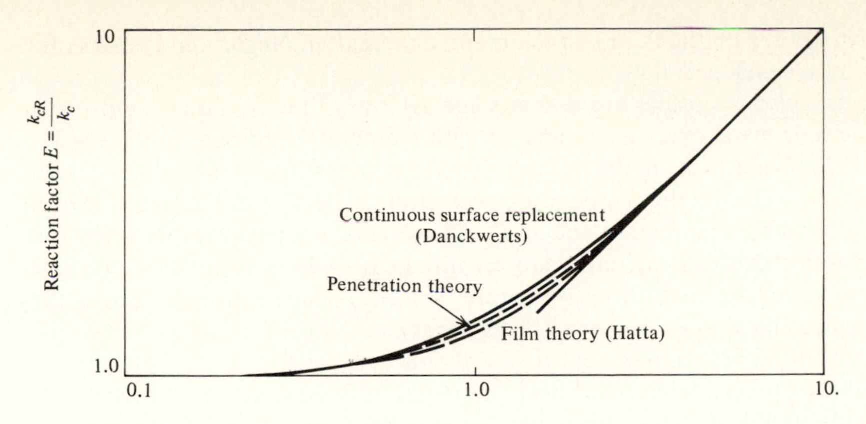

Figure 7-11 Comparison of three theories of interfacial mass transfer accompanied by first-order irreversible chemical reaction. (*From T. K. Sherwood, R. L. Pigford, and C. R. Wilke, "Mass Transfer," McGraw-Hill, New York, 1975*).

Heat-Momentum Transport at Unsteady State with Internal Generation

Example 7-4 A nuclear engineer is interested in predicting $\dot{Q}_{Lt}(t)$, the net rate of heat transport to the cooling surfaces during the startup of a nuclear reactor, consisting of a solid annular fuel element of outer radius R_2 cooled both at its periphery and internally, the latter by a tube of radius R_1 which penetrates the fuel element (Fig. 7-12). Assume that the cooling surfaces at R_1 and R_2 are maintained at a constant temperature T_0 throughout the length of the reactor at all $t > 0$. Initially the fuel element is also at temperature T_0, but at time $t = 0$ the nuclear reaction begins generating heat uniformly throughout the fuel element at a rate Φ_N (energy per unit volume and unit time) which is assumed constant at all times and locations. (*a*) Show that this problem has a momentum-transport analogy. (*b*) Show how it could be used to find $\dot{Q}_{Lt}(t)$ in the reactor if one had a solution for the momentum case.

SOLUTION (*a*) An analogous problem would be the following. An incompressible newtonian fluid is contained at $t < 0$ in a long vertical annulus of inner radius R_1, outer radius R_2, and length L. At time $t = 0$ a dynamic-pressure drop $\mathcal{P}_0 - \mathcal{P}_L$ is

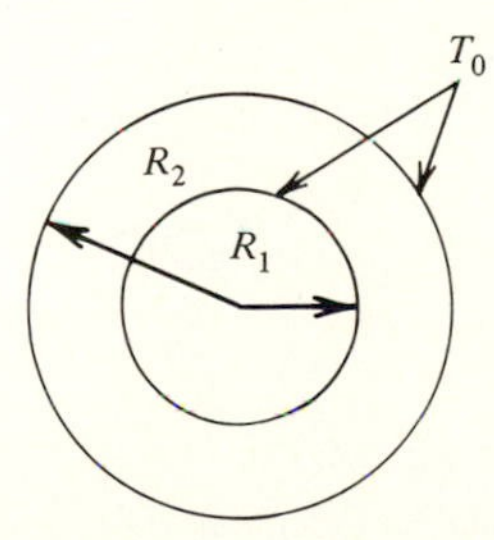

Figure 7-12 Annular fuel element.

suddenly imposed on the fluid and maintained thereafter. Neglect end effects and assume fully developed flow.

Typical velocity profiles at $t = 0$ at some arbitrary time t_1 and at steady state or infinite time are shown and compared with temperature profiles in the reactor in Fig. 7-13. Note that in the reactor (conduit) the initial generation of heat (momentum) will raise the temperature (velocity) above T_0 (zero velocity) except at the walls, where it is maintained at T_0. This results in a temperature (velocity) gradient which causes heat (momentum) transport to each wall. As more heat (momentum) is generated, the temperature (velocity) inside continues to rise, the gradients become steeper, and even more heat (momentum) is transported to the walls. This process continues until the gradients become so steep that the transport to the walls equals the total heat (momentum) generated. At this point we have steady state and the accumulation is zero. The force balance in the z direction is

$$\sigma_{rz}(2\pi rL)\Big|_r - \sigma_{rz}(2\pi rL)\Big|_{r+\Delta r} + (p_0 - p_L)(2\pi r\,\Delta r) + \frac{\rho g_z}{g_c}(2\pi r\,\Delta r\,L) = \frac{\Delta(\rho v_z 2\pi r\,\Delta r\,L)}{g_c\,\Delta t}$$

The momentum balance is

$$\tau_{rz}(2\pi rL)\Big|_r - \tau_{rz}(2\pi rL)\Big|_{r+\Delta r} + (p_0 - p_L)(2\pi r\,\Delta r)g_c + \rho g_z(2\pi r\,\Delta r\,L) = \frac{\Delta(\rho v_z 2\pi r\,\Delta r\,L)}{\Delta t}$$

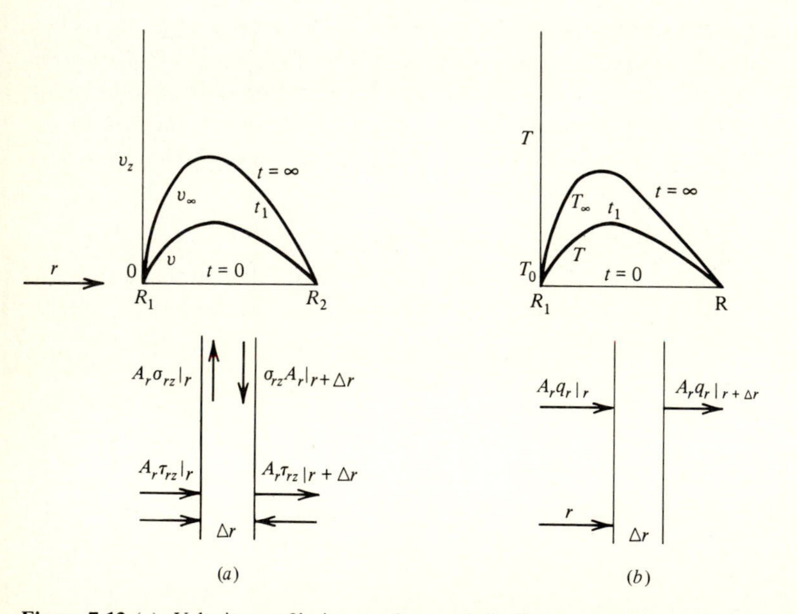

Figure 7-13 (*a*) Velocity profile in annular conduit; (*b*) temperature profile in fuel element.

Dividing by $2\pi r\,\Delta r\,L$ and letting $\Delta r \to 0$ and $\Delta t \to 0$ gives

$$-\frac{\partial(\tau_{rz}r)}{r\,\partial r} + \frac{g_c(p_0 - p_L) + \rho g_z L}{L} = \frac{\partial(\rho v_z)}{\partial t} \tag{7-81}$$

Momentum transport — Momentum generation — Momentum accumulation

Using $\tau_{rz} = -\mu(\partial v_z/\partial r)$, we have

$$-\mu\frac{\partial(r\,\partial v_z/\partial r)}{r\,\partial r} + \Phi_M = \rho\frac{\partial v_z}{\partial t} \tag{7-82}$$

where the momentum generation per unit time and unit volume is

$$\Phi_M \equiv \frac{g_c(\mathcal{P}_0 - \mathcal{P}_L)}{L} \equiv \frac{(p_0 - p_L + \rho g_L/g_c)g_c}{L} \tag{7-83}$$

The energy balance is

$$-\frac{\partial(q_r r)}{r\,\partial r} + \Phi_H = \frac{\partial(\rho\hat{C}_p T)}{\partial t} \tag{7-84}$$

Heat transport — Heat generation — Energy accumulation

or

$$k\frac{\partial(r\,\partial T/\partial r)}{r\,\partial r} + \Phi_H = \rho\hat{C}_p\frac{\partial T}{\partial t} \tag{7-85}$$

These equations can be put in dimensionless form by choosing characteristic lengths, times, and velocities or temperatures. As a characteristic length we can choose either R_1 or R_2; we arbitrarily choose R_2, letting $\eta \equiv r/R_2$. From previous experience we choose R_2^2/ν as a characteristic time for momentum transport, letting $s_m \equiv \nu t/R_2^2$. The characteristic velocity will temporarily be designated as v_c and we let $\phi_M\ (\equiv v_z/v_c)$ be the dimensionless velocity. Then if the momentum equation is multiplied by $R_2^2/\mu v_c$, we obtain

$$\frac{\partial(\eta\,\partial\phi_M/\partial\eta)}{\eta\,\partial\eta} + \frac{\Phi_M R_2^2}{\mu v_c} = \frac{\partial\phi_M}{\partial s_M}$$

For convenience we can define v_c so that the second term is unity, or

$$v_c \equiv \frac{\Phi_M R_2^2}{\mu}$$

For the heat-transport case, since we must have the same dimensionless boundary conditions at the walls, we must use $\phi_H \equiv (T - T_0)/T_c$, and in order to obtain the same differential equation we must use $T_c \equiv \Phi_H R_2^2/k$. Then for either case

$$\frac{\partial(\eta\,\partial\phi/\partial\eta)}{\eta\,\partial\eta} + 1 = \frac{\partial\phi}{\partial s} \tag{7-85a}$$

Table 7-3 Table of conditions

Condition	η	s	v_z	ϕ_M	T	ϕ_H	ϕ_∞	$\phi_t \equiv \phi_\infty - \phi$
Boundary 1	κ	≥ 0	0	0	T_0	0	0	0
Boundary 2	1	≥ 0	0	0	T_0	0	0	0
Initial	$\kappa \geq \eta \geq 1$	0	0	0	T_0	0	ϕ_∞	ϕ_∞

Since there is a steady-state solution, which we will call $\phi_\infty(\eta)$, it is convenient to define a transient solution as

$$\phi_t(s, \eta) \equiv \phi_\infty(\eta) - \phi(s, \eta) \tag{7-85b}$$

If we let $R_1/R_2 \equiv \kappa$, the boundary and initial conditions become as shown in Table 7-3. Since the dimensionless differential equations and the dimensionless initial and boundary conditions are identical for each case, there is an analogy between the two, so that the solution for one case can be used for the other.

(*b*) To find $\dot{Q}_{Lt}(t)$ we need to find the flux at each wall, multiply by the wall area, and then add the results—with due regard to the sign convention (Sec. 1-19). Since n_r is negative at $r = R_1$ and positive at $r = R_2$ (or since heat must be lost at *each* wall and q_r is negative at R_1 and positive at R_2), we obtain

$$-\dot{Q}_t \equiv \dot{Q}_{Lt} = -q_r(2\pi rL)\Big|_{r=R_1} + q_r(2\pi rL)\Big|_{r=R_2}$$

where
$$q_r = -k\frac{\partial T}{\partial r} = -\frac{kT_c}{R_2}\frac{\partial \phi}{\partial \eta}$$

Then
$$\dot{Q}_{Lt} = 2\pi LkT_c\left[\kappa\frac{\partial \phi}{\partial \eta}\Big|_\kappa - \frac{\partial \phi}{\partial \eta}\Big|_{n=1}\right] \tag{7-86}$$

For the analogous momentum-transport case the partial differential equation [the second of Eqs. (7-24)], the initial condition and the boundary conditions in dimensionless form will be identical to those for the energy case, or

$$\phi_M(\eta, s_M; \kappa) \equiv \phi_H(\eta, s_H; \kappa) \equiv \phi(\eta, s; \kappa)$$

Substituting Eq. (7-85*b*) into Eq. (7-86) gives

$$\dot{Q}_{Lt} = 2\pi LkT_c\left[\kappa\left(\frac{d\phi_\infty}{d\eta}\Big|_\kappa - \frac{\partial \phi_t}{\partial \eta}\Big|_\kappa\right) - \left(\frac{d\phi_\infty}{d\eta}\Big|_1 - \frac{\partial \phi_t}{\partial \eta}\Big|_1\right)\right]$$

The steady-state solution, as obtained in Chap. 3, is

$$\phi_\infty = \frac{1}{4}\left(1 - \eta^2 - \frac{1-\kappa^2}{\ln \kappa}\ln \eta\right)$$

The transient solution is of the form

$$\phi_t = \sum_{n=0}^{\infty} A_n X_n(\eta) S_n(s) \qquad \text{where} \qquad S_n \equiv e^{-\beta_n s}$$

Then
$$\frac{\partial \phi_t}{\partial \eta} = \sum_{n=0}^{\infty} A_n S_n(s) \frac{dX_n}{d\eta}$$

For the steady case $d\phi_\infty/d\eta$ is found by differentiating ϕ_∞. It is evaluated at $\eta = \kappa$ and $\eta = 1$. The details are left to the student. For the transient part, one must differentiate a known solution for $X_n(\eta; \beta_n)$ and evaluate it at $\eta = \kappa$ and $\eta = 1$. The solution can be obtained in terms of Bessel functions. □

Example 7-5: Reactor with multiple cooling tubes In Chap. 1 we have shown that if a physical situation is such that a mathematical analogy exists between two transport processes and if an analytical solution is available in dimensionless form for one process, that solution can be used to obtain a desired physical quantity for another process. But if an analytical solution is not available, it may still be possible to carry out an experimental program in which certain quantities are measured for one transport process; then the analogy is used to obtain the desired physical quantities for another transport process for which it may be too costly, too hazardous, or too time-consuming to carry out experiments.

In this example a nuclear engineer is interested in predicting $|\dot{Q}(t)|$, the net rate of heat transport to the cooling surfaces, during the startup of a nuclear reactor (Fig. 7-14) which consists of a solid cylindrical fuel element (of radius R_2) that is cooled both at its periphery and internally by tubes (each of radius R_1) which penetrate the fuel element. For simplicity we assume that all cooling surfaces are maintained at a constant temperature T_0 throughout the length of the reactor and at all $t > 0$. Initially the fuel element is also at temperature T_0, but at time $t = 0$ the nuclear reaction begins generating heat uniformly throughout the fuel element at a rate Φ_N (energy per unit volume and unit time) which is assumed constant at all times and locations. The nuclear engineer suggests that this problem has a momentum-transport analogy; thus one could carry out an experiment in which one measures, as a function of time, the average velocity of a fluid $v_a(t) \equiv V(t)$ flowing under a pressure drop $p_0 - p_L$ through a cylindrical conduit pierced with tubes.

Derive the necessary equations to show how one could theoretically calculate $|\dot{Q}(t)|$ in the reactor from data on $V(t)$ obtained in the flow experiment. Describe how the experiment could be carried out and state what quantities, if any, are to be the same in each case. (Note that it is not considered feasible to measure point unsteady-state velocity profiles or to measure drag forces at the walls, but the average velocity can be measured easily.)

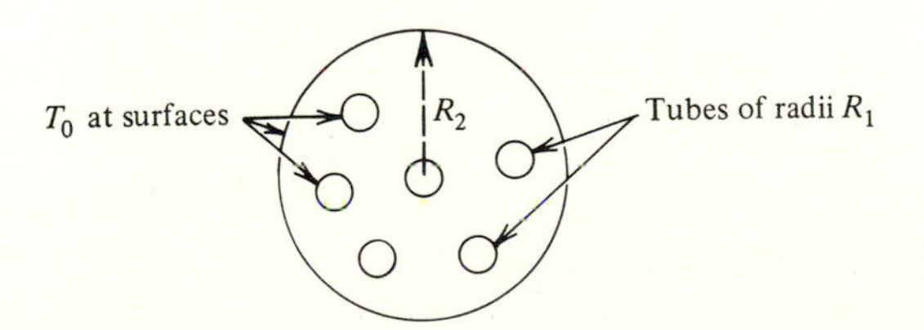

Figure 7-14 Reactor with internal cooling tubes.

SOLUTION We note that $|\dot{Q}(t)|$ is *not* analogous to $V(t)$ but *is* analogous to $|\dot{P}_z(t)|$, the net rate of momentum transport at *all* the surfaces. However, $|\dot{P}_z(t)| \equiv |F_z g_c|$ can be related to $V(t)$ by a macroscopic momentum balance

$$\underset{\substack{\text{Input}\\\text{rate}}}{0} - \underset{\substack{\text{Output}\\\text{rate}}}{|F_z g_c|} + \underset{\substack{\text{Gener-}\\\text{ation}\\\text{rate}}}{\Phi_M \mathcal{V}_M} = \underset{\substack{\text{Accumu-}\\\text{lation}\\\text{rate}}}{\rho \mathcal{V} \frac{dV}{dt}}$$

where $\mathcal{V}_M = A_M L_M =$ volume of momentum system
$A_M =$ cross-sectional *open* area
$\Phi_M = (\mathcal{P}_0 - \mathcal{P}_L) g_c / L =$ momentum generation per unit volume and unit time

Similarly for energy transport, the macroscopic balance is

$$0 - |\dot{Q}(t)| + \Phi_H \mathcal{V}_H = \rho \mathcal{V}_H \widehat{C}_p \frac{d\langle T \rangle}{dt}$$

These can be made dimensionless by using the variables defined in Example 7-3:

$$s_M \equiv \frac{\nu t}{R_2^2} \qquad \phi_M \equiv \frac{v_z}{v_c} \qquad v_c \equiv \frac{\Phi_M R_2^2}{\mu L} \qquad \kappa \equiv \frac{R_1}{R_2}$$

$$s_H \equiv \frac{\alpha t}{R_2^2} \qquad \phi_H \equiv \frac{T'}{T_c} \qquad T_c \equiv \frac{\Phi_H R_2^2}{k} \qquad T' \equiv T - T_0$$

Then the dimensionless macroscopic balances are

$$0 - \frac{|F_z| g_c}{\Phi_M \mathcal{V}_M} + 1 = \frac{d\langle \phi_M \rangle}{ds_M} \qquad \text{and} \qquad \frac{-|\dot{Q}|}{\Phi_H \mathcal{V}_H} + 1 = \frac{d\langle \phi_H \rangle}{ds_H}$$

We note that

$$\langle \phi \rangle \equiv \frac{\iint \phi \, dA}{\iint dA} = \langle \phi \rangle (s; \kappa)$$

Thus, if we make $\kappa \equiv R_1/R_2$ the same (not necessarily $R_1 = R_2$) in the momentum experiment as in the reactor, $\langle \phi \rangle$ will depend only on s and $\langle \phi_M \rangle$ versus s_M will be the same curve as $\langle \phi_H \rangle$ versus s_H (see Fig. 7-15). To obtain this curve we obtain $V(t) = \dot{\mathcal{V}}/A_M$ by measuring $\dot{\mathcal{V}}$, the volumetric rate of flow through the cross section A_M. We also measure the pressure drop (held constant) from which we can calculate Φ_M and v_c and, for any time t, get s_M and $\langle \phi_M \rangle$. We then plot $\langle \phi_M \rangle$ versus s_M and measure slopes to get

$$\frac{d\langle \phi_M \rangle}{ds_M} \equiv \frac{d\langle \phi_H \rangle}{ds_H} = 1 - \frac{|\dot{Q}|}{s_H \mathcal{V}_H}$$

whence
$$|\dot{Q}(t)| = \Phi_H \mathcal{V}_H \left(1 - \frac{d\langle \phi_M \rangle}{ds_M}\right)$$

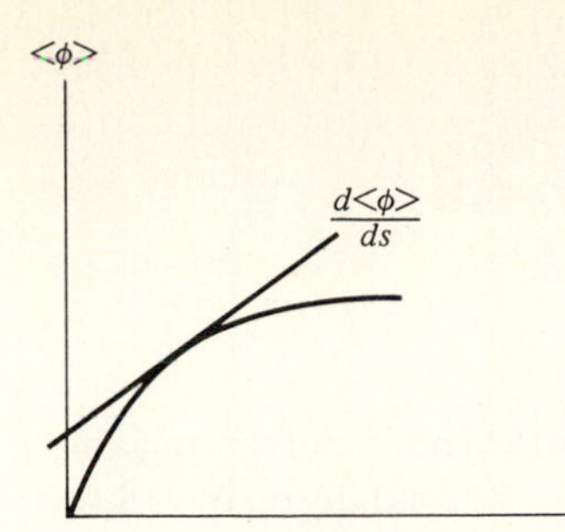

Figure 7-15 Evaluation of $d\langle\phi\rangle/ds$.

Note that we do *not* need to have the same diffusivities in each system; that is, $\alpha \neq \nu$. This is important since α will be very much larger than ν. However, we *do* need to be able to measure $V(t)$ accurately at sufficient times to cover the range of values of $s_H = s_M$ for which $|\dot{Q}(t)|$ is sought. Although the plot of ϕ_M versus s_M is identical to a plot of ϕ_H versus s_H, at the same dimensionless times the dimension*al* times will be related by

$$t_H = t_M \frac{\nu}{\alpha} \left[\frac{(R_2)_H}{(R_2)_M} \right]^2$$

Thus if we take a momentum data point after 100*s* (say), if† $\nu/\alpha \approx 10^{-6}$, and if $(R_2)_H = (R_2)_M$, the value of t_H that corresponds to that value of $s_M = s_H$ will be 10^{-4} s. This is desirable since we would like to have the $|\dot{Q}(t)|$ values at very short time intervals. □

The actual design of nuclear reactors is a very involved process, and the above example is given to illustrate the use of analogies rather than to propose a design method.

APPENDIX 7-1 SOLUTION OF HEAT-DIFFUSION EQUATION WITH SIDES AT DIFFERENT TEMPERATURES

We try to reduce

$$\alpha \frac{\partial^2 \theta_t}{\partial x^2} = \frac{\partial \theta_t}{\partial t} \tag{A7-1-1}$$

to two ordinary differential equations that can be solved by the usual methods. We postulate that *a* solution to Eq. (A7-1-1) can be written as the product of two functions, one of which, $F(x)$, depends on x only and the other of which, $G(t)$, depends on t only; i.e., we let

$$\theta_t(x, t) = F(x)G(t) \tag{A7-1-2}$$

†See Ref. 1, p. 810, for properties of nuclear fuels.

We now substitute Eq. (A7-1-2) into Eq. (A7-1-1) by taking the appropriate derivatives. Let us start with the left-hand side of Eq. (A7-1-1). Since the partial derivative with respect to t is taken at constant x, $F(x)$ in Eq. (A7-1-2) is treated as a constant. Then

$$\frac{\partial \theta_t}{\partial t} = F(x)\frac{dG(t)}{dt} = F(x)G'(t) \tag{A7-1-3}$$

where the prime means differentiation with respect to the independent variable. Since only one independent variable is involved, we use the total derivative. Similarly

$$\frac{\partial^2 \theta_t}{\partial x^2} = \frac{d^2F(x)}{dx^2}G(t) = F''(x)G(t) \tag{A7-1-4}$$

Then if we substitute into Eq. (A7-1-1), we find

$$\alpha F''(x)G(t) = F(x)G'(t)$$

If we now divide both sides by $\alpha F(x)G(t)$, the left-hand side will be a function of x only and the right-hand side will be a function of t only. The only way this can be so is for *both* sides to equal a constant, which we call $-\gamma^2$ (for reasons to be apparent later). Then

$$\frac{F''(x)}{F(x)} = \frac{G'(t)}{\alpha G(t)} = \text{const} \equiv -\gamma^2 \tag{A7-1-5}$$

Each equality gives us an ordinary differential equation. They are

$$G'(t) + \alpha\gamma^2 G(t) = 0 \tag{A7-1-6}$$

and

$$F''(x) + \gamma^2 F(x) = 0 \tag{A7-1-7}$$

The solutions to these equations can be written (see Appendix 1-2) as

$$G(t) = c_1 e^{-\gamma^2 \alpha t} \tag{A7-1-8}$$

and

$$F(x) = c_2 \cos \gamma x + c_3 \sin \gamma x \tag{A7-1-9}$$

where we have used the form in Eq. (A7-1-9) appropriate for γ a real number. To show that γ is real, we use the condition that $\theta_t \to 0$ as $t \to \infty$. In order for this to hold, $G(t)$ must go to 0 as $t \to \infty$ (according to Eq. A7-1-2). But if γ were imaginary, Eq. (A7-1-8) would predict $G(t) \to \infty$. Hence γ must be real, and that is why it is convenient to choose the constant in Eq. (A7-1-5) as $-\gamma^2$.

We now evaluate the constant c_2 in Eq. (A7-1-9). If we apply the boundary condition that $\theta_t = 0$ at $x = 0$, then

$$0 = F(0) = c_2 \cos 0 + c_3 \sin 0 = c_2 \tag{A7-1-10}$$

or

$$C_2 = 0$$

To evaluate c_3 we will use the boundary condition that $\theta_t = 0$ at $x = L_x$, or

$$F(L_x) = 0 = c_3 \sin \gamma L_x$$

Now c_3 cannot be zero; if it were, θ_t would not vary with x at all since c_2 is zero. Therefore

$$\sin \gamma L_x = 0 \tag{A7-1-11}$$

We know that the sine function is zero at intervals of π and that there are an infinite number of these roots. We call the nth root $\gamma_n L_x$, where

$$\gamma_n L_x = n\pi \qquad n = 1, 2, 3, \ldots \tag{A7-1-12}$$

The roots given by γ_n are called *eigenvalues* or *characteristic values*. We have shown that there is *a* solution of Eq. (A7-1-1) for every value of n of the form

$$\theta_{tn}(x,t) = F_n(\gamma_n, x)G_n(\gamma_n, t) \tag{A7-1-13}$$

or

$$\theta_{tn} = (c_3 \sin \gamma_n x)[c_1 \exp(-\gamma_n^2 \alpha t)] = A_n(\sin \gamma_n x)e^{-\gamma_n^2 \alpha t} \tag{A7-1-14}$$

where $A_n = c_1 c_3$.

It is a property of linear differential equations that any linear combination of solutions is also a solution. This property also holds for an infinite sum of all the solutions; i.e.,

$$\theta_t = \sum_{n=1}^{\infty} \theta_{tn} = \sum_{n=1}^{\infty} A_n F_n G_n$$

$$= \sum_{n=1}^{\infty} A_n(\sin \gamma_n x) \exp(-\gamma_n^2 \alpha t) \tag{A7-1-15}$$

is also a solution. In order to find the constants A_n we use the condition that $\theta_t = \theta_\infty = 1 - x/L_x$, at $t = 0$. Since $G_n(0) = 1.0$ at $t = 0$, Eq. (A7-1-15) becomes

$$\theta_t(x, 0) = \theta_\infty = \left(1 - \frac{x}{L_x}\right) = \sum_{n=1}^{\infty} A_n F_n$$

$$= \sum_{n=1}^{\infty} A_n \sin \gamma_n x \tag{A7-1-16}$$

This means that the A_n must be determined so that Eq. (A7-1-16) will hold. The method by which this is accomplished is called an expansion of the function $\theta_\infty(x)$ into a Fourier sine series with coefficients A_n. It is based upon the fact that if $\gamma_m = m\pi/L_x$,

$$\int_0^{L_x} \sin \gamma_n x \sin \gamma_m x \, dx = \begin{cases} 0 & \text{if } n \neq m \quad \text{(A7-1-17)} \\ \dfrac{L_x}{2} & \text{if } n = m \quad \text{(A7-1-18)} \end{cases}$$

(The student should verify the above integrals by carrying out the integrations.)

This property of the functions $F_n \equiv \sin \gamma_n x$ and $F_m \equiv \sin \gamma_m x$ is called the *orthogonality condition*. In order to use it, we multiply both sides of Eq. (A7-1-16) by F_m and integrate from 0 to L_x. Then

$$\int_0^{L_x} \left(1 - \frac{x}{L_x}\right) F_m(x)\, dx = \int_0^{L_x} \sum_{n=1}^{\infty} A_n F_n F_n \, dx$$

If we reverse the order of integration and summation in the right-hand side (RHS), we get

$$\text{RHS} = \sum_{n=1}^{\infty} A_n \int_0^{L_x} F_n F_m \, dx = 0 + \cdots + A_m \int_0^{L_x} F_m^2 \, dx + \cdots + 0$$

because of Eq. (A7-1-17). Also because of Eq. (A7-1-18),

$$\text{RHS} = \frac{A_m L_x}{2} \tag{A7-1-19}$$

If we evaluate the left-hand side (LHS) by usual procedures, we obtain

$$\text{LHS} = \frac{1}{\gamma_m}$$

Equating sides gives

$$A_m = \frac{2}{\gamma_m L_x} = \frac{2}{m\pi}$$

To obtain A_n we merely replace m with n

$$A_n = \frac{2}{\gamma_n L_x} = \frac{2}{n\pi}$$

The solution for θ_t is then

$$\theta_t = \sum_{n=1}^{\infty} A_n F_n G_n = \sum_{n=1}^{\infty} \frac{2}{n\pi} \left(\sin \frac{n\pi x}{L_x}\right) \exp\left(\frac{-n^2\pi^2\alpha t}{L_x^2}\right) \tag{A7-1-20}$$

and the complete solution is

$$\theta \equiv \frac{T - T_2}{T_1 - T_2} = 1 - \frac{x}{L_x} - \sum_{n=1}^{\infty} \frac{2}{n\pi} \left(\sin \frac{n\pi x}{L_x}\right) \exp\left(\frac{-n^2\pi^2\alpha t}{L_x^2}\right) \tag{A7-1-21}$$

APPENDIX 7-2 SOLUTION TO HEAT-DIFFUSION EQUATION WITH BOTH SIDES AT SAME TEMPERATURE

The partial differential equation for heat transport in a slab (originally at temperature T_0 and then cooled on both sides $x = \pm a$ by maintaining these surfaces at $T = T_1$) was found to be

$$\alpha \frac{\partial^2 T}{\partial x^2} = \frac{\partial T}{\partial t} \quad \text{(A7-2-1)}$$

When the dimensionless variables

$$\eta = \frac{x}{a} \qquad s \equiv \frac{\alpha t}{a^2} \qquad \theta \equiv \frac{T - T_1}{T_0 - T_1} \quad \text{(A7-2-2)}$$

are used, this becomes

$$\frac{\partial^2 \theta}{\partial \eta^2} = \frac{\partial \theta}{\partial s} \quad \text{(A7-2-3)}$$

The boundary conditions are given in Table A7-2-1. Note that in this case, the *steady*-state solution is $\theta_\infty = 0$.

Following the procedure used in Appendix 7-1, we solve this equation by separation of variables. We let

$$\theta = X(\eta)S(s) \quad \text{(A7-2-4)}$$

Substituting into the partial differential equation and dividing by $X(\eta)S(s)$ gives

$$\frac{X''(\eta)S(s)}{X(\eta)S(s)} = \frac{X(\eta)S'(s)}{X(\eta)S(s)} = -\gamma^2 \quad \text{(A7-2-5)}$$

where γ^2 must be a constant (an eigenvalue), the significance of which is to be determined later. Each of the above equalities gives an ordinary differential equation. These are

$$S' + \gamma^2 S = 0 \qquad \text{and} \qquad X'' + \gamma^2 X = 0 \quad \text{(A7-2-6)}$$

Their respective solutions are

$$S = c_1 e^{-\gamma^2 s} \qquad \text{and} \qquad X = c_2 \cos \gamma\eta + c_3 \sin \gamma\eta \quad \text{(A7-2-7)}$$

From the second boundary condition, γ must be real (otherwise θ would approach ∞ as $t \rightarrow \infty$ instead of approaching zero). Thus the choice of $-\gamma^2$ is convenient.

The boundary conditions indicate that $c_3 = 0$ and

$$\cos \gamma = 0 \quad \text{(A7-2-8)}$$

For any variable u, a plot of $\cos u$ versus u would indicate that there are roots at $\pi/2$, $3\pi/2$, and $5\pi/2$, that is, at $(n + \frac{1}{2})\pi$. Then we can let

$$\gamma_n = (n + \tfrac{1}{2})\pi \quad \text{(A7-2-9)}$$

Table A7-2-1 Table of conditions

Condition	t	x	T	s	η	θ
Initial	0	$\neq \pm a$	T_0	0	$\neq +1$	1
Boundary 1	≥ 0	$-a$	T_1	≥ 0	-1	0
Boundary 2	≥ 0	$+a$	T_1	≥ 0	$+1$	0

represent the nth eigenvalue and let

$$X_n(\gamma_n \eta) \equiv \cos \gamma_n \eta \tag{A7-2-10}$$

represent an *eigenfunction*.

Then if we consolidate $c_1 c_2$ into a single constant A_n and let $S_n = \exp(-\gamma_n^2 s)$,

$$\theta_n = A_n X_n S_n \tag{A7-2-11}$$

which is the solution corresponding to the nth eigenvalue. Since any linear combination of solutions is also a solution, and since there are an infinite number of eigenvalues, we can postulate that the general solution is

$$\theta = \sum_{n=0}^{\infty} A_n X_n(\eta) S_n(s) \tag{A7-2-12}$$

and that the constants A_n must be determined to fit the boundary condition at $t = 0$, that is,

$$\theta(\eta, 0) = 1 = \sum_{n=0}^{\infty} A_n X_n \tag{A7-2-13}$$

As in the previous example, we can multiply both sides of this equation by X_m ($\equiv \cos \gamma_m \eta \, d\eta$) and integrate over all η. Then

$$\int_{-1}^{1} X_m \, d\eta = \int_{-1}^{1} \sum_{n=0}^{\infty} A_n X_n X_m \, d\eta = \sum_{n=0}^{\infty} A_n \int_{-1}^{1} X_n X_m \, d\eta \tag{A7-2-14}$$

By integration it can be shown that

$$\int_{-1}^{1} X_n X_m \, d\eta = \begin{cases} 0 & \text{if } n \neq m \\ \left. \dfrac{\frac{1}{2}(m + \frac{1}{2})\pi\eta + \frac{1}{4}\sin(m + \frac{1}{2})2\pi\eta}{(m + \frac{1}{2})\pi} \right|_{-1}^{+1} & \text{if } n = m \end{cases}$$

Therefore only the term $n = m$ remains as a result of this orthogonality condition.

Then if we integrate the right-hand side of Eq. (A7-2-14), we get

$$\left. \frac{\sin(m + \frac{1}{2})\pi\eta}{(m + \frac{1}{2})\pi} \right|_{-1}^{+1} = A_m \left. \frac{\frac{1}{2}(m + \frac{1}{2})\pi\eta + \frac{1}{4}\sin(m + \frac{1}{2})2\pi\eta}{(m + \frac{1}{2})\pi} \right|_{-1}^{+1} \tag{A7-2-15}$$

Solving for A_m gives

$$A_m = \frac{2(-1)^m}{(m + \frac{1}{2})\pi} \tag{A7-2-16}$$

Substituting into Eq. (A7-2-12) gives the solution

$$\theta = 2 \sum_{n=0}^{\infty} \frac{(-1)^n}{(n + \frac{1}{2})\pi} \exp\left[-\frac{(n + \frac{1}{2})^2 \pi^2 \alpha t}{a^2}\right] \cos(n + \tfrac{1}{2})\frac{\pi x}{a} \tag{A7-2-17}$$

In terms of dimensionless variables this becomes

$$\theta = \sum_{n=0}^{\infty} A_n \exp(-\gamma_n^2 s) \cos \gamma_n \eta \tag{A7-2-18}$$

where

$$A_n = \frac{2(-1)^n}{(n+\frac{1}{2})\pi} \qquad \eta \equiv \frac{x}{a}$$

$$\gamma_n = (n+\tfrac{1}{2})\pi \quad s_n = \exp(-\gamma_n^2 s) \quad s \equiv \frac{\alpha t}{a^2}$$

APPENDIX 7-3 COMBINATION-OF-VARIABLES METHOD

In Sec. 7-9 we discussed the case of transport into a semi-infinite medium and the use of a similarity transformation

$$\xi \equiv \frac{x}{\delta(t)} \tag{A7-3-1}$$

to transform the partial differential equation (7-7)

$$\alpha \frac{\partial^2 \theta}{\partial x^2} = \frac{\partial \theta}{\partial t} \qquad \text{where } \theta \equiv \frac{T - T_2}{T_1 - T_2} \tag{A7-3-2}$$

with initial and boundary conditions

$$T \begin{cases} \to T_2 & \text{as } x \to \infty & \text{(A7-3-3)} \\ = T_2 & \text{at } t = 0,\ x \geq 0 & \text{(A7-3-4)} \\ = T_1 & \text{at } x = 0,\ t \geq 0 & \text{(A7-3-5)} \end{cases}$$

into an ordinary differential equation of the form $\theta(\xi)$. We now illustrate the solution by the method of combination of variables, which is often used for problems of this type.

First, in order to determine whether the similarity transformation is possible, we must substitute the definition Eq. (A7-3-1) into the partial differential equation (A7-3-2) (see above). Thus, by the chain rule of differentiation,

$$\frac{\partial \theta}{\partial t} = \frac{d\theta}{d\xi}\frac{\partial \xi}{\partial t} = \frac{d\theta}{d\xi}\left(-\frac{x}{\delta^2}\frac{d\delta}{dt}\right)$$

$$\frac{\partial \theta}{\partial x} = \frac{d\theta}{d\xi}\frac{\partial \xi}{\partial x} = \frac{d\theta}{d\xi}\frac{1}{\delta}$$

$$\frac{\partial^2 \theta}{\partial x^2} = \left[\frac{d}{d\xi}\left(\frac{\partial \theta}{\partial x}\right)\right]\frac{\partial \xi}{\partial x} = \left(\frac{d^2\theta}{d\xi^2}\frac{1}{\delta}\right)\frac{1}{\delta} = \frac{1}{\delta^2}\frac{d^2\theta}{d\xi^2}$$

Then Eq. (A7-3-2) becomes

$$\frac{\alpha}{\delta^2}\frac{d^2\theta}{d\xi^2} = -\frac{x}{\delta^2}\frac{d\theta}{d\xi}\frac{d\delta}{dt}$$

or, using the definition of ξ to eliminate x,

$$\frac{d^2\theta}{d\xi^2} + \frac{\xi\delta}{\alpha}\frac{d\delta}{dt}\frac{d\theta}{d\xi} = 0 \tag{A7-3-6}$$

This equation still contains t, but it will not if we take

$$\frac{\delta}{\alpha}\frac{d\delta}{dt} = \text{const}$$

For convenience later we let the constant be 2. Then since $\delta = 0$ at $t = 0$ (from physical principles), we get

$$\int_0^\delta \delta\, d\delta = 2\alpha \int_0^t dt \qquad \text{or} \qquad \delta^2 = 4\alpha t$$

and

$$\delta = \sqrt{4\alpha t} \tag{A7-3-7}$$

The ordinary differential equation (A7-3-6) then becomes

$$\frac{d^2\theta}{d\xi^2} + 2\xi\frac{d\theta}{d\xi} = 0 \tag{A7-3-8}$$

The initial condition (A7-3-3) and the boundary condition (A7-3-4) then become the single condition

$$\theta = 0 \qquad \text{at } \xi = \infty \tag{A7-3-9}$$

and the third condition becomes

$$\theta = 1 \qquad \text{at } \xi = 0 \tag{A7-3-10}$$

Thus we can reduce the three conditions to the above two, making the method possible. To solve Eq. (A7-3-8) we let $\phi \equiv d\theta/d\xi$, so that

$$\frac{d\phi}{d\xi} + 2\xi\phi = 0 \qquad \text{or} \qquad \frac{d\phi}{\phi} = -2\xi\, d\xi \qquad \text{or} \qquad \phi = c_1 e^{-\xi^2} = \frac{d\theta}{d\xi}$$

Then

$$\int_1^\theta d\theta = c_1 \int_0^\xi e^{-\xi^2}\, d\xi \tag{A7-3-11}$$

and

$$\int_1^0 d\theta = c_1 \int_0^\infty e^{-\xi^2}\, d\xi \tag{A7-3-12}$$

Dividing Eq. (A7-3-11) by Eq. (A7-3-12) and rearranging gives

$$\theta = 1 - \frac{\int_0^\xi e^{-\xi^2}\, d\xi}{\int_0^\infty e^{-\xi^2}\, d\xi} = 1 - \frac{2}{\sqrt{\pi}}\int_0^\xi e^{-\xi^2}\, d\xi \equiv 1 - \operatorname{erf}\xi \equiv \operatorname{erfc}\xi \tag{A7-3-13}$$

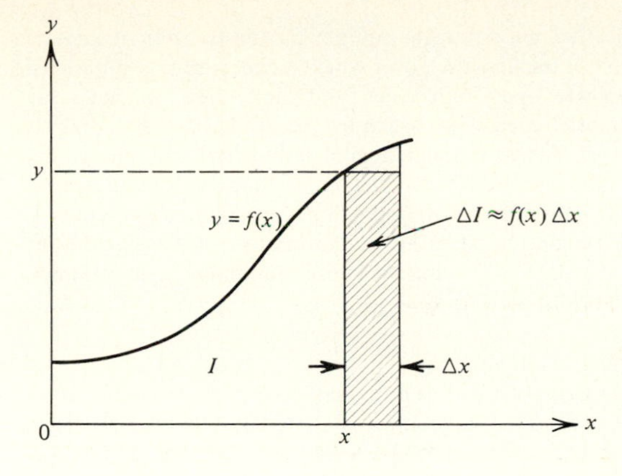

Figure A7-4-1 Derivative of integral with variable upper limit.

APPENDIX 7-4 DERIVATIVE OF AN INTEGRAL WITH VARIABLE UPPER LIMIT

Consider the integral

$$I = \int_0^x f(x)\,dx$$

which is the area under the curve of $f(x)$ between $x = 0$ and $x = x$ (see Fig. A7-4-1). Suppose we wish to find dI/dx. The derivative is then

$$\frac{dI(x)}{dx} = \lim_{\Delta x \to 0} \frac{I(x + \Delta x) - I(x)}{\Delta x} = \lim_{\Delta x \to 0} \frac{\displaystyle\int_x^{x+\Delta x} f(x)\,dx}{\Delta x}$$

since the numerator is the rectangular area between x and $x + \Delta x$ or $\Delta I = f(x)\,\Delta x$, and

$$\frac{dI(x)}{dx} = \lim_{\Delta x \to 0} \frac{f(x)\,\Delta x}{\Delta x} = f(x)$$

or

$$\frac{d}{dx} \int_0^x f(x)\,dx = f(x) \tag{A7-4-1}$$

Derivative of a Definite Integral

If both upper and lower limits are independent of x, the derivative dI/dx is zero since I depends only on the limits.

PROBLEMS

7-1 Carry out the detailed steps to derive Eqs. (7-23) and (7-24)

7-2 Derive equations similar to the first part of Eqs. (7-23) and the second of Eqs. (7-24) for a cylindrical geometry.

7-3 In absorption columns, extractors, and other mass-transfer equipment, the transfer of mass is assumed to take place through a stagnant film of thickness L_x, over which a concentration difference $C_{A1} - C_{A2}$ exists. The rate of mass transfer per unit time and unit area is given by $N_A = k_c(C_{A1} - C_{A2})$, where k_c is the mass-transfer coefficient, which for the steady state is D_{AB}/L_x. It is desired to estimate the unsteady-state transfer of mass through the film and the effect of time on the mass-transfer coefficient. For this purpose, consider a simplified model for leaching a substance A from a solid by a solvent B. At $t < 0$ the concentration throughout the liquid film is everywhere equal to C_{A2}. At all times $t > 0$ at the surface (which may be considered flat) the concentration of A is equal to its solubility C_{A1}, while in the liquid stream at the edge of the film the concentration is C_{A2} at all times. Since the solution is dilute, bulk flow can be neglected. The steady-state concentration profile will be linear.

(*a*) Obtain a differential equation for the molar flux N_{Ax}.

(*b*) Write a general expression for N_{Ax} in terms of diffusive flux and flux resulting from bulk flow. Write Fick's law of diffusion and derive a differential equation in C_A, stating assumptions made.

(*c*) Using an analogy to the heat-transport case, write the solution for θ_∞, θ_t, and θ. Let $\theta = (C_A - C_{A2})(C_{A1} - C_{A2})$. Prepare a table of all conditions on C_A, θ, θ_∞, and θ_t at various values of t and x.

(*d*) Derive an equation for N_{Ax} at $x = 0$ and for k_c.

7-4 In a lubrication system an oil is confined between two flat and parallel metal surfaces with area A. The distance between the plates L_y is small compared with the other dimensions. At time $t < 0$ the fluid is at rest. At $t > 0$, the lower plate is pulled in the x direction with a constant velocity V while the upper plate is held stationary. It requires 2 s to achieve an average velocity (in the region between the plates) of $\langle v \rangle$.

(*a*) Write a differential equation for momentum flux and velocity.

(*b*) Define dimensionless variables and write the velocity equation in dimensionless form.

(*c*) If the plate spacing were doubled and the velocity of the lower plate remained the same, how long would it take to achieve the same average velocity as with the smaller plate spacing?

(*d*) Instead of changing the plate spacing the kinematic viscosity of the fluid is increased 50 percent. What time will be required to achieve the same average velocity after 2 s as with the original fluid?

7-5 A thin sheet is being dried on *both* sides by exposure to nearly dry air. Originally ($t = 0$) the sheet has a uniform concentration of water vapor C_{A0}. At $t > 0$ both sides are suddenly exposed to air with a lower concentration C_{A1} of water vapor, which can be assumed constant at all later times. Let $x = +a$ at one side and $x = -a$ at the other side; the midpoint, or centerline, is $x = 0$. Assume that the length L and the width W are so large with respect to the thickness $2a$ that mass transfer (which is assumed to occur by diffusion and bulk flow) will be in the x direction only. After an infinite length of time, the sheet is assumed to be at unknown concentration $C_{A\infty} = C_{A1}$.

(*a*) Sketch and label curves of C_A versus x at $t = 0$, $t = t$, and $t = \infty$.

(*b*) Derive a partial differential equation for the molar flux N_{Ax}.

(*c*) Write the flux law and use it, neglecting bulk flux, to obtain the diffusion equation.

(*d*) Write the above equation and the initial conditions and boundary conditions in dimensionless form.

(*e*) Determine whether this problem is analogous to the problem in Example 7-2, and if so write the solution.

7-6 A rubber sheet initially at 288 K is cured by exposing both sides of the sheet to hot air at 422 K. Calculate the time required for the centerline of the sheet to reach 400 K (use charts) if the thickness of the sheet is 5.08 mm; $\rho = 6408\ \mathrm{kg/m^3}$; $\widehat{C}_p = 3.347\ \mathrm{kJ/kg \cdot K}$; $k = 0.1592\ \mathrm{W/m \cdot K}$.

7-7 A roast is usually considered ready when the center has reached some minimum temperature T_M. Assuming that it takes 1 h to cook a 4-kg roast, use dimensional analysis to estimate the cooking time for a 7.5-kg roast. Assume the roast is a sphere. List all other assumptions.

7-8 Using an incremental volume, compare the derivation of the partial differential equations for one-dimensional unsteady-state transport of energy, momentum, and moles A by listing the following in parallel columns: transport rate in, transport rate out, net transport rate, accumulation rate, differential equation for flux, flux law, differential equation for potential, dimensionless dependent variable, dimensionless time.

7-9 (*a*) A long cylinder 7.62×10^{-2} m in diameter is being cooled by blowing air over it at high velocity. If the initial temperature of the cylinder is 722 K and the air temperature is 300 K, calculate the time for the temperature at the center of the cylinder to fall to 333 K. Assume that α is 1.032×10^{-5} m²/s and that there is negligible surface resistance.

(*b*) Repeat if the diameter is doubled.

7-10 In the naval stores industry, rosin is leached out of wood chips with various organic solvents such as gasoline. Assume that $D_{AB} = 2.58 \times 10^{-7}$ m²/s, that the rosin concentration in gasoline is negligible, and that the initial rosin concentration in the chip was 160 kg/m³. Neglecting surface resistance, calculate the leaching time to remove 95 percent of the rosin from (*a*) 1.27-mm-thick chips and (*b*) 0.63-mm-thick chips. *Note:* Assume the chips are a slab; but since the Gurney-Lurie chart (Fig. 7-3) does not give a curve for the average concentration in the slab, and since it is needed to find the total resin content, you must integrate point values. This can be done either numerically using values from the chart or analytically using the analytical solution, Eq. (7-36). The latter is much more accurate.

7-11 A steam pipe of length L and inner radius R_1 is insulated to a total radius of R_2. Initially the pipe and insulation are at the outside surface temperature T_2. At time $t = 0$ steam has filled the pipe and the inner surface of the insulation is held at T_1. (Neglect the period during which the metal wall is being heated up and assume that temperature T_1 is achieved instantly at $t = 0$ and that both T_1 and T_2 are constant.) Determine whether the log-mean area is still valid at unsteady state by carrying out the following steps:

(*a*) Derive partial differential equations for (1) heat flux and (2) temperature.

(*b*) Write the equation in dimensionless form by using the variables $\phi \equiv (T - T_2)/(T_1 - T_2)$, $\xi \equiv r/R_1$, and $s \equiv \alpha t/R_1^2$.

(*c*) Prepare a table of boundary and initial conditions for the dimensional and dimensionless variables.

(*d*) Expressing the solution in terms of a steady-state term ϕ_∞ and a transient term ϕ_t, prepare a sketch showing ϕ, ϕ_∞, and ϕ_t versus position ξ at typical dimensionless times. Label curves.

(*e*) Carry out the solution of the steady-state equation, making use of the boundary conditions.

(*f*) Write the boundary conditions for the transient equation.

(*g*) Outline the method of solution for the unsteady-state case; i.e., show that

$$\phi = \phi_\infty - \sum_{n=1}^{\infty} A_n X_n(\beta_n \xi) S_n(s; \beta_n) \tag{P7-11-1}$$

where

$$S_n = \exp(-\beta_n^2 s) \tag{P7-11-2}$$

β_n and X_n are respectively eigenvalues and eigenfunctions, and the A_n are constants. (You need not find the β_n, X_n, or A_n.)

(*h*) If the mean area is defined by

$$\dot{Q}_1 \equiv k A_m \frac{T_1 - T_2}{R_2 - R_1} \tag{P7-11-3}$$

where the heat flow $\dot{Q}_1$ can be calculated at R_1 from

$$\dot{Q}_1 = A_r q_r \Big|_{R_1} \tag{P7-11-4}$$

show clearly how to obtain A_m at unsteady state and whether it is the same as for steady state. Assume that the analytical solution is known.

7-12 An engineer in your company is working on a die for coating wires by pulling the wire of radius R_1 through a coating fluid contained in an annulus of outer radius R_2. You explain that this problem may be analogous in both the steady and unsteady cases to heat transport in an annulus (Prob. 7-11).

(*a*) State the physical conditions for which this problem is analogous to Prob. 7-11.

(*b*) By analogy write the differential equations for momentum flux and velocity. (No formal derivation required.)

(*c*) Tabulate the boundary conditions on velocity, time, and position.

(*d*) Give definitions for the dimensionless quantities ϕ, s, and ζ that will yield the same dimensionless differential equation and boundary and initial conditions for the momentum-transport problem as for the heat-transport problem.

(*e*) Write an expression analogous to Eq. (P7-11-4) and state how the solution [Eq. (P7-11-1)] to the heat case could be used in the momentum case to find the force required to pull the wire if A_n, β_n, and X_n were known.

7-13 Describe an analogy to mass transport and state how Eqs. (P7-11-1) and (P7-11-2) would be used so as to be applicable to that case.

7-14 Show by the following problems that the results obtained in Prob. 7-11 can be used without finding a complete analytical solution to the problem if the form of the solution is as given but X_n, β_n, and A_n are not known. Assume that only one term in the series is important at large times.

(*a*) When $T_1 = 1100°F$ and $T_2 = 100°F$, $R_1 = 2$ in, and $R_2 = 4$ in, the temperature midway between R_1 and R_2 is found to be 500°F after 1 h and 510°F after 1.5 h. (1) If both dimensions are doubled, how long will it take to obtain the same temperature at the midpoint? (2) If both dimensions are increased 50 percent, what will the temperature be after 1 h? The insulation has a thermal diffusivity of 0.01 ft^2/h.

(*b*) Assume that the above information is known for the thermal system and that R_2/R_1 is the same for both the heat-transfer system and the wire-coating system (Prob. 7-12). If the kinematic viscosity of the fluid and the diffusivity of the insulation are known, is it possible to obtain the velocity of the fluid at the midpoint ($\zeta = 1.5$) as a function of time? If not, why not; if so, how? Use H to refer to the heat-transfer system and M to refer to the momentum system. Assume that *no* data are available and no measurements can be made on the momentum system.

7-15 Refer to the slab in Prob. 1-2. Plot the rate of heat flow into the slab versus time for various small values of time. Also plot the penetration depths $\sqrt{\pi\alpha t}$ and $4\sqrt{\alpha t}$ and the temperatures at the penetration depths.

REFERENCES

1. Bonilla, C. F.: "Nuclear Engineering," McGraw-Hill, New York, 1957.
2. Carberry, J. C.: *Trans. Inst. Chem. Eng.*, **80**:75 (1981).
3. Carslaw, H. S., and J. C. Jaeger: "Conduction of Heat in Solids," 2d ed., Oxford University Press, London, 1959.
4. "CRC Handbook of Tables for Mathematics," 3d ed., Chemical Rubber Co., Cleveland, 1967.
5. Danckwerts, P. V.: *Ind. Eng. Chem.*, **43**:1460 (1951).
6. Gurney, H. P., and J. Lurie: *Ind. Eng. Chem.*, **15**:1170 (1923).
7. Higbie, R.: *Trans. AIChE J.*, **31**:365 (1935).
8. Scriven, L. E.: *Chem. Eng. Educ.*, **2**:150 (1968); **3**:26, 94 (1969).
9. Sherwood, T. K., R. L. Pigford, and C. R. Wilke: "Mass Transfer," McGraw-Hill, New York, 1975.
10. Toor, H. L., and J. M. Marchello: *AIChE J.*, **4**:97 (1958); *Ind. Eng. Chem. Fund.*, **2**:8 (1963).

PART TWO

MULTIDIMENSIONAL TRANSPORT

INTRODUCTION TO PART TWO

In Part One we used an incremental balance over an element of volume to develop a differential equation for one-dimensional transport of heat, momentum, or mass. For each physical situation, e.g., heat transport in a slab, Couette flow, diffusion with bulk flow, we showed how to derive the differential equation appropriate to the processes. This procedure was valuable in illustrating the meaning of the terms involved and as a general procedure it is also useful in simple physical cases; but in more complex problems, such as in multidimensional transport, it becomes tedious and repetitious. For these complex cases it is better to use previously derived general differential balance equations which can then be simplified for specific physical applications. These general differential equations apply at an arbitrary point in the medium and do not involve the boundary conditions. The general equations can be written (1) as partial differential equations in the flux components or in the potentials (T, C_A, v_x, etc.) and (2) as vector-tensor equations in the flux vectors or tensors or the potentials. The general partial differential equations for various coordinate systems are tabulated in Tables I-6, I-7, I-8 and the vector equations in Tables I-12, I-13, I-14, of this Introduction.

The vector equations have the advantage of being compact, in that several terms can be combined into one, and of being general, in that a vector equation applies to any coordinate system. The use of vectors also enhances our understanding of the physical processes occurring in the three-dimensional world in which we live, and vector notation is widely used in the literature. For that reason it will be emphasized in this part of the text. For instructors who do not wish to delve into vector notation at this time, a parallel treatment which does not use vector notation is provided.

Paths through Part Two

There are three main paths through Part Two, each at a different level. At the advanced level, all derivations in the text and Appendixes can be included and selected advanced parts of Chaps. 9 to 11. At an intermediate level the basic derivations and examples in Chaps. 8 to 11 can be studied with sections so indicated by a footnote omitted on first reading, along with selected advanced parts of Chaps. 9 to 11. The shortest path, suited to an elementary course, is to forgo all or part of the discussion in Chaps. 8 and 9 of vectors and tensors and instead move directly to the examples listed in the accompanying table, which illustrate the simplification and application of the general equations as given in the tables (see Summary of Procedures following). A recommended modification of the above would be to include Secs. 8-1, 8-3, 8-5, 8-8, 8-12, and 8-18. One may also include the derivations (not requiring vector notation) in Secs. 8-22, 9-3, and 10-1. A final option would be to include some of the basic material on vector operations in Chap. 8, e.g., Secs. 8-2, 8-4, 8-6, 8-7, 8-11, 8-23, 8-26, and 8-28.

Example	Section	Title
8-1	8-1	Heat conduction in a solid
8-3	8-5	Energy balance in a solid
8-5	8-23	Application of continuity equation to pipe flow
9-1	9-11	Flow between parallel plates (equation of motion)
9-3	9-19	Slit flow (Navier-Stokes equation)
9-4		Cylindrical-tube flow
9-5		Torque in a viscometer
9-8	9-18	Flow in entrance region between parallel plates
10-1	10-4	Viscous heating in an annulus
10-2	10-5	Laminar-flow heat transfer with constant wall temperature
10-3	10-6	Forced-convection laminar-flow mass transfer
10-4	10-7	Heat transfer at entrance to a pipe
10-5		Flat velocity profile—falling-film mass transfer
10-6	10-9	Turbulent-flow heat transfer in a pipe
10-7	10-11	Turbulent diffusion in the atmosphere
10-8	10-12	Tubular-reactor design: two-dimensional model
10-9	10-14	Free convection from a vertical flat plate

Summary of Procedures for the Derivation of General Equations

For cartesian coordinates the general equations are derived by making a balance over an element $\Delta\mathcal{V} = \Delta x\, \Delta y\, \Delta z$. Such derivations are carried out for mass and for moles of A in Sec. 8-22, for momentum in Sec. 9-3, and for total energy in Sec. 10-1. An energy equation in terms of temperature is derived in Chap. 10.

For curvilinear coordinate systems, e.g., cylindrical, an incremental balance over an element of volume in that system can similarly be used to derive mass-balance equations, one for the entire fluid and another for species A. This method cannot be used for momentum transport, however, because momentum is a vector, for which changes in *direction* must be considered as well as changes in magnitude. The momentum-balance equations in curvilinear coordinates can be obtained by transformation of the equations in cartesian coordinates into the curvilinear coordinate system. One method is to use the transformation equations, for example, $x = r\cos\theta$, $y = r\sin\theta$, $z = z$, and the chain rule of *differentiation,* but this method is tedious and is not used in this text.

A second method is to write the cartesian component equations in vector notation ($\mathbf{v}$ rather than v_x) (see Table I-12) and then use various methods to expand the individual terms into other coordinate systems. These procedures are described in Chaps. 8 and 9, and the expansions are given in the tables. This method is based upon the fact that a vector equation in vector notation applies to *all* coordinate systems. Therefore an equation that is written in vector notation can be transformed into any other system. This method has several advantages: vector notation is much more compact, and the use of vector notation can enhance our understanding of the physical processes involved. The physical meaning of various vector operations is discussed in Chap. 8.

Another approach is to derive the equations of change directly in vector form using an element of volume of arbitrary shape. This derivation is carried out in Appendix 10-2.

How to Use the General Equations

To use the component equations in Tables I-6, I-7, I-8 one need not be skilled in vector and tensor analysis. One merely needs to know how to simplify the general equations by eliminating terms that are zero or negligible for the case at hand; e.g., for unsteady-state conditions $\partial/\partial t = 0$, and one strikes out all partial derivatives with respect to time. For no change in the y direction, $\partial/\partial y = 0$, and so on.

In some cases this method is awkward and cumbersome compared with the use of the vector equations in Table I-12, in which one can immediately delete such vector terms as $\nabla \cdot \mathbf{v}$, $\mathbf{v} \cdot \nabla\mathbf{v}$, etc., which are zero or negligible. Since these terms each consist of several terms when they are expanded, a number of terms are thus eliminated at once. Then the other terms in the vector equation can be expanded into components using the list of vector operations in the tables.

Example of Use of the Tables for a Problem in Part One

As an example of the use of the equations in the tables, consider unsteady-state mass transport in one direction (Sec. 7-3). The component equation (from part A of Table I-6) is

$$\underset{(1)}{\frac{\partial C_A}{\partial t}} + \underset{(2)}{v_x \frac{\partial C_A}{\partial x}} + \underset{(3)}{v_y \frac{\partial C_A}{\partial y}} + \underset{(4)}{v_z \frac{\partial C_A}{\partial z}} + C_A\left(\underset{(5)}{\frac{\partial v_x}{\partial x}} + \underset{(6)}{\frac{\partial v_y}{\partial y}} + \underset{(7)}{\frac{\partial v_z}{\partial z}}\right)$$

$$= -\underset{(8)}{\frac{\partial J_{Ax}}{\partial x}} - \underset{(9)}{\frac{\partial J_{Ay}}{\partial y}} - \underset{(10)}{\frac{\partial J_{Az}}{\partial z}} + \underset{(11)}{R_A}$$

Since the system is not at steady state, term (1) is not zero. If it is assumed that there is no bulk flow, $v_x = 0 = v_y = v_z$, thus eliminating terms (2) to (7). For diffusion in the x direction only terms (9) and (10) are zero, and for no reaction term (11) is zero. Then

$$\frac{\partial C_A}{\partial t} = -\frac{\partial J_{Ax}}{\partial x}$$

which is the first of Eqs. (7-23).

If we assume constant ρ and D_{AB}, part B of Table I-6 gives

$$\frac{\partial C_A}{\partial t} = D_{AB}\frac{\partial^2 C_A}{\partial x^2}$$

which is identical to the second of Eqs. (7-23).

Vector Equations†

Now let us use the vector equation† in part A of Table I-12

$$\underset{(1)}{\frac{\partial C_A}{\partial t}} + \underset{(2)}{\mathbf{v}\cdot\nabla C_A} + \underset{(3)}{C_A\nabla\cdot\mathbf{v}} = \underset{(4)}{-\nabla\cdot\mathbf{J}_A} + \underset{(5)}{R_A}$$

If bulk flow is neglected, terms (2) and (3) can be eliminated without having to expand them into components. Then

$$\frac{\partial C_A}{\partial t} + \nabla\cdot\mathbf{J}_A = 0$$

Substituting $\nabla\cdot\mathbf{J}_A$ from Table I-3 (or part A of Table I-6), we get

$$\frac{\partial C_A}{\partial t} + \frac{\partial J_{Ax}}{\partial x} + \frac{\partial J_{Ay}}{\partial y} + \frac{\partial J_{Az}}{\partial z} = 0$$

which for diffusion in the x direction only reduces to

$$\frac{\partial C_A}{\partial t} + \frac{\partial J_{Ax}}{\partial x} = 0$$

†May be deleted in nonvector approach.

Table I-1 Transformation Equations

Cylindrical coordinates

System defined by the transformation equations

$$x = r\cos\theta \qquad y = r\sin\theta \qquad z = z$$

In index notation, if $u_1 \equiv r,\ u_2 \equiv \theta,\ u_3 \equiv z$,

$$x_1 = u_1 \cos u_2 \qquad x_2 = u_1 \sin u_2 \qquad x_3 = u_3$$

A point P can be represented as either $P(x, y, z)$ or $P(r, \theta, z)$:

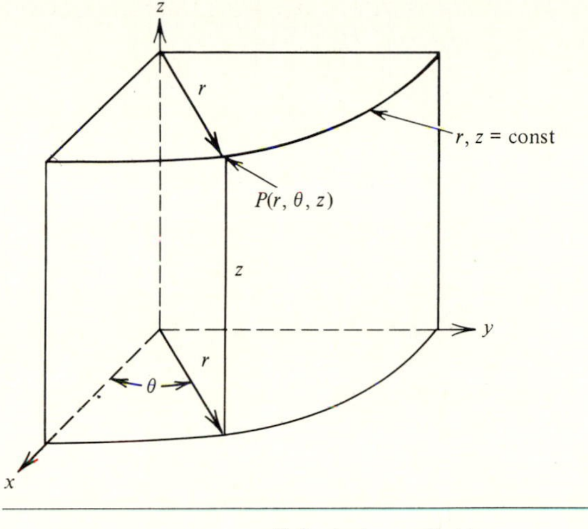

Spherical coordinates

System defined by the transformation equations

$$x = r\sin\theta\cos\phi \qquad y = r\sin\theta\sin\phi \qquad z = r\cos\theta$$

A point $P(x, y, z)$ can be also represented by $P(r, \theta, \phi)$:

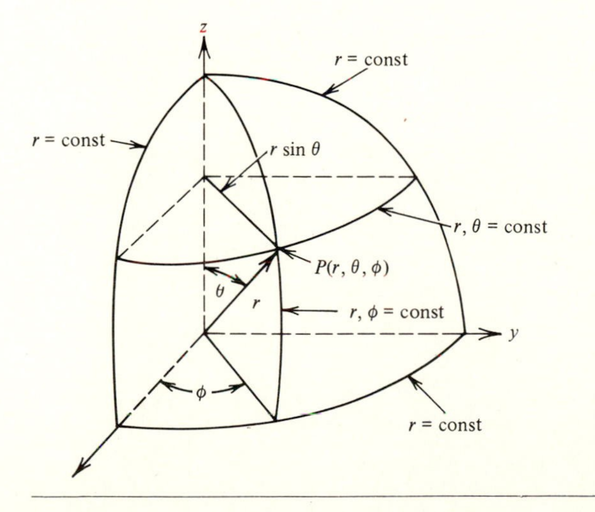

Table I-2 Basic vector operations

Vector $\mathbf{q}$	Gradient of a scalar ∇T	Divergence of a vector $\nabla \cdot \mathbf{q}$	Laplacian of a scalar $\nabla^2 T$
Cartesian coordinates			
$q_x \boldsymbol{\delta}_x + q_y \boldsymbol{\delta}_y + q_z \boldsymbol{\delta}_z$ (8-8)	$\frac{\partial T}{\partial x} \boldsymbol{\delta}_x + \frac{\partial T}{\partial y} \boldsymbol{\delta}_y + \frac{\partial T}{\partial z} \boldsymbol{\delta}_z$ (8-13)	$\frac{\partial q_x}{\partial x} + \frac{\partial q_y}{\partial y} + \frac{\partial q_z}{\partial z}$ (8-28)	$\frac{\partial^2 T}{\partial x^2} + \frac{\partial^2 T}{\partial y^2} + \frac{\partial^2 T}{\partial z^2}$ (8-30)
Cylindrical coordinates			
$q_r \boldsymbol{\delta}_r + q_\theta \boldsymbol{\delta}_\theta + q_z \boldsymbol{\delta}_z$ (8-14b)	$\frac{\partial T}{\partial r} \boldsymbol{\delta}_r + \frac{\partial T}{r\,\partial \theta} \boldsymbol{\delta}_\theta + \frac{\partial T}{\partial z} \boldsymbol{\delta}_z$ (8-19)	$\frac{\partial\left(r \frac{\partial T}{\partial r}\right)}{r\,\partial r} + \frac{\partial^2 T}{r^2\,\partial \theta^2} + \frac{\partial^2 T}{\partial z^2}$ (8-98b)	$\frac{\partial (r q_r)}{r\,\partial r} + \frac{\partial q_\theta}{r\,\partial \theta} + \frac{\partial q_z}{\partial z}$ (8-98)
Spherical coordinates			
$q_r \boldsymbol{\delta}_r + q_\theta \boldsymbol{\delta}_\theta + q_\phi \boldsymbol{\delta}_\phi$	$\frac{\partial T}{\partial r} \boldsymbol{\delta}_r + \frac{\partial T}{r\,\partial \theta} \boldsymbol{\delta}_\theta + \frac{1}{r \sin\theta} \frac{\partial T}{\partial \phi} \boldsymbol{\delta}_\phi$	$\frac{1}{r^2} \frac{\partial}{\partial r}(r^2 q_r) + \frac{1}{r \sin\theta} \frac{\partial}{\partial \theta}(q_\theta \sin\theta) + \frac{1}{r \sin\theta} \frac{\partial q_\phi}{\partial \phi}$	$\frac{\partial}{r^2\,\partial r}\left(r^2 \frac{\partial T}{\partial r}\right) + \frac{1}{r^2 \sin\theta} \frac{\partial\left(\sin\theta \frac{\partial T}{\partial \theta}\right)}{\partial \theta} + \frac{1}{r^2 \sin^2\theta} \frac{\partial^2 T}{\partial \phi^2}$

Table I-3 Expansion of the term $\mathbf{v}\cdot\nabla T$ appearing in the substantial derivative of a scalar T

$$\frac{DT}{Dt}\equiv\frac{\partial T}{\partial t}+\mathbf{v}\cdot\nabla T$$

Rectangular	Cylindrical
$v_x\frac{\partial T}{\partial x}+v_y\frac{\partial T}{\partial y}+v_z\frac{\partial T}{\partial z}$	$v_r\frac{\partial T}{\partial r}+v_\theta\frac{\partial T}{r\,\partial\theta}+\frac{\partial v_z}{\partial z}$
Spherical	**General orthogonal**
$v_r\frac{\partial T}{\partial r}+v_\theta\frac{\partial T}{r\,\partial\theta}+\frac{v_\phi}{r\sin\theta}\frac{\partial T}{\partial\phi}$	$\sum_m v_m\frac{\partial T}{h_m\,\partial u_m}$ where $h_m^2=\sum_i\left(\frac{\partial x_i}{\partial u_m}\right)^2$

Table I-4 Terms appearing in equation of motion†

$$\underset{(1)}{\rho\frac{D\mathbf{v}}{Dt}}\equiv\rho\left(\underset{(2)}{\frac{\partial\mathbf{v}}{\partial t}}+\underset{(3)}{\mathbf{v}\cdot\nabla\mathbf{v}}\right)=\underset{(4)}{-\nabla\cdot\underline{\tau}}\underset{(5)}{-\nabla p}+\underset{(6)}{\rho\mathbf{g}}\tag{9-34}$$

Cartesian

$$\rho\left(\frac{\partial v_i}{\partial t}+\sum_j v_j\frac{\partial v_i}{\partial x_j}\right)=-\sum_j\frac{\partial\tau_{ji}}{\partial x_j}-\frac{\partial p}{\partial x_i}+\rho g_i$$

Orthogonal

$$\rho\left(\frac{\partial v_u}{\partial t}+[\![(\mathbf{v}\cdot\nabla\mathbf{v})_u]\!]\right)=-[\![(\nabla\cdot\underline{\tau})_u]\!]-[\![(\nabla p)_u]\!]+\rho g_u$$

Cylindrical

$$[\![(\mathbf{v}\cdot\nabla\mathbf{v})_r]\!]=v_r\frac{\partial v_r}{\partial r}+v_\theta\frac{\partial v_r}{r\,\partial\theta}+v_z\frac{\partial v_r}{\partial z}-\frac{v_\theta^2}{r}$$

$$[\![(\nabla\cdot\underline{\tau})_r]\!]=\frac{\partial(r\tau_{rr})}{\partial r}+\frac{1}{r}\frac{\partial\tau_{\theta r}}{\partial\theta}+\frac{\partial\tau_{zr}}{\partial z}-\frac{\tau_{\theta\theta}}{r}$$

$$[\![(\nabla p)_r]\!]=\frac{\partial p}{\partial r}$$

$$[\![(\mathbf{v}\cdot\nabla\mathbf{v})_\theta]\!]=v_r\frac{\partial v_\theta}{\partial r}+\frac{v_\theta}{r}\frac{\partial v_\theta}{\partial\theta}+v_z\frac{\partial v_\theta}{\partial z}+\frac{v_r v_\theta}{r}$$

$$[\![(\nabla\cdot\underline{\tau})_\theta]\!]=\frac{1}{r^2}\frac{\partial}{\partial r}(r^2\tau_{r\theta})+\frac{1}{r}\frac{\partial\tau_{\theta\theta}}{\partial\theta}+\frac{\partial\tau_{z\theta}}{\partial z}$$

$$[\![(\nabla p)_\theta]\!]=\frac{\partial p}{r\,\partial\theta}$$

†Brackets [[ ]] are used to denote a scalar component.

Table I-4 Terms appearing in equation of motion (*cont.*)

$$[\![(\mathbf{v}\cdot\nabla\mathbf{v})_z]\!] = v_r \frac{\partial v_z}{\partial r} + v_\theta \frac{\partial v_z}{r\,\partial\theta} + v_z \frac{\partial v_z}{\partial z}$$

$$[\![(\nabla\cdot\underline{\tau})_z]\!] = \frac{1}{r}\frac{\partial(r\tau_{rz})}{\partial r} + \frac{1}{r}\frac{\partial\tau_{\theta z}}{\partial\theta} + \frac{\partial\tau_{zz}}{\partial z}$$

$$[\![(\nabla p)_z]\!] = \frac{\partial p}{\partial z}$$

Spherical

$$[\![(\mathbf{v}\cdot\nabla\mathbf{v})_r]\!] = v_r \frac{\partial v_r}{\partial r} + \frac{v_\theta}{r}\frac{\partial v_r}{\partial\theta} + \frac{v_\phi}{r\sin\theta}\frac{\partial v_r}{\partial\phi} - \frac{v_\theta^2 + v_\phi^2}{r}$$

$$[\![(\nabla\cdot\underline{\tau})_r]\!] = \frac{1}{r^2}\frac{\partial}{\partial r}(r^2\tau_{rr}) + \frac{1}{r\sin\theta}\frac{\partial}{\partial\theta}(\tau_{r\theta}\sin\theta) + \frac{1}{r\sin\theta}\frac{\partial\tau_{\phi r}}{\partial\phi} - \frac{\tau_{\theta\theta} + \tau_{\phi\phi}}{r}$$

$$[\![(\nabla p)_r]\!] = \frac{\partial p}{\partial r}$$

$$[\![(\mathbf{v}\cdot\nabla\mathbf{v})_\theta]\!] = v_r \frac{\partial v_\theta}{\partial r} + \frac{v_\theta}{r}\frac{\partial v_\theta}{\partial\theta} + \frac{v_\phi}{r\sin\theta}\frac{\partial v_\theta}{\partial\phi} + \frac{v_r v_\theta}{r} - \frac{v_\phi^2\cot\theta}{r}$$

$$[\![(\nabla\cdot\underline{\tau})_\theta]\!] = \frac{1}{r^2}\frac{\partial}{\partial r}(r^2\tau_{r\theta}) + \frac{1}{r\sin\theta}\frac{\partial}{\partial\theta}(\tau_{\theta\theta}\sin\theta) + \frac{1}{r\sin\theta}\frac{\partial\tau_{\phi\theta}}{\partial\phi} + \frac{\tau_{r\theta}}{r} - \frac{\cot\theta}{r}\tau_{\phi\phi}$$

$$[\![(\nabla p)_\theta]\!] = \frac{1}{r}\frac{\partial p}{\partial\theta}$$

$$[\![(\mathbf{v}\cdot\nabla\mathbf{v})_\phi]\!] = v_r \frac{\partial v_\phi}{\partial r} + \frac{v_\theta}{r}\frac{\partial v_\phi}{\partial\theta} + \frac{v_\phi}{r\sin\theta}\frac{\partial v_\phi}{\partial\phi} + \frac{v_\phi v_r}{r} + \frac{v_\theta v_\phi}{r}\cot\theta$$

$$[\![(\nabla\cdot\underline{\tau})_\phi]\!] = \frac{1}{r^2}\frac{\partial}{\partial r}(r^2\tau_{r\phi}) + \frac{1}{r}\frac{\partial\tau_{\theta\phi}}{\partial\theta} + \frac{1}{r\sin\theta}\frac{\partial\tau_{\phi\phi}}{\partial\phi} + \frac{\tau_{r\phi}}{r} + \frac{2\cot\theta}{r}\tau_{\theta\phi}$$

$$[\![(\nabla p)_\phi]\!] = \frac{1}{r\sin\theta}\frac{\partial p}{\partial\phi}$$

Table I-5 Terms appearing in Navier-Stokes equations (constant ρ, μ)†

$$\rho \frac{D\mathbf{v}}{Dt} \equiv \rho\left(\frac{\partial \mathbf{v}}{\partial t} + \mathbf{v}\cdot\nabla\mathbf{v}\right) = \mu\,\nabla^2\mathbf{v} - \nabla p + \rho\mathbf{g}$$

(1) (2) (3) (4) (5) (6)

Cartesian

$$[\![(\nabla^2\mathbf{v})_j]\!] = \sum_i \frac{\partial^2 v_j}{\partial x_i}$$

Cylindrical

$$[\![(\nabla^2\mathbf{v})_r]\!] = \frac{\partial}{\partial r}\left[\frac{1}{r}\frac{\partial}{\partial r}(rv_r)\right] + \frac{1}{r^2}\frac{\partial^2 v_r}{\partial\theta^2} - \frac{2}{r^2}\frac{\partial v_\theta}{\partial\theta} + \frac{\partial^2 v_r}{\partial z^2}$$

$$[\![(\nabla^2\mathbf{v})_\theta]\!] = \frac{\partial}{\partial r}\left[\frac{1}{r}\frac{\partial}{\partial r}(rv_\theta)\right] + \frac{1}{r^2}\frac{\partial^2 v_\theta}{\partial\theta^2} + \frac{2}{r^2}\frac{\partial v_r}{\partial\theta} + \frac{\partial^2 v_\theta}{\partial z^2}$$

$$[\![(\nabla^2\mathbf{v})_z]\!] = \frac{1}{r}\frac{\partial}{\partial r}\left(r\frac{\partial v_z}{\partial r}\right) + \frac{1}{r^2}\frac{\partial^2 v_z}{\partial\theta^2} + \frac{\partial^2 v_z}{\partial z^2}$$

Spherical

$$[\![(\nabla^2\mathbf{v})_r]\!] = \dot{\nabla}^2 v_r - \frac{2}{r^2}v_r - \frac{2}{r^2}\frac{\partial v_\theta}{\partial\theta} - \frac{2}{r^2}v_\theta\cot\theta - \frac{2}{r^2\sin\theta}\frac{\partial v_\phi}{\partial\phi}$$

where

$$\nabla^2 \equiv \frac{1}{r^2}\frac{\partial}{\partial r}\left(r^2\frac{\partial}{\partial r}\right) + \frac{1}{r^2\sin\theta}\frac{\partial}{\partial\theta}\left(\sin\theta\frac{\partial}{\partial\theta}\right) + \frac{1}{r^2\sin^2\theta}\frac{\partial^2}{\partial\phi^2}$$

$$[\![(\nabla^2\mathbf{v})_\theta]\!] = \nabla^2 v_\theta + \frac{2}{r^2}\frac{\partial v_r}{\partial\theta} - \frac{v_\theta}{r^2\sin^2\theta} - \frac{2\cos\theta}{r^2\sin^2\theta}\frac{\partial v_\phi}{\partial\phi}$$

$$[\![(\nabla^2\mathbf{v})_\phi]\!] = \nabla^2 v_\phi - \frac{v_\phi}{r^2\sin^2\theta} + \frac{2}{r^2\sin\theta}\frac{\partial v_r}{\partial\phi} + \frac{2\cos\theta}{r^2\sin^2\theta}\frac{\partial v_\theta}{\partial\phi}$$

†Since all terms except (4) are identical to those in the EOM, only $[\![(\nabla^2\mathbf{v})_u]\!]$ will be given. Brackets $[\![\ ]\!]$ are used to denote a scalar component.

Table I-6 Equations of change in cartesian coordinates

A. In terms of fluxes

Continuity

$$\frac{\partial\rho}{\partial t} = -\frac{\partial}{\partial x}(\rho v_x) - \frac{\partial}{\partial y}(\rho v_y) - \frac{\partial}{\partial z}(\rho v_z)$$

Moles of species A

$$\frac{\partial C_A}{\partial t} + v_x\frac{\partial C_A}{\partial x} + v_y\frac{\partial C_A}{\partial y} + v_z\frac{\partial C_A}{\partial z} + C_A\left(\frac{\partial v_x}{\partial x} + \frac{\partial v_y}{\partial y} + \frac{\partial v_z}{\partial z}\right) = -\frac{\partial J_{Ax}}{\partial x} - \frac{\partial J_{Ay}}{\partial y} - \frac{\partial J_{Az}}{\partial z} + R_A$$

x momentum

$$\rho\left(\frac{\partial v_x}{\partial t} + v_x\frac{\partial v_x}{\partial x} + v_y\frac{\partial v_x}{\partial y} + v_z\frac{\partial v_x}{\partial z}\right) = -\frac{\partial\tau_{xx}}{\partial x} - \frac{\partial\tau_{yx}}{\partial y} - \frac{\partial\tau_{zx}}{\partial z} - \frac{\partial p}{\partial x} + \rho g_x$$

Table I-6 Equations of change in cartesian coordinates (*cont.*)

A. In terms of fluxes

y momentum

$$\rho\left(\frac{\partial v_y}{\partial t} + v_x \frac{\partial v_y}{\partial x} + v_y \frac{\partial v_y}{\partial y} + v_z \frac{\partial v_y}{\partial z}\right) = -\frac{\partial \tau_{xy}}{\partial x} - \frac{\partial \tau_{yy}}{\partial y} - \frac{\partial \tau_{zy}}{\partial z} - \frac{\partial p}{\partial y} + \rho g_y$$

z momentum

$$\rho\left(\frac{\partial v_z}{\partial t} + v_x \frac{\partial v_z}{\partial x} + v_y \frac{\partial v_z}{\partial y} + v_z \frac{\partial v_z}{\partial z}\right) = -\frac{\partial \tau_{xz}}{\partial x} - \frac{\partial \tau_{yz}}{\partial y} - \frac{\partial \tau_{zz}}{\partial z} - \frac{\partial p}{\partial z} + \rho g_z$$

Energy†

$$\rho \hat{C}_v\left(\frac{\partial T}{\partial t} + v_x \frac{\partial T}{\partial x} + v_y \frac{\partial T}{\partial y} + v_z \frac{\partial T}{\partial z}\right) \equiv -\frac{\partial q_x}{\partial x} - \frac{\partial q_y}{\partial y} - \frac{\partial q_z}{\partial z} + \Phi_c + \Phi_v + \Phi_H$$

B. In terms of potentials for constant ρ, μ, D_{AB}, and k

Continuity

$$\frac{\partial v_x}{\partial x} + \frac{\partial v_y}{\partial y} + \frac{\partial v_z}{\partial z} = 0$$

Continuity (species A)

$$\frac{\partial C_A}{\partial t} + v_x \frac{\partial C_A}{\partial x} + v_y \frac{\partial C_A}{\partial y} + v_z \frac{\partial C_A}{\partial z} = D_{AB}\left(\frac{\partial^2 C_A}{\partial x^2} + \frac{\partial^2 C_A}{\partial y^2} + \frac{\partial^2 C_A}{\partial z^2}\right) + R_A$$

x momentum

$$\rho\left(\frac{\partial v_x}{\partial t} + v_x \frac{\partial v_x}{\partial x} + v_y \frac{\partial v_x}{\partial y} + v_z \frac{\partial v_x}{\partial z}\right) = \mu\left(\frac{\partial^2 v_x}{\partial x^2} + \frac{\partial^2 v_x}{\partial y^2} + \frac{\partial^2 v_x}{\partial z^2}\right) - \frac{\partial p}{\partial x} + \rho g_x$$

y momentum

$$\rho\left(\frac{\partial v_y}{\partial t} + v_x \frac{\partial v_y}{\partial x} + v_y \frac{\partial v_y}{\partial y} + v_z \frac{\partial v_y}{\partial z}\right) = \mu\left(\frac{\partial^2 v_y}{\partial x^2} + \frac{\partial^2 v_y}{\partial y^2} + \frac{\partial^2 y_y}{\partial z^2}\right) - \frac{\partial p}{\partial y} + \rho g_y$$

z momentum

$$\rho\left(\frac{\partial v_z}{\partial t} + v_x \frac{\partial v_z}{\partial x} + v_y \frac{\partial v_z}{\partial y} + v_z \frac{\partial v_z}{\partial z}\right) = \mu\left(\frac{\partial^2 v_z}{\partial x^2} + \frac{\partial^2 v_z}{\partial y^2} + \frac{\partial^2 y_z}{\partial z^2}\right) - \frac{\partial p}{\partial z} + \rho g_z$$

Energy†

$$\rho \hat{C}_v\left(\frac{\partial T}{\partial t} + v_x \frac{\partial T}{\partial x} + v_y \frac{\partial T}{\partial y} + v_z \frac{\partial T}{\partial z}\right) = k\left(\frac{\partial^2 T}{\partial x^2} + \frac{\partial^2 T}{\partial y^2} + \frac{\partial^2 T}{\partial z^2}\right) + \Phi_c + \Phi_v + \Phi_H$$

† The viscous dissipation Φ_v, is given in Tables I-10 and I-11. For incompressible fluids

$$\Phi_c \equiv -T\left(\frac{\partial p}{\partial T}\right)_\rho \left(\frac{\partial v_x}{\partial x} + \frac{\partial v_y}{\partial y} + \frac{\partial v_z}{\partial z}\right) = 0$$

Φ_H = rate of heat generation per unit volume by other sources.

Table I-7 Equations of change in cylindrical coordinates

A. In terms of fluxes

Continuity
$$\frac{\partial \rho}{\partial t} = -\frac{\partial(\rho r v_r)}{r\,\partial r} - \frac{\partial(\rho v_\theta)}{r\,\partial\theta} - \frac{\partial(\rho v_z)}{\partial z}$$

Species A
$$-\frac{\partial(rJ_{Ar})}{r\,\partial r} - \frac{\partial J_{A\theta}}{r\,\partial\theta} - \frac{\partial J_{Az}}{\partial z} + R_A = \frac{\partial C_A}{\partial t} + v_r\frac{\partial C_A}{\partial r} + v_\theta\frac{\partial C_A}{r\,\partial\theta} + v_z\frac{\partial C_A}{\partial z} + C_A\left[\frac{\partial(rv_r)}{r\,\partial r} + \frac{\partial v_\theta}{r\,\partial\theta} + \frac{\partial v_z}{\partial z}\right]$$

r momentum
$$\rho\left(\frac{\partial v_r}{\partial t} + v_r\frac{\partial v_r}{\partial r} + \frac{v_\theta}{r}\frac{\partial v_r}{\partial\theta} + v_z\frac{\partial v_r}{\partial z} - \frac{v_\theta^2}{r}\right) = -\frac{\partial p}{\partial r} - \left[\frac{\partial(r\tau_{rr})}{r\,\partial r} + \frac{\partial\tau_{\theta r}}{r\,\partial\theta} - \frac{\tau_{\theta\theta}}{r} + \frac{\partial\tau_{zr}}{\partial z}\right] + \rho g_r$$

θ momentum
$$\rho\left(\frac{\partial v_\theta}{\partial t} + v_r\frac{\partial v_\theta}{\partial r} + \frac{v_\theta}{r}\frac{\partial v_\theta}{\partial\theta} + v_z\frac{\partial v_\theta}{\partial z} + \frac{v_r v_\theta}{r}\right) = -\frac{\partial p}{r\,\partial\theta} - \left[\frac{\partial(\tau_{r\theta}r^2)}{r^2\,\partial r} + \frac{\partial\tau_{\theta\theta}}{r\,\partial\theta} + \frac{\partial\tau_{z\theta}}{\partial z}\right] + \rho g_\theta$$

z momentum
$$\rho\left(\frac{\partial v_z}{\partial t} + v_r\frac{\partial v_z}{\partial r} + v_\theta\frac{\partial v_z}{r\,\partial\theta} + v_z\frac{\partial v_z}{\partial z}\right) = -\frac{\partial p}{\partial z} - \left[\frac{\partial(r\tau_{rz})}{r\,\partial r} + \frac{\partial\tau_{\theta z}}{r\,\partial\theta} + \frac{\partial\tau_{zz}}{\partial z}\right] + \rho g_z$$

Energy†
$$\rho\hat{C}_v\left(\frac{\partial T}{\partial T} + v_r\frac{\partial T}{\partial r} + \frac{v_\theta}{r}\frac{\partial T}{\partial\theta} + v_z\frac{\partial T}{\partial z}\right) = -\frac{\partial(rq_r)}{r\,\partial r} - \frac{\partial q_\theta}{r\,\partial\theta} - \frac{\partial q_z}{\partial z} + \Phi_c + \Phi_v + \Phi_H$$

B. In terms of potentials for constant k, μ, ρ, and D_{AB}

Continuity
$$\frac{\partial(rv_r)}{r\,\partial r} + \frac{\partial v_\theta}{r\,\partial\theta} + \frac{\partial v_z}{\partial z} = 0$$

Species A
$$\frac{\partial C_A}{\partial t} + v_r\frac{\partial C_A}{\partial r} + v_\theta\frac{\partial C_A}{r\,\partial\theta} + v_z\frac{\partial C_A}{\partial z} = D_{AB}\left[\frac{\partial\left(r\frac{\partial C_A}{\partial r}\right)}{r\,\partial r} + \frac{\partial^2 C_A}{r^2\,\partial\theta^2} + \frac{\partial^2 C_A}{\partial z^2}\right] + R_A$$

r momentum
$$\rho\left(\frac{\partial v_r}{\partial t} + v_r\frac{\partial v_r}{\partial r} + v_\theta\frac{\partial v_r}{r\,\partial\theta} + v_z\frac{\partial v_r}{\partial z} - \frac{v_\theta^2}{r}\right) = \mu\left\{\frac{\partial}{\partial r}\left[\frac{\partial(rv_r)}{r\,\partial r}\right] + \frac{\partial^2 v_r}{r^2\,\partial\theta^2} - \frac{2}{r^2}\frac{\partial v_\theta}{\partial\theta} + \frac{\partial^2 v_r}{\partial z^2}\right\} - \frac{\partial p}{\partial r} + \rho g_r$$

θ momentum
$$\rho\left(\frac{\partial v_\theta}{\partial t} + v_r\frac{\partial v_\theta}{\partial r} + \frac{v_\theta}{r}\frac{\partial v_\theta}{\partial\theta} + v_z\frac{\partial v_\theta}{\partial z} + \frac{v_r v_\theta}{r}\right) = \mu\left\{\frac{\partial}{\partial r}\left[\frac{\partial(rv_\theta)}{r\,\partial r}\right] + \frac{\partial^2 v_\theta}{r^2\,\partial\theta^2} + \frac{2}{r^2}\frac{\partial v_r}{\partial\theta} + \frac{\partial^2 v_\theta}{\partial z^2}\right\} - \frac{\partial p}{r\,\partial\theta} + \rho g_\theta$$

Table I-7 Equations of change in cylindrical coordinates (*cont.*)

B. In terms of potentials for constant k, μ, ρ, and D_{AB}

z momentum

$$\rho\left(\frac{\partial v_z}{\partial t} + v_r\frac{\partial v_z}{\partial r} + v_\theta\frac{\partial v_z}{r\,\partial\theta} + v_z\frac{\partial v_z}{\partial z}\right) = \mu\left[\frac{\partial\left(r\frac{\partial v_z}{\partial r}\right)}{r\,\partial r} + \frac{\partial^2 v_z}{r^2\,\partial\theta^2} + \frac{\partial^2 v_z}{\partial z^2}\right] - \frac{\partial p}{\partial z} + \rho g_z$$

Energy†

$$\rho\hat{C}_v\left(\frac{\partial T}{\partial t} + v_r\frac{\partial T}{\partial r} + v_\theta\frac{\partial T}{r\,\partial\theta} + v_z\frac{\partial T}{\partial z}\right) = k\left[\frac{\partial\left(r\frac{\partial T}{\partial r}\right)}{r\,\partial r} + \frac{\partial^2 T}{r^2\,\partial\theta^2} + \frac{\partial^2 T}{\partial z^2}\right] + \Phi_c + \Phi_v + \Phi_H$$

†The viscous dissipation Φ_v is given in Tables I-10 and I-11. For incompressible fluids

$$\Phi_c \equiv -T\left(\frac{\partial p}{\partial T}\right)_\rho\left[\frac{\partial(rv_r)}{r\,\partial r} + \frac{\partial v_\theta}{r\,\partial\theta} + \frac{\partial v_z}{\partial z}\right] = 0$$

Φ_H = rate of heat generation per unit volume by other sources.

Table I-8 Equations of change in spherical coordinates

A. In terms of fluxes

Continuity

$$\frac{\partial\rho}{\partial t} = -\frac{1}{r^2}\frac{\partial(\rho r^2 v_r)}{\partial r} - \frac{1}{r\sin\theta}\frac{\partial(\rho v_\theta\sin\theta)}{\partial\theta} - \frac{1}{r\sin\theta}\frac{\partial(\rho v_\phi)}{\partial\phi}$$

Species A

$$\frac{\partial C_A}{\partial t} + v_r\frac{\partial C_A}{\partial r} + v_\theta\frac{\partial C_A}{r\,\partial\theta} + v_\phi\frac{\partial C_A}{r\sin\theta\,\partial\phi} + C_A\left[\frac{\partial(r^2 v_r)}{r^2\,\partial r} + \frac{\partial(v_\theta\sin\theta)}{r\sin\theta\,\partial\theta} + \frac{\partial v_\phi}{r\sin\theta\,\partial\phi}\right]$$

$$= -\frac{\partial(r^2 J_{Ar})}{r^2\,\partial r} - \frac{\partial(J_{A\theta}\sin\theta)}{r\sin\theta\,\partial\theta} - \frac{\partial J_{A\phi}}{r\sin\theta\,\partial\phi} + R_A$$

r momentum

$$\rho\left(\frac{\partial v_r}{\partial t} + v_r\frac{\partial v_r}{\partial r} + \frac{v_\theta}{r}\frac{\partial v_r}{\partial\theta} + \frac{v_\phi}{r\sin\theta}\frac{\partial v_r}{\partial\phi} - \frac{v_\theta^2 + v_\phi^2}{r}\right)$$

$$= -\frac{\partial p}{\partial r} + \rho g_r - \frac{1}{r^2}\frac{\partial}{\partial r}(r^2\tau_{rr}) - \frac{1}{r\sin\theta}\frac{\partial}{\partial\theta}(\tau_{r\theta}\sin\theta) - \frac{1}{r\sin\theta}\frac{\partial\tau_{r\phi}}{\partial\phi} - \frac{\tau_{\theta\theta} + \tau_{\phi\phi}}{r}$$

θ momentum

$$\rho\left(\frac{\partial v_\theta}{\partial t} + v_r\frac{\partial v_\theta}{\partial r} + \frac{v_\theta}{r}\frac{\partial v_\theta}{\partial\theta} + \frac{v_\phi}{r\sin\theta}\frac{\partial v_\theta}{\partial\phi} + \frac{v_r v_\theta}{r} - \frac{v_\phi^2\cot\theta}{r}\right)$$

$$= -\frac{\partial p}{r\,\partial\theta} + \rho g_\theta - \frac{1}{r^2}\frac{\partial}{\partial r}(r^2\tau_{r\theta}) - \frac{1}{r\sin\theta}\frac{\partial}{\partial\theta}(\tau_{\theta\theta}\sin\theta) - \frac{1}{r\sin\theta}\frac{\partial\tau_{\phi\theta}}{\partial\phi} - \frac{\tau_{r\theta}}{r} + \frac{\cot\theta}{r}\tau_{\phi\phi}$$

ϕ momentum

$$\rho\left(\frac{\partial v_\phi}{\partial t} + v_r\frac{\partial v_\phi}{\partial r} + \frac{v_\theta}{r}\frac{\partial v_\phi}{\partial\theta} + \frac{v_\phi}{r\sin\theta}\frac{\partial v_\phi}{\partial\phi} + \frac{v_\phi v_r}{r} + \frac{v_\theta v_\phi}{r}\cot\theta\right)$$

$$= -\frac{1}{r\sin\theta}\frac{\partial p}{\partial\phi} + \rho g_\phi - \frac{1}{r^2}\frac{\partial}{\partial r}(r^2\tau_{r\phi}) - \frac{1}{r}\frac{\partial\tau_{\theta\phi}}{\partial\theta} - \frac{1}{r\sin\theta}\frac{\partial\tau_{\phi\phi}}{\partial\phi} - \frac{\tau_{r\phi}}{r} - \frac{2\cot\theta}{r}\tau_{\theta\phi}$$

Table I-8 Equations of change in spherical coordinates (*cont.*)

A. In terms of flux

Energy†

$$\rho \widehat{C}_v \left(\frac{\partial T}{\partial t} + v_r \frac{\partial T}{\partial r} + v_\theta \frac{\partial T}{r\,\partial \theta} + \frac{v_\phi}{r \sin\theta} \frac{\partial T}{\partial \phi} \right) = -\frac{\partial (r^2 q_r)}{r^2\,\partial r} - \frac{\partial (q_\theta \sin\theta)}{r \sin\theta\,\partial\theta} - \frac{\partial q_\phi}{r\sin\theta\,\partial\phi} + \Phi_c + \Phi_v + \Phi_H$$

B. In terms of potentials‡ for constant ρ, μ, D_{AB}, and k

Continuity

$$\frac{\partial (r^2 v_r)}{r^2\,\partial r} + \frac{\partial (v_\theta \sin\theta)}{r\sin\theta\,\partial\theta} + \frac{\partial v_\phi}{r\sin\theta\,\partial\phi} = 0$$

Species A

$$\frac{\partial C_A}{\partial t} + v_r \frac{\partial C_A}{\partial r} + v_\theta \frac{\partial C_A}{r\,\partial\theta} + v_\phi \frac{\partial C_A}{r\sin\theta\,\partial\phi} = D_{AB}\nabla^2 C_A + R_A$$

r momentum

$$\rho\left(\frac{\partial v_r}{\partial t} + v_r\frac{\partial v_r}{\partial r} + \frac{v_\theta}{r}\frac{\partial v_r}{\partial\theta} + \frac{v_\phi}{r\sin\theta}\frac{\partial v_r}{\partial\phi} - \frac{v_\theta^2 + v_\phi^2}{r}\right) = -\frac{\partial p}{\partial r} + \rho g_r + \mu\left(\nabla^2 v_r - \frac{2}{r^2}v_r - \frac{2}{r^2}\frac{\partial v_\theta}{\partial\theta} - \frac{2}{r^2}v_\theta\cot\theta - \frac{2}{r^2\sin\theta}\frac{\partial v_\phi}{\partial\phi}\right)$$

θ momentum

$$\rho\left(\frac{\partial v_\theta}{\partial t} + v_r\frac{\partial v_\theta}{\partial r} + \frac{v_\theta}{r}\frac{\partial v_\theta}{\partial\theta} + \frac{v_\phi}{r\sin\theta}\frac{\partial v_\theta}{\partial\phi} + \frac{v_r v_\theta}{r} - \frac{v_\phi^2\cot\theta}{r}\right) = -\frac{1}{r}\frac{\partial p}{\partial\theta} + \rho g_\theta + \mu\left(\nabla^2 v_\theta + \frac{2}{r^2}\frac{\partial v_r}{\partial\theta} - \frac{v_\theta}{r^2\sin^2\theta} - \frac{2\cos\theta}{r^2\sin^2\theta}\frac{\partial v_\phi}{\partial\phi}\right)$$

ϕ momentum

$$\rho\left(\frac{\partial v_\phi}{\partial t} + v_r\frac{\partial v_\phi}{\partial r} + \frac{v_\theta}{r}\frac{\partial v_\phi}{\partial\theta} + \frac{v_\phi}{r\sin\theta}\frac{\partial v_\phi}{\partial\phi} + \frac{v_\phi v_r}{r} + \frac{v_\theta v_\phi}{r}\cot\theta\right) = -\frac{1}{r\sin\theta}\frac{\partial p}{\partial\phi} + \rho g_\phi + \mu\left(\nabla^2 v_\phi - \frac{v_\phi}{r^2\sin^2\theta} + \frac{2}{r^2\sin\theta}\frac{\partial v_r}{\partial\phi} + \frac{2\cos\theta}{r^2\sin^2\theta}\frac{\partial v_\theta}{\partial\phi}\right)$$

Energy†

$$\rho\widehat{C}_v\left(\frac{\partial T}{\partial t} + v_r\frac{\partial T}{\partial r} + v_\theta\frac{\partial T}{r\,\partial\theta} + \frac{v_\phi}{r\sin\theta}\frac{\partial T}{\partial\phi}\right) = k\nabla^2 T + \Phi_v + \Phi_H$$

†The viscous dissipation Φ_v is given by Tables I-10 and I-11. For incompressible fluids

$$\Phi_c \equiv -T\left(\frac{\partial p}{\partial T}\right)_\rho\left[\frac{\partial(r^2 v_r)}{r^2\,\partial r} + \frac{\partial(v_\theta\sin\theta)}{r\sin\theta\,\partial\theta} + \frac{\partial v_\phi}{r\sin\theta\,\partial\phi}\right] = 0$$

Φ_H = rate of heat generation per unit volume by other sources.

‡The laplacian operator ∇^2 is used, where

$$\nabla^2 \equiv \frac{\partial\left(r^2\frac{\partial}{\partial r}\right)}{r^2\,\partial r} + \frac{\partial\left(\sin\theta\frac{\partial}{\partial\theta}\right)}{r^2\sin\theta\,\partial\theta} + \frac{\partial^2}{r^2\sin^2\theta\,\partial\phi^2}$$

Table I-9 Stress and rate-of-deformation tensors in various coordinates

If newtonian,

$$\tau_{ij} = -\mu\Delta_{ij} - \tfrac{2}{3}\mu\delta_{ij}\nabla\cdot\boldsymbol{v}$$

For $\nabla\cdot\mathbf{v}$, see Table I-2. If incompressible, $\nabla\cdot\mathbf{v} = 0$ and

$$\tau_{ij} = -\mu\Delta_{ij} \qquad \Delta_{ij} = [\![(\nabla\mathbf{v})_{ij}]\!] + [\![(\nabla\mathbf{v})_{ji}]\!]$$

Cartesian

$$\Delta_{xy} = \frac{\partial v_x}{\partial y} + \frac{\partial v_y}{\partial x} \qquad \Delta_{ij} = \Delta_{ji} = \frac{\partial v_i}{\partial x_j} + \frac{\partial v_j}{\partial x_i}$$

Cylindrical

$$\Delta_{rr} = 2\frac{\partial v_r}{\partial r} \qquad \Delta_{\theta\theta} = 2\frac{\partial v_\theta}{r\,\partial\theta} \qquad \Delta_{zz} = 2\frac{\partial v_z}{\partial z}$$

$$\Delta_{r\theta} = \Delta_{\theta r} = r\frac{\partial}{\partial r}\left(\frac{v_\theta}{r}\right) + \frac{1}{r}\frac{\partial v_r}{\partial\theta}$$

$$\Delta_{\theta z} = \Delta_{z\theta} = \frac{\partial v_\theta}{\partial z} + \frac{1}{r}\frac{\partial v_z}{\partial\theta} \qquad \Delta_{rz} = \Delta_{zr} = \frac{\partial v_z}{\partial r} + \frac{\partial v_r}{\partial z}$$

Spherical

$$\Delta_{rr} = 2\frac{\partial v_r}{\partial r} \qquad \Delta_{\theta\theta} = 2\left(\frac{1}{r}\frac{\partial v_\theta}{\partial\theta} + \frac{v_r}{r}\right)$$

$$\Delta_{\phi\phi} = 2\left(\frac{1}{r\sin\theta}\frac{\partial v_\phi}{\partial\phi} + \frac{v_r}{r} + \frac{v_\theta\cot\phi}{r}\right)$$

$$\Delta_{r\theta} = \Delta_{\theta r} = r\frac{\partial}{\partial r}\left(\frac{v_\theta}{r}\right) + \frac{1}{r}\frac{\partial v_r}{\partial\theta}$$

$$\Delta_{\theta\phi} = \Delta_{\phi\theta} = \frac{\sin\theta}{r}\frac{\partial}{\partial\theta}\left(\frac{v_\phi}{\sin\theta}\right) + \frac{1}{r\sin\theta}\frac{\partial v_\theta}{\partial\phi}$$

$$\Delta_{\phi r} = \Delta_{r\phi} = \frac{1}{r\sin\theta}\frac{\partial v_r}{\partial\phi} + r\frac{\partial}{\partial r}\left(\frac{v_\phi}{r}\right)$$

Table I-10 Expressions for viscous-dissipation term Φ_v

Rectangular

$$\Phi_v = \left(\tau_{xx}\frac{\partial v_x}{\partial x} + \tau_{yy}\frac{\partial v_y}{\partial y} + \tau_{zz}\frac{\partial v_z}{\partial z}\right) + \left[\tau_{xy}\left(\frac{\partial v_x}{\partial y} + \frac{\partial v_y}{\partial x}\right) + \tau_{xz}\left(\frac{\partial v_x}{\partial z} + \frac{\partial v_z}{\partial x}\right) + \tau_{yz}\left(\frac{\partial v_y}{\partial z} + \frac{\partial v_z}{\partial y}\right)\right]$$

Cylindrical

$$\Phi_v = \left[\tau_{rr}\frac{\partial v_r}{\partial r} + \tau_{\theta\theta}\frac{1}{r}\left(\frac{\partial v_\theta}{\partial \theta} + v_r\right) + \tau_{zz}\frac{\partial v_z}{\partial z}\right]$$
$$+ \left\{\tau_{r\theta}\left[r\frac{\partial}{\partial r}\left(\frac{v_\theta}{r}\right) + \frac{1}{r}\frac{\partial v_r}{\partial \theta}\right] + \tau_{rz}\left(\frac{\partial v_z}{\partial r} + \frac{\partial v_r}{\partial z}\right) + \tau_{\theta z}\left(\frac{1}{r}\frac{\partial v_z}{\partial \theta} + \frac{\partial v_\theta}{\partial z}\right)\right\}$$

Spherical

$$\Phi_v = \left[\tau_{rr}\frac{\partial v_r}{\partial r} + \tau_{\theta\theta}\left(\frac{1}{r}\frac{\partial v_\theta}{\partial \theta} + \frac{v_r}{r}\right) + \tau_{\phi\phi}\left(\frac{1}{r\sin\theta}\frac{\partial v_\phi}{\partial \phi} + \frac{v_r}{r} + \frac{v_\theta\cot\theta}{r}\right)\right]$$
$$+ \left[\tau_{r\theta}\left(\frac{\partial v_\theta}{\partial r} + \frac{1}{r}\frac{\partial v_r}{\partial \theta} - \frac{v_\theta}{r}\right) + \tau_{r\phi}\left(\frac{\partial v_\phi}{\partial r} + \frac{1}{r\sin\theta}\frac{\partial v_r}{\partial \phi} - \frac{v_\phi}{r}\right)\right.$$
$$\left. + \tau_{\theta\phi}\left(\frac{1}{r}\frac{\partial v_\phi}{\partial \theta} + \frac{1}{r\sin\theta}\frac{\partial v_\theta}{\partial \phi} - \frac{\cot\theta}{r}v_\phi\right)\right]$$

Table I-11 Expressions† for viscous-dissipation function $\phi_v \equiv \Phi_v/\mu$

Rectangular

$$\phi_v = 2\left[\left(\frac{\partial v_x}{\partial x}\right)^2 + \left(\frac{\partial v_y}{\partial y}\right)^2 + \left(\frac{\partial v_z}{\partial z}\right)^2\right]$$
$$+ \left(\frac{\partial v_y}{\partial x} + \frac{\partial v_x}{\partial y}\right)^2 + \left(\frac{\partial v_z}{\partial y} + \frac{\partial v_y}{\partial z}\right)^2 + \left(\frac{\partial v_x}{\partial z} + \frac{\partial v_z}{\partial x}\right)^2 - \tfrac{2}{3}(\nabla \cdot \mathbf{v})^2$$

Cylindrical

$$\phi_v = 2\left[\left(\frac{\partial v_r}{\partial r}\right)^2 + \left(\frac{1}{r}\frac{\partial v_\theta}{\partial \theta} + \frac{v_r}{r}\right)^2 + \left(\frac{\partial v_z}{\partial z}\right)^2\right] + \left[r\frac{\partial}{\partial r}\left(\frac{v_\theta}{r}\right) + \frac{1}{r}\frac{\partial v_r}{\partial \theta}\right]^2$$
$$+ \left(\frac{1}{r}\frac{\partial v_z}{\partial \theta} + \frac{\partial v_\theta}{\partial z}\right)^2 + \left(\frac{\partial v_r}{\partial z} + \frac{\partial v_z}{\partial r}\right)^2 - \tfrac{2}{3}(\nabla \cdot \mathbf{v})^2$$

Spherical

$$\phi_v = 2\left[\left(\frac{\partial v_r}{\partial r}\right)^2 + \left(\frac{1}{r}\frac{\partial v_\theta}{\partial \theta} + \frac{v_r}{r}\right)^2 + \left(\frac{1}{r\sin\theta}\frac{\partial v_\phi}{\partial \phi} + \frac{v_r}{r} + \frac{v_\theta\cot\theta}{r}\right)^2\right]$$
$$+ \left[r\frac{\partial}{\partial r}\left(\frac{v_\theta}{r}\right) + \frac{1}{r}\frac{\partial v_r}{\partial \theta}\right]^2 + \left[\frac{\sin\theta}{r}\frac{\partial}{\partial \theta}\left(\frac{v_\phi}{\sin\theta}\right) + \frac{1}{r\sin\theta}\frac{\partial v_\theta}{\partial \phi}\right]^2$$
$$+ \left[\frac{1}{r\sin\theta}\frac{\partial v_r}{\partial \phi} + r\frac{\partial}{\partial r}\left(\frac{v_\phi}{r}\right)\right]^2 - \tfrac{2}{3}(\nabla \cdot \mathbf{v})^2$$

†For a newtonian fluid $\mu\phi_v = \Phi_v$.

Table I-12 Equations of change in lagrangian vector form†

A. In terms of fluxes

Momentum	$\rho \frac{D\mathbf{v}}{Dt} = -\nabla \cdot \underline{\tau} - \nabla p + \rho \mathbf{g}$
Energy	$\rho \widehat{C}_v \frac{DT}{Dt} = -\nabla \cdot \mathbf{q} - T\left(\frac{\partial p}{\partial T}\right)_\rho \nabla \cdot \mathbf{v} - \underline{\tau}:\nabla \mathbf{v} + \Phi_H$
Continuity	$\frac{D\rho}{Dt} = -\rho \nabla \cdot \mathbf{v}$
Moles of A	$\frac{DC_A}{Dt} = -\nabla \cdot \mathbf{J}_A - C_A \nabla \cdot \mathbf{v} + R_A$ where $\frac{D}{Dt} \equiv \frac{\partial}{\partial t} + \mathbf{v} \cdot \nabla$

B. In terms of potentials at constant k, ρ, μ, and D_{AB}

$\nabla \cdot v = 0$ $\quad -\nabla \cdot \underline{\tau} = \mu \nabla^2 \mathbf{v}$ $\quad -\nabla \cdot \mathbf{q} = k \nabla^2 T$ $\quad -\nabla \cdot \mathbf{J}_A = D_{AB} \nabla^2 C_A$

C. Transformation to eulerian form

If $\widehat{G}$ = any quantity ($\mathbf{v}$, ω_A, or T), the identity

$$\rho \frac{D\widehat{G}}{Dt} \equiv \frac{\partial(\rho \widehat{G} \mathbf{v})}{\partial t} + \nabla \cdot \rho \widehat{G} \mathbf{v}$$

can be used to convert from lagrangian into eulerian form

†For DT/Dt and $D\mathbf{v}/Dt$, see Tables I-3 and I-4; for $\nabla \cdot \mathbf{q}$, $\nabla \cdot \mathbf{v}$, or $\nabla \cdot \mathbf{J}_A$, see Table I-2; and for $\underline{\tau}:\nabla \mathbf{v}$ see Tables I-10 and I-11. For $\nabla^2 v$ see Table I-4. For $\nabla^2 T$, $\nabla^2 C_A$, see Table I-2.

Table I-13 Equations of change for multicomponent mixtures

$$\frac{\partial \rho_i}{\partial t} + \nabla \cdot \rho_i \mathbf{v} = -\nabla \cdot \mathbf{j}_i + r_i \qquad \frac{\partial C_i}{\partial t} = -\nabla \cdot (\mathbf{J}_i + C_i \mathbf{v}) + R_i$$

$$\frac{DC_i}{Dt} + C_i \nabla \cdot \mathbf{v} = -\nabla \cdot \mathbf{J}_i + R_i \qquad \rho \frac{D\omega_i}{Dt} = -\nabla \cdot \mathbf{j}_i + r_i$$

$$\frac{D\rho}{Dt} = -\rho \nabla \cdot \mathbf{v} \qquad \frac{\partial \rho}{\partial t} = -\nabla \cdot \rho \mathbf{v} \qquad \rho \frac{D\widehat{G}}{Dt} \equiv \frac{\partial(\rho \widehat{G})}{\partial t} + \nabla \cdot \rho \widehat{G} \mathbf{v}$$

$$\rho \frac{D\mathbf{v}}{Dt} \equiv \frac{\partial(\rho \mathbf{v})}{\partial t} + \nabla \cdot \rho \mathbf{v}\mathbf{v} = -\nabla \cdot \underline{\tau} - \nabla p + \sum \rho_i \widehat{\mathbf{F}}_i$$

$$\rho \frac{D(\widehat{U} + \widehat{K})}{Dt} = -\nabla \cdot \mathbf{q} - \nabla \cdot p\mathbf{v} - \nabla \cdot \underline{\tau} \cdot \mathbf{v} + \sum \rho_i \widehat{\mathbf{F}}_i \cdot \mathbf{v}_i$$

$$\rho \frac{D\widehat{K}}{Dt} = -\mathbf{v} \cdot \nabla \cdot \underline{\tau} - \mathbf{v} \cdot \nabla p + \sum \rho_i \mathbf{v} \cdot \widehat{\mathbf{F}}_i$$

$$\rho \frac{D\widehat{U}}{Dt} = -\nabla \cdot \mathbf{q} - \underline{\tau} : \nabla \mathbf{v} - p \nabla \cdot \mathbf{v} + \sum \mathbf{j}_i \cdot \widehat{\mathbf{F}}_i$$

$$\rho \frac{D\widehat{H}}{Dt} = -\nabla \cdot \mathbf{q} - \underline{\tau} : \nabla \mathbf{v} + \frac{Dp}{Dt} + \sum \mathbf{j}_i \cdot \widehat{\mathbf{F}}_i$$

$$\rho \frac{D\widehat{H}}{Dt} = \rho \widehat{C}_p \frac{DT}{Dt} + (1 - \beta T) \frac{Dp}{Dt} + \sum \left(\frac{\bar{H}_i}{\mu_i} \right) \frac{D\omega_i}{Dt}$$

where $\beta \equiv \rho \left[\frac{\partial(1/\rho)}{\partial T} \right]_p$ and $\beta = \begin{cases} \frac{1}{T} & \text{for ideal gas} \\ 0 & \text{for incompressible fluid} \end{cases}$

$$\boxed{\rho \widehat{C}_p \frac{DT}{Dt} = -\nabla \cdot \mathbf{q}^{(c)} - \sum \mathbf{J}_i \cdot \nabla \bar{H}_i - \sum \bar{H}_i R_i - \underline{\tau} : \nabla \mathbf{v} + \beta T \frac{Dp}{Dt} + \sum \mathbf{j}_i \cdot \widehat{\mathbf{F}}_i}$$

$-\Sigma \bar{H}_i R_i$ is the enthalpy change due to reaction per unit time and unit volume

$$-\sum_i \bar{H}_i R_i = -\sum_j (\nabla H_R)_j \tilde{R}_j = \Phi_R$$

If the terms $\Sigma \mathbf{J}_i \cdot \nabla H_i$, $\beta T\, Dp/Dt$, and $\underline{\tau} : \nabla \mathbf{v}$, are neglected, we get

$$\frac{\partial(\rho \widehat{C}_p T)}{\partial t} + \nabla \cdot \rho \widehat{C}_p T \mathbf{v} = -\nabla \cdot \mathbf{q}^{(c)} + \Phi_R$$

Table I-14 Flux laws

	Laminar fluxes (neglect coupling)	
Mass	$\mathbf{J}_A = -D_{AB}\nabla C_A$ binary, constant ρ	
Energy	$\mathbf{q} = -k\nabla T$	
Momentum	$\underline{\boldsymbol{\tau}} = -\mu\underline{\boldsymbol{\Delta}} = -\mu[\nabla\mathbf{v} + (\nabla\mathbf{v})^T]$ newtonian	
	Turbulent fluxes	
	Anisotropic	Isotropic
Mass	$\mathbf{J}_A^{(t)} = -\underline{\mathbf{D}}_{AB}^{(t)}\cdot\nabla\bar{C}_A$	$\mathbf{J}_A^{(t)} = -D_{AB}^{(t)}\nabla\bar{C}_A$
Energy	$\mathbf{q}^{(t)} = -\underline{\mathbf{k}}^{(t)}\cdot\nabla\bar{T}$	$\mathbf{q}^{(t)} = -k^{(t)}\nabla\bar{T}$
Momentum	$\tau_{ij}^{(t)} = -\sum_k\sum_l \mu_{ijkl}^{(t)}\bar{\Delta}_{lk}$	$\underline{\boldsymbol{\tau}}^{(t)} = -\mu^{(t)}\bar{\underline{\boldsymbol{\Delta}}}$
	Total fluxes	
Mass	$\bar{\mathbf{J}}_A = \overline{\mathbf{J}_A^{(l)} + \mathbf{J}_A^{(t)}} = -(D_{AB}\underline{\boldsymbol{\delta}} + \underline{\mathbf{D}}_{AB}^{(t)})\cdot\nabla\bar{C}_A = -\underline{\mathbf{E}}\cdot\nabla\bar{C}_A$	
Energy	$\bar{\mathbf{q}} = \overline{\mathbf{q}^{(l)} + \mathbf{q}^{(t)}} = -(k\underline{\boldsymbol{\delta}} + \underline{\mathbf{k}}^{(t)})\cdot\nabla\bar{T} = -\underline{\mathbf{K}}\cdot\nabla\bar{T}$	
Momentum	$\bar{\tau}_{ij} = \overline{\tau_{ij}^{(l)} + \tau_{ij}^{(t)}} = -\sum_k\sum_l(\mu\delta_{ijkl} + \mu_{ijkl}^{(t)})\bar{\Delta}_{lk} = -\sum_k\sum_l \epsilon_{ijkl}\bar{\Delta}_{lk}$	
	Conservation laws for turbulent flow	
	To apply the conservation laws to turbulent flow one interprets the flux as the total flux (laminar plus turbulent) and concentration, velocity, and temperature as time-averaged quantities	
	Total transport properties	
	$\underline{\mathbf{E}} = D_{AB}^{(l)}\underline{\boldsymbol{\delta}} + D_{AB}^{(t)}$ $\quad\underline{\mathbf{K}} = k\underline{\boldsymbol{\delta}} + \underline{\mathbf{k}}^{(t)}$ $\quad\underline{\boldsymbol{\delta}}$ = unit tensor	
	$\epsilon_{ijkl} = \mu\delta_{ijkl} + \mu_{ijkl}^{(t)}$ $\quad\delta_{ijkl}$ = unit tensor	

CHAPTER

EIGHT

HEAT AND MASS TRANSPORT IN MORE THAN ONE DIRECTION

Now that we have studied the application of the basic flux laws and conservation equations to unidirectional transport, we can easily extend our treatment to the transport of heat and mass in three dimensions. To do so efficiently, however, requires that we make use of vector notation, which we will take up, as needed, for a particular application. Our emphasis will be on the physical meaning of vector operations and not on precise definitions and proofs. A parallel treatment which does not use vector notation will be provided.

Since momentum flux is a tensor while heat and mass flux are vectors, the subject of three-dimensional momentum transport is much more complex and will be left to the next chapter. Heat and mass transport can conveniently be treated together because the flux laws and conservation equations are often analogous when properties such as $\widehat{C}_p$, k, μ, D_{AB}, and ρ can be assumed constant.

8-1 ENERGY TRANSPORT IN THREE DIMENSIONS*

In order to illustrate the principles involved in three-dimensional energy transport, we will use a specific physical situation similar to that described in Example 7-2 but extended to three dimensions. The basic differential equations and flux laws developed will apply generally to problems in heat conduction when there is no convection or internal generation.

*This section does not require use of vector notation.

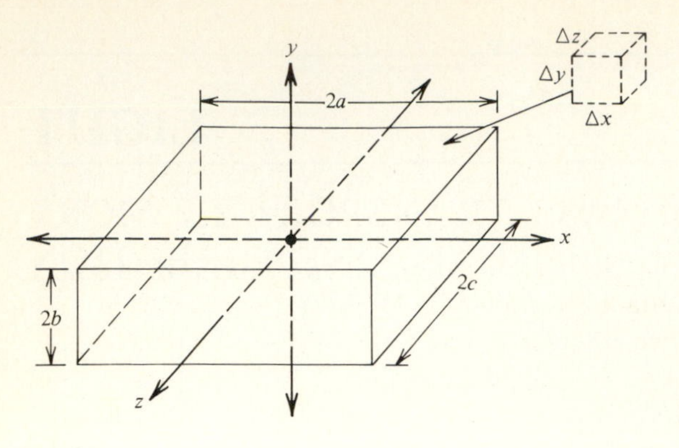

Figure 8-1 Dimensions of solid.

Example 8-1: Heat conduction in a solid Consider a parallelepiped, a brick-shaped solid, that is originally at a high temperature T_0 throughout at all time $t < 0$ (see Fig. 8-1). At time $t = 0$ all six sides are suddenly subjected to a lower temperature T_1 and maintained at that temperature at all subsequent times. Describe what happens physically in the solid.

Solution Since $T_0 > T_1$, heat will be transported from the inside to all six faces, resulting in a cooling of the brick eventually to the surface temperature $T = T_1$. We will take the origin ($x = 0, y = 0, z = 0$) as the center of the solid whose sides are given by $x = \pm a, y = \pm b, z = \pm c$. It is apparent that the temperature in the solid depends upon position (x, y, z) and upon time t. Thus if we hold y, z, and t constant at arbitrary y_1, z_1, and t_1, we can plot T versus x (see Fig. 8-2).

Now consider an arbitrary point $x = x_1$. Since there is a temperature gradient in the x direction, we would expect that at that point there would be a heat flux $\mathbf{q}_x$ in the x direction. Similarly, a plot of T versus y at constant x, z, and t would indicate that there is a gradient in the y direction and a heat flux $\mathbf{q}_y$. The same would be true of $\mathbf{q}_z$. □

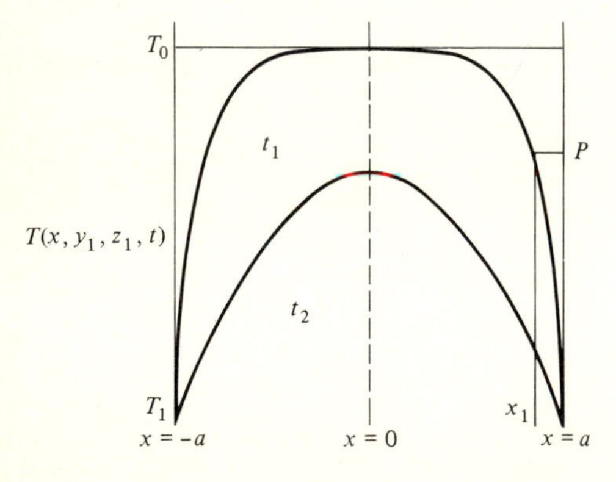

Figure 8-2 Temperature profiles at y_1, z_1.

8-2 VECTORS†

The above leads us to believe that the heat flux **q** is a vector since we must specify not only its magnitude but also its direction. We will show later that it is a vector. Actually $\mathbf{q}_x$, $\mathbf{q}_y$, and $\mathbf{q}_z$ are *vector components* of the vector **q**, the total flux. We must add them to get **q**, but we cannot simply add their magnitudes because that would not take into consideration the fact that each has a different direction. It also would not give us the resultant direction of **q**. We now digress a bit to study vector addition in general as well as some of the general properties of vectors.

We have indicated that a vector has both magnitude and direction while a scalar has only magnitude. As an example, suppose we wanted to measure the temperature and velocity at a point in a pipe through which a fluid is flowing. When we measure the temperature (using a thermometer or thermocouple), we find that no matter how we orient the instrument, we still get the same reading. Thus temperature does not have direction and is a scalar; so is concentration, and so is pressure. However, when we measure velocity (using, say, a pitot tube or an anemometer), we obtain different results depending upon how we orient the instrument, i.e., whether we point the instrument in the direction of flow, perpendicular to it, or at some angle to the flow. We therefore call velocity a vector since it has direction.

Graphically, a vector can be drawn on a sheet of paper as a line of specified length and direction, called a *directed line segment*. One end is called the head and the other the tail. We usually deal with *free vectors,* which are equal if and only if they have the same magnitude and direction, regardless of their point of origin. Free vectors can be translated up or down or from right to left, as opposed to *bound vectors,* which have a fixed point of origin.

Vector Addition and Subtraction

Let **u** and **v** be vectors pointing in different directions and having different magnitudes

$$u \equiv |\mathbf{u}| \qquad \text{and} \qquad v = |\mathbf{v}| \tag{8-1}$$

where the vertical rules indicate "magnitude‡ of." The sum of **u** and **v** can be written

$$\mathbf{u} + \mathbf{v} = \mathbf{R} \tag{8-2}$$

where **R** is called the *resultant vector* or simply the resultant. We wish to find the direction and magnitude of **R**. We first illustrate how to find **R** graphically and then show how to calculate it. As indicated in Fig. 8-3*a*, we can add **u** and **v** by

†For a nonvector approach, the reader may proceed to Secs. 8-3, 8-5, 8-8, 8-12, 8-18, 8-22.

‡We designate vectors by boldface type. Scalars and magnitudes of vectors are in lightface type.

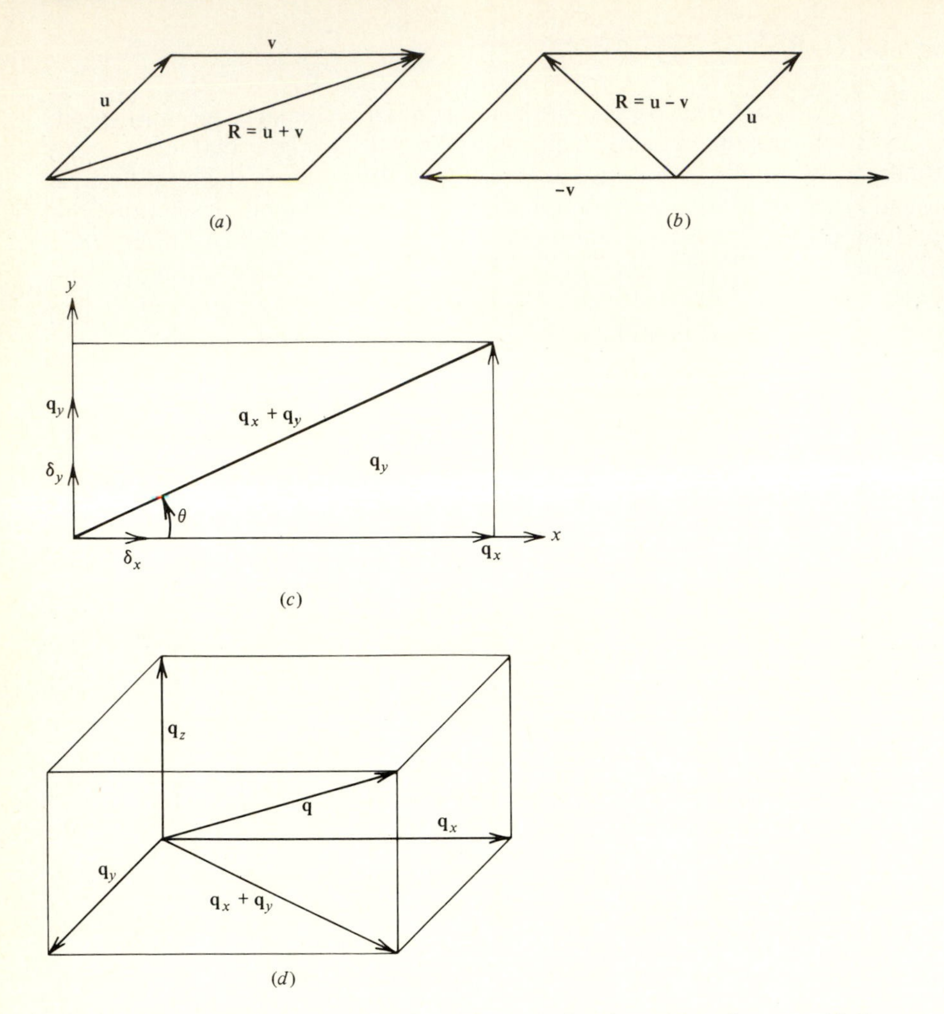

Figure 8-3 (*a*) Addition and (*b*) subtraction of vector. (*c*) Resultant vector of $\mathbf{q}_x + \mathbf{q}_y$. (*d*) Components of vector **q**.

placing the tail of **v** at the head of **u** and drawing a line from the tail of **u** to the head of **v**. This line is the resultant vector **R**. However, we can also obtain **R** by placing the tail of **u** at the head of **v** and drawing **R** from the tail of **v** to the head of **u**. Thus, addition is said to be *commutative* or

$$\mathbf{u} + \mathbf{v} = \mathbf{v} + \mathbf{u} = \mathbf{R}$$

Since **R** forms the diagonal of a parallelogram, vector addition is said to follow the parallelogram law.

To illustrate vector subtraction we define $\mathbf{w} = -\mathbf{v}$ as a vector with the same magnitude as **v** but with opposite direction. We first draw **w** by reversing the direction of **v** (Fig. 8-3*b*) and then add it to **u** as before.

Application to Heat Flux

In the case of the brick-shaped solid, if $\mathbf{q}_x$ and $\mathbf{q}_y$ are vectors, the sum $\mathbf{q}_x + \mathbf{q}_y = \mathbf{R}'$ is the resultant vector, which appears in Fig. 8-3*c* as the diagonal of a rectangle in the *xy* plane with sides $\mathbf{q}_x$ and $\mathbf{q}_y$. The magnitude of the resultant is

$$|R'| = \sqrt{q_x^2 + q_y^2} \tag{8-3}$$

and the angle θ between the vectors is given by

$$\theta = \arctan \frac{q_y}{q_x} \tag{8-4}$$

If we add $\mathbf{q}_z$ to $\mathbf{q}_x + \mathbf{q}_y$, we obtain the heat-flux vector

$$\mathbf{q} = \mathbf{q}_x + \mathbf{q}_y + \mathbf{q}_z \tag{8-5}$$

which is the diagonal of the parallelepiped formed by $\mathbf{q}_x$, $\mathbf{q}_y$, and $\mathbf{q}_z$ (Fig. 8-3*d*); these are called the *vector components* of $\mathbf{q}$. The magnitude of $\mathbf{q}$ is

$$q \equiv |\mathbf{q}| = \sqrt{q_x^2 + q_y^2 + q_z^2} \tag{8-6}$$

Each vector component of $\mathbf{q}$, such as $\mathbf{q}_x$, can be written as a product of its *scalar component* q_x (which gives its magnitude and sign) and a unit vector $\boldsymbol{\delta}_x$, which indicates its direction, e.g.,

$$\mathbf{q}_x = q_x \boldsymbol{\delta}_x \tag{8-7}$$

where $\boldsymbol{\delta}_x$ is pointed in the direction of the $+x$ axis.

If we substitute Eq. (8-7) and similar equations for q_y and q_z into Eq. (8-5), we obtain $\mathbf{q}$ in terms of its cartesian scalar components and unit vectors

$$\mathbf{q} = q_x \boldsymbol{\delta}_x + q_y \boldsymbol{\delta}_y + q_z \boldsymbol{\delta}_z \tag{8-8}$$

More generally, for any vector $\mathbf{v}$,

$$\mathbf{v} = \mathbf{v}_x + \mathbf{v}_y + \mathbf{v}_z = v_x \boldsymbol{\delta}_x + v_y \boldsymbol{\delta}_y + v_z \boldsymbol{\delta}_z \tag{8-9}$$

Note that we do not actually have to draw vectors to add them; instead we can add their components. Thus

$$\mathbf{u} + \mathbf{v} = (u_x + v_x)\boldsymbol{\delta}_x + (u_y + v_y)\boldsymbol{\delta}_y + (u_z + v_z)\boldsymbol{\delta}_z$$

since $\qquad \mathbf{R} = R_x \boldsymbol{\delta}_x + R_y \boldsymbol{\delta}_y + R_z \boldsymbol{\delta}_z \qquad R_x = u_x + v_x \qquad$ etc.

Remember that a scalar component of a vector such as q_x may be negative as well as positive but that $\boldsymbol{\delta}_x$ is always positive.

8-3 FOURIER'S LAW IN THREE DIMENSIONS*

The heat flux in the x direction at any point in a medium can be expressed in terms of Fourier's law

*This section does not require use of vector notation.

$$q_x \equiv -k \lim_{\Delta x \to 0} \frac{T(x + \Delta x, y, z, t) - T(x, y, z, t)}{\Delta x} = -k \frac{\partial T}{\partial x} \qquad \text{(8-10}a\text{)}$$

where y, z, and t are held constant. Similarly, the heat flux in the y direction is

$$q_y = -k \frac{\partial T}{\partial y} \qquad \text{(8-10}b\text{)}$$

at constant x, z, and t and the heat flux in the z direction at constant x, y, t is

$$q_z = -k \frac{\partial T}{\partial z} \qquad \text{(8-10}c\text{)}$$

Therefore, at any arbitrary point (x, y, z) the heat flux in each direction is given by Eqs. (8-10).

8-4 FOURIER'S LAW IN VECTOR FORM

In Eqs. (8-10) the flux laws are written in terms of the scalar components. Substituting them into Eq. (8-8) gives

$$\mathbf{q} = -k \frac{\partial T}{\partial x} \boldsymbol{\delta}_x - k \frac{\partial T}{\partial y} \boldsymbol{\delta}_y - k \frac{\partial T}{\partial z} \boldsymbol{\delta}_z \qquad \text{(8-11)}$$

which can be written in operator form as

$$\mathbf{q} = -k \left(\boldsymbol{\delta}_x \frac{\partial}{\partial x} + \boldsymbol{\delta}_y \frac{\partial}{\partial y} + \boldsymbol{\delta}_z \frac{\partial}{\partial z} \right) T$$

It is convenient to use the shorthand notation

$$\mathbf{q} = -k \nabla T \qquad \text{Fourier's law in vector form} \qquad \text{(8-12)}$$

where ∇ is the *del operator,* defined in cartesian coordinates as

$$\nabla \equiv \boldsymbol{\delta}_x \frac{\partial}{\partial x} + \boldsymbol{\delta}_y \frac{\partial}{\partial y} + \boldsymbol{\delta}_z \frac{\partial}{\partial z} \qquad \text{(8-13)}$$

Equation (8-12) is an equation for Fourier's law in *Gibbs vector notation,* also called *vector form*. It holds for *any* coordinate system, assuming that the medium is *isotropic,* i.e., that the conductivity is the same no matter what the direction of transport (or how the coordinate axes are rotated). An example of a nonisotropic medium is wood, which has different conductivities along and against the grain (see Chap. 10).

Example 8-2: Heat flux in cylindrical coordinates Find expressions for $\mathbf{q}$ in cylindrical coordinates (r, θ, z) (see Fig. I-1 in Table I-1).

SOLUTION The vector $\mathbf{q}$ can be written in terms of its *vector* components

$$\mathbf{q} = \mathbf{q}_r + \mathbf{q}_\theta + \mathbf{q}_z \qquad \text{(8-14}a\text{)}$$

or in terms of its *scalar* components q_r, q_θ, q_z

$$\mathbf{q} = q_r \boldsymbol{\delta}_r + q_\theta \boldsymbol{\delta}_\theta + q_z \boldsymbol{\delta}_z \tag{8-14b}$$

where $\boldsymbol{\delta}_r$, $\boldsymbol{\delta}_\theta$, and $\boldsymbol{\delta}_z$ are unit vectors pointing in the r, θ, and z directions, respectively. In general [see Eq. (1-39)] if u is an arbitrary coordinate (x, y, r, θ, etc.) and s_u is a distance along the coordinate, the flux in the u direction is

$$q_u = -k \lim_{\Delta s_u \to 0} \frac{\Delta T}{\Delta s_u} = -k \frac{\partial T}{\partial s_u} = -k \frac{\partial T}{h_u \, \partial u} \tag{8-15}$$

where h_u is called a *scale factor* and is defined by

$$ds_u \equiv h_u \, du \tag{8-16}$$

For cylindrical coordinates we can determine from the geometry (see Fig. 1-8) that $h_r = 1$, $h_\theta = r$, and $h_z = 1$. For more complex systems, generalized methods discussed in Appendix 8-7 must be used to get h_u.

The flux laws are then

$$q_r = -k \frac{\partial T}{\partial r} \qquad q_z = -k \frac{\partial T}{\partial z}$$
$$q_\theta = -k \lim_{\Delta s_\theta \to 0} \frac{T(r, \theta + \Delta\theta, z) - T(r, \theta, z)}{\Delta s_\theta} = -k \lim_{\Delta\theta \to 0} \frac{\Delta T}{r \, \Delta\theta} = -k \frac{\partial T}{r \, \partial\theta} \tag{8-17}$$

Using the expressions obtained above for q_r, q_θ, and q_z, we obtain

$$\mathbf{q} = -k \frac{\partial T}{\partial r} \boldsymbol{\delta}_r - k \frac{\partial T}{r \, \partial\theta} \boldsymbol{\delta}_\theta - k \frac{\partial T}{\partial z} \boldsymbol{\delta}_z = -k \left(\boldsymbol{\delta}_r \frac{\partial}{\partial r} + \boldsymbol{\delta}_\theta \frac{\partial}{r \, \partial\theta} + \boldsymbol{\delta}_z \frac{\partial}{\partial z} \right) T \tag{8-18}$$

Comparing this equation with the vector equation (8-12) for Fourier's law indicates that in *cylindrical* coordinates the operator ∇ is given by

$$\nabla \equiv \boldsymbol{\delta}_r \frac{\partial}{\partial r} + \boldsymbol{\delta}_\theta \frac{\partial}{r \, \partial\theta} + \boldsymbol{\delta}_z \frac{\partial}{\partial z} \tag{8-19}$$

Note that while a vector expression [such as Eq. (8-12) for Fourier's law] holds for *any* coordinate system, the expansion into components of vectors (such as $\mathbf{q}$) or vector operators (such as ∇) differs for various coordinate systems. The above procedure then can be used to expand ∇ into *any* orthogonal coordinate system (see Sec. 8-21). □

8-5 ENERGY BALANCE IN THREE DIMENSIONS*

Example 8.3: Energy balance in a solid Consider any point (x, y, z) in the solid in Example 8.1 (Fig. 8-1). Obtain a partial differential equation for the temperature distribution.

*This section does not require use of vector notation.

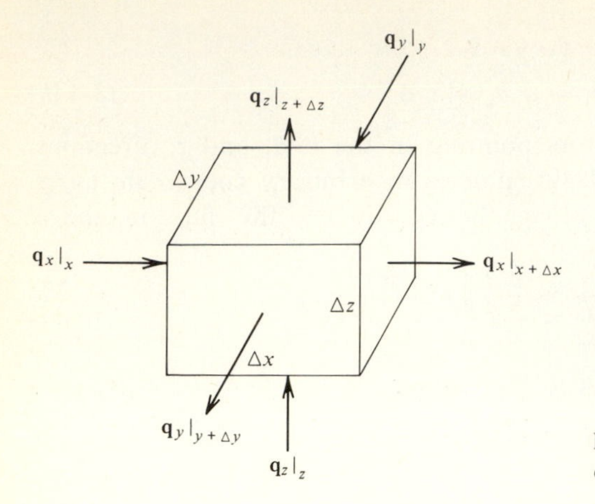

Figure 8-4 Heat fluxes into and out of element of volume.

Solution We can now apply the conservation law to an element of volume $\Delta\mathcal{V} = \Delta x\,\Delta y\,\Delta z$. Since there is no internal generation, the energy balance is (see Fig. 8-4)

	Input	*Output*	*Accumulation*
x faces:	$(q_x\,\Delta A_x)\Big|_x$	$-\,(q_x\,\Delta A_x)\Big|_{x+\Delta x}$	
y faces:	$+(q_y\,\Delta A_y)\Big|_y$	$-\,(q_y\,\Delta A_y)\Big|_{y+\Delta y}$	$= \dfrac{\Delta(\widehat{C}_p T\rho\,\Delta\mathcal{V})}{\Delta t}$
z faces:	$+(q_z\,\Delta A_z)\Big|_z$	$-\,(q_z\,\Delta A_z)\Big|_{z+\Delta z}$	

Using the difference operator Δ_x for the x direction,

$$-\Delta_x(q_x\,\Delta A_x) \equiv (q_x\,\Delta A_x)\Big|_x - (q_x\,\Delta A_x)\Big|_{x+\Delta x} \qquad \text{etc.}$$

letting $\qquad \Delta A_x = \Delta y\,\Delta z \qquad \Delta A_y = \Delta x\,\Delta z \qquad \Delta A_z = \Delta x\,\Delta y$

and dividing by the volume $\Delta x\,\Delta y\,\Delta z$, we obtain

$$-\frac{\Delta_x(q_x\,\Delta y\,\Delta z)}{\Delta x\,\Delta y\,\Delta z} - \frac{\Delta_y(q_y\,\Delta x\,\Delta z)}{\Delta x\,\Delta y\,\Delta z} - \frac{\Delta_z(q_z\,\Delta x\,\Delta y)}{\Delta x\,\Delta y\,\Delta z} = \frac{\Delta(\rho\widehat{C}_p T)}{\Delta t}$$

Since we are dealing with cartesian coordinates, we can cancel the area terms and take limits with respect to each independent variable while the others are held constant

$$-\lim_{\Delta x\to 0}\frac{\Delta q_x}{\Delta x} - \lim_{\Delta y\to 0}\frac{\Delta q_y}{\Delta y} - \lim_{\Delta z\to 0}\frac{\Delta q_z}{\Delta z} = \lim_{\Delta t\to 0}\frac{\Delta(\rho\widehat{C}_p T)}{\Delta t}$$

Then from the *definition* of the partial derivative (Appendix 1-1)

$$-\frac{\partial q_x}{\partial x} - \frac{\partial q_y}{\partial y} - \frac{\partial q_z}{\partial z} = \frac{\partial(\rho \widehat{C}_p T)}{\partial t} = \rho \widehat{C}_p \frac{\partial T}{\partial t} \tag{8-20}$$

If we substitute Fourier's-law expressions for the flux components (8-10) and assume constant k, we obtain an equation for three-dimensional energy transport

$$k\left(\frac{\partial^2 T}{\partial x^2} + \frac{\partial^2 T}{\partial y^2} + \frac{\partial^2 T}{\partial z^2}\right) = \rho \widehat{C}_p \frac{\partial T}{\partial t} \tag{8-21}$$

□

In the next section we show how Eqs. (8-14) and (8-15) can be written in vector notation. That section may be skipped in a nonvector approach.

8-6 VECTOR OPERATIONS

Scalar or Dot Product of Two Vectors

Since vectors contain information on both magnitude and direction, multiplication of vectors can take various forms. We illustrate one form, the dot or scalar product, by considering the work done in moving a blackboard eraser along its tray a distance **s** by the action of a constant force **F**. Assume that the eraser is guided by retaining walls so that it can move only in the x direction. Hence **s** is measured along the x axis. Now suppose that a constant force F is applied in the x direction (see Fig. 8-5a) and the eraser moves from position 1 to position 2. Then the work done is equal to the force in the x direction times the distance. For this case, if $F \equiv |\mathbf{F}|$ and $s \equiv |\mathbf{s}|$,

$$W = Fs \tag{8-22}$$

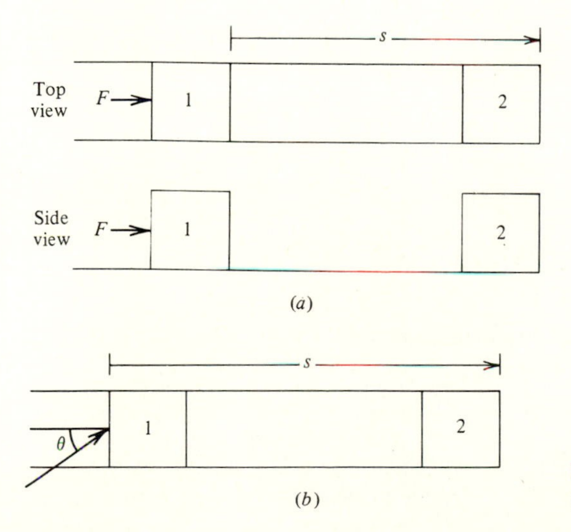

Figure 8-5 Force applied (a) directly and (b) at angle θ.

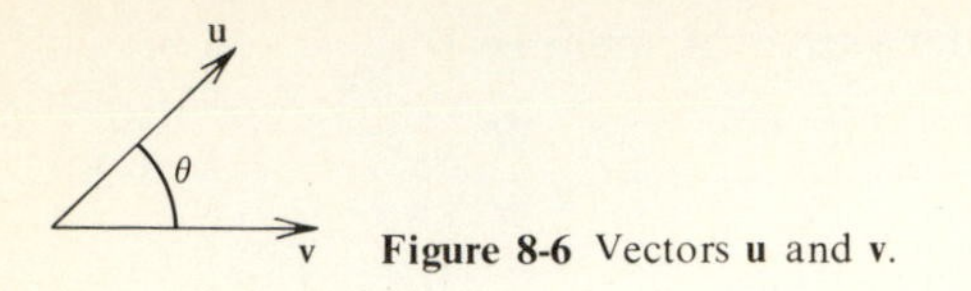

Figure 8-6 Vectors **u** and **v**.

Now suppose we apply the force obliquely, say at an angle θ with the x axis, as in Fig. 8-5*b*. If $\theta = \pi/2$, the force will be at right angles to the eraser and we will not be able to move it; that is, $W = 0$. In general, we can see that for any angle θ, only the x component of the force will be effective in doing work, or

$$W = (F \cos \theta)s \tag{8-23}$$

Note that at $\theta = 0$ this reduces to the case described in Eq. (8-22).

Both force **F** and distance **s** have direction as well as magnitude, and they are both vectors. Equation (8-23) is a form of multiplication of two vectors called the *dot product* or *scalar product* or *inner product* and can be written

$$W = \mathbf{F} \cdot \mathbf{s} = (F \cos \theta)(s) = Fs \cos \theta \tag{8-24}$$

where W is a scalar, which is the work done. In this case the dot-product operation is not merely a mathematical abstraction but is also an expression of a physical idea.

In *general,* if **u** and **v** are vectors, their dot product is a scalar given by

$$\mathbf{u} \cdot \mathbf{v} = uv \cos \theta \tag{8-25}$$

where θ is the angle between **u** and **v** (see Fig. 8-6). From the above we can see that if **u** and **v** are perpendicular, $\theta = \pi/2$, $\cos \theta = 0$, and the dot product is zero. This result has important significance for the unit vectors, for which

$$\boldsymbol{\delta}_x \cdot \boldsymbol{\delta}_y = 0 \qquad \boldsymbol{\delta}_x \cdot \boldsymbol{\delta}_z = 0 \tag{8-26}$$

and similarly for other such products. However,

$$\boldsymbol{\delta}_x \cdot \boldsymbol{\delta}_x = 1(1) \cos 0 = 1 \tag{8-26a}$$

and similarly for other products of a unit vector with itself. Then

$$\begin{aligned} \mathbf{u} \cdot \mathbf{v} &= (u_x \boldsymbol{\delta}_x + u_y \boldsymbol{\delta}_y + u_z \boldsymbol{\delta}_z) \cdot (v_x \boldsymbol{\delta}_x + v_y \boldsymbol{\delta}_y + v_z \boldsymbol{\delta}_z) \\ &= u_x v_x + u_y v_y + u_z v_z \end{aligned} \tag{8-27}$$

The Operation $\nabla \cdot \mathbf{q}$

In cartesian coordinates the operation $\nabla \cdot \mathbf{q}$ is given by

$$\nabla \cdot \mathbf{q} = \frac{\partial q_x}{\partial x} + \frac{\partial q_y}{\partial y} + \frac{\partial q_z}{\partial z} \tag{8-28}$$

In order to show this, we write the operator ∇ and the vector **q** in terms of cartesian coordinates and carry out the indicated operations term by term. Then, using

Eqs. (8-13) and (8-8), we obtain

$$\nabla \cdot \mathbf{q} \equiv \left(\boldsymbol{\delta}_x \frac{\partial}{\partial x} + \boldsymbol{\delta}_y \frac{\partial}{\partial y} + \boldsymbol{\delta}_z \frac{\partial}{\partial z}\right) \cdot (q_x \boldsymbol{\delta}_x + q_y \boldsymbol{\delta}_y + q_z \boldsymbol{\delta}_z)$$

The first terms in each parenthesis give

$$\left(\boldsymbol{\delta}_x \frac{\partial}{\partial x}\right) \cdot q_x \boldsymbol{\delta}_x = \frac{\partial q_x}{\partial x} \boldsymbol{\delta}_x \cdot \boldsymbol{\delta}_x + q_x \boldsymbol{\delta}_x \cdot \frac{\partial \boldsymbol{\delta}_x}{\partial x} = \frac{\partial q_x}{\partial x}(1) + q_x(0) = \frac{\partial q_x}{\partial x}$$

since the dot product of a unit vector with itself is unity and since the unit vector $\boldsymbol{\delta}_x$ is independent of x. The next term is

$$\left(\boldsymbol{\delta}_x \frac{\partial}{\partial x}\right) \cdot q_y \boldsymbol{\delta}_y = \boldsymbol{\delta}_x \cdot \boldsymbol{\delta}_y \frac{\partial q_y}{\partial x} + q_y \boldsymbol{\delta}_x \cdot \frac{\partial \boldsymbol{\delta}_y}{\partial x} = 0 + 0 = 0$$

since the unit vectors are perpendicular and $\boldsymbol{\delta}_y$ does not depend upon x. Proceeding in the same manner with other terms results in the desired equation (8-28) for $\nabla \cdot \mathbf{q}$.

Laplacian Operator

The laplacian operator ∇^2 (del second) is defined by

$$\nabla^2 \equiv \nabla \cdot \nabla \tag{8-29}$$

In cartesian coordinates it is given as

$$\nabla^2 = \frac{\partial^2}{\partial x^2} + \frac{\partial^2}{\partial y^2} + \frac{\partial^2}{\partial z^2} \tag{8-30}$$

In order to obtain this result we use the expression for ∇ [Eq. (8-13)] and carry out the operation

$$\nabla^2 \equiv \nabla \cdot \nabla \equiv \left(\boldsymbol{\delta}_x \frac{\partial}{\partial x} + \boldsymbol{\delta}_y \frac{\partial}{\partial y} + \boldsymbol{\delta}_z \frac{\partial}{\partial z}\right) \cdot \left(\boldsymbol{\delta}_x \frac{\partial}{\partial x} + \boldsymbol{\delta}_y \frac{\partial}{\partial y} + \boldsymbol{\delta}_z \frac{\partial}{\partial z}\right)$$

The first term is

$$(\boldsymbol{\delta}_x \cdot \boldsymbol{\delta}_x) \frac{\partial^2}{\partial x^2} + \boldsymbol{\delta}_x \cdot \frac{\partial \boldsymbol{\delta}_x}{\partial x} \frac{\partial}{\partial x} = \frac{\partial^2}{\partial x^2} + 0$$

Since $\boldsymbol{\delta}_x \cdot \boldsymbol{\delta}_y = 0$ and $\boldsymbol{\delta}_x \cdot \boldsymbol{\delta}_z = 0$, and since the derivatives of the unit vectors are zero, the second and third terms are zero. Similar arguments give the desired result.

8-7 CONDUCTION EQUATION IN VECTOR NOTATION

When Eq. (8-30) for the laplacian operator in cartesian coordinates is substituted, Eq. (8-21) becomes

$$\alpha \nabla^2 T = \frac{\partial T}{\partial t} \tag{8-31}$$

Equation (8-20) can also be written in vector form using Eq. (8-28). Thus

$$-\nabla \cdot \mathbf{q} = \rho \widehat{C}_p \frac{\partial T}{\partial t} \tag{8-32}$$

where $\nabla \cdot \mathbf{q}$ is the *divergence* of the vector **q**. Although Eq. (8-32) holds for any coordinate system, Eq. (8-28) applies only to cartesian coordinates. Expressions for other coordinate systems are given in Appendix B and will be derived later. The reader who does not intend to study vector operations at this time may look upon $\nabla \cdot \mathbf{q}$ and $\nabla^2 T$ as a form of mathematical shorthand. Note that we could also obtain Eq. (8-31) by substituting Eq. (8-12) into Eq. (8-32) and using Eq. (8-29). Thus

$$-\nabla \cdot \mathbf{q} = +\nabla \cdot (k\, \nabla T) = k \nabla \cdot \nabla T = k\, \nabla^2 T \tag{8-33}$$

8-8 GENERAL EQUATION FOR HEAT CONDUCTION*

In the previous illustration we discussed a solid being cooled from a temperature T_0 by maintaining the temperature of its surfaces at a lower constant temperature T_1. We derived the partial differential equations for the flux [(8-20) and (8-32)] and for the temperature [(8-21) and (8-31)] in cartesian coordinates and in vector form. In doing so it was actually not necessary to use the boundary conditions (temperatures T_1 or T_0 or dimensions a, b, c) at all to obtain the differential equations. Of course it was through the boundary conditions that we observed that a physical situation existed in which heat would be transported in all three directions; but three-dimensional transport could have resulted from other boundary conditions than the ones used; e.g., (1) the flux could have been specified at one or more of the surfaces in terms of a heat-transfer coefficient; (2) the temperature on each surface might have been different; or (3) the temperature or flux might have varied with the location on each surface or with time. The point is that in *all* these cases both the flux differential equation and the partial differential equation for temperature would have been identical to Eqs. (8-20), (8-21), (8-31), and (8-32). This result means that we can use Eq. (8-32) as a *general* flux equation for heat conduction in a solid (without internal generation) and then supply the flux law and boundary conditions. Then if the conditions of the problem indicate that there is no heat conduction in, say, the y direction, one would merely drop the term involving q_y. Likewise, if the physical condition indicates that there is no heat conduction in the z direction either, one would also eliminate the term involving q_z. The resulting equation would be Eqs. (7-4) and (7-5), which we derived for heat conduction in one direction in slabs with different boundary conditions.

Similarly for cases in which the thermal conductivity is constant, Eq. (8-21) is a *general* partial differential equation for the temperature distribution; it can be

*This section does not require use of vector notation.

simplified so as to apply to individual cases. These equations could also be used if the problem were one in *steady-state* transport by setting the partial derivative with respect to time equal to zero. The general equations are also independent of the actual physical *direction* of heat transport. For example, suppose in the previous illustration that the solid were being heated instead of being cooled ($T_1 > T_0$). Our *sign conventions* would tell us that q_x is the symbol for a positive flux in the positive x direction. In this case we would discover upon applying the boundary conditions (and perhaps making numerical calculations) that q_x is negative, meaning physically that heat flow is *actually* in the negative x direction, or from a high temperature at the surface to a lower temperature inside.

This result means that *it is actually unnecessary* to use an incremental balance to derive a differential equation for each and every case that arises. Instead we can derive general equations that apply to a variety of cases and simplify them so that they apply to the specific physical situation under consideration. In the next chapter we will derive a very general equation for momentum transport and use it to derive a very general equation for energy transport.

8-9 TEMPERATURE DISTRIBUTION IN A SOLID*

We now solve Eq. (8-21), the partial differential equation for the temperature in a brick-shaped solid. As in the one-dimensional case (Example 7-1) we define the dimensionless temperature (fraction unaccomplished temperature change), but in order to avoid confusion with the angle θ we use

$$T^* \equiv \frac{T - T_1}{T_0 - T_1} \tag{8-34}$$

Then

$$T = T_1 + (T_0 - T_1)T^* \tag{8-35}$$

and

$$\alpha \left(\frac{\partial^2 T^*}{\partial x^2} + \frac{\partial^2 T^*}{\partial y^2} + \frac{\partial^2 T^*}{\partial z^2} \right) = \frac{\partial T^*}{\partial t} \tag{8-36}$$

Although ordinarily we would now employ dimensionless variables such as $\eta_x \equiv x/a$, $\eta_y \equiv y/b$, $\eta_z \equiv z/c$, and $s \equiv \alpha t/a^2$, at this point it is more instructive and more convenient *not* to use dimensionless time and distance variables. We will obtain the solution for $T^*(x, y, z, t)$ by means of the *product rule,* also called *Newman's rule* (Appendix 8-1). This principle states that if T_x^*, T_y^*, and T_z^* are the solutions of

$$\alpha \frac{\partial^2 T_x^*}{\partial x^2} = \frac{\partial T_x^*}{\partial t} \qquad \alpha \frac{\partial^2 T_y^*}{\partial y^2} = \frac{\partial T_y^*}{\partial t} \qquad \alpha \frac{\partial^2 T_z^*}{\partial z^2} = \frac{\partial T_z^*}{\partial t}$$

then the solution of the equation

$$\alpha \left(\frac{\partial^2 T^*}{\partial x^2} + \frac{\partial^2 T^*}{\partial y^2} + \frac{\partial^2 T^*}{\partial z^2} \right) = \frac{\partial T^*}{\partial t}$$

*This section does not require use of vector notation.

is (see Appendix 8-1)

$$T^*(x, y, z, t) = T_x^*(x, t)T_y^*(y, t)T_z^*(z, t) \tag{8-37}$$

provided that the boundary conditions on T_x^*, T_y^*, and T_z^* are compatible with those on T^*. We use Table 8-1 to determine that the boundary conditions are indeed compatible; that is, $T^* = T_x^* T_y^* T_z^*$ at all boundaries. For transport in the x direction only, we have obtained the solution (see Appendix 7-2)

$$T_x^* = \sum_{n=0}^{\infty} A_n X_n S_n \qquad n = 0, 1, 2, \ldots$$

where
$$X_n = \cos \gamma_n \eta_x = \cos\left[(n + \tfrac{1}{2})\pi \frac{x}{a}\right]$$

$$S_n = \exp\left[-(n + \tfrac{1}{2})^2 \frac{\pi^2 \alpha t}{a^2}\right] \qquad A_n = \frac{2(-1)^n}{(n + \frac{1}{2})\pi}$$

If we define a dimension*al* eigenvalue,

$$\lambda_n \equiv \frac{(n + \frac{1}{2})\pi}{a} \equiv \frac{\gamma_n}{a} \tag{8-38}$$

we get

$$X_n = \cos \lambda_n x \qquad S_n = \exp(-\lambda_n^2 \alpha t) \qquad A_n = \frac{2(-1)^n}{\lambda_n a} \tag{8-39}$$

and
$$T_x^* = \sum_{n=0} A_n (\cos \lambda_n x) e^{-\lambda_n^2 \alpha t} \tag{8-40}$$

We now note that the form of the solution for T_y^* and T_z^* would be the same except for having to replace the distance a with the distance b in the case of T_y^*. Of course, it is also necessary to use a different summation index in each case. We will sum over m in the case of T_x^*, n in the case of T_y^*, and p in the case of T_z^*. To avoid double subscripts we will use the subscript m to mean not only the mth term in the series but also the fact that the length scale is a; subscript n to mean that it is b, and p to mean that it is c. Then

$$T_x^* = \sum_{m=0}^{\infty} A_m X_m(x) S_m(t) \tag{8-41}$$

Table 8-1 Initial and boundary conditions

Condition	t	x	y	z	T	T^*	T_x^*	T_y^*	T_z^*
1	0	†	†	†	T_0	1	1	1	1
2, 3	>0	$\pm a$	All	All	T_1	0	0	T_y^*	T_z^*
4, 5	>0	All	$\pm b$	All	T_1	0	T_x^*	0	T_z^*
6, 7	>0	All	All	$\pm c$	T_1	0	T_x^*	T_y^*	0

† All except on surface.

where
$$A_m = \frac{2(-1)^m}{(m+\frac{1}{2})\pi} \qquad X_m = \cos\lambda_m x$$
$$\lambda_m = \frac{(m+\frac{1}{2})\pi}{a} \equiv \frac{\gamma_m}{a} \qquad S_m = e^{-\lambda_m^2 \alpha t} \tag{8-42}$$

Similarly
$$T_y^* = \sum_{n=0}^{\infty} A_n X_n(y) S_n(t) \tag{8-43}$$

where
$$X_n = \cos\lambda_n y \qquad \lambda_n \equiv \frac{(n+\frac{1}{2})\pi}{b} = \frac{\gamma_n}{b} \tag{8-44}$$

and
$$T_z^* = \sum_{p=0}^{\infty} A_p X_p(z) S_p(t) \tag{8-45}$$

where
$$X_p = \cos\lambda_p z, \qquad \lambda_p = \frac{(p+\frac{1}{2})\pi}{c} = \frac{\gamma_p}{c} \tag{8-46}$$

By the product rule (Appendix 8-1)
$$T^*(x, y, z, t) = T_x^*(x, t)T_y^*(y, t)T_z^*(z, t)$$
$$T^* = \sum_{m=0}^{\infty} A_m X_m S_m \sum_{n=0}^{\infty} A_n X_n S_n \sum_{p=0}^{\infty} A_p X_p S_p \tag{8-47}$$
$$T^* = \sum_{m=0}^{\infty}\sum_{n=0}^{\infty}\sum_{p=0}^{\infty} A_m A_n A_p X_m X_n X_p S_m S_n S_p \tag{8-48}$$
$$T^* = \sum_{m=0}^{\infty}\sum_{n=0}^{\infty}\sum_{p=0}^{\infty} \frac{8(-1)^{m+n+p}\cos\left[(m+\frac{1}{2})\frac{\pi x}{a}\right]\cos\left[(n+\frac{1}{2})\frac{\pi y}{b}\right]\cos\left[(p+\frac{1}{2})\frac{\pi z}{c}\right]}{(m+\frac{1}{2})(n+\frac{1}{2})(p+\frac{1}{2})\pi^3}$$
$$\exp\left[\frac{(m+\frac{1}{2})^2\pi^2\alpha t}{a^2}\right]\exp\left[\frac{-(n+\frac{1}{2})^2\pi^2\alpha t}{b^2}\right]\exp\left[\frac{-(p+\frac{1}{2})^2\pi^2\alpha t}{c^2}\right] \tag{8-48a}$$

8-10 CALCULATION OF TOTAL TRANSPORT OF HEAT*†

The design of engineering equipment usually requires a knowledge of the total rate of transport of heat or mass across the boundaries of the system under consideration. For example, in designing a heat exchanger we are interested in finding the total rate of heat transport to or from a fluid that is being heated or cooled; in designing a gas absorber we are interested in the total rate of transport of the solute from a gas phase to a liquid phase. In the case of one-dimensional transport

*This section does not require use of vector notation.

†This section may be omitted on first reading.

we have previously illustrated the calculation of the total transport of heat from systems such as a thin slab (Chaps. 1 and 7) and a nuclear reactor (Chaps. 3 and 7) by calculating the fluxes at the boundaries and multiplying them by the areas. We have also calculated the average potential (temperature or concentration) and have used a macroscopic balance to relate it to the net transport at the boundaries [Eq. (7-27)]. It is now our purpose to extend these procedures to three-dimensional problems, and we illustrate the method by considering the simple case of heat transport from the bricklike solid.

Total Rate of Heat Transport in Terms of Flux

Heat will be transferred at each face from the brick, and the total heat loss will be the sum of the heat losses from each face. Consider first the surface at $x = +a$ (see Fig. 8-7). At an arbitrary point $P(a, y, z)$ on the surface the heat flux will be

$$q_x \bigg|_{x=a} = -k \frac{\partial T}{\partial x}\bigg|_{x=a} \tag{8-49}$$

Therefore, in order to get the heat flux at P we need to know the temperature gradient at $x = a$. This can be obtained by differentiating an equation for T as a function of x, y, z, and t.

We first qualitatively examine how the temperature $T(x, y, z, t)$ varies with position inside the brick. Let us consider its variation with x at a constant value of y and z. The intersection of the two planes $y = 0$ and $z = 0$ represents a line (the x axis) through the center of the brick. The variation of temperature with x along this line will depend upon time. At $t = 0$ the temperature is constant at T_0 except at $x = \pm a$, where it is T_1 (see Fig. 8-8). At an arbitrary time t_1 the region near the surfaces $x = \pm a$ has cooled, but no noticeable heat loss has occurred near the center. At $t = t_2$ heat has been lost from the center as well, and at $t = \infty$ the entire brick has been cooled to T_1. If we were to plot T versus x at a value of

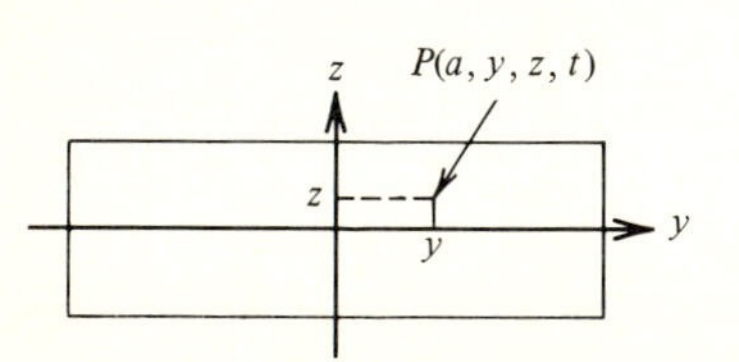

Figure 8-7 Point on $x = +a$ surface.

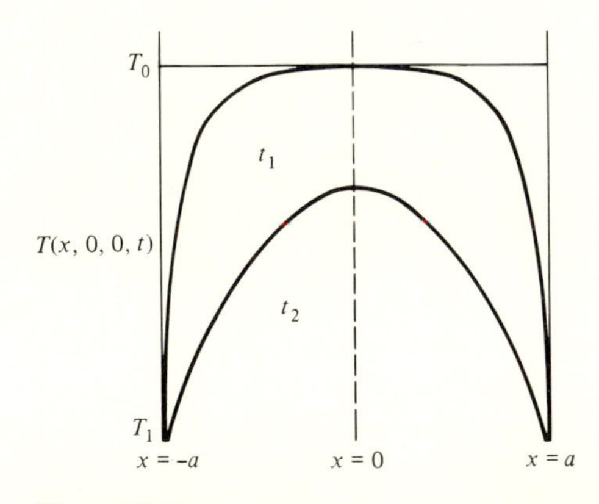

Figure 8-8 Temperature versus x at $y = 0$ and $z = 0$.

y or z different from zero, the general appearance of the curves at arbitrary values of t would be the same; in fact the curves at $t = 0$ and $t = \infty$ would be identical to the case where $y = 0 = z$.

Since, at any t, $T = T(x, y, z)$, the temperature gradient $\partial T/\partial x$ will also depend on x, y, z and the gradient at $x = a$ will depend on y and z, that is, on the location on the surface. Hence the heat flux will also vary with location on the surface. Since the flux varies, we *cannot* obtain the heat transport by multiplying the flux q_x by the *total* area of the face $A_x = 4bc$. Instead, we must consider an element of surface area $\Delta A_x = \Delta y \, \Delta z$. If ΔA_x is small, the rate of heat loss through the area is approximately (see Fig. 8-9a)

$$\Delta \dot{Q}_{L(x)} = q_x \, \Delta A_x \tag{8-50}$$

where (x) refers to the x face. If $\Delta A_x \to 0$, we get

$$d\dot{Q}_{L(x)} = q_x \, dA_x \tag{8-50a}$$

The *total* heat-loss rate through the entire area A_x is obtained by integrating over A_x

$$\dot{Q}_{Lt}\bigg|_{x=a} = \iint_{A_x} (q_x \, dA_x)\bigg|_{x=a} = \int_{-c}^{c} \int_{-b}^{b} q_x\bigg|_{x=a} dy \, dz \tag{8-51a}$$

Now let us consider the face $x = -a$. We note from Fig. 8-2 that the slope of the T-versus-x curve is positive at $x = -a$ whereas it is negative at $x = +a$; hence the heat flux will be negative at $x = -a$ and positive at $x = +a$. Since we know from the physical situation that heat will be lost at *both* faces, the total heat-loss rate from the x faces will be

$$\dot{Q}_{Lt(x)} = \left(\int q_x \, dA_x\right)\bigg|_{x=a} - \left(\int q_x \, dA_x\right)\bigg|_{x=-a} \tag{8-51b}$$

and there would be similar expressions for the other faces. The total heat-loss rate is

$$\dot{Q}_{Lt} \equiv \dot{Q}_{L(x)} + \dot{Q}_{L(y)} + \dot{Q}_{L(z)} \tag{8-51c}$$

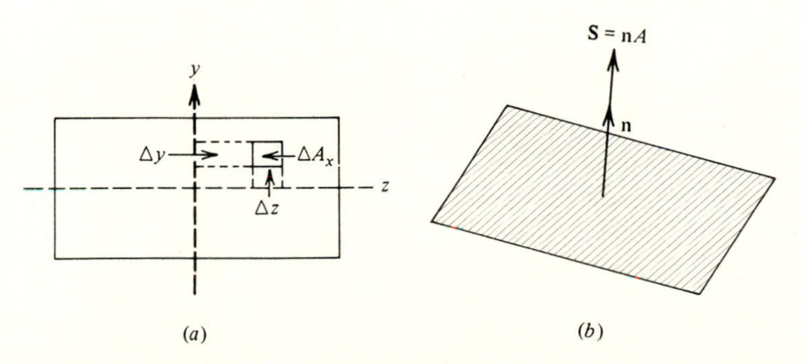

Figure 8-9 (a) Element of area on $x = a$ face. (b) Surface and unit normal vectors.

8-11 HEAT FLOW RATE IN TERMS OF FLUX VECTOR

If we use vector notation, we can write the above expression in more convenient form. To do so we must make a distinction between an area and a surface. If we think of a sheet of paper, we know that it has two sides or surfaces, each with the same area; we can distinguish between the two sides by constructing unit normal vectors so that they point *outward* from the sheet in opposite directions. It is apparent that a surface **S** has a direction given by its *outwardly* directed unit normal **n** as well as magnitude given by its area A. Thus (Fig. 8-9*b*) **S** is a vector given by

$$\mathbf{S} = \mathbf{n}A \tag{8-52a}$$

where

$$A \equiv |\mathbf{S}| \tag{8-52b}$$

is its magnitude. For a surface perpendicular to the x axis

$$\mathbf{S}_x = \mathbf{n}_x A_x \tag{8-53a}$$

where

$$\mathbf{n}_x = n_x \boldsymbol{\delta}_x \qquad \text{and} \qquad n_x = \mathbf{n}_x \cdot \boldsymbol{\delta}_x = \cos\theta \tag{8-53b}$$

Then $n_x = +1$ if $\mathbf{n}_x$ and $\boldsymbol{\delta}_x$ point in the same direction ($\theta = 0$), and $n_x = -1$ if $\mathbf{n}_x$ and $\boldsymbol{\delta}_x$ point in the opposite directions ($\theta = \pi$).

Then for the brick (Fig. 8-10) the first term in Eq. (8-51*b*) can be written

$$(\mathbf{q}_x \cdot d\mathbf{S}_x)\Big|_{x=a} = (q_x \boldsymbol{\delta}_x \cdot \mathbf{n}_x\, dA_x)\Big|_{x=a} = (q_x\, dA_x \boldsymbol{\delta}_x \cdot \mathbf{n}_x)\Big|_{x=a} = (q_x\, dA_x)\Big|_{x=a} \tag{8-54}$$

and the second term is

$$(\mathbf{q}_x \cdot d\mathbf{S}_x)\Big|_{x=-a} = (q_x\, dA_x \boldsymbol{\delta}_x \cdot \mathbf{n}_x)\Big|_{x=-a} = (-q_x\, dA_x)\Big|_{x=-a}$$

Then we can write Eq. (8-51*b*) in the form

$$\dot{Q}_{L(x)} = \left(\iint \mathbf{q}_x \cdot d\mathbf{S}_x\right)\Big|_{x=a} + \left(\iint \mathbf{q}_x \cdot d\mathbf{S}_x\right)\Big|_{x=-a} \tag{8-55}$$

or

$$\dot{Q}_{L(x)} = \int_{\mathbf{S}_x}\!\!\int \mathbf{q}_x \cdot d\mathbf{S}_x \tag{8-56}$$

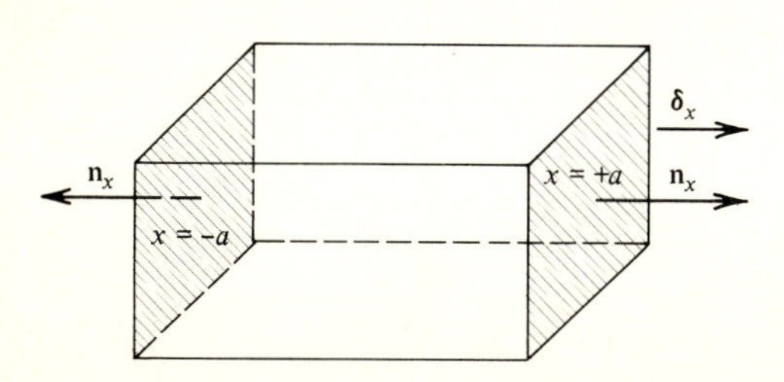

Figure 8-10 Unit normals at $x = a$ and $x = -a$.

where the integration is to be taken over *both* the face $x = a$ and the face $x = -a$.

We can readily extend this treatment to include the $y = \pm b$ and the $z = \pm c$ faces. The total heat loss is then

$$\dot{Q}_{Lt} = \iint_{A_x} \mathbf{q}_x \cdot d\mathbf{S}_x + \iint_{A_x} \mathbf{q}_y \cdot d\mathbf{S}_y + \iint_{A_x} \mathbf{q}_z \cdot d\mathbf{S}_z \tag{8-57}$$

or

$$\dot{Q}_L = \iint \mathbf{q} \cdot d\mathbf{S} = \iint \mathbf{q} \cdot \mathbf{n}\, dA \tag{8-58}$$

In terms of the total rate of heat gained $\dot{Q}_{Gt}$ we have

$$\boxed{\dot{Q}_{Gt} = -\iint \mathbf{q} \cdot \mathbf{n}\, dA} \tag{8-59}$$

Although derived for cartesian coordinates, Eq. (8-59) is *general* and it can be applied to all coordinate systems (as proved later) by writing the vectors $\mathbf{q}$ and $d\mathbf{S}$ in that system. In cartesian coordinates we would expand Eq. (8-59) into Eq. (8-57) and then substitute (8-51*b*) and Fourier's law

$$\begin{aligned} \dot{Q}_{L(x)} &\equiv \int_{-c}^{c}\int_{-b}^{b} \mathbf{q}_x \cdot d\mathbf{S}_x = \int_{-c}^{c}\int_{-b}^{b}\left(q_x\Big|_{x=a} - q_x\Big|_{x=-a}\right) dy\,dz \\ &= 2\int_{-c}^{c}\int_{-b}^{b} q_x\Big|_{x=a}\, dy\,dz \text{ (by antisymmetry)} \\ &= 2\int_{-c}^{c}\int_{-b}^{b}\left(-k\frac{\partial T}{\partial x}\Big|_{x=a}\right) dy\,dz \end{aligned} \tag{8-60}$$

and similar equations for $\dot{Q}_{L(y)}$ and $Q_{L(z)}$. Thus we now merely need to (1) substitute an equation for the temperature distribution, (2) take the specified derivatives, (3) evaluate them at the boundary, and (4) carry out successively the indicated integrations over the area.

8-12 EVALUATION OF HEAT-LOSS RATE*†

The total rate of heat loss is given either by Eq. (8-51*b*) or by expanding Eqs. (8-58) and (8-59). Substituting the definition of T^* [Eqs. (8-34) and (8-37)] and recalling that T_y^* and T_z^* do not depend on x gives

$$\dot{Q}_{L(x)} = -2k\int_{-c}^{c}\int_{-b}^{b} (T_0 - T_1)\frac{\partial T_x^*}{\partial x}\Big|_{x=a} T_y^* T_z^*\, dy\,dz \tag{8-61}$$

Using the expression for T_x^* [Eq. (8-41)] gives

$$\frac{\partial T_x^*}{\partial x}\Big|_{x=a} = \sum_{m=0}^{\infty} A_m(-\lambda_m)(\sin \lambda_m a)S_m = -\sum_{m=0}^{\infty}\frac{2}{a}\exp(-\lambda_m^2 \alpha t) \tag{8-62}$$

*This section does not require use of vector notation.

†Although the details of this section may be omitted on first reading, the procedure should be understood and the results will be used later.

since, by Eq. (8-38),

$$\sin \lambda_m a = \sin (m + \tfrac{1}{2})\pi = (-1)^m$$

Also

$$\int_{-b}^{b} T_y^* \, dy = \sum_{n=0}^{\infty} A_n S_n \left(2 \int_0^b \cos \lambda_n y \, dy \right)$$

$$= 2 \sum_{n=0}^{\infty} \frac{A_n S_n}{\lambda_n} \sin \lambda_n y \Big|_0^b = \sum_{n=0}^{\infty} \frac{4b}{[(n + \frac{1}{2})\pi]^2} S_n \tag{8-63}$$

Using a similar expression for the integral over z and recognizing that the summations are independent so that the products of the sums can be written as a triple summation gives

$$\dot{Q}_{L(x)} = \frac{64kbc(T_0 - T_1)}{a} \sum_{m=0}^{\infty} \sum_{n=0}^{\infty} \sum_{p=0}^{\infty} \frac{S_n S_m S_p}{[(n + \frac{1}{2})\pi]^2 [(p + \frac{1}{2})\pi]^2} \tag{8-64}$$

Using similar expressions for $\dot{Q}_{L(y)}$ and $\dot{Q}_{L(z)}$ and writing $\dot{Q}_{Lt}$ in terms of a dimensionless heat flow gives

$$Q_{Lt}^* \equiv \frac{\dot{Q}_{Lt}}{k(T_0 - T_1)a} = \frac{64bc}{\pi^4 a^2} \left[\sum_m S_m \sum_n \frac{S_n}{(n + \frac{1}{2})^2} \sum_p \frac{S_p}{(p + \frac{1}{2})^2} \right.$$

$$\left. + \frac{a^2}{b^2} \sum_n S_n \sum_m \frac{S_m}{(m + \frac{1}{2})^2} \sum_p \frac{S_p}{(p + \frac{1}{2})^2} + \frac{a^2}{c^2} \sum_p S_p \sum_m \frac{S_m}{(m + \frac{1}{2})^2} \sum_n \frac{S_n}{(n + \frac{1}{2})^2} \right] \tag{8-65}$$

For large *times*, only the term $m = 0 = n = p$ is important. Using the third and fourth of Eqs. (8-42) for S_n, S_m, and S_p gives

$$Q_{Lt}^* = \frac{1024bc}{\pi^4 a^2} \left\{ \exp\left[-\frac{\alpha \pi^2 t}{4a^2} \left(1 + \frac{a^2}{b^2} + \frac{b^2}{c^2} \right) \right] \right\} \left(1 + \frac{a^2}{b^2} + \frac{a^2}{c^2} \right) \tag{8-66}$$

Average Temperature

From Eq. (A1-4-3).

$$\langle T^* \rangle \equiv \frac{1}{\mathcal{V}} \iiint T^* \, d\mathcal{V}$$

or

$$\langle T^* \rangle = \frac{\int_{-c}^{c} \int_{-b}^{b} \int_{-a}^{a} T^* \, dx \, dy \, dz}{\int_{-c}^{c} \int_{-b}^{b} \int_{-a}^{a} dx \, dy \, dz} \tag{8-67}$$

Substituting for T^* [Eq. (8-48)] and integrating gives

$$\langle T^* \rangle = \frac{8}{\pi^6} \sum_{m=0}^{\infty} \frac{S_m}{(m+\frac{1}{2})^2} \sum_{n=0}^{\infty} \frac{S_n}{(n+\frac{1}{2})^2} \sum_{p=0}^{\infty} \frac{S_p}{(n+\frac{1}{2})^2} \tag{8-68}$$

For $m = n = p = 0$ (larger times) and $S_0 = \exp(-\alpha\pi^2 t/4a^2)$ this becomes

$$\langle T^* \rangle = \frac{8}{\pi^6}(64) \exp\left[\left(-\frac{\alpha\pi^2 t}{4}\right)\left(\frac{1}{a^2} + \frac{1}{b^2} + \frac{1}{c^2}\right)\right] \tag{8-69}$$

If we let $B_m \equiv (m + \frac{1}{2})^{-2}$, etc., Eq. (8-68) becomes (8-70)

$$\langle T^* \rangle = \frac{8}{\pi^6} \sum_{m=0}^{\infty} B_m S_m \sum_{n=0}^{\infty} B_n S_n \sum_{p=0}^{\infty} B_p S_p \tag{8-71}$$

Relation between Rate of Heat Loss and Average Temperature

For *any* object the rate of heat loss to the surroundings is equal to the rate of decrease of the enthalpy of the body, which in turn is proportional to the rate of decrease of the average temperature with time. Thus

$$-\dot{Q}_t = \dot{Q}_{Lt} = -\rho\widehat{C}_p \frac{d}{dt} \iiint T\, d\mathcal{V} = -\rho\widehat{C}_p \mathcal{V} \frac{d\langle T \rangle}{dt} \tag{8-72}$$

We can relate $d\langle T^* \rangle/dt$ to Q^*_{Lt} and make it dimensionless by multiplying by $8bc/\alpha$,

$$\frac{8bc\rho\widehat{C}_p}{k}\frac{d\langle T^* \rangle}{dt} \equiv \frac{d\langle T^* \rangle}{ds_H} \qquad \text{where} \qquad s_H \equiv \frac{\alpha t}{8bc} = \frac{\alpha t a}{\mathcal{V}} \tag{8-73}$$

If

$$\dot{Q}^*_{Lt} \equiv \frac{\dot{Q}_{Lt}}{k(T_0 - T_1)a}$$

then the *macroscopic dimensionless balance* is

$$\frac{d\langle T^* \rangle}{ds_H} = -\dot{Q}^*_{Lt} \tag{8-74}$$

or in *dimensional* form

$$\left(\rho\widehat{C}_p \frac{d\langle T \rangle}{dt}\right)(8abc) = -\dot{Q}_{Lt} \tag{8-75}$$

8-13 GAUSS DIVERGENCE THEOREM†

The Gauss divergence theorem is

$$\iiint_{\mathcal{V}} \nabla \cdot \mathbf{q}\, d\mathcal{V} = \iint_A \mathbf{q} \cdot \mathbf{n}\, dA = \iint_{\mathbf{S}} \mathbf{q} \cdot d\mathbf{S} \tag{8-76}$$

†This section may be omitted on first reading.

where $d\mathcal{V}$ is a volume element with surface vector $d\mathbf{S}$ whose outwardly directed normal is $\mathbf{n}$. We will illustrate this theorem using the case of the cooling of the brick whose sides are bounded by $x = \pm a$, $y = \pm b$, and $z = \pm c$. Consider the left-hand side of the equation and use the expression for $\nabla \cdot \mathbf{q}$ in cartesian coordinates

$$\iiint \nabla \cdot \mathbf{q}\, d\mathcal{V} \equiv \int_{-c}^{c}\int_{-b}^{b}\int_{-a}^{a} \left(\frac{\partial q_x}{\partial x} + \frac{\partial q_y}{\partial y} + \frac{\partial q_z}{\partial z}\right) dx\, dy\, dz$$

The integral of the first term under the integral sign can be written

$$\iint \left(\int_{-a}^{a} \frac{\partial q_x}{\partial x} dx\right) dA_x = \iint_{A_x} \left(q_x\Big|_{x=a} - q_x\Big|_{x=-a}\right) dA_x$$
$$= \iint_{A_x} \mathbf{q}_x \cdot d\mathbf{S}_x = \iint_{A_x} \mathbf{q}_x \boldsymbol{\delta}_x \cdot \mathbf{n}_x\, dA_x$$

Using similar expressions for the other terms, we have

$$\iiint \nabla \cdot \mathbf{q}\, d\mathcal{V} = \iint \mathbf{q}_x \cdot d\mathbf{S}_x + \iint \mathbf{q}_y \cdot d\mathbf{S}_y + \iint \mathbf{q}_z \cdot d\mathbf{S}_z \tag{8-77}$$

which is the same as the right-hand side of Eq. (8-76). For a solid being cooled

$$-\dot{Q}_t = \dot{Q}_{Lt} = \iint \mathbf{q} \cdot d\mathbf{S} = \iiint \nabla \cdot \mathbf{q}\, d\mathcal{V}$$

But since

$$-\nabla \cdot \mathbf{q} = \rho \widehat{C}_p \frac{\partial T}{\partial t}$$

for a solid of *any* shape,

$$-\dot{Q}_{Lt} = -\iiint \nabla \cdot \mathbf{q}\, d\mathcal{V} = \rho \widehat{C}_p \iiint \frac{\partial T}{\partial t} d\mathcal{V} = \rho \widehat{C}_p \mathcal{V} \frac{d\langle T \rangle}{dt} \tag{8-78}$$

Thus one obtains the same macroscopic balance in a more formal manner. The Gauss divergence theorem is quite useful in deriving such equations. *Since the Gauss divergence theorem applies to a volume element of any size, it can be used for any arbitrary element of volume within a medium or for the body as a whole.*

8-14 GENERAL EQUATION FOR THE TOTAL RATE OF HEAT TRANSPORT†

For a differential area dA on any element of surface $d\mathbf{S} = \mathbf{n}\, dA$ of the brick-shaped solid, the rate of heat gain is

$$d\dot{Q}_G = -\mathbf{n} \cdot \mathbf{q}\, dA \tag{8-79}$$

where $\mathbf{n}$ is the outwardly directed unit normal. The total rate of heat loss $\dot{Q}_{Lt}$ or

†This section may be omitted on first reading.

heat gain $\dot{Q}_t$ is then obtained from Eq. (8-58)

$$\dot{Q}_{Lt} = \iint \mathbf{q} \cdot \mathbf{n}\, dA \equiv -\dot{Q}_t$$

where the integral is taken over the entire surface (all six faces).

It is now appropriate to show that Eqs. (8-58) and (8-79) are general, i.e., hold for an arbitrary internal element, for any coordinate system, or for any angle θ between **q** and **n.** To show this let us first recognize that there is no difference in principle between applying Eq. (8-58) to a solid whose boundaries are $x = \pm a$, $y = \pm b$, $z = \pm c$ and applying it to an arbitrary *interior* element bounded by planes $x = \pm x_1, y = \pm y_1, z = \pm z_1$. Therefore, Eq. (8-58) *does* hold for an arbitrary internal element if it holds for the entire volume. Secondly, Eqs. (8-58) and Eq. (8-79) involve the dot product of two vectors [Eq. (8-25)] and therefore tacitly assume that **q** is a vector. So in the long run their generality depends upon whether **q** is a vector. If it *is* a vector, they will apply for any element of *any* shape for *any* orientation of the axes or for *any* coordinate system. We show that **q** *is* a vector in Appendix 8-4.

8-15 HEAT SCALAR†

Since **q** is a vector, the dot product of **q** with **n** will give, in accordance with the general equation (8-25) for the dot product, a scalar

$$\mathbf{n} \cdot \mathbf{q} = q \cos\theta \equiv q_n \tag{8-80}$$

for any value of the angle θ between **q** and **n** (see Fig. 8-11). The proof of Eq. (8-80) is given in Appendix 8-4. We will call‡ q_n the *heat scalar.* Since q_n is the projection of **q** onto **n,** Eq. (8-80) indicates that only the component of **q** in the direction of **n** contributes to the heat flow rate. If we multiply Eq. (8-80) by the differential element of area dA whose normal is **n,** we get the heat-loss rate

$$d\dot{Q}_L = q_n\, dA = \mathbf{q} \cdot \mathbf{n}\, dA = \mathbf{q} \cdot d\mathbf{S} = -d\dot{Q}_G \tag{8-81}$$

†This section may be omitted on first reading.

‡Some texts, especially in physics call q_n the flux, in which case the vector **q** is called the flux density.[3]

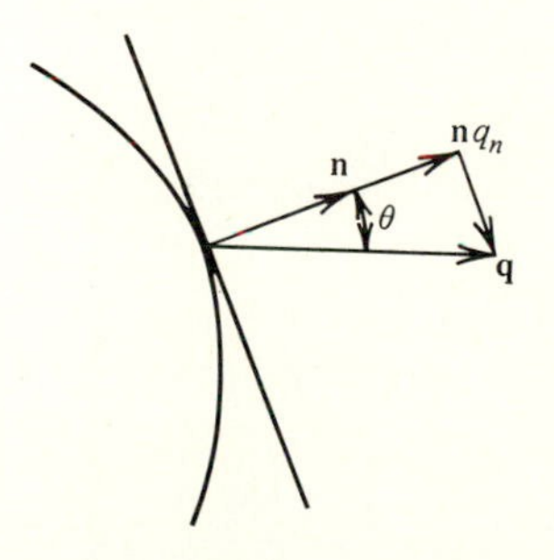

Figure 8-11 Projection of **q** onto ***n***.

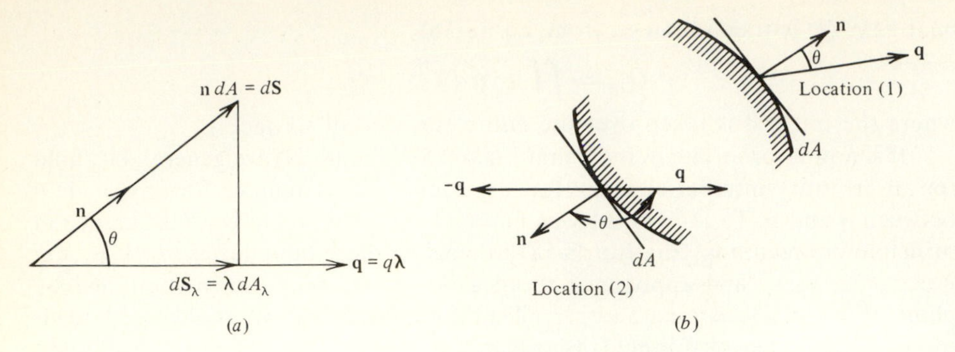

Figure 8-12 (*a*) Projection of differential area element onto **q**. (*b*) Vectors **n** and **q** at locations (1) and (2).

where $d\mathbf{S}$ is an element of surface whose magnitude is dA and normal is **n.** Note that if the heat flux at the surface is positive, the temperature gradient is negative and heat is leaving the system; that is, $d\dot{Q}_L > 0$, $d\dot{Q}_G < 0$. If $\boldsymbol{\lambda}$ is a unit vector in the direction of $\mathbf{q} \equiv q\boldsymbol{\lambda}$, taking the dot product of **q** and $d\mathbf{S}$ gives

$$d\dot{Q}_G = -\mathbf{q} \cdot d\mathbf{S} = -|\mathbf{q}|\,|d\mathbf{S}| \cos\theta = -q \cos\theta\, dA = -q\, dA_\lambda \tag{8-82}$$

where $dA_\lambda \equiv dA \cos\theta$ is the projection of the area $d\mathbf{S}$ onto **q** (see Fig. 8-12*a*). Note that the result is the same whether we project **q** onto **n** to get q_n or $d\mathbf{S}$ onto **q** to get dA_λ. If we integrate over the entire area of a control volume, we get an equation for $\dot{Q}_G$ [Eq. (8-59)].

Physical Picture of Heat Scalar

Now let us get a physical picture of $\mathbf{q} \cdot \mathbf{n}$ and q_n. In Fig. 8-12*b* we depict **n** and **q** at two different locations (1) and (2) on the surface. At (1), $|\theta| < \pi/2$; $\mathbf{n} \cdot \mathbf{q} = \cos\theta > 0$. Then q_n and $d\dot{Q}_L$ are greater than zero, meaning that heat is leaving the system. At (2), however, $|\theta| > \pi/2$, $\cos\theta < 0$, and q_n and $d\dot{Q}_L$ are negative, meaning that heat is entering the system. However, if at (2) we reverse the direction of ∇T and hence that of **q**, $d\dot{Q}_L > 0$, meaning that heat is leaving the system. Equation (8-58) then merely adds up, with due regard for sign, all the heat flows through all the elements of area.

8-16 GENERAL DEFINITION OF HEAT FLUX†

If we let $\mathbf{q} \equiv q\boldsymbol{\lambda}$, where $\boldsymbol{\lambda}$ is a unit vector in the direction of the heat flux, the magnitude of **q** is given by

$$q \equiv \left| \frac{d\dot{Q}_G}{\cos\theta\, dA} \right| = \left| \frac{d\dot{Q}_G}{dA_\lambda} \right| \tag{8-83}$$

†This section may be omitted on first reading.

where θ is the angle between $\boldsymbol{\lambda}$ (or $\mathbf{q}$) and $\mathbf{n}$ and $dA_\lambda \equiv dA \cos\theta$ is an element of area whose norm is in the direction of $\mathbf{q} = q\boldsymbol{\lambda}$.

8-17 UNSTEADY-STATE CONDUCTION IN A SOLID OF ARBITRARY SHAPE†

Let us now consider a solid of arbitrary shape (brick, sphere, cylinder, ellipsoid, etc, or irregular). The solid is originally at T_0, and for $t > 0$ its surface is maintained at a uniform temperature T_1. Since there is no generation, the net input rate will equal the accumulation rate and Eqs. (8-31) and (8-32)

$$\alpha \nabla^2 T = \frac{\partial T}{\partial t} \qquad \text{and} \qquad -\nabla \cdot \mathbf{q} = \rho \widehat{C}_p \frac{\partial T}{\partial t}$$

will apply at any point in the solid. In addition, the total rate of heat gain will be given by

$$-\dot{Q}_{Lt} = \dot{Q}_t = -\iiint \mathbf{q} \cdot \mathbf{n}\, dA = \rho \widehat{C}_p \mathcal{V} \frac{d\langle T\rangle}{dt} \tag{8-72}$$

where the average temperature (see Appendix 1-4) is

$$\langle T\rangle = \frac{1}{\mathcal{V}} \iiint T\, d\mathcal{V}$$

The dimensionless temperature is

$$T^* \equiv \frac{T - T_1}{T_0 - T_1}$$

Usually we are interested in finding $\dot{Q}_{Lt}$ and/or $\langle T\rangle$ as functions of time. For the brick we obtained an analytical solution [Eqs. (8-48) and (8-49)], which can be written

$$T^* = \sum_{m=0}^{\infty} \sum_{n=0}^{\infty} \sum_{p=0}^{\infty} \Phi_{mnp}(x, y, z) S_{mnp}(t)$$

where

$$\Phi_{mnp} \equiv X_m(x) X_n(y) X_p(z) \qquad S_{mnp} \equiv \exp\left(-\frac{\alpha t \gamma_{mnp}^2}{a^2}\right)$$

$$\gamma_{mnp}^2 = \gamma_m^2 + \left(\frac{\gamma_n a}{b}\right)^2 + \left(\frac{\gamma_p a}{c}\right)^2$$

$$\gamma_m = (m + \tfrac{1}{2})\pi \qquad \gamma_n = (n + \tfrac{1}{2})\pi \qquad \gamma_p = (p + \tfrac{1}{2})\pi$$

†This section may be omitted on first reading.

For sufficiently long times, only the first term ($m = 0$, $n = 0$, $p = 0$) is important (see Sec. 7-7) and

$$T^* = \Phi_{000}(x, y, z)S_{000}(t)$$
$$= \Phi_{000}(x, y, z) \exp\left\{-\frac{\alpha t}{a^2}\frac{\pi^2}{4}\left[1 + \left(\frac{a}{b}\right)^2 + \left(\frac{a}{c}\right)^2\right]\right\}$$
$$= \Phi \exp\left(\frac{-t}{t_c}\right)$$

where t_c is a time constant and $\Phi_{000} \equiv \Phi$ for brevity. This equation indicates that, at large enough times, T^* can be separated into the product of a function of the space coordinates and a function of time, the latter being exponential. This leads us to believe that such a separation might also hold for other shapes. For a sphere or finite cylinder such can be shown to be the case since analytical solutions can be obtained which give the form of the spatial function Φ and the time constant t_c (see the Problems).

Now let us consider shapes for which an analytical solution *cannot* be obtained. If we let

$$T^* = \Phi(\text{space})S(t)$$

and substitute into Eq. (8-31), we get

$$\alpha \nabla^2\Phi\, S(t) = \Phi S'(t) \qquad \text{where } S' \equiv \frac{dS}{dt}$$

Dividing by $S\Phi$ gives

$$\frac{(\alpha \nabla^2\Phi)S}{S\Phi} = \frac{S'\Phi}{S\Phi} \qquad \text{or} \qquad \frac{\alpha \nabla^2\Phi}{\Phi} = \frac{S'(t)}{S(t)}$$

which equals a constant since the left-hand side is not a function of time and the right-hand side is not a function of space. The constant, for reasons to be apparent later, is taken as $-\beta^2$. Then

$$\alpha \nabla^2\Phi + \beta^2\Phi = 0 \qquad \text{and} \qquad \frac{dS}{S} = -\beta^2\, dt$$

or
$$S = e^{-\beta^2 t} \qquad \text{and} \qquad T^* = \Phi e^{-t/t_c} \qquad \text{for } t_c = \beta^{-2}$$

In order for T^* to approach zero monotonically as $t \to \infty$, the constant $-\beta^2$ must be negative and β must be real (see Appendixes 7-1 and 7-2).

Since Φ is not specified, the above result means that, for large times, the separation of T^* into a product of Φ(space) and $\exp(-t/t_c)$ is valid for any arbitrary shape. Thus, even when the analytical solution is not known, if we could take data on T at a fixed point (say the center), we could obtain Φ at that point and t_c from a plot of ln T^* versus t, since at large times straight lines should result with intercepts Φ and slope t_c.

Actually, it is not necessary to use interior temperatures. Instead the average temperature can be used since

$$\langle T^* \rangle = \frac{1}{\mathcal{V}} \int \Phi S(t)\, d\mathcal{V} = \langle \Phi \rangle S = \langle \Phi \rangle \exp\left(-\frac{t}{t_c}\right)$$

Then a plot of ln $\langle T^* \rangle$ versus t at large times will give $\langle \Phi \rangle$ from the intercept and t_c from the slope. In order to find the rate of heat gain we can use Eq. (8-72)

$$\dot{Q}_t = \rho \widehat{C}_p \mathcal{V} \frac{d\langle T \rangle}{dt} = \rho \widehat{C}_p \mathcal{V}(T_0 - T_1) \frac{d\langle T^* \rangle}{dt}$$

For large enough times, this becomes

$$\dot{Q}_t = \rho \widehat{C}_p \mathcal{V}(T_0 - T_1) \langle \Phi \rangle \frac{1}{t_c} \exp\left(-\frac{t}{t_c}\right)$$

8-18 EXTERNAL AND INTERNAL RESISTANCES*†

In the above development it is assumed that all the resistance was internal and that there was no external resistance or, more generally, that the Biot number $\text{Bi} \equiv hL_c/k$ [Eq. (A7-44)] is large. If that were not so, the boundary condition at the surface would be replaced by one involving Eq. (7-38)

$$q_s = h(T_1 - T_a)$$

where T_a is the ambient temperature and q_s is the heat flux at the surface as given, say, by Fourier's law.

On the other hand, if the Biot number is small enough (low ratio of internal to external resistance), the above development is unnecessary and the solid can be treated as if it were at a uniform temperature [see Eq. (7-43) and Fig. 8-13].

*This section does not require use of vector notation.

†This section may be omitted on first reading.

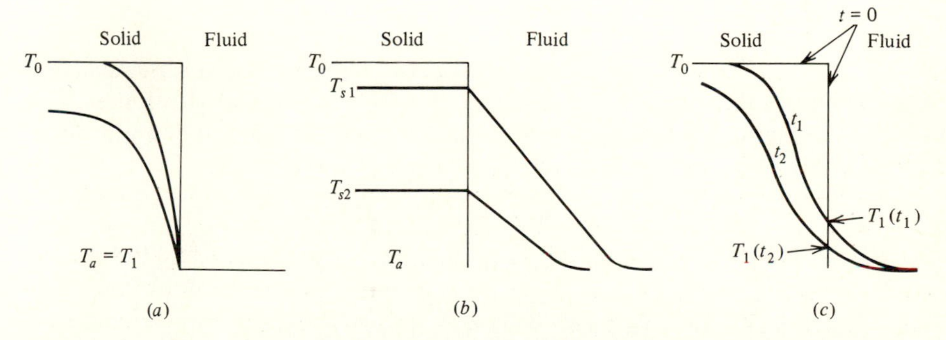

Figure 8-13 Temperature profiles for various conditions of Biot number: (*a*) no external resistance, Bi large; (*b*) no internal resistance, Bi small; (*c*) both internal and external resistance important.

8-19 ANALOGY TO MASS TRANSPORT IN THREE DIMENSIONS

Drying a porous solid can often be treated as a diffusional process for which Fick's law holds. It can be extended to three dimensions by a procedure similar to that used for heat conduction. Thus, for constant ρ and D_{AB}, the molar diffusive flux with respect to the mass-average velocity is given by

$$J_{Ax} = -D_{AB}\frac{\partial C_A}{\partial x} \qquad \text{or} \qquad \mathbf{J}_A = -D_{AB}\nabla C_A \tag{8-84}$$

where C_A is the molar concentration of water in the solid and D_{AB} is the diffusivity (see the Problems). If bulk flow can be neglected, a molar balance in A over a volume element $\Delta\mathcal{V}$ gives

$$-\frac{\partial J_{Ax}}{\partial x} - \frac{\partial J_{Ay}}{\partial y} - \frac{\partial J_{Az}}{\partial z} \equiv -\nabla\cdot\mathbf{J}_A = \frac{\partial C_A}{\partial t} \tag{8-85a}$$

Inserting Eq. (8-84) gives

$$D_{AB}\left(\frac{\partial^2 C_A}{\partial x^2} + \frac{\partial^2 C_A}{\partial y^2} + \frac{\partial^2 C_A}{\partial z^2}\right) \equiv D_{AB}\nabla^2 C_A = \frac{\partial C_A}{\partial t} \tag{8-85b}$$

For a solid of arbitrary shape the molar rate of transport of A from a surface element $d\mathbf{S}$ is

$$d\mathcal{W}_{AL} = -d\mathcal{W}_{AG} = \mathbf{J}_A\cdot d\mathbf{S} \tag{8-86a}$$

and the total rate of transport is

$$\mathcal{W}_{At} = -\iint_{\mathbf{S}} \mathbf{J}_A\cdot d\mathbf{S} \tag{8-86b}$$

The analogy to the heat-transport case should be noted (see the Problems).

8-20 DEFINITION AND PHYSICAL MEANING OF VECTOR OPERATIONS†

We have seen in Chap. 1 that the flux law can be interpreted as a statement that heat flows downhill; i.e., if the temperature gradient $\partial T/\partial x$ is negative, there will be a positive flow of heat in the $+x$ direction, and vice versa. It is now appropriate to seek a physical interpretation for the vector equation

$$\mathbf{q} = -k\,\nabla T \tag{8-12}$$

and for the gradient, which, in cartesian coordinates [Eq. (8-13)], is

$$\nabla T \equiv \boldsymbol{\delta}_x\frac{\partial T}{\partial x} + \boldsymbol{\delta}_y\frac{\partial T}{\partial y} + \boldsymbol{\delta}_z\frac{\partial T}{\partial z}$$

†This section may be omitted on first reading.

or, from Eq. (A8-2-5),

$$\nabla T = \sum_i \boldsymbol{\delta}_i \frac{\partial T}{\partial x_i}$$

First we note that ∇T has both magnitude and direction and therefore has the physical characteristics which we have associated with a vector. If we assume for the moment that $\nabla T \equiv \mathbf{G}$, a vector, we can see that it has components

$$\mathbf{G} = \mathbf{G}_x + \mathbf{G}_y + \mathbf{G}_z \tag{8-87}$$

where

$$\mathbf{G}_x = \frac{\partial T}{\partial x} \boldsymbol{\delta}_x \quad \text{etc.} \tag{8-88}$$

To get another physical picture of the gradient we consider, for simplicity, an isotropic medium in which $T = T(x, y)$ or $\partial T/\partial z = 0$. For example, we might take a slab that is infinitely long in the z direction and originally at T_0 (see Fig. 8-14). At $t \geq 0$ its faces $x = \pm a$ and $y = \pm b$ are maintained at a lower temperature $T = T_1$. Neglecting end effects, there will be transport only in the x and y directions, and so we can draw contours of constant T versus x and y at a cross section $z = z$ (see Fig. 8-15). If we think in terms of a three-dimensional surface $T(x, y)$, where T is the vertical axis, the lines of constant T represent the intersection of planes of $T =$ const with the surface $T(x, y)$. Now it is apparent that at an arbitrary location $P(x, y)$ the temperature will be decreasing in the $+x$ direction at constant y (path 1) or in the $+y$ direction at constant x (path 2). Hence there will be components of the gradient $\mathbf{G}_x$ and $\mathbf{G}_y$ wherever there are temperature gradients in these directions.

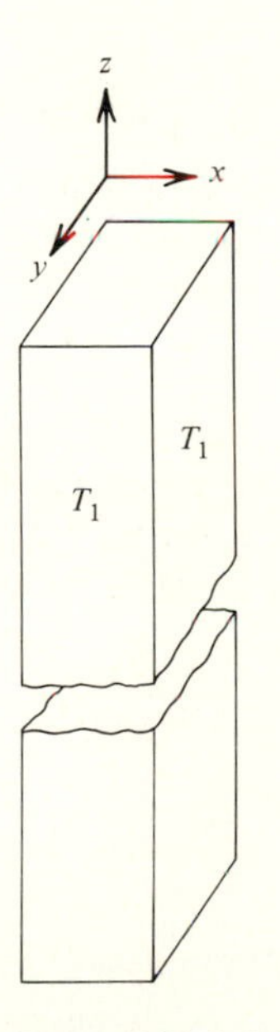

Figure 8-14 Infinitely long slab.

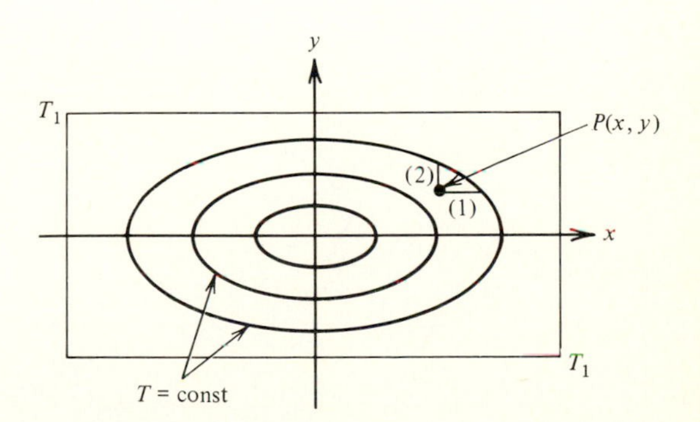

Figure 8-15 Temperature contours.

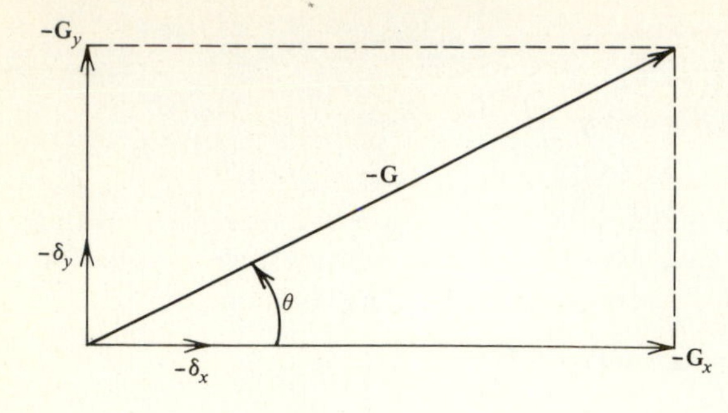

Figure 8-16 Vector components of temperature gradient.

These components can be resolved into a resultant vector $\mathbf{G} = \mathbf{G}_x + \mathbf{G}_y$. Now the scalar component of the gradient G_x represents the slope of the T-versus-x (at constant y) curve or the steepness of the "hill" [on the $T(x, y)$ surface] that is experienced by moving in the x direction at constant y. A similar statement can be made for G_y. The direction of the resultant vector $\mathbf{G}$ will be that given by the angle $\theta = \arctan G_y/G_x$ (see Fig. 8-16). We now develop some mathematical tools before proceeding further.

Physical Meaning of the Chain Rule of Differentiation

Let us examine how the temperature at an arbitrary point (x_a, y_a, z_a) changes along some arbitrary path Δs during which x changes by Δx and y changes by Δy (see Fig. 8-17). The total change in temperature ΔT can be presumed to consist of the sum of the change that occurs over Δx at constant $y = y_a$ (which we will call $\Delta_x T$) and the change that occurs over Δy at constant $x = x_a$ (which we will call $\Delta_y T$) or

$$\Delta T = \Delta_x T + \Delta_y T \tag{8-89}$$

In other words, we are treating the $T(x, y)$ surface as a hill with T representing the elevation, x representing an easterly direction, and y representing a direction to

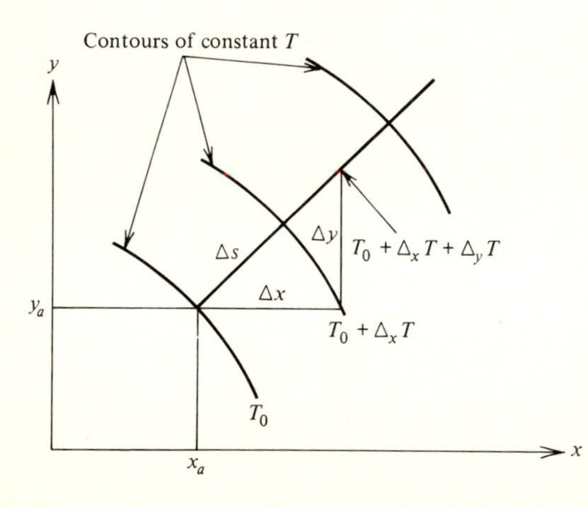

Figure 8-17 Changes of T with changes in Δx and Δy.

the north, and we are saying that if we go east Δx units, the change in elevation will be $\Delta_x T$ and if we go north Δy units, the added change in elevation will be $\Delta_y T$ units. Then the total change will be the sum of the two partial changes in elevation.

If we know G_x, the gradient in the x direction at (x_a, y_a), we can assume that it is constant over a small distance Δx and

$$\Delta_x T \approx \left(\frac{\partial T}{\partial x}\right)\bigg|_{x_a, y_a} \Delta x \equiv G_x \bigg|_{x_a, y_a} \Delta x$$

Similarly
$$\Delta_y T \approx \frac{\partial T}{\partial y}\bigg|_{x_a+\Delta x, y_a} \Delta y \equiv G_y \bigg|_{x_a+\Delta x, y_a} \Delta y$$

Then
$$\Delta T = \frac{\partial T}{\partial x}\bigg|_{x_a, y_a} \Delta x + \frac{\partial T}{\partial y}\bigg|_{x_a+\Delta x, y_a} \Delta y$$

As we let Δx and $\Delta y \to 0$, this becomes

$$dT = \frac{\partial T}{\partial x} dx + \frac{\partial T}{\partial y} dy = G_x\, dx + G_y\, dy \tag{8-90}$$

since, as $\Delta x \to 0$,

$$G_y \bigg|_{x_a+\Delta x, y_a} \to G_y \bigg|_{x_a, y_a} = G_y$$

Similarly we obtain G_x.

Physical Meaning of Gradient of a Scalar[1]

By an application of the chain rule, the rate of increase of T along an arbitrary path $\Delta s \to 0$ (see Fig. 8-18) will be

$$\frac{dT}{ds} = \frac{\partial T}{\partial x}\frac{dx}{ds} + \frac{\partial T}{\partial y}\frac{dy}{ds} \tag{8-91}$$

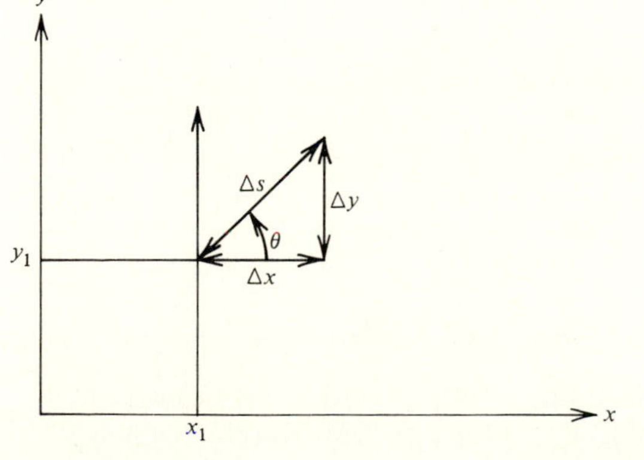

Figure 8-18 Changes in temperature for arbitrary path.

which can be generalized as

$$\frac{dT}{ds} = \sum_i G_i \frac{dx_i}{ds} \tag{8-91a}$$

This looks like the dot product of a vector **G**, whose scalar components are G_i, with another vector $d\mathbf{x}/ds$, whose scalar components are dx_i/dt [Eq. (A8-2-4)]. To find it let us differentiate the *position* vector [Eq. (A8-2-3)] to obtain the *displacement* vector

$$d\mathbf{x} = dx\,\boldsymbol{\delta}_x + dy\,\boldsymbol{\delta}_y + dz\,\boldsymbol{\delta}_z = \sum_i dx_i\,\boldsymbol{\delta}_i \tag{8-91b}$$

whose magnitude is ds. Since

$$\frac{d\mathbf{x}}{ds} = \sum_i \frac{dx_i}{ds}\boldsymbol{\delta}_i \tag{8-91c}$$

we can write Eq. (8-91a) as a dot product [see (Eq. A8-2-4)]

$$\frac{dT}{ds} = G \cdot \frac{d\mathbf{x}}{ds} = \sum_i G_i\,\boldsymbol{\delta}_i \cdot \sum_i \frac{ds_i}{ds}\boldsymbol{\delta}_i = \sum_i G_i \frac{dx_i}{ds}$$

Using Eq. (8-25) gives

$$\frac{dT}{ds} = |\mathbf{G}| \cos\theta$$

since $|d\mathbf{x}| = ds$. The angle between **G** and $d\mathbf{x}$ is θ.

If $\theta = \pi/2$, **G** and $d\mathbf{x}$ are perpendicular. This means that $dT/ds = 0$ since $\mathbf{G} \cdot d\mathbf{x} = 0$; therefore we are moving along a path of constant T (an isotherm), i.e., along a contour line. If $\theta = 0$, $\cos\theta = 1$, which is its maximum value. Therefore **G** is a vector whose magnitude is

$$\mathbf{G} = \left(\frac{dT}{ds}\right)_{\max}$$

which is the path of *steepest ascent*. But if $\theta = 0$, **G** and $d\mathbf{x}$ are in the same direction and are *normal* to the contour lines. Therefore **G** is

1. A vector because its dot product with respect to another vector gives a scalar
2. Normal to the contour
3. The slope of steepest ascent

Physical Meaning of Fourier's Law in Three Dimensions

Since Fourier's law tells us that the heat flux is proportional to the negative of the gradient, we can see that the vectors $\mathbf{q}_x$, $\mathbf{q}_y$, and $\mathbf{q}_z$ will represent the tendency for heat to flow downhill against the gradients in the x, y, and z directions. Furthermore, the vector flux $\mathbf{q}$ will represent the *resultant* flow of heat in the direction of steepest descent down the temperature hill.

Fourier's law in vector form consequently tells us that for an isotropic system (1) the flux of heat is in the direction opposite the temperature gradient ∇T; that is, heat flows downhill, and (2) the magnitude of the heat flux is directly proportional to the magnitude of the temperature gradient; the steeper the hill, the faster the rate of heat transport will be.

A word about coordinate systems is now in order. We note that coordinate systems (cartesian, cylindrical, etc.) are an arbitrary concept that are specified by the investigator. The physical quantities $\mathbf{q}$, ∇T (or T itself) are independent of whatever coordinate system is used to describe them. In other words, the heat flux is a vector that describes a certain physical phenomenon (heat flow per area), and the temperature gradient is a vector that describes a physical condition (nonequilibrium) at a point in space. The flux law is a statement of the relationship between these two quantities. When written in vector form, it expresses this natural law in a manner that does not depend on the coordinate system chosen. This also means that we can select any coordinate system we wish and express vector quantities in terms of their components in that system. The choice of coordinate systems usually is determined by the physical boundary conditions of the problem.

Definition and Physical Meaning of the Divergence of a Vector

The operation $\nabla \cdot \mathbf{q}$ for cartesian coordinates [Eq. (8-28)] was found by carrying out an energy balance over an incremental element of volume $\Delta \mathcal{V}$. Reviewing this process in terms of index notation, we write the incremental balance as

$$\underbrace{-\sum_i \Delta_i(q_i\,\Delta A_i)}_{\text{Net input}} = \underbrace{\rho \widehat{C}_p\,\Delta \mathcal{V}\,\frac{\Delta T}{\Delta t}}_{\text{Accumulation}}$$

Dividing by $\Delta \mathcal{V}$ and letting $\Delta \mathcal{V} \to 0$ and $\Delta t \to 0$ gives

$$-\lim_{\Delta \mathcal{V}\to 0} \sum_i \frac{\Delta_i(q_i\,\Delta A_i)}{\Delta \mathcal{V}} = \rho \widehat{C}_p \lim_{\Delta t \to 0} \frac{\Delta T}{\Delta t} = \rho \widehat{C}_p \frac{\partial T}{\partial t} \tag{8-92}$$

The left-hand side is physically the net rate of *in*flow of heat per unit volume; its negative is called the *divergence* of the vector $\mathbf{q}$. Thus

$$\operatorname{div} \mathbf{q} \equiv \lim_{\Delta \mathcal{V}\to 0} \sum_i \frac{\Delta_i(q_i\,\Delta A_i)}{\Delta \mathcal{V}} \tag{8-93}$$

In general the divergence of a vector $\mathbf{q}$ is defined physically as *the net rate of outflow per unit of volume of a quantity whose flux is* $\mathbf{q}$. Here the quantity that is flowing is heat, but it could also be mass or moles of species A, etc.

In vector notation, Eq. (8-93) can be written

$$\operatorname{div} \mathbf{q} \equiv \lim_{\Delta \mathcal{V} \to 0} \frac{\iint \mathbf{q} \cdot d\mathbf{S}}{\Delta \mathcal{V}} \tag{8-94}$$

where the integral is taken over *all* surfaces of the incremental volume $\Delta\mathcal{V}$. Note that Eq. (8-94) is a form of the Gauss divergence theorem (8-76) applied to an increment of volume over which div **q** varies slightly. Note also that Eq. (8-94) applies to any coordinate system since it is a vector equation.

Let us now find an expression for the divergence for cartesian coordinates. From Eq. (8-94), proceeding as before but using index notation, we let $\Delta\mathcal{V} = \Delta x_i \, \Delta A_i$; that is, $\Delta\mathcal{V} = \Delta x \, \Delta A_x = \Delta x \, (\Delta y \, \Delta z)$, but also $\Delta\mathcal{V} = \Delta y \, \Delta A_y = \Delta y \, (\Delta x \, \Delta z)$, etc. Then, since ΔA_i is constant,

$$\operatorname{div} \mathbf{q} \equiv \lim_{\Delta x_i \to 0} \sum_i \frac{\Delta_i (q_i \, \Delta A_i)}{\Delta x_i \, \Delta A_i} = \lim_{\Delta x_i \to 0} \sum_i \frac{\Delta_i q_i}{\Delta x_i} = \sum_i \frac{\partial q_i}{\partial x_i} \tag{8-95}$$

the latter coming from the definition of the partial derivative (see Appendix 1-1).

Previously we showed that the operator $\nabla \cdot \mathbf{q}$ in cartesian coordinates gives Eq. (8-28), which can be written in index notation as

$$\nabla \cdot \mathbf{q} = \sum_i \frac{\partial q_i}{\partial x_i} \tag{8-96}$$

Thus the physical quantity which we have so far called the divergence is the same as the vector operation $\nabla \cdot \mathbf{q}$, or

$$\operatorname{div} \mathbf{q} \equiv \nabla \cdot \mathbf{q} \tag{8-97}$$

Combining Eqs. (8-97) and (8-93) and substituting into Eq. (8-92) gives

$$-\nabla \cdot \mathbf{q} = \rho \hat{C}_p \frac{\partial T}{\partial t} \tag{8-32}$$

Since Eq. (8-97) is in general vector notation, it applies to all coordinate systems. However, the expressions used to expand each side into the scalar components of **q** do depend upon the coordinate system.

Equation (8-97) tells us that there are two ways to expand the divergence into components, either by using Eq. (8-93), which is equivalent to using an incremental balance, or by using the right-hand side of Eq. (8-97), which requires expanding ∇ and **q** into components and taking the dot product. We have carried out the latter for cartesian components [Eq. (8-28)] and will discuss the procedure for curvilinear coordinates (Appendix 8-5). The forms of $\nabla \cdot \mathbf{q}$ for various coordinates are given in Table I-1.

8-21 VECTOR OPERATIONS IN CURVILINEAR COORDINATES†

Let us now find expressions for various vector operations in cylindrical coordinates to illustrate the procedures and equations needed for other orthogonal coordinate systems, i.e., systems in which the coordinate axes are mutually perpendicular.

Divergence The cylindrical coordinate system is defined and illustrated in Table I-1. In Fig. 8-19 we show an incremental element of volume in this system formed by increasing r by Δr, θ by $\Delta\theta$, and z by Δz. The volume of the element is obtained by multiplying together the elements of length along each coordinate, or (see Sec. 1-9)

$$\Delta\mathcal{V} = \Delta s_\theta \, \Delta s_r \, \Delta s_z$$

where $$\Delta s_\theta = h_\theta \, \Delta\theta = r \, \Delta\theta$$

is the length of an arc made as $\Delta\theta$ increases at constant r and z (see Example 8-2) and where $h_\theta = r$ is the scale factor, as defined by Eq. (8-16), that converts an angular change $\Delta\theta$ into a change in distance Δs_θ. For the other coordinates $h_r = 1 = h_z$. Then

$$\Delta s_r = \Delta r \qquad \text{and} \qquad \Delta s_z = \Delta z$$

†This section may be omitted on first reading.

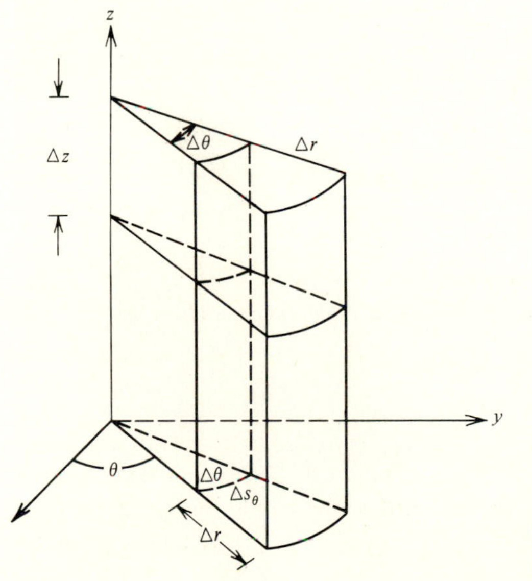

Figure 8-19 Incremental element of volume in cylindrical coordinates.

The volume is then

$$\Delta\mathcal{V} = r\,\Delta\theta\,\Delta r\,\Delta z$$

and the areas of the faces are

$$\Delta A_r = \Delta s_\theta\,\Delta s_z = r\,\Delta\theta\,\Delta z$$

$$\Delta A_\theta = \Delta s_r\,\Delta s_z = \Delta r\,\Delta z$$

$$\Delta A_z = \Delta s_\theta\,\Delta s_r = r\,\Delta\theta\,\Delta r$$

Suppose we had unsteady heat conduction occurring in a finite cylinder, with conditions similar to those occurring in the brick in Fig. 8-1. In general there still would be a heat flux in the x, y, and z directions, and the differential equations derived for an element of volume $\Delta x\,\Delta y\,\Delta z$ would still apply. However, the boundary conditions would be inconvenient and the equations difficult to solve. Therefore it would be preferable to use cylindrical coordinates. Then the total flux would be given by Eqs. (8-14) as

$$\mathbf{q} = \mathbf{q}_r + \mathbf{q}_\theta + \mathbf{q}_z = q_r\,\boldsymbol{\delta}_r + q_\theta\,\boldsymbol{\delta}_\theta + q_z\,\boldsymbol{\delta}_z$$

In index notation (see Appendix 8-2) these become

$$\mathbf{q} = \sum_n \mathbf{q}_n = \sum_n q_n\,\boldsymbol{\delta}_n$$

Now the *net* rate of heat transport into the element is equal to the rate in less the rate out of each face, or

$$-\Delta_r(q_r\,\Delta A_r) - \Delta_\theta(q_\theta\,\Delta A_\theta) - \Delta_z(q_z\,\Delta A_z) = -\sum_n \Delta_n(q_n\,\Delta A_n)$$

After division by $\Delta\mathcal{V} = \Delta s_n\,\Delta A_n = h_n\,\Delta u_n\,\Delta A_n$ when the limit $\Delta\mathcal{V} \to 0$ is taken, this gives the negative of the divergence as defined by Eq. (8-94). Then changing signs gives

$$\operatorname{div}\mathbf{q} = \lim_{\Delta\mathcal{V}\to 0} \sum_n \frac{\Delta_n(q_n\,\Delta A_n)}{h_n\,\Delta(u_n\,\Delta A_n)} = \nabla\cdot\mathbf{q}$$

since Eq. (8-94) holds in *any* coordinate system.

Substituting for the areas and scale factors gives

$$\Delta A_r = \Delta s_\theta\,\Delta s_z = r\,\Delta\theta\,\Delta z$$

$$\Delta A_\theta = \Delta s_r\,\Delta s_z = \Delta r\,\Delta z$$

$$\Delta A_z = \Delta s_\theta\,\Delta s_r = r\,\Delta\theta\,\Delta r$$

Then

$$\operatorname{div}\mathbf{q} = \lim_{\Delta r\to 0}\frac{\Delta_r(q_r r\,\Delta\theta\,\Delta z)}{r\,\Delta\theta\,\Delta r\,\Delta z} + \lim_{\Delta\theta\to 0}\frac{\Delta_\theta(q_\theta\,\Delta r\,\Delta z)}{r\,\Delta\theta\,\Delta r\,\Delta z} + \lim_{\Delta z\to 0}\frac{\Delta_z(q_z r\,\Delta\theta\,\Delta r)}{r\,\Delta\theta\,\Delta r\,\Delta z}$$

If we cancel constant terms in numerator and denominator and use the definition of a partial derivative, we get

$$\operatorname{div} \mathbf{q} = \frac{\partial(q_r r)}{r\,\partial r} + \frac{\partial q_\theta}{r\,\partial \theta} + \frac{\partial q_z}{\partial z} \tag{8-98}$$

This equation can also be found by carrying out the operation $\nabla \cdot \mathbf{q}$ in Sec. 8-2. Expressions for $\nabla \cdot \mathbf{q}$ are given in Table I-2.

Laplacian operator in cylindrical coordinates Since $\mathbf{q} = -k\,\nabla T$,

$$\nabla \cdot \mathbf{q} = -k\,\nabla \cdot \nabla T = -k\,\nabla^2 T \qquad \text{if } k = \text{const}$$

Then the laplacian is

$$\nabla^2 T = -\frac{1}{k}\nabla \cdot \mathbf{q} \tag{8-98a}$$

Equation (8-98) for $\nabla \cdot \mathbf{q}$ is

$$\nabla \cdot \mathbf{q} \equiv \frac{\partial(r q_r)}{r\,\partial r} + \frac{\partial q_\theta}{r\,\partial \theta} + \frac{\partial q_z}{\partial z}$$

If we substitute flux laws (8-17) for each component into this expression, we get

$$\nabla \cdot \mathbf{q} = -k\left[\frac{\partial(r\,\partial T/\partial r)}{r\,\partial r} + \frac{\partial^2 T}{r^2\,\partial \theta^2} + \frac{\partial^2 T}{\partial z^2}\right]$$

Dividing by $-k$ and using Eq. (8-98*a*) gives $\nabla^2 T$, from which

$$\nabla^2 \equiv \frac{\partial(r\,\partial/\partial r)}{r\,\partial r} + \frac{\partial^2}{r^2\,\partial \theta^2} + \frac{\partial^2}{\partial r^2}$$

where ∇^2 is the laplacian operator.

Generalized Orthogonal Coordinates

The procedures used for cylindrical coordinates can be extended to a generalized orthogonal (perpendicular) coordinate system (see Fig. 8-20). We label the coordinates† u_1, u_2, u_3; for example, for cylindrical coordinates $u_1 = r$, $u_2 = \theta$, $u_3 = z$. In any direction, the incremental length Δs_m along a coordinate axis is given by the product of the incremental change in the coordinate Δu_m and a scale factor h_m or $\Delta s_m = h_m\,\Delta u_m$. For example, in cylindrical coordinates, if $u_2 = \theta$, $h_2 = r$ and $\Delta s_\theta = \Delta s_2 = h_2\,\Delta u_2 = r\,\Delta\theta$. A vector $\mathbf{q}$ can be written

$$\mathbf{q} = q_1\,\boldsymbol{\delta}_1 + q_2\,\boldsymbol{\delta}_2 + q_3\,\boldsymbol{\delta}_3 = \sum_{m=1} q_m\,\boldsymbol{\delta}_m \tag{8-99}$$

†In this text as a mnemonic we use x_1, x_2, x_3 and subscripts i, j, k, etc., for cartesian coordinates; we will use u_1, u_2, and u_3 with subscripts m, n, p, etc., for curvilinear coordinates (see Fig. 8-21 and Appendix 8-2).

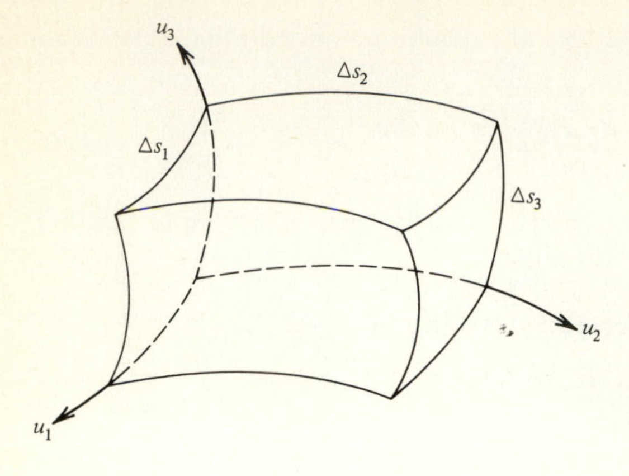

Figure 8-20 Generalized orthogonal coordinates.

where $\boldsymbol{\delta}_1$, $\boldsymbol{\delta}_2$, and $\boldsymbol{\delta}_3$ are unit vectors in the u_1, u_2, and u_3 directions and q_1, q_2, and q_3 are scalar components.

Example 8-4: Generalized coordinate system (*a*) Using an arbitrary element of volume $h_1\,\Delta u_1\,h_2\,\Delta u_2\,h_3\,\Delta u_3$, derive an equation for unsteady-state heat conduction in a solid. (*b*) Obtain expressions for (1) the divergence of a vector $\mathbf{q}$

$$\nabla\cdot\mathbf{q} = \frac{1}{h}\sum_{m=1}^{3}\frac{\partial}{\partial u_m}\frac{q_m h}{h_m} \tag{8-100}$$

(2) the gradient of a scalar T

$$\nabla T = \sum_{m=1}^{3}\boldsymbol{\delta}_m\frac{\partial T}{h_m\,\partial u_m} \tag{8-101}$$

and (3) the laplacian of a scalar T

$$\nabla^2 T = \frac{1}{h}\sum_{m=1}^{3}\frac{\partial}{\partial u_m}\left(\frac{h}{h_m^2}\frac{\partial T}{\partial u_m}\right) \tag{8-102}$$

SOLUTION (*a*) Net transport ($\Delta A_1 = \Delta s_2\,\Delta s_3$, etc.)

u_1 face: $\quad -\Delta_1(q_1\,\Delta A_1) = -\Delta_1(q_1 h_2\,\Delta u_2\,h_3\,\Delta u_3)$

u_2 face: $\quad -\Delta_2(q_2\,\Delta A_2) = -\Delta_2(q_2 h_1\,\Delta u_1\,h_3\,\Delta u_3)$

u_3 face: $\quad -\Delta_3(q_3\,\Delta A_3) = -\Delta_3(q_3 h_1\,\Delta u_1\,h_2\,\Delta u_2)$

$$\text{Accumulation} = \frac{\Delta}{\Delta t}[\rho\widehat{C}_p(T - T_0)\,\Delta\mathcal{V}]$$

where $\quad \Delta\mathcal{V} = h_1 h_2 h_3\,\Delta u_1\,\Delta u_2\,\Delta u_3 = h\,\Delta u_1\,\Delta u_2\,\Delta u_3$

and

$$h \equiv h_1 h_2 h_3 \tag{8-103}$$

Divide by $\Delta\mathcal{V}$ and let all Δ's $\to 0$, giving

$$-\frac{\partial(q_1 h_2 h_3)}{h\,\partial u_1} - \frac{\partial(q_2 h_3 h_1)}{h\,\partial u_2} - \frac{\partial(q_3 h_1 h_2)}{h\,\partial u_3} = \rho \widehat{C}_p \frac{\partial T}{\partial t}$$

(*b*) The divergence is the negative of the left-hand side, or

$$\nabla \cdot \mathbf{q} = \frac{1}{h} \sum_{m=1}^{3} \frac{\partial}{\partial u_m} \frac{q_m h}{h_m}$$

$$\mathbf{q} = -k\,\nabla T = -k\left(\boldsymbol{\delta}_1 \frac{\partial}{h_1\,\partial u_1} + \boldsymbol{\delta}_2 \frac{\partial}{h_2\,\partial u_2} + \boldsymbol{\delta}_3 \frac{\partial}{h_3\,\partial u_3}\right)T$$

Then the gradient of T is

$$\nabla T = \sum_{m=1}^{3} \boldsymbol{\delta}_m \frac{\partial T}{h_m\,\partial u_m}$$

Substituting $q_m = -k\,\partial T/h_m\,\partial u_m$ into expressions for $\nabla \cdot \mathbf{q}$ gives the laplacian of T

$$\nabla^2 T = \frac{1}{h}\left[\frac{\partial\left(\dfrac{h_2 h_3}{h_1}\dfrac{\partial T}{\partial u_1}\right)}{\partial u_1} + \frac{\partial\left(\dfrac{h_1 h_3}{h_2}\dfrac{\partial T}{\partial u_2}\right)}{\partial u_2} + \frac{\partial\left(\dfrac{h_1 h_2}{h_3}\dfrac{\partial T}{\partial u_3}\right)}{\partial u_3}\right]$$

or

$$\nabla^2 T = \frac{1}{h} \sum_{m=1}^{3} \frac{\partial}{\partial u_m}\left(\frac{h}{h_m^2}\frac{\partial T}{\partial u_m}\right)$$ □

8-22 EQUATIONS OF CONTINUITY*

Consider a compressible fluid in motion with velocity components in all three directions. We wish to derive an equation in cartesian coordinates relating the velocity components v_x, v_y, and v_z to each other. To do so we write a mass balance over an element of volume $\Delta\mathcal{V} = \Delta x\,\Delta y\,\Delta z$ located at a fixed arbitrary point in space (Fig. 8-21). This is called the eulerian point of view. In Chap. 5 we defined the total mass flux in the x direction as $n_{tx} \equiv \rho v_x$, where n_{tx} is the total mass flux of the fluid in the x direction. The net mass rate of flow of the *mixture* through the incremental area $\Delta A_x = \Delta y\,\Delta z$ at $x = x$ is then

$$(n_{tx}\,\Delta A_x)\Big|_x - (n_{tx}\,\Delta A_x)\Big|_{x+\Delta x} = -\Delta_x(n_{tx}\,\Delta A_x)$$

Taking into consideration all six faces, the mass balance is

$$-\Delta_x(n_{tx}\,\Delta A_x) - \Delta y(n_{ty}\,\Delta A_y) - \Delta_z(n_{tz}\,\Delta A_z) = \frac{\Delta(\rho\,\Delta\mathcal{V})}{\Delta t}$$

*This section does not require use of vector notation.

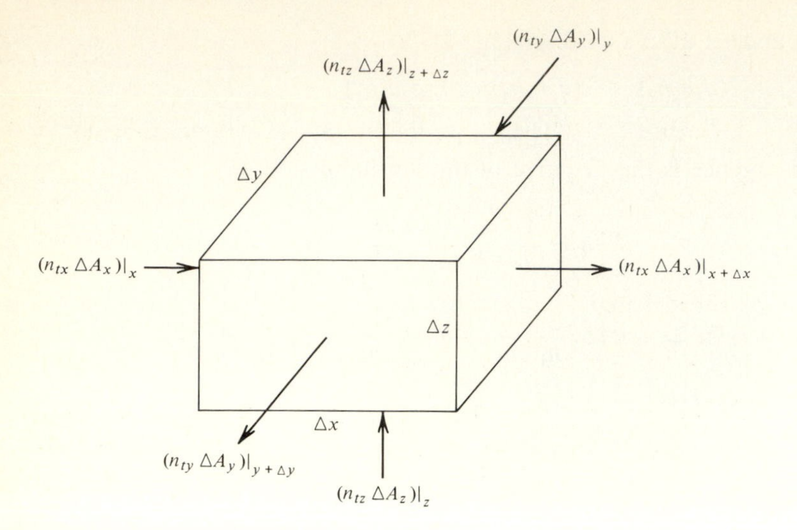

Figure 8-21 Mass flows into and out of element of volume.

Dividing by $\Delta\mathcal{V} = \Delta x\,\Delta y\,\Delta z$ and taking limits as $\Delta x \to 0$, $\Delta y \to 0$, $\Delta z \to 0$, and $\Delta t \to 0$ gives

$$-\frac{\partial n_{tx}}{\partial x} - \frac{\partial n_{ty}}{\partial y} - \frac{\partial n_{tz}}{\partial z} = \frac{\partial \rho}{\partial t}$$

or

$$-\frac{\partial(\rho v_x)}{\partial x} - \frac{\partial(\rho v_y)}{\partial y} - \frac{\partial(\rho v_z)}{\partial z} = \frac{\partial \rho}{\partial t} \tag{8-104}$$

The above equation is called the *equation of continuity* (EOC). For an incompressible fluid, ρ is constant and the EOC becomes

$$-\frac{\partial v_x}{\partial x} - \frac{\partial v_y}{\partial y} - \frac{\partial v_z}{\partial z} = 0 \tag{8-104a}$$

These equations are tabulated in Table I-6. It is convenient to write them in *index notation* (see Appendix 8-2) by letting $x \equiv x_1$, $y \equiv x_2$, $z \equiv z_2$, $v_x \equiv v_1$, $v_y \equiv v_2$, $v_z \equiv v_3$. Then

$$-\sum_{i=1}^{3} \frac{\partial(\rho v_i)}{\partial x_i} = \frac{\partial \rho}{\partial t} \qquad \text{and} \qquad \sum_{i=1}^{3} \frac{\partial v_i}{\partial x_i} = 0 \tag{8-104b}$$

Customarily $\sum_i$ is used as an abbreviation for $\sum_{i=1}^{3}$.

Equation of Change for Species A

By means of a mass balance for a single species A over an element of volume at a fixed point, one obtains an equation of continuity for species A

$$\frac{\partial \mathcal{R}_A}{\partial t} = -\frac{\partial n_{Ax}}{\partial x} - \frac{\partial n_{Ay}}{\partial y} - \frac{\partial n_{Az}}{\partial z} + r_A \tag{8-105}$$

where r_A is the production of A in mass A per unit time and unit volume. Or in index notation

$$-\sum_i \frac{\partial n_{Ai}}{\partial x_i} + r_A = \frac{\partial \mathcal{R}_A}{\partial t} \tag{8-105a}$$

In Chap. 5 we showed that

$$n_{Ax} = j_{Ax} + \rho_A v_x \qquad \text{or} \qquad n_{Ai} = j_{Ai} + \rho_A v_i$$

where j_{Ai} is the mass diffusive flux of A in the x direction with respect to the mass-average velocity v_i, $\rho_A v_i$ is the bulk flux of A, and n_{Ai} is the total flux of A. This gives

$$-\sum_i \frac{\partial j_{Ai}}{\partial x_i} - \sum_i \frac{\partial(\rho_A v_i)}{\partial x_i} + r_A = \frac{\partial \rho_A}{\partial t} \tag{8-105b}$$

The second term in the above can be differentiated as follows:

$$\sum_i \frac{\partial(\rho_A v_i)}{\partial x_i} = \rho_A \sum_i \frac{\partial v_i}{\partial x_i} + \sum_i v_i \frac{\partial \rho_A}{\partial x_i}$$

Then the balance can be written

$$-\sum_i \frac{\partial j_{Ai}}{\partial x_i} + r_A = \frac{\partial \rho_A}{\partial t} + \sum v_i \frac{\partial \rho_A}{\partial x_i} + \rho_A \sum \frac{\partial v_i}{\partial x_i} \tag{8-105c}$$

Dividing by molecular weight M_A gives the molar balance

$$-\sum_i \frac{\partial J_{Ai}}{\partial x_i} + R_A = \frac{\partial C_A}{\partial t} + \sum_i v_i \frac{\partial C_A}{\partial x_i} + C_A \sum \frac{\partial v_i}{\partial x_i} \tag{8-105d}$$

The first two terms on the right-hand side are

$$\frac{DC_A}{Dt} \equiv \frac{\partial C_A}{\partial t} + \sum_i v_i \frac{\partial C_A}{\partial x_i} \tag{8-106}$$

where DC_A/Dt is called the *substantial* or *material derivative* of C_A. This derivative is widely used in fluid mechanics. Its physical meaning is discussed in detail later, but briefly it represents the change in C_A with time that occurs when an observer moves with the velocity of the stream. The first term represents the change that occurs at a fixed point, and the second represents the change due to the motion. A balance equation written in terms of D/Dt is said to be in the *lagrangian* form, to distinguish it from an equation in *eulerian* form, which describes changes that occur at a fixed point in space.

8-23 EQUATIONS OF CONTINUITY IN VECTOR NOTATION

Again consider a compressible fluid in motion with velocity components in all three directions. At any arbitrary point ($x = x_1, y = x_2, z = x_3$) in the fluid, the velocity vector is

$$\mathbf{v} = v_x \boldsymbol{\delta}_x + v_y \boldsymbol{\delta}_y + v_z \boldsymbol{\delta}_z$$

or (Appendix 8-2)

$$\mathbf{v} = \sum_i v_i \boldsymbol{\delta}_i$$

We can define the total mass flux vector $\mathbf{n}_t$ at any point (see Chap. 5) as

$$\rho \mathbf{v} \equiv \mathbf{n}_t \equiv \sum_i \rho \mathbf{v}_i \boldsymbol{\delta}_i = \sum_i n_{ti} \boldsymbol{\delta}_i$$

Note that to multiply a vector $\mathbf{v}$ by a scalar ρ is to multiply each component by the scalar. We can make a mass balance on an element of volume $\Delta \mathcal{V} = \Delta x \, \Delta y \, \Delta z$ (see Fig. 8-21). The element is fixed in space, and the boundaries form an imaginary cage of fixed volume through which the fluid is flowing. This approach is called the *eulerian point of view* (as opposed to the lagrangian point of view, in which the element moves with the stream velocity). The incremental balance and the procedure are therefore identical to that used for heat transport in the brick-shaped solid except that n_{tx} replaces q_x, etc., and $\rho \, \Delta \mathcal{V}$ is the accumulation instead of $\rho \widehat{C}_p T \, \Delta \mathcal{V}$. Therefore the incremental final equations can be written down by analogy by merely replacing $\mathbf{q}$ with $\mathbf{n}_t$. Equating net input to accumulation, dividing by $\Delta \mathcal{V} \equiv \Delta x_i \, \Delta A_i$, and taking limits gives

$$-\lim_{\Delta \mathcal{V} \to 0} \sum_i \frac{\Delta_i (n_{ti} \, \Delta A_i)}{\Delta \mathcal{V}} = \lim_{\Delta t \to 0} \frac{\Delta \rho}{\Delta t} = \frac{\partial \rho}{\partial t}$$

According to the second of Eqs. (A8-2-5), the left-hand side is the negative of the divergence of the vector $\mathbf{n}_t$, which can be written

$$\operatorname{div} \mathbf{n}_t = \sum_i \frac{\partial n_{ti}}{\partial x_i} \equiv \nabla \cdot \mathbf{n}_t \tag{8-107}$$

Thus the mass-balance equation in vector form is

$$-\nabla \cdot \mathbf{n}_t = \frac{\partial \rho}{\partial t} \tag{8-108}$$

Net rate of inflow of mass per unit volume — Rate of accumulation of mass per unit volume

Since $\mathbf{n}_t = \rho \mathbf{v}$, this becomes

$$-\nabla \cdot \rho \mathbf{v} = \frac{\partial \rho}{\partial t} \tag{8-109}$$

or, in terms of cartesian components,†

$$-\frac{\partial(\rho v_x)}{\partial x} - \frac{\partial(\rho v_y)}{\partial y} - \frac{\partial(\rho v_z)}{\partial z} = \frac{\partial \rho}{\partial t} \tag{8-110}$$

If the fluid is incompressible, ρ is constant and Eq. (8-110) becomes

$$\frac{\partial v_x}{\partial x} + \frac{\partial v_y}{\partial y} + \frac{\partial v_z}{\partial z} = 0 \tag{8-111}$$

or

$$\nabla \cdot \mathbf{v} = 0 \tag{8-112}$$

The above are all forms of the *equation of continuity,* which tells us that for an incompressible fluid a change, say, in a velocity component v_x in the x direction must be compensated for by a change of another velocity component v_y in the y direction and/or v_z in the z direction.

Example 8-5: Application of continuity equation to pipe flow Consider incompressible fluid flow in a pipe of circular cross section. (*a*) Express the equation of continuity in cylindrical coordinates (1) using the general equations in Tables I-6 through I-14 and (2)* using the vector equations and Table I-2; (*b*)* simplify the EOC for flow in a pipe.

SOLUTION (*a*) (1) From Part B of Table I-7

$$\frac{\partial(rv_r)}{r\,\partial r} + \frac{\partial v_\theta}{r\,\partial \theta} + \frac{\partial v_z}{\partial z} = 0$$

(2) From Part B of Table I-12, for an incompressible fluid

$$\nabla \cdot \mathbf{v} = 0$$

From Table I-2 we find that $\nabla \cdot \mathbf{v}$ in cylindrical coordinates is as given above.

(*b*) Because of angular symmetry we can assume that $\partial v_\theta/\partial\theta = 0$. In the fully developed flow region, $\partial v_z/\partial z = 0$ and hence

$$\frac{d(rv_r)}{r\,dr} = 0 \qquad \text{or} \qquad rv_r = \text{const}$$

Since at the wall $v_r = 0$ and $r \neq \infty$, the constant must be zero or

$$v_r = 0$$

Now consider the entrance region to a pipe, where $\partial v_z/\partial z$ is not constant. We can see that if $\partial v_z/\partial z$ is positive, $\partial(rv_r)/\partial r$ must be negative and vice versa. Hence the equation of continuity permits us to obtain a qualitative if not quantitative picture of how the velocity components are changing with position. We analyze the problem of entrance-region flow more fully in Chap. 9. □

†Expressions for the divergence in other coordinate systems are given in Table I-2.
*These parts do not require use of vector notation.

Equation of Continuity of a Species A

Consider a fluid in which a homogeneous chemical reaction is generating a species A at a rate r_A (mass per unit volume and unit time). The fluid is in motion with mass-average velocity defined as [Eq. (5-5)]

$$\mathbf{v} \equiv \frac{\mathbf{n}_t}{\rho} \tag{8-113}$$

where

$$\mathbf{n}_t = \mathbf{n}_A + \mathbf{n}_B + \mathbf{n}_C + \cdots \tag{8-114}$$

or

$$\mathbf{n}_t = \rho_A \mathbf{v}_A + \rho_B \mathbf{v}_B + \rho_C \mathbf{v}_C + \cdots \tag{8-115}$$

is the total flux with respect to a fixed axis of all the species A, B, C, etc., and

$$\rho = \rho_A + \rho_B + \rho_C + \cdots \tag{8-116}$$

is the density of the mixture. In Eq. (5-10) we showed that

$$\underset{\substack{\text{Total}\\\text{mass}\\\text{flux}}}{\mathbf{n}_A} = \underset{\substack{\text{Diffu-}\\\text{sive}\\\text{flux}}}{\mathbf{j}_A} + \underset{\substack{\text{Bulk}\\\text{flux}}}{\rho_A \mathbf{v}} \tag{8-117}$$

Applying the balance equation for a species A to an element of volume $\Delta\mathcal{V}$, dividing by $\Delta\mathcal{V}$, and taking limits gives

$$-\lim_{\Delta\mathcal{V}\to 0} \frac{\sum_i \Delta_i (j_{Ai}\,\Delta A_i)}{\Delta\mathcal{V}} - \lim_{\Delta\mathcal{V}\to 0} \frac{\sum_i \Delta_i (\rho_A v_i\,\Delta A_i)}{\Delta\mathcal{V}} + r_A = \frac{\partial \rho_A}{\partial t}$$

Using the definition of the divergence [Eq. (8-93)] gives

$$-\nabla \cdot \mathbf{j}_A - \nabla \cdot \rho_A \mathbf{v} + r_A = \frac{\partial \rho_A}{\partial t} \tag{8-118}$$

which is the equation of change or equation of continuity for species A in terms of the mass diffusive flux with respect to the mass-average velocity. To express it in terms of molar units we divide each term by M_A, the molecular weight. Then

$$-\nabla \cdot \mathbf{J}_A - \nabla \cdot (C_A \mathbf{v}) + R_A = \frac{\partial C_A}{\partial t} \tag{8-119}$$

In terms of its components, the molar-diffusive-flux vector $\mathbf{J}_A$ is

$$J_A = \sum_i J_{Ai}\,\boldsymbol{\delta}_i \tag{8-120}$$

If ρ is constant, the flux law for the x_i direction is

$$J_{Ai} = -D_{AB} \frac{\partial C_A}{\partial x_i} \tag{8-121}$$

Substituting Eq. (8-121) into Eq. (8-120) gives

$$\mathbf{J}_A = -D_{AB} \sum_i \boldsymbol{\delta}_i \frac{\partial C_A}{\partial x_i} \equiv -D_{AB} \nabla C_A \tag{8-122}$$

where we have used Eq. (8-13) for ∇. Equation (8-122) is *Fick's law in three dimensions for constant* ρ.

Substituting the flux law (8-121) into the expression for $\nabla \cdot \mathbf{J}_A$ gives

$$D_{AB} \nabla^2 C_A - \nabla \cdot C_A \mathbf{v} + R_A = \frac{\partial C_A}{\partial t} \qquad D_{AB} = \text{const} \tag{8-123}$$

Since $\mathbf{N}_A = \mathbf{J}_A + C_A \mathbf{v}$, Eq. (8-119) becomes

$$-\nabla \cdot \mathbf{N}_A + R_A = \frac{\partial C_A}{\partial t} \tag{8-124}$$

8-24 HEAT TRANSPORT WITH BULK FLOW

We now proceed, as we did for mass transport, to derive equations of change for heat flux and for temperature in a flowing, reacting incompressible fluid. In order to do so we can again use an element of volume $\Delta \mathcal{V}$, but in this case we write the balance equation for energy. Although this should include work done by the fluid pressure and the viscous stresses, which cause a heating of the fluid by internal friction, rigorous treatment including these terms is left to Chap. 10 because it requires the use of the general momentum balance.

In this chapter we will assume not only that the fluid density is constant but also that its specific heat, thermal conductivity, and other properties are constant. This permits us to write an energy balance over the element of volume as

Net rate of heat transport by diffusion	+	Net rate of change in enthalpy due to bulk flow	+	Rate of generation of heat	=	Rate of accumulation of enthalpy

The rate of generation of heat due to all sources (chemical, electrical, nuclear, or fluid friction) is expressed as Φ_H J/s · m^3.

The enthalpy per unit volume of the fluid (or "enthalpy concentration") is taken as $\rho \hat{C}_p T$. If we note that $\rho \hat{C}_p T$ corresponds to ρ_A, the total energy flux in the x direction can then be written analogously to the total mass flux [Eq. (8-118)] as the sum of the heat flux and the bulk flux [Eqs. (5-42)]

$$e_x = q_x + \rho \hat{C}_p T v_x \tag{8-125}$$

After dividing by $\Delta \mathcal{V}$ and taking limits, we obtain

$$\underset{\text{Diffusion}}{-\nabla \cdot \mathbf{q}} \underset{\text{Bulk flow}}{- \nabla \cdot \rho \hat{C}_p T \mathbf{v}} \underset{\text{Generation}}{+ \Phi_H} = \underset{\text{Accumulation}}{\frac{\partial(\rho \hat{C}_p T)}{\partial t}} \tag{8-126}$$

Table 8-2 Flux laws (constant properties)

	Laminar	Turbulent (isotropic)	Total	Eq. no.
Heat	$\mathbf{q}^{(l)} = -k \nabla T$	$\overline{\mathbf{q}}^{(t)} = -k^{(t)} \nabla \overline{T}$	$\overline{\mathbf{q}} = -(k + k^{(t)}) \nabla \overline{T}$	(8-128)
Mass	$\mathbf{J}_{\mathrm{A}}^{(l)} = -D_{\mathrm{AB}} \nabla C_{\mathrm{A}}$	$\overline{\mathbf{J}_{\mathrm{A}}}^{(t)} = -D_{\mathrm{AB}}^{(t)} \nabla \overline{C_{\mathrm{A}}}$	$\overline{\mathbf{J}_{\mathrm{A}}} = -(D_{\mathrm{AB}} + D_{\mathrm{AB}}^{(t)}) \nabla \overline{C_{\mathrm{A}}}$	(8-129)

which is analogous to Eq. (8-119) for species A. If we substitute the flux law [Eq. (8-12)] and assume constant k this becomes

$$k \nabla^2 T - \nabla \cdot \rho \widehat{C}_p T \mathbf{v} + \Phi_H = \frac{\partial(\rho \widehat{C}_p T)}{\partial t} \tag{8-127}$$

For application of Eq. (8-127) see Examples 10-2, 10-4, and 10-9.

8-25 TURBULENT FLOW†

For turbulent flow we can use the balance equations above if we replace $\mathbf{v}$, C_{A}, T, $\mathbf{q}$, and $\mathbf{J}_{\mathrm{A}}$ with the time-averaged values $\overline{\mathbf{v}}$, $\overline{C_{\mathrm{A}}}$, $\overline{T}$, $\overline{\mathbf{q}}$, and $\overline{\mathbf{J}_{\mathrm{A}}}$, where $\overline{\mathbf{q}}$ and $\overline{\mathbf{J}_{\mathrm{A}}}$ are the time-averaged total fluxes, i.e., the sum of the laminar and turbulent contributions. These fluxes in vector form are summarized in Table 8-2. A more thorough discussion of turbulent transport is given in Chap. 10.

8-26 EULERIAN AND LAGRANGIAN POINTS OF VIEW†

Let us consider a pollutant, such as carbon monoxide, whose concentration C_{A} in an urban atmosphere is to be monitored. Since there are various sources of carbon monoxide, and since wind currents and diffusion will disperse it away from its sources, the concentration will generally vary with lateral position x, horizontal position y, and vertical position z. Also, at a given location, changes in automobile traffic during the day will result in changes in concentration with time. So in general $C_{\mathrm{A}} = C_{\mathrm{A}}(x, y, z, t)$.

We know from mathematics that a differential change in concentration can be expressed as the sum of the changes in concentration due to changes in each of the independent variables x, y, z and t, or [see Eq. (8-90)]

$$dC_{\mathrm{A}} = \frac{\partial C_{\mathrm{A}}}{\partial t} dt + \frac{\partial C_{\mathrm{A}}}{\partial x} dx + \frac{\partial C_{\mathrm{A}}}{\partial y} dy + \frac{\partial C_{\mathrm{A}}}{\partial z} dz$$

so that the rate of change with time will be given by the *total* derivative

$$\underset{(3)}{\frac{dC_{\mathrm{A}}}{dt}} = \underset{(1)}{\frac{\partial C_{\mathrm{A}}}{\partial t}} + \underset{(2)}{\frac{\partial C_{\mathrm{A}}}{\partial x} \frac{dx}{dt}} + \frac{\partial C_{\mathrm{A}}}{\partial y} \frac{dy}{dt} + \frac{\partial C_{\mathrm{A}}}{\partial z} \frac{dz}{dt} \tag{8-130}$$

†This section may be omitted on first reading.

or
$$\frac{dC_A}{dt} = \frac{\partial C_A}{\partial t} + \sum_i \frac{\partial C_A}{\partial x_i}\frac{dx_i}{dt} \tag{8-131}$$

Let us now examine what each term means.

Term (1) If we were to measure the variation in the concentration of CO with time at a point above the top of a tall tower, i.e., at a *fixed* point (x, y, z), we could obtain $\partial C_A/\partial t$ by plotting C_A versus t and measuring the slope. This derivative $\partial C_A/\partial t$ is called the *partial* time derivative and is defined in Appendix 1-1. We would describe our observation as being made from the *eulerian point of view* since it is made at a fixed point in space (see Fig. 8-22).

Term (2) If there were a row of towers, and if we were to measure C_A at the same elevation z_1 on each simultaneously, we could then plot C_A versus x, where x is the distance along the row (see Fig. 8-23). Then the slope would give $\partial C_A/\partial x$ since t, y, and z are held constant.

Term (3) If we were to fly through the atmosphere in an airplane while taking CO samples, we would be measuring directly dC_A/dt along our path and dx/dt, dy/dt, and dz/dt (or dx_i/dt) would represent components of the velocity of the airplane. Note that these values of dC_A/dt would have to be measured at locations in space different from the values in (1) and (2)—unless we hopped from tower to tower on a helicopter. In Eq. (8-130), however, all terms must be evaluated at the same location and time.

Term (4) [See Eq. (8-132).] Suppose instead of flying in an airplane, we were to glide through the atmosphere with the velocity of the wind. Then $dx_i/dt = v_i$ is the x_i component of the velocity of the wind. For this special case, in which the observer is moving with the velocity of the stream, we call the total time derivative the *substantial* derivative and use the special symbol D/Dt. Then†

†The expansion of $\mathbf{v}\cdot\nabla$ is carried out by a method discussed in Appendix 8-3; see Table I-3.

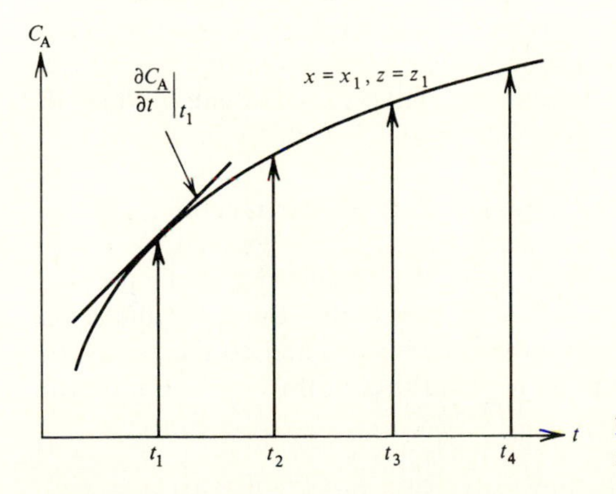

Figure 8-22 Evaluation of partial time derivative.

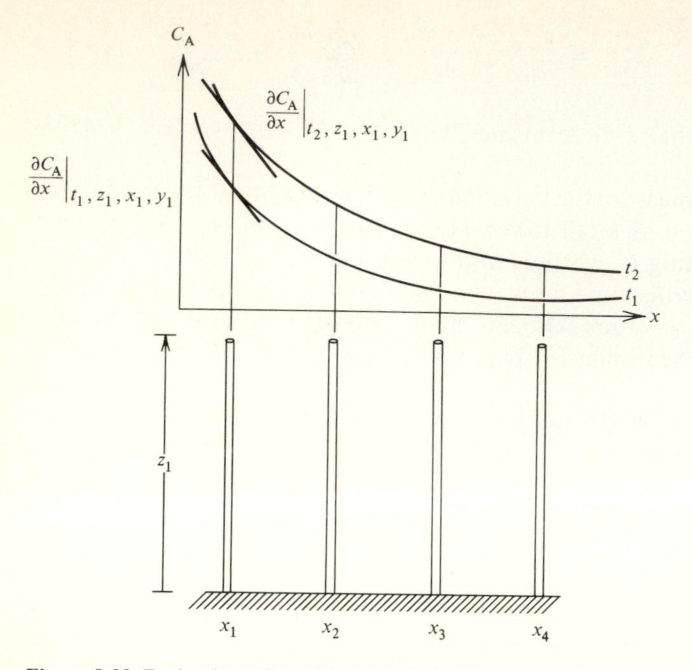

Figure 8-23 Evaluation of partial space derivative.

$$\underset{(4)}{\frac{DC_A}{Dt}} = \frac{\partial C_A}{\partial t} + \sum_i v_i \frac{\partial C_A}{\partial x_i} = \frac{\partial C_A}{\partial t} + \mathbf{v} \cdot \nabla C_A \tag{8-132}$$

Mathematically, the substantial (or material) derivative is

$$\frac{D}{Dt} \equiv \frac{\partial}{\partial t} + \mathbf{v} \cdot \nabla \tag{8-133}$$

In summary, *all* time derivatives can be found by measuring C_A versus t and finding the limit as $\Delta t \to 0$ of $\Delta C_A/\Delta t$. However, for the partial derivative, the limit is evaluated at a fixed point (x, y, z) in space (measured from a fixed set of axes); for the total derivative it is evaluated while the observer is moving with an arbitrary velocity; and for the substantial derivative it is evaluated while the observer is moving with the same velocity as the fluid.

Example 8-6: Flow in the entrance of a laminar-flow tubular reactor In the entrance region of a tubular reactor the velocity profile is not yet fully developed and therefore v_z varies with axial distance z until the familiar parabolic profile is achieved. If the entering profile is flat, $v_z = \langle v_z \rangle \equiv V$, the velocity at the center will increase until it reaches $2V$. Assume that there is a center core in which the profile is flat, that is, v_z does not vary with r, and that in this core the following curve-fit equation holds

$$v_z = v_F(z) = V\left[2 - \left(1 - \frac{z}{L_e}\right)^2\right]$$

where L_e is an entrance length. (See Example 9-19 for a related problem.) (*a*) Use the equation of continuity to find $v_r(r, z)$, assuming that the fluid is incompressible, that the flow is at steady state, and that there is axial symmetry. (*b*) If the concentration of product A in the above reaction is given approximately by

$$C_A = C_0 e^{\beta z}[(1 - r^{*2})]^2$$

where $r^* \equiv r/R$, R is the radius of tube, and β and C_0 are constants, obtain an expression for the substantial time derivative of C_A at any r and z.

SOLUTION (*a*) For incompressible fluids $\rho = \text{const}$, $\nabla \cdot \mathbf{v} = 0$. Due to symmetry,

$$\frac{\partial v_\theta}{\partial \theta} = 0$$

Expanding $\nabla \cdot \mathbf{v}$ into cylindrical coordinates (Table I-7) gives

$$\frac{\partial(rv_r)}{r\,\partial r} = -\frac{\partial v_z}{\partial z} = -\frac{\partial}{\partial z}\left\{V\left[2 - \left(1 - \frac{z}{L_e}\right)^2\right]\right\}$$

$$= +2V\left(1 - \frac{z}{L_e}\right)\left(-\frac{1}{L_e}\right) = -\frac{2V}{L_e}\left(1 - \frac{z}{L_e}\right)$$

Since $rv_r = 0$ at $r = 0$, integrating at constant z gives

$$\int_0^{rv_r} d(rv_r) = rv_r = -\frac{2V}{L_e}\left(1 - \frac{z}{L_e}\right)\int_0^r r\,dr = -\frac{V}{L_e}\left(1 - \frac{z}{L_e}\right)r^2$$

or

$$v_r = -\frac{V}{L_e}\left(1 - \frac{z}{L_e}\right)r$$

(*b*) By definition [Eq. (8-133)]

$$\frac{DC_A}{Dt} = \frac{\partial C_A}{\partial t} + \mathbf{v} \cdot \nabla C_A$$

$$\frac{\partial C_A}{\partial t} = 0 \qquad \text{at steady state}$$

$$\mathbf{v} \cdot \nabla C_A = (v_r\,\boldsymbol{\delta}_r + v_\theta\,\boldsymbol{\delta}_\theta + v_z\,\boldsymbol{\delta}_z) \cdot \left(\boldsymbol{\delta}_r \frac{\partial C_A}{\partial r} + \boldsymbol{\delta}_\theta \frac{\partial C_A}{r\,\partial \theta} + \boldsymbol{\delta}_z \frac{\partial C_A}{\partial z}\right)$$

$$= v_r \frac{\partial C_A}{\partial r} + v_\theta \frac{\partial C_A}{r\,\partial \theta} + v_z \frac{\partial C_A}{\partial z} = v_r \frac{\partial C_A}{\partial r} + v_z \frac{\partial C_A}{\partial z}$$

Since $\mathbf{v} \cdot \nabla C_A \neq 0$, $DC_A/Dt \neq 0$ even at steady state.

We note that v_r is negative, which means that there will be a flow inward from the wall to the center. This causes the velocity v_z at the center to increase with z

($\partial v_z/\partial z > 0$) until it reaches $2V$ at L_e. Since both the velocity components and the concentration vary with r and z, the particle paths will not be straight lines (as in fully developed flow) and DC_A/Dt will vary with position. If we were to choose an initial location at an arbitrary r_1 and z_1, DC_A/Dt would be the slope of the concentration-versus-travel-time curve that would be experienced by an observer moving with the velocity of the fluid. In order to obtain $C_A(r, z)$ from these data one would need to know the path of the particle of fluid originating at r_1 and z_1. This path could be obtained by integrating the velocity components.

Since in this case we have equations for C_A, v_r, and v_z, we can calculate DC_A/Dt. Thus

$$\frac{\partial C_A}{\partial r} = \frac{\partial C_A}{\partial r^*}\frac{1}{R} = 2C_0 e^{\beta z}(1 - r^{*2})(-2r^*)\frac{1}{R} = -\frac{4C_0}{R}(1 - r^{*2})r^* e^{\beta z}$$

$$\frac{\partial C_A}{\partial \theta} = 0 \text{ (symmetry)} \qquad \frac{\partial C_A}{\partial z} = C_0 \beta e^{\beta z}(1 - r^{*2})^2$$

$$\frac{DC_A}{Dt} = -\frac{4C_0}{R}(1 - r^{*2})r^* e^{\beta z} v_r + C_0 \beta e^{\beta z}[(1 - r^{*2})]^2 v_z$$

$$= -\frac{4C_0 e^{\beta z} V}{L_e}\left(1 - \frac{z}{L_e}\right) r^{*2}(1 - r^{*2}) + C_0 \beta e^{\beta z} V\left[2 - \left(1 - \frac{z}{L_e}\right)^2\right](1 - r^{*2})^2$$

Note that DC_A/Dt varies with position and, at unsteady state, would vary with time.

By a similar method one can evaluate the substantial derivative of a velocity component Dv_r/Dt

$$\frac{Dv_r}{Dt} \equiv \frac{\partial v_r}{\partial t} + \mathbf{v} \cdot \nabla v_r$$

and similarly Dv_z/Dt. But for the velocity vector $\mathbf{v}$

$$\frac{D\mathbf{v}}{Dt} \equiv \frac{\partial \mathbf{v}}{\partial t} + \mathbf{v} \cdot \nabla \mathbf{v}$$

where $D\mathbf{v}/Dt$ is a *lagrangian* acceleration. However, the so-called inertial term $\mathbf{v} \cdot \nabla \mathbf{v}$ requires special attention for curvilinear coordinates, as we demonstrate in Chap. 9. □

8-27 EQUATION OF CONTINUITY: LAGRANGIAN POINT OF VIEW†

Consider an incremental element of fluid of *constant mass* (see Figs. 8-24 and 8-25) moving with the stream velocity. If the fluid is compressible, the volume will be variable since the density varies from one position to another. For that reason we will use the symbol $\delta\mathcal{V}$ to represent an incremental element of volume rather

†This section may be omitted on first reading.

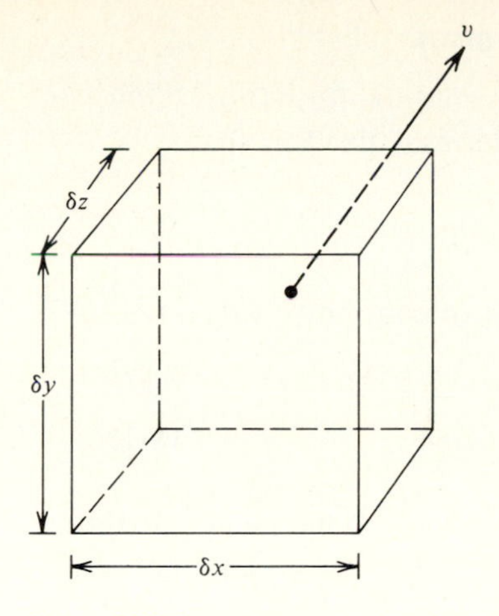

Figure 8-24 Lagrangian element of volume.

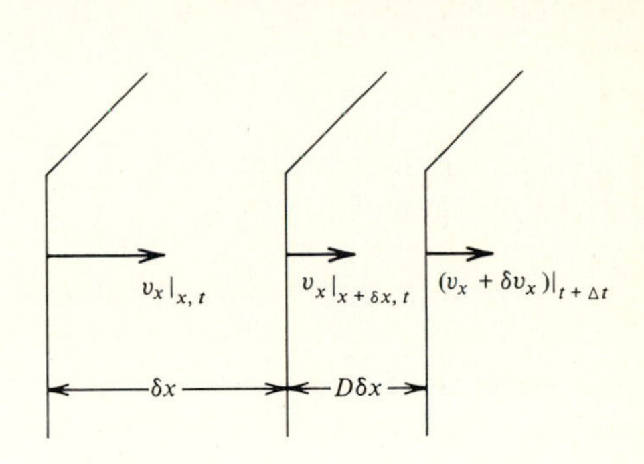

Figure 8-25 Expansion of one face of element.

than $\Delta\mathcal{V}$. In cartesian coordinates

$$\delta\mathcal{V} = \delta x\, \delta y\, \delta z \tag{8-134}$$

where δx, δy, and δz are increments of length. Since the mass of the element $\rho\,\delta\mathcal{V}$ is constant†

$$\frac{D(\rho\,\delta\mathcal{V})}{Dt} = 0 \qquad \text{or} \qquad \delta\mathcal{V}\frac{D\rho}{Dt} = -\rho\frac{D\delta\mathcal{V}}{Dt} \tag{8-135}$$

But by Eq. (8-134)

$$\frac{D\delta\mathcal{V}}{Dt} = \frac{D\delta x}{Dt}\delta y\,\delta z + \frac{D\delta y}{Dt}\delta x\,\delta z + \frac{D\delta z}{Dt}\delta x\,\delta y \tag{8-135a}$$

Now $D\delta x/Dt$, which is the time rate of increase of the length δx, can be related to the difference between the velocity of the face at x and the face at $x + \delta x$; in other words, the extension of the length δx is reflected in the difference in the velocities of the two faces, or δv_x. Then

$$\frac{D\delta x}{Dt} = \delta v_x = \frac{\partial v_x}{\partial x}\delta x$$

If we write similar expressions for the other axes and substitute into Eq. (8-135*a*), we get

$$\delta\mathcal{V}\frac{D\rho}{Dt} = -\rho\left(\frac{\partial v_x}{\partial x} + \frac{\partial v_y}{\partial y} + \frac{\partial v_z}{\partial z}\right)\delta x\,\delta y\,\delta z$$

or

$$\frac{D\rho}{Dt} = -\rho\nabla\cdot\mathbf{v} \tag{8-136}$$

which is the equation of continuity in lagrangian form.

†The increment δx should not be confused with the unit vector $\boldsymbol{\delta}_x$.

Relation between Eulerian and Lagrangian Forms

Previously we derived the equation of continuity in eulerian form [Eq. (8-109)] by using an element of constant volume and variable mass fixed in space

$$\frac{\partial \rho}{\partial t} = -\nabla \cdot \rho \mathbf{v} \qquad \text{eulerian} \tag{8-109}$$

This equation must be equivalent to the equation of continuity which we have derived in lagrangian form

$$\frac{D\rho}{Dt} = -\rho \nabla \cdot \mathbf{v} \qquad \text{lagrangian} \tag{8-136}$$

for an element of volume of constant mass and variable volume moving with the velocity of the stream. We can now relate the two forms of the equation of continuity. Starting with the lagrangian form (8-136), we apply the definition of the substantial derivative (8-133) to its left-hand side

$$\frac{D\rho}{Dt} \equiv \frac{\partial \rho}{\partial t} + \mathbf{v} \cdot \nabla \rho = -\rho \nabla \cdot \mathbf{v} \tag{8-137}$$

which can be written

$$\frac{\partial \rho}{\partial t} = -\rho \nabla \cdot \mathbf{v} - \mathbf{v} \cdot \nabla \rho \tag{8-138}$$

Comparing Eqs. (8-138) and (8-109) indicates that the latter can be obtained from the former by using the following identity (derived in Appendix 8-3)

$$\nabla \cdot \rho \mathbf{v} \equiv \rho \nabla \cdot \mathbf{v} + \mathbf{v} \cdot \nabla \rho \tag{8-139}$$

If we substitute Eq. (8-139) into Eq. (8-138), we obtain

$$\frac{\partial \rho}{\partial t} = -\nabla \cdot \rho \mathbf{v} \tag{8-109}$$

the eulerian form of the equation of continuity. Thus the eulerian and lagrangian forms are shown to be mathematically equivalent—as is necessary since they both are statements of the same physical law, the conservation of mass.

8-28 ENERGY AND MASS EQUATIONS IN LAGRANGIAN FORM (CONSTANT PROPERTIES)

By means of an energy balance made over an incremental element fixed in space the following energy equation for constant ρ was obtained in eulerian form

$$\frac{\partial(\rho \widehat{C}_p T)}{\partial t} = -\nabla \cdot \mathbf{q} - \nabla \cdot \rho \widehat{C}_p T \mathbf{v} + \Phi_H \tag{8-126}$$

where Φ_H is the total generation rate, per unit volume, of heat due to chemical, electrical, or nuclear sources as well as to heat generated internally by fluid friction. (An expression for the latter will be developed in Chap. 10, where a more rigorous energy equation will be derived.)

To derive the energy equation in lagrangian form we would use an element of variable volume $\delta \mathcal{V} = \delta x \, \delta y \, \delta z$ and constant mass $\rho \, \delta \mathcal{V}$ (see Fig. 8-24). *Since the element is moving with the velocity of the stream, there would be* no *net bulk flow*. The energy balance would be

$$\frac{D(\rho \, \delta \mathcal{V} \, \hat{C}_p T)}{Dt} = \rho \, \delta \mathcal{V} \frac{D(\hat{C}_p T)}{Dt} = (q_x \, \delta y \, \delta z)\Big|_x - (q_x \, \delta y \, \delta z)\Big|_{x+\Delta x}$$

$$+ \text{ similar terms for } y \text{ and } z \text{ directions } + \Phi_H \, \delta \mathcal{V} \qquad (8\text{-}140)$$

Upon dividing by $\delta \mathcal{V}$ and taking the limit as $\delta \mathcal{V} \to 0$ the *lagrangian* form of the energy equation for constant ρ is obtained

$$\rho \frac{D(\hat{C}_p T)}{Dt} = -\nabla \cdot \mathbf{q} + \Phi_H \qquad (8\text{-}141)$$

Transformation to Eulerian Form

To convert Eq. (8-141) back to eulerian form, we use the definition of the substantial derivative [Eq. (8-133)] in Eq. (8-141)

$$\rho \frac{D(\hat{C}_p T)}{Dt} \equiv \rho \frac{\partial(\hat{C}_p T)}{\partial t} + \rho \mathbf{v} \cdot \nabla \hat{C}_p T = -\nabla \cdot \mathbf{q} + \Phi_H \qquad (8\text{-}142)$$

Treating $\rho \hat{C}_p T$ as the product $\rho \hat{C}_p T$ gives

$$\frac{\partial(\rho \hat{C}_p T)}{\partial t} = \rho \frac{\partial(\hat{C}_p T)}{\partial t} + \hat{C}_p T \frac{\partial \rho}{\partial t}$$

Then using the above and the equation of continuity in eulerian form (8-109) gives

$$\frac{\partial(\rho \hat{C}_p T)}{\partial t} - \hat{C}_p T(-\nabla \cdot \rho \mathbf{v}) + \rho \mathbf{v} \cdot \nabla \hat{C}_p T = -\nabla \cdot \mathbf{q} + \Phi_H$$

Comparing with Eq. (8-126) indicates that we need to use the identity

$$\nabla \cdot \rho \hat{C}_p T \mathbf{v} \equiv \rho \mathbf{v} \cdot \nabla \hat{C}_p T + \hat{C}_p T \nabla \cdot \rho \mathbf{v} \qquad (8\text{-}143)$$

which is derived by the procedure discussed in Appendix 8-3. By a similar method one can show that, for any property $\hat{G}$,

$$\boxed{\rho \frac{D\hat{G}}{Dt} \equiv \frac{\partial(\rho \hat{G})}{\partial t} + \nabla \cdot \rho \hat{G} \mathbf{v}} \qquad (8\text{-}144)$$

This equation provides a useful method of making transformations between lagrangian and eulerian forms.

Example 8-7: Transformation from eulerian to lagrangian form Use Eq. (8-144) to transform Eq. (8-126) to the lagrangian form.

SOLUTION In order for Eq. (8-126) to be in the same form as Eq. (8-144) we must let $\widehat{G} \equiv \widehat{C}_p T$. Then Eq. (8-126) becomes

$$\frac{\partial(\rho\widehat{G})}{\partial t} = -\nabla \cdot \mathbf{q} - \nabla \cdot \rho\widehat{G}\mathbf{v} + \Phi_H$$

Using Eq. (8-144) gives

$$\rho \frac{D(\widehat{C}_p T)}{Dt} = -\nabla \cdot \mathbf{q} + \Phi_H$$

which agrees with Eq. (8-141). □

Likewise we can transform the equation of continuity for species A [Eq. (8-119)] to the lagrangian form (see the Problems)

$$\rho \frac{D\omega_A}{Dt} = -\nabla \cdot \mathbf{j}_A + r_A \tag{8-145}$$

In terms of molar concentration, this is

$$\rho \frac{DC_A}{Dt} + C_A \nabla \cdot \mathbf{v} = -\nabla \cdot \mathbf{J}_A + R_A \tag{8-146}$$

For constant ρ and D_{AB} this gives

$$\frac{DC_A}{Dt} \equiv \frac{\partial C_A}{\partial t} + \mathbf{v} \cdot \nabla C_A = D_{AB} \nabla^2 C_A + R_A \tag{8-147}$$

which is given for various coordinate systems in Tables I-7, I-8, and I-9.

APPENDIX 8-1 PRODUCT RULE

If T_x^*, T_y^*, and T_z^* are the solutions of

$$\alpha \frac{\partial^2 T_x^*}{\partial x^2} = \frac{\partial T_x^*}{\partial t} \qquad \alpha \frac{\partial^2 T_y^*}{\partial y^2} = \frac{\partial T_y^*}{\partial t} \qquad \alpha \frac{\partial^2 T_z^*}{\partial z^2} = \frac{\partial T_z^*}{\partial t} \tag{A8-1-1}$$

then the solution of the equation

$$\alpha \left(\frac{\partial^2 T^*}{\partial x^2} + \frac{\partial^2 T^*}{\partial y^2} + \frac{\partial^2 T^*}{\partial z^2} \right) = \frac{\partial T^*}{\partial t} \tag{A8-1-2}$$

is

$$T^*(x, y, z, t) = T_x^*(x, t) T_y^*(y, t) T_z^*(z, t) \tag{A8-1-3}$$

provided that the boundary conditions on T_x^*, T_y^*, and T_z^* are compatible with those on T^*; that is, they must also satisfy Eq. (A8-1-3).

Proof for Two Dimensions

We must show that

$$T^*(x, y, t) = T_x^*(x, t)T_y^*(y, t) \tag{A8-1-4}$$

is a solution of

$$\alpha\left(\frac{\partial^2 T^*}{\partial x^2} + \frac{\partial^2 T^*}{\partial y^2}\right) = \frac{\partial T^*}{\partial t} \tag{A8-1-5}$$

if T_x^* and T_y^* are solutions of

$$\alpha \frac{\partial^2 T_x^*}{\partial x^2} = \frac{\partial T_x^*}{\partial t} \qquad \text{and} \qquad \alpha \frac{\partial^2 T_y^*}{\partial y^2} = \frac{\partial T_y^*}{\partial t} \tag{A8-1-6}$$

Substituting Eq. (A8-1-4) into (A8-1-5) gives

$$\alpha\left(\frac{\partial^2 T_x^*}{\partial x^2} T_y^* + \frac{\partial^2 T_y^*}{\partial y^2} T_x^*\right) = \frac{\partial T_x^*}{\partial t} T_y^* + \frac{\partial T_y^*}{\partial t} T_x^* \tag{A8-1-7}$$

Using Eqs. (A8-1-6), we have

$$T_y^*\left(\alpha \frac{\partial^2 T_x^*}{\partial x^2} - \frac{\partial T_x^*}{\partial t}\right) + T_x^*\left(\alpha \frac{\partial^2 T_y^*}{\partial y^2} - \frac{\partial T_y^*}{\partial t}\right) = 0 \tag{A8-1-8}$$

Since the terms in the parentheses are zero by virtue of Eqs. (A8-1-6), the product rule satisfies the equation. Of course, the boundary conditions must also satisfy the product-rule equation.

APPENDIX 8-2 INDEX NOTATION

In manipulating vector equations in three dimensions it is convenient to number the coordinate axes (see Fig. A8-2-1) by letting

$$x \equiv x_1 \qquad y \equiv x_2 \qquad z \equiv x_3 \tag{A8-2-1}$$

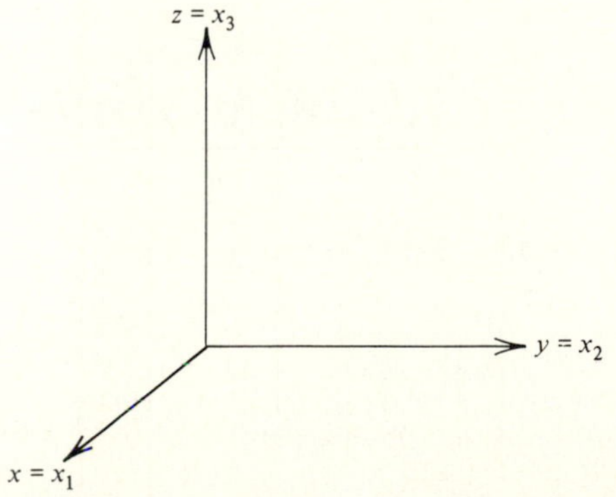

$z = x_3$

$y = x_2$

$x = x_1$

Figure A8-2-1 Index notation.

If we similarly number the vector components of a vector $\mathbf{q}$ [Eq. (8-5)], we can conveniently write them as a summation†

$$\mathbf{q} = \mathbf{q}_1 + \mathbf{q}_2 + \mathbf{q}_3 \equiv \sum_{i=1}^{3} \mathbf{q}_i = \sum_i \mathbf{q}_i$$

where $\sum_i$ is an abbreviation for $\sum_{i=1}^{3}$. In terms of scalar components, this becomes

$$\mathbf{q} = \sum_{i=1}^{3} q_i \boldsymbol{\delta}_i \tag{A8-2-2}$$

which is equivalent to Eq. (8-8).

It is often convenient to define the *position* vector $\mathbf{x}$ (Table I-1) by

$$\mathbf{x} = x\boldsymbol{\delta}_x + y\boldsymbol{\delta}_y + z\boldsymbol{\delta}_z = \sum_i x_i \boldsymbol{\delta}_i \tag{A8-2-3}$$

The dot product can be written

$$\mathbf{u} \cdot \mathbf{v} = \sum_i u_i \boldsymbol{\delta}_i \cdot \sum_j u_j \boldsymbol{\delta}_j = \sum_i \sum_j u_i v_j \boldsymbol{\delta}_i \cdot \boldsymbol{\delta}_j$$

Since $\boldsymbol{\delta}_i \cdot \boldsymbol{\delta}_j$ is zero if $i \neq j$ and unity if $i = j$,

$$\mathbf{u} \cdot \mathbf{v} = \sum_i u_i v_i \tag{A8-2-4}$$

The gradient, the divergence, the laplacian, and the lagrangian convective terms are then

$$\nabla T = \sum_i \boldsymbol{\delta}_i \frac{\partial T}{\partial x_i} \qquad \nabla \cdot \mathbf{q} = \sum_i \frac{\partial q_i}{\partial x_i} \qquad \nabla^2 T = \sum_i \frac{\partial^2 T}{\partial x_i^2}$$

$$\mathbf{v} \cdot \nabla T = \sum_i v_i \frac{\partial T}{\partial x_i} \qquad \text{and} \qquad \mathbf{v} \cdot \nabla \mathbf{v} = \sum_i \sum_j v_j \frac{\partial v_i}{\partial v_j} \tag{A8-2-5}$$

APPENDIX 8-3 PROOF OF VECTOR IDENTITIES

Suppose we wish to prove a vector identity such as Eq. (8-139)

$$\rho \nabla \cdot \mathbf{v} + \mathbf{v} \cdot \nabla \rho \equiv \nabla \cdot \rho \mathbf{v} \tag{A8-3-1}$$

†To be consistent with Eqs. (8-5) and (8-9) we should use the notation q_{x_1} instead of q_1 but there is seldom any ambiguity in the use of q_1, and it avoids the need for sub-subscripts.

We label the coordinate axes by letting $x = x_1, y = x_2, z = x_3$ (see Fig. A8-2-1); then

$$\nabla \equiv \boldsymbol{\delta}_1 \frac{\partial}{\partial x_1} + \boldsymbol{\delta}_2 \frac{\partial}{\partial x_2} + \boldsymbol{\delta}_3 \frac{\partial}{\partial x_3} = \sum_{i=1}^{3} \boldsymbol{\delta}_i \frac{\partial}{\partial x_1}$$

$$\mathbf{v} = \sum_{i=1}^{3} v_i \boldsymbol{\delta}_i \qquad \rho\mathbf{v} = \sum_{i=1}^{3} \rho v_i \boldsymbol{\delta}_i$$

$$\nabla \cdot \rho\mathbf{v} = \left(\sum_{i=1}^{3} \boldsymbol{\delta}_i \frac{\partial}{\partial x_i}\right) \cdot \left(\sum_{k=1}^{3} \rho v_k \boldsymbol{\delta}_k\right)$$

For generality we must use a different dummy summation index k for the second summation. Since, for cartesian coordinates, the unit vectors do not depend upon position, $\partial \boldsymbol{\delta}_k / \partial x_i = 0$, these summations can be written

$$\sum_{i=1}^{3} \sum_{k=1}^{3} \boldsymbol{\delta}_i \cdot \boldsymbol{\delta}_k \frac{\partial}{\partial x_i} (\rho v_k)$$

The dot product of the unit vectors $\boldsymbol{\delta}_i \cdot \boldsymbol{\delta}_k$ will be unity if $i = k$, but since the coordinates are orthogonal (perpendicular), it will be zero if $i \neq k$. Therefore we can let

$$\boldsymbol{\delta}_i \cdot \boldsymbol{\delta}_k = \delta_{ik} \tag{A8-3-2a}$$

where δ_{ik} is called the *Kronecker delta*. It has the property

$$\delta_{ik} = \begin{cases} 0 & \text{when } i \neq k \qquad \text{(A8-3-2b)} \\ 1 & \text{when } i = k \qquad \text{(A8-3-2c)} \end{cases}$$

Then

$$\nabla \cdot \rho\mathbf{v} = \sum_{i=1}^{3} \frac{\partial(\rho v_i)}{\partial x_i} = \sum_{i=1}^{3} \rho \frac{\partial v_i}{\partial x_i} + \sum_{i=1}^{3} v_i \frac{\partial \rho}{\partial x_i} \tag{A8-3-3}$$

But

$$\begin{aligned} \rho\nabla \cdot \mathbf{v} &= \rho\left(\sum_{i=1}^{3} \boldsymbol{\delta}_i \frac{\partial}{\partial x_i}\right) \cdot \left(\sum_{k=1}^{3} v_k \boldsymbol{\delta}_k\right) = \rho \sum_i \sum_k \boldsymbol{\delta}_i \cdot \boldsymbol{\delta}_k \frac{\partial v_k}{\partial x_i} \\ &= \rho \sum_i \sum_k \delta_{ik} \frac{\partial v_k}{\partial x_i} = \rho \sum_i \frac{\partial v_i}{\partial x_i} \end{aligned} \tag{A8-3-4}$$

Similarly we can show that

$$\mathbf{v} \cdot \nabla\rho = \left(\sum_i v_i \boldsymbol{\delta}_i\right) \cdot \left(\sum_k \boldsymbol{\delta}_k \frac{\partial \rho}{\partial x_k}\right) = \sum_i v_i \frac{\partial \rho}{\partial x_i} \tag{A8-3-5}$$

Substitution of Eqs. (A8-3-3) to (A8-3-5) into Eq. (A8-3-1) gives

$$\rho \sum_{i=1}^{3} \frac{\partial v_i}{\partial x_i} + \sum_{i=1}^{3} v_i \frac{\partial \rho}{\partial x_i} = \sum_{i=1}^{3} \rho \frac{\partial v_i}{\partial x_i} + \sum_{i=1}^{3} v_i \frac{\partial \rho}{\partial v_i}$$

which shows that Eq. (8-139) is an identity.

APPENDIX 8-4 HEAT FLUX AS A VECTOR: ENERGY TETRAHEDRON

We wish to establish that

$$-d\dot{Q}_G = \mathbf{q} \cdot \mathbf{n}\, d\mathbf{A} = \mathbf{q} \cdot d\mathbf{S} \tag{8-79}$$

applies not only to the brick-shaped solid (for which **n** and **q** are oriented along the cartesian coordinate axes) but is quite general and applies to any orientation of the unit vector **n** with respect to **q.** The equation will be general if **q** is a vector since then its dot product with another vector $d\mathbf{S}$ will result in a scalar, as the definition of the dot product [Eq. (8-25)] indicates.

Consider a solid for which $T = T(x, y, z, t)$. Let us cut a plane through the solid making arbitrary angles with the coordinate axes; then at a point on the plane let us draw a tetrahedral element of sides ΔA_x, ΔA_y, ΔA_z and of slant face ΔA_n (Fig. A8-4-1). Let us try to relate the rate of heat flow passing through the area ΔA_n on the slant plane to the rate of heat flow passing through the other faces. We can then compare this result with the expansion of the right-hand side of Eq. (8-58) into a cartesian coordinate system.

To simplify the geometry we restrict ourselves first to the two-dimensional case and assume the solid to be so long that $q_z = 0$. Then instead of a tetrahedron

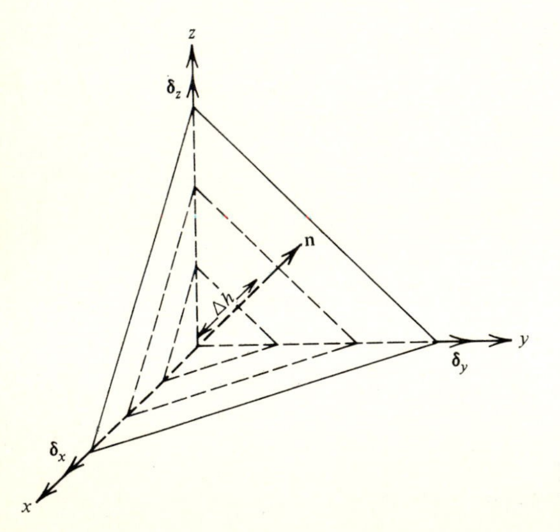

Figure A8-4-1 Tetrahedral elements of volume.

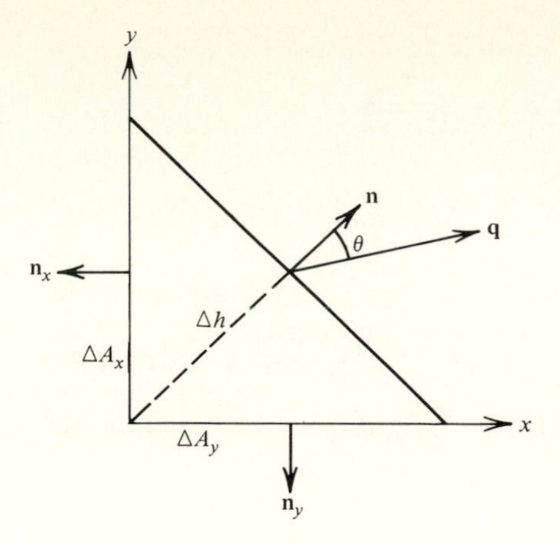

Figure A8-4-2 Element in two dimensions.

we have a wedge (Fig. A8-4-2) of sides $\Delta A_x = L\,\Delta y$ and $\Delta A_y = L\,\Delta x$ and length L in z direction. We now construct a normal to the slant area ΔA_n. We wish to relate the flux through ΔA_n to the flux through the sides as the slant face moves toward the origin or as $\Delta\mathcal{V} \to 0$, where

$$\Delta\mathcal{V} = \frac{\Delta h}{2}\frac{\Delta A_n}{L}$$

and Δh is the height of the triangle. Making an energy balance gives

$$q_x\,\Delta A_x + q_y\,\Delta A_y - q_n\,\Delta A_n = \frac{\Delta(\rho\widehat{C}_p T\,\Delta\mathcal{V})}{\Delta t}$$

where q_n is the rate of flow through A_n per unit area.

Now ΔA_x is the projection of ΔA_n on the x axis, or

$$\Delta A_x = \Delta A_n \cos\theta_{(x)}$$

where $\theta_{(x)}$ is the angle between $\mathbf{n}$ and $\boldsymbol{\delta}_x$. Then by the definition of the dot product [Eq. (8-25)]

$$\Delta A_x = \mathbf{n}\cdot\boldsymbol{\delta}_x\,\Delta A_n \tag{A8-4-1}$$

Likewise

$$\Delta A_y = \mathbf{n}\cdot\boldsymbol{\delta}_y\,\Delta A_n \tag{A8-4-2}$$

Dividing by ΔA_n and letting $\Delta\mathcal{V} \to 0$ gives

$$q_x\,\mathbf{n}\cdot\boldsymbol{\delta}_x + q_y\,\mathbf{n}\cdot\boldsymbol{\delta}_y - q_n = \lim_{\substack{\Delta\mathcal{V}\to 0\\ \text{or } \Delta h\to 0}} \frac{\Delta h}{2}\frac{\Delta(\rho\widehat{C}_p T)}{\Delta t} = 0$$

i.e., the accumulation term disappears since the volume-to-area ratio approaches zero. Then

$$\mathbf{n}\cdot(q_x\boldsymbol{\delta}_x + q_y\boldsymbol{\delta}_y) - q_n = 0$$

The term in parentheses is a vector if $\mathbf{q}$ is a vector.

Then for three dimensions, using Fig. A8-4-1,

$$\mathbf{n} \cdot (q_x \boldsymbol{\delta}_x + q_y \boldsymbol{\delta}_y + q_z \boldsymbol{\delta}_z) - q_n = 0$$

or

$$q_n = \mathbf{n} \cdot \mathbf{q} \tag{A8-4-3}$$

We will call q_n the *heat scalar* (for reasons to be apparent in Chap. 9, where a stress vector $\boldsymbol{\sigma}_{(n)}$ is defined).

Since **n** is arbitrary, this result tells us that no matter what the direction of **n**, we can obtain the rate of heat loss through dA_n from the equation

$$d\dot{Q}_L = q_n \, dA_n = \mathbf{n} \cdot \mathbf{q} \, dA_n = \sum_i q_i \, dA_i \tag{A8-4-4}$$

which says that the rate of heat loss through the slant face equals the sum of the heat flows in the direction of the coordinate axes. This result also establishes **q** as a vector since its dot product with the vector **n** gives the scalar q_n no matter what the coordinate system or the direction of **n** in space. For a proof that if $\mathbf{n} \cdot \mathbf{q} = q_n$, a scalar, **q** is a vector, see the Problems and Appendix 8-7. More general treatments are available.[3]

APPENDIX 8-5 THE OPERATION $\nabla \cdot \mathbf{q}$ IN CURVILINEAR COORDINATES

Suppose we wish to find $\nabla \cdot \mathbf{q}$ in cylindrical coordinates. In Sec. 8-1 we indicated that Fourier's law can be written in general for an arbitrary coordinate u [Eq. (8-15)]. In index notation

$$q_n = -k \frac{\partial T}{h_n \, \partial u_n} \tag{A8-5-1}$$

where h_n is the scale factor that converts the coordinate into a length. (For cylindrical coordinates, $h_1 = 1$, $h_2 = r$, $h_3 = 1$; $u_1 = r$, $u_2 = \theta$, $u_3 = z$). Then if we write

$$\mathbf{q} = \sum_n q_n \boldsymbol{\delta}_n = -k \sum_n \frac{\partial T}{h_n \, \partial u_n} \boldsymbol{\delta}_n = -k \, \nabla T$$

we can observe that the operator ∇ is given by

$$\nabla \equiv \sum_n \boldsymbol{\delta}_n \frac{\partial}{h_n \, \partial y_n}$$

To obtain $\nabla \cdot \mathbf{q}$ we take the dot product of ∇ and **q** as indicated but, for generality, we change the summation index on ∇ to m. Then

$$\nabla \cdot \mathbf{q} = \left(\sum_m \boldsymbol{\delta}_m \frac{\partial}{h_n\, \partial u_m}\right) \cdot \left(\sum_n q_n \boldsymbol{\delta}_n\right)$$

$$= \sum_m \sum_n \left(\boldsymbol{\delta}_m \cdot \boldsymbol{\delta}_n \frac{\partial q_n}{h_m\, \partial u_m} + q_n \boldsymbol{\delta}_m \frac{\partial \boldsymbol{\delta}_n}{h_m\, \partial u_m}\right) \qquad \text{(A8-5-2)}$$

The first term contains the product $\boldsymbol{\delta}_m \cdot \boldsymbol{\delta}_n = \cos\theta$, which is unity if $m = n$ ($\theta = \pi/2$) and zero if $m \neq n$ ($\theta = 0$). In general

$$\boldsymbol{\delta}_m \cdot \boldsymbol{\delta}_n = \delta_{mn} \qquad \text{where } \delta_{mn} = \begin{cases} 1 & m = n \\ 0 & m \neq n \end{cases} \qquad \text{(A8-5-3)}$$

(δ is the Kronecker delta). Then only the $m = n$ term in the first summation is nonzero, giving

$$\sum_n \frac{\partial q_n}{h_n\, \partial u_n}$$

The second summation term in Eq. (A8-5-2) contains $\partial \boldsymbol{\delta}_m / \partial u_n$, which is zero for cartesian coordinates but not necessarily zero for curvilinear coordinates because a unit vector can change direction as a coordinate changes. For example, in cylindrical coordinates, $\partial \boldsymbol{\delta}_r / \partial \theta \neq 0$, as can be seen from the diagram (Fig. A8-5-1) showing the unit vector $\boldsymbol{\delta}_r$ at location θ and at $\theta + d\theta$, where $d\theta$ is an infinitesimal angle (exaggerated) approaching zero as a limit. We note that while the *magnitude* of $\boldsymbol{\delta}_r$ is always the same, its *direction* at θ and at $\theta + d\theta$ is *not* the same. The difference between $\boldsymbol{\delta}_r + d\boldsymbol{\delta}_r$ and $\boldsymbol{\delta}_r$ is a vector $d\boldsymbol{\delta}_r$ which will be in the θ direction. By similar triangles we see that its magnitude will be

$$|d\boldsymbol{\delta}_r| = d\theta\, |\boldsymbol{\delta}_r| = d\theta$$

Then $$d\boldsymbol{\delta}_r = \text{magnitude} \times \text{direction} = d\theta\, \boldsymbol{\delta}_\theta$$

or $$\boxed{\frac{d\boldsymbol{\delta}_r}{d\theta} = \boldsymbol{\delta}_\theta} \qquad \text{(A8-5-4)}$$

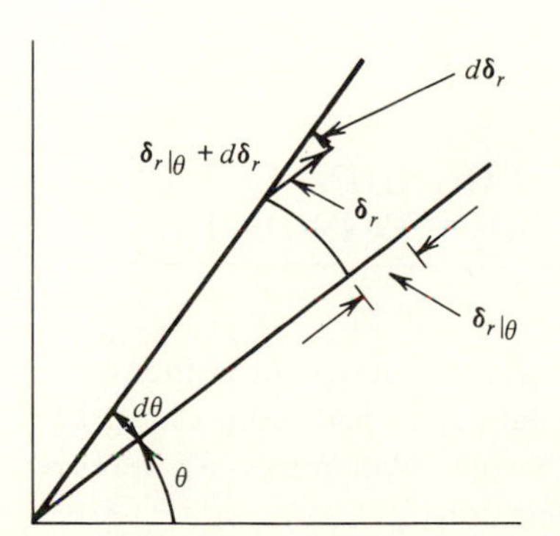

Figure A8-5-1 Variation of $\boldsymbol{\delta}_r$ with θ.

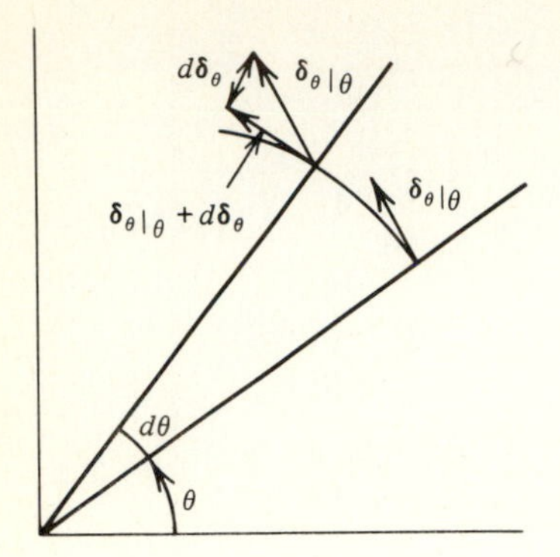

Figure A8-5-2 Variation of $\boldsymbol{\delta}_\theta$ with θ.

A similar diagram (Fig. A8-5-2) shows that $\partial\boldsymbol{\delta}_\theta/\partial\theta \neq 0$. Although $\boldsymbol{\delta}_\theta$ at θ and at $\theta + d\theta$ have the same magnitude, the directions have changed since $\boldsymbol{\delta}_\theta$ is always tangent to a constant r curve. The change in $\boldsymbol{\delta}_\theta$ or $d\boldsymbol{\delta}_\theta$ is a vector in the direction $-\boldsymbol{\delta}_r$, as seen by forming a triangle with sides $\boldsymbol{\delta}_\theta$ and $\boldsymbol{\delta}_\theta + d\boldsymbol{\delta}_\theta$. By similar triangles the angle between these vectors will be $d\theta$. The magnitude of the resultant $d\boldsymbol{\delta}_\theta$ will be $\sin d\theta \approx d\theta$ since the sine of a small angle is approximately equal to the angle. Then

$$d\boldsymbol{\delta}_\theta = -d\theta\, \boldsymbol{\delta}_r$$

or

$$\boxed{\frac{d\boldsymbol{\delta}_\theta}{d\theta} = -\boldsymbol{\delta}_r} \qquad \text{(A8-5-5)}$$

These results lead to

$$\nabla \cdot \mathbf{q} = \frac{\partial q_r}{\partial r} + \frac{\partial q_\theta}{r\,\partial\theta} + \frac{\partial q_z}{\partial z} + \frac{q_r}{r} = \frac{\partial(rq_r)}{r\,\partial r} + \frac{\partial q_\theta}{r\,\partial\theta} + \frac{\partial q_z}{\partial z} \qquad \text{(A8-5-6)}$$

When geometric methods cannot be used to find the derivative of the unit vectors, an analytical method can be used, as described in Appendix 8-7 [Eq. (A8-7-36)].

APPENDIX 8-6 TRANSFORMATION OF COORDINATES USING GEOMETRIC CONSIDERATIONS

It is sometimes necessary to transform from cartesian coordinates into other coordinate systems, such as cylindrical, spherical, prolate spherical, ellipsoidal, etc. Although there are general methods for carrying this out (Appendix 8-7), here we illustrate a procedure based on geometric considerations.

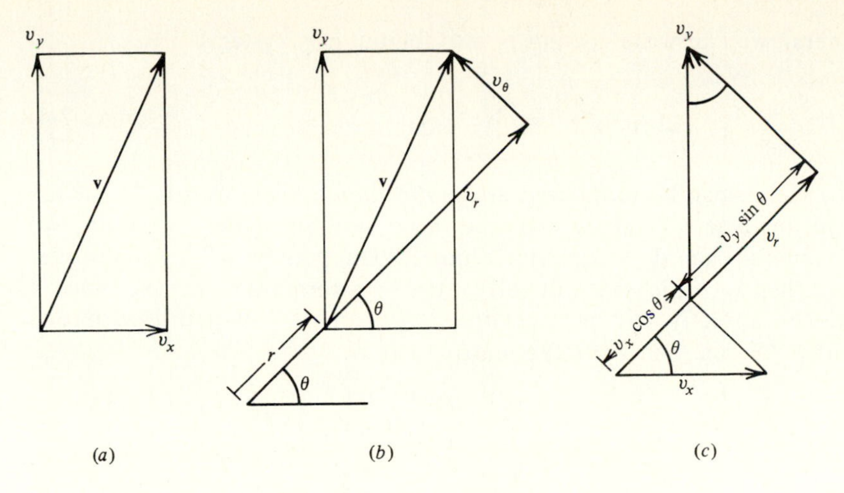

Figure A8-6-1 Geometric relation between vector components: (*a*) vector $\mathbf{v}$ in cartesian coordinates; (*b*) vector $\mathbf{v}$ in polar coordinates; (*c*) component v_r in terms of v_x and v_y.

Vectors As an example, consider a cylindrical coordinate system. At an arbitrary point (x, y, z) or (r, θ, z) we can alternatively write (see Fig. A8-6-1)

$$\mathbf{v} = \mathbf{v}_x + \mathbf{v}_y + \mathbf{v}_z = \mathbf{v}_r + \mathbf{v}_\theta + \mathbf{v}_z \tag{A8-6-1}$$

or

$$\mathbf{v} = v_x \boldsymbol{\delta}_x + v_y \boldsymbol{\delta}_y + v_z \boldsymbol{\delta}_z \tag{A8-6-2}$$

and

$$\mathbf{v} = v_r \boldsymbol{\delta}_r + v_\theta \boldsymbol{\delta}_\theta + v_z \boldsymbol{\delta}_z \tag{A8-6-3}$$

where $\boldsymbol{\delta}_r$ and $\boldsymbol{\delta}_\theta$ are unit vectors in the r and θ direction.

Let us relate the components v_r and v_θ to v_x and v_z. From the diagram, geometric considerations (Fig. A8-6-1*c*) indicate that v_r is the sum of the projections of v_x and v_y on a line in the r direction or

$$v_r = v_x \cos\theta + v_y \sin\theta \qquad \text{and} \qquad v_\theta = -v_x \sin\theta + v_y \cos\theta \tag{A8-6-4}$$

APPENDIX 8-7 GENERAL ORTHOGONAL ANALYTICAL TRANSFORMATION OF COORDINATES

If we consider cylindrical coordinates as an example, and if we label the cartesian coordinates $x_1 \equiv x$, $x_2 \equiv y$, $x_3 \equiv z$ and the cylindrical coordinates $u_1 \equiv r$, $u_2 \equiv \theta$, $u_3 \equiv z$, we can write the transformation equations (A8-6-4) in terms of coefficients of v_x and v_y as

$$v_r = (\cos\theta)v_x + (\sin\theta)v_y = L_{rx}v_x + L_{ry}v_y \tag{A8-7-1}$$

$$v_\theta = (-\sin\theta)v_x + (\cos\theta)v_y = L_{\theta x}v_x + L_{\theta y}v_y \tag{A8-7-2}$$

$$v_z = v_z = L_{zz}v_z \tag{A8-7-3}$$

In general, we can write† (using primes for the new system)

$$v'_m(u_1, u_2, u_3) = \sum_{i=1}^{3} L_{mi} v_i(x_1, x_2, x_3) \tag{A8-7-4}$$

The L_{mi} terms can be recognized as the *direction cosines;* that is, L_{mi} is the cosine of the angle between the u_m axis and the x_i axis. Since the L_{mi} require two subscripts, one for each direction, they represent components of a second-order tensor (see Chap. 1) which we will call $\underline{\mathbf{L}}$, the transformation matrix.‡ Such a quantity can be correctly written as a *matrix,* i.e., as an array of quantities in rows and columns. We can then write the matrix of $\underline{\mathbf{L}}$ as

$$\underline{\mathbf{L}} = \begin{bmatrix} L_{rx} & L_{ry} & L_{rz} \\ L_{\theta x} & L_{\theta y} & L_{\theta z} \\ L_{zx} & L_{zy} & L_{zz} \end{bmatrix} = \begin{bmatrix} \cos\theta & \sin\theta & 0 \\ -\sin\theta & \cos\theta & 0 \\ 0 & 0 & 1 \end{bmatrix} = \begin{bmatrix} L_{11} & L_{12} & L_{13} \\ L_{21} & L_{22} & L_{23} \\ L_{31} & L_{32} & L_{33} \end{bmatrix} \tag{A8-7-5}$$

We have seen how to obtain the L_{mi} from geometrical considerations, but for some coordinate systems it is not convenient to use such a method and it is better to use the mathematical procedure which follows.

Determination of Scale Factors

First let us develop an equation for the scale factor h_m, defined by Eq. (8-16). In index notation

$$ds_m = h_m\, du_m \tag{A8-7-6}$$

where ds_m is an element of length along the coordinate u_m. For example, in cylindrical coordinates, if u_m is the angle θ, we have shown that $ds_\theta = r\,d\theta$ and therefore $h_m = h_\theta = r$. The other scale factors are $h_r = 1$ and $h_z = 1$. Also, for cartesian coordinates all $h_i = 1$.

Now let us consider an element of length ds at some arbitrary position in space (see Fig. A8-7-1). In cartesian coordinates the pythagorean formula gives

†Subscripts i, j, and k are reserved for cartesian coordinates, x_1, x_2, and x_3.

‡$\underline{\mathbf{L}}$ is also a second-order tensor which is designated by boldface and one underline.

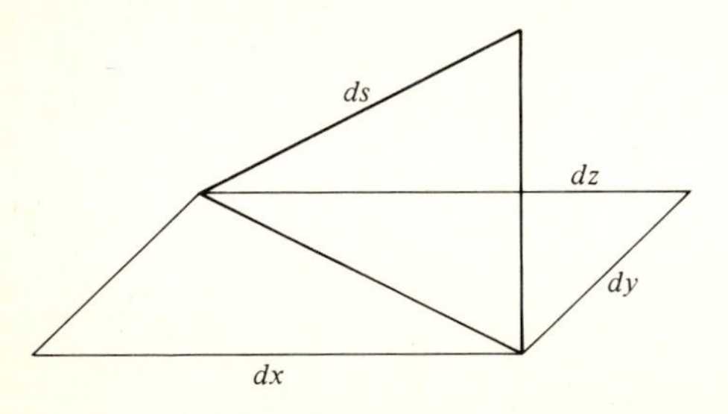

Figure A8-7-1 Element of length *ds*.

$$(ds)^2 = (dx)^2 + (dy)^2 + (dz)^2 = \sum_i (dx_i)^2 \tag{A8-7-7}$$

A coordinate system is usually defined by a set of transformation equations

$$x_i = x_i(u_1, u_2, u_3) \tag{A8-7-8}$$

For example, for cylindrical coordinates

$$x_1 \equiv x = r\cos\theta = u_1 \cos u_2 \tag{A8-7-9}$$

$$x_2 \equiv y = r\sin\theta = u_1 \sin u_2 \tag{A8-7-10}$$

$$x_3 \equiv z = z \tag{A8-7-11}$$

Although in this special case $x_i = x_i(u_1, u_2)$ only, in general, the coordinate x_i (and similarly an element dx_i) can be expressed as a function of each of the coordinates, as in Eq. (A8-7-8). Then by the chain rule of differentiation [see Eq. (8-90)] the differentials of the coordinates are given by

$$dx_i = \sum_m \frac{\partial x_i}{\partial u_m} du_m \tag{A8-7-12}$$

Substituting into Eq. (A8-7-7) gives

$$(ds)^2 = \sum_i \left[\sum_m \left(\frac{\partial x_i}{\partial u_m}\right) du_m\right]^2 \tag{A8-7-13}$$

Since we are dealing with orthogonal coordinates, we can also write

$$(ds)^2 = \sum_m h_m^2\, du_m^2 \tag{A8-7-14}$$

Equating coefficients of the differentials in (A8-7-14) and (A8-7-13) leads to

$$h_m^2 = \sum_i \left(\frac{\partial x_i}{\partial u_m}\right)^2 \tag{A8-7-15}$$

Calculation of Transformation Matrix

If we cannot use a geometrical method, we proceed as follows. Let an element of length along the u_m coordinate be $ds_m = h_m\, du_m$. Using the chain rule gives

$$ds_m = h_m\, du_m = h_m \sum_i \frac{\partial u_m}{\partial x_i} dx_i \tag{A8-7-16}$$

But by the transformation law [Eq. (A8-7-4)]

$$ds_m = \sum_i L_{mi}\, dx_i \tag{A8-7-17}$$

Then by comparing coefficients of dx_i

$$L_{mi} = \frac{h_m \, \partial u_m}{\partial x_i} \tag{A8-7-18}$$

Substituting Eq. (A8-7-18) into Eq. (A8-7-9), we get the transformation law for a vector

$$v'_m(u_1, u_2, u_3) = \sum_i h_m \frac{\partial u_m}{\partial x_i} v_i(x_1, x_2, x_3) \tag{A8-7-19}$$

or, in vector notation,

$$\mathbf{v}' = \underline{\mathbf{L}} \cdot \mathbf{v} \tag{A8-7-20}$$

If the inverse transformation is

$$v_i = \sum_m P_{im} v'_m \tag{A8-7-21}$$

or

$$\mathbf{v} = \underline{\mathbf{P}} \cdot \mathbf{v}' \tag{A8-7-22}$$

by transforming $\mathbf{v}' = \underline{\mathbf{L}} \cdot \mathbf{v}$ we get

$$\underline{\mathbf{P}} \cdot \mathbf{v}' = \mathbf{v} = \underline{\mathbf{P}} \cdot \underline{\mathbf{L}} \cdot \mathbf{v} \tag{A8-7-23}$$

$$\underline{\mathbf{P}} \cdot \underline{\mathbf{L}} = \underline{\boldsymbol{\delta}} \tag{A8-7-24}$$

$$\underline{\mathbf{P}} = \underline{\mathbf{L}}^{-1} = \text{inverse of } \underline{\mathbf{L}} \tag{A8-7-25}$$

For an orthogonal coordinate system, it can be shown that the inverse equals the transpose $\underline{\mathbf{L}}^T$, or

$$\underline{\mathbf{L}}^{-1} = \underline{\mathbf{L}}^T \qquad \text{or} \qquad L^{-1}_{im} = L_{mi} = L^T_{im} \tag{A8-7-26}$$

Then

$$\underline{\mathbf{P}} = \underline{\mathbf{L}}^T \tag{A8-7-27}$$

or

$$\boxed{P_{im} = L^T_{im}} \tag{A8-7-28}$$

where $\underline{\mathbf{L}}^T$ is the transpose of $\underline{\mathbf{L}}$ obtained by interchanging subscripts in L_{mi} or rows and columns in Eq. (A8-7-5). Thus

$$L^T_{x\theta} = L_{\theta x} = -\sin\theta$$

Then for cylindrical coordinates

$$\underline{\mathbf{L}}^{-1} = \underline{\mathbf{P}} = \begin{bmatrix} P_{xr} & P_{x\theta} & P_{xz} \\ P_{yr} & P_{y\theta} & P_{yz} \\ P_{zr} & P_{z\theta} & P_{zz} \end{bmatrix} = \begin{bmatrix} \cos\theta & -\sin\theta & 0 \\ \sin\theta & \cos\theta & 0 \\ 0 & 0 & 1 \end{bmatrix} = \underline{\mathbf{L}}^T \tag{A8-7-29}$$

Note that $P_{im} = L_{mi} = (L^T)_{im}$; for example, $P_{12} = L_{21} = L^T_{12}$, or

$$P_{x\theta} = L_{\theta x} = -\sin\theta$$

Differentiation of Unit Vectors

The derivative of a unit vector such as $\partial \boldsymbol{\delta}_m / \partial u_n$ can be found as follows. Let the displacement vector *along* a coordinate m be

$$d\mathbf{x}_m = ds_m \, \boldsymbol{\delta}_m = h_m \, du_m \, \boldsymbol{\delta}_m \tag{A8-7-30}$$

Also

$$d\mathbf{x}_m = \frac{\partial \mathbf{x}}{\partial u_m} du_m \tag{A8-7-31}$$

Solving for $\boldsymbol{\delta}_m$ after equating (A8-7-30) and (A8-7-31) gives

$$\boldsymbol{\delta}_m = \frac{\partial \mathbf{x}}{h_m \, \partial u_m} \tag{A8-7-32}$$

But

$$d\mathbf{x} = \sum_i dx_i \, \boldsymbol{\delta}_i \tag{A8-7-33}$$

or by the chain rule

$$\frac{\partial \mathbf{x}}{\partial u_m} = \sum_i \frac{\partial x_i}{\partial u_m} \boldsymbol{\delta}_i \tag{A8-7-34}$$

Then substituting into Eq. (A8-7-30) gives

$$\boldsymbol{\delta}_m = \sum_i \frac{\partial x_i}{h_m \, \partial u_m} \boldsymbol{\delta}_i = \sum L_{im}^{-1} \, \boldsymbol{\delta}_i \tag{A8-7-35}$$

and differentiating leads to

$$\frac{\partial \, \boldsymbol{\delta}_m}{\partial u_n} = \sum_i \frac{\partial}{\partial u_n} \left(\frac{\partial x_i}{h_m \, \partial u_m} \right) \boldsymbol{\delta}_i = \sum_i \frac{\partial}{\partial u_n} (L_{im}^{-1}) \boldsymbol{\delta}_i \tag{A8-7-36}$$

where L_{im}^{-1} is the inverse transformation matrix. It equals the transpose of L_{im}.

Transformation of Tensors

Let $\mathbf{G} = \underline{\boldsymbol{\tau}} \cdot \mathbf{v}$ be a vector formed by multiplication of a tensor $\underline{\boldsymbol{\tau}}$ with a vector $\mathbf{v}$. Transform $\mathbf{G}$ to the curvilinear system $G'(u_n, u_m, u_p)$. Then

$$\mathbf{G}' = \underline{\mathbf{L}} \cdot \mathbf{G} = \underline{\mathbf{L}} \cdot \underline{\boldsymbol{\tau}} \cdot \mathbf{v} = \underline{\mathbf{L}} \cdot \underline{\boldsymbol{\tau}} \cdot (\underline{\mathbf{L}}^{-1} \cdot \underline{\mathbf{L}}) \cdot \mathbf{v} \tag{A8-7-37}$$

(since $\underline{\mathbf{L}}^{-1} \cdot \underline{\mathbf{L}} = \underline{\boldsymbol{\delta}}$ can be placed in any position). Now the vector in the new system is

$$\underline{\mathbf{L}} \cdot \mathbf{v} = \mathbf{v}' \tag{A8-7-38}$$

and

$$\mathbf{G}' = \underline{\boldsymbol{\tau}}' \cdot \mathbf{v}' \tag{A8-7-39}$$

must hold for the new system. Hence

$$\underline{\boldsymbol{\tau}}' = \underline{\mathbf{L}} \cdot \underline{\boldsymbol{\tau}} \cdot \underline{\mathbf{L}}^{-1} \qquad \text{or} \qquad \tau_{mn} = \sum_i \sum_j L_{mi} \tau_{ij} L_{nj}^{-1} \tag{A8-7-40}$$

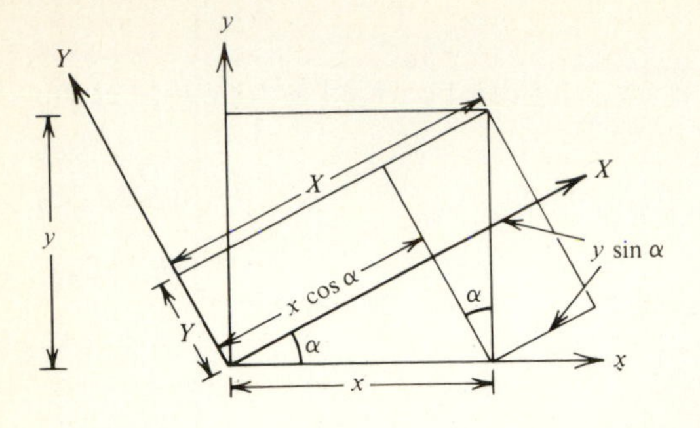

Figure A8-7-2 Transformation of coordinates upon rotation of axes by angle α.

We have indicated previously that Eq. (A8-7-19) is a mathematical *definition* of a vector; i.e., a vector is a quantity that obeys the transformation law given by Eq. (A8-7-19). By the same token a *second-order tensor* is defined as a quantity that obeys the transformation law Eq. (A8-7-40) or

$$\tau'_{mn}(u_1, u_2, u_3) = \sum_i \sum_j h_m \frac{\partial u_m}{\partial x_i} \tau_{ij} h_n \frac{\partial u_n}{\partial x_j} \tag{A8-7-41}$$

$$\tau'_{mn} = \sum_i \sum_j L_{mi} \tau_{ij} L^{-1}_{nj} \tag{A8-7-42}$$

or

$$\underline{\tau}' = \underline{\mathbf{L}} \cdot \underline{\tau} \cdot \underline{\mathbf{L}}^{-1} \tag{A8-7-43}$$

where $\underline{\mathbf{L}}^{-1}$ is the inverse of $\underline{\mathbf{L}}$.

Transformation under Rotation of Axes

Consider the position vector $\mathbf{x}$ defined in the coordinate system x, y, z as

$$\mathbf{x} = x\boldsymbol{\delta}_x + y\boldsymbol{\delta}_y + z\boldsymbol{\delta}_z \tag{A8-7-44}$$

Now suppose we define a new system in which the coordinate axes are rotated over an angle α (see Fig. A8-7-2) and the new coordinates are X, Y, Z. How are the scalar components under the old system related to those under the new system?

For simplicity we consider the two-dimensional case. We can do this geometrically or analytically. Using the diagram (Fig. A8-7-2), we get

$$X = x\cos\alpha + y\sin\alpha \qquad Y = x(-\sin\alpha) + y\cos\alpha \tag{A8-7-45}$$

or

$$x = X\cos\alpha - Y\sin\alpha \qquad y = X\sin\alpha + Y\cos\alpha \tag{A8-7-46}$$

PROBLEMS

8-1 A ceramic brick at 400 K is cooled by subjecting it to a constant surface temperature of 311 K. The brick is 12 by 8 by 8 in and has a thermal diffusivity of 5.16×10^{-7} m^2/s and a thermal conductivity of 0.642 W/m · K.

(*a*) Using the brick dimensions given, calculate the heat loss in joules per hour from the brick.

(*b*) Calculate the temperature after 5 h at the center ($x = 0, y = 0, z = 0$) and at the location $x = 3$ in, $y = \frac{1}{2}$ in, $z = 1.5$ in. Compare your answers with those obtained from the Gurney-Lurie charts using the product rule.

(*c*) Calculate the rate of heat loss in joules per hour from the brick after 5 h. (Use only one term in the series.)

(*d*) Calculate the average temperature and its rate of change with time.

(*e*) Compare your result in part (*c*) with that in part (*d*) by using a macroscopic balance.

8-2 Prove the product-rule principle for the three-dimensional case (see Appendix 8-1).

8-3 Show whether the product rule holds in cylindrical coordinates, illustrating with a sketch.

8-4 (*a*) By dropping out terms that are zero, simplify the general heat-conduction equations (8-31) and (8-32) so as to apply them to the problem of heat transport in a slab discussed in Sec. 7-1. Compare your results with the equations derived in that section.

(*b*) Repeat for the steady-state case discussed in Chap. 1.

8-5 Consider the drying of a brick-shaped solid that is originally at moisture concentration C_{A0} throughout. At $t = 0$ the entire surface is exposed to air. Assume that the moisture content at the surface is maintained at concentration C_{A1} at all $t > 0$, that a diffusion mechanism is responsible for transport of the water vapor to the surface, and that the final concentration of the solid (at $t = \infty$) is C_{A1}. Using the same coordinate system as in heat transport from a brick-shaped object, show that this problem is analogous by carrying out the following:

(*a*) Sketch curves of concentration C_A versus x at constant y and z for various times.

(*b*) Write Fick's-law expressions for J_{Ax}, J_{Ay}, and J_{Az}, assuming constant ρ.

(*c*) Combine the above to obtain a Fick's-law expression in vector form, defining the vector operator ∇ in cartesian coordinates.

(*d*) Neglecting bulk flow, use an incremental mass balance to derive a three-dimensional equation of change for species A in terms of the flux components. Express the result in vector notation.

(*e*) Use the flux law to obtain a three-dimensional equation for C_A; then express the result in vector notation, defining the laplacian operator.

(*f*) Compare your results with those of the heat-transport case. State which equations are analogous. Why are the vector operators the same in each case? Why haven't we used any boundary conditions?

(*g*) Define a dimensionless concentration and write the boundary and initial conditions. Then, by analogy to Sec. 8-3, write the solution.

(*h*) Write an equation for $\mathcal{W}_A$, the total molar transport of A.

(*i*) Write an equation for the rate of change of mean concentration $\langle C_A \rangle$ with time.

(*j*) Show that $\mathcal{W}_A$ and $d\langle C_A \rangle/dt$ can be related by a macroscopic balance.

8-6 Apply the Gauss divergence theorem to Prob. 8-5 by starting with the answer in part (*h*) and using it and the result in part (*d*) to relate the net molar rate of transport of A to the change in average concentration with respect to time.

8-7 A circular cylinder, e.g., a can of your favorite beverage, has a radius R and length L, where $L/R \neq \infty$. It is initially at uniform temperature T_0 and is to be cooled by subjecting it to a uniform surface temperature T_1. We wish to find the temperature distribution in the can, its average temperature, and the rate of heat loss from the can. The external resistance and the resistance of the metal wall are neglected. Let the z axis be taken as the axis of the can, with $z = -L/2$ at the base and $z = +L/2$ at the top. Neglect free convection.

(*a*) Assuming that T_1 is constant, derive a partial differential equation for the temperature.

(*b*) Letting $\phi \equiv (T - T_1)/(T_0 - T_1)$ be the dimensionless temperature, determine whether the product rule holds by writing the partial differential equation and the boundary and initial conditions in terms of ϕ. If it holds, express ϕ in terms of one-dimensional solutions for an infinite cylinder and a slab.

(*c*) If $R = 64$ mm, $L = 116$ mm, $T_0 = 300$ K, $T_1 = 275$ K, and α is that of water, describe how you would use the Gurnie-Lurie charts to find the temperature at the center, the average temperature, and the rate of heat loss as a function of time (for large times).

(*d*) The solution of the differential equation for a cylinder of radius R and no axial conduction (infinite length) is

$$\phi \equiv \frac{T - T_1}{T_0 - T_1} = \sum_{n=1}^{\infty} A_n J_0 \frac{\gamma_n r}{R} \exp\left(-\gamma_n^2 \frac{\alpha t}{R^2}\right)$$

where $A_n = 2/\gamma_n J_1(\gamma_n)$, γ_n is the nth root of $J_0(\gamma_n) = 0$, and J_0 and J_1 are, respectively, the Bessel functions of order 0 and 1. The solution of the differential equation for heat conduction in a slab of thickness $2a$ is

$$\phi \equiv \frac{T - T_1}{T_0 - T_1} = \sum_{n=1}^{\infty} A_n \cos \lambda_n x \exp(-\lambda_n^2 \alpha t)$$

where $\lambda_n = (n + \frac{1}{2})\pi/a$ and $A_n = 2(-1)^n/(n + \frac{1}{2}\pi)$. Use the above to write the solution for the finite cylinder.

(*e*) If $\gamma_1 = 2.40$, $J_1(\gamma_1) = 0.511$, $J_0(0) = 1.0$, and $dJ_2(\gamma_1 r)/dr = \gamma_1 J_1(\gamma r)$, use the above solution to work part (*c*) assuming large times.

8-8 A long solid cylinder is of hexagonal cross section with each side of length B. It is initially at 21°C throughout. At $t = 0$ the solid is placed in a medium which is at a temperature of 3°C. Measurement after 20 min shows that the center temperature cools to 5°C. Another cylinder of the same material but with each side of length $B/2$ is initially at the same temperature and is placed in a medium at the same temperature. How long will it take for the second cylinder to reach the same center temperature? Assume that surface resistance is negligible and that no analytical solution is available.

8-9 A cylindrical solid of length L_D is of hexagonal cross section with each side of length B_D. It is to be dried by exposure of its surfaces at $t > 0$ to a constant concentration C_{A1}, where C_A is the concentration of water vapor. At $t = 0$ the concentration of water vapor in the pellet is C_{A0}. It is suggested that one can calculate the average concentration in the pellet as a function of time if experiments can be carried out on heat transfer in a hexagonal cylinder of side B_H and length L_H in which the total heat loss $\dot{Q}_{Lt}$ is measured as a function of time. Assume that you do *not* have an analytical solution, that external or surface resistance is negligible in each case, and that Fick's law applies. If the experiment can be done, state whether $\langle C_A \rangle$ can be calculated as a function of time from $\dot{Q}_{Lt}$ as a function of time. If it cannot, state why not. If it can, state why and how.

8-10 A coliseum is shaped in the form of an enclosed hemispherical dome. The surface temperature of the dome is $T_s(\theta, \phi)$. Radiation from the sun means that the temperature of the surface T_s will not generally be uniform. The floor temperature inside the dome is $T_F(r, \phi)$. As an illustration of problems in spherical coordinates, show how one can obtain the rate of heat gain and the change of average temperature in the dome with time if the air-conditioning equipment is turned off. As a first approximation neglect free or forced convection.

(*a*) Write equations showing how $\dot{Q}_t(t)$, the net rate of heat transport into the dome, can be found from a knowledge of the temperature distribution in the dome $T(r, \theta, \phi, t)$, giving all integration limits or points of evaluation. Assume that near the ground $T > T_F$; near the roof $T < T_s$.

(*b*) Relate $\dot{Q}_t$ to the average internal temperature $\langle T \rangle$ by means of a macroscopic balance, defining $\langle T \rangle$.

(*c*) For simplicity assume that the surface temperature T_s is constant (cloudy day) and equal to T_1 and that there is no heat transport at the floor to the ground (perfect insulation). Let T_0 be the temperature at time $t = 0$ within the dome (assume uniform) when the air conditioning is turned off. Assuming an incompressible fluid, obtain a partial differential equation for temperature and write it in dimensionless form.

(*d*) Obtain the solution to the above problem. *Note:* Let $[(T - T_1)/(T_0 + T_1)](r/R) \equiv \phi$ to transform the partial differential equation into a familiar form (see Appendix 7-2).

(*e*) Devise a mass-transfer analogy to part (*c*).

8-11 (*a*) Derive the equation of continuity for a species A in spherical coordinates in terms of molar flux with respect to mass-average velocity.

(*b*) For constant ρ write Fick's law of diffusion for each spherical coordinate.

(*c*) Using the results in part (*b*), express part (*a*) in terms of C_A.

8-12 Find $\nabla \cdot \mathbf{q}$ and $\nabla^2 T$ in spherical coordinates (see Examples 8-2 and 8-4).

8-13 Find $\nabla^2 T$ in cylindrical coordinates by differentiation of the unit vectors using the method described in Appendix 8-3.

8-14 Using the method of Appendix 8-7, find $\nabla \cdot \mathbf{q}$ and $\nabla^2 T$ for a prolate spheroidal coordinate system

$$x = a \sinh \eta \sin \theta \cos \psi \qquad y = a \sinh \eta \sin \theta \sin \psi \qquad z = a \cosh \eta \cos \theta$$

8-15 For constant properties derive the energy equation in lagrangian form in terms of T, identifying all terms.

8-16 (*a*) If $\widehat{G}$ is any property per unit mass, show that

$$\frac{\partial(\widehat{G}\rho)}{\partial t} + \nabla \cdot \rho \widehat{G}\mathbf{v} \equiv \rho \frac{D\widehat{G}}{Dt}$$

Use the definition of D/Dt and the equation of continuity.

(*b*) Use the result in part (*a*) to convert both the energy equation and the equation of change for species A into lagrangian form.

8-17 Prove that if $\mathbf{n} \cdot \mathbf{q} = q_n$, a scalar, $\mathbf{q}$ is a vector.

8-18 Carry out the analytical solution to obtain the equation for unsteady-state heat transport in an infinite cylinder as given in Prob. 8-12.

8-19 Consider unsteady-state heat transport in a solid with an external heat-transfer coefficient h. Obtain analytical solutions for (*a*) a slab and (*b*) a cylinder.

8-20 Show that $\mathbf{n} \cdot \mathbf{j}_{An} = J_{An}$, where j_{An} is the mass scalar.

8-21 Derive Eq. (8-145) from Eq. (8-118).

REFERENCES

1. Feynman, R. P., R. B. Leighton, and M. Sands: "The Lectures on Physics," vols. I and II, Addison-Wesley, Reading, Mass., 1965.
2. Scriven, L. E., Lecture Notes on Essentials of Fluid Dynamics, Chemical Engineering Department, University of Minnesota, 1966.
3. Stokes, V. K., and D. Ramkrishna: *Chem. Eng. Educ.*, **16**:82 (1982).

CHAPTER

NINE

MULTIDIMENSIONAL MOMENTUM TRANSPORT

9-1 OBJECTIVE

In Chap. 1 we indicated that the momentum conservation equations and flux laws derived there were valid only for flow and momentum transport in one direction. We also usually assumed the fluid to be incompressible. In this chapter we treat the more general case of three-dimensional flow and transport, and we no longer make the assumption that the fluid is incompressible.

Our immediate objective is to derive a *general* partial differential equation for the momentum flux. This procedure differs from that followed in Part One in that previously we derived a differential equation for various *specific* cases, e.g., Couette flow and flow in a pipe or annulus. Instead we derive a differential equation for a very general case and use it for any application or set of boundary conditions that may occur in practice. All we need do is to drop those terms which are not important in the application involved; then we will have the desired differential equation without having to go through the steps of making an incremental balance. The general equation we will derive is called the *equation of motion* (EOM).

For generality we will consider a fluid in unsteady-state laminar three-dimensional flow that is subject to pressure forces and gravity. First we will derive the

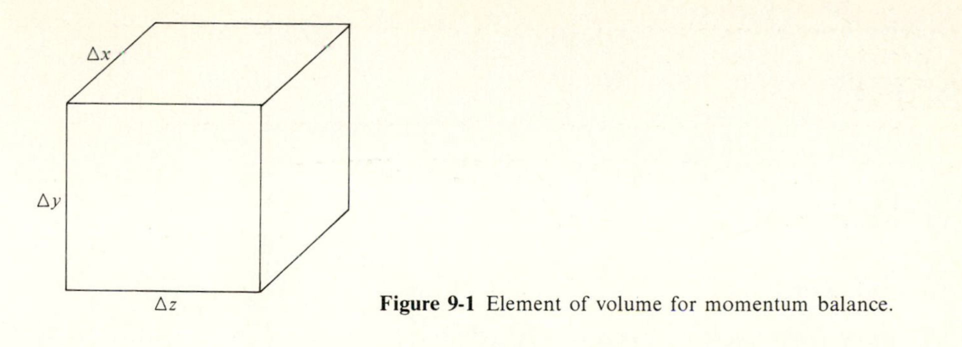

Figure 9-1 Element of volume for momentum balance.

equation of motion in the x direction, express it in three-dimensional vector form, and show how to write it in any orthogonal coordinate system. In doing so we will illustrate certain physical concepts that are involved in using cylindrical or other curvilinear coordinates. Finally, we will show how to use the equations in specific cases.

We will make our momentum balance over an incremental volume element $\Delta \mathcal{V}$ (see Fig. 9-1) that is fixed in space (eulerian point of view). At an arbitrary x, y, z, the balance law for momentum is applicable and can be stated as

$$\begin{matrix}\text{Rate of}\\ \text{accumulation}\\ \text{of momentum}\end{matrix} = \begin{matrix}g_c \times \text{pressure}\\ \text{and gravity}\\ \text{forces acting}\\ \text{on element}\end{matrix} + \begin{matrix}\text{Net diffusive}\\ \text{flow of momentum}\\ \text{into element}\end{matrix} + \begin{matrix}\text{Net bulk flow}\\ \text{of momentum into}\\ \text{the element}\end{matrix} \tag{9-1}$$

9-2 FORCES ACTING ON ELEMENT OF VOLUME†

Let us now consider the forces acting on the element of volume $\Delta \mathcal{V}$. As indicated in Chap. 3, these forces can be classified into *body* forces $\Delta \mathbf{F}_g$ (such as gravity) and *surface* forces (such as pressure $\Delta \mathbf{F}_P$ and viscous stress $\Delta \mathbf{F}_v$). Thus

$$\Delta \mathbf{F} = \Delta \mathbf{F}_g + \Delta \mathbf{F}_P + \Delta \mathbf{F}_v$$

The student may have already learned in physics courses that the surface forces acting on a body can be described as shear forces, which act tangentially, and compressive and tensile forces, both of which act normally. Similarly, the viscous forces that act on an element of fluid may likewise be classified as shear forces and normal forces. As in the case of a solid body, it is convenient to express these forces as stresses, i.e., as forces per unit area. The stresses include the shear stresses, which act tangentially to the specified area, and the normal stresses, which act perpendicular to the specified area.

† On first reading, the detailed discussion in Sec. 9-2 on individual terms may be skipped and the reader may proceed directly to Sec. 9-3 and Eq. (9-26), for which a derivation not involving Gibbs vector-tensor notation is given (Table 9-1).

Gravity Force

The gravity force in the x direction is

$$\Delta F_{gx} = \frac{\rho g_x}{g_c} \Delta \mathcal{V} \qquad \text{or} \qquad \Delta F_{gi} = \frac{\rho g_i \, \Delta \mathcal{V}}{g_c} \tag{9-2}$$

in the x_i direction.

Pressure Force

Whereas viscous forces occur only when a fluid is in motion, a pressure force exists also when the fluid is stationary. We have previously indicated that a pressure force can arise due to the bombardment of molecules against a wall, resulting in an exchange of their momentum equal to the force required to keep the wall stationary. A pressure force also arises on an arbitrary plane within the fluid at $x = x$ due to the exchange of momentum resulting from the collision of the molecules on one side (region A) of the plane with those on the opposite side (region B) (see Fig. 9-2). Then

$$g_c \mathbf{F}_{px}^A \Big|_x = \dot{\mathbf{P}}_{px}^A \Big|_x = -\dot{\mathbf{P}}_{px}^B \Big|_x = -g_c \mathbf{F}_{px}^B \Big|_x$$

as could be seen by first taking region A, with outward normal $\mathbf{n}^A$, as the interior and then taking region B, with outward normal $\mathbf{n}^B$, as the interior. If the positive axis goes from left to right,

$$\mathbf{n}^A = +\boldsymbol{\delta}_x \qquad \text{and} \qquad \mathbf{n}^B = -\boldsymbol{\delta}_x \qquad \text{or} \qquad \mathbf{n}^A = -\mathbf{n}^B$$

If the pressure force $\mathbf{F}_{px}^A$ is acting in the $+x$ direction, it is acting in a direction opposite to $\mathbf{n}^A$ and we can write

$$\mathbf{F}_{px}^A = -\mathbf{n}^A p A_x \qquad \text{and} \qquad \mathbf{F}_{px}^B = -\mathbf{n}^B p A_x$$

assuming p does not vary over A_x.

Suppose we change the orientation of the area by rotating it so that it makes an angle θ with the x axis. Since neither the random molecular motion, i.e., average molecular speed, nor the fluid density changes as θ changes, the pressure will

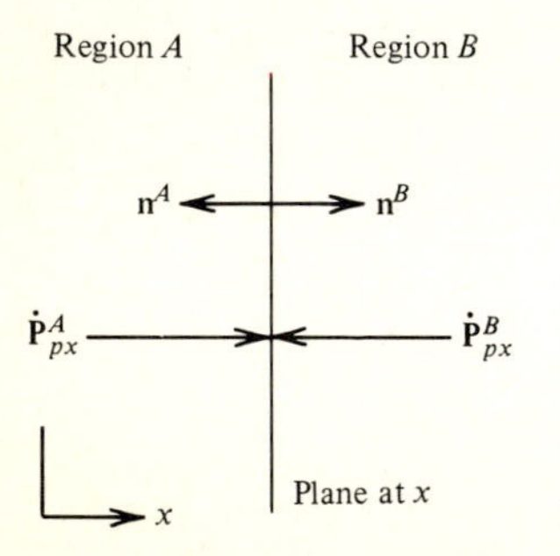

Figure 9-2 Pressure forces at a plane $x = x$.

not change and the same will be true for any orientation with respect to the y or z axis. Thus pressure is a scalar and acts isotropically on an element. The pressure force therefore would always be in the direction opposite **n** or

$$\mathbf{F}_p = -\mathbf{n}pA \tag{9-3}$$

if p does not vary over the area. Otherwise

$$d\mathbf{F}_p = -\mathbf{n}p\,dA = -p\,d\mathbf{S} \tag{9-4}$$

For the x direction the force due to pressure on area ΔA_x is

$$\Delta F_{px} = -(n_x p\,\Delta A_x)\Big|_x - (n_x p\,\Delta A_x)\Big|_{x+\Delta x} = \left(p\Big|_x - p\Big|_{x+\Delta x}\right)\Delta A_x = -\Delta_x p\,\Delta A_x$$

or, in index notation,

$$\Delta F_{pi} = -\Delta_i p\,\Delta A_i \tag{9-5}$$

Viscous Forces

In Chap. 1 we considered a Couette flow in which a fluid between two horizontal parallel plates each of area A_x was maintained in motion by pulling the lower plate with a force F_y (or with shear stress $\sigma_{xy} = F_y/A_x$) (see Fig. 9-3). This model was used as a basis for a sign convention and resulted in a positive shear stress σ_{xy} at $x = 0$, a negative velocity gradient, and a positive momentum flux τ_{xy} in accordance with

$$\sigma_{xy} g_c = \tau_{xy} = -\mu \frac{\partial v_y}{\partial x} \tag{1-48}$$

Since σ_{xy} did not vary over A_x, we found that the viscous force acting *on* the face of an element of volume $A_x \Delta x$ at either x or $x + \Delta x$ is given by Eq. (1-74)

$$F_{vy} = -n_x A_x \sigma_{xy}$$

where n_x, the outwardly directed unit normal to A_x, has the values (Secs. 1-7, 1-18, and 1-19)

$$n_x = \begin{cases} -1 & \text{at } x = x \\ +1 & \text{at } x = x + \Delta x \end{cases}$$

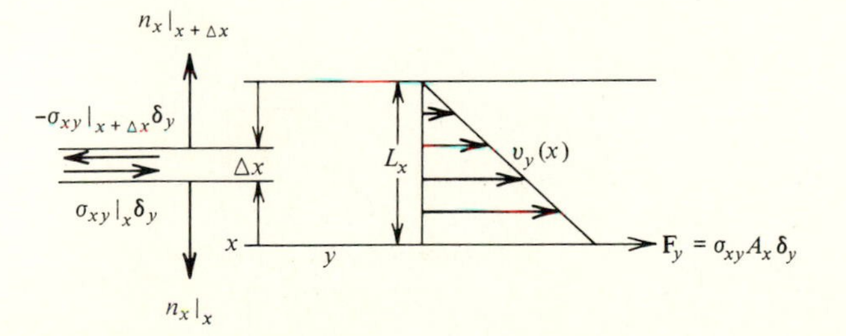

Figure 9-3 Incremental element for Couette flow.

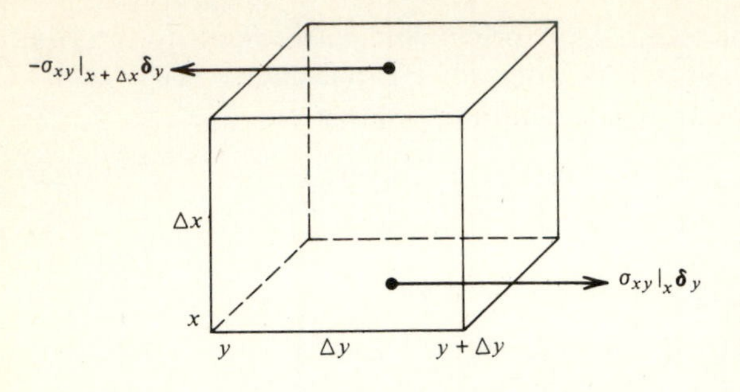

Figure 9-4 Shear stress acting on x faces of element.

We will extend the same convention to three-dimensional flow. For an x face of the element of volume $\Delta x\,\Delta y\,\Delta z$, the viscous force acting in the y direction on an area ΔA_x is then

$$\Delta\mathbf{F}_{vy}^{(x)} \equiv \Delta F_{vy}^{(x)}\boldsymbol{\delta}_y = -n_x\,\Delta A_x\,\sigma_{xy}\boldsymbol{\delta}_y \tag{9-6}$$

where (x) is used to indicate an x face. This gives

$$-n_x\sigma_{xy}\boldsymbol{\delta}_y = \frac{\Delta\mathbf{F}_{vy}^{(x)}}{\Delta A_x} \tag{9-7}$$

In Fig. 9-4 $\sigma_{xy}\boldsymbol{\delta}_y$ at x and $-\sigma_{xy}\boldsymbol{\delta}_y$ at $x+\Delta x$ are sketched showing their opposite directions. Note that the sign of the scalar component σ_{xy} does not itself change over Δx, as do the directions of $\Delta\mathbf{F}_{vy}^{(x)}$ and the sign of n_x. Similar equations can be written for σ_{xz} and for shear stresses on the other faces.

Normal Viscous Force

The normal forces acting on an x face are given by an equation of the same form as Eq. (9-6) but with x replacing y

$$\Delta\mathbf{F}_{vx}^{(x)} \equiv \Delta F_{vx}^{(x)}\,\boldsymbol{\delta}_x = -n_x\sigma_{xx}\,\Delta A_x\,\boldsymbol{\delta}_x \tag{9-8}$$

As shown in Fig. 9-5, if σ_{xx} is positive, $-n_x\sigma_{xx}\,\boldsymbol{\delta}_x$ is acting in the $+x$ direction at x and in the $-x$ direction at $x+\Delta x$. Thus compressive forces are considered positive and tensile forces are considered negative. Although an opposite convention is used in most fluid mechanics texts, the present convention allows analogous flux-law equations to be written for one-dimensional heat, momentum, and

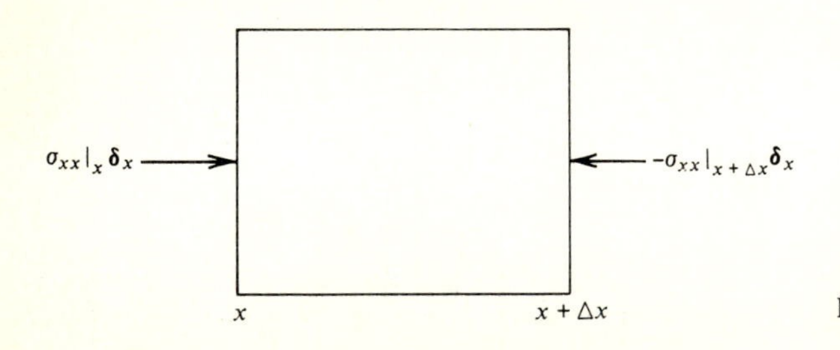

Figure 9-5 Normal stresses.

mass transport. We will later make the transition to the fluid mechanics convention.

The total normal stress on an area ΔA_x would then be

$$\Delta \mathbf{F}_x^{(x)} = \Delta \mathbf{F}_{vx}^{(x)} + \Delta \mathbf{F}_{px}^{(x)} = -n_x \boldsymbol{\delta}_x \, \Delta A_x \, \sigma_{xx} - n_x \boldsymbol{\delta}_x p \, \Delta A_x$$
$$= -\mathbf{n}_x \, \Delta A x (\sigma_{xx} + p) \tag{9-9}$$

Viscous-Stress Vector

In general, on an area ΔA_x at $x = x$ there will be components of the viscous force in the x, y, and z directions. The resultant viscous-force vector then can be expressed in terms of these components by means of the general equation (8-9)

$$\Delta \mathbf{F}_v^{(x)} = \Delta \mathbf{F}_{vx}^{(x)} + \Delta \mathbf{F}_{vy}^{(x)} + \Delta \mathbf{F}_{vz}^{(x)} = \Delta F_{vx}^{(x)} \boldsymbol{\delta}_x + \Delta F_{vy}^{(x)} \boldsymbol{\delta}_y + \Delta F_{vz}^{(x)} \boldsymbol{\delta}_z \tag{9-10}$$

If we divide by ΔA_x and substitute equations of the form of (9-7) and (9-8) (with $n_x|_x = -1$), we get

$$\frac{\Delta \mathbf{F}_v^{(x)}}{\Delta A_x} = \sigma_{xx} \boldsymbol{\delta}_x + \sigma_{xy} \boldsymbol{\delta}_y + \sigma_{xz} \boldsymbol{\delta}_z \equiv \boldsymbol{\sigma}^{(x)} \tag{9-11}$$

where $\boldsymbol{\sigma}^{(x)}$ = *resultant viscous-stress vector*
σ_{xx} = scalar component of normal stress
σ_{xy}, σ_{yz} = scalar components of shear stress

Note that $\boldsymbol{\sigma}^{(x)}$ is *not* normal to the x face but is a resultant vector with components $\sigma_{xx}\boldsymbol{\delta}_x$, $\sigma_{xy}\boldsymbol{\delta}_y$, and $\sigma_{xz}\boldsymbol{\delta}_z$. These vector components are illustrated in Fig. 9-6 for the $x = x$ face.

Stress Tensor

If we considered the faces perpendicular to the y and z axes, we could write expressions similar to Eq. (9-11) for $\boldsymbol{\sigma}^{(y)}$ and $\boldsymbol{\sigma}^{(z)}$. We can therefore see intuitively that to describe the stress we need to specify not only magnitude but also *two* directions, the first indicating the orientation of the area under consideration and

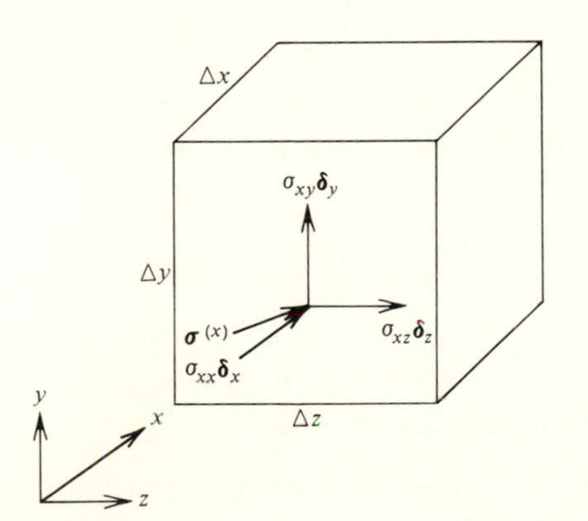

Figure 9-6 Components of resultant vector $\boldsymbol{\sigma}^{(x)}$ on x face of area ΔA_x; $\boldsymbol{\delta}_x$, $\boldsymbol{\delta}_y$, and $\boldsymbol{\delta}_z$ are unit vectors.

the second indicating the direction of the component. Such a property is characteristic of a second-order tensor or, in common parlance, a tensor. Strictly speaking, a vector is a first-order tensor because one direction is specified and a scalar is a zero-order tensor.

Mathematically a second-order tensor can be defined as a quantity that obeys a certain transformation law [Eq. (A8-7-40)] or as a quantity which gives another vector when it is dot-multiplied by a vector. Thus if $\mathbf{a} \cdot \underline{\mathbf{B}} = \mathbf{c}$ (Appendix 8-7), where $\mathbf{a}$ and $\mathbf{c}$ are vectors, then $\underline{\mathbf{B}}$ is a second-order tensor. We will show later that $\underline{\boldsymbol{\sigma}}$ is a tensor. Note that boldface with an underscore is used for a tensor.

Just as a vector can be expanded into its scalar components and unit vectors, so can a tensor of second order be expanded into its vector components and unit vectors. Thus the viscous-stress tensor is

$$\underline{\boldsymbol{\sigma}} = \boldsymbol{\delta}_x \boldsymbol{\sigma}^{(x)} + \boldsymbol{\delta}_y \boldsymbol{\sigma}^{(y)} + \boldsymbol{\delta}_z \boldsymbol{\sigma}^{(z)} \tag{9-12}$$

Then a second-order tensor can be expanded into scalar components σ_{ij} and *unit dyads* $\boldsymbol{\delta}_i \boldsymbol{\delta}_j$, which consist of two unit vectors written in juxtaposition with no dot or cross between them. This combination is called a *dyadic product*. Substituting Eq. (9-7) and similar equations for $\boldsymbol{\sigma}^{(y)}$ and for $\boldsymbol{\sigma}^{(z)}$ gives

$$\begin{aligned}
\underline{\boldsymbol{\sigma}} = {} & \sigma_{xx}\boldsymbol{\delta}_x\boldsymbol{\delta}_x + \sigma_{xy}\boldsymbol{\delta}_x\boldsymbol{\delta}_y + \sigma_{xz}\boldsymbol{\delta}_x\boldsymbol{\delta}_z \\
& \qquad\qquad (x \text{ face}) \\
& + \sigma_{yx}\boldsymbol{\delta}_y\boldsymbol{\delta}_x + \sigma_{yy}\boldsymbol{\delta}_y\boldsymbol{\delta}_y + \sigma_{yz}\boldsymbol{\delta}_y\boldsymbol{\delta}_z \\
& \qquad\qquad (y \text{ face}) \\
& + \sigma_{zx}\boldsymbol{\delta}_z\boldsymbol{\delta}_x + \sigma_{zy}\boldsymbol{\delta}_z\boldsymbol{\delta}_y + \sigma_{zz}\boldsymbol{\delta}_z\boldsymbol{\delta}_z \\
& \qquad\qquad (z \text{ face})
\end{aligned} \tag{9-13}$$

We can write this compactly as

$$\underline{\boldsymbol{\sigma}} = \sum_{i=1} \sum_{j=1} \sigma_{ij} \boldsymbol{\delta}_i \boldsymbol{\delta}_j \tag{9-14}$$

It is convenient to write $\underline{\boldsymbol{\sigma}}$ as a 3×3 matrix.

$$\underline{\boldsymbol{\sigma}} = \begin{bmatrix} \sigma_{xx} & \sigma_{xy} & \sigma_{xz} \\ \sigma_{yx} & \sigma_{yy} & \sigma_{yz} \\ \sigma_{zx} & \sigma_{zy} & \sigma_{zz} \end{bmatrix} = \begin{bmatrix} \sigma_{11} & \sigma_{12} & \sigma_{13} \\ \sigma_{21} & \sigma_{22} & \sigma_{23} \\ \sigma_{31} & \sigma_{32} & \sigma_{33} \end{bmatrix} \tag{9-15}$$

Note that the σ_{ij} are scalar components not only of the tensor $\underline{\boldsymbol{\sigma}}$ but also of the vector $\boldsymbol{\sigma}^{(i)}$.

Momentum-Flux Tensor

In Chap. 1 we indicated that a shear stress, say σ_{xy}, is not only a force per unit area but also corresponds to a flux of y momentum in the x direction. If τ_{xy} is the momentum flux, then in accordance with Newton's law of motion,

$$\tau_{xy} = g_c \sigma_{xy} \tag{9-16}$$

where g_c, the gravitational constant, is a conversion factor between force and momentum rate units. For an absolute system of units the unit of force is defined so that $g_c = 1.0$ and is dimensionless; but for the English gravitational system $g_c = 32.2\ (\text{lbm} \cdot \text{ft/s}^2)/\text{lbf}$. Since for three-dimensional flow *each* component of $\underline{\boldsymbol{\sigma}}$ and $\underline{\boldsymbol{\tau}}$ will follow the relationship above, the *momentum-flux tensor* can be obtained from Eq. (9-14) as

$$\underline{\boldsymbol{\tau}} = g_c \underline{\boldsymbol{\sigma}} = \sum_{i=1}^{3} \sum_{j=1}^{3} \tau_{ij} \boldsymbol{\delta}_i \boldsymbol{\delta}_j \tag{9-17}$$

where τ_{ij} is the ijth component of the momentum-flux tensor.

As stated earlier, our goal is to write a momentum rate balance over the element $\Delta\mathcal{V}$. We will do so by first writing an x-momentum rate balance, which we will add vectorially to the y- and z-momentum balances. In order to obtain an x-momentum balance we must add the flows of x momentum through each of the six faces. As indicated previously, an x-momentum rate balance is equivalent to a balance of forces acting in the x direction. Let us now try to picture these forces and momentum flows.

In Fig. 9-7 we show the viscous forces per unit area acting in the x direction on the various faces; e.g., on the y faces we show†

$$\frac{\Delta \mathbf{F}_{vx}^{(y)}}{\Delta A_y} = -n_y \sigma_{yx} \boldsymbol{\delta}_y = -n_y \boldsymbol{\sigma}_{yx} \tag{9-18}$$

Thus at y, where $n_y = -1$, we show $\sigma_{yx}\boldsymbol{\delta}_x$, whereas at $y + \Delta y$, where $n_y = +1$, we show $-\sigma_{yx}\boldsymbol{\delta}_x$. For each of the viscous forces acting in the x direction in Fig. 9-7

† For simplicity, we do not show unit vector $\boldsymbol{\delta}_y$ in Fig. 9-7.

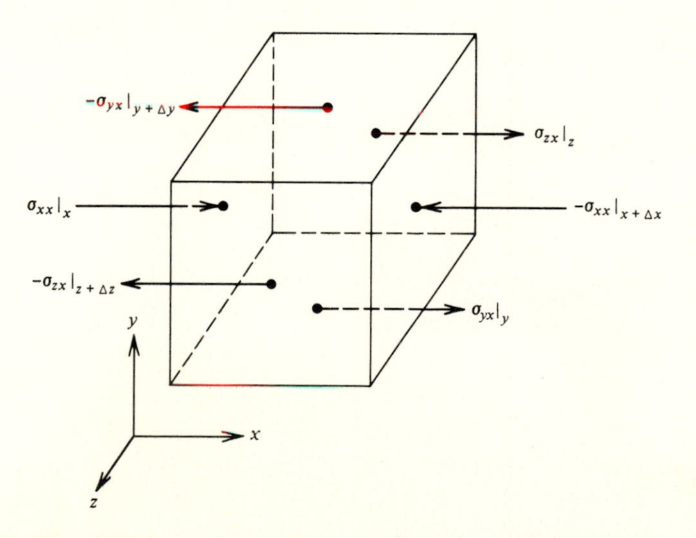

Figure 9-7 Viscous stresses in the $\pm x$ direction on volume element.

there is a corresponding rate of transport of x momentum; e.g., for the y faces it is

$$\Delta \dot{P}_{vy}^{(y)} = \Delta F_{vx}^{(y)} g_c = -n_y \, \Delta A_y \, \tau_{yx} \tag{9-19}$$

For a similar geometry and coordinate system, the analogous equation for the rate of gain of heat by an element due to a heat flux q_y through an area ΔA_y would be (Secs. 1-7 and 8-2)

$$\Delta \dot{Q}_G^{(y)} = -n_y \, \Delta A_y \, q_y \tag{9-20}$$

For this case we would sketch arrows pointing in the y direction depicting the vector $\mathbf{q}_y = q_y \boldsymbol{\delta}_y$ at y and at $y + \Delta y$. Similarly, for x-momentum transport in the y direction we sketch arrows pointing in the y direction showing the vector $\tau_{yx}\boldsymbol{\delta}_y = \boldsymbol{\tau}_{yx}$ at y and at $y + \Delta y$ (Fig. 9-8).† These vectors correspond to the faces shown in Fig. 9-7. These two figures illustrate that for a second-order tensor two directions are involved, only one of which is emphasized in each figure.

We follow the convention adopted in Chap. 1 that the momentum flux τ_{xy} is positive if the flow is in the $+x$ direction. The flow is then *into* the element at x and *out* of the element at $x + \Delta x$. Remember that the arrows in Fig. 9-8 indicate the direction of *transport* whereas those in Fig. 9-7 show the direction of *forces* acting *on* the system. Note that since n_x changes sign, the arrows on the forces and flows have the same direction as x but opposite directions at $x + \Delta x$.

† For simplicity, we do not show the unit vectors in Fig. 9-8; however these are in the direction of the arrows.

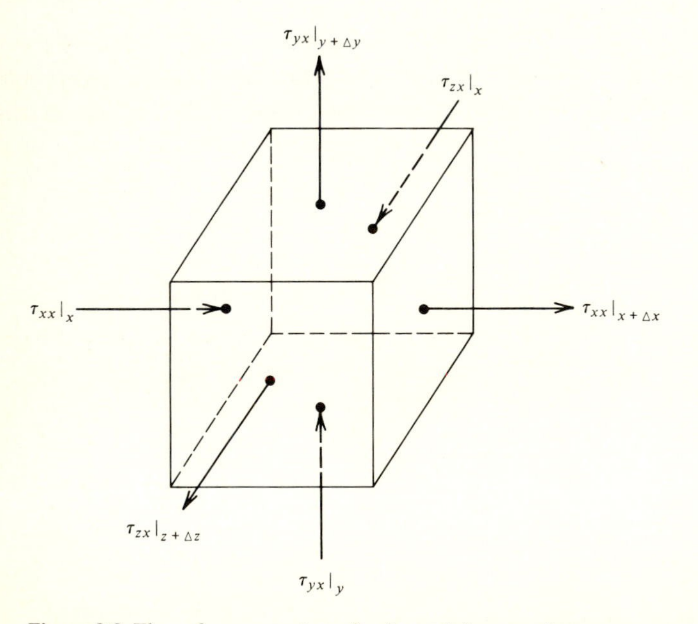

Figure 9-8 Flux of x momentum. $\boldsymbol{\delta}_x$, $\boldsymbol{\delta}_y$, and $\boldsymbol{\delta}_z$ are unit vectors.

The net force in the x direction is obtained by adding the contribution of all six faces. Thus for the y faces we have

$$\Delta F_{vx}^{(y)} = -(n_y \, \Delta A_y \, \sigma_{yx})\Big|_y - (n_y \, \Delta A_y \, \sigma_{yx})\Big|_{y+\Delta y}$$

$$= (\Delta A_y \, \sigma_{yx})\Big|_y - (\Delta A_y \, \sigma_{yx})\Big|_{y+\Delta y} = -\Delta_y (\Delta A_y \, \sigma_{yx}) \qquad (9\text{-}21)$$

At this point it is convenient to use index notation, replacing y with j and x with i, as follows

$$\Delta F_{vi}^{(j)} = -\Delta_j(\Delta A_j \, \sigma_{ji}) \qquad (9\text{-}22)$$

Summing over all faces gives

$$\Delta F_{vi} = -\sum_j \Delta_j(\Delta A_j \, \sigma_{ji}) \qquad (9\text{-}23)$$

9-3 MOMENTUM BALANCE†

Equation (9-1) for the element $\Delta \mathcal{V}$ can be written for the x direction as

$$\Delta \dot{P}_{ax} = \Delta \dot{P}_{vx} + \Delta \dot{P}_{px} + \Delta \dot{P}_{gx} + \Delta \dot{P}_{bx}$$

or for the i direction as

$$\Delta \dot{P}_{acx} = \Delta \dot{P}_{vi} + \Delta \dot{P}_{pi} + \Delta \dot{P}_{gi} + \Delta \dot{P}_{bi} \qquad (9\text{-}24)$$

where $\Delta \dot{P}_{acx}$, the acceleration or accumulation term, is obtained as in Chap. 7 by multiplying the x_i momentum per unit volume by the volume of the increment and taking the difference over an increment of time Δt (see Table 9-1).

As shown in Fig. 9-8, for a face ΔA_y the flow of x momentum into the element at y is $(\tau_{yx} \, \Delta A_y)|_y$ and the flow out at $y + \Delta y$ is $(\tau_{yx} \, \Delta A_y)|_{y+\Delta y}$. The net inward flow is then

$$\Delta \dot{P}_{vx}^{(y)} = -\Delta_y(\tau_{yx} \, \Delta A_y) \equiv (\tau_{yx} \, \Delta A_y)\Big|_y - (\tau_{yx} \, \Delta A_y)\Big|_{y+\Delta y}$$

with similar terms for other faces (see Table 9-1). In index notation, for the j faces

$$\Delta \dot{P}_{vi}^{(j)} = -\Delta_j(\Delta A_j \, \tau_{ji})$$

The total for the i direction is the sum over all faces or

$$\Delta \dot{P}_{vi} = -\sum_j \Delta_j(\Delta A_j \, \tau_{ji})$$

†Does not require knowledge of vector notation.

Table 9-1 *x***-momentum balance, eulerian view (see Figs. 9-1 and 9-8)**

	Net diffusive-transport rate	+	Net bulk-transport rate	+	Generation rate due to: Pressure	+	Gravity	=	Rate of accumulation
x faces:	$-\Delta_x(\tau_{xx}\,\Delta A_x)$		$-\Delta_x[(v_x\,\Delta A_x)\rho v_x]$		$-\Delta_x(p\,\Delta A_x)g_c$	+	$\rho g_x\,\Delta x\,\Delta y\,\Delta z$	=	$\dfrac{\Delta_t(\rho v_x\,\Delta x\,\Delta y\,\Delta z)}{\Delta t}$
y faces:	$-\Delta_y(\tau_{yx}\,\Delta A_y)$		$-\Delta_y[(v_y\,\Delta A_y)\rho v_x]$						
z faces:	$-\Delta_z(\tau_{zx}\,\Delta A_z)$		$-\Delta_z[(v_z\,\Delta A_z)\rho v_x]$						

Substituting $\Delta A_x = \Delta y\,\Delta z$, etc., dividing by volume $\Delta x\,\Delta y\,\Delta z$, and taking limits gives

$$\lim_{\substack{\Delta x\to 0\\ \Delta y\to 0\\ \Delta z\to 0}} \left\{ \frac{-\Delta(\tau_{xx}\,\Delta y\,\Delta z)}{\Delta x\,\Delta y\,\Delta z} - \frac{\Delta_x(v_x\,\Delta y\,\Delta z\,\rho v_x)}{\Delta x\,\Delta y\,\Delta z} - \frac{\Delta_y(\tau_{yx}\,\Delta x\,\Delta z)}{\Delta x\,\Delta y\,\Delta z} - \frac{\Delta_y(v_y\,\Delta x\,\Delta z\,\rho v_x)}{\Delta x\,\Delta y\,\Delta z} - \frac{\Delta_z(\tau_{zx}\,\Delta x\,\Delta y)}{\Delta x\,\Delta y\,\Delta z} - \frac{\Delta_z(v_z\,\Delta x\,\Delta y\,\rho v_x)}{\Delta x\,\Delta y\,\Delta z} - \frac{\Delta_x(p\,\Delta y\,\Delta z)g_c}{\Delta x\,\Delta y\,\Delta z} \right\} + \rho g_x = \lim_{\Delta t\to 0} \frac{\Delta_t(\rho v_x)}{\Delta t}$$

Cancelling terms and using the definition of the partial derivative gives Eq. (9-26), the x component of the EOM

$$\frac{\partial \tau_{xx}}{\partial x} - \frac{\partial \tau_{yx}}{\partial y} - \frac{\partial \tau_{zx}}{\partial z} - \frac{\partial(\rho v_x v_x)}{\partial x} - \frac{\partial(\rho v_y v_x)}{\partial y} - \frac{\partial(\rho v_z v_x)}{\partial z} - \frac{\partial p}{\partial x}g_c + \rho g_x = \frac{\partial(\rho v_x)}{\partial t}$$

The y and z components can be found by interchanging x and y and x and z, respectively.

The rate of momentum generated by the pressure force on the x face is

$$\Delta \dot{P}_{px} = g_c \left(p \Big|_x - p \Big|_{x+\Delta x} \right) \Delta A_x = -\Delta_x p \, \Delta A_x g_c$$

or, in index notation

$$\Delta \dot{P}_{pi} = -g_c \Delta_i p \, \Delta A_i$$

The momentum rate due to gravity force is

$$\Delta \dot{P}_{gx} = \rho g_x \, \Delta \mathcal{V} \qquad \text{or} \qquad \Delta \dot{P}_{gi} = \rho g_i \, \Delta \mathcal{V}$$

The bulk-flow term $\Delta \dot{P}_{bi}$ includes expressions for the bulk transport of x_i momentum into and out of each face. As we showed in Chaps. 1 and 5, the quantity ρv_x is the "concentration" of x momentum at a point with the units x momentum per unit volume. Since the volumetric rate of flow into a face at $x = x$ is $(v_x \, \Delta A_x)|_x$, the x momentum is being transported by bulk flow through the face at x at the rate

$$\frac{\text{Volume}}{\text{Time}} \times \frac{x \text{ momentum}}{\text{vol}} = (v_x \, \Delta A_x)(\rho v_x) \Big|_x$$

The *net* rate of transport of x momentum by bulk flow in the x direction is then

$$\Delta \dot{P}_{bx}^{(x)} \equiv (v_x \, \Delta A_x \, \rho v_x) \Big|_x - (v_x \, \Delta A_x \, \rho v_x) \Big|_{x+\Delta x} = -\Delta_x (v_x \, \Delta A_x \, \rho v_x)$$

However, since volumetric flow also occurs in the y direction at the rate $v_y \, \Delta A_y$, we also need to include the term $-\Delta_y (v_y \, \Delta A_y \, \rho v_x)$ and a similar term for the z direction. In general then

$$\Delta \dot{P}_{bi} = -\sum_j \Delta_j (v_j \, \Delta A_j \, \rho v_i)$$

Inserting the various terms into Eq. (9-24), dividing by $\Delta \mathcal{V} = \Delta x_j \, \Delta A_j = \Delta x_i \, \Delta A_i$, and taking limits gives

$$\lim_{\Delta t \to 0} \frac{\Delta(\rho v_i \, \Delta \mathcal{V})}{\Delta \mathcal{V} \, \Delta t} = - \lim_{\substack{\Delta x_j \to 0 \\ \Delta \mathcal{V} \to 0}} \sum_j \left[\frac{\Delta_j (\Delta A_j \tau_{ji})}{\Delta x_j \, \Delta A_j} + \frac{\Delta_j (v_j \, \Delta A_j \rho v_i)}{\Delta x_j \, \Delta A_j} \right] + \lim_{\Delta x_i \to 0} \left(\frac{g_c \, \Delta_i p \, \Delta A_i}{\Delta x_i \, \Delta A_i} \right) + \rho g_i$$

Using the definition of the partial derivative gives

$$\frac{\partial(\rho v_i)}{\partial t} = -\sum_j \frac{\partial \tau_{ji}}{\partial x_j} - \sum_j \frac{\partial(\rho v_j v_i)}{\partial x_j} - \frac{\partial p}{\partial x_i} g_c + \rho g_i \tag{9-25}$$

For the x component, this becomes

$$\frac{\partial(\rho v_x)}{\partial t} = -\frac{\partial \tau_{xx}}{\partial x} - \frac{\partial \tau_{yx}}{\partial y} - \frac{\partial \tau_{zx}}{\partial z} - \frac{\partial(\rho v_x v_x)}{\partial x} - \frac{\partial(\rho v_y v_x)}{\partial y} - \frac{\partial(\rho v_z v_x)}{\partial z} - \frac{\partial p}{\partial x} g_c + \rho g_x \tag{9-26}$$

Similar equations can be obtained for the y and z components. Equation (9-25) is in eulerian form. It can be put into lagrangian form, as we did for moles of species A in Sec. 8-22, by using the substantial derivative [Eq. (8-106)], defined as

$$\frac{Dv_i}{Dt} \equiv \frac{\partial v_i}{\partial t} + \sum_j v_j \frac{\partial v_i}{\partial x_j} \tag{9-26a}$$

Combining the accumulation and bulk-flow terms in Eq. (9-25) and differentiating each as products gives the identity

$$\frac{\partial(\rho v_i)}{\partial t} + \sum_j \frac{\partial(\rho v_j v_i)}{\partial x_j} \equiv \rho \frac{\partial v_i}{\partial t} + v_i \frac{\partial \rho}{\partial t} + \sum_j v_i \frac{\partial(\rho v_j)}{\partial x_j} + \sum_j \rho v_j \frac{\partial v_i}{\partial x_j} \tag{9-26b}$$

Since v_i can be removed from inside the Σ_j in the third term on the right, the second and third terms cancel each other according to the EOC [Eq. (8-104b)]. Multiplying Eq. (9-26a) by ρ and combining with Eqs. (9-26b) and (9-25) gives the equation of motion in lagrangian form

$$\rho \frac{Dv_i}{Dt} = -\sum_j \frac{\partial \tau_{ji}}{\partial x_j} - \frac{\partial p}{\partial x_i} g_c + \rho g_i \tag{9-26c}$$

Note that Eq. (9-26c) does not hold for curvilinear coordinates. Expressions for it in various coordinate systems are given in part A of Tables I-6 to I-7. The method of obtaining them requires the use of vector notation, discussed later.

9-4 VECTOR FORM OF EQUATION OF MOTION

The diffusion term in Eq. (9-26c) is

$$-\sum_j \frac{\partial \tau_{ji}}{\partial x_j} \qquad \text{or} \qquad -\frac{\partial \tau_{xx}}{\partial x} - \frac{\partial \tau_{yx}}{\partial y} - \frac{\partial \tau_{zx}}{\partial z}$$

It is similar to the divergence of a vector, which in Eqs. (8-28) and (8-96) was a scalar.

$$\nabla \cdot \mathbf{q} = \sum_j \frac{\partial q_j}{\partial x_j}$$

But, since $\underline{\boldsymbol{\tau}}$ is a tensor, it is now useful to obtain an expression for $\nabla \cdot \underline{\boldsymbol{\tau}}$ from our previous definitions for ∇ and $\underline{\boldsymbol{\tau}}$ [Eqs. (8-13) and (9-17)]. Using them, we have

$$\nabla \cdot \underline{\boldsymbol{\tau}} = \left(\sum_k \boldsymbol{\delta}_k \frac{\partial}{\partial x_k}\right) \cdot \left(\sum_j \sum_i \tau_{ji} \boldsymbol{\delta}_j \boldsymbol{\delta}_i\right) = \sum_i \sum_j \sum_k \frac{\partial \tau_{ji}}{\partial x_k} \boldsymbol{\delta}_k \cdot \boldsymbol{\delta}_j \boldsymbol{\delta}_i \tag{9-27}$$

since the derivatives of the unit vectors in cartesian coordinates are zero. We now must take the dot product of the *adjacent* unit vectors. From Appendix 8-3

$$\boldsymbol{\delta}_k \cdot \boldsymbol{\delta}_j = \delta_{kj}$$

where δ_{kj} is the Kronecker delta. Since this is zero for $k \neq j$, one summation in Eq. (9-27) disappears; i.e., we can select either k or j. However, $\boldsymbol{\delta}_i$ remains, indicating that

$$\nabla \cdot \underline{\tau} = \sum_i \sum_j \frac{\partial \tau_{ji}}{\partial x_j} \boldsymbol{\delta}_i = \sum_i [\![(\nabla \cdot \underline{\tau})_i]\!] \, \boldsymbol{\delta}_i \equiv \sum_i (\nabla \cdot \underline{\tau})_i \tag{9-28}$$

is a vector with its scalar component in the i direction given by†

$$[\![(\nabla \cdot \underline{\tau})_i]\!] \equiv \sum_j \frac{\partial \tau_{ji}}{\partial x_j} \tag{9-29}$$

and vector components given by $(\nabla \cdot \underline{\tau})_i$. By a similar method we can show that the bulk-flow term can be written

$$\sum_j \frac{\partial(\rho v_j v_i)}{\partial x_j} \equiv \nabla \cdot \rho \mathbf{v} v_i \tag{9-30}$$

Eulerian Point of View

If we let $x_i = x$, the equation of motion can be written for the x direction as

$$-[\![(\nabla \cdot \underline{\tau})_x]\!] - \nabla \cdot \rho \mathbf{v} v_x - (\nabla p)_x g_c + \rho g_x = \frac{\partial(\rho v_x)}{\partial t} \tag{9-31}$$

Similar equations can be written for the y or z directions by letting $i = y$ or z. When the x equation is multiplied by $\boldsymbol{\delta}_x$, the y equation by $\boldsymbol{\delta}_y$, and the z equation by $\boldsymbol{\delta}_z$, we obtain three vector components which can be added to obtain a vector in space—or we could merely multiply Eq. (9-25) by $\boldsymbol{\delta}_i$ and sum over i. In either case we obtain the equation of motion in *Gibbs vector notation,* eulerian point of view

$$-\nabla \cdot \underline{\tau} - \nabla \cdot \rho \mathbf{v}\mathbf{v} - g_c \nabla p + \rho \mathbf{g} = \frac{\partial(\rho \mathbf{v})}{\partial t} \qquad \text{eulerian} \tag{9-32}$$

Lagrangian Point of View

The lagrangian form of the equation of motion can be obtained either by (1) writing a momentum balance over an element of volume moving with the velocity of the stream or (2) converting the eulerian form to the lagrangian form

† We will use [[ ]] to indicate a scalar component.

by using the definition of the substantial derivative and of the equation of continuity. The second method requires one to prove the vector identity

$$\nabla \cdot \rho \mathbf{v}\mathbf{v} \equiv \rho \mathbf{v} \cdot \nabla \mathbf{v} + \mathbf{v} \nabla \cdot \rho \mathbf{v}$$

The proof is carried out by writing an expression for each of the vectors (∇ and $\mathbf{v}$) in terms of summations. The dot products are then taken as in Appendix 8-3; the Kronecker delta is made use of, and the differentiation operations are carried out (see Appendix 8-3). The result is

$$-\nabla \cdot \underline{\boldsymbol{\tau}} - g_c \nabla p + \rho \mathbf{g} = \frac{\rho \, D\mathbf{v}}{Dt} \qquad \text{lagrangian} \tag{9-33}$$

9-5 ORDER OF A VECTOR EXPRESSION

A vector identity should be checked to see whether all its terms have the same order. To find the order of a term, the following rule is useful: *Find the sum of the orders of all vectors and tensors and subtract twice the number of dots.* For example, the order of $\nabla \cdot \rho \mathbf{v}\mathbf{v}$ is $1 + 1 + 1 - 2 = 1$; i.e., it is a vector. The order of $\nabla \cdot \underline{\boldsymbol{\tau}}$ is $1 + 2 - 2 = 1$, and it also is a vector. The order of $\underline{\boldsymbol{\sigma}} : \nabla \mathbf{v}$ (to appear later) is $2 + 1 + 1 - 2(2) = 0$; it is a scalar.

9-6 USE OF SI UNITS†

The equation of motion is usually written with $g_c = 1.0$ since the scientific community universally uses an absolute system of units. *From now on we adopt this simplification.* Then

$$p g_c = p \qquad \sigma_{ij} g_c = \tau_{ij} \tag{9-34}$$

9-7 SUMMATION CONVENTION‡

Using cartesian tensor notation, the equation of motion can be written for a component i as

$$\rho \frac{Dv_i}{Dt} = -\sum_j \frac{\partial \tau_{ji}}{\partial x_j} - \frac{\partial p}{\partial x_i} + \rho g_i \tag{9-35}$$

Sometimes $p_{,i}$ is used for $\partial p / \partial x_i$ and $\tau_{ji,j}$ is used for $\Sigma_j \, \partial \tau_{ji} / \partial x_j$. The convention is that in general a subscripted comma followed by an index refers to differentiation

† Does not require knowledge of vector notation.

‡ This section may be omitted on first reading.

with respect to the coordinate with that index and a repeated index means a summation with respect to that index. Hence in $\tau_{ji,j}$ we sum over j. Thus

$$\rho \frac{Dv_i}{Dt} = -\tau_{ji,j} - p_{,i} + \rho g_i \tag{9-36}$$

A disadvantage of this notation is that when we do *not* wish to indicate summation by a repeated index, we must say so.

9-8 TOTAL-STRESS VECTOR†

In fluid mechanics it is convenient to define a *total-stress vector* by the equation

$$\mathbf{t}_{(n)} \equiv \frac{d\mathbf{F}}{dA} \tag{9-37}$$

where

$$d\mathbf{F} = d\mathbf{F}_v + d\mathbf{F}_p \tag{9-38}$$

is the total (viscous plus pressure) resultant force acting on a differential element of area dA whose outwardly directed normal is $\mathbf{n}$. [The subscript (n) refers to the area whose normal is $\mathbf{n}$ and *not* to the direction of $\mathbf{t}_{(n)}$, which, in general, is not normal to the area.] Since the normal $\mathbf{n}$ may be oriented in space in various ways, it is desirable to try to relate $\mathbf{t}_{(n)}$ to the cartesian components of the stress tensor and to $\mathbf{n}$. It is also desirable to generalize the previous approach, which was based on an element $\Delta\mathcal{V}$ in cartesian coordinates. This can be done by a method similar to that used in Chap. 8 for heat transport by making, in this case, a momentum balance over a tetrahedral element of volume formed by a slant face ΔA_n and three faces normal to each of the coordinate axes (see Fig. A8-4-1). When this is done, one finds that in the limit as $\Delta\mathcal{V} \to 0$, the accumulation and body-force terms (which depend on the volume) become negligible compared with the surface-force terms that depend on the area, and (see Appendix 9-1)‡

$$\mathbf{t}_{(n)} = \mathbf{n} \cdot \underline{\mathbf{T}} = \sum_i \sum_j n_i T_{ij} \boldsymbol{\delta}_j \tag{9-39}$$

where

$$\underline{\mathbf{T}} = \sum_i \boldsymbol{\delta}_i \mathbf{t}_{(i)} \tag{9-40}$$

is called the total-stress tensor and $\mathbf{t}_{(i)}$ is the total-stress vector on a face perpendicular to the ith axis. Equation (9-39) shows that $\mathbf{t}_{(n)}$ is linear in the total-stress components T_{ij} and in $\mathbf{n}$ and that one need only specify $\mathbf{n}$ to describe the stress acting on any plane at a point in the fluid. Equation (9-39) also shows that $\underline{\mathbf{T}}$ is a second-order tensor since when it is dot-multiplied by a vector, another vector

† This section may be omitted on first reading.

‡ In fluid mechanics, the transpose T_{ji} of T_{ij} is used; i.e., the first subscript on the stress vector is the direction of the force and the second refers to the face. However, usually $\underline{\mathbf{T}}$ is symmetrical and equal to its transpose, so the result is the same.

results. Equation (9-40) also shows that $\underline{\mathbf{T}}$ is a second-order tensor since it is the dyadic product of two vectors. The consequence is that Eq. (9-39) is applicable to any orientation in space, to any coordinate system, or to any body.

If we multiply Eq. (9-39) by dA, we obtain $d\mathbf{F} = \mathbf{t}_{(n)}\, dA$. Let us define

$$\boldsymbol{\sigma}_{(n)} \equiv -\frac{d\mathbf{F}_v}{dA} \tag{9-41}$$

as the *viscous-stress vector*. If we substitute Eqs. (9-37), (9-41), and (9-6), into (9-38), we get

$$d\mathbf{F} = \mathbf{t}_{(n)}\, dA = -\boldsymbol{\sigma}_{(n)}\, dA - p\mathbf{n}\, dA \tag{9-42}$$

Then

$$\mathbf{t}_{(n)} = -\boldsymbol{\sigma}_{(n)} - p\mathbf{n} \tag{9-43}$$

Now let us relate $\underline{\mathbf{T}}$ to $\underline{\boldsymbol{\sigma}}$. They differ not only in sign but also in that $\underline{\mathbf{T}}$ includes the effect of pressure, which contributes only to the normal stress, as in Eq. (9-13), and not to the shear stress. Furthermore $\underline{\mathbf{T}}$ is a tensor and p is a scalar and cannot be added directly to $\underline{\boldsymbol{\sigma}}$ or subtracted from $\underline{\mathbf{T}}$ except as indicated below. When Eq. (9-13) is written in index notation, the normal components of $\underline{\mathbf{T}}$ are given by

$$T_{ii} = -p - \sigma_{ii} \tag{9-44}$$

Since the off-diagonal components would be unaffected by pressure, they become

$$T_{ij} = -\sigma_{ij} \qquad i \neq j \tag{9-45}$$

The *general* equation is then

$$T_{ij} = -p\delta_{ij} - \sigma_{ij} \tag{9-46}$$

where δ_{ij} is the Kronecker delta (see Appendix 8-3).

In Gibbs vector notation, this becomes

$$\underline{\mathbf{T}} = -p\underline{\boldsymbol{\delta}} - \underline{\boldsymbol{\sigma}} \tag{9-46a}$$

where $\underline{\boldsymbol{\delta}}$ is the unit tensor whose components are δ_{ij}. In matrix form $\underline{\boldsymbol{\delta}}$ is defined as

$$\underline{\boldsymbol{\delta}} = \begin{bmatrix} 1 & 0 & 0 \\ 0 & 1 & 0 \\ 0 & 0 & 1 \end{bmatrix} \tag{9-47}$$

We can then obtain $\underline{\mathbf{T}}$ by matrix addition

$$\begin{bmatrix} T_{xx} & T_{xy} & T_{xz} \\ T_{yx} & T_{yy} & T_{yz} \\ T_{zx} & T_{zy} & T_{zz} \end{bmatrix} = -p\begin{bmatrix} 1 & 0 & 0 \\ 0 & 1 & 0 \\ 0 & 0 & 1 \end{bmatrix} - \begin{bmatrix} \sigma_{xx} & \sigma_{xy} & \sigma_{xz} \\ \sigma_{yx} & \sigma_{yy} & \sigma_{yz} \\ \sigma_{zx} & \sigma_{zy} & \sigma_{zz} \end{bmatrix} = -\begin{bmatrix} p+\sigma_{xx} & \sigma_{xy} & \sigma_{xz} \\ \sigma_{yx} & p+\sigma_{yy} & \sigma_{yz} \\ \sigma_{zx} & \sigma_{zy} & p+\sigma_{zz} \end{bmatrix} \tag{9-48}$$

We now relate $\boldsymbol{\sigma}_{(n)}$ to $\underline{\boldsymbol{\sigma}}$. Solving Eq. (9-38) for $d\mathbf{F}_v$ gives

$$d\mathbf{F}_v = d\mathbf{F} - d\mathbf{F}_p \tag{9-48a}$$

Combining Eqs. (9-37) and (9-39) gives

$$dF = \mathbf{n} \cdot \underline{\mathbf{T}}\, dA \tag{9-48b}$$

We can rewrite Eq. (9-4) as

$$d\mathbf{F}_p = -\mathbf{n} \cdot \underline{\boldsymbol{\delta}} p\, dA \tag{9-49}$$

since $\mathbf{n} \cdot \underline{\boldsymbol{\delta}} = \mathbf{n}$. Using Eq. (9-41) for $d\mathbf{F}_v$, (9-48b) for $d\mathbf{F}$, and (9-49) for $d\mathbf{F}_p$ and substituting into (9-48a) gives

$$-\boldsymbol{\sigma}_{(n)}\, dA = \mathbf{n} \cdot \underline{\mathbf{T}}\, dA + \mathbf{n} \cdot \underline{\boldsymbol{\delta}} p\, dA$$

Using Eq. (9-46a) gives

$$\boldsymbol{\sigma}_{(n)} = \mathbf{n} \cdot \underline{\boldsymbol{\sigma}} \tag{9-50}$$

which means that $\underline{\boldsymbol{\sigma}}$ is a tensor. In terms of $\underline{\mathbf{T}}$, the EOM is

$$\rho \frac{D\mathbf{v}}{dt} = \nabla \cdot \underline{\mathbf{T}} + \rho \mathbf{g}$$

Total Force

The total force acting on a finite volume of fluid is obtained by integrating Eq. (9-37) over the surface area of the element and using Eq. (9-39)

$$\mathbf{F} = \mathbf{F}_v + \mathbf{F}_p = \iint \mathbf{t}_{(n)}\, dA = \iint \mathbf{n} \cdot \underline{\mathbf{T}}\, dA \tag{9-50a}$$

or, using Eq. (9-46a),

$$\mathbf{F} = -\iint \mathbf{n} \cdot \underline{\boldsymbol{\sigma}}\, dA - \iint p\mathbf{n}\, dA \tag{9-50c}$$

9-9 CURVILINEAR COORDINATE SYSTEMS†‡

Equation of Motion in Cylindrical Coordinates

When the EOM is expressed in Gibbs vector notation [Eqs. (9-32) and (9-33)], it applies to any coordinate system but the form of the components of each term will differ for various coordinate systems. In Appendix 8-5 we showed that a term such as $\nabla \cdot \mathbf{q}$ (which occurs in the energy equation) can be obtained for cylindrical coordinates by using graphical methods to find the derivatives of the unit vectors.

†The equation of motion in cartesian, cylindrical, and spherical coordinates is given in Tables I-6 to I-8. The mathematical principles involved in deriving this equation are illustrated for cylindrical coordinates below and in Appendixes 8-5 to 8-7. On first reading this part may be omitted, and the reader may skip to the applications (Secs. 9-11 and 9-14). Readers interested in general transformation methods for arbitrary orthogonal coordinate systems are referred to Appendix 8-7. For further information see Ref. 1 or 18.

‡May be omitted on first reading.

We now use this method to obtain the acceleration term $\rho(D\mathbf{v}/Dt)$ and the term $\nabla \cdot \underline{\tau}$ in cylindrical coordinates.

Velocity and acceleration The acceleration of a fluid is defined as the time rate of change of its velocity $\mathbf{v}$

$$\mathbf{a} \equiv \frac{D\mathbf{v}}{Dt} \tag{9-51}$$

The velocity is defined at the time rate of change of travel distance

$$\mathbf{v} = \frac{D\mathbf{x}}{Dt} \tag{9-52}$$

where $\mathbf{x}$ is called the *position* vector,† which, in cartesian coordinates, is defined (see Fig. 9-9) as

$$\mathbf{x} = x\boldsymbol{\delta}_x + y\boldsymbol{\delta}_y + z\boldsymbol{\delta}_z \tag{9-53}$$

Then by differentiation

$$\mathbf{v} = \frac{Dx}{Dt}\boldsymbol{\delta}_x + \frac{Dy}{Dt}\boldsymbol{\delta}_y + \frac{Dz}{Dt}\boldsymbol{\delta}_z \tag{9-54}$$

Since $\mathbf{v}$ is a vector, it can be expressed in terms of its components as

$$\mathbf{v} = v_x\boldsymbol{\delta}_x + v_y\boldsymbol{\delta}_y + v_z\boldsymbol{\delta}_z \tag{9-55}$$

Comparing coefficients of Eqs. (9-54) and (9-55), we get

$$v_x \equiv \frac{Dx}{Dt} \qquad v_y \equiv \frac{Dy}{Dt} \qquad v_z \equiv \frac{Dz}{Dt} \tag{9-56}$$

† Note that $|\mathbf{x}| = (x^2 + y^2 + z^2)^{1/2}$ does not have the magnitude of $r = (x^2 + y^2)^{1/2}$ in a cylindrical coordinate system but does have the magnitude of the r used in a spherical coordinate system.

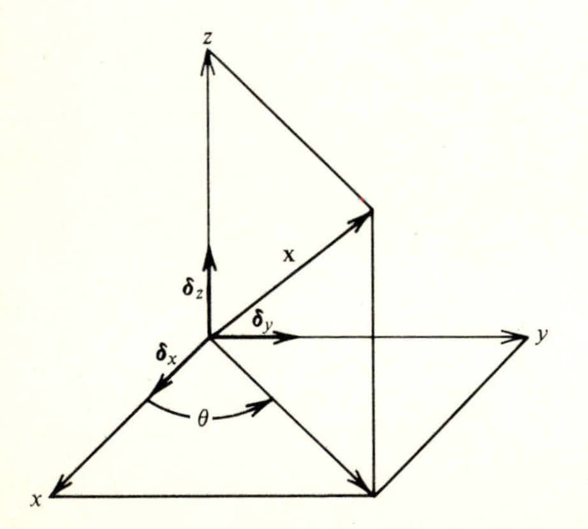

Figure 9-9 Position vector.

In *cylindrical* coordinates, the velocity vector is

$$\mathbf{v} = v_r\boldsymbol{\delta}_r + v_\theta\boldsymbol{\delta}_\theta + v_z\boldsymbol{\delta}_z \tag{9-57}$$

The acceleration is then

$$\frac{D\mathbf{v}}{Dt} = \frac{Dv_r}{Dt}\boldsymbol{\delta}_r + v_r\frac{D\boldsymbol{\delta}_r}{Dt} + \frac{Dv_\theta}{Dt}\boldsymbol{\delta}_\theta + v_\theta\frac{D\boldsymbol{\delta}_\theta}{Dt} + \frac{Dv_z}{Dt}\boldsymbol{\delta}_z + v_z\frac{D\boldsymbol{\delta}_z}{Dt} \tag{9-58}$$

In Appendix 8-5 we showed that $\boldsymbol{\delta}_r$ and $\boldsymbol{\delta}_\theta$ depend on θ. Using these results [Eqs. (A8-5-4) and (A8-5-5)], and the chain rule we obtain

$$\frac{D\boldsymbol{\delta}_r}{Dt} = \frac{d\boldsymbol{\delta}_r}{d\theta}\frac{D\theta}{Dt} = \boldsymbol{\delta}_\theta\frac{D\theta}{Dt} \qquad \frac{D\boldsymbol{\delta}_\theta}{Dt} = \frac{d\boldsymbol{\delta}_\theta}{d\theta}\frac{D\theta}{Dt} = -\boldsymbol{\delta}_r\frac{D\theta}{Dt} \tag{9-59}$$

The velocity component in the θ direction is given by the time rate of change of distance in that direction; thus

$$v_\theta = \frac{Ds_\theta}{Dt} = r\frac{D\theta}{Dt} \tag{9-60}$$

Combining Eqs. (9-58) to (9-60) and collecting terms, we obtain

$$\mathbf{a} \equiv \frac{d\mathbf{v}}{Dt} = \left(\frac{Dv_r}{Dt} - \frac{v_\theta^2}{r}\right)\boldsymbol{\delta}_r + \left(\frac{Dv_\theta}{Dt} + \frac{v_r v_\theta}{r}\right)\boldsymbol{\delta}_\theta + \frac{Dv_z}{Dt}\boldsymbol{\delta}_z \tag{9-61}$$

Since $\mathbf{a}$ is a vector,

$$\mathbf{a} = a_r\boldsymbol{\delta}_r + a_\theta\boldsymbol{\delta}_\theta + a_z\boldsymbol{\delta}_z \tag{9-62}$$

Equating coefficients in Eqs. (9-61) and (9-62) gives

$$a_r = \frac{Dv_r}{Dt} - \frac{v_\theta^2}{r} \qquad a_\theta = \frac{Dv_\theta}{Dt} + \frac{v_r v_\theta}{r} \qquad a_z = \frac{Dv_z}{Dt} \tag{9-63}$$

The term $-v_\theta^2/r$ is the contribution due to *centrifugal force;* it is an effective force in the r direction. The term $v_r v_\theta/r$ is sometimes called a *Coriolis acceleration;*†[7] it produces an effective force in the θ direction, as will be discussed next.

Coriolis and Centrifugal Terms in Polar Coordinates

In order to understand the acceleration terms a_θ and a_r, let us use polar coordinates. We define the position vector as

$$\mathbf{r} = r\boldsymbol{\delta}_r(\theta) \tag{9-64}$$

Refer to Fig. 9-10, which compares the velocity $\mathbf{v}$ at arbitrary $\mathbf{r}$ and θ with the velocity $\mathbf{v} + d\mathbf{v}$ at $\mathbf{r} + d\mathbf{r}$ and $\theta + d\theta$. In order to emphasize the Coriolis and centrifugal accelerations, we have assumed that the *magnitude* of $\mathbf{v}$ does not

† For a more conventional definition, see next subsection.

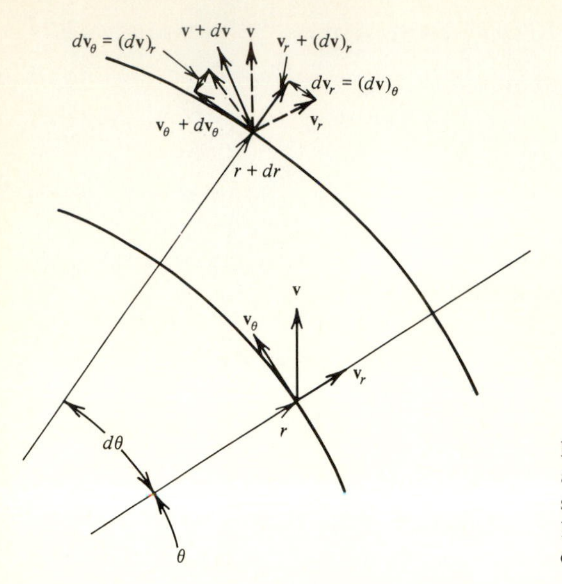

Figure 9-10 Velocity-vector components at r and θ and at $r + dr$ and $\theta + d\theta$, showing origin of acceleration terms. Note that $|\mathbf{v}|$ does not change but $\mathbf{v}$ changes to $\mathbf{v} + d\mathbf{v}$.

change but that its direction does change (due to a change in θ). Then

$$d\mathbf{v} = d\mathbf{v}_\theta + d\mathbf{v}_r \equiv d(v_\theta \boldsymbol{\delta}_\theta) + d(v_r \boldsymbol{\delta}_r) = v_\theta \, d\boldsymbol{\delta}_\theta + v_r \, d\boldsymbol{\delta}_r \tag{9-65}$$

since $|\mathbf{v}|$ is constant.

Using Eqs. (A8-5-4) and (A8-5-5), we have

$$d\mathbf{v} = v_\theta(-\boldsymbol{\delta}_r) \, d\theta + v_r(\boldsymbol{\delta}_\theta) \, d\theta \tag{9-66a}$$

Also in terms of its scalar components

$$d\mathbf{v} = [\![d\mathbf{v}_r]\!]\boldsymbol{\delta}_r + [\![d\mathbf{v}_\theta]\!]\boldsymbol{\delta}_\theta \tag{9-66b}$$

Then equating coefficients of the unit vectors in the above gives

$$[\![d\mathbf{v}_r]\!] = -v_\theta \, d\theta \qquad \text{and} \qquad [\![d\mathbf{v}_\theta]\!] = v_r \, d\theta \tag{9-66c}$$

Also $$d\mathbf{v} \equiv (\mathbf{v} + d\mathbf{v}) - \mathbf{v} = (d\mathbf{v})_r + (d\mathbf{v})_\theta = [\![d\mathbf{v}_r]\!]\boldsymbol{\delta}_r + [\![d\mathbf{v}_\theta]\!]\boldsymbol{\delta}_\theta \tag{9-66d}$$

Now the change in the vector component $\mathbf{v}_r$, from Eqs. (9-66), (9-66a), and (9-66b), is

$$d\mathbf{v}_r = [\![d\mathbf{v}_r]\!]\boldsymbol{\delta}_r = -v_\theta \, d\theta \, \boldsymbol{\delta}_r = d(v_\theta \boldsymbol{\delta}_\theta) = (d\mathbf{v})_\theta \tag{9-66e}$$

which is the θ component of the resultant vector $d\mathbf{v}$. Similarly

$$d\mathbf{v}_\theta = [\![d\mathbf{v}_\theta]\!]\boldsymbol{\delta}_\theta = v_r \, d\theta \, \boldsymbol{\delta}_\theta = d(v_r \boldsymbol{\delta}_r) = (d\mathbf{v})_r \tag{9-66f}$$

The Coriolis term $(d\mathbf{v})_\theta = d\mathbf{v}_r$ is directed in the θ direction and occurs because $\boldsymbol{\delta}_r$ is changing with θ, resulting in a component $(d\mathbf{v})_\theta$—even though v_θ itself does not change—whereas the centrifugal term $(d\mathbf{v})_r$ occurs due to a change in $\boldsymbol{\delta}_\theta$ with θ. If v_θ *is* changing and we let $v_\theta = \omega r$, where ω is the magnitude of the angular velocity, the total acceleration in the θ direction [as given in the second of Eqs. (9-63)] can be written

$$a_\theta = \frac{Dv_\theta}{Dt} + \frac{v_r v_\theta}{r} = \frac{D(\omega r)}{Dt} + v_r \omega = \frac{D\omega}{Dt} r + \frac{\omega\, Dr}{Dt} + v_r \omega = \frac{D\omega}{Dt} r + 2v_r \omega \tag{9-67}$$

In this form the term $2v_r\omega$ is called the Coriolis acceleration.†

Gradient of Pressure

Since pressure is a scalar, we can use Eq. (8-19) to obtain

$$\nabla p = \boldsymbol{\delta}_r \frac{\partial p}{\partial r} + \boldsymbol{\delta}_\theta \frac{\partial p}{r\, \partial \theta} + \boldsymbol{\delta}_z \frac{\partial p}{\partial z} \tag{9-68}$$

Then $$[\![(\nabla p)_r]\!] = \frac{\partial p}{\partial r} \qquad [\![(\nabla p)_\theta]\!] = \frac{\partial p}{r\, \partial \theta} \qquad [\![(\nabla p)_z]\!] = \frac{\partial p}{\partial z} \tag{9-69}$$

Divergence of Momentum Flux

Using Eqs. (8-19) and (9-17), we obtain

$$\nabla \cdot \underline{\tau} = \left(\boldsymbol{\delta}_r \frac{\partial}{\partial r} + \boldsymbol{\delta}_\theta \frac{\partial}{r\, \partial \theta} + \boldsymbol{\delta}_z \frac{\partial}{\partial z}\right) \cdot \left(\sum_m \sum_n \boldsymbol{\delta}_m \boldsymbol{\delta}_n \tau_{mn}\right) \tag{9-70}$$

where m and n refer to cylindrical coordinates r, θ, and z. Now $\partial \boldsymbol{\delta}_m / \partial r = 0 = \partial \boldsymbol{\delta}_m / \partial z$ for all m, but $\partial \boldsymbol{\delta}_r / \partial \theta = \boldsymbol{\delta}_\theta$ and $\partial \boldsymbol{\delta}_\theta / \partial \theta = -\boldsymbol{\delta}_r$ (see Appendix 8-5). Hence

$$\left(\boldsymbol{\delta}_r \frac{\partial}{\partial r}\right) \cdot \left(\sum_m \sum_n \boldsymbol{\delta}_m \boldsymbol{\delta}_n \tau_{mn}\right) = \sum_m \sum_n \boldsymbol{\delta}_r \cdot \boldsymbol{\delta}_m \boldsymbol{\delta}_n \frac{\partial \tau_{mn}}{\partial r} = \sum \boldsymbol{\delta}_n \frac{\partial \tau_{rn}}{\partial r}$$

† The transformation of a lagrangian acceleration to cylindrical coordinates is comparable to a transformation from a fixed cartesian set of axes (a *frame of reference*) to a set of axes rotating with angular velocity $\boldsymbol{\omega}$. Newton's law of motion applies only to an inertial frame, i.e., to a set of axes fixed in space. Thus an effective tangential or Coriolis force is experienced when one tries to walk in a straight line from the center of a carousel (or a rotating disk) to the edge. The rotation of the disk will produce an effective acceleration in the θ direction that must be compensated for by an effective force if one is to maintain a straight-line path as viewed from a fixed coordinate system on the ground. A similar effect occurs in a centrifugal pump (see Chap. 11), when a fluid flows from the center to the periphery of a rotating disk with radial velocity v_r (due to the flow) and tangential velocity v_θ (due to the rotation of the disk). A Coriolis effect also occurs in the earth's atmosphere, where the rotation of the earth tends to produce winds (cyclones) that move in a counterclockwise direction in the northern hemisphere and clockwise in the southern hemisphere (anticyclones).

Texts on mechanics give the expression $2\boldsymbol{\omega} \times \mathbf{v}$ for the Coriolis acceleration, where $\boldsymbol{\omega}$ is the angular velocity of a rotating coordinate system and $\mathbf{v}$ is the velocity of a particle relative to the moving axes. In cylindrical coordinates the θ component of the Coriolis acceleration is then $2\omega v_r \boldsymbol{\delta}_\theta$, which corresponds to the second term in Eq. (9-67).

since $\qquad \boldsymbol{\delta}_r \cdot \boldsymbol{\delta}_m = 0 \qquad m \neq r$

Similarly $\qquad \left(\boldsymbol{\delta}_z \frac{\partial}{\partial z}\right) \cdot \left(\sum_m \sum_n \boldsymbol{\delta}_m \boldsymbol{\delta}_n \tau_{mn}\right) = \sum_n \boldsymbol{\delta}_n \frac{\partial \tau_{zn}}{\partial z}$

But

$$\left(\boldsymbol{\delta}_\theta \frac{\partial}{r\,\partial\theta}\right) \cdot \left(\sum_m \sum_n \boldsymbol{\delta}_m \boldsymbol{\delta}_n \tau_{mn}\right) = \sum_n \boldsymbol{\delta}_n \frac{\partial \tau_{\theta n}}{r\,\partial\theta} + \sum_n \frac{\partial \boldsymbol{\delta}_n}{r\,\partial\theta}\tau_{\theta n} + \sum_m \sum_n \boldsymbol{\delta}_\theta \cdot \frac{\partial \boldsymbol{\delta}_m}{r\,\partial\theta} \boldsymbol{\delta}_n \tau_{mn}$$

Now $\qquad \sum_n \frac{\partial \boldsymbol{\delta}_n}{r\,\partial\theta}\tau_{\theta n} = \frac{1}{r}\left(\boldsymbol{\delta}_\theta \tau_{\theta r} - \boldsymbol{\delta}_r \tau_{\theta\theta}\right)$

and

$$\boldsymbol{\delta}_\theta \cdot \left(\sum_m \frac{\partial \boldsymbol{\delta}_m}{r\,\partial\theta} \sum_n \boldsymbol{\delta}_n \tau_{mn}\right) = \boldsymbol{\delta}_\theta \cdot \frac{1}{r}\boldsymbol{\delta}_\theta \left(\boldsymbol{\delta}_r \tau_{rr} + \boldsymbol{\delta}_\theta \tau_{r\theta} + \boldsymbol{\delta}_z \tau_{rz}\right)$$

$$-\boldsymbol{\delta}_\theta \cdot \frac{1}{r}\boldsymbol{\delta}_r(\boldsymbol{\delta}_r \tau_{\theta r} + \boldsymbol{\delta}_\theta \tau_{\theta\theta} + \boldsymbol{\delta}_z \tau_{\theta z}) + 0$$

Using $\boldsymbol{\delta}_\theta \cdot \boldsymbol{\delta}_r = 0$, we can collect terms having the same unit vectors

$$\nabla \cdot \underline{\boldsymbol{\tau}} = \boldsymbol{\delta}_r[\![(\nabla \cdot \underline{\boldsymbol{\tau}})_r]\!] + \boldsymbol{\delta}_\theta[\![(\nabla \cdot \underline{\boldsymbol{\tau}})_\theta]\!] + \boldsymbol{\delta}_z[\![(\nabla \cdot \underline{\boldsymbol{\tau}})_z]\!] \tag{9-71}$$

or $\qquad (\nabla \cdot \underline{\boldsymbol{\tau}})_r = \boldsymbol{\delta}_r\left(\frac{\partial \tau_{rr}}{\partial r} + \frac{\partial \tau_{zr}}{\partial z} + \frac{\partial \tau_{\theta r}}{r\,\partial\theta} + \frac{\tau_{rr}}{r} - \frac{\tau_{\theta\theta}}{r}\right)$

$$[\![(\nabla \cdot \underline{\boldsymbol{\tau}})_r]\!] = \frac{\partial(r\tau_{rr})}{r\,\partial r} + \frac{\partial \tau_{zr}}{\partial z} + \frac{\partial \tau_{\theta r}}{r\,\partial\theta} - \frac{\tau_{\theta\theta}}{r} \tag{9-72}$$

$$(\nabla \cdot \underline{\boldsymbol{\tau}})_\theta = \boldsymbol{\delta}_\theta\left(\frac{\partial \tau_{r\theta}}{\partial r} + \frac{\partial \tau_{z\theta}}{\partial z} + \frac{\partial \tau_{\theta\theta}}{r\,\partial\theta} + \frac{\tau_{\theta r}}{r} + \frac{\tau_{r\theta}}{r}\right)$$

If $\tau_{\theta r} = \tau_{r\theta}$, we get

$$[\![(\nabla \cdot \underline{\boldsymbol{\tau}})_\theta]\!] = \frac{\partial(\tau_{r\theta} r^2)}{r^2\,\partial r} + \frac{\partial \tau_{\theta\theta}}{r\,\partial\theta} + \frac{\partial \tau_{z\theta}}{\partial z} \tag{9-73}$$

$$[\![(\nabla \cdot \underline{\boldsymbol{\tau}})_z]\!] = \frac{\partial \tau_{rz}}{\partial r} + \frac{\partial \tau_{zz}}{\partial z} + \frac{\partial \tau_{\theta z}}{r\,\partial\theta} + \frac{\tau_{rz}}{r} \tag{9-74}$$

Components of Equation of Motion in Cylindrical Coordinates

We can now substitute the above results into the general equation (taking $g_c = 1$)

$$\rho\mathbf{a} \equiv \rho\frac{D\mathbf{v}}{Dt} \equiv \rho\left(\frac{\partial \mathbf{v}}{\partial t} + \mathbf{v} \cdot \nabla\mathbf{v}\right) = -\nabla p - \nabla \cdot \underline{\boldsymbol{\tau}} + \rho\mathbf{g}$$

The r component is

$$\rho\left(\frac{Dv_r}{Dt} - \frac{v_\theta^2}{r}\right) \equiv \rho\left(\frac{\partial v_r}{\partial t} + v_r\frac{\partial v_r}{\partial r} + \frac{v_\theta}{r}\frac{\partial v_r}{\partial \theta} + v_z\frac{\partial v_r}{\partial z} - \frac{v_\theta^2}{r}\right)$$

$$= -\frac{\partial p}{\partial r} - \left[\frac{\partial(r\tau_{rr})}{r\,\partial r} + \frac{\partial \tau_{r\theta}}{r\,\partial \theta} - \frac{\tau_{\theta\theta}}{r} + \frac{\partial \tau_{rz}}{\partial z}\right] + \rho g_r \quad (9\text{-}75)$$

The z component is

$$\rho\left(\frac{\partial v_z}{\partial t} + \mathbf{v}\cdot\nabla v_z\right) \equiv \rho\left(\frac{\partial v_z}{\partial t} + v_r\frac{\partial v_z}{\partial r} + v_\theta\frac{\partial v_z}{r\,\partial \theta} + v_z\frac{\partial v_z}{\partial z}\right)$$

$$= -\frac{\partial p}{\partial z} - \left(\frac{\partial(\tau_{rz}r)}{r\,\partial r} + \frac{\partial \tau_{\theta z}}{r\,\partial \theta} + \frac{\partial \tau_{zz}}{\partial z}\right) + \rho g_z \quad (9\text{-}76)$$

The θ component, in terms of linear momentum [from Eq. (9-61), the first of Eqs. (9-63) and (9-69), and from (9-73)]

$$\rho\left(\frac{Dv_\theta}{Dt} + \frac{v_r v_\theta}{r}\right) \equiv \rho\left(\frac{\partial v_\theta}{\partial t} + v_r\frac{\partial v_\theta}{\partial r} + \frac{v_\theta}{r}\frac{\partial v_\theta}{\partial \theta} + v_z\frac{\partial v_\theta}{\partial z} + \frac{v_r v_\theta}{r}\right)$$

$$= -\frac{\partial p}{r\,\partial \theta} - \left[\frac{\partial(\tau_{r\theta}r^2)}{r^2\,\partial r} + \frac{\partial \tau_{\theta\theta}}{r\,\partial \theta} + \frac{\partial \tau_{z\theta}}{\partial z}\right] + \rho g_\theta \quad (9\text{-}77)$$

which has the units ($g_c = 1$) of

$$\frac{\text{Momentum}}{\text{Volume} \times \text{time}} = \frac{\text{force}}{\text{volume}}$$

If we multiply both sides by the lever arm r, we get (see Appendix 9-2)

$$\frac{\text{Angular momentum}}{\text{Volume} \times \text{time}} = \frac{\text{torque}}{\text{volume}}$$

The left-hand side can then be shown by differentiation to be

$$\frac{D(rv_\theta)}{Dt} \equiv r\frac{Dv_\theta}{Dt} + v_\theta\frac{Dr}{Dt} = r\left(\frac{Dv_\theta}{Dt} + \frac{v_\theta v_r}{r}\right) \quad (9\text{-}78)$$

since $Dr/Dt \equiv v_r$.

If Eq. (9-77) is multiplied by r, we obtain an angular-momentum equation which will be discussed in the next section

$$\rho\frac{D(rv_\theta)}{Dt} = -\frac{\partial p}{\partial \theta} - r[\![(\nabla \cdot \underline{\tau})_\theta]\!] + \rho g_\theta r \quad (9\text{-}79)$$

Rate of change of angular momentum	Net pressure torque	Net viscous torque	Gravity torque

9-10 ANGULAR-MOMENTUM EQUATION†

Assuming that the stress tensor is symmetric (for proof see Appendix 9-3), we can obtain an equation of change for angular momentum by taking the cross product of the position vector $\mathbf{x}$ and the equation of motion. We proceed as we did in deriving the law of conservation of angular momentum. Starting with the lagrangian form of the equation of motion (9-34), we obtain

$$\mathbf{x} \times \rho \frac{D\mathbf{v}}{Dt} = \frac{\rho D(\mathbf{x} \times \mathbf{v})}{Dt} = -\mathbf{x} \times (\nabla \cdot \underline{\tau}) - \mathbf{x} \times \nabla p + \rho \mathbf{x} \times \mathbf{g} \tag{9-80}$$

If we use a cylindrical coordinate system, the z component of $\mathbf{x} \times \mathbf{v}$, as given by Eq. (A9-2-2), is $rv_\theta \boldsymbol{\delta}_z$; this is the angular momentum per unit mass of an element of fluid moving with velocity v_θ at a distance r from the z axis. Using similar reasoning on the other terms, we obtain an equation for the z component of the angular-momentum equation which agrees with Eq. (9-79). Note that Eq. (9-79) is the same as r times the θ component of the linear-momentum equation (9-77).

In eulerian form Eq. (9-80) (for all but highly nonpolar fluids) is

$$\frac{\partial}{\partial t}(\mathbf{x} \times \rho \mathbf{v}) = -\nabla \cdot (\mathbf{x} \times \rho \mathbf{v}\mathbf{v}) - \nabla \cdot (x \times \underline{\tau}) - \nabla \cdot (x \times p\underline{\boldsymbol{\delta}}) + \mathbf{x} \times \rho \mathbf{g} \tag{9-81}$$

9-11 APPLICATION OF THE EQUATION OF MOTION‡

Although an understanding of the derivation of the equation of motion is an important part of its successful application, our ultimate objective, of course, is its application. The customary procedure is to simplify either the general vector equation (9-34) or its components so that they apply to a given physical situation; e.g., if the flow is at steady state, all $\partial/\partial t$ terms will be eliminated.

Usually one chooses a coordinate system on the basis of the boundary conditions and simplifies the equation of continuity (EOC) and equation of motion (EOM) for each component (see Tables I-6 to I-8).

Another procedure is first to simplify the EOM in Gibbs vector notation [Eq. (9-34)] and then to expand the remaining vector terms into their components in the appropriate coordinate system. For cartesian, cylindrical, or spherical coordinates the general expansions in Tables I-2 to I-5 can be used for this purpose. For other coordinate systems, the methods described in Sec. 9-9 or in Appendix 8-7 for transforming coordinates may be used.

In many cases the EOM will provide information on the shear-stress distribution, but it is not adequate for deriving equations for the velocity profiles. To obtain them it is necessary to have a flux law or *constitutive equation,* i.e., a

† This section may be omitted on first reading.
‡ Does not require knowledge of vector notation.

relationship between the shear stress and velocity gradients. Newton's law of viscosity, which holds for newtonian fluids, is an example of a constitutive equation. Its general form, which is more complicated than the flux law developed in Chap. 1, will be taken up later. Its substitution into the EOM results in the Navier-Stokes equations (Tables I-5 to I-8).

In the following examples we will discuss applications of the EOM. In each case Table I-9 is to be referred to as well as part A of Tables I-6 to I-8 (or Tables I-4 and I-12).

Example 9-1: Flow between parallel plates (equation of motion) Consider a slit flow between two wide plates, separated by a narrow gap of width $2Y$ (see Fig. 9-11). The width of the plates is W, where $W/Y \to \infty$. The flow is steady state, laminar, and fully developed, and the fluid is assumed to be incompressible. We wish to determine which components of $\underline{\tau}$ are zero and how τ_{yx} varies with position in the gap, with pressure gradient, and with Y.

SOLUTION At steady state the equation of continuity (part B of Table I-6) for an incompressible fluid (ρ = const) is

$$\frac{\partial v_x}{\partial x} + \frac{\partial v_y}{\partial y} + \frac{\partial v_z}{\partial z} = 0$$

Since there is no flow in the z direction, $v_z = 0$; and since the flow is fully developed,

$$\frac{\partial v_x}{\partial x} = 0$$

Then $\partial v_y/\partial y = 0$ or v_y = const, since it cannot depend upon x (the flow is fully developed) or on z (W/Y is very large). The no-slip condition indicates that $v_y = 0$ at the walls, and thus the constant must be zero and v_y is zero everywhere. The velocity field is therefore described by

$$v_x = v_x\,(y \text{ only}) \qquad v_y = 0 = v_z \tag{9-82}$$

and may be envisioned as a series of planes, all perpendicular to the y axis, which slide past each other with varying relative velocities.

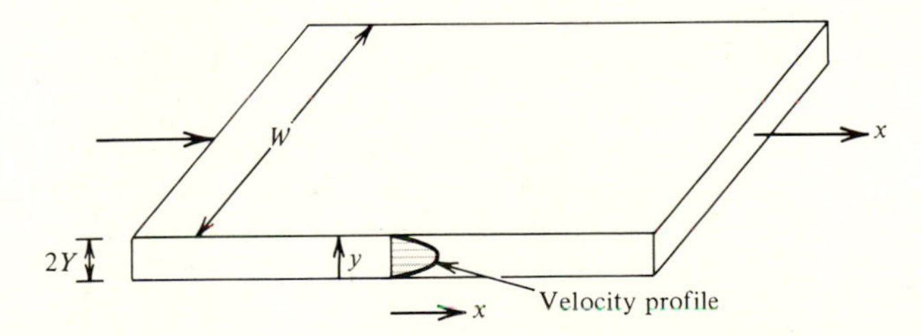

Figure 9-11 Slit flow between two parallel plates.

It is apparent therefore that at a given y there is a shear stress σ_{yx} acting in the x direction on an area A_y. In Appendix 9-3 we show by an angular-momentum balance that $\underline{\tau}$ is symmetrical. Therefore, if there is a τ_{yx}, there is also a shear stress τ_{xy} $(= \tau_{yx})$ acting in the y direction on an area A_x perpendicular to the x axis. Thus we see that if there is a velocity gradient dv_x/dy, there will be not only a shear stress τ_{yx} but also a shear stress τ_{xy}. Likewise we would expect that if there were a velocity gradient dv_y/dx (instead of dv_x/dy), shear-stress components τ_{xy} as well as τ_{yx} would again appear. *This tells us that* τ_{yx} *is not only the flux of x momentum in the y direction but also the flux of y momentum in the x direction.*

In general, then, we would expect both τ_{xy} and τ_{yx} to depend upon *both* $\partial v_y/\partial x$ and $\partial v_x/\partial y$. An equation which meets this requirement is

$$\tau_{xy} = -\mu \, \Delta_{xy} = \tau_{yx} \tag{9-83}$$

where

$$\Delta_{xy} = \frac{\partial v_x}{\partial y} + \frac{\partial v_y}{\partial x} = \Delta_{yx} \tag{9-84}$$

is called the *rate-of-deformation tensor*. In index notation, for a newtonian incompressible fluid we can write

$$\tau_{ij} = -\mu \, \Delta_{ij} = -\mu \left(\frac{\partial v_i}{\partial x_j} + \frac{\partial v_j}{\partial x_i} \right) \tag{9-85}$$

For a nonnewtonian fluid, τ_{xy} may depend upon Δ_{xy} in a nonlinear manner, as in the power-law model discussed in Chap. 1.

Also, for a newtonian incompressible fluid, the normal stress is

$$\tau_{xx} = -\mu \left(\frac{\partial v_x}{\partial x} + \frac{\partial v_x}{\partial x} \right) = -2\mu \left(\frac{\partial v_x}{\partial x} \right) \tag{9-86}$$

For a compressible fluid an additional term is needed (Appendix 9-5).

Similar expressions hold for other normal components τ_{yy} and τ_{zz}. However, for a class of nonnewtonian fluids, τ_{xx} may depend upon $\partial v_y/\partial x$, say, as well as upon $\partial v_x/\partial x$. Strictly speaking, therefore, it is necessary to know something about the relationship between $\underline{\tau}$ and $\underline{\Delta}$ (called a constitutive equation) before one can state which components of $\underline{\tau}$ are zero for a given "flow."

In the present example we can readily see that because $v_y = 0 = v_z$ and $v_x = v_x(y)$ only, Δ_{xx}, Δ_{yy}, Δ_{zz}, Δ_{xz}, Δ_{zx}, Δ_{zy}, and Δ_{yz} are all zero for a newtonian fluid. Thus the corresponding components of $\underline{\tau}$ will also be zero. But for a nonnewtonian fluid the normal components τ_{xx} and τ_{yy} may not be zero (they may depend on dv_x/dy, which is not zero). However, in order to obtain the velocity profile, we need shear stresses such as $\tau_{xy} = \tau_{yx}$ which are zero if Δ_{xy} is zero (no flow). For newtonian fluids the matrix of the components of $\underline{\tau}$ is then

$$\underline{\tau} = \begin{bmatrix} 0 & \tau_{xy} & 0 \\ \tau_{yx}(= \tau_{xy}) & 0 & 0 \\ 0 & 0 & 0 \end{bmatrix}$$

The x component of the EOM is obtained from either Table I-6, part A, or Table I-4.

$$\rho\left(\frac{\partial v_x}{\partial t}+v_x\frac{\partial v_x}{\partial x}+v_y\frac{\partial v_x}{\partial y}+v_z\frac{\partial v_x}{\partial z}\right)=\rho g_x-\frac{\partial p}{\partial x}-\frac{\partial \tau_{xx}}{\partial x}-\frac{\partial \tau_{yx}}{\partial y}-\frac{\partial \tau_{zx}}{\partial z}$$

Note that

$$\frac{\partial v_x}{\partial t}=0 \qquad \text{steady state}$$

$$\frac{\partial v_x}{\partial x}=0 \qquad v_x=v_x(y) \qquad \text{fully developed flow}$$

$$\frac{\partial \tau_{xx}}{\partial x}=0 \qquad \tau_{xx}\neq\tau_{xx}(x)$$

Assuming that gravity acts normal to the plates and using our knowledge of the velocity and stress fields, we find that the x component reduces to

$$\frac{-\partial p}{\partial x}=\frac{\partial \tau_{yx}}{\partial y} \tag{9-87}$$

The y component of the EOM is

$$\rho\left(\frac{\partial v_y}{\partial t}+v_x\frac{\partial v_y}{\partial x}+v_y\frac{\partial v_y}{\partial y}+v_z\frac{\partial v_y}{\partial z}\right)=\rho g_y-\frac{\partial p}{\partial y}-\frac{\partial \tau_{xy}}{\partial x}-\frac{\partial \tau_{yy}}{\partial y}-\frac{\partial \tau_{zy}}{\partial z}$$

Since $\partial\tau_{xy}/\partial x=\partial[-\mu(\partial v_x/\partial y)]/\partial x=-\mu\partial[(\partial v_x/\partial x)]/\partial y=0$, this simplifies to

$$0=\rho g_y-\frac{\partial p}{\partial y}-\frac{\partial \tau_{yy}}{\partial y} \tag{9-88}$$

It is now desirable to investigate the variation of the pressure with y and x. Differentiating Eq. (9-88) with respect to x, we have

$$0=0-\frac{\partial^2 p}{\partial x\,\partial y}$$

But we know from calculus that we can reverse the order of differentiation to obtain

$$\frac{\partial}{\partial y}\left(\frac{\partial p}{\partial x}\right)=0 \tag{9-89}$$

Thus Eq. (9-89) shows that $\partial p/\partial x$ is independent of y. On the other hand, $\partial\tau_{yx}/\partial y$ in Eq. (9-87) is at most a function of y, as a result of the exclusive dependence of τ_{yx} on the velocity gradient which is independent of x. Thus Eq. (9-87) implies that

$$\frac{\partial p}{\partial x} = \text{const} \equiv G_p \tag{9-90}$$

where for convenience we use G_p for the pressure gradient.

Since G_p is constant, we can integrate Eq. (9-90) between $p(x = 0) \equiv p_0$ and $p(x = L) \equiv p_L$ to obtain

$$G_p = \frac{p_L - p_0}{L} = \frac{-\Delta p}{L}$$

The symmetry of the region below the midplane ($y < Y$) with that above the midplane ($y > Y$) leads us to expect the velocity gradient at the midplane to be zero and the momentum flux to be zero also. Equation (9-87) can then be integrated between limits to give

$$\int_0^{\tau_{yx}} d\tau_{yx} = \frac{-\Delta p}{L} \int_Y^y dy$$

or

$$\tau_{yx} = \frac{-\Delta p}{L}(y - Y) \tag{9-91}$$

The magnitude of shear stress is a maximum at the walls and decreases linearly to zero at the center of the slit. It is also directly proportional to the pressure gradient.

Note that Eq. (9-91) was derived independently of the nature of the substance flowing through the slit and holds equally well for water, asphalt, or molasses. It is also clear that this equation is not very useful by itself, since it cannot tell us how much fluid will flow for a given pressure drop or the pressure drop required for a given flow rate. In order to solve this problem (as we noted earlier) τ_{yx} must be related to the velocity field. This will be done later. □

9-12 RELATION BETWEEN STRESS, DEFORMATION RATE, AND VORTICITY TENSORS†‡

In Chap. 1 we indicated that Newton's law of viscosity states that the shear stress (or momentum flux) is proportional to the rate of strain (or rate of deformation). Thus

$$\tau_{xy} = -\mu \, \Delta_y \tag{9-92}$$

where

$$\Delta_{xy} \equiv \frac{\partial v_x}{\partial y} + \frac{\partial v_y}{\partial x} \tag{9-93}$$

† This section may be omitted on first reading although certain defining equations will be used later.

‡ Combines vector and nonvector approaches.

is the rate-of-deformation tensor (or rate-of-strain tensor). In this section we verify the above expression for Δ_{xy} and attempt to gain insight into its physical meaning.

Deformation, Translation, and Rotation

What do we mean physically by fluid deformation or strain? When two fluid elements at different points move to another location, has the fluid undergone deformation or translation or rotation about an axis? How do we distinguish between deformation, translation, and rotation? These are some of the questions we must answer for a full understanding of fluid behavior in general and Newton's law of viscosity in particular.

When a force (or stress) is applied to an elastic solid, it deforms and its deformation is measured in terms of the strain, which is a *deformation per unit length*. Initially we can see that the deformation results in the displacement of a particle of the material at a point in the medium. However, if every particle were displaced the same amount in the same linear direction, there would be no deformation but only *translation*. Hence the deformation of a medium must depend upon the *relative* displacement of two particles. Likewise, if we had pure rotation, the two particles would be moving with the same angular velocity $\omega = v_\theta/r$ (like specks of dust at different radial locations on a phonograph record); while there would be no deformation, there would be displacement. However, if the angular velocity of the two particles were different, we would have *deformation* since the particles would be moving relative to each other.

Rate of Strain

An *elastic solid* is a material that bounces back like a glob of gelatin when subjected to a stress; i.e., the stress depends upon the strain. But a *stockesian fluid* is a material that deforms continuously. Thus, the stress depends upon the *rate* of strain, the deformation per unit time per unit length or the *relative* motion between two points in a fluid.

Let us then consider the motion in a fluid at a point R relative to a point O (see Fig. 9-12). If the velocity in the x direction at O is v_x and that at R is $v_x + dv_x(x, y, z)$, the *relative* motion in the x direction is dv_x. Now since $v_x = v_x(x, y, z)$, the chain rule of differentiation tells us that

$$dv_x = \frac{\partial v_x}{\partial x} dx + \frac{\partial v_x}{\partial y} dy + \frac{\partial v_x}{\partial z} dz \tag{9-94}$$

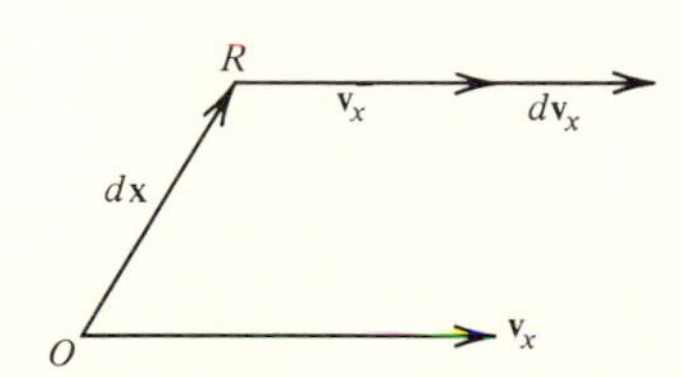

Figure 9-12 Relative motion in x direction of a fluid element.

Similar equations can be written for the y and z directions. In general, the relative velocity in the jth direction is

$$dv_j = \sum_i \frac{\partial v_j}{\partial x_i} dx_i \tag{9-95}$$

In Gibbs vector notation, if $\mathbf{x}$ is the position vector defined by Eq. (9-53), then

$$d\mathbf{v} = d\mathbf{x} \cdot \nabla \mathbf{v} \tag{9-96}$$

where $\nabla \mathbf{v}$ and $\partial v_j / \partial x_i$ are, respectively, the velocity-gradient tensor and its ij component.

In general, part of this motion may be due to deformation and part to rotation. In order to separate these contributions, let us consider an element (Fig. 9-13) of size $\Delta x \, \Delta y \, \Delta z$ with the points O, R, P, and Q initially at the corners. Assuming that v_y does not vary with y and that v_x does not vary with x, we let the velocity components at O be v_x and v_y, those at P be $v_x + (\partial v_x / \partial y) \, \Delta y$ and v_y, and those at Q be v_x and $v_y + (\partial v_y / \partial x) \, \Delta x$. After a time dt (see Fig. 9-14) point Q has moved to loction Q' and point P to location P', due to the shear stresses σ_{yx} and σ_{xy}. Relative to O, the *linear* movement of point P to P' in time dt can be expressed as $(\partial v_x / \partial y) \, \Delta y \, dt$ and that of point Q to Q' as $(\partial v_y / \partial x) \, \Delta x \, dt$. Now we can see that each linear motion corresponds to an angular movement measured by the angles α_{xy} and α_{yx}. For small angles the rate of an angular movement of α_{xy} (in radians per second) would be given by dividing the linear movement by the radius Δy, or

$$\frac{d\alpha_{xy}}{dt} = \frac{(\partial v_x / \partial y) \, \Delta y}{\Delta y} = \frac{\partial v_x}{\partial y} \tag{9-97}$$

Similarly

$$\frac{d\alpha_{yx}}{dt} = \frac{(\partial v_y / \partial x) \, \Delta x}{\Delta x} = \frac{\partial v_y}{\partial x} \tag{9-98}$$

If these rates are added (Fig. 9-15*a*), we have a *total deformation* rate of

$$\frac{d\alpha_{xy}}{dt} + \frac{d\alpha_{yx}}{dt} = \frac{\partial v_x}{\partial y} + \frac{\partial v_y}{\partial x} \equiv \Delta_{xy} \tag{9-99}$$

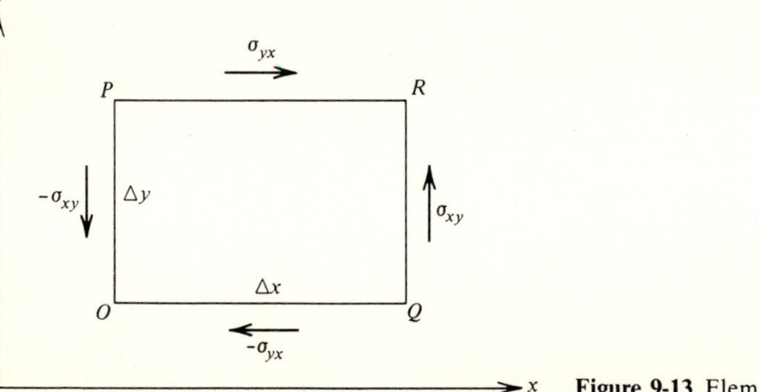

Figure 9-13 Element before shear.

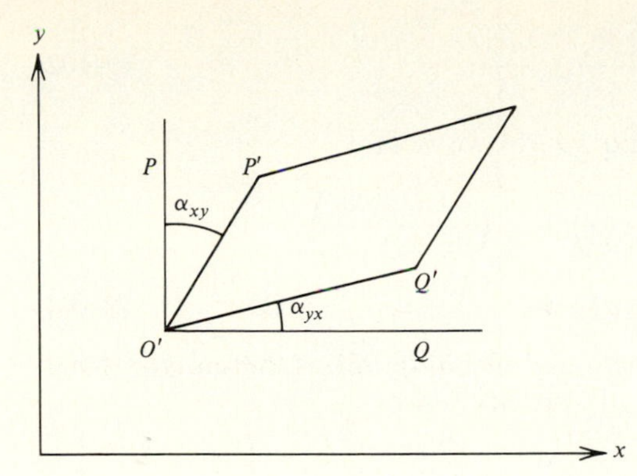

Figure 9-14 Fluid element after shear.

But if these rates are subtracted, we get a rotation about the z axis at an average angular velocity

$$\omega_z = \frac{1}{2}\left(\frac{\partial v_y}{\partial x} - \frac{\partial v_x}{\partial y}\right) \tag{9-100}$$

where ω_z represents a component of the average angular velocity of a fluid element in rotation (see Fig. 9-15*b*).

Vorticity

Equation (9-93) or (9-99) gives expressions for the xy component of the rate-of-deformation tensor. This relation resulted because the effects of stress on each face were additive (Fig. 9-15*a*); but if the effects had been subtracted, we would have obtained a *rotation* of the fluid element (Fig. 9-15*b*). To describe this rotation we define a *vorticity tensor* by

$$\Omega_{xy} = \frac{\partial v_y}{\partial x} - \frac{\partial v_x}{\partial y} \tag{9-101}$$

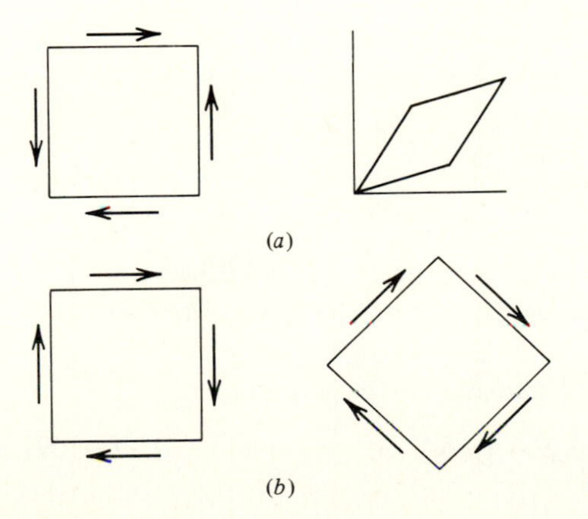

Figure 9-15 (*a*) Shear; (*b*) rotation.

or, in general,
$$\Omega_{ji} = \frac{\partial v_i}{\partial x_j} - \frac{\partial v_j}{\partial x_i} \tag{9-102}$$

By adding and subtracting terms in Eq. (9-95) we obtain

$$\frac{\partial v_i}{\partial x_j} = \frac{1}{2}\left(\frac{\partial v_i}{\partial x_j} + \frac{\partial v_j}{\partial x_i}\right) + \frac{1}{2}\left(\frac{\partial v_i}{\partial x_j} - \frac{\partial v_j}{\partial x_i}\right) \tag{9-103}$$

or
$$[\![(\nabla \mathbf{v})_{ji}]\!] = \tfrac{1}{2}(\Delta_{ji} + \Omega_{ji}) \quad \text{or} \quad \nabla \mathbf{v} = \tfrac{1}{2}(\underline{\boldsymbol{\Delta}} + \underline{\boldsymbol{\Omega}}) \tag{9-104}$$

The *velocity-gradient tensor* $\partial v_j/\partial x_i$, obtained in Eq. (9-95), is therefore one-half the sum of the rate-of-deformation and vorticity tensors.

Newton's Law of Viscosity

Since Newton's law of viscosity says that the stress is proportional to the rate of strain, it can be written

$$-\sigma_{yx} g_c \equiv \sigma_{yx} g_c \equiv -\tau_{yx} = \mu\, \Delta_{yx} = \mu \left(\frac{\partial v_y}{\partial x} + \frac{\partial v_x}{\partial y}\right) \tag{9-105}$$

or, more generally, for cartesian coordinates

$$\sigma_{ij} g_c = \tau_{ij} = -\mu\, \Delta_{ij} = -\mu \left(\frac{\partial v_i}{\partial x_j} + \frac{\partial v_j}{\partial x_i}\right) \tag{9-106}$$

Note that $\Delta_{yx} = \Delta_{xy}$ and hence $\sigma_{yx} = \sigma_{xy}$. In other words, $\Delta_{ij} = \Delta_{ji}$, $\tau_{ij} = \tau_{ji}$, and $\sigma_{ij} = \sigma_{ji}$. Tensors which have the above property are said to be *symmetrical tensors*. The shear stress can be shown more rigorously to be symmetrical for nonpolar fluids.[1,20]

If the fluid is *incompressible,* Eq. (9-106) also holds for $i = j$. For example, the normal stress in the x direction on the x face (Appendix A9-5) is

$$\tau_{xx} = \sigma_{xx} g_c = -2\mu \frac{\partial v_x}{\partial x} \tag{9-107}$$

This stress corresponds to a compression or tension (note the factor 2). For a *compressible* fluid, a fluid element may change in volume upon compression or tension, and to account for this fact another term must be added to Eq. (9-106)

$$\tau_{ii} = \sigma_{ii} g_c = -2\mu \frac{\partial v_i}{\partial x_j} - \lambda \sum_{i=1}^{3} \frac{\partial v_j}{\partial x_i} \tag{9-108}$$

where λ is called the *coefficient of bulk viscosity* or the *second viscosity coefficient.* It is difficult to measure and is usually taken as $-\frac{2}{3}\mu$ on the basis of a theoretical argument put forward by Stokes.

In Gibbs notation, Newton's law of viscosity can be written

$$\underline{\boldsymbol{\sigma}} g_c = \underline{\boldsymbol{\tau}} = -\mu\, \underline{\boldsymbol{\Delta}} + \tfrac{2}{3}\mu\, \underline{\boldsymbol{\delta}}\, \nabla \cdot \mathbf{v} \tag{9-109}$$

where $\underline{\boldsymbol{\delta}}$ is the unit tensor (its diagonal terms are unity, and its off-diagonal terms are zero) [see (Eq. 9-47)]. The rate-of-deformation tensor is

$$\underline{\boldsymbol{\Delta}} = (\nabla \mathbf{v}) + (\nabla \mathbf{v})^T \tag{9-110}$$

where $\nabla \mathbf{v}$, the *velocity-gradient tensor,* is the dyad [Eqs. (9-95) and (9-96)]

$$\nabla \mathbf{v} = \sum_i \sum_j \frac{\partial v_j}{\partial x_i} \boldsymbol{\delta}_i \boldsymbol{\delta}_j = \sum_i \sum_j [\![\nabla \mathbf{v}_{ij}]\!] \boldsymbol{\delta}_i \boldsymbol{\delta}_j \tag{9-111}$$

and $(\nabla \mathbf{v})^T$ is the transpose of $\nabla \mathbf{v}$, obtained by interchanging rows and columns when $\nabla \mathbf{v}$ is written as a matrix. For an incompressible fluid, $\nabla \cdot \mathbf{v} = 0$ and Newton's law in Gibbs notation [Eq. (9-109)] is

$$\underline{\boldsymbol{\tau}} = -\mu \, \underline{\boldsymbol{\Delta}} = -\mu[\nabla \mathbf{v} + (\nabla \mathbf{v})^T] \tag{9-112}$$

Vorticity Vector

Let us define the *vorticity vector* as

$$2\boldsymbol{\omega} = \nabla \times \mathbf{v} \tag{9-113}$$

In general the cross product of two vectors **u** and **v** yields a vector (Appendix 9-2) perpendicular to the plane of both **u** and **v**. If the vector **u** is the operator ∇, this becomes the determinant

$$2\boldsymbol{\omega} = \nabla \times \mathbf{v} = \begin{vmatrix} \boldsymbol{\delta}_1 & \boldsymbol{\delta}_2 & \boldsymbol{\delta}_3 \\ \dfrac{\partial}{\partial x_1} & \dfrac{\partial}{\partial x_2} & \dfrac{\partial}{\partial x_3} \\ v_1 & v_2 & v_3 \end{vmatrix} \tag{9-114}$$

Now consider the components of the vorticity tensor [Eq. (9-102)]

$$\Omega_{ji} = \frac{\partial v_i}{\partial x_j} - \frac{\partial v_j}{\partial x_i}$$

Then
$$\Omega_{12} = \frac{\partial v_2}{\partial x_1} - \frac{\partial v_1}{\partial x_2} = -\Omega_{21}$$

But by expansion of Eq. (9-114)

$$2\omega_3 = \frac{\partial v_2}{\partial x_1} - \frac{\partial v_1}{\partial x_2} = \Omega_{12} = -\Omega_{21}$$

The vector $\boldsymbol{\omega}$ is thus seen to represent the components of $\underline{\boldsymbol{\Omega}}$ as follows

$$\boldsymbol{\omega} = \begin{bmatrix} \omega_1 \\ \omega_2 \\ \omega_3 \end{bmatrix} \qquad \tfrac{1}{2}\underline{\boldsymbol{\Omega}} = \begin{bmatrix} 0 & \omega_3 & -\omega_2 \\ -\omega_3 & 0 & \omega_2 \\ \omega_2 & -\omega_1 & 0 \end{bmatrix} \tag{9-115}$$

Irrotational Flow; Velocity Potential

A flow for which $\nabla \times \mathbf{v} = 0$ is said to be an *irrotational flow*. For such flows, it is possible to treat the velocity as a quantity that results from a *velocity potential* ϕ in the same manner as heat flow results from a temperature potential. The velocity potential ϕ is defined by

$$v_x = -\frac{\partial \phi}{\partial x} \qquad \text{and} \qquad v_y = -\frac{\partial \phi}{\partial y} \tag{9-116}$$

(see Sec. 9-19 for an example).

Distinction between Deformation and Rotation: A General Treatment

Consider the points P and Q separated by the distance $|d\mathbf{x}| \equiv ds$ (Fig. 9-16). Point P is moving with velocity $\mathbf{v}$ and point Q is moving with velocity $\mathbf{v} + d\mathbf{v}$. Substituting the second of Eqs. (9-104) into Eq. (9-96) gives

$$d\mathbf{v} = \tfrac{1}{2}\underline{\mathbf{\Delta}} \cdot d\mathbf{x} + \tfrac{1}{2}\underline{\mathbf{\Omega}} \cdot d\mathbf{x} \tag{9-117}$$

which represents the relative velocity between the points in terms of the velocity attributed to the deformation tensor and that attributed to the vorticity tensor.

Now let us show that the first term is the only part of the motion that contributes to the deformation (in other words that the name we have given $\underline{\mathbf{\Delta}}$ is not a misnomer). The rate of strain can be expressed in terms of the time rate of change of the distance ds per unit length, or

$$\frac{1}{ds}\frac{D(ds)}{Dt}$$

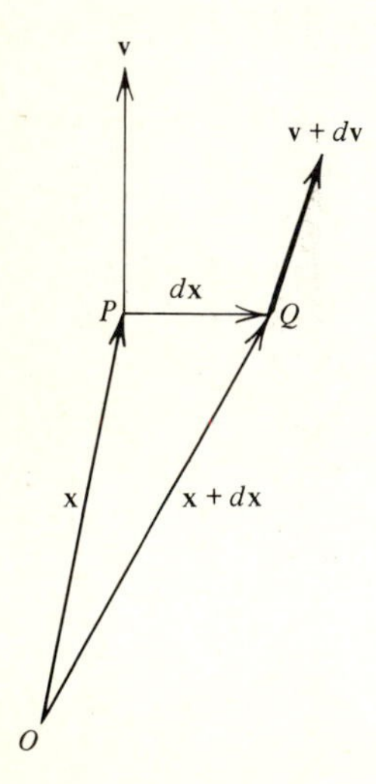

Figure 9-16 Relative movement of two points.

The time rate of change of the distance between the two points P and Q will depend upon the relative velocity between the points, but in general the relative-velocity vector $d\mathbf{v}$ will *not* be in the direction of the vector distance $d\mathbf{x}$ between the points P and Q, although it *may* have a component in that direction. To find that component we first need to define a unit vector in the direction pointing from P to Q as $\boldsymbol{\lambda} \equiv d\mathbf{x}/ds$, where ds is the magnitude of $d\mathbf{x}$. The component of $d\mathbf{v}$ in the direction of $\boldsymbol{\lambda}$ is $d\mathbf{v} \cdot \boldsymbol{\lambda}$ and is $D(ds)/Dt$, the time rate of change of the distance between the points.

Then using Eqs. (9-96) and (9-117) gives

$$\frac{D}{Dt}(ds) = (d\mathbf{v}) \cdot \boldsymbol{\lambda} = d\mathbf{x} \cdot \nabla\mathbf{v} \cdot \boldsymbol{\lambda} = \frac{1}{2} d\mathbf{v} \cdot \underline{\boldsymbol{\Delta}} \cdot \boldsymbol{\lambda} + \frac{1}{2} d\mathbf{x} \cdot \underline{\boldsymbol{\Omega}} \cdot \boldsymbol{\lambda} \qquad (9\text{-}118)$$

If we divide both sides by ds, we get the rate of strain, or

$$\frac{1}{ds}\frac{D(ds)}{Dt} = \frac{1}{2}(\boldsymbol{\lambda} \cdot \underline{\boldsymbol{\Delta}} \cdot \boldsymbol{\lambda} + \boldsymbol{\lambda} \cdot \Omega \cdot \boldsymbol{\lambda}) \qquad (9\text{-}119)$$

The second term can be shown to be zero by writing it in terms of components and using the summation convention.† Therefore it does not actually contribute to the deformation but represents a rigid-body rotation. Then we can write Eq. (9-118) as

$$\frac{1}{ds}\frac{D(ds)}{Dt} = \frac{1}{2}\boldsymbol{\lambda} \cdot \underline{\boldsymbol{\Delta}} \cdot \boldsymbol{\lambda} \qquad (9\text{-}120)$$

which shows that the fluid undergoes strain or deformation if the relative *distance* between two material points is changing. It can also be written

$$\frac{D}{Dt}(ds)^2 = d\mathbf{x} \cdot \underline{\boldsymbol{\Delta}} \cdot d\mathbf{x} \qquad (9\text{-}121)$$

Deformation Tensor in Curvilinear Coordinates‡

In cylindrical or spherical coordinates the expression for $\underline{\boldsymbol{\Delta}}$ could in principle be obtained by transforming the cartesian coordinate expression into the appropriate coordinate system. For example, to find $\Delta_{r\theta}$ in cylindrical coordinates, one could start with Δ_{xy} and transform the velocity terms v_x and v_y into v_r and v_θ. Then the terms $\partial/\partial x$, $\partial/\partial y$, and $\partial/\partial z$ would be transformed into $\partial/\partial r$, $\partial/\partial \theta$, and $\partial/\partial z$, by means of the chain rule. The resulting differentiations would then be carried out to obtain the final result.

An easier method would be to write ∇ and $\mathbf{v}$ in terms of the unit vectors and carry out the operations of differentiation and summation. From Eq. (9-110)

$$\underline{\boldsymbol{\Delta}} = \nabla\mathbf{v} + (\nabla\mathbf{v})^T$$

† Let $\lambda_i\Omega_{ij}\lambda_j = \frac{1}{2}(\lambda_i\Omega_{ij}\lambda_j + \lambda_j\Omega_{ji}\lambda_i) = \frac{1}{2}(\Omega_{ij} + \Omega_{ji})\lambda_i\lambda_j = 0$, since $\Omega_{ij} = -\Omega_{ji}$ (skew-symmetric).
‡ Expressions for $\underline{\boldsymbol{\Delta}}$ in various coordinate systems are given in Table I-9.

where

$$\nabla \mathbf{v} = \left(\sum_m \boldsymbol{\delta}_m \frac{\partial}{\partial u_m}\right)\left(\sum_n \boldsymbol{\delta}_n v_n\right) = \sum_m \sum_n \left(\boldsymbol{\delta}_m v_n \frac{\partial \boldsymbol{\delta}_n}{\partial u_m} + \boldsymbol{\delta}_m \boldsymbol{\delta}_n \frac{\partial v_n}{\partial u_m}\right)$$

$$= \sum_m \sum_n [\![\nabla \mathbf{v}]\!]_{mn} \boldsymbol{\delta}_m \boldsymbol{\delta}_n \qquad (9\text{-}122)$$

The $\partial \boldsymbol{\delta}_n / \partial u_m$ can be found from Eq. (A8-7-36) or from geometrical considerations (see Appendix 8-5). To find a given component, such as the $r\theta$ component, one first collects all terms in $\nabla \mathbf{v}$ that are coefficients of the unit dyad $\boldsymbol{\delta}_r \boldsymbol{\delta}_\theta$. To find $(\nabla \mathbf{v})^T$ the transpose of the $r\theta$ component is needed; this would be the θr component; hence all coefficients of $\boldsymbol{\delta}_\theta \boldsymbol{\delta}_r$ would be obtained.

Carrying out these operations gives

$$[\![(\nabla \mathbf{v})_{r\theta}]\!] = \frac{\partial v_\theta}{\partial r} \qquad (9\text{-}123)$$

and

$$[\![(\nabla \mathbf{v})_{\theta r}]\!] = \frac{\partial v_r}{r\,\partial \theta} - \frac{v_\theta}{r} \qquad (9\text{-}124)$$

Then

$$\Delta_{r\theta} = [\![(\nabla \mathbf{v})_{r\theta}]\!] + [\![(\nabla \mathbf{v})_{\theta r}]\!] \qquad (9\text{-}125)$$

and

$$\Delta_{r\theta} = r \frac{\partial (v_\theta / r)}{\partial r} + \frac{\partial v_r}{r\,\partial \theta} \qquad (9\text{-}126)$$

so that for a newtonian fluid

$$\tau_{r\theta} = -\mu \left[r \frac{\partial (v_\theta / r)}{\partial r} + \frac{\partial v_r}{r\,\partial \theta} \right] \qquad (9\text{-}127)$$

Similar expressions can be found for other components and other coordinate systems.

Physical Interpretation of $\Delta_{r\theta}$

A physical interpretation of the expression for $\tau_{r\theta}$ can be obtained (see Fig. 9-17) by considering the tangential laminar flow of an incompressible fluid between two vertical coaxial cylinders with the outer cylinder rotating and the inner one remaining stationary, as shown in Fig. 9-17. Since $v_r = 0$ [Eq. (9-127)],

$$\tau_{r\theta} = -\mu r \frac{\partial (v_\theta / r)}{\partial r} \qquad (9\text{-}128)$$

where v_θ / r represents the angular velocity of an element of fluid at radial position r. Since there is a change in *angular* velocity with respect to r, deformation occurs. But if *both* cylinders were rotating with the *same* angular velocity, each element of fluid would have the same angular velocity. Hence the *relative* distance between fluid elements would be the same and there would be no deformation.

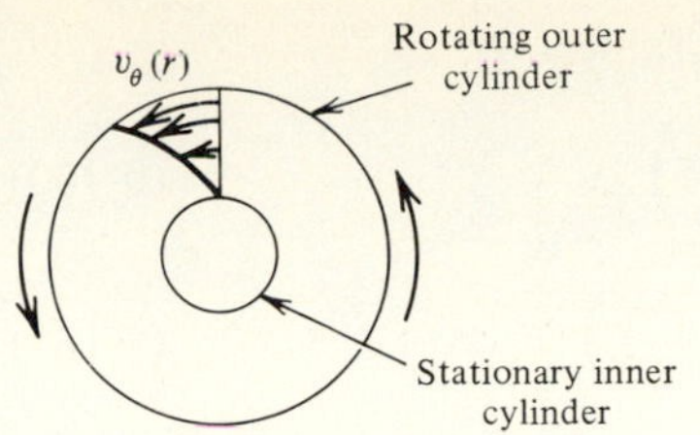

Figure 9-17 Tangential flow between concentric cylinders.

Example 9-2 Find $(\nabla \mathbf{v})_{r\theta}$, $(\nabla \mathbf{v})_{\theta r}$, and $\Delta_{r\theta}$ for cylindrical coordinates, using the method of differentiation of unit vectors.

SOLUTION Using expressions for ∇ and $\mathbf{v}$, we obtain

$$\nabla \mathbf{v} = \left(\boldsymbol{\delta}_r \frac{\partial}{\partial r} + \boldsymbol{\delta}_\theta \frac{\partial}{r\,\partial\theta} + \boldsymbol{\delta}_z \frac{\partial}{\partial z}\right)(\boldsymbol{\delta}_r v_r + \boldsymbol{\delta}_\theta v_\theta + \boldsymbol{\delta}_z v_z)$$

$$= \boldsymbol{\delta}_r\boldsymbol{\delta}_r \frac{\partial v_r}{\partial r} + \boldsymbol{\delta}_r\boldsymbol{\delta}_\theta \frac{\partial v_\theta}{\partial r} + \boldsymbol{\delta}_r\boldsymbol{\delta}_z \frac{\partial v_z}{\partial r} + \boldsymbol{\delta}_\theta\boldsymbol{\delta}_r \frac{\partial v_r}{r\,\partial\theta} + \boldsymbol{\delta}_\theta \frac{\partial \boldsymbol{\delta}_r}{r\,\partial\theta} v_r + \boldsymbol{\delta}_\theta\boldsymbol{\delta}_\theta \frac{\partial v_\theta}{r\,\partial\theta}$$

$$+ \boldsymbol{\delta}_\theta \frac{\partial \boldsymbol{\delta}_\theta}{r\,\partial\theta} v_\theta + \boldsymbol{\delta}_\theta\boldsymbol{\delta}_z \frac{\partial v_z}{r\,\partial\theta} + \boldsymbol{\delta}_z\boldsymbol{\delta}_m \text{ terms}$$

Using $\partial\boldsymbol{\delta}_r/\partial\theta = \boldsymbol{\delta}_\theta$ and $\partial\boldsymbol{\delta}_\theta/\partial\theta = -\boldsymbol{\delta}_r$ (Appendix 8-5) and collecting the coefficients of $\boldsymbol{\delta}_r\boldsymbol{\delta}_\theta$ and $\boldsymbol{\delta}_\theta\boldsymbol{\delta}_r$ separately gives, respectively,

$$[\![(\nabla \mathbf{v})_{r\theta}]\!] = \frac{\partial v_\theta}{\partial r} \qquad \text{and} \qquad [\![(\nabla \mathbf{v})_{\theta r}]\!] = \frac{\partial v_r}{r\,\partial\theta} - \frac{v_\theta}{r}$$

Then
$$\Delta_{r\theta} \equiv \frac{\partial v_\theta}{\partial r} + \frac{\partial v_r}{r\,\partial\theta} - \frac{v_\theta}{r} = r\frac{\partial(v_\theta/r)}{\partial r} + \frac{\partial v_r}{r\,\partial\theta} \qquad \square$$

9-13 NAVIER-STOKES EQUATIONS†

Now that we have derived the equation of motion in terms of the momentum flux and the expression for the flux law for a newtonian fluid, the next logical step is to derive an equation of motion in terms of velocity. In doing so we will assume that both density and viscosity are constant. The resulting set of equations (one for each component) is called the *Navier-Stokes equations*.

Let us consider the x component of the equation of motion in lagrangian form [from Eq. (9-34)]

$$\rho \frac{Dv_x}{Dt} = -[\![(\nabla \cdot \underline{\boldsymbol{\tau}})_x]\!] - \frac{\partial p}{\partial x} + \rho g_x \tag{9-129}$$

†Combines vector and nonvector approach.

The components of $\underline{\tau}$ are [from Eq. (9-106) and Table I-9]

$$\tau_{xx} = -\mu\left(\frac{\partial v_x}{\partial x} + \frac{\partial v_x}{\partial x}\right) \tag{9-130}$$

$$\tau_{yx} = \tau_{xy} = -\mu\left(\frac{\partial v_x}{\partial y} + \frac{\partial v_y}{\partial x}\right) \tag{9-131}$$

and

$$\tau_{zx} = -\mu\left(\frac{\partial v_x}{\partial z} + \frac{\partial v_z}{\partial x}\right) \tag{9-132}$$

If we expand $[\![(\nabla \cdot \underline{\tau})_x]\!]$ [see Eq. (9-27) or Table I-4] and substitute the above, we get

$$\begin{aligned}
[\![(\nabla \cdot \underline{\tau})_x]\!] &= \frac{\partial \tau_{xx}}{\partial x} + \frac{\partial \tau_{yx}}{\partial y} + \frac{\partial \tau_{zx}}{\partial z} \\
&= \frac{\partial}{\partial x}\left[-\mu\left(\frac{\partial v_x}{\partial x} + \frac{\partial v_x}{\partial x}\right)\right] + \frac{\partial}{\partial y}\left[-\mu\left(\frac{\partial v_x}{\partial y} + \frac{\partial v_y}{\partial x}\right)\right] + \frac{\partial}{\partial z}\left[-\mu\left(\frac{\partial v_x}{\partial z} + \frac{\partial v_z}{\partial x}\right)\right] \\
&= -\mu\left[\left(\frac{\partial^2 v_x}{\partial x^2} + \frac{\partial^2 v_x}{\partial y^2} + \frac{\partial^2 v_x}{\partial z^2}\right) + \frac{\partial}{\partial x}\left(\frac{\partial v_x}{\partial x} + \frac{\partial v_y}{\partial y} + \frac{\partial v_z}{\partial z}\right)\right]
\end{aligned} \tag{9-133}$$

where the last term is zero because

$$\nabla \cdot \mathbf{v} \equiv \frac{\partial v_x}{\partial x} + \frac{\partial v_y}{\partial y} + \frac{\partial v_z}{\partial z} = 0$$

for an incompressible fluid. Using the definition of the laplacian of a scalar [Table I-2 or Eq. (8-30)] gives

$$[\![(\nabla \cdot \tau)_x]\!] = -\mu\, \nabla^2 v_x \tag{9-134}$$

and the x component of Eq. (9-129) is

$$\boxed{\rho \frac{Dv_x}{Dt} = \mu\, \nabla^2 v_x - \frac{\partial p}{\partial x} + \rho g_x} \tag{9-135}$$

If we multiply the x component by $\boldsymbol{\delta}_x$, the y component by $\boldsymbol{\delta}_y$, and the z component by $\boldsymbol{\delta}_z$ and sum, we get the vector equation

$$\boxed{\rho \frac{D\mathbf{v}}{Dt} = \mu\, \nabla^2 \mathbf{v} - \nabla p + \rho \mathbf{g}} \tag{9-136}$$

The substantial derivative is given as Eq. (8-133)

$$\rho \frac{D\mathbf{v}}{Dt} \equiv \rho\left(\frac{\partial \mathbf{v}}{\partial t} + \mathbf{v} \cdot \nabla \mathbf{v}\right) \qquad \text{or} \qquad \rho \frac{Dv_i}{Dt} = \rho\left(\frac{\partial v_i}{\partial t} + \sum_j v_j \frac{\partial v_i}{\partial x_j}\right) \tag{9-137}$$

and the x component is

$$\rho \frac{Dv_x}{Dt} \equiv \rho\left(\frac{\partial v_x}{\partial t} + v_x \frac{\partial v_x}{\partial x} + v_y \frac{\partial v_x}{\partial y} + v_z \frac{\partial v_x}{\partial z}\right) \tag{9-138a}$$

The Navier-Stokes equations for various coordinate systems are given in part B of Tables I-6 to I-8 or in Table I-5.

9-14 APPLICATION OF THE NAVIER-STOKES EQUATION TO ONE-DIMENSIONAL FLOW: GENERAL SOLUTION PROCEDURES†

The first step in dealing with fluid-flow problems is usually to expand the equation of continuity into components and simplify by eliminating negligible terms (see Tables I-6 to I-8). This may give a relationship between velocity components (or show that one of the components is zero). The next step depends upon the nature of the problem. If only velocity profiles are desired, the following direct procedure can be used (for newtonian fluids with constant μ and ρ):

1. Expand the Navier-Stokes equation into its components, using the appropriate coordinate system (use part B of Tables I-6 to I-8 or Table I-5).
2. Then simplify by eliminating negligible terms.
3. Determine upon what independent variables the pressure may or may not depend.
4. Solve for the velocity distribution using the boundary condition.

This procedure leads to a second-order equation for velocity.

If shear stresses and velocities are both required, it may be preferable to use the following procedure:

1. Determine which components of the momentum-flux (shear-stress) tensor are nonzero (Table I-9 may be used).
2. Expand the equation of motion (in terms of flux) into its components in the appropriate coordinate system (part A of Tables I-6 to I-8 or Table I-4 may be used).
3. Solve the equation of motion for the flux if possible. (It may be necessary to obtain velocities first.)
4. Substitute the appropriate flux-law expression (Table I-9 if the fluid is newtonian).
5. Solve for the velocity distribution.

†Combines vector and nonvector approach.

In the following examples we apply the Navier-Stokes equation to a one-dimensional flow problem to obtain results of immediate practical importance and to accustom the reader to some of the various mathematical techniques and simplifications which have been developed.

Example 9-3: Slit flow (Navier-Stokes equation) This is the same example we looked at previously in a study of the shear-stress distribution (Example 9-1). We obtained

$$\tau_{xy} = \frac{p(0) - p(L)}{L}(Y - y) \qquad \begin{aligned} v_x &= v_x(y) \\ v_y &= 0 \\ v_z &= 0 \end{aligned}$$

For this flow, the x and y components of the Navier-Stokes equations (see part B of Table I-6 or Table I-4) reduce to

$$\underset{x \text{ component}}{-\frac{\partial p}{\partial x} + \frac{d^2 v_x}{dy^2} = 0} \qquad \underset{y \text{ component}}{-\frac{\partial p}{\partial y} + \rho g_y = 0} \tag{9-138b}$$

Differentiating the second of Eqs. (9-138*b*) with respect to x and reversing the order of differentiation of p, we find that $\partial p/\partial x$ is independent of y. It is also independent of z since the plates are assumed to be infinite in width, and thus dp/dx is at most a function of x. Differentiating the first of Eqs. (9-138*b*) gives

$$\frac{\partial}{\partial x}\left(\frac{\partial p}{\partial x}\right) = \mu \frac{d^2}{dy^2}\left(\frac{\partial v_x}{\partial x}\right) = 0$$

since v_x depends only on y. Then

$$-\mu \frac{d^2 v_x}{dy^2} = -\frac{dp}{dx} = \text{const} = G_p = \frac{p_0 - p_L}{L} \tag{9-138c}$$

We now integrate twice to obtain the velocity profile

$$v_x = -\frac{G_p y^2}{2\mu} + C_1 y + C_2$$

Using the no-slip boundary conditions $v_x = 0$ at $y = 0$ and at $y = 2Y$, this becomes

$$v_x = \frac{G_p Y^2}{2\mu}\left[\frac{2y}{Y} - \left(\frac{y}{Y}\right)^2\right]$$

or, from Eq. (9-138*c*),

$$v_x = \frac{p_0 - p_L}{2\mu L} Y^2 \left(\frac{2y}{Y} - \frac{y^2}{Y^2}\right) \tag{9-139}$$

□

Example 9-4: Cylindrical-tube flow Derive an expression for the velocity distribution for the fully developed laminar steady-state vertical pipe flow of an incompressible newtonian fluid (see Chap. 3).

SOLUTION At steady state $\partial/\partial t = 0$. From symmetry $\partial v_\theta/\partial\theta = 0$ and $v_\theta = 0$. Also, $\partial v_z/\partial z = 0$ for fully developed flow. For an incompressible fluid, ρ is constant. Then the continuity equation (Table I-7) leads to $v_r = 0$ (see Example 8-5). The Navier-Stokes equations (Table I-5 or part B of Table I-7) become

$$0 = 0 - \frac{\partial p}{\partial r} + 0 \qquad 0 = -\mu\left[\frac{1}{r}\frac{\partial}{\partial r}\left(r\frac{\partial v_z}{\partial r}\right)\right] + \rho g_z - \frac{\partial p}{\partial z}$$

r direction $\qquad$ z direction

In the θ direction all terms are zero, and hence

$$\frac{-\partial p}{\partial z} + \rho g_z = \frac{-dp}{dz} + \rho g_z = \text{const} \equiv \frac{\mathcal{P}_0 - \mathcal{P}_L}{L} \equiv \Phi_M$$

where $\mathcal{P}$ is the dynamic pressure (defined in Chap. 3) and Φ_M is the momentum generation per unit volume. Then

$$\frac{-\mu}{r}\frac{d}{dr}\left(r\frac{dv_z}{dr}\right) = \Phi_M$$

Integrating twice, first letting the velocity gradient be zero at the center and then using $v_z = 0$ at $r = R$, gives

$$v_z = \frac{(\mathcal{P}_0 - \mathcal{P}_L)R^2}{4\mu L}\left(1 - \frac{r^2}{R^2}\right) \tag{9-140}$$

□

Example 9-5: Torque in a viscometer We wish to calculate the torque in a viscometer, which can be modeled as two concentric cylinders of length L, the outer cylinder of radius R_2 rotating at an angular velocity ω_0 and the inner cylinder of radius R_1 stationary. The incompressible fluid inside flows tangentially (in the θ direction only) in steady-state laminar flow. The cylinders are vertical and long (see Fig. 9-17). (*a*) Use the tables to write simplified expressions for the r, θ, and z components of the equation of motion (in terms of the fluxes). Justify the elimination of any terms. Interpret each equation physically in terms of forces, torques, etc. (*b*) Solve for the shear stress and torque distribution in terms of the torque at the inner cylinder Υ_0 and radial position r. (*c*) Write boundary conditions *on velocity*. (*d*) Assuming the fluid is newtonian, find the components of $\mathbf{\Delta}$ and solve for the velocity distribution $v_\theta(r)$ and the torque Υ_0 in terms of $R_1/R_2 \equiv \kappa$, ω_0, R_2, μ, and L.

SOLUTION (*a*) At steady state $\partial/\partial t = 0$. For a very long cylinder, we can assume that end effects are negligible. Then

$$v_z = 0 \quad \text{and} \quad \frac{\partial v_z}{\partial z} = 0$$

in the EOC. Also

$$\frac{\partial v_\theta}{\partial \theta} = 0 \text{ (by symmetry)} \quad \text{and} \quad v_\theta = v_\theta(r)$$

The EOC then gives (part B of Table I-7)

$$\frac{\partial (r v_r)}{r\, \partial r} = 0$$

which gives $\qquad v_r r = \text{const} = 0 \quad \text{or} \quad v_r = 0$

From Table I-9 for a newtonian fluid, $\Delta_{rr} = 0 = \Delta_{\theta\theta} = \Delta_{zz}$, and since $\tau_{ij} = -\mu\Delta_{ij}$, the stress tensor has these components

$$\underline{\tau} = \begin{bmatrix} \tau_{rr} & \tau_{r\theta} & \tau_{rz} \\ \tau_{\theta r} & \tau_{\theta\theta} & \tau_{\theta z} \\ \tau_{zr} & \tau_{z\theta} & \tau_{zz} \end{bmatrix} = \begin{bmatrix} 0 & \tau_{r\theta} & 0 \\ \tau_{\theta r} & 0 & 0 \\ 0 & 0 & 0 \end{bmatrix}$$

The EOM for cylindrical coordinates is given in part A of Table I-7 or Table I-4:

z component:
$$0 = \frac{-\partial p}{\partial z} + \rho g_z \tag{1}$$

Thus the pressure force in the z direction is balanced by the gravity force in the z direction

r component:
$$-\rho \frac{v_\theta^2}{r} = \frac{-\partial p}{\partial r} \tag{2}$$

Thus the centrifugal acceleration is balanced by a net radial pressure gradient

θ component:
$$0 = \frac{d(r^2 \tau_{r\theta})}{r^2\, dr} \tag{3}$$

Multiplying Eq. (3) by $2\pi L r^2$ gives

$$0 = \frac{d}{dr}[r(2\pi r L \tau_{r\theta})] \tag{4}$$

The viscous force acting in the θ direction at $r = r$ is

$$\tau_{r\theta}(2\pi r L) = F_{v\theta} \tag{5}$$

and
$$r F_{v\theta} = \text{viscous torque about } z \text{ axis} \equiv \Upsilon_z \tag{6}$$

Equations (4) to (6) indicate that

$$\frac{d\Upsilon_z}{dr} = 0 \tag{7}$$

This equation can also be derived from an angular momentum balance in which the net torque acting on an element $\Delta \mathcal{V} = 2\pi r L \, \Delta r$ is set equal to zero and divided by $\Delta \mathcal{V}$ while $\Delta r \to 0$.

(*b*) Integrating Eq. (7) gives the *torque* distribution

$$\Upsilon_z = \text{const} = \Upsilon_0 \tag{8}$$

Using Eqs. (5) to (7) gives

$$\Upsilon_0 = \tau_{r\theta}(2\pi r^2 L)$$

and the *flux* distribution is

$$\tau_{r\theta} = \frac{\Upsilon_0}{2\pi r^2 L} \tag{9}$$

(*c*)

$$v_\theta = \begin{cases} 0 & \text{at } r = R_1 \quad (10) \\ \omega_0 R_2 & \text{at } r = R_2 \quad (11) \end{cases} \qquad \text{no slip}$$

(*d*) Only $\Delta_{r\theta} = \Delta_{\theta r}$ will be nonzero. With substitution from Example 9-2, Eq. (9-126) gives

$$\Delta_{r\theta} = \Delta_{\theta r} = \frac{\partial v_\theta}{\partial r} + \left(\frac{\partial v_r}{r\,\partial\theta} - \frac{v_\theta}{r}\right) = \frac{r d(v_\theta/r)}{dr}$$

Then, from Newton's law,

$$\tau_{r\theta} = -\mu \Delta_{r\theta} = -\mu r \frac{d(v_\theta/r)}{dr} \tag{12}$$

To find $v_\theta(r)$ we need to equate (9) and (12) and integrate using boundary condition (10)

$$\frac{\Upsilon_0}{2\pi L\mu}\int_{R_1}^{r}\frac{dr}{r^3} = \int_0^{v_\theta/r} d\frac{v_\theta}{r} \qquad \frac{\Upsilon_0}{4\pi L\mu}\left(\frac{1}{R_1^2} - \frac{1}{r^2}\right) = \frac{v_\theta}{r} \tag{13}$$

Using boundary condition (11) and letting $v_\theta/r = \omega$ gives

$$\frac{\Upsilon_0}{4\pi L\mu}\left(\frac{1}{R_1^2} - \frac{1}{R_2^2}\right) = \omega_0 \tag{14}$$

Eliminating Υ_0 in Eqs. (13) and (14) gives

$$\frac{v_\theta}{r} = \omega_0 \frac{(1/R_1^2) - 1/r^2}{(1/R_1^2) - 1/R_2^2} \tag{15}$$

and

$$v_\theta = \frac{r\omega_0 R_2[(r/\kappa R_2) - \kappa R_2/r]}{(1/\kappa) - \kappa} \qquad \text{if } \kappa \equiv \frac{R_1}{R_2}$$

Solving Eq. (14) for Υ_0 gives

$$\Upsilon_0 = \frac{4\pi\mu L\omega_0}{\kappa^2 R_2^2}(1 - \kappa^2)$$

□

9-15 NONNEWTONIAN FLUIDS†

Generalized Newtonian Fluid

For nonnewtonian incompressible fluids, Newton's law of viscosity (9-112) can be modified to read

$$\underline{\tau} = -\eta \underline{\Delta} \tag{9-141}$$

where η is an effective viscosity that varies with the velocity gradient. Equation (9-141) is the constitutive equation for a *generalized newtonian fluid.* An example is the power-law fluid, where

$$\eta = m|\sqrt{\tfrac{1}{2}(\underline{\Delta}:\underline{\Delta})}|^{n-1} \tag{9-142}$$

where m and n are constants for a given fluid, and where

$$\underline{\Delta}:\underline{\Delta} = \sum_i \sum_j \Delta_{ij}\boldsymbol{\delta}_i\boldsymbol{\delta}_j : \sum_k \sum_l \Delta_{kl}\boldsymbol{\delta}_k\boldsymbol{\delta}_l$$

$$= \sum_i \sum_j \sum_k \sum_l \Delta_{ij}\Delta_{kl}\boldsymbol{\delta}_i\boldsymbol{\delta}_j : \boldsymbol{\delta}_k\boldsymbol{\delta}_l$$

To take the double dot product we first dot-multiply the adjacent unit vectors

$$\boldsymbol{\delta}_j \cdot \boldsymbol{\delta}_k = \delta_{jk} = \begin{cases} 1 & \text{if } j = k \\ 0 & \text{if } j \neq k \end{cases}$$

Then we dot-multiply the remaining unit vectors

$$\boldsymbol{\delta}_i \cdot \boldsymbol{\delta}_l = \delta_{il} = \begin{cases} 1 & \text{if } i = l \\ 0 & \text{if } i \neq l \end{cases}$$

Thus we need obtain only $j = k$ and $l = i$, and

$$\underline{\Delta}:\underline{\Delta} = \sum_i \sum_j \Delta_{ij}\Delta_{ji} + \sum_i \sum_j \Delta_{ij}^2 \tag{9-143}$$

Then

$$\eta = m \left| \left(\frac{1}{2} \sum_i \sum_j \Delta_{ij}^2 \right)^{1/2} \right|^{n-1} \tag{9-144}$$

Example 9-6: Couette flow of a power-law fluid Find an expression for $\underline{\tau}$ for a simple Couette flow for which

$$v_x = v_x(y) \text{ only} \qquad \text{and} \qquad v_y = 0 = v_z$$

assuming that the power-law equation (9-142) holds.

†Sections 9-15 through 9-21 can be delayed until after Secs. 10-1 to 10-5 have been studied.

SOLUTION Let $\dot{\gamma} \equiv dv_x/dy$; then

$$\underline{\Delta} = \begin{bmatrix} 0 & \dot{\gamma} & 0 \\ \dot{\gamma} & 0 & 0 \\ 0 & 0 & 0 \end{bmatrix}$$

$$\Delta_{xy} = \Delta_{yx} = \dot{\gamma} \qquad \Delta_{xz} = 0 = \Delta_{zx} \equiv \Delta_{xx} = \Delta_{yy} = \Delta_{zz}$$

and

$$\eta = m|[\tfrac{1}{2}(\Delta_{12}^2 + \Delta_{12}^2)]^{1/2}|^{n-1} = m|\dot{\gamma}|^{n-1}$$

Then

$$\tau_{yx} = -m|\dot{\gamma}|^{n-1}\dot{\gamma} \tag{9-145}$$

which agrees with our previous expression (2-48). □

Other Constitutive Equations

The Williamson model [Eq. (2-50)] can be generalized to

$$\mu_A \equiv \eta = \eta_\infty + \frac{\eta_0 - \eta_\infty}{(1 + |\dot{\gamma}|/\alpha_1)^{\alpha_2}} \qquad \alpha_1, \alpha_2 = \text{const}$$

The Bingham-plastic model [Eq. (2-51)] becomes

$$\eta = \eta_0 + \frac{\tau_0}{|\dot{\gamma}|} \qquad \text{for } \underline{\tau}:\underline{\tau} > 2\tau_0^2$$

$$\underline{\Delta} = 0 \qquad \text{for } \underline{\tau}:\underline{\tau} < 2\tau_0^2$$

Viscoelastic Fluids†

Viscoelasticity is defined as behavior somewhere between that of a viscous fluid and an elastic solid. Thus, viscoelastic fluids exhibit such phenomena as recoil, die swell, and stress relaxation (see Fig. 9-18*a* to *c*). Many polymer solutions and molten polymers exhibit viscoelastic behavior. For example, die swell is a real problem in most extrusion processes and must be accounted for empirically. In order to describe viscoelastic behavior mathematically complex expressions for the momentum flux are required. The development of such viscoelastic constitutive equations forms one of the major areas of research in the science of rheology.

The *Maxwell model* can be constructed by using mechanical models, such as combinations of springs and dashpots. The behavior of an ideal elastic spring is described (Fig. 9-18*d*) by $F = Gx$, and the behavior of a purely viscous dashpot (Fig. 9-18*e*) by $F = \mu\, dx/dt$. If we combine the two mechanical elements in series, we have (Fig. 9-18*f*)

$$x = x_1 + x_2 \qquad \dot{x} = \dot{x}_1 + \dot{x}_2 = \frac{\dot{F}}{G} + \frac{F}{\mu} \qquad F + \frac{\mu}{G}\dot{F} = \mu\dot{x}$$

Assuming that this model can adequately describe a viscoelastic fluid, with the momentum flux τ_{yx} replacing F and the velocity gradient $\partial v_z/\partial y$ replacing $\dot{x}$,

†Contributed by Prof. R. Gordon.

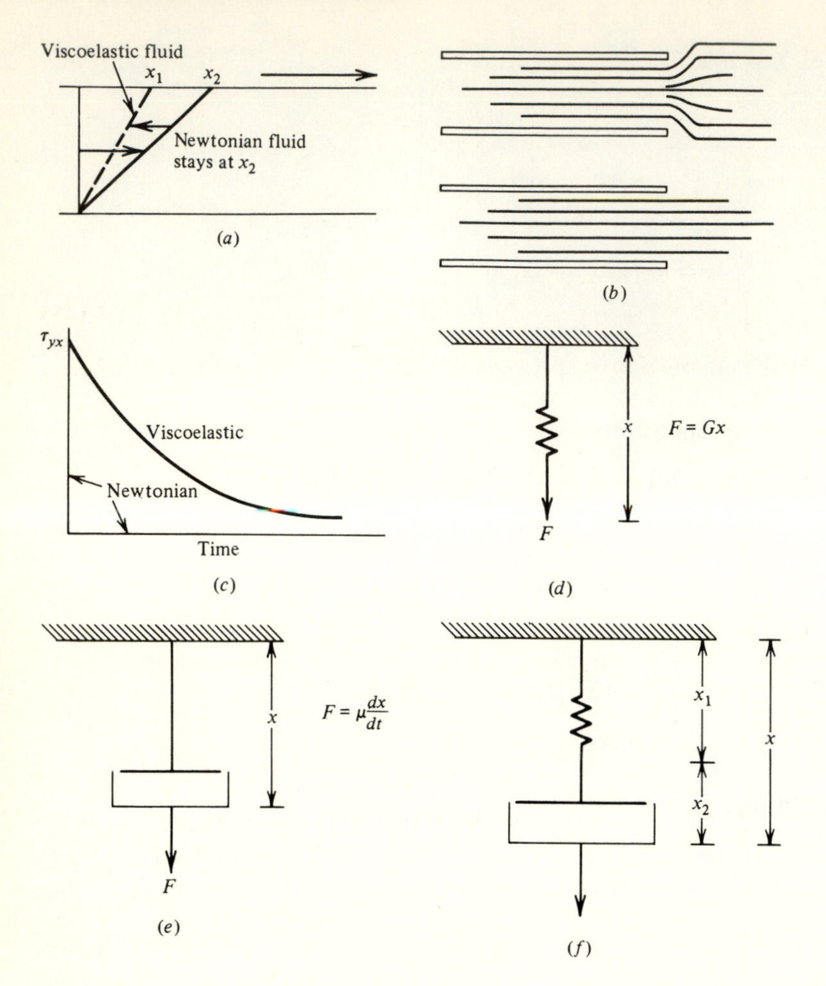

Figure 9-18 Typical behavior exhibited by viscoelastic fluids. (*a*) Recoil: fluid contained between two flat plates. Top plate suddenly moved to position x_2 and released. Plate springs back to position x_1. (*b*) Die swell: fluid leaves end of tube and swells to much greater size than tube. (*c*) Stress relaxation: fluid is being sheared between two flat plates, and motion is suddenly stopped. Stresses decay as in figure. (*d*) Ideal elastic spring. (*e*) Purely viscous dashpot. (*f*) Spring and dashpot in series (Maxwell model).

we have

$$\tau_{yx} + \theta \dot{\tau}_{yx} = -\mu \frac{\partial v_x}{\partial y} \tag{9-146}$$

where $\theta = \mu/G$ has dimensions of time and is referred to as the *relaxation time*. Equation (9-146) is known as the Maxwell model and was first introduced by James Clerk Maxwell (the father of electromagnetic theory) in 1867. This expression represents the simplest example of a viscoelastic fluid.

When one attempts to generalize Eq. (9-146) to an arbitrary velocity field, difficulties arise in the definition of the time derivative of $\dot{\tau}_{yx}$. These difficulties are associated with the fact that the momentum flux should not depend on *bulk rotations* in the fluid but only on relative displacements of fluid particles from each other. This is a major problem in developing rheological constitutive equations, and the principle that such equations be independent of bulk rotations is known as the *principle of material-frame indifference.* A generalization of Eq. (9-146) which satisfies this principle is the Criminale-Ericsen-Filberg (CEF) equation[5]

$$\underline{\tau} = -\eta \underline{\Delta} - \tfrac{1}{2}(\theta + 2\beta)\{\underline{\Delta} \cdot \underline{\Delta}\} + \tfrac{1}{2}\theta \frac{\mathfrak{D}\underline{\Delta}}{\mathfrak{D}t} \tag{9-147}$$

where η, θ, and β are measurable so-called *material functions* of the velocity gradients and $\mathfrak{D}\underline{\Delta}/\mathfrak{D}t$ is the *Jaumann* or *corotational derivative* of $\underline{\Delta}$, given by

$$\frac{\mathfrak{D}\underline{\Delta}}{\mathfrak{D}t} \equiv \frac{D\underline{\Delta}}{Dt} + \tfrac{1}{2}[(\underline{\Omega} \cdot \underline{\Delta}) - (\underline{\Delta} \cdot \underline{\Omega})] \tag{9-147a}$$

where $D\underline{\Delta}/Dt$ is the substantial derivative given by

$$\frac{D\underline{\Delta}}{Dt} \equiv \frac{\partial \underline{\Delta}}{\partial t} + v \cdot \nabla \underline{\Delta}$$

For Couette flow, $\dot{\gamma} = |dv_x/dx|$, and this equation gives[6]

$$\underline{\Omega} = \begin{bmatrix} 0 & -\dot{\gamma} & 0 \\ \dot{\gamma} & 0 & 0 \\ 0 & 0 & 0 \end{bmatrix} \quad \text{and} \quad \begin{aligned} \tau_{xy} = \tau_{yx} &= -\eta\dot{\gamma} \\ \tau_{xx} - \tau_{yy} &= -\theta\dot{\gamma}^2 \\ \tau_{yy} - \tau_{zz} &= -\beta\dot{\gamma}^2 \end{aligned}$$

where $\tau_{xx} - \tau_{yy}$ and $\tau_{yy} - \tau_{zz}$ are called the *primary* and *secondary normal-stress differences* respectively.

A special case is the *Maxwell model,*

$$\tau_{ij} = -\mu\Delta_{ij} - \frac{\mu}{G}\frac{\partial \tau_{ij}}{\partial t} \qquad \mu, G = \text{const} \tag{9-148}$$

In terms of the Jaumann derivative

$$\tau_{ij} = -\eta\dot{\gamma}_{ij} - \lambda \frac{\mathfrak{D}\tau_{ij}}{\mathfrak{D}t} \tag{9-148a}$$

which is the same as Eq. (9-147) if

$$\theta = -2\beta = -2\lambda$$

9-16 DIMENSIONAL ANALYSIS OF NAVIER-STOKES EQUATION†

Let us carry out a dimensional analysis of the Navier-Stokes equation

$$\rho\left(\frac{\partial \mathbf{v}}{\partial t} + \mathbf{v} \cdot \nabla \mathbf{v}\right) = \mu \nabla^2 \mathbf{v} - \nabla p + \rho \mathbf{g}$$

†May be omitted on first reading.

We let

$$\mathbf{v}^* \equiv \frac{\mathbf{v}}{v_{ch}} \qquad \nabla^* = D_{ch}\nabla \qquad p^* = \frac{p}{p_{ch}} \qquad t^* = \frac{t}{t_{ch}} \qquad g^* = \frac{\mathbf{g}}{g} \tag{9-149}$$

where v_{ch}, D_{ch}, p_{ch}, and t_{ch} are characteristic velocity, length, pressure, and time scales. Then

$$\frac{\rho v_{ch}}{t_{ch}}\frac{\partial \mathbf{v}^*}{\partial t^*} + \frac{\rho v_{ch}^2}{D_{ch}}(\mathbf{v}^* \cdot \nabla^*\mathbf{v}^*) = \frac{\mu}{D_{ch}^2}\nabla^{*2}\mathbf{v}^* - \frac{p_{ch}}{D_{ch}}\nabla^* p^* + \rho g \mathbf{g}^*$$

Multiplying by $D_{ch}^2/\mu v_{ch}$ gives

$$\frac{D_{ch}^2\rho}{t_{ch}}\frac{\partial \mathbf{v}^*}{\partial t^*} + \frac{D_{ch}\rho v_{ch}}{\mu}\mathbf{v}^* \cdot \nabla^*\mathbf{v}^* = \nabla^{*2}\mathbf{v}^* - \frac{p_{ch}D_{ch}}{\mu v_{ch}}\nabla^* p^* + \frac{D_{ch}^2\rho g}{\mu v_{ch}}\mathbf{g}^*$$

or

$$\text{Re}\left[(\text{Sr})^{-1}\frac{\partial \mathbf{v}^*}{\partial t^*} + \mathbf{v}^* \cdot \nabla^*\mathbf{v}^*\right] = \nabla^{*2}\mathbf{v}^* - \text{Re}\frac{p_{ch}}{\rho v_{ch}^2}\nabla^* p^* + \text{Re}\frac{D_{ch}g}{v_{ch}^2}\mathbf{g}^* \tag{9-150}$$

where

$$\text{Re} \equiv \frac{D_{ch}\rho v_{ch}}{\mu} = \text{Reynolds number} \qquad \text{Sr} \equiv \frac{v_{ch}t_{ch}}{D_{ch}} = \text{Strouhal number}$$

Also, we let

$$\frac{v_{ch}^2}{D_{ch}g} \equiv \text{Fr} = \text{Froude number}$$

which is important for wave motion and in free surface flows. If we let $p_{ch} = \rho v_{ch}^2$ (as in Chap. 4),

$$\text{Re}\left(\frac{1}{\text{Sr}}\frac{\partial \mathbf{v}^*}{\partial t^*} + \mathbf{v}^* \cdot \nabla^*\mathbf{v}^*\right) = \nabla^{*2}\mathbf{v}^* - \text{Re}\,\nabla^* p^* + \frac{\text{Re}}{\text{Fr}}\mathbf{g}^* \tag{9-151}$$

For steady flows $\partial \mathbf{v}^*/\partial t^* = 0$. Also, at very low Re, the left-hand side approaches zero and the equation becomes linear. This is called the *creeping-flow limit*. This very slow flow is realized not only at low velocities but also for fluids of high viscosity (polymers) and for small-diameter tubes. An example of creeping flow will be discussed later.

At the other extreme, for large Re, the viscous forces become negligible ($\mu \to 0$), and we have an inviscid fluid, for which solutions can often be obtained. This condition does not hold near solid surfaces but can be used outside the boundary layer, the layer of fluid near a surface. Inside the boundary layer, the inertial terms $\mathbf{v}^* \cdot \nabla^*\mathbf{v}^*$ cannot be zero. This case is discussed later.

For *fully developed* steady laminar flow in a duct of constant cross section, the inertial terms are zero and the Navier-Stokes equation becomes

$$0 = \mu\,\nabla^2\mathbf{v} - \nabla\mathcal{P}$$

where $\mathcal{P}$ is the dynamic pressure. For entrance-region flow the inertial term $\mathbf{v}^* \cdot \nabla^* \mathbf{v}^*$ is not negligible, and the equation becomes nonlinear.

9-17 BOUNDARY-LAYER THEORY†‡

It is well known that when an airplane flies through the atmosphere, there is appreciable air resistance that must be accounted for in the design of the plane. Similarly, when a fluid flows past any object (or any solid surface), there is resistance to the flow due to the drag or frictional forces at the surface. This resistance depends on the fluid viscosity, but in the early days of fluid mechanics (or hydrodynamics) the fluid was assumed to be without viscosity, or *inviscid,* making analytical solutions for flow around various objects possible. Although these solutions were useful in predicting flow patterns not near a surface, they were less useful near the surface because they predicted zero drag as a consequence of the neglect of viscous friction.

The existence of a developing boundary layer near a surface was postulated early in the century by Prandtl, who solved the Navier-Stokes equation in the boundary layer. The inviscid-flow solution then became useful for determining the flow distribution in the region outside the boundary layer. It was then matched to the solution within the boundary layer.

General Equations for the Boundary Layer

As a general model consider flow past an object such as an airplane wing (air foil) (see Fig. 9-19*a*). The fluid approaches the solid at an angle of incidence θ with a free-stream velocity v_∞. At the solid surface the velocity must be zero (due to the no-slip boundary condition), but it will increase rapidly in the boundary layer and

†May be omitted on first reading.
‡Does not require knowledge of vector notation.

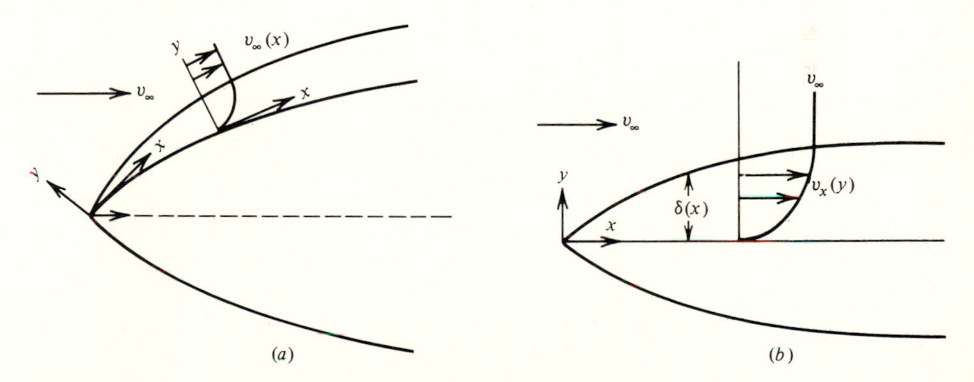

Figure 9-19 (*a*) Model for laminar-boundary-layer flow. (*b*) Flow past flat plate.

will approach the free-stream velocity v_∞. In the general case v_∞ will then vary with downstream distance x but is assumed independent of lateral position y. Thus we divide the overall flow into two regions: (1) an inner region, called the *boundary layer,* in which the velocity increases rapidly, and (2) an outer region, called the *free stream,* in which the velocity profile is assumed flat. In the latter region, since there is no lateral velocity gradient $\partial v_\infty/\partial y = 0$, viscous effects are not important and the fluid can be treated as being *inviscid.*

Since $\partial v_z/\partial z = 0$, the EOC can be written (part B of Table I-6) as

$$\frac{\partial v_x}{\partial x} = -\frac{\partial v_y}{\partial y}$$

Therefore, since near the surface the velocity v_x is dropping suddenly from v_∞ to nearly zero, $\partial v_x/\partial x$ is negative and $\partial v_y/\partial y$ must be positive; that is, v_y increases with y. Then, since $v_y = 0$ at the surface, it must be positive for $y > 0$, meaning that there is a flow *away* from the wall. As a consequence, *both* terms in the EOC are important, and the flow is two-dimensional. This means that two components of the EOM must be written and made dimensionless in such a manner that it is apparent which terms can be neglected.

The procedure for doing so (called *scaling and ordering*) is described in Appendix 9-7. With this procedure the y component of the Navier-Stokes equation is found to be simply

$$\frac{\partial p}{\partial y} = 0$$

and the x component in the boundary layer is then

$$\rho v_x \frac{\partial v_x}{\partial x} + \rho v_y \frac{\partial v_x}{\partial y} - \frac{dp}{dx} + \mu \frac{\partial^2 v_x}{\partial y^2} \tag{9-152}$$

In the free stream, $v_x = v_\infty$, and the second term on the left-hand side is small compared with the first; then for the free stream the x component becomes

$$\rho v_\infty \frac{dv_\infty}{dx} = -\frac{dp}{dx} \tag{9-153}$$

Combining Eqs. (9-152) and (9-153) gives

$$\rho\left(v_x \frac{\partial v_x}{\partial x} + v_y \frac{\partial v_x}{\partial y} - v_\infty \frac{dv_\infty}{dx}\right) = \mu \frac{\partial^2 v_x}{\partial y^2} \tag{9-154}$$

We now take up two special cases: (1) flow past a flat plate and (2) flow in the entrance of a duct.

Example 9-7: Flow past a flat plate Consider steady-state flow past a flat plate of total length L and width W with zero angle of incidence (see Fig. 9-19b). We wish to find the drag force at the wall and the drag coefficient. For this case, the free-stream velocity v_∞ is taken independent of x and the axial pressure gradient can be neglected. Then Eq. (9-154) becomes

$$\rho\left(v_x \frac{\partial v_x}{\partial x} + v_y \frac{\partial v_x}{\partial y}\right) = \mu \frac{\partial^2 v_x}{\partial y^2} \tag{9-155}$$

In Appendix 9-7 we show that if the inertial terms are to remain, the length scale in the y direction should be of the order of magnitude of $\sqrt{\nu L/v_\infty}$, where L is the length scale in the x direction. If we note that L/v_∞ is the time required for the fluid to flow a distance L with velocity v_∞, this compares with the characteristic time we obtained for the accelerating plate in Sec. 7-9, for which we successfully used the similarity transformation $\eta = y/\sqrt{\nu t}$ to obtain a solution by a combination-of-variables method (Appendix 7-3).

Therefore let us try a solution of the form

$$v_x^* = v_x^*(\eta) \tag{9-156}$$

where

$$v^* \equiv \frac{v_x}{v_\infty} \tag{9-157}$$

and

$$\eta \equiv \frac{y}{\sqrt{\nu x/v_\infty}} = \frac{y}{\delta} \tag{9-157a}$$

It is convenient to let

$$\frac{df}{d\eta} \equiv f' \equiv v_x^*(\eta) \tag{9-158}$$

Then

$$\frac{\partial v_x}{\partial x} \qquad \frac{\partial v_y}{\partial y} \qquad \text{and} \qquad \frac{\partial^2 v_x}{\partial x^2}$$

can be transformed into derivatives of f with respect to η (see Sec. 7-7). The lateral velocity in Eq. (9-155) is found from the EOC as

$$v_y = -\int_0^y \frac{\partial v_x}{\partial x}\, dy \tag{9-159}$$

Using the rules for differentiation (as in Chap. 7), these substitutions after much manipulation give

$$ff'' + 2f''' = 0 \tag{9-160}$$

which is called the *Blasius equation*. The boundary conditions are found as follows.

At the wall, where $y = 0$ and $\eta = 0$, there is no slip and therefore $f'(0) = 0$. Since we defined f' as the velocity, $f(0)$ is arbitrary and we can let $f(0) = 0$. At the leading edge, where $x = 0$, we should have $v_x = v_\infty$. But $x = 0$ corresponds to $\eta = \infty$, and so does $y = \infty$. This means that in order to be consistent, we should use $f'(\infty) = 1$. Thus $v_x \to v_\infty$ as $y \to \infty$. Since this approach to v_∞ is asymptotic, it is conventional to define the effective boundary thickness δ_e as the distance at which $v_x = 0.99v_\infty$ or $f' = 0.99$.

The solution of this set of equations was carried out[8] by means of an infinite-series solution and later more thoroughly by Howarth.[14] The results are plotted and tabulated by Schlicting,[19] and are found to agree well with data. From them one obtains (for later use)

$$\left.\frac{dv^*}{d\eta}\right|_0 = f''(0) = 0.332$$

Also, at $f' = 0.99$, one obtains $\eta = 5$. Then since

$$\eta \equiv \frac{y}{\sqrt{\nu x/v_\infty}}$$

we can find the boundary-layer thickness by letting $\eta = 5$ and $y = \delta_e$, or

$$\delta_e = 5\sqrt{\frac{\nu x}{v_\infty}} = 5x\,(\mathrm{Re})_x^{-1/2}$$

where $(\mathrm{Re})_x \equiv v_\infty x/\nu$ is a length Reynolds number. This shows that the thickness of the boundary layer increases with $x^{1/2}$.

The local drag stress at the wall at x is given by

$$\left.\tau_{yx}\right|_{y=0} = -\mu\left.\frac{\partial v_x}{\partial y}\right|_{y=0} = -\mu\left(\left.\frac{dv_x^*}{d\eta}\right|_{\eta=0}\left.\frac{\partial\eta}{\partial y}\right|_{y=0}\right)v_\infty \tag{9-161}$$

Using $(dv^*/d\eta)|_{\eta=0} \equiv f''(0) = 0.332$ and $(\partial\eta/\partial y)|_{y=0} \equiv 1/\delta$, we get

$$\left.\tau_{yx}\right|_{y=0} \equiv f''(0)\frac{v_\infty\mu}{\delta} = -\frac{0.332\mu^{1/2}v_\infty}{\sqrt{x/\rho v_\infty}} \tag{9-162}$$

The total drag force on the entire plate (both sides) is then

$$F_{Dt} = \int_0^{A_w}\tau_w\,dA_w = 2\int_0^L \left.\tau_{yx}\right|_{y=0} W\,dX \tag{9-163}$$

since *both* sides contribute to the drag and $dA_w = 2W\,dX$. We can express F_{Dt} also in terms of an average drag coefficient C_{fL}

$$F_{Dt} \equiv \tfrac{1}{2}A_w C_{fL}\rho v_\infty^2 \tag{9-164}$$

where C_{fL} is defined in the same way as a friction factor [Eq. (4-9)]. Then the average drag coefficient over L is given by Eqs. (9-162) to (9-164) as

$$C_{fL} \equiv \frac{F_{Dt}/A_w}{\frac{1}{2}\rho v_\infty^2} = \frac{2\int_0^L \left.\tau_{yx}\right|_{y=0} W\,dX}{\frac{1}{2}\rho v_\infty^2(2WL)} = \frac{2(0.332)v_\infty^{3/2}(\mu\rho)^{1/2}}{\rho v_\infty^2 L}\int_0^L \frac{dx}{x^{1/2}}$$

$$= \frac{0.664L^{1/2}v_\infty^{3/2}(\mu\rho)^{1/2}}{\frac{1}{2}v_\infty^2 L} = 1.328\left(\frac{\mu}{v_\infty\rho L}\right)^{1/2}$$

$$= 1.328(\mathrm{Re})_L^{-1/2} \tag{9-164a}$$

□

9-18 FLOW IN ENTRANCE REGION: INTEGRAL EQUATION†

Example 9-8: Flow in entrance region between parallel plates Consider steady-state flow of an incompressible newtonian fluid in the entrance region between parallel plates separated by a distance $2Y$, where Y is small compared with the width W and length L (see Fig. 9-20). Let the main direction of flow be the x direction, and let y be the distance from the lower plate. Assume that there is no flow in the z direction or *no variation with z*. Find the length of the entrance region L_e.

SOLUTION Assume that the fluid enters with a flat velocity profile $v_x = v_a$ (see Fig. 9-20). Since the velocity drops suddenly from v_a to zero at the wall, the velocity gradient is infinite and there is an infinite rate of momentum transport to the wall, leading to a subsequent reduction of velocity v_x near the wall. However, near the center the fluid can be assumed to continue to have a flat velocity, which we call v_F (rather than v_∞) since the flow is confined. Since we are at steady state, the mass rate of flow in kilograms per second past every cross section of area A_{cs} must be the same, or

$$w\bigg|_{x=0} = w\bigg|_{x=x} = \rho v_a A_{cs}$$

Then, since the fluid is incompressible, ρ is constant; and since A_{cs} is also constant, the average velocity v_a must be the same at each axial location x. Therefore, as the fluid moves downstream, v_F must become larger in order to compensate for the decrease in axial velocity near the wall caused by momentum transport. This means (1) that there must be flow away from the wall toward the centerline (or $v_y > 0$), as would also be revealed by using the continuity equation (8-110) near the wall (where v_x decreases with x), and (2) the condition that $v_y = 0$ obtains at the wall. Thus as we move farther from the wall, the region in which the velocity can be considered flat will decrease, until it disappears at $x = L_e$.

We designate the center region, in which the velocity profile is nearly flat, as the *core region* and the region near the wall as the boundary layer. Since the velocity gradient is flat in the core, we can assume that the fluid there is inviscid;

†Does not require knowledge of vector notation.

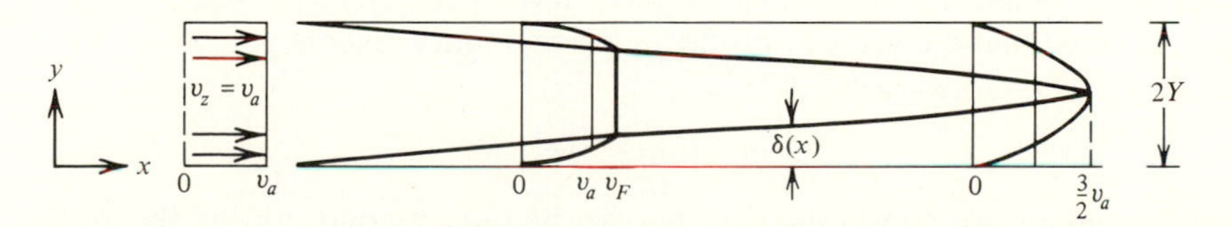

Figure 9-20 Flow in entrance region of a slit showing velocity distribution at $x = 0$, $x = x$, and $x = L_e$.

i.e., it has zero viscosity. The plates are assumed to be horizontal, and gravity is negligible. Also, axial-momentum diffusion is neglected in accordance with our scaling procedure.

We let the thickness of the boundary layer be $\delta(x)$. As indicated in Fig. 9-20, $\delta = 0$ at $x = 0$ and $\delta = Y$ at the entrance length $x = L_e$. At $x > L_e$ the flow becomes fully developed; that is, v_x no longer depends on x. In the fully developed region the velocity profile has been found (in Example 9-3) to be parabolic and is given by

$$v_x = \frac{\mathcal{P}_0 - \mathcal{P}_L}{2\mu L} Y^2 \left(\frac{2y}{Y} - \frac{y^2}{Y^2}\right) \tag{9-165}$$

where y is the distance from the lower plate, and

$$\frac{\mathcal{P}_0 - \mathcal{P}_L}{L} \equiv \frac{p_0 - p_L + \rho g_x L}{L}$$

The average velocity in the *fully developed region* can be obtained in terms of the pressure drop by integrating over the cross section [see Eq. (3-10)]. It is

$$v_a = \frac{\mathcal{P}_0 - \mathcal{P}_L}{3\mu L} Y^2 = \tfrac{2}{3} v_{\max} \tag{9-166}$$

Equation of motion The x component of the Navier-Stokes equation is

$$\rho \left(v_y \frac{\partial v_x}{\partial y} + v_x \frac{\partial v_x}{\partial x}\right) = \mu \frac{\partial^2 v_x}{\partial y^2} - \frac{\partial p}{\partial x} \tag{9-167}$$

In the *core region,* since there is no viscous transport, $\tau_{yx} = 0$, $\partial v_x / \partial y = 0$, and $v_x = v_F(x)$. Then Eq. (9-167) becomes

$$\rho v_F \frac{dv_F}{dx} = -\left(\frac{dp}{dx}\right)_F \tag{9-168}$$

Later we will use the above equation by assuming that the pressure gradient is the same in the boundary layer as in the core. In the *boundary layer* we can simplify the Navier-Stokes equation to obtain

$$\rho \left(v_y \frac{\partial v_x}{\partial y} + v_x \frac{\partial v_x}{\partial x}\right) = \mu \frac{\partial^2 v_x}{\partial y^2} - \left(\frac{dp}{dx}\right)_F \tag{9-169}$$

Velocity profile in boundary layer Since the fully developed velocity profile is parabolic, it is assumed that the velocity profile in the boundary layer is approximately parabolic also. Therefore we let

$$v_x = my + ny^2 + v_0 \tag{9-170}$$

where m, n, and v_0 are constants. The no-slip boundary condition at the wall indicates that $v_0 = 0$. We can determine m and n from the boundary conditions at

$y = \delta$, where, for the solutions to match,

$$v_x \Big|_{\delta} = v_F \tag{9-171}$$

and for there to be no momentum transport at the core

$$\frac{\partial v_x}{\partial y}\Big|_{y=\delta} = 0 \tag{9-172}$$

These conditions then give

$$n = -\frac{v_F}{\delta^2} \qquad \text{and} \qquad m = \frac{2v_F}{\delta} \tag{9-173}$$

and the velocity profile is

$$v_x = 2v_F \frac{y}{\delta} - v_F \left(\frac{y}{\delta}\right)^2 \tag{9-174}$$

Material balance We can relate the variation of v_F with x to the variation of δ with x by means of a material balance. The total mass entering at $x = 0$ must equal the total mass passing a cross section at arbitrary $x = x_1$; or if W is the width of each plate, the mass rate of flow w is

$$w = \rho v_a W(2Y) = \int \rho v_x \, dA_x = 2\int_0^Y \rho v_x W \, dy = 2\rho\left(\int_0^{\delta(x)} v_x W \, dy + \int_{\delta(x)}^{Y} v_F W \, dy\right)$$

Then since ρ and W are constant, they will cancel out. Substituting Eq. (9-174), integrating, and rearranging gives

$$\frac{\delta}{Y} = 3\left(1 - \frac{v_a}{v_F}\right) \tag{9-175}$$

Von Kármán integral equation We now integrate the Navier-Stokes equation (9-169) over the boundary layer to obtain a macroscopic or integral equation

$$\underset{(1)}{\rho \int_0^\delta v_y \frac{\partial v_x}{\partial y} dy} + \underset{(2)}{\rho \int_0^\delta v_x \frac{\partial v_x}{\partial x} dy} = \underset{(3)}{-\int_0^\delta \left(\frac{dp}{dx}\right)_F dy} + \underset{(4)}{\mu \int_0^\delta \frac{\partial^2 v_x}{\partial y^2} dy} \tag{9-176}$$

Terms (1) and (2) are integrated by using the equation of continuity and Eq. (9-171). When Eq. (9-168) is substituted into term (3), term (4) integrated, and Eq. (9-172) used, the final result is

$$\rho\left[\underset{(5)}{\frac{d}{dx}\int_0^\delta v_x(v_F - v_x)\, dy} + \underset{(6)}{\frac{dv_F}{dx}\int_0^\delta (v_F - v_x)\, dy}\right] = \underset{(7)}{-\mu \frac{\partial v_x}{\partial y}\Big|_{y=0}} \tag{9-177}$$

called the *von Kármán integral equation.*

Equation for v_F We now substitute the equation for the velocity distribution (9-174) into (9-177). Calculating each term separately and using Eq. (9-175) gives

$$0.3 \frac{dv_F}{dx} Y^2(9v_F - 7v_a)\frac{1}{v_F^2}\frac{\rho}{\mu}(v_F - v_a) = 0 \tag{9-178}$$

We now define the dimensionless variables

$$\zeta \equiv \frac{16(x/D_{\text{eq}})}{(\text{Re})_{\text{eq}}} \qquad (\text{Re})_{\text{eq}} \equiv \frac{D_{\text{eq}} v_a \rho}{\mu} \qquad D_{\text{eq}} \equiv 4Y$$
$$v_F^* \equiv \frac{v_F}{v_a} \qquad \delta^* \equiv \frac{\delta}{Y} \tag{9-179}$$

and obtain

$$0.3(9v_F^* - 7)(v_F^* - 1)\frac{1}{v_F^{*2}} dv_F^* = d\zeta \tag{9-180}$$

If this equation is integrated starting at $\zeta = 0$, where $v_F^* = 1$, we get

$$\zeta = 0.3\left[9(v_F^* - 1) - 16 \ln v_F^* - 7\left(\frac{1}{v_F^*} - 1\right)\right] \tag{9-181}$$

To find ζ_e we need to recall that at $x = L_e$ the flow is fully developed and that v_F will correspond to the maximum velocity in the fully developed profile [Eq. (9-165); see Fig. 9-21]. Thus by Eq. (9-181) $\zeta_e = 0.104$ is obtained as the value of ζ at which $v_F^* = \frac{3}{2}$. Using the definition of ζ, we then obtain

$$\frac{L_e}{D_{\text{eq}}} = 0.00055(\text{Re})_{\text{eq}} \tag{9-182}$$

It is of interest to compare the above result with that obtained from the previous example of flow past a flat plate. In that case we set $\zeta = Y$ at $x = L_e$

$$Y = 5\sqrt{\frac{L_e \nu}{v_\infty}} = 5\sqrt{\frac{\nu}{4Yv_\infty}}\sqrt{4YL_e} = \frac{5}{\sqrt{Re}}\sqrt{4YL_e}$$

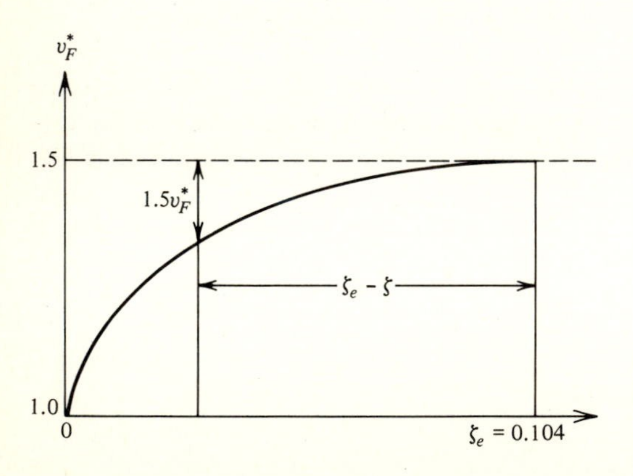

Figure 9-21 Determination of entrance length as point where boundary-layer thickness equals the half thickness.

or $$Y^2\,\mathrm{Re} = 25(4YL_e) \qquad \text{or} \qquad \frac{L_e}{D_{\mathrm{eq}}} \equiv \frac{L_e}{4Y} = \frac{\mathrm{Re}}{100}$$

Pressure drop The equation of motion in the core was shown to be

$$\rho v_F \frac{dv_F}{dx} = -\frac{dp}{dx} \qquad \text{or} \qquad \frac{\rho}{2}\frac{d}{dx}(v_F^2) = -\frac{dp}{dx}$$

and $$-(p - p_0) = \frac{\rho}{2}(v_F^2 - v_a^2)$$

at $z = L_e$:

$$p = p_e \qquad \text{and} \qquad v_F = \frac{3v_a}{2}$$

and at $z = 0$:

$$p = p_0 \qquad \text{and} \qquad v_F = v_a$$

Substituting, we obtain the entrance-region pressure drop

$$p_0 - p_e = \frac{\rho}{2}\left(\frac{9}{4} - 1\right)v_a^2 = \tfrac{5}{8}\rho v_a^2 \tag{9-183}$$

□

Flow in Entrance Region of a Pipe

The entrance length for a pipe of diameter D is found[17] to be

$$\frac{L_e}{D} = 0.0575\,\mathrm{Re} \tag{9-184}$$

The experimental value is 0.055 Re. For a pipe, the entrance pressure drop is

$$p_0 - p_e = \tfrac{3}{2}\rho V^2 \tag{9-185}$$

9-19 STREAM FUNCTION AND VELOCITY POTENTIAL†

The equation of continuity provides a means for interrelating the components of velocity for a given flow and makes it possible, for the special case of two-dimensional flow of an incompressible fluid, to express one component of the velocity in terms of the other. For example, in cartesian coordinates, the equation of continuity is from Eq. (8-111),

$$\frac{\partial v_x}{\partial x} + \frac{\partial v_y}{\partial y} = 0$$

† May be omitted on first reading.

This equation suggests that for two-dimensional flows the solution procedure can be simplified by making use of the equation of continuity. To accomplish this, the *stream function* ψ is utilized. Suppose we let

$$v_x = -\frac{\partial \psi}{\partial y} \qquad \text{and} \qquad v_y = \frac{\partial \psi}{\partial x} \tag{9-186}$$

Substituting into the equation of continuity gives

$$\frac{\partial^2 \psi}{\partial x\, \partial y} - \frac{\partial^2 \psi}{\partial y\, \partial x} = 0$$

and therefore ψ not only satisfies the equation of continuity but is also an exact differential.

Now let us show that Eqs. (9-186) are the consequence of a more general definition by considering the two-dimensional flow in Fig. 9-22. We will assume that there is no variation in velocity with z and that the depth of the channel in the z direction is L. The flow is bounded by areas A_1 and A_2 and by two streamlines ψ_1 and ψ_2 where *a streamline is defined as a curve across which no flow occurs*. Thus if $\mathbf{n}$ is an outwardly directed normal to the curve, the volumetric rate of flow from left to right across an element of length ds along a streamline ψ is

$$d\dot{\mathcal{V}} = \mathbf{v} \cdot \mathbf{n}\, dA = \mathbf{v} \cdot \mathbf{n} L\, ds = 0 \tag{9-187}$$

Note that this equation is similar to Eq. (8-79), which we derived for the rate of heat flow.

For a two-dimensional incompressible flow in cartesian coordinates it is convenient to deal with the volumetric rate of flow per *unit depth* and to define the stream function ψ by

$$\frac{d\dot{\mathcal{V}}}{L} \equiv d\psi = \mathbf{v} \cdot \mathbf{n}\, ds \tag{9-188}$$

where ds is along an arbitrary path. Thus Eq. (9-188) indicates that along a streamline $\mathbf{n} \cdot \mathbf{v} = 0$ or $d\psi = 0$ and ψ is constant.

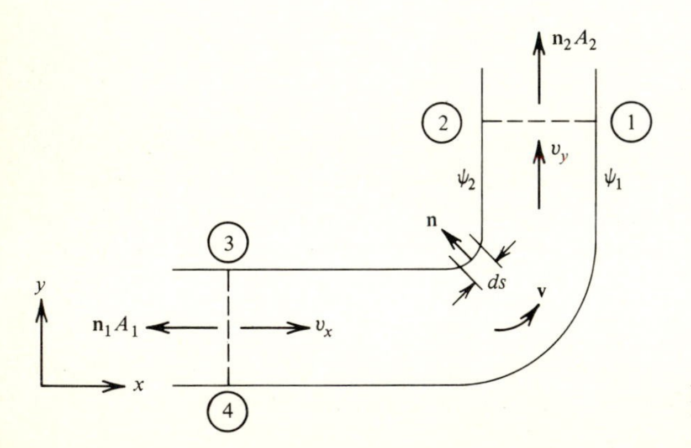

Figure 9-22 Two-dimensional flow between streamlines ψ_1 and ψ_2.

Now consider the total or net rate of flow into and out of the system 1-2-3-4 shown in Fig. 9-22. It is apparent that since there is no flow across the streamline, the magnitude of the flow in at A_1 must equal the magnitude of the flow out at A_2; that is,

$$|\dot{\mathcal{V}}_1| - |\dot{\mathcal{V}}_2| = 0$$

since $$\boldsymbol{\delta}_x \cdot \mathbf{n}_x \Big|_{A_1} = -1 \qquad \text{and} \qquad \boldsymbol{\delta}_y \cdot \mathbf{n}_y \Big|_{A_2} = 1$$

This can be written

$$\int_{A_2} \mathbf{v}_y \cdot \mathbf{n}_y \, dA_y + \int_{A_1} \mathbf{v}_x \cdot \mathbf{n}_x \, dA_x = 0 \tag{9-189}$$

Since $\mathbf{n} \cdot \mathbf{v} = 0$ along the streamline, the total *net* volume rate of flow is

$$\dot{\mathcal{V}}_t = \int_{A_t} d\dot{\mathcal{V}}_t = \int_{A_t} \mathbf{v} \cdot \mathbf{n} \, dA = 0 \tag{9-190}$$

where A_t is the total area of all boundaries.†

For the two-dimensional flow depicted in Fig. 9-22 we can divide Eq. (9-190) by the depth L and obtain the line integral

$$\int_C d\psi = \int_C \mathbf{v} \cdot \mathbf{n} \, ds = 0 \tag{9-191}$$

where n is the *outwardly* directed normal to ds and C is a closed path 1-2-3-4. Thus

$$\int_C d\psi = \int_{1\to2} d\psi + \int_{2\to3} d\psi + \int_{3\to4} d\psi + \int_{4\to1} d\psi = \int_{1\to2} \mathbf{v} \cdot \mathbf{n} \, ds + \int_{3\to4} \mathbf{v} \cdot \mathbf{n} \, ds$$

Since the equation of continuity holds at all points and since the divergence theorem holds for *any* volume of fluid, Eq. (9-191) holds for *any* closed path; in particular it applies to any path within the system as well as to a path along its boundaries. For example, consider the arbitrary path in Fig. 9-23*a*. If we start at point P_1 and move counterclockwise to P_2 over path C_1 and then back to P_1 over path C_2,

$$\int_{C_1} \mathbf{v} \cdot \mathbf{n} \, ds + \int_{C_2} \mathbf{v} \cdot \mathbf{n} \, ds = 0 \tag{9-192}$$

This tells us that the flow per unit depth across path C_1 has the same magnitude as the flow across path C_2. But since P_1 and P_2 are arbitrary, Eq. (9-191) indicates that the integral along an arbitrary *closed* path is zero and that ψ is an *exact* differential. This means that along an open path any change $\Delta\psi \equiv \int d\psi$ is independent of the path.

† This equation is *general* for any incompressible flow because the Gauss divergence theorem tells us that $\int \mathbf{v} \cdot \mathbf{n} \, dA = \int \nabla \cdot \mathbf{v} \, d\mathcal{V} = 0$ (since, by the equation of continuity, $\nabla \cdot \mathbf{v} = 0$).

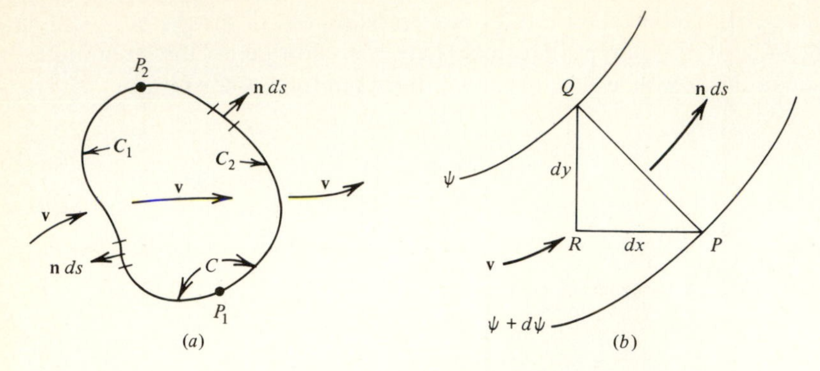

Figure 9-23 Two-dimensional flow: (*a*) arbitrary path and (*b*) flow between streamlines ψ and $\psi + d\psi$.

If two points P and Q separated by a distance ds are *not* on the same streamline, the stream function will change from ψ to $\psi + d\psi$, as shown in Fig. 9-23*b*. If we go along ds from point P to point Q,

$$(d\psi)_{PQ} = (\mathbf{v} \cdot \mathbf{n}\, ds)_{PQ}$$

where $\mathbf{n}$ is the normal to ds. But if we return by the paths QR at constant x and RP at constant y, we obtain

$$(d\psi)_{QRP} = (d\psi)_{QR} + (d\psi)_{RP}$$

or

$$(d\psi)_{QRP} = v_x \boldsymbol{\delta}_x \cdot \mathbf{n}_x\, dy + v_y \boldsymbol{\delta}_y \cdot \mathbf{n}_y\, dx = v_x\, dy - v_y\, dx$$

Now for the *total* path $PQRP$, $d\psi = 0$; so

$$(d\psi)_{PQ} = -(d\psi)_{QRP} = -v_x\, dy + v_y\, dx \tag{9-193}$$

But since ψ is an exact differential, from the calculus,

$$d\psi = \frac{\partial \psi}{\partial x} dx + \frac{\partial \psi}{\partial y} dy \tag{9-194}$$

Then by comparison with Eq. (9-193)†

$$v_x = -\frac{\partial \psi}{\partial y} \qquad \text{and} \qquad v_y = \frac{\partial \psi}{\partial x} \tag{9-195}$$

For *cylindrical* coordinates, a similar procedure results in the following equations for the cylindrical coordinate stream function ψ_c (see Fig. 9-24*a*)

$$v_r = -\frac{\partial \psi_c}{r\, \partial \theta} \qquad \text{and} \qquad v_\theta = \frac{\partial \psi_c}{\partial r} \tag{9-196}$$

For *spherical* coordinates (Fig. 9-24*b*) a stream function is defined for the case of *axisymmetric* flow, i.e., flow symmetrical about the z axis and for which $\partial/\partial\phi = 0$.

† Since we are dealing only with *changes* in ψ, the sign convention is arbitrary. If we had taken $d\psi$ in Fig. 9-23*b* to be of opposite sign, the opposite signs would have appeared in Eq. (9-195).

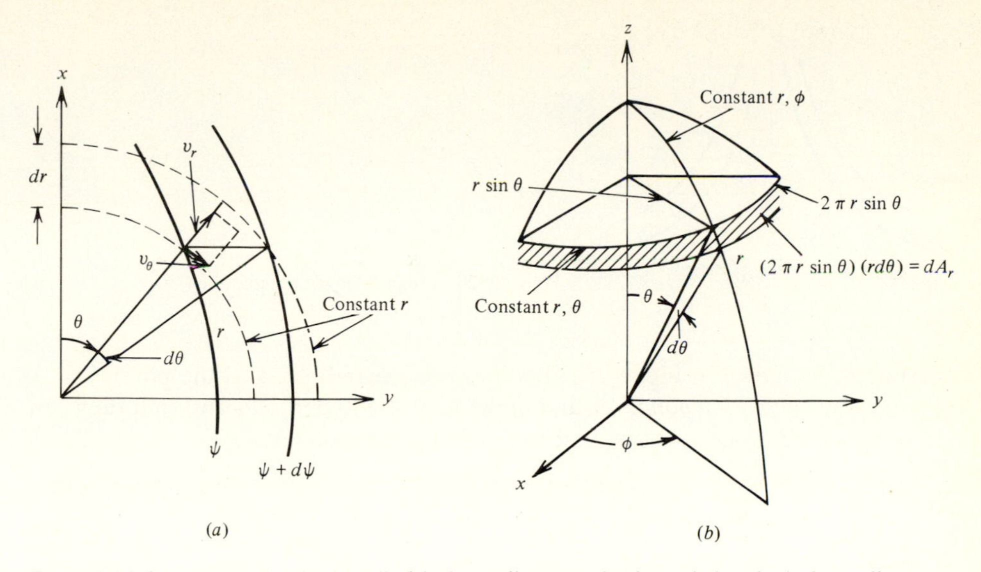

Figure 9-24 Stream function in (*a*) cylindrical coordinates at fixed z and (*b*) spherical coordinates at fixed ϕ.

However, since the integral over all ϕ is 2π (as opposed to a length L), the stream function is defined by dividing the volume rate of flow by 2π. Then

$$d\psi_s = \frac{\mathbf{v} \cdot \mathbf{n}\, dA}{2\pi} \tag{9-197}$$

or
$$d\psi_s = \mathbf{v}_\theta \cdot \mathbf{n}_\theta \frac{dA_\theta}{2\pi} + \mathbf{v}_r \cdot \mathbf{n}_r \frac{dA_r}{2\pi} = v_\theta \frac{dA_\theta}{2\pi} - v_r \frac{dA_r}{2\pi}$$

Since, from Fig. 9-24*b* (see Secs. 1-9, 1-29, and Examples 8-2 and 8-4),

$$dA_\theta = 2\pi r \sin\theta\, dr \qquad \text{and} \qquad dA_r = 2\pi r^2 \sin\theta\, d\theta \tag{9-198}$$

we have

$$d\psi_s = v_\theta r \sin\theta\, dr - v_r r^2 \sin\theta\, d\theta \tag{9-199}$$

Also
$$d\psi_s = \frac{\partial \psi_s}{\partial r} dr + \frac{\partial \psi_s}{\partial \theta} d\theta \tag{9-200}$$

Comparing Eqs. (9-199) and (9-200) gives

$$v_r = -\frac{1}{r^2 \sin\theta} \frac{\partial \psi_s}{\partial \theta} \qquad \text{and} \qquad v_\theta = \frac{1}{r \sin\theta} \frac{\partial \psi_s}{\partial r} \tag{9-201}$$

Generalized equations for arbitrary orthogonal coordinates are given by Goldstein.[11]

Example 9-9: Stagnation flow Consider the *stagnation flow* that occurs, for example, when a jet of water flows toward a plane surface and is forced to change

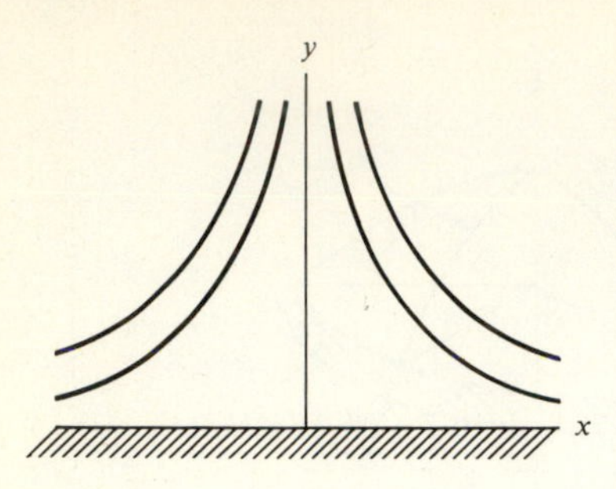

Figure 9-25 Streamlines for stagnation flow.

directions, as shown in Fig. 9-25. The lines are streamlines, and the point $x = 0$, $y = 0$ is the *stagnation point*. Assuming the lines are hyperbolas and that they can be represented by $\psi = -cxy$, find the velocity components.

SOLUTION

$$v_x = -\frac{\partial \psi}{\partial y} = cx \qquad \text{and} \qquad v_y = \frac{\partial \psi}{\partial x} = -cy$$

which correctly gives $v_y = 0$ at $y = 0$ and $v_x = 0$ at $x = 0$. It gives $v_y = 0 = v_x$ at the point ($x = 0$, $y = 0$), the *stagnation point*. It also describes the general flow pattern except near the plane $y = 0$, where it fails to give $v_x = 0$. This region is treated by Batchelor.[2] □

Potential Flow

At a sufficiently large distance from the solid boundary the effects of fluid viscosity can often be neglected, in which case the fluid is said to be *inviscid*. If, in addition, its density is constant, it is said to be an *ideal* fluid. A *potential* flow is the flow of an ideal fluid such that the flow is *irrotational,* or

$$\nabla \times \mathbf{v} = 0 \tag{9-202}$$

For such a flow, according to Eq. (9-113),

$$2\boldsymbol{\omega} = \nabla \times \mathbf{v} = 0$$

and so the vorticity vector $\boldsymbol{\omega} = 0$.

If there exists a velocity potential ϕ such that

$$\mathbf{v} = -\nabla\phi \tag{9-203}$$

then

$$\nabla \times \mathbf{v} = -\nabla \times \nabla\phi = 0$$

because $\nabla\phi$ and ∇ are in the same direction ($\boldsymbol{\delta}_x \times \boldsymbol{\delta}_x = 0$, etc.). Therefore if the flow is irrotational, ϕ will exist, and

$$v_x = -\frac{\partial \phi}{\partial x} \qquad \text{and} \qquad v_y = -\frac{\partial \phi}{\partial y} \tag{9-116}$$

follow. Equating their right-hand sides to the corresponding equations (9-186) in terms of ψ gives

$$\frac{\partial \phi}{\partial x} = \frac{\partial \psi}{\partial y} \qquad \text{and} \qquad \frac{\partial \phi}{\partial y} = -\frac{\partial \psi}{\partial x} \tag{9-204}$$

These are called *the Cauchy-Riemann equations,* for which solutions are available.[9]

By differentiation and substituting into the equation of continuity, it can be shown that both ϕ and ψ satisfy the two-dimensional laplacian equation

$$\nabla^2\phi = 0 \qquad \text{and} \qquad \nabla^2\psi = 0$$

Lines of constant ϕ are equipotential lines and are perpendicular to the streamline of constant ψ. Both can be found by a procedure using complex variables, called *conformal mapping.*[7]

Example 9-10: Velocity potential for stagnation flow Find ϕ for stagnation flow for which (Example 9-9)

$$v_x = cx \qquad \text{and} \qquad v_y = -cy$$

SOLUTION From Eq. (9-116)

$$\frac{\partial \phi}{\partial x} = -cx \qquad \text{and} \qquad \frac{\partial \phi}{\partial y} = cy \tag{9-205}$$

Integrating gives

$$\phi = -2cx^2 + f_1(y) \qquad \phi = 2cy^2 + f_2(x)$$

Then let

$$\phi = -2c(x^2 - y^2)$$

□

Use of the Stream Function to Solve Navier-Stokes Equations

We now illustrate how the stream function can be used, at least in principle, to solve the Navier-Stokes equations in two dimensions. If we assume the Reynolds number to be very low (creeping flow), we can neglect the term $\mathbf{v} \cdot \nabla \mathbf{v}$ in the expansion

$$\rho \frac{D\mathbf{v}}{Dt} \equiv \rho \left(\frac{\partial \mathbf{v}}{\partial t} + \mathbf{v} \cdot \nabla \mathbf{v} \right) \tag{9-206}$$

This term (commonly called the *inertial term*) is nonlinear and makes the solution of the Navier-Stokes equations quite difficult when it must be included.

If we further impose the restrictions of steady state, we have $D\mathbf{v}/Dt = 0$. Finally for two-dimensional flow, we can let $v_z = 0$ and $\partial/\partial z = 0$. The x and y components of the equation of motion are then

$$\frac{\partial p}{\partial x} = \mu \nabla^2 v_x + \rho g_x \tag{9-207}$$

and
$$\frac{\partial p}{\partial y} = \mu \nabla^2 v_y + \rho g_y \tag{9-208}$$

We now try to eliminate the pressure by making use of the fact that

$$\frac{\partial^2 p}{\partial y\, \partial x} - \frac{\partial^2 p}{\partial x\, \partial y} = 0 \tag{9-209}$$

Therefore we can differentiate Eq. (9-207) with respect to y and Eq. (9-208) with respect to x and substitute into Eq. (9-209). At the same time we let $v_x = -\partial\psi/\partial y$ and $v_y = \partial\psi/\partial x$; then we get

$$\frac{\partial}{\partial y}\left(\frac{\partial p}{\partial x}\right) = \mu \frac{\partial}{\partial y}\left[\frac{\partial^2}{\partial x^2}\left(-\frac{\partial \psi}{\partial y}\right) + \frac{\partial^2}{\partial y^2}\left(-\frac{\partial \psi}{\partial y}\right)\right] = -\mu\left(\frac{\partial^4 \psi}{\partial x^2\, \partial y^2} + \frac{\partial^4 \psi}{\partial y^4}\right) \tag{9-210}$$

and
$$\frac{\partial}{\partial x}\left(\frac{\partial p}{\partial y}\right) = \mu \frac{\partial}{\partial x}\left[\frac{\partial^2}{\partial x^2}\left(\frac{\partial \psi}{\partial x}\right) + \frac{\partial^2}{\partial y^2}\left(\frac{\partial \psi}{\partial x}\right)\right] = \mu\left(\frac{\partial^4 \psi}{\partial x^4} + \frac{\partial^4 \psi}{\partial y^2\, \partial x^2}\right) \tag{9-211}$$

Then substituting the second form of Eqs. (9-210) and (9-211) into Eq. (9-209) gives

$$\frac{\partial^4 \psi}{\partial x^4} + 2\frac{\partial^4 \psi}{\partial y^2\, \partial x^2} + \frac{\partial^4 \psi}{\partial y^4} = 0 \tag{9-212}$$

called the *biharmonic equation*. It can be written

$$\nabla^4 \psi = 0 \qquad \text{where } \nabla^4 \equiv \nabla^2 \nabla^2 \tag{9-213}$$

We have consequently converted two second-order partial differential equations in $v_x(x, y)$, $v_y(x, y)$, and $p(x, y)$ into one partial differential equation in terms of $\psi(x, y)$. The procedure then would be:

1. Using the applicable boundary conditions, solve the equation for $\psi(y, x)$.
2. Differentiate ψ with respect to y at constant x to find v_x.
3. Differentiate ψ with respect to x at constant y to find v_y.
4. Substitute into Eqs. (9-207) and (9-208) to find $\partial p/\partial x$ and $\partial p/\partial y$ as a function of x and y. Then integrate to find $p(x, y)$.

We illustrate this procedure for the important application of creeping flow past a sphere.

9-20 FLOW PAST SUBMERGED OBJECTS†

Many engineering problems involve the flow of a fluid past a submerged object, e.g., packed catalytic reactors, absorption columns, pebble heat exchangers. Of

†May be omitted on first reading.

course, in these cases an individual object is not isolated but surrounded by numerous others and each affects the flow pattern of the other. Nevertheless, if the problem of flow past a single particle is solved, one can gain insight into the more complicated problems of flow through a bed of particles. Such knowledge can then be used, for example, to obtain equations for the friction factor or pressure drop in packed reactors or absorbers.

Actually, the problem of flow past a particle is important not only in itself but also because it is mathematically equivalent to the problem of a particle *falling* through a fluid. In either case we are dealing with the velocity of the fluid with respect to the *surface* of the particle (where the no-slip condition may hold); it is therefore immaterial whether it is the fluid or the sphere that is moving with respect to an external reference (see Fig. 9-26).

The problem of the falling particle is important in the design of sedimentation equipment, particle classifiers, dust collectors, demisters, extractors, and numerous other applications. In some of the above cases one is dealing with a solid pellet *falling* through a gas or liquid while in others a liquid drop is falling through another liquid (extractor). In still another application a gas bubble is rising through a liquid (as in boiling). When the particle is a fluid rather than a solid, circulation patterns exist inside and have been calculated by Hadamard.[12] However, in this example we will assume that there is no internal motion.

Example 9-11: Creeping flow around a sphere A fluid is approaching a sphere of radius r with an approach velocity v_∞; that is, at an infinite distance away from the sphere (see Fig. 9-26) the fluid has a constant velocity in the z direction equal to v_∞, where v_∞ is small enough for the particle Reynolds number $\text{Re} \equiv 2Rv_\infty/\nu$ to be much less than one. (*a*) Simplify the Navier-Stokes equations in spherical coordinates. (*b*) Write the boundary conditions. (*c*) Describe a method of solution

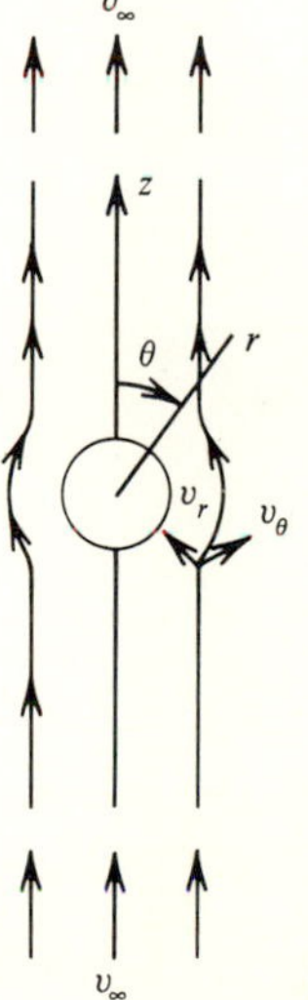

Figure 9-26 Flow past a sphere.

that can be used to obtain the velocity components $v_r(r, \theta)$ and $v_z(r, \theta)$ and the pressure distribution $p(r, \theta)$.

SOLUTION We will use a spherical coordinate system with $r = 0$ and $z = 0$ at the center of the sphere. For creeping flow $\mathbf{v} \cdot \nabla \mathbf{v} = 0$, and for steady state $\partial/\partial t = 0$; therefore $D\mathbf{v}/Dt = 0$. The assumption of axial symmetry means that $\partial/\partial\phi = 0$ and $v_\phi = 0$, where ϕ is an angle of rotation about the z axis (see Fig. 9-25). Consequently, only the θ and r components of the equation of motion (or Navier-Stokes equations) are needed. They can be written (Table I-5)

$$\frac{\partial p}{r\,\partial\theta} = \mu(\nabla^2\mathbf{v})_\theta + \rho g_\theta \qquad \frac{\partial p}{\partial r} = \mu(\nabla^2\mathbf{v})_r + \rho g_r \tag{9-214}$$

[The expressions for $\nabla^2\mathbf{v}$ in spherical coordinates are given in Table I-5 and need not be rewritten for our purposes, but it should be noted that $(\nabla^2\mathbf{v})_\theta \neq \nabla^2 v_\theta \boldsymbol{\delta}_\theta$, etc.]

We can substitute for the stream function [see Eqs. (9-201)]

$$v_r = \frac{1}{r^2 \sin\theta} \frac{\partial\psi_s}{\partial\theta} \qquad v_\theta = \frac{1}{r \sin\theta} \frac{\partial\psi_s}{\partial r}$$

Carrying out the operations of differentiating the θ equation with respect to r and r equation with respect to θ and equating the second derivatives of the pressure gives[11]

$$E^4\psi_s = 0 \tag{9-215}$$

where the operator E^4 is defined as E^2E^2 and the operator E^2 is

$$E^2 \equiv \frac{\partial^2}{\partial r^2} - \frac{\sin\theta}{r^2} \frac{\partial}{\partial\theta} \left(\frac{1}{\sin\theta} \frac{\partial}{\partial\theta} \right) \tag{9-216}$$

Equation (9-215) must be solved subject to the boundary conditions

At $r = R$: $\qquad v_r = 0 \quad \text{and} \quad v_\theta = 0 \qquad$ (9-217)

At $r = \infty$: $\qquad v = v_z = v_\infty \qquad$ (9-218)

The method of solution of the above equations is described in Refs. 17 and 7; another approach is given in Ref. 13.

The last boundary condition can be written in terms of the stream function as follows. We note (Fig. 9-27) that if $\mathbf{v} = v_\infty \boldsymbol{\delta}_z$,

$$v_r = \boldsymbol{\delta}_r \cdot \mathbf{v} = v_\infty \cos\theta = -\frac{1}{r^2 \sin\theta} \frac{\partial\phi_s}{\partial\theta} \tag{9-219}$$

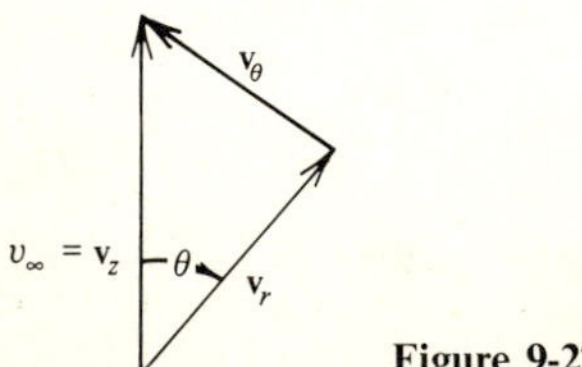

Figure 9-27 Velocity components.

Integrating at constant r gives

$$\psi_s + f_1(\theta) = \int d\psi_s = \int v_\infty r^2 \sin\theta \cos\theta \, d\theta = -\tfrac{1}{2} v_\infty r^2 \sin^2\theta \qquad (9\text{-}220)$$

where $f_1(\theta)$ is a function of θ. But

$$v_\theta = v_\infty \sin\theta = \frac{1}{r \sin\theta} \frac{\partial \psi_s}{\partial r} \qquad (9\text{-}221)$$

Integrating at constant θ gives

$$\psi_s + f_2(r) - \tfrac{1}{2} v_\infty r^2 \sin^2\theta \qquad (9\text{-}222)$$

By cross differentiation $f_1(\theta) = 0$ and $f_2(r) = 0$. Then

$$\psi_s = -\tfrac{1}{2} v_\infty r^2 \sin^2\theta \qquad (9\text{-}223)$$

This suggests a solution to the differential equation of the form

$$\psi_s = f(r) \sin^2\theta \qquad (9\text{-}224)$$

Substituting into Eqs. (9-215) and (9-216), we get

$$\left(\frac{d^2}{dr^2} - \frac{2}{r^2}\right)\left(\frac{d^2}{dr^2} - \frac{2}{r^2}\right) f(r) = 0 \qquad (9\text{-}225)$$

If we assume $f(r) = Cr^n$, we get solutions for $n = -1, 1, 2,$ and 4. Then we take the sum of these as a general solution

$$f(r) = \frac{C_1}{r} + C_2 r + C_3 r^2 + C_4 r^4 \qquad (9\text{-}226)$$

By the third boundary condition,

$$C_4 = 0 \qquad C_3 = -\tfrac{1}{2} v_\infty \qquad (9\text{-}227)$$

Then

$$\psi_s = \left(\frac{C_1}{r} + C_2 r - \tfrac{1}{2} v_\infty r^2\right) \sin^2\theta \qquad (9\text{-}228)$$

Using the boundary conditions at $r = R$, we can evaluate the constants

$$C_1 = -\tfrac{1}{4} v_\infty R^3 \qquad \text{and} \qquad C_2 = \tfrac{3}{4} v_\infty R$$

The solutions for the velocities are then obtained from Eqs. (9-201)

$$\frac{v_r}{v_\infty} = \left[1 - \frac{3}{2}\frac{R}{r} + \frac{1}{2}\left(\frac{R}{2}\right)^3\right] \cos\theta \qquad (9\text{-}229)$$

$$\frac{v_\theta}{v_\infty} = -\left[1 - \frac{3}{4}\frac{R}{r} - \frac{1}{r}\left(\frac{R}{r}\right)^3\right] \sin\theta \qquad (9\text{-}230)$$

The pressure distribution is obtained by substituting them in the Navier-Stokes equations and integrating. The result is

$$p = p_0 - \rho g z - \frac{3}{2}\frac{\mu v_\infty}{R}\left(\frac{R}{r}\right)^2 \cos\theta \qquad (9\text{-}231)$$

where p_0 is the pressure at $z = 0$, $r \to \infty$, and $z = r\cos\theta$. Note that the maximum pressure is at the front, $\theta = \pi$. □

9-21 TOTAL FORCE ON AN OBJECT†

The total viscous plus pressure force **F** exerted by the surroundings on the surface of an element of *fluid* of arbitrary shape with surface area A and outwardly directed normal **n** is given by Eqs. (9-50*a*) and (9-50*b*),

$$\mathbf{F} = \mathbf{F}_v + \mathbf{F}_p$$

where the viscous force and pressure force are, respectively,

$$\mathbf{F}_y = -\iint_A \mathbf{n} \cdot \underline{\tau}\, dA \qquad \text{and} \qquad \mathbf{F}_p = -\iint p\mathbf{n}\, dA$$

If the surroundings are fixed solid surfaces or fluid-fluid interfaces, these forces will exert a drag or frictional force on the fluid. This drag force will generally be opposite to the direction of flow of the fluid and therefore will be negative if the axes are chosen so that the velocity is positive (see Fig. 9-28). For this reason the drag force is often taken as the magnitude of the component of **F** in the direction of the main flow. Thus if the flow is in the z direction,

$$F_D \equiv |\mathbf{F}_z|$$

Often we wish to find $\mathbf{F}_f$, the force of the solid (the surroundings) on the fluid. By Newton's third law it is $\mathbf{F}_f = -\mathbf{F}$. The total force acting on a submerged object is therefore

$$\mathbf{F}_f = \mathbf{F}_{vf} + \mathbf{F}_{pf}$$

or

$$\mathbf{F}_f = \iint \mathbf{n} \cdot \underline{\tau}\, dA + \iint p\mathbf{n}\, dA \tag{9-232}$$

Force Acting on a Sphere

In the case of a sphere, we must evaluate the terms in Eq. (9-232) at $r = R$ and use an element of area (see Fig. 9-29)

$$dA_r\Big|_R = (R\sin\theta\, d\phi)(R\, d\theta)$$

Since the main flow is in the z direction, we are interested in the z component of the drag force of the solid on the fluid, or $\mathbf{F}_f \cdot \boldsymbol{\delta}_z$. Note that this will be negative since the drag acts in the direction opposite the flow; i.e., it opposes the tendency to flow. Then, since the *outward* normal to the system is $\mathbf{n} = \mathbf{n}_r = -\boldsymbol{\delta}_r$, we obtain

$$\mathbf{n}_r \cdot \underline{\tau} = -(\boldsymbol{\delta}_r \cdot \underline{\tau}) = \tau_{rr}\boldsymbol{\delta}_r + \tau_{r\theta}\boldsymbol{\delta}_\theta + \tau_{r\phi}\boldsymbol{\delta}_\phi$$

†May be omitted on first reading.

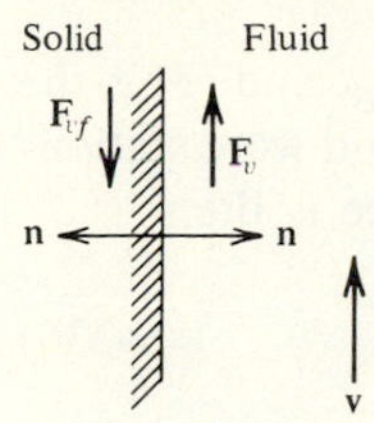

Figure 9-28 Viscous force and velocity vectors for flow past a surface.

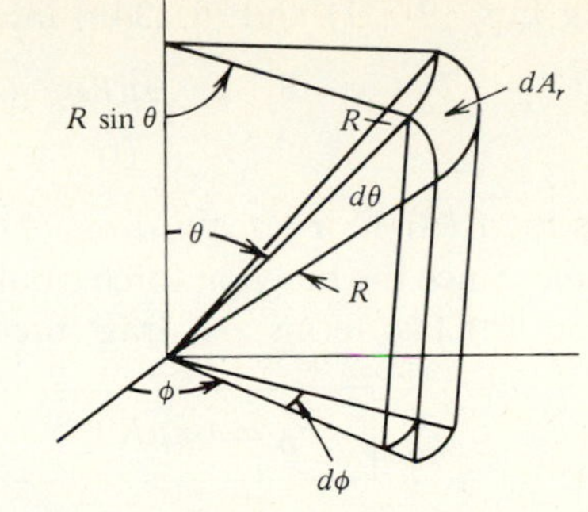

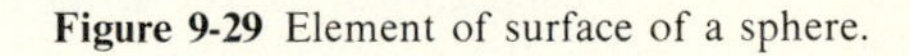

Figure 9-29 Element of surface of a sphere.

Using Table I-9 and letting $r = R$, we obtain

$$\tau_{rr}\Big|_R = -2\mu \frac{\partial v_r}{\partial r}\Big|_R = -2\mu v_\infty \left(\frac{3}{2}\frac{R}{r^2} - \frac{3}{2}\frac{R^3}{r^4}\right)\Big|_R = 0$$

Similarly
$$\tau_{r\phi}\Big|_R = 0$$

Since we want the z component (see Fig. 9-29),

$$\mathbf{n}_r \cdot \underline{\boldsymbol{\tau}}\Big|_R \cdot \boldsymbol{\delta}_z = -(\boldsymbol{\delta}_r \cdot \underline{\boldsymbol{\tau}})\Big|_R \cdot \boldsymbol{\delta}_z = -\tau_{r\theta}\Big|_R \boldsymbol{\delta}_\theta \cdot \boldsymbol{\delta}_z = \tau_{r\theta}\Big|_R \sin\theta$$

Also
$$p\mathbf{n}_r \cdot \boldsymbol{\delta}_z = -p\boldsymbol{\delta}_r \cdot \boldsymbol{\delta}_z = -p\cos\theta$$

Then

$$-F_{fz} = \int_{\phi=0}^{2\pi}\int_{\theta=0}^{\pi} -(\tau_{r\theta}\sin\theta\, dA_r)\Big|_{r=R} + \int_{\phi=0}^{2\pi}\int_0^{\pi}(p\cos\theta\, dA_r)\Big|_{r=R} \qquad \text{(9-233)}$$

Note that because of axial symmetry there is no variation in the ϕ direction and

$$\int_0^{2\pi} f\, d\phi = 2\pi f$$

where f is any $f(r, \theta)$.

To obtain $\tau_{r\theta}$ we must use Table I-9 for spherical coordinates

$$\tau_{r\theta} = -\mu\left[r\frac{\partial}{\partial r}\left(\frac{v_\theta}{r}\right) + \frac{1}{r}\frac{\partial v_r}{\partial \theta}\right] \qquad \text{(9-234a)}$$

If we substitute for the velocity components and let $r = R$, the result is

$$\tau_{r\theta}\Big|_{r=R} = \frac{3}{2}\frac{\mu v_\infty}{R}\left(\frac{R}{r}\right)^4\Big|_{r=R}\sin\theta = \frac{3}{2}\frac{\mu v_\infty}{R}\sin\theta \qquad \text{(9-234b)}$$

Substituting Eqs. (9-231) and (9-234*b*) into Eq. (9-233) and integrating, we get

$$F_D \equiv |-\mathbf{F}_z| = \underset{(1)}{4\pi\mu R v_\infty} + \underset{(2)}{2\pi\mu R v_\infty} + \underset{(3)}{\tfrac{4}{3}\pi R^3 \rho g} \tag{9-235}$$

where (1) is the friction or viscous drag, (2) is called the *form drag,* and (3) is the buoyant force. Since the buoyant force would also appear if the fluid were stationary, only the first two terms are drag forces. The total drag force is then

$$\boxed{F_D = 6\pi\mu R v_\infty \qquad \text{Stokes' law}} \tag{9-236}$$

which is widely used to obtain the drag on falling objects.

9-22 DRAG COEFFICIENT†

We can now obtain the friction factor or drag coefficient for a sphere. The *general* definition of friction factor [Eq. (4-6)] is

$$f = \frac{F_D/A_{\text{ch}}}{\frac{1}{2}\rho v_{\text{ch}}} \tag{9-237}$$

where F_D = total drag force
A_{ch} = characteristic area
$\frac{1}{2}\rho v_{\text{ch}}^2$ = kinetic energy per unit volume
v_{ch} = characteristic velocity

†Does not require knowledge of vector notation.

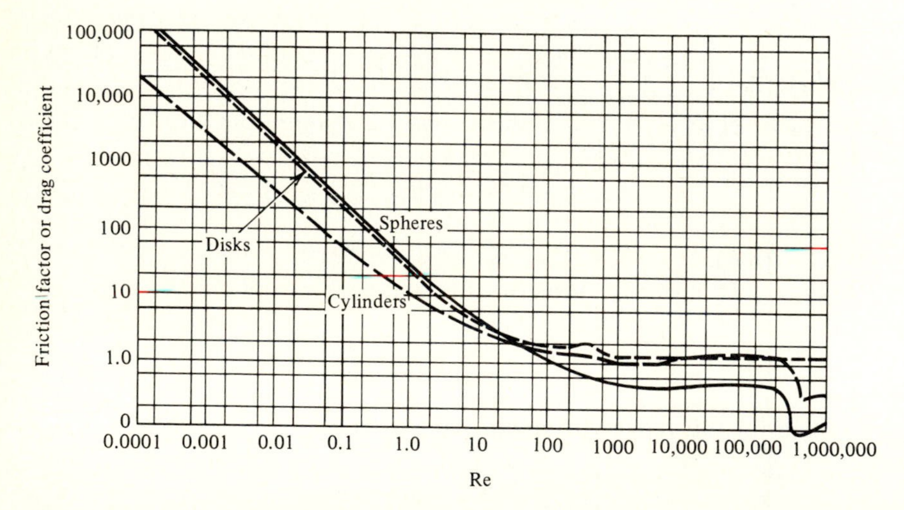

Figure 9-30 Drag coefficients for spheres, disks, and cylinders. [*From C. E. Lapple and R. Shepherd, Ind. Eng. Chem.,* **32:***605 (1940).*]

For flow past a submerged object we take A_{ch} as the *projected* area and v_{ch} as the approach velocity v_∞. Then for the sphere, if we substitute Stokes' law for F_D [Eq. (9-236)] we get

$$\boxed{f = \frac{24}{\text{Re}}} \tag{9-238}$$

where $\text{Re} \equiv 2Rv_\infty \rho/\mu$. This equation (and Stokes' law) hold for $\text{Re} < 0.1$.

At higher velocities, including *turbulent* flow, the curve of friction factor versus Re has been determined experimentally (see Fig. 9-30). A turbulent boundary layer is formed, and the flow eventually separates from the surface, forming vortices. At large values of Re, the friction factor is nearly constant and equal to 0.44. This region is called the *Newton's-law region.*

9-23 FLOW THROUGH PACKED COLUMNS AND POROUS MEDIA

Absorption columns, packed-bed tubular reactors, and pebble heat exchangers consist of a pipe packed with pellets of spherical, cylindrical, or irregular shapes. The calculation of the pressure drop of the fluid flowing through such columns is an important part of the design of the column and associated pumps and piping. Instead of flow past a single sphere or other solid, the problem becomes one of flow past an assemblage of particles. Since the velocity distribution around a given particle is influenced by the flow past each neighboring particle, the problem is much more complex than that of flow past a single sphere. A similar problem exists for flow through a porous medium like that involved in petroleum recovery operations.

A common procedure used in such problems is to substitute for the actual medium an idealized model for which the equation of motion can be solved or for which empirical equations are available. For example, for laminar flow one can assume that the fluid flows through a bundle of channels of some effective diameter D_e or hydraulic radius

$$R_H = \frac{D_e}{4} \tag{9-239}$$

In Chap. 4 we defined the hydraulic radius as

$$R_H = \frac{\text{cross-sectional area}}{\text{perimeter of wetted surface}} \tag{9-240}$$

Since a packed bed consists of void regions through which the fluid flows and solid regions of packing, there will be channels of various cross sections within the packed bed; thus it is more convenient to define an average hydraulic radius based on an element of bed volume of length L. Then within this volume

$$R_H = \frac{\text{volume void region}}{\text{area of packing surface}}$$

The volume of the void region can be expressed in terms of the void fraction, or

$$\epsilon = \frac{\text{volume of voids}}{\text{volume of total bed}} \tag{9-241}$$

The surface area of the packing is usually expressed in terms of the quantity

$$a = \frac{\text{area of packing}}{\text{volume of bed}} \tag{9-242}$$

Then
$$R_H = \frac{\text{volume voids/volume bed}}{\text{area of packing/volume bed}} = \frac{\epsilon}{a} \tag{9-243}$$

It is convenient to let

$$a = S_0(1 - \epsilon) \tag{9-244}$$

where $1 - \epsilon$ is the volume of solid per volume of bed and S_0, called the specific surface, is the area of particle per volume of particle or A_p/V_p. For a sphere of pellet diameter D_p

$$S_0 = \frac{\pi D_p^2}{\pi D_p^3/6} = \frac{6}{D_p} \tag{9-245}$$

Then
$$a = \frac{6(1 - \epsilon)}{D_p} \quad \text{and} \quad R_H = \frac{\epsilon D_p}{6(1 - \epsilon)} \tag{9-246}$$

For nonspherical particles an equivalent diameter is used, as defined by Eq. (9-245), or $D_p = 6/S_0 = 6V_p/A_p$.

For laminar flow of a newtonian fluid through a pipe we derived the Hagen-Poiseuille equation (3-29) for a pipe of diameter D as

$$\frac{\mathcal{P}_0 - \mathcal{P}_L}{L} \equiv \frac{p_0 - p_L + \rho g L}{L} = \frac{32\mu v_a}{D^2} \tag{9-247}$$

where v_a is the average velocity over the flow across section. For a packed bed it is convenient to use the *superficial velocity* v_0, which is defined as the volumetric flow rate $\dot{\mathcal{V}}_v$ through the column divided by the total cross-sectional area of the column A_t, or

$$v_0 \equiv \frac{\dot{\mathcal{V}}_v}{A_t} \tag{9-248}$$

If the column were not packed, the average velocity would be the superficial velocity, but because of the packing, the *actual* average velocity within the voids v_v is greater. If we let A_v be the void area on a total cross section A_t, a mass balance gives

$$\dot{\mathcal{V}}_v = v_0 A_t = v_v A_v \tag{9-249}$$

If we assume that the *area* void fraction A_v/A_t is equal to the *volume* void fraction ϵ, the above equation gives

$$v_v = \frac{v_0 A_t}{A_v} = \frac{v_0}{\epsilon} \tag{9-250}$$

This relationship is called the *du Puit assumption.*

It is now usually assumed that the Hagen-Poiseuille equation holds if we replace v_a with v_0/ϵ and D with an equivalent diameter equal to $4R_H$ (see Chap. 4). Substituting into Eqs. (9-247), (9-250), and the second of (9-246) for v_a and R_H, we get

$$\frac{\mathcal{P}_0 - \mathcal{P}_L}{L} = \frac{32\mu(v_0/\epsilon)}{4\epsilon D_p/6(1-\epsilon)^2} = \frac{72\mu v_0(1-\epsilon)^2}{D_p^2 \epsilon^3} \tag{9-251}$$

In order to correct for the fact that the actual path meanders through the bed and also for errors in the use of the hydraulic-radius concept, a constant larger than 72 must be used. Experimental data indicate that a value of 150 is appropriate. Then

$$\frac{\mathcal{P}_0 - \mathcal{P}_L}{L} = \frac{150\mu v_0(1-\epsilon)^2}{D_p^2 \epsilon^3} \tag{9-252}$$

The friction factor for the bed can be defined by Eq. (4-14) as

$$f \equiv \frac{\dfrac{\mathcal{P}_0 - \mathcal{P}_L}{L} R_H}{\frac{1}{2}\rho v_v^2} \equiv \frac{\dfrac{\mathcal{P}_0 - \mathcal{P}_L}{L} \dfrac{\epsilon D_p}{6(1-\epsilon)}}{\frac{1}{2}\rho (v_0/\epsilon)^2} \equiv \frac{(\mathcal{P}_0 - \mathcal{P}_L)\epsilon^3 D_p}{3\rho v_0^2 (1-\epsilon) L} \tag{9-253}$$

Substituting for $(\mathcal{P}_0 - \mathcal{P}_L)/L$ [Eq. (9-252)] gives

$$f = \frac{150\mu v_0(1-\epsilon)^2}{D_p^2 \epsilon^3} \frac{\epsilon^3 D_p}{3\rho v_0^2(1-\epsilon)} = \frac{50\mu(1-\epsilon)}{D_p \rho v_0} \tag{9-254}$$

We now define a Reynolds number for the bed in terms of hydraulic radius and v_v

$$(\mathrm{Re})_H \equiv \frac{4R_H \rho v_v}{\mu} \equiv \frac{4\epsilon D_p}{6(1-\epsilon)} \frac{\rho v_0/\epsilon}{\mu} \equiv \frac{2D_p \rho v_0}{3(1-\epsilon)\mu} \tag{9-255}$$

Substituting into Eq. (9-254), we obtain

$$f = \frac{33}{(\mathrm{Re})_H} \qquad \text{laminar flow} \tag{9-256}$$

This equation and Eq. (9-254) are called the *Blake-Kozeny equation.* It is valid for void fractions less than 0.5 and for $(\mathrm{Re})_H < 7$. It is plotted in Fig. 9-31, but to simplify numerical calculations the plot is actually of $f' \equiv 3f$ versus $\mathrm{Re}' \equiv \frac{3}{2}(\mathrm{Re})_H$. (Note that the mass velocity $G_0 = \rho v_0$ is used.) In terms of

$$f' \equiv \frac{(\mathcal{P}_0 - \mathcal{P}_L)\epsilon^3 D_p}{\rho v_0^2(1-\epsilon)L} \qquad \text{and} \qquad \mathrm{Re}' \equiv \frac{D_p \rho v_0}{\mu(1-\epsilon)} \tag{9-257}$$

Eq. (9-256) is

$$f' = \frac{150}{\mathrm{Re}'} \tag{9-258}$$

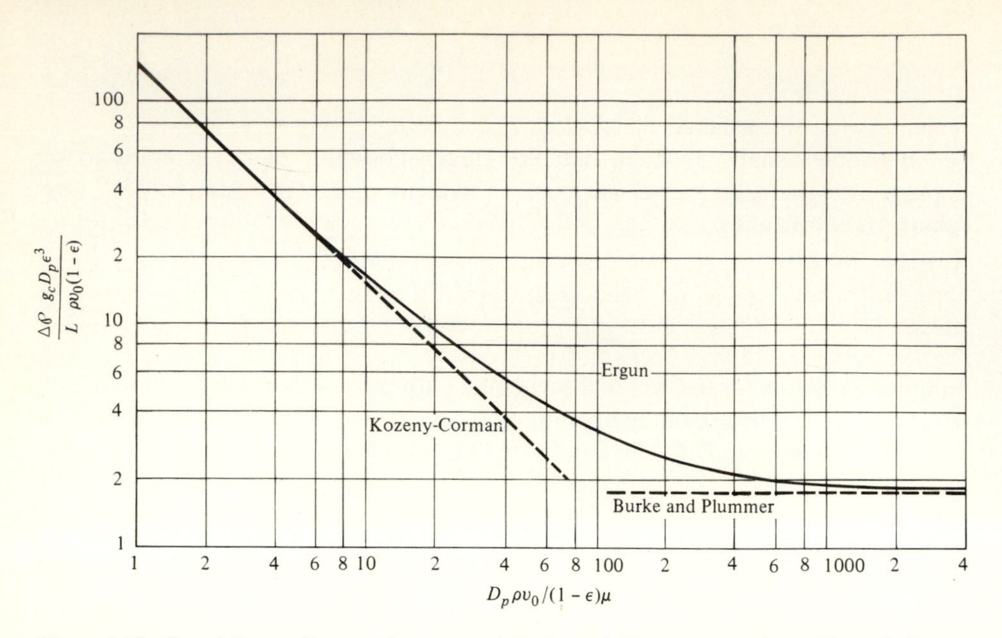

Figure 9-31 Plot of Ergun, Kozeny-Carman, and Burke and Plummer equations. [*From S. Ergun, Chem. Eng. Prog.*, **48:**89 (1952).]

In the *turbulent region*, for flow past a sphere we recall that f approaches a constant at high Re. Thus we let

$$f' \equiv 3f \equiv \frac{(\mathcal{P}_0 - \mathcal{P}_L) D_p \epsilon^3}{\rho v_0^2 (1-\epsilon) L} = \text{const} = f_2 \tag{9-259}$$

Experimental data indicate that the best value of the constant is $f_2 = 1.75$. This is called the *Blake-Plummer equation*.

In the *intermediate* region Ergun assumed that the effects are additive. That is, with subscripts L and T for laminar and turbulent

$$\left(\frac{\mathcal{P}_0 - \mathrm{P}_L}{L}\right)_t = \left(\frac{\mathcal{P}_0 - \mathcal{P}_L}{L}\right)_L + \left(\frac{\mathcal{P}_0 - \mathcal{P}_L}{L}\right)_T$$

$$= \frac{150 \mu v_0 (1-\epsilon)^2}{\epsilon^3 D_p^2} + \frac{1.75 \rho v_0^2 (1-\epsilon)}{D_p \epsilon^3} \tag{9-260}$$

In terms of friction factor this becomes

$$3f \equiv f' = \frac{150}{\mathrm{Re}'} + 1.75 \tag{9-261}$$

called the *Ergun equation*. It is plotted in Fig. 9-31.

The Ergun equation gives reasonably good results for spheres or other particles with an average porosity of about 0.44 but is not as accurate for higher porosities. Fahien and Schriver[10] tested it against experimental data for particles of widely varying porosities and over a large range of Re′ by writing the equation in the form

$$f' \operatorname{Re}' \equiv \phi = f_1 + f_2 \operatorname{Re}' \tag{9-262}$$

and plotting experimental data in the form of ϕ versus Re′. These data should give a straight line of intercept $f_1 = 150$ and slope $f_2 = 1.75$ if the Ergun equation is correct. Instead, the curves through the data in the turbulent region extrapolated to give an intercept of f_{1T} and the data in the laminar region gave a different intercept, f_{1L}. Also, the values of f_{1L}, f_{1T}, and f_2 were not constants (as predicted by the Ergun equation) but functions of porosity. They were correlated by the empirical equations

$$f_{1L} = \frac{136}{(1-\epsilon)^{0.38}} \qquad f_{1T} = \frac{29}{(1-\epsilon)^{1.45}\epsilon^2} \qquad f_2 = \frac{1.87\epsilon^{0.75}}{(1-\epsilon)^{0.26}} \tag{9-263}$$

Then for turbulent and laminar flow respectively one could write

$$\phi_T \equiv f'_T \operatorname{Re}' = f_{1T} + f_2 \operatorname{Re}' \tag{9-264}$$

and

$$\phi_L \equiv f'_L \operatorname{Re}' = f_{1L} \tag{9-265}$$

Finally, Fahien and Schriver modified the Ergun equation by taking into account the probability that the flow at an arbitrary overall bed Reynolds number would be turbulent in some regions and laminar in other regions. The data indicate that these effects are not simply additive (as indicated by the Ergun equation). Letting q be the laminar fraction of the flow and $1 - q$ the turbulent fraction, it was assumed that the overall effect is given by

$$\phi \equiv f' \operatorname{Re}' = q\phi_L + (1-q)\phi_T = qf_{1L} + (1-q)(f_{1T} + f_2 \operatorname{Re}') \tag{9-266}$$

and further that the turbulent condition was approached asymptotically according to

$$q = \exp(-C_0 \operatorname{Re}') \tag{9-267}$$

The parameter C_0 was found to be approximately equal to f_2/f_{1T} and was correlated by the equation

$$C_0 = \frac{\epsilon^2(1-\epsilon)}{12.6} \tag{9-268}$$

It was also found that $C_0 \operatorname{Re}' \approx \operatorname{Re}^*$, where Re^* is defined later.

A new *total* friction factor f^* was then defined by

$$f^* \equiv \frac{\phi}{\operatorname{Re}' f_2} \equiv \frac{f'}{f_2} \tag{9-269}$$

Using the expression for ϕ from Eq. (9-266) gives

$$f^* = \frac{qf_{1L}}{\operatorname{Re}' f_2} + \frac{(1-q)(f_{1T} + f_2 \operatorname{Re}')}{\operatorname{Re}' f_2} \tag{9-270}$$

If we define laminar-flow and turbulent-flow Reynolds numbers as

$$(\mathrm{Re}^*)_L \equiv \frac{\mathrm{Re}' f_2}{f_{1L}} \qquad \text{and} \qquad (\mathrm{Re}^*)_T \equiv \frac{f_2\,\mathrm{Re}'}{f_{1T}} \tag{9-271}$$

and use the definitions in the first of Eqs. (9-257) and in (9-269), then Eq. (9-270) can be written

$$\frac{(\mathcal{P}_0 - \mathcal{P}_L)\epsilon^3 d}{\rho v_0^2(1-\epsilon)Lf_2} \equiv f^* = \frac{q}{(\mathrm{Re}^*)_L} + (1-q)\left[1 + \frac{1}{(\mathrm{Re}^*)_T}\right]$$

$$= (1-q) + \left[\frac{q}{(\mathrm{Re}^*)_L} + \frac{1-q}{(\mathrm{Re}^*)_T}\right] = (1-q) + \frac{1}{\mathrm{Re}^*} \tag{9-272}$$

where

$$\mathrm{Re}^* \equiv \left[\frac{q}{(\mathrm{Re}^*)_L} + \frac{1-q}{(\mathrm{Re}^*)_T}\right]^{-1} \tag{9-273}$$

is a weighted overall Reynolds number. Note that Eq. (9-273) can be written as

$$\underset{\left(\substack{\text{Total}\\ \text{resist-}\\ \text{ance}}\right)}{\frac{v_0}{\mathrm{Re}^*}} = \underset{\left(\substack{\text{Frac-}\\ \text{tion}\\ \text{lam-}\\ \text{inar}}\right)}{q} \times \underset{\left(\substack{\text{Lam-}\\ \text{inar}\\ \text{resist-}\\ \text{ance}}\right)}{\frac{v_0}{(\mathrm{Re}^*)_L}} + \underset{\left(\substack{\text{Frac-}\\ \text{tion}\\ \text{tur-}\\ \text{bulent}}\right)}{(1-q)} \underset{\left(\substack{\text{Tur-}\\ \text{bulent}\\ \text{resist-}\\ \text{ance}}\right)}{\frac{v_0}{(\mathrm{Re}^*)_T}} \tag{9-274}$$

The idea of additive resistance is strengthened if for laminar- and turbulent-flow friction factors we write

$$f_L^* = \frac{1}{(\mathrm{Re}^*)_L} \qquad \text{and} \qquad f_T^* = 1 + [(\mathrm{Re}^*)_T]^{-1} \tag{9-275}$$

Then the first line of Eq. (9-272) can be written as

$$f^* = qf_L^* + (1-q)f_T^* \tag{9-276}$$

Based on the empirical correlations for f_{1L}, f_{1T}, f_2, and C_0, Fig. 9-32 is a plot of f^* versus Re^* as predicted by the first line of Eq. (9-272).

Darcy's Law

For flow through porous media, (e.g., in oil recovery operations, it is customary to use *Darcy's law*

$$v_0 = K\frac{-\Delta\mathcal{P}}{L} \tag{9-277}$$

where K is called the *permeability* of the porous media and $\mathcal{P}$ is the dynamic pressure. Comparison with Eq. (9-252) indicates that

$$K \equiv \frac{D_p^2\epsilon^3}{150(1-\epsilon)^2} \tag{9-278}$$

In order to eliminate the effect of the fluid a *specific* permeability k is often used,

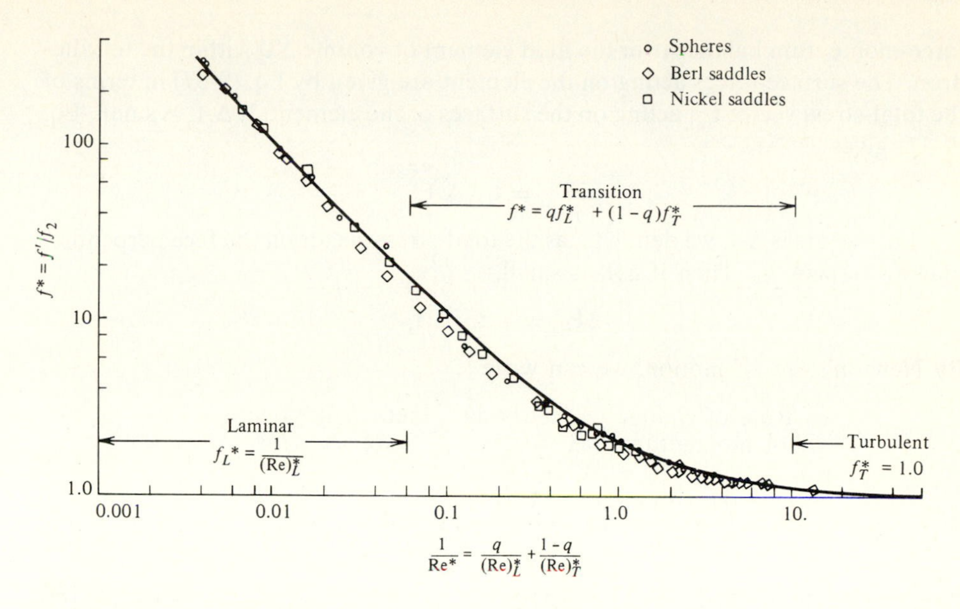

Figure 9-32 Plot of Eq. (9-272).

as defined by

$$v_0 = \frac{g_c k}{\mu} \frac{(-\Delta \mathcal{P})}{L} \tag{9-279}$$

where k has the units of (length)^2 and is a measure of the cross section for the flow.

An often used equation for the specific permeability is based on the Kozeny-Carmen equation, which includes a *tortuosity factor* L/L_e to account for the meandering path of the flow, a *Kozeny constant* k_0 to account for the shape of the conduit, and a hydraulic radius, defined previously. The equation for the specific permeability is then

$$k = \left[\frac{\epsilon D_p}{6(1-\epsilon)}\right]^2 \left(\frac{L}{L_e}\right)^2 \frac{\epsilon}{k_0} \tag{9-280}$$

Usually k_0 is taken as 2.5 and L_e/L as $\sqrt{2}$. Then

$$k = \frac{\epsilon^3 D_p^2}{180(1-\epsilon)^2} \tag{9-281}$$

APPENDIX 9-1 STRESS TETRAHEDRON

Consider a tetrahedron formed of areas $\Delta A_x \equiv \Delta A_1$, $\Delta A_y \equiv \Delta A_2$, and $\Delta A_z \equiv \Delta A_3$ (each perpendicular to the axis subscripted) and of another arbitrary area ΔA_n (with outwardly directed normal $\mathbf{n}$) (see Fig. A8-4-1). It is our purpose to write a force-momentum balance over the fluid element of volume $\Delta\mathcal{V}$ within the tetrahedron. The surface forces acting on the element are given by Eq. (9-37) in terms of the total-stress vector $\mathbf{t}_{(n)}$ acting on the surfaces of the element. If ΔA_n is small, Eq. (9-37) gives

$$\Delta\mathbf{F}_{(n)} = \mathbf{t}_{(n)}\,\Delta A_n$$

For the areas ΔA_i we define $\mathbf{t}_{(i)}$ as the total-stress vector on the face perpendicular to the axis x_i. Then if ΔA_i is small,

$$\Delta\mathbf{F}_{(i)} = -\mathbf{t}_{(i)}\,\Delta A_i$$

By Newton's law of motion, we can write

$$\begin{matrix}\text{Rate of change}\\ \text{of momentum}\end{matrix} = \left(\begin{matrix}\text{net body}\\ \text{forces}\end{matrix} + \begin{matrix}\text{net surface}\\ \text{forces}\end{matrix}\right)g_c$$

$$\frac{D(\rho\mathbf{v}\,\Delta\mathcal{V})}{g_c\,Dt} = \frac{\rho\,\Delta\mathcal{V}\,\mathbf{g}}{g_c} - \sum_{i=1}^{3}\Delta A_i\,\mathbf{t}_{(i)} + \mathbf{t}_{(n)}\,\Delta A_n$$

Now we note that if Δh is a characteristic length, then $\Delta\mathcal{V} \propto (\Delta h)^3$ and $\Delta A \propto (\Delta h)^2$. Dividing by ΔA_n and letting $\Delta A_n \to 0$, gives $\Delta\mathcal{V}/\Delta A_n \to 0$ and

$$0 = 0 - \lim_{\Delta A_n \to 0} \sum \frac{\Delta A_i}{\Delta A_n}\mathbf{t}_{(i)} + \mathbf{t}_{(n)} = 0$$

This equation is called the *principle of local stress equilibrium*. It can also be written

$$\lim_{\Delta h \to 0} \frac{\iint \mathbf{t}_{(n)}\,dA}{(\Delta h)^2} = 0 \qquad \text{(A9-1-1)}$$

By projection of ΔA_n onto coordinate planes $\Delta A_i = \mathbf{n}\cdot\boldsymbol{\delta}_i\,\Delta A_n$, so that

$$\mathbf{t}_{(n)} = \sum_i \mathbf{n}\cdot\boldsymbol{\delta}_i\,\mathbf{t}_{(i)} = \mathbf{n}\cdot\sum_i \boldsymbol{\delta}_i\,\mathbf{t}_{(i)} \equiv \mathbf{n}\cdot\underline{\mathbf{T}}$$

or

$$\mathbf{t}_{(n)} = \mathbf{n}\cdot\underline{\mathbf{T}} \qquad \text{where} \qquad \underline{\mathbf{T}} \equiv \sum_i \boldsymbol{\delta}_i\,\mathbf{t}_{(i)} \qquad \text{(A9-1-2)}$$

It is important to note that $\mathbf{t}_{(i)}$, in general, is *not* a vector in the direction of axis x_i but is instead the resultant $\Sigma_j\, T_{ij}\boldsymbol{\delta}_j$. We note that since $\boldsymbol{\delta}_i$ and $\mathbf{t}_i$ are vectors, $\underline{\mathbf{T}}$ must be a second-order tensor; it is called the *total-stress tensor*. It can be expressed in terms of its cartesian components as

$$T = \sum_{ij} T_{ij}\boldsymbol{\delta}_i\boldsymbol{\delta}_j \qquad \text{(A9-1-3)}$$

Equation (A9-1-2) is an important equation in that it shows that the total stress acting on an area is linear in the normal to the area and in the components of $\underline{\mathbf{T}}$. Thus at any point in the fluid we can describe the stress acting on any plane through the point by specifying the normal $\mathbf{n}$ to the plane.

Although we obtained a similar result previously for a cube, we have now shown that this result applies to any shape since the orientation of the slant face ΔA_n is arbitrary. Also, since it is expressed in tensor notation, it applies to any coordinate system.

APPENDIX 9-2 CROSS PRODUCT OR VECTOR PRODUCT

We have already seen that we can multiply vectors by means of the dot-product (or scalar-product) operation. We illustrated that concept with the physical example of the dot product of the force and distance vectors to obtain work, a scalar. But there is yet another way to multiply vectors, namely the cross or vector product, and it has a still different physical meaning.

To illustrate this operation, consider a force $F = |\mathbf{F}|$ acting at P at a distance r from a pivot point O, as shown in Fig. A9-2-1*a*. We might think of a door of width r hinged at point O. We wish to calculate the torque used to open the door. If we push on the door in a direction perpendicular to the door, we obtain the maximum torque, which has the magnitude

$$\Upsilon = rF = \text{lever arm} \times \text{force}$$

Now suppose we apply the force at an angle θ to the perpendicular (as shown in Fig. A9-2-1*b*); then the magnitude of the torque is

$$\Upsilon = rF \sin\theta$$

where θ is the angle between the vectors $\mathbf{r}$ and $\mathbf{F}$. Thus for maximum torque the vectors $\mathbf{F}$ and $\mathbf{r}$ are perpendicular, or $\sin\theta = 1$. But for zero torque, $\mathbf{F}$ and $\mathbf{r}$ are collinear and $\sin\theta = 0$. We have measured this torque about an axis through point O and perpendicular to the plane of the paper (the z axis), but in general a torque can be measured about other axes (x or y) also. The axis then gives it a

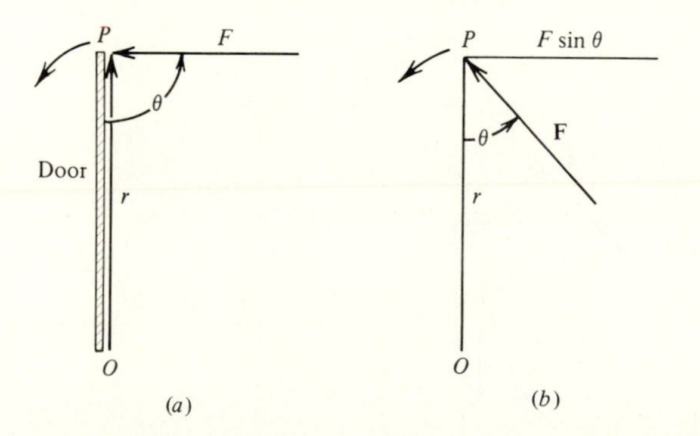

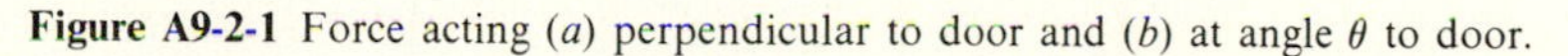
Figure A9-2-1 Force acting (*a*) perpendicular to door and (*b*) at angle θ to door.

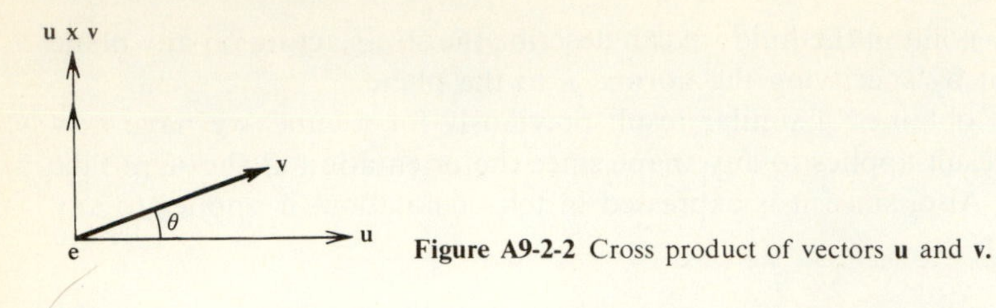

Figure A9-2-2 Cross product of vectors **u** and **v**.

direction. Thus, torque is a vector quantity, and in this example we can represent its direction by a unit vector **e** measured perpendicular to the paper. We will arbitrarily define the positive direction as being upward from the plane of the paper. If we place **F** and **r** in a head-to-head position at point O (see Fig. A9-2-2), we can visualize this upward motion by the advance of a right-hand screw at point O, moving from **r** to **F**, or counterclockwise. Then we can write this operation as

$$\boldsymbol{\Upsilon} = \mathbf{r} \times \mathbf{F} = rF \sin\theta\, \mathbf{e} \tag{A9-2-1}$$

In general for two vectors **u** and **v** the cross (or vector) product is *defined* as

$$\boxed{\mathbf{u} \times \mathbf{v} \equiv uv \sin\theta\, \mathbf{e}} \tag{A9-2-2}$$

where θ is the angle between the vectors and **e** is a unit vector in the direction perpendicular to the plane of **u** and **v** and with sign given by the right-hand rule: If we turn a right-hand screw counterclockwise from **u** to **v**, **e** will be positive if it advances away from the plane of **u** and **v** and negative if it moves toward **u** and **v**.

If we apply this definition to the unit vectors, we get

$$\boldsymbol{\delta}_x \times \boldsymbol{\delta}_y = \boldsymbol{\delta}_z \qquad \boldsymbol{\delta}_z \times \boldsymbol{\delta}_x = \boldsymbol{\delta}_y \qquad \boldsymbol{\delta}_y \times \boldsymbol{\delta}_z = \boldsymbol{\delta}_x \qquad \boldsymbol{\delta}_y \times \boldsymbol{\delta}_x = -\boldsymbol{\delta}_z \tag{A9-2-3}$$

and so forth. These equations define what is called a right-hand coordinate system (Fig. A9-2-3*a*). For a left-hand system the signs on the right-hand side would change and $\boldsymbol{\delta}_z$ would point downward (Fig. A9-2-3*b*).

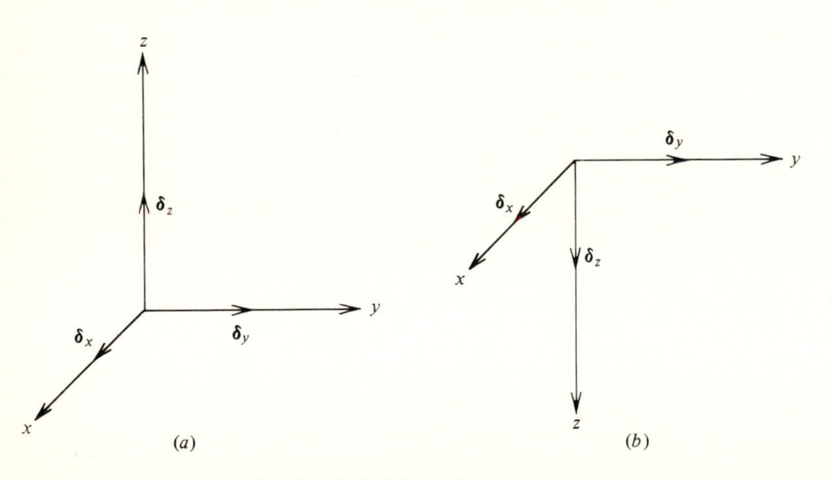

Figure A9-2-3 (*a*) Right-hand and (*b*) left-hand system.

The vector product of two vectors **u** and **v** can be written in terms of components as

$$\mathbf{u} \times \mathbf{v} = \left(\sum_i u_i \boldsymbol{\delta}_i\right) \times \left(\sum_j v_j \boldsymbol{\delta}_j\right)$$

Since $\boldsymbol{\delta}_1 \times \boldsymbol{\delta}_2 = \boldsymbol{\delta}_3$ and $\boldsymbol{\delta}_2 \times \boldsymbol{\delta}_1 = -\boldsymbol{\delta}_3$, we can write

$$(\mathbf{u} \times \mathbf{v})_3 = (u_1 v_2 - u_2 v_1)\boldsymbol{\delta}_3 \tag{A9-2-4}$$

or

$$\mathbf{u} \times \mathbf{v} = \sum_i \sum_j \sum_k \epsilon_{ijk} u_i v_j \boldsymbol{\delta}_k \tag{A9-2-5}$$

where ϵ_{ijk} is the *alternating tensor,* defined as

$$\epsilon_{ijk} = \begin{cases} 0 & \text{if any two } i, j, k \text{ are the same} \\ 1 & \text{if } i, j, k \text{ is an even permutation of 1, 2, 3} \\ -1 & \text{if } i, j, k \text{ is an odd permutation of 1, 2, 3} \end{cases} \tag{A9-2-6}$$

We can also express $\mathbf{u} \times \mathbf{v}$ as a determinant

$$\mathbf{u} \times \mathbf{v} = \begin{vmatrix} \boldsymbol{\delta}_1 & \boldsymbol{\delta}_2 & \boldsymbol{\delta}_3 \\ u_1 & u_2 & u_3 \\ v_1 & v_2 & v_3 \end{vmatrix} \tag{A9-2-7}$$

It is apparent that we can multiply either term by a scalar as follows

$$\rho\mathbf{u} \times \mathbf{v} \equiv \mathbf{u} \times \rho\mathbf{v} \equiv (\mathbf{u} \times \mathbf{v})\rho$$

To take the derivative $d(\mathbf{u} \times \mathbf{v})/dt$ we first find

$$\Delta(\mathbf{u} \times \mathbf{v}) = (\mathbf{u} + \Delta\mathbf{u}) \times (\mathbf{v} + \Delta\mathbf{v}) - \mathbf{u} \times \mathbf{v}$$

$$= \mathbf{u} \times \Delta\mathbf{v} + \Delta\mathbf{u} \times \mathbf{v} + \Delta\mathbf{u} \times \Delta\mathbf{v}$$

Then if we divide by Δt and let $\Delta t \to 0$,

$$\boxed{\frac{d}{dt}(\mathbf{u} \times \mathbf{v}) = \mathbf{u} \times \frac{d\mathbf{v}}{dt} + \frac{d\mathbf{u}}{dt} \times \mathbf{v}} \tag{A9-2-8}$$

Since the cross product is not commutative, we cannot reverse the order of the operation.

APPENDIX 9-3 ANGULAR-MOMENTUM BALANCE

Newton's law of motion can be written ($g_c = 1.0$)

$$\mathbf{F} = \frac{d(m\mathbf{v})}{dt}$$

If we take the cross product [see Eq. (A9-2-2)] of this equation with respect to the position vector **x** [Eq. (9-53)], we get

$$\mathbf{x} \times \mathbf{F} = \mathbf{x} \times \frac{d(m\mathbf{v})}{dt} \tag{A9-3-1}$$

or

$$\mathbf{x} \times \mathbf{F} = \frac{d(\mathbf{x} \times m\mathbf{v})}{dt} - \underbrace{\frac{d\mathbf{x}}{dt} \times m\mathbf{v}}_{(1)} = \frac{d(\mathbf{x} \times m\mathbf{v})}{dt} \tag{A9-3-2}$$

Since $d\mathbf{x}/dt = \mathbf{v}$ and $\mathbf{v}$ and $d\mathbf{x}/dt$ are in the same direction, term (1) is zero.

We define the torque $\boldsymbol{\Upsilon}$ acting on a body of mass m and the angular momentum $\boldsymbol{\Gamma}$ as, respectively,

$$\boldsymbol{\Upsilon} \equiv \mathbf{x} \times \mathbf{F} \qquad \text{and} \qquad \boldsymbol{\Gamma} \equiv m\mathbf{x} \times \mathbf{v} \tag{A9-3-3}$$

Combining Eqs. (A9-3-2) and (A9-3-3) gives

$$\boldsymbol{\Upsilon} = \frac{d\boldsymbol{\Gamma}}{dt} \tag{A9-3-4}$$

i.e., torque is proportional to the time rate of change of angular momentum.

For a fluid, we would write an angular-momentum counterpart to the linear-momentum balance as

$$\begin{matrix}\text{Rate of change} \\ \text{of angular} \\ \text{momentum}\end{matrix} = \begin{matrix}\text{Net torque} \\ \text{(due to pressure} \\ \text{and body forces)}\end{matrix} + \begin{matrix}\text{Net diffusive} \\ \text{flow of angular} \\ \text{momentum}\end{matrix} + \begin{matrix}\text{Net bulk} \\ \text{flow of angular} \\ \text{momentum}\end{matrix} \tag{A9-3-5}$$

APPENDIX 9-4 SYMMETRY OF THE STRESS TENSOR

Consider an element of fluid of volume $\Delta \mathcal{V} = \Delta x \, \Delta y \, \Delta z$ located at an arbitrary position (x, y, z) in a fluid. In general the fluid will be subject to shear forces acting on each face (see Fig. A9-4-1, in which, for simplicity, only the xy components are shown). The element must satisfy an angular-momentum balance taken clockwise about the z axis at the origin. Using Eq. (A9-3-5), we have

$$g_c \sum \text{torques} + \begin{matrix}\text{Net rate of flow of} \\ \text{angular momentum}\end{matrix} = \begin{matrix}\text{Rate of accumulation of} \\ \text{angular momentum}\end{matrix}$$

By definition, torque is given by Eq. (A9-2-1)

$$\boldsymbol{\Upsilon} = \mathbf{x} \times \mathbf{F} = |\mathbf{x}|\,|\mathbf{F}| \sin\theta \, \mathbf{e}$$

where θ is the angle between **x** and **F** and **e** is a unit vector normal to both **x** and **F**.

A torque about the z axis is in general [Eq. (A9-2-4)]

$$\Upsilon_z = xF_y - yF_x$$

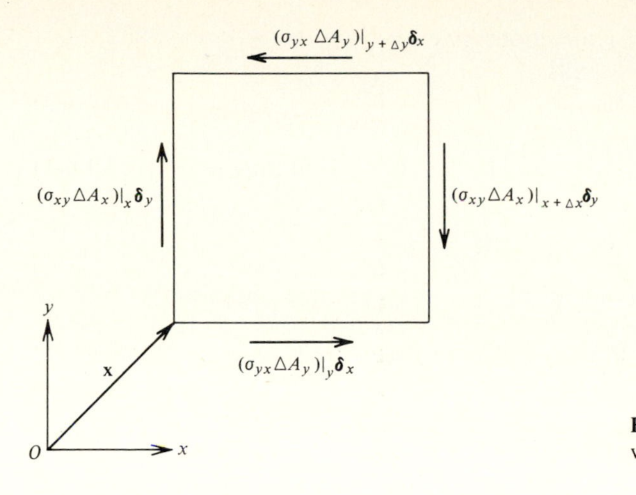

Figure A9-4-1 Torques acting on volume element.

since $\theta = \pi/2$ and $\mathbf{e}$ is a vector in the $+z$ direction. Applying this to the element in question, we note that the forces acting on the element may consist not only of viscous forces but also pressure forces, gravity forces, or other body forces as well. For simplicity, however, at this time we consider only viscous forces and assume that there is no net bulk flow of angular momentum. (As an exercise, the student can show later that our result does not depend on this assumption.)

The viscous force acting on the area $\Delta A_y = \Delta x\,\Delta z$ of the element at $y = y$ is then $(\sigma_{yx}\,\Delta A_y)|_y$, and the torque is $(y\sigma_{yx}\,\Delta A_y)\big|_y$. Writing similar terms at $y + \Delta y$, x, and $x + \Delta x$ gives

$$-(y\sigma_{yx}\,\Delta A_y)\Big|_y + (y\sigma_{yx}\,\Delta A_y)\Big|_{y+\Delta y} + (x\sigma_{xy}\,\Delta A_x)\Big|_x - (x\sigma_{xy}\,\Delta A_y)\Big|_{x+\Delta x}$$
$$= \frac{\Delta}{\Delta t}[\rho\,\Delta\mathcal{V}\,(xv_y - yv_x)]$$

where the angular momentum per unit mass of the element is taken as [Eq. (A9-3-3)]

$$[\![(\mathbf{x}\times\mathbf{v})_z]\!] = xv_y - yv_x$$

Dividing by $\Delta\mathcal{V} = \Delta x\,\Delta y\,\Delta z$, letting $\Delta A_x = \Delta y\,\Delta z$, and taking limits as $\Delta x \to 0$, $\Delta y \to 0$, and $\Delta t \to 0$ gives

$$\frac{\partial(y\sigma_{yx})}{\partial y} - \frac{\partial(x\sigma_{xy})}{\partial x} = \frac{\partial}{\partial t}[\rho(v_y x - v_x y)] \qquad \text{(A9-4-1)}$$

or

$$y\frac{\partial\sigma_{yx}}{\partial y} - x\frac{\partial\sigma_{xy}}{\partial x} - \sigma_{yx} + \sigma_{xy} = x\frac{\partial(\rho v_y)}{\partial t} - y\frac{\partial(\rho v_x)}{\partial t}$$

Force balances in the x and y directions give

$$-\frac{\partial \sigma_{yx}}{\partial y} = \frac{\partial}{\partial t}(\rho v_x) \qquad \text{and} \qquad -\frac{\partial \sigma_{xy}}{\partial x} = \frac{\partial}{\partial t}(\rho v_y) \tag{A9-4-2}$$

Multiplying the first by y and the second by $-x$ and substituting into Eq. (A9-4-1) gives

$$-\sigma_{yx} + \sigma_{xy} = 0$$

Similarly, by taking balances about the x and y axes one obtains

$$\sigma_{yz} = \sigma_{zy} \qquad \text{and} \qquad \sigma_{xz} = \sigma_{zx}$$

or, in general,

$$\sigma_{ij} = \sigma_{ji} \qquad \text{and} \qquad \tau_{ij} = \tau_{ji} \qquad \text{or} \qquad \underline{\tau} = \underline{\tau}^T \tag{A9-4-4}$$

where $\underline{\tau}^T$ is the transpose of $\underline{\tau}$. (A *transpose* of a matrix is obtained by interchanging rows and columns.) A tensor that is equal to its transpose is called a *symmetric tensor*. Obviously the shear-stress tensor $\underline{\boldsymbol{\sigma}}$ and the total-stress tensor $\underline{\mathbf{T}}$ are also symmetric.

The above derivation applies only to a nonpolar fluid. Certain nonnewtonian fluids are highly polar and capable of transmitting internal stress couples and body torques which must be included. When they are, the stress tensor is not symmetric. However, the effect is believed to be small, and the stress tensor is always taken to be symmetric. Therefore, the angular-momentum equation can be obtained by taking the cross product of the position vector with the equation of motion.

Equation of Change for Angular Momentum

If all torque (viscous, gravity, and pressure) as well as bulk-flow terms are considered, a general angular-momentum equation can be derived by the method above using Eq. (A9-3-5). This gives[4]

$$\frac{\partial}{\partial t}(\mathbf{r} \times \rho \mathbf{v}) = -\nabla \cdot (\mathbf{r} \times \rho \mathbf{v}\mathbf{v})^T - \nabla \cdot (\mathbf{r} \times \underline{\tau})^T - \nabla \cdot (\mathbf{r} \times p\underline{\boldsymbol{\delta}})^T + \mathbf{r} \times \rho \mathbf{g} \tag{A9-4-5}$$

APPENDIX 9-5 NEWTON'S LAW FOR COMPRESSIBLE FLUIDS

For a compressible fluid an additional term must be added to the flux law to account for the *dilatation* of the fluid element, i.e., the deformation occurring due to the simultaneous change in length of its sides such that there is a change in the *volume* of the element. (Previously the volume was kept constant.) More generally, during time dt this relative deformation per unit volume of element $\Delta x\,\Delta y\,\Delta z$ would be

$$\frac{\left(dx + \frac{\partial v_x}{\partial x}dx\,dt\right)\left(dy + \frac{\partial v_y}{\partial y}dy\,dt\right)\left(dz + \frac{\partial v_z}{\partial z}dz\,dt\right) - dx\,dy\,dz\,dt}{dx\,dy\,dz\,dt}$$

$$= \frac{\partial v_x}{\partial x} + \frac{\partial v_y}{\partial y} + \frac{\partial v_z}{\partial z} \equiv \nabla \cdot \mathbf{v}$$

For compressible fluids the stress tensor can then be written

$$\underline{\tau} = -\mu\underline{\boldsymbol{\Delta}} - \lambda\nabla \cdot \mathbf{v}\,\underline{\boldsymbol{\delta}} \tag{A9-5-1}$$

where λ is the coefficient of *bulk viscosity,* usually taken as $-\frac{2}{3}\mu$.

APPENDIX 9-6 NORMAL STRESSES

By an analysis of the angular-deformation rate we have arrived at the expression (9-106) for the shear stress τ_{ij} for a newtonian fluid. Previously we indicated that for an incompressible fluid this equation also gives the normal stress τ_{ii} by setting $i = j$. But it is not clear that such a generalization is valid since it is not obvious that an angular deformation is involved in a compression. As illustrated by the before and after views in Fig. A9-6-1 an angular deformation can be shown to occur by rotating through an angle β, the *axes* of the simple shear case discussed

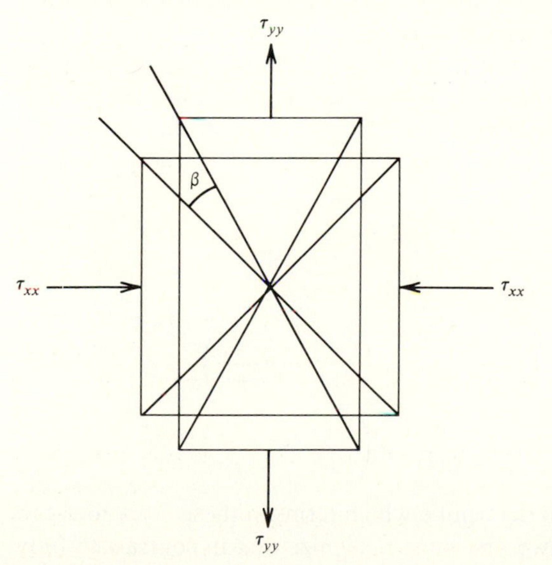

Figure A9-6-1 Deformation due to normal stress.

previously. As shown in Fig. A9-6-1, the effect of a compression in the xy coordinate system is equivalent to that of the pure shears indicated in the $x'y'$ system. Since such an analysis becomes involved,[3] it will not be carried out at this time. Instead we note that both shear and normal deformations are apparently a manifestation of the same basic tensor relationship $\underline{\tau} = \underline{\tau}(\underline{\Delta})$ and that for isotropic incompressible fluids

$$\underline{\tau} = -\eta \underline{\Delta} \tag{A9-6-1}$$

where η will not be constant but may depend upon the rate of deformation. The above expression describes a *generalized newtonian fluid.*

APPENDIX 9-7 SCALING AND ORDERING

In Appendix 4-1 we discussed the nondimensionalization of a differential equation in order to find the dimensionless groups involved. In doing so we found that there is some choice in the selection of characteristic dimensions or scales; e.g., for laminar flow in a pipe, the characteristic velocity can be defined either as the average velocity v_a or as $v_c \equiv (\mathcal{P}_0 - \mathcal{P}_L)g_c/R^2L$. In this example it was not of major importance which characteristic velocity was used since eventually we arrived at the same result, namely that the friction factor is a function of Reynolds number.

In more complex cases, however, a special kind of dimensional analysis called *scaling and ordering* is used. In this method the goal is to determine not only the characteristic parameters or scales but also to find the relative importance of each term. This ordering is done so that, for limiting values of the variables and parameters, the equations can be simplified and solved.

The subject of scaling is thoroughly discussed by Lin and Segal,[16] and the reader is referred to that excellent source for more detail. For our purposes we will illustrate the procedure for the case of a laminar-boundary-layer flow past a flat plate, as discussed in Example 9-7. The EOC is

$$\frac{\partial v_x}{\partial x} + \frac{\partial v_y}{\partial y} = 0$$

and the Navier-Stokes equations are

$$\rho\left(v_x \frac{\partial v_x}{\partial x} + v_y \frac{\partial v_x}{\partial y}\right) = -\frac{\partial p}{\partial x} - \mu\left(\frac{\partial^2 v_x}{\partial x^2} + \frac{\partial^2 v_x}{\partial y^2}\right)$$

and

$$\rho\left(v_x \frac{\partial v_y}{\partial x} + v_y \frac{\partial v_y}{\partial y}\right) = -\frac{\partial p}{\partial y} - \mu\left(\frac{\partial^2 v_y}{\partial x^2} + \frac{\partial^2 v_y}{\partial y^2}\right)$$

We will use scaling and ordering to determine which terms in these equations can be neglected. But before doing so we first must recognize that in contrast to fully developed flow in a pipe, there is now more than one length scale involved, namely the length of the plate and the thickness of the boundary layer. As we will see later there is a velocity scale for both the axial and the lateral components of the velocity.

The procedure is to choose as scales the maximum values of the variables involved, so that the dimensionless variables will vary from zero to unity. Thus for the axial velocity we choose v_∞ as the characteristic velocity (or velocity scale) and let

$$v_x^* \equiv \frac{v_x}{v_\infty}$$

Since the axial coordinate x varies from 0 to L, we let

$$x^* \equiv \frac{x}{L}$$

For a derivative the procedure is to choose scales that would make it unity in dimensionless form. For example, let

$$\frac{L}{v_\infty}\frac{\partial v_x}{\partial x} = \frac{\partial v_x^*}{\partial x^*} \approx \frac{1-0}{1-0} \approx 1$$

where we have used the maximum values minus the minimum value of each variable.

For the y direction, the length scale is some measure of the thickness of the developing boundary layer. Since v_x is approaching v_∞ as y increases, we will let the boundary-layer thickness $\delta(x)$ be the distance at which $v_x/v_\infty \approx 1$. Then we let

$$y^* \equiv \frac{y}{\delta}$$

We do not as yet have a characteristic velocity scale in the y direction; so we will call it V. Then

$$v_y^* \equiv \frac{v_y}{V}$$

Similarly we define

$$p^* \equiv \frac{p}{p_c}$$

Since we have flow in two directions, the EOC becomes

$$\frac{v_\infty \delta}{VL}\frac{\partial v_x^*}{\partial x^*} + \frac{\partial v_y^*}{\partial y^*} = 0$$

Now each velocity gradient is on the order of magnitude of unity. Thus in order to make each term roughly equal in importance, we must let $V \equiv v_\infty \delta / L$. Then since $\delta \ll L$, we must also have $V \ll v_\infty$.

Using the above dimensionless variables in the x component of the EOM gives

$$\frac{\rho v_\infty \delta^2}{\mu L}\left(v_x^* \frac{\partial v_x^*}{\partial x^*} + v_y^* \frac{\partial v_x^*}{\partial y^*}\right) = -\frac{p_c \delta^2}{\mu v_\infty L}\frac{\partial p^*}{\partial x^*} + \left[\left(\frac{\delta}{L}\right)^2 \frac{\partial^2 v_x^*}{\partial x^{*2}} + \frac{\partial^2 v_x^*}{\partial y^{*2}}\right]$$

$$\text{(1)} \qquad\qquad \text{(2)} \qquad\qquad \text{(3)} \qquad\qquad \text{(4)} \qquad \text{(A9-7-1)}$$

In the viscous- or diffusive-transport terms (3) and (4) we recall that both $\partial^2 v_x^*/\partial x^{*2}$ and $\partial^2 v_y^*/\partial y^{*2}$ were made to be of the order of magnitude of unity. Therefore since $\delta/L \ll 1$, term (3), representing axial diffusion, is negligible compared with (4).

On the other hand, both the inertia term (1) and the viscous-diffusion term (4) must be of nearly equal importance. To make them so, the group $\rho v_\infty \delta^2/\mu L$ must be of the order of unity. Setting it equal to 1 and solving for δ gives

$$\delta = \sqrt{\frac{\mu L}{\rho v_\infty}} = L\,(\text{Re})_L^{-1/2} \qquad \text{(A9-7-2)}$$

where

$$(\text{Re})_L \equiv \frac{L v_\infty \rho}{\mu} \qquad \text{(A9-7-3)}$$

is a length Reynolds number. Thus the boundary-layer thickness decreases as $(\text{Re})_L$ increases until turbulent flow occurs.

Similarly, we can make the coefficient of $\partial p^*/\partial x^*$ in term (2) equal to unity if we let

$$p_c \equiv \frac{\mu v_\infty L}{\delta^2} = \frac{\mu v_\infty L}{\mu L/\rho v_\infty} = \rho v_\infty^2$$

This is twice the kinetic energy per unit volume, which is the characteristic pressure we used in Chap. 4 for fully developed laminar flow in a pipe.

Expressing the y component of the Navier-Stokes equation in dimensionless form gives

$$\left(\frac{\delta}{L}\right)^2 \left(v_y^* \frac{\partial v_y^*}{\partial x^*} + v_y^* \frac{\partial v_y^*}{\partial x^*}\right) = \left(\frac{\delta}{L}\right)^2 \left\{\left(\frac{\delta}{L}\right)^2 \frac{\partial^2 v_y^*}{\partial x^{*2}} + \frac{\partial^2 v_y^*}{\partial y^{*2}}\right\}\left(-\frac{\partial p^*}{\partial y^*}\right)$$

Since $\delta/L \ll 1$, and since the terms in the braces are on the order of unity, this gives

$$\frac{\partial p^*}{\partial y^*} \approx 0$$

or there is no lateral pressure gradient. As a result

$$p^* = p^*(x^*)$$

only. Then we can assume that

$$\left.\frac{\partial p^*}{\partial x^*}\right|_{\text{bdy layer}} = \left.\frac{\partial p^*}{\partial x^*}\right|_{\text{outer layer}} = \frac{dp^*}{dx^*} \qquad \text{(A9-7-4)}$$

The x component is then

$$v_x^* \frac{\partial v_x^*}{\partial x^*} + v_y^* \frac{\partial v_x^*}{\partial y^*} = -\frac{dp^*}{dx^*} + \frac{\partial^2 v_x^*}{\partial y^{*2}} \tag{A9-7-5}$$

PROBLEMS

9-1 (*a*) Show that

$$\nabla \cdot \rho \mathbf{v}\mathbf{v} \equiv \rho \mathbf{v} \cdot \nabla \mathbf{v} + \mathbf{v} \nabla \cdot \rho \mathbf{v}$$

(*b*) Convert the equation of motion in eulerian form into lagrangian form using the result in part (*a*).

(*c*) Repeat part (*b*) but use the identity

$$\rho \frac{D\hat{G}}{Dt} \equiv \frac{\partial(\rho \hat{G})}{\partial t} + \nabla \cdot \rho \hat{G} \mathbf{v} \tag{8-144}$$

9-2 (*a*) Simplify the EOC and the EOM for steady-state fully developed flow of an incompressible newtonian fluid in a circular pipe.

(*b*) Derive an equation for the shear-stress distribution.

9-3 Explain how the z component of the angular-momentum equation is related to the θ component of the linear-momentum equation.

9-4 Distinguish (*a*) between $\mathbf{v} \cdot \nabla$, $\nabla \cdot \mathbf{v}$, $\nabla \mathbf{v}$, and $(\nabla \mathbf{v})^T$ and (*b*) between $\mathbf{v} \cdot \underline{\boldsymbol{\tau}}$ and $\underline{\boldsymbol{\tau}} \cdot \mathbf{v}$.

9-5 For newtonian Couette flow, write expressions for the components of $\underline{\boldsymbol{\sigma}}$, $\underline{\boldsymbol{\tau}}$, $\underline{\mathbf{T}}$, $\boldsymbol{\sigma}_{(n)}$, $\mathbf{t}_{(n)}$, $\mathbf{F}_p$, and $\mathbf{F}_v$, on each plate.

9-6 Show that if $\mathbf{n} \cdot \underline{\mathbf{T}} = \mathbf{t}_{(n)}$, a vector, then $\underline{\mathbf{T}}$ is a second-order tensor.

9-7 Find expressions for the $r\theta$ components of $\nabla \mathbf{v}$, $\underline{\boldsymbol{\Delta}}$, and $\underline{\boldsymbol{\Omega}}$ in spherical coordinates.

9-8 A newtonian incompressible fluid is flowing under a pressure difference in the z direction through a vertical pipe with an elliptical cross section of major semiaxis A and minor semiaxis B. Assume steady state and fully developed flow. Use orthogonal elliptic cylinder coordinates defined by

$$x = a \cos \theta \cosh \eta \qquad y = a \sin \theta \sinh \eta \qquad z = z$$

where a = const is the focal length.

(*a*) Find $(\nabla \cdot \underline{\boldsymbol{\tau}})_z$ and $\nabla^2 v_z$.

(*b*) Write the simplified z components of the Navier-Stokes equations.

(*c*) Show that curves of constant η are confocal ellipses.

9-9 Show that the Poisson equation $\mu \nabla^2 v_z = \phi_M$ obtained in Prob. 9-8 is satisfied by

$$v_z = \frac{\phi_M}{2\mu} \frac{A^2 B^2}{A^2 + B^2} \left(1 - \frac{x^2}{A^2} - \frac{y^2}{B^2}\right)$$

9-10 (*a*) Find the average velocity in Prob. 9-9. *Hint:* Use a nonorthogonal coordinate system ρ, θ, z, where $\eta \equiv N$ is constant and $\rho \equiv a$ varies:

$$x = \rho\alpha \cos \theta \qquad y = \rho\beta \sin \theta \qquad z = z$$

where $\alpha \equiv \cosh N$, $\beta \equiv \sinh N \equiv \sqrt{1 - \alpha^2}$, R = value of ρ at wall $= A/\alpha = B/\beta$. Thus curves of constant ρ are concentric ellipses. It can be shown by vector identities that

$$dA_z = \alpha\beta\rho \, d\rho$$

(*b*) Relate the friction factor to the Reynolds number.

9-11 Consider fully developed steady flow under a pressure difference of an incompressible newtonian fluid through a pipe of arbitrary cross section (see Appendix 4-2).

(*a*) Show that $\mu \nabla^2 v_z \equiv \phi_M$.

(*b*) Write the above in dimensionless form, letting $v^* \equiv v_z/v_{\rm ch}$ and $\nabla^* \equiv R_{\rm ch}\nabla$. Then define appropriate characteristic values $R_{\rm ch}$ and $v_{\rm ch}$.

(*c*) Show that $f\,{\rm Re} = C$, where C is a constant depending on shape.

9-12 A newtonian fluid of constant density and viscosity is in a long cylindrical container of radius R. The container is rotating about its own axis with angular velocity ω. The cylinder axis is vertical. It is desired to find an equation $z_s(r)$ for the free upper surface. Assume that neither the air above the fluid nor the bottom surface of the cylinder exerts a viscous drag and that the pressure everywhere on the free surface is $p = p_0$. The low point on the free surface is taken to be $z_s = z_0$.

(*a*) Write the stress tensor as a matrix, showing which terms are zero.

(*b*) Use the tables to write simplified expressions for the r, θ, and z components of the equation of motion (in terms of the fluxes). Justify your elimination of any terms. Interpret each equation physically in terms of forces, torques, etc.

(*c*) Substitute the flux equation for $\tau_{r\theta}$ into the θ component of the EOM. Compare the resulting expression with the simplified θ component of the Navier-Stokes equation (Table I-6).

(*d*) Write boundary conditions *on velocity*.

(*e*) Solve for the velocity distribution $v_\theta(r)$, using the boundary conditions.

(*f*) Find $\partial p/\partial r$ in terms of r.

(*g*) Find $p(r, z)$ and $z_s(r)$.

(*h*) Obtain $\Delta_{r\theta}$ versus r and interpret the result.

(*i*) Obtain $\Omega_{r\theta}$ versus r and interpret.

9-13 Consider the steady-state laminar flow under pressure of a nonnewtonian incompressible fluid between the narrow slit formed by two horizontal parallel plates of width W and length L and separated by a distance $2H$. Let y = distance from the lower plate, z = distance in the direction of flow, and x = distance perpendicular to y and z. Assume that flow is *fully developed* and that v_z depends only on y.

(*a*) Write equations for *each* component of the EOM, eliminating *all* unnecessary terms and justifying each.

(*b*) Obtain the shear-stress distribution (in terms of needed constants).

(*c*) Show all components of the rate-of-deformation tensor $\underline{\Delta}$ as a matrix in terms of $\dot{\gamma}$.

(*d*) Obtain expressions for each component of the shear stress $\underline{\tau}$ in terms of the gradient $\dot{\gamma} \equiv (\frac{1}{2}\underline{\Delta}:\underline{\Delta})^{1/2} \equiv dv_z/dy$ for a CEF fluid [Eq. (9-147)].

9-14 For unsteady Couette flow (see Prob. 7-4) of a Maxwell fluid, write expressions for $\underline{\Delta}$, $\underline{\sigma}$, $\underline{\tau}$, and $\underline{T}$. Assume that the fluid is at rest at $t = 0$ and that the upper plate is suddenly pulled with a constant velocity.

9-15 An incompressible newtonian fluid is flowing in steady-state laminar flow through a pipe. The fluid enters the pipe at $z = 0$ with uniform velocity $v_z(r) = V$. At a distance $z = L_e$ from the entrance, the fully developed (parabolic) velocity profile is obtained. At an arbitrary location $z_0 < L_e$ it is assumed that there is a central core region in which the velocity profile is flat, or $v_z = v_F(z)$ only. Outside this inner region (and especially near the wall) the velocity profile increases rapidly with distance from the wall. This outer region near the wall is the boundary layer, and $\delta(z)$ is the boundary-layer thickness. At the entrance $\delta = 0$; and at $z = L_e$, $\delta = R$ is the radius of the pipe. Neglect axial diffusion.

(*a*) Assuming angular symmetry, simplify the equation of continuity in both the boundary layer and the core regions.

(*b*) Write simplified expressions for the components of the EOM in the wall region and core region.

(*c*) Write simplified expressions for the Navier-Stokes equation in the wall region and core region.

(*d*) Relate δ to v_F by a mass balance.

(*e*) Find $p_0 - p_L$ in the entrance.

(*f*) Describe how to get L_e.

9-16 Consider the case of creeping flow around a sphere (Example 9-6).

(*a*) Verify the expressions for the velocity components v_r and v_θ in terms of the stream function ψ_s.

(*b*) Draw sketches showing how v_r, v_θ, p, and $\tau_{r\theta}$ vary with $-\theta$ at a fixed value of r/R.

(*c*) Carry out in detail all the steps in the solution of Example 9-6 (you need not find $E^4\psi_s$).

9-17 A spherical object with a volume of 1 ft^3 and a specific gravity of 2.80 is found to have a terminal velocity of 10 ft in an oil with density 58 lbm/ft^3.

(*a*) What is the viscosity of the oil?

(*b*) If the sphere were suspended in a stream of this oil flowing at 20 ft/s, what drag force would be exerted on the sphere?

REFERENCES

1. Aris, R.: "Vectors, Tensors, and Basic Equations of Fluid Mechanics," Prentice-Hall, Englewood Cliffs, N.J., 1962.
2. Batchelor, G. K.: "An Introduction to Fluid Dynamics," Cambridge University Press, New York, 1967.
3. Bennett, C. O., and J. E. Myers: "Momentum, Heat, and Mass Transfer," 2d ed., McGraw-Hill, New York, 1974.
4. Bird, R. B.: *Chem. Eng. Prog. Symp. Ser.* 4, 1965.
5. Bird, R. B.: *AIChE Cont. Educ. Ser.* 4, 1969.
6. Bird, R. B., R. C. Armstrong, and O. Hunsaker: "Dynamics of Polymeric Liquids," vol. 1, Wiley, New York, 1977.
7. Bird, R. B., W. E. Stewart, and E. N. Lightfoot: "Transport Phenomena," Wiley, New York, 1960.
8. Blasius, H.: *Angew. Math. Phys.,* **56**:1 (1908).
9. Churchill, R. V., J. W. Brown, and R. V. Verheys: "Complex Variables and Applications," McGraw-Hill, New York, 1974.
10. Fahien, R. W., and C. B. Schriver: Paper presented at Denver meeting of AIChE, 1961.
11. Goldstein, S.: "Modern Developments in Fluid Dynamics," Oxford University Press, London, 1939; reprinted by Dover, New York, 1965.
12. Hadamard, J.: *Comptes Rendus,* **152**:1735 (1911).
13. Happel, J., and H. Brenner: "Low Reynolds Number Hydrodynamics," Prentice-Hall, Englewood Cliffs, N.J., 1965.
14. Howarth, L.: *Proc. R. Soc. Lond.,* **A269**:539 (1938).
15. Langhaar, H. L.: *Trans. ASME,* **A64**:55 (1942).
16. Lin, C. C., and L. A. Segal: "Mathematics Applied to Deterministic Problems in the Material Sciences," Macmillan, New York, 1974.
17. Milne-Thompson, L. M.: "Theoretical Hydrodynamics," Macmillan, New York, 1955.
18. Moon, P., and D. E. Spencer: "Field Theory for Engineers," Van Nostrand, Princeton, N.J., 1961.
19. Schlichting, H.: "Bounday Layer Theory," 7th ed., McGraw-Hill, New York, 1968.
20. Whitaker, S.: "Introduction to Fluid Mechanics," Prentice-Hall, Englewood Cliffs, N.J., 1968.

CHAPTER

TEN

EQUATIONS OF CHANGE

10-1 TOTAL-ENERGY EQUATION†

In this chapter we will derive an energy-balance equation that applies to both compressible and incompressible fluids. We begin with the first law of thermodynamics, which can be written

Rate of change of internal and kinetic energy of system	=	Net rate of entry of internal and kinetic energy into system by bulk flow and diffusion	+	Rate work is being done by surroundings on system	+	Rate of production of thermal energy within system due to transformation from other forms of energy	(10-1)

The production term is actually not necessary for nonnuclear processes since the energy is conserved; i.e., there is no net *creation* of energy; however, there may be a transformation of energy such as the conversion of work into thermal energy (friction into heat). The potential energy is not included in the left-hand side with the internal and kinetic energies because it is to be treated as a work term. We will apply Eq. (10-1) to an arbitrary element of volume of a fluid $\Delta\mathcal{V} = \Delta x\, \Delta y\, \Delta z$. A nonvector derivation is summarized in Table 10-1. A detailed discussion of the individual terms follows.

†Does not require knowledge of vector notation.

Table 10-1 Derivation of total-energy equation

Face	Heat diffusion	+	Bulk flow	+	Viscous work	+	Pressure work	+
x	$-\Delta_x(q_x\,\Delta A_x)$		$-\Delta_x[\rho(\hat{U}+\hat{K})\,\Delta A_x\,v_x]$		$-\Delta_x[\Delta A_x\,(\tau_{xx}v_x+\tau_{xy}v_y+\tau_{xz}v_z)]$		$-\Delta_x(pv_x\,\Delta A_x)$	
y	$-\Delta_y(q_y\,\Delta A_y)$		$-\Delta_y[\rho(\hat{U}+\hat{K})\,\Delta A_y\,v_y]$		$-\Delta_y[\Delta A_y\,(\tau_{yx}v_x+\tau_{yy}v_y+\tau_{yz}v_z)]$		$-\Delta_y(pv_y\,\Delta A_y)$	
z	$-\Delta_z(q_z\,\Delta A_z)$		$-\Delta_z[\rho(\hat{U}+\hat{K})\,\Delta A_z\,v_z]$		$-\Delta_z[\Delta A_z\,(\tau_{zx}v_x+\tau_{zy}v_y+\tau_{zz}v_z)]$		$-\Delta_z(pv_z\,\Delta A_z)$	

Gravity work	+	Generation	=	Accumulation
$\rho\,\Delta\mathcal{V}\,(g_x v_x + g_y v_y + g_z v_z)$		$\Phi_H\Delta\mathcal{V}$	=	$\dfrac{-\Delta[\rho(\hat{U}+\hat{K})\,\Delta\mathcal{V}]}{\Delta t}$

Dividing by $\Delta\mathcal{V} = \Delta x\,\Delta y\,\Delta z$ and letting all $\Delta \to 0$ gives

$$-\frac{\partial q_x}{\partial x}-\frac{\partial q_y}{\partial y}-\frac{\partial q_z}{\partial z}-\frac{\partial[\rho(\hat{U}+\hat{K})v_x]}{\partial x}-\frac{\partial[\rho(\hat{U}+\hat{K})v_y]}{\partial y}-\frac{\partial[\rho(\hat{U}+\hat{K})v_z]}{\partial z}-\frac{\partial}{\partial x}(\tau_{xx}v_x+\tau_{xy}v_y+\tau_{xz}v_z)-\frac{\partial}{\partial y}(\tau_{yx}v_x+\tau_{yy}v_y+\tau_{yz}v_z)$$

$$-\frac{\partial}{\partial z}(\tau_{zx}v_x+\tau_{zy}v_y+\tau_{zz}v_z)-\frac{\partial(pv_x)}{\partial x}-\frac{\partial(pv_y)}{\partial y}-\frac{\partial(pv_z)}{\partial z}+\rho g_x v_x+\rho g_y v_y+\rho g_z v_z+\Phi_H=\frac{\partial[\rho(\hat{U}+\hat{K})]}{\partial t} \quad (10\text{-}1a)$$

The rate of accumulation of internal and kinetic energy in the element is

$$\frac{\Delta[\rho(\widehat{U} + \widehat{K})\,\Delta\mathcal{V}]}{\Delta t} \tag{10-2}$$

where $\widehat{U}$ is the internal energy per unit mass and $\widehat{K} \equiv \frac{1}{2}v^2$ is the kinetic energy per unit mass. We first consider the work-rate terms since they were not included in our development in Chap. 8.

Recall that the work done by a constant force of magnitude F in moving an object a distance s in the direction of the force is $W = Fs$ [Eq. (8-22)]. Suppose that the force is not constant; then the differential work done by a force F_x in moving an object a distance dx is

$$dW = F_x\,dx \tag{10-3}$$

The *rate* at which work is being done is

$$\dot{W} \equiv \frac{dW}{dt} = F_x \frac{dx}{dt} \tag{10-4}$$

where dx/dt is the velocity v_x of the body. Now consider an element $\Delta\mathcal{V}$ of fluid being acted on by a force ΔF_x in the x direction (only) at a point where the fluid velocity is v_x. Then the rate at which work is being done on the element is

$$\Delta\dot{W} = \Delta F_x\,v_x \tag{10-4a}$$

If in particular the force is the force due to gravity,

$$\Delta F_x = \frac{\rho\,\Delta\mathcal{V}\,g_x}{g_c} \tag{10-5}$$

where we take $g_c = 1$. If the axes are oriented so that there are components of gravity and of velocity in the y and z directions as well as the x direction,

$$\Delta\dot{W}_g = \Delta F_{gx}\,v_x + \Delta F_{gy}\,v_y + \Delta F_{gz}\,v_z = \rho\,\Delta\mathcal{V}\,(g_x v_x + g_y v_y + g_z v_z) \tag{10-5a}$$

If the force is a surface force, such as pressure, acting on an area A_x at $x = x$, the incremental force is

$$\Delta F_{px}\Big|_x = (p\,\Delta A_x)\Big|_x$$

and the incremental rate of gain of energy by the system is

$$\Delta\dot{W}_G\Big|_x = (p\,\Delta A_x\,v_x)\Big|_x$$

At $x + \Delta x$ the force would be given by a similar term and the net or total rate of gain would be

$$\Delta\dot{W}_t = -\Delta_x(p\,\Delta A_x\,v_x) \equiv (p\,\Delta A_x\,v_x)\Big|_x - (p\,\Delta A_x\,v_x)\Big|_{x+\Delta x}$$

The total rate for all six surfaces is the sum of similar terms for each face, or

$$-\sum_i (\Delta_i p\, \Delta A_i v_i)$$

The viscous-work terms are obtained by multiplying the components of the viscous force acting on each face by the velocity acting in the same direction. These force components are expressed in terms of the viscous stress. For example, on the x face at $x = x$ the viscous force acting in the y direction is $\Delta A_x\, \tau_{xy}$ and the work rate due to that component is $\Delta A_x\, \tau_{xy} v_y$. The total work rate for the x_i face at x_i is the sum of all components, or

$$\Delta A_i \left(\sum_j \tau_{ij} v_j\right)\Bigg|_{x_i}$$

As in the case of the force due to pressure, the net work rate for the x face is the difference between the values of the above at x_i and at $x_i + \Delta x_i$. Similar terms for each face are added to give the total work rate

$$-\sum_i \Delta_i \left[\Delta A_i \left(\sum_j \tau_{ij} v_j\right)\right]$$

The diffusion terms are obtained in the same manner as in Chap. 8

$$-\sum_i \Delta_i (q_i\, \Delta A_i)$$

In the case of bulk flow we include both internal and kinetic energy. For an area ΔA_i the volumetric flow rate into the element at x_i is $v_i\, \Delta A_i|_{x_i}$, and the internal energy per unit volume of the fluid is $\rho \widehat{U}$. The bulk flow of internal energy (energy per unit time) is then given by $(v_i\, \Delta A_i\, \rho \widehat{U})|_{x_i}$, and the bulk flow of the internal and kinetic energy at x_i is $[v_i\, \Delta A_i\, \rho(\widehat{U} + \widehat{K})]|_{x_i}$. There are similar terms at each of the other faces. The total is then

$$-\sum_i \Delta_i [\rho \widehat{U} + \widehat{K})\, \Delta A_i\, v_i]$$

For a single component substance, thermodynamics tells us that the internal energy $\widehat{U}$ will depend upon only two state variables, which are usually taken as temperature and specific volume, but for a multicomponent mixture (Appendix 10-5) the internal energy is also a function of composition. Thus the heat given off by chemical reaction can be expressed in terms of the difference in the internal energy (or, more customarily, the difference in enthalpy) of products and reactants. It is often more convenient, however, to let the internal-heat-generation rate be expressed (as in Chaps. 7 and 8) as part of Φ_H, the heat-generation rate per unit volume due to all sources (except friction), e.g., chemical reaction, nuclear sources, or electrical heating. If this is done, $\widehat{U}$ is considered a function of temperature and volume and not of composition. The heat due to internal friction, called the *viscous dissipation,* will *not* be included in Φ_H because an expression for it will be derived.

When all terms in Eq. (10-1) are divided by $\Delta \mathcal{V}$ and limits taken as $\Delta x_i \to 0$ and as $\Delta t \to 0$ (Table 10-1), a partial differential equation for the total-energy balance is obtained [Eq. (10-1*a*)].

Index Notation

The total-energy balance in Table 10-1 can be expressed in index notation

$$\frac{\partial[\rho(\widehat{U}+\widehat{K})]}{\partial t} = -\sum_i \frac{\partial q_i}{\partial x_i} - \sum_i \frac{\partial[\rho(\widehat{U}+\widehat{K})v_i]}{\partial x_i} - \sum_i \frac{\partial\left(\sum_j \tau_{ij} v_j\right)}{\partial x_i}$$

$$-\sum_i \frac{\partial(p v_i)}{\partial x_i} + \sum_i \rho g_i v_i + \Phi_H \tag{10-6}$$

Vector Notation†

Equation (10-6) can be written in Gibbs vector-tensor notation as

$$\frac{\partial}{\partial t}[\rho(\widehat{U}+\widehat{K})] = -\nabla\cdot\mathbf{q} - \nabla\cdot[\rho(U+K)\mathbf{v}] - \nabla\cdot\underline{\boldsymbol{\tau}}\cdot\mathbf{v} - \nabla\cdot p\mathbf{v} + \rho\mathbf{g}\cdot\mathbf{v} + \Phi_H \tag{10-7}$$

The first two terms on the right-hand side are divergences obtained by the methods of Chap. 8.

The next three terms are work-rate terms. Note that in vector notation the rate at which work is being done by a force $\mathbf{F}$ is

$$\dot{W} = \mathbf{F}\cdot\frac{d\mathbf{x}}{dt} = \mathbf{F}\cdot\mathbf{v}$$

where $\mathbf{x}$ is the position vector and $\mathbf{v}$ is the velocity. Thus the work done by the gravity force acting on an element of fluid volume $\Delta\mathcal{V}$ is

$$\Delta \dot{W}_g = \rho\,\Delta\mathcal{V}\,\mathbf{g}\cdot\mathbf{v}$$

†Although we use vector notation in the following development, the reader can see by the comparison between Eqs. (10-6) and (10-7) that vector notation can easily be translated into index notation or cartesian component notation by means of Tables I-2 to I-8. For this reason we no longer provide a parallel treatment in index notation. It must be emphasized, however, that index notation generally cannot be transformed into curvilinear coordinates by a mere change in subscripts; for example, $\nabla\cdot\mathbf{q} = \sum_i \frac{\partial q_i}{\partial x_i} \neq \sum_n \frac{\partial q_n}{\partial u_n}$, where u_n is a curvilinear coordinate. (See Appendix 8-3.)

which in index notation is

$$\Delta \dot{W}_g = \rho\, \Delta \mathcal{V} \sum_i g_i v_i$$

Dividing by the volume gives

$$\frac{\Delta \dot{W}_g}{\Delta \mathcal{V}} = \rho \sum_i g_i v_i = \rho \mathbf{g} \cdot \mathbf{v}$$

Similarly the rate at which work is being done by the pressure force per unit volume is

$$\lim_{\Delta \mathcal{V} \to 0} \frac{\Delta \dot{W}_p}{\Delta \mathcal{V}} = -\sum_i \frac{\partial (p v_i)}{\partial x} = -\nabla \cdot p\mathbf{v}$$

To find the work done by the viscous forces let us expand $\underline{\boldsymbol{\tau}} \cdot \mathbf{v}$. Using the same methods and the summation expressions for $\underline{\boldsymbol{\tau}}$ [Eq. (9-17)] and $\mathbf{v}$ [Eq. (8-9)], we obtain

$$\underline{\boldsymbol{\tau}} \cdot \mathbf{v} = \sum_i \sum_j \tau_{ij} v_j \boldsymbol{\delta}_i$$

Then

$$\nabla \cdot \underline{\boldsymbol{\tau}} \cdot \mathbf{v} = \left(\sum_k \boldsymbol{\delta}_k \frac{\partial}{\partial x_k} \right) \cdot \left(\sum_i \sum_j \tau_{ij} v_j \boldsymbol{\delta}_i \right) = \sum_i \sum_j \frac{\partial (\tau_{ij} v_j)}{\partial x_i} \tag{10-8}$$

This result agrees with that given in Eq. (10-6).

Other Forms of Total-Energy Equation

Since we were using an element fixed in space, Eq. (10-7) is in the *eulerian* form. To put it into lagrangian form we can either use the definition of D/Dt and the equation of continuity or use the identity [see Eq. (8-133)]

$$\rho \frac{D\widehat{G}}{Dt} \equiv \frac{\partial (\rho \widehat{G})}{\partial t} + \nabla \cdot \rho \widehat{G} v = \frac{\partial (\rho \widehat{G})}{\partial t} + \sum_i \frac{\partial (\rho \widehat{G} v_i)}{\partial x_i}$$

Here we let $\widehat{G} = \widehat{U} + \widehat{K}$ and

$$\boxed{\frac{\rho D(\widehat{U} + \widehat{K})}{Dt} = -\nabla \cdot \mathbf{q} - \nabla \cdot p\mathbf{v} - \nabla \cdot \underline{\boldsymbol{\tau}} \cdot \mathbf{v} + \rho \mathbf{g} \cdot \mathbf{v} + \Phi_H} \tag{10-9}$$

An equation of change can also be written in terms of the total energy $\widehat{E} = \widehat{U} + \widehat{K} + \widehat{\psi}$, where $\widehat{\psi}$ is the potential energy per unit mass. To do so, we need to add $\rho(D\widehat{\psi}/Dt)$ to both sides of Eq. (10-9). But by the definition of D/Dt,

$$\rho \frac{D\widehat{\psi}}{Dt} = \rho \frac{\partial \widehat{\psi}}{\partial t} + \rho \mathbf{v} \cdot \nabla \widehat{\psi} = \rho \mathbf{v} \cdot \nabla \widehat{\psi} \tag{10-10}$$

since ψ is independent of time (for terrestial problems). However, the gravitational force per unit mass $\mathbf{g}$ can be expressed as the gradient of $\widehat{\psi}$, or

$$\mathbf{g} = -\nabla \widehat{\psi}$$

For example, if $\mathbf{g}$ acts downward along the z axis,

$$\mathbf{g} = -g\boldsymbol{\delta}_z = -\frac{d\widehat{\psi}}{dz}\boldsymbol{\delta}_z$$

and $\widehat{\psi} - \widehat{\psi}_0 = g(z - z_0)$, where z_0 is a reference plane. Thus an increase in potential energy is equal to the increase in elevation times the gravitational acceleration. Then

$$\frac{D\widehat{\psi}}{Dt} = -\rho \mathbf{v} \cdot \mathbf{g} \tag{10-12}$$

We note that this cancels the similar term in Eq. (10-9), and therefore we can write the total-energy equation in the form

$$\rho \frac{D\widehat{E}}{Dt} \equiv \frac{\rho D(\widehat{U} + \widehat{K} + \widehat{\psi})}{Dt} = -\nabla \cdot \mathbf{q} - \nabla \cdot p\mathbf{v} - \nabla \cdot \underline{\boldsymbol{\tau}} \cdot \mathbf{v} + \Phi_H \tag{10-13}$$

10-2 MECHANICAL- AND INTERNAL-ENERGY EQUATIONS

The total-energy equation includes the kinetic energy per unit mass $v^2/2$ and various work terms. It is desirable to be able to eliminate the kinetic energy so that an equation for the internal energy can be developed. We can accomplish this by recalling that $\mathbf{v} \cdot \mathbf{v} = v^2$ and that

$$\rho \frac{D(v^2/2)}{Dt} = \frac{D(\mathbf{v} \cdot \mathbf{v})}{2Dt} = \frac{\rho}{2}\left(\mathbf{v} \cdot \frac{D\mathbf{v}}{Dt} + \mathbf{v} \cdot \frac{D\mathbf{v}}{Dt}\right) = \rho \mathbf{v} \cdot \frac{D\mathbf{v}}{Dt} \tag{10-14}$$

Hence we merely need to write the equation of motion in terms of $\rho(D\mathbf{v}/Dt)$ and take the dot product of it by $\mathbf{v}$.

The equation of motion in lagrangian form [Eq. (9-33)] is

$$\rho \frac{D\mathbf{v}}{Dt} = -\nabla \cdot \underline{\boldsymbol{\tau}} - \nabla p + \rho \mathbf{g} \tag{10-15}$$

If we form $\mathbf{v} \cdot \rho(D\mathbf{v}/Dt)$, we get

$$\rho \frac{D(v^2/2)}{Dt} = -\mathbf{v} \cdot \nabla \cdot \underline{\boldsymbol{\tau}} - \mathbf{v} \cdot \nabla p + \rho \mathbf{g} \cdot \mathbf{v} \tag{10-16}$$

Substituting the mechanical-energy equation (10-16) into Eq. (10-9) and using the identities

$$\nabla \cdot \underline{\tau} \cdot \mathbf{v} \equiv \underline{\tau} : \nabla \mathbf{v} + \mathbf{v} \cdot \nabla \cdot \underline{\tau} \tag{10-17}$$

and†

$$\nabla \cdot p\mathbf{v} \equiv p \nabla \cdot \mathbf{v} + \mathbf{v} \cdot \nabla p \tag{10-18}$$

gives

$$\rho \frac{D\hat{U}}{Dt} = -\nabla \cdot \mathbf{q} - \underline{\tau} : \nabla \mathbf{v} - p \nabla \cdot \mathbf{v} + \Phi_H \tag{10-19}$$

If we substitute Eq. (10-17) into Eq. (10-16), the term $\Phi_v \equiv -\underline{\tau} : \nabla \mathbf{v}$ also appears in the mechanical-energy balance but with the opposite sign. It represents the conversion of mechanical energy into heat energy as a result of the work done by the viscous stresses. Since it depends on squared terms, it is always positive and corresponds to positive entropy production, i.e., an irreversible process. Thus it represents the work lost through fluid friction and is called the *viscous dissipation* (see Appendix 10-5). It is important in polymer extrusion, in lubrication, and in high-speed flows. Expressions for Φ_v for various coordinate systems are given in Table I-10.

For a newtonian incompressible fluid, we can show that

$$\Phi_v \equiv -\underline{\tau} : \nabla \mathbf{v} = \frac{\mu}{2} \underline{\mathbf{\Delta}} : \underline{\mathbf{\Delta}}$$

as follows. We first recognize that we can interchange i and j in Eq. (10-17b) without changing the result. In vector notation

$$\underline{\tau} : \nabla \mathbf{v} \equiv \underline{\tau}^T : (\nabla \mathbf{v})^T \tag{10-20}$$

† In order to prove Eq. (10-17) we first consider the left-hand side. Writing it as a summation and differentiating, we obtain

$$\nabla \cdot \underline{\tau} \cdot \mathbf{v} \equiv \sum_i \sum_j \frac{\partial(\tau_{ij} v_j)}{\partial x_i} = \sum_i \sum_j \tau_{ij} \frac{\partial v_j}{\partial x_i} + \sum_i \sum_j v_j \frac{\partial \tau_{ij}}{\partial x_i} \tag{10-17a}$$

To obtain the first term on the right-hand side we use expressions for $\underline{\tau}$ [Eq. (9-17)] and $\nabla \mathbf{v}$ [Eq. (9-111)]

$$\underline{\tau} : \nabla \mathbf{v} \equiv \Big(\sum_i \sum_j \tau_{ij} \boldsymbol{\delta}_i \boldsymbol{\delta}_j \Big) : \Big(\sum_k \sum_l \frac{\partial v_l}{\partial x_k} \boldsymbol{\delta}_k \boldsymbol{\delta}_l \Big) = \sum_i \sum_j \sum_k \sum_l \tau_{ij} \frac{\partial v_l}{\partial x_k} \boldsymbol{\delta}_i \boldsymbol{\delta}_j : \boldsymbol{\delta}_k \boldsymbol{\delta}_l$$

The dyadic double dot product is obtained by first taking the dot product of the two *inner* unit vectors and then of the two *outer* unit vectors, as discussed in Chap. 9. Thus

$$\underline{\tau} : \nabla \mathbf{v} \equiv \sum_i \sum_j \tau_{ij} \frac{\partial v_i}{\partial x_j} \tag{10-17b}$$

Similarly

$$\mathbf{v} \cdot \nabla \cdot \underline{\tau} \equiv \sum_i \sum_j v_j \frac{\partial \tau_{ij}}{\partial x_i} \tag{10-17c}$$

Assuming symmetrical $\underline{\tau}$ in Eq. (10-17b) and adding it to Eq. (10-17c) gives Eq. (10-17a).

Then we can write the identity

$$\underline{\tau}:\nabla\mathbf{v} \equiv \tfrac{1}{2}(\underline{\tau}:\nabla\mathbf{v}) + \tfrac{1}{2}[\underline{\tau}^T:(\nabla\mathbf{v})^T]$$

Since $\underline{\tau}$ is equal to its transpose $\underline{\tau}^T$ (Appendix 9-4), we have

$$\underline{\tau}:\nabla\mathbf{v} = \tfrac{1}{2}\underline{\tau}:[\nabla\mathbf{v} + (\nabla\mathbf{v})^T] \equiv \tfrac{1}{2}\underline{\tau}:\underline{\mathbf{\Delta}} = -\frac{\mu}{2}(\underline{\mathbf{\Delta}}:\underline{\mathbf{\Delta}})$$

It is convenient to define a term ϕ_v by the expression

$$\Phi_v \equiv \underline{\tau}:\nabla\mathbf{v} \equiv \mu\phi_v \tag{10-21}$$

Then for an incompressible newtonian fluid

$$\phi_v \equiv \tfrac{1}{2}(\underline{\mathbf{\Delta}}:\underline{\mathbf{\Delta}}) = \frac{1}{2}\sum_i\sum_j\left(\frac{\partial v_i}{\partial x_j} + \frac{\partial v_j}{\partial x_i}\right)^2 \tag{10-22}$$

Expressions for various coordinate systems are given in Table I-11.

In contrast to ϕ_v, which is always positive, the term $p\nabla\cdot\mathbf{v}$, which also appears in the mechanical-energy equation, can be either positive or negative. Hence it represents reversible work, that of compression, and the resulting internal generation of heat. For an incompressible fluid it would be zero.

10-3 ENERGY EQUATION IN TERMS OF ENTHALPY AND TEMPERATURE

In engineering calculations it is convenient to express the energy equation in terms of enthalpy. To do so we make use of the thermodynamic definition

$$\widehat{H} = \widehat{U} + \frac{p}{\rho} \tag{10-23}$$

Then

$$\rho\frac{D\widehat{H}}{Dt} = \rho\left[\frac{D\widehat{U}}{Dt} + \frac{1}{\rho}\frac{Dp}{Dt} + p\frac{D(1/\rho)}{Dt}\right] \tag{10-24}$$

But

$$\frac{D(1/\rho)}{Dt} = -\frac{1}{\rho^2}\frac{D\rho}{Dt} \tag{10-25}$$

where we have used the continuity equation $D\rho/Dt = -\rho\nabla\cdot\mathbf{v}$. Then

$$\boxed{\rho\frac{D\widehat{H}}{Dt} = -\nabla\cdot\mathbf{q} - \underline{\tau}:\nabla\mathbf{v} + \frac{Dp}{Dt} + \Phi_H} \tag{10-26}$$

From the phase rule we know that the enthalpy of a pure component substance will depend upon only two independent variables. If we take them as T and P, we can write

$$d\hat{H} = \left(\frac{\partial \hat{H}}{\partial T}\right)_p dT + \left(\frac{\partial \hat{H}}{\partial p}\right)_T dp \qquad (10\text{-}27)$$

where $\left(\frac{\partial \hat{H}}{\partial T}\right)_p \equiv \hat{C}_p$ and $\left(\frac{\partial \hat{H}}{\partial p}\right)_T = \frac{1}{\rho} - T\left[\frac{\partial(1/\rho)}{\partial T}\right]_p$

If we let

$$\beta \equiv \rho\left[\frac{\partial(1/\rho)}{\partial T}\right]_p = \text{coefficient of volume expansion} \qquad (10\text{-}28)$$

then

$$\rho \frac{D\hat{H}}{Dt} = \rho\hat{C}_p \frac{DT}{Dt} + (1 - \beta T)\frac{Dp}{Dt} \qquad (10\text{-}29)$$

Substituting into the enthalpy equation (10-26) and using Eqs. (8-12) and (8-30) gives

$$\boxed{\rho\hat{C}_p \frac{DT}{Dt} = -\nabla \cdot \mathbf{q} - \underline{\boldsymbol{\tau}}:\nabla\mathbf{v} + \beta T \frac{Dp}{Dt} + \Phi_H} \qquad (10\text{-}30)$$

For an ideal gas $\beta = T^{-1}$, and for an incompressible fluid $\beta = 0$.

If instead we choose to write an internal-energy equation in terms of temperature, from thermodynamics we obtain (for a pure substance)

$$d\hat{U} = \hat{C}_v\, dT + \left[-T\left(\frac{\partial p}{\partial T}\right)_\rho\right] d\frac{1}{\rho} \qquad (10\text{-}31)$$

Substituting into the internal-energy equation (10-19) and using the equation of continuity gives†

$$\boxed{\rho\hat{C}_v \frac{DT}{Dt} = -\nabla \cdot \mathbf{q} - \underline{\boldsymbol{\tau}}:\nabla\mathbf{v} - T\left(\frac{\partial p}{\partial T}\right)_\rho \nabla \cdot \mathbf{v} + \Phi_H} \qquad (10\text{-}32)$$

Since, for an incompressible fluid, $\nabla \cdot \mathbf{v} = 0$ and $\hat{C}_p - \hat{C}_v = 0$, Eq. 10-32 becomes [provided that $T(\partial p/\partial T)_\rho \neq \infty$]

$$\boxed{\rho\hat{C}_v \frac{DT}{Dt} = \rho\hat{C}_p \frac{DT}{Dt} = -\nabla \cdot \mathbf{q} - \underline{\boldsymbol{\tau}}:\nabla\mathbf{v} + \Phi_H \quad \text{incompressible fluid}} \qquad (10\text{-}33)$$

For solids, $\mathbf{v} = 0$, and we immediately obtain

$$\boxed{\rho\hat{C}_p \frac{\partial T}{\partial t} = -\nabla \cdot \mathbf{q} + \Phi_H} \qquad (10\text{-}34)$$

†Equation (10-32) in various coordinate systems is given in Tables I-6, I-7, and I-8.

10-4 APPLICATIONS OF THE ENERGY EQUATION

The energy equation can be writen in various forms and coordinate systems, as shown in Tables I-6 to I-8 with $\Phi_H = 0$. For a given application they can be simplified by elimination of unneeded terms, as was done for the equation of motion and the Navier-Stokes equations. Note that these equations do *not* include internal-energy generation due to chemical reaction, nuclear effects, or electrical heating but *do* include heat generation due to viscous dissipation as $\underline{\boldsymbol{\tau}}:\nabla\mathbf{v}$.

Example 10-1: Viscous heating in an annulus In Example 9-5 the following velocity distribution [Eq. (15)] in a viscometer was obtained

$$v_\theta = \frac{\omega_0 r(r^{-2} - R_1^{-2})}{R_2^{-2} - R_1^{-2}}$$

We now wish to obtain the temperature distribution due to viscous heating in the same system. Let the temperature at R_1 be T_1 and at R_2 be T_2. Assume the fluid is incompressible. (*a*) Find the viscous-dissipation function ϕ_v versus radial position. (*b*) Either (1) simplify the equations in Tables I-6 to I-8, or (2) simplify the energy equation in Table I-12 using Tables I-2 to I-5 to obtain DT/Dt, $\nabla \cdot \mathbf{q}$, or $\nabla^2 T$ and simplify them to apply to this problem. (*c*) Letting $\theta \equiv (T - T_1)/(T_2 - T_1)$, $\xi \equiv r/R_2$, and $\text{Br} \equiv \mu\omega_0^2 R_2^2/[k(T_2 - T_1)]$ = Brinkman number, write the differential equation in dimensionless form. (*d*) Use the results above and the solution to Example 9-5 to obtain an equation for dimensionless temperature.

SOLUTION (*a*) From Table I-11, since $v_r = 0 = v_z$, $\partial/\partial\theta = 0 = \partial/\partial z$,

$$\phi_v = \left[r\frac{\partial(v_\theta/r)}{\partial r}\right]^2$$

From Example 9-5, Eq. (15),

$$\frac{v_\theta}{r} = \frac{\omega_0(r^{-2} - R_1^{-2})}{R_2^{-2} - R_1^{-2}}$$

Then
$$\frac{d(v_\theta/r)}{dr} = \frac{\omega_0(-2)r^{-3}}{R_2^{-2} - R_1^{-2}} = \frac{2\omega_0 R_2^2 r^{-3}}{1 - (R_1/R_2)^2}$$

and
$$\phi_v = \left[\frac{2\omega_0 R_2^2 \kappa^2}{r^2(1 - \kappa^2)}\right]^2 \tag{10-34a}$$

(*b*) (1) Using Table I-7B

$$0 = k\frac{d(r\,dT/dr)}{r\,dr} + \mu\phi_v = k\frac{d(r\,dT/dr)}{r\,dr} + \left(\frac{2\omega_0 R_2^2 r^2 \mu}{1 - \kappa^2}\frac{1}{r^2}\right)^2 \tag{10-34b}$$

(2) The energy equation (10-33) (or Table I-12), assuming $\widehat{C}_v \approx \widehat{C}_p$, is

$$\rho \widehat{C}_p \left(\frac{\partial T}{\partial t} + \mathbf{v} \cdot \nabla T \right) = \rho \widehat{C}_p \frac{DT}{Dt} = -\nabla \cdot \mathbf{q} - \underline{\tau} : \nabla \mathbf{v}$$

$$\rho \widehat{C}_p \mathbf{v} \cdot \nabla T = k \nabla^2 T + \mu \phi_v$$

For cylindrical coordinates, Table I-3 gives

$$\mathbf{v} \cdot \nabla T = v_r \frac{\partial T}{\partial r} + v_\theta \frac{\partial T}{r \partial \theta} + v_z \frac{\partial T}{\partial z} = 0 + 0 + 0$$

since $v_r = 0 = v_z$ and $\partial T/\partial \theta = 0$. Using Table I-2 for $\nabla^2 T$, letting $\partial T/\partial z = 0$, we get the same result, Eq. (10-34*b*).

(*c*) In dimensionless form Eq. (10-34) becomes

$$\frac{1}{\xi} \frac{d(\xi \, d\theta/d\xi)}{d\xi} = -\frac{4 \, \mathrm{Br} \, \kappa^4}{\xi^2 (1 - \kappa^2)^2}$$

(*d*) Integrating twice and using the boundary conditions gives

$$\theta = \left[(Y + 1) - \frac{Y}{\xi^2} \right] - \left[(Y + 1) - \frac{Y}{\kappa^2} \right] \frac{\ln \xi}{\ln \kappa}$$

where
$$Y \equiv \mathrm{Br} \frac{\kappa^4}{(1 - \kappa^2)^2}$$

□

10-5 APPLICATION TO DESIGN OF HEAT EXCHANGERS

In its simplest form a heat exchanger consists of a pipe through which the fluid flows while being heated or cooled by means of transport to or from the walls (Figs. 1-1 and 4-3*a*). More complex shell-and-tube exchangers have tubes (*passes*) that loop back and forth through a shell, which itself may have buffers to divert the flow up and down across the tubes. We now return to the heat exchanger discussed in Sec. 4-3 (Fig. 4-3*a*), but with $T_w \equiv T_R > T_0$ (Fig. 10-1).

To design a heat exchanger one must relate the total heat transport $\dot{Q}_w$ to the length of the exchanger L. A macroscopic balance over the exchanger gives [Eq. (4-23)]

$$\dot{Q}_w \approx w \widehat{C}_p [T_b(0) - T_b(L)] \tag{10-35}$$

where (see Appendix 4-3)

$$T_b(z) \equiv \langle T \rangle \equiv \frac{\iint T(r, z) v_z(r) \, dA_z}{\iint v_z \, dA_z} = \frac{\int_0^R T(r, z) v_z(r) 2\pi r \, dr}{\int_0^R v_z 2\pi r \, dr} \tag{10-36}$$

and $T_{bL} \equiv T_b(L)$ and $T_{b0} \equiv T_b(0)$ are bulk mean temperatures at $z = L$ and $z = 0$, respectively. Then in order to find $T(r, z)$ and $v_z(r)$ we would need to solve the

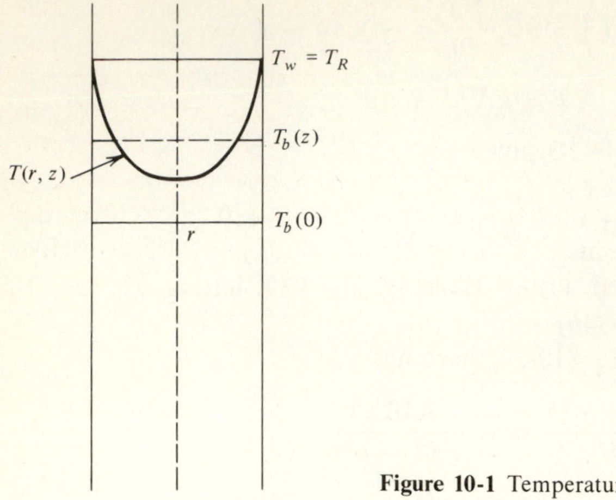

Figure 10-1 Temperature profile in a heat exchanger.

equations of change for T and v_z, which would be substituted into the above to obtain T_b and $\dot{Q}_w$ at any z including $z = L$.

The above is called the transport-phenomena approach. The classical approach would be to relate $\dot{Q}_w$ to T_b through Eq. (4-26g),

$$\dot{Q}_w = \langle h \rangle ZL \langle T_w - T_b \rangle_{\mathrm{LM}}$$

where Z = perimeter
$\langle T_w - T_b \rangle_{\mathrm{LM}}$ = log-mean temperature difference
$\langle h \rangle$ = average heat-transfer coefficient (usually found from dimensionless correlation.)

If the transport-phenomena approach is used to find T_{bL}, this result can then be used to find $\langle h \rangle$ (as we will show). Thus a theoretical equation for $\langle h \rangle$ can be derived. When the equations cannot be solved, we will show how dimensional analysis can be used to determine which dimensionless variables should be varied in an experimental program and correlated with the Nusselt number.

Example 10-2: Laminar-flow heat transfer with constant wall temperature An incompressible newtonian fluid is flowing upward through a cylindrical tube in fully developed laminar flow. The wall is at temperature T_R, and the fluid enters with uniform temperature T_0. Assume that T_R is constant. (See Fig. 10-2.) (a) Simplify

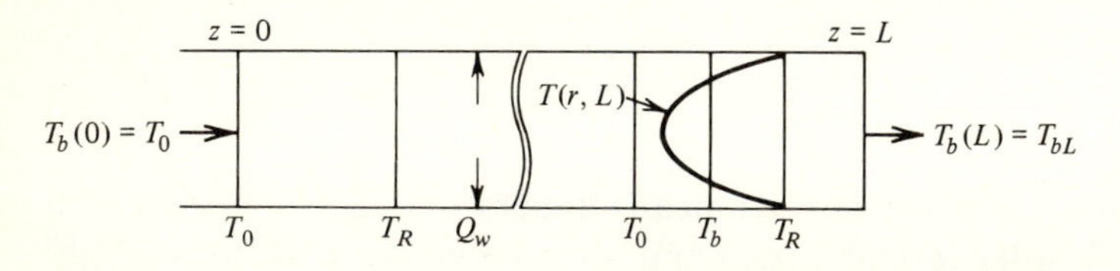

Figure 10-2 Heat exchanger with constant wall temperature $T_w \equiv T_R$.

the velocity and temperature equations of change to apply to this problem. Neglect axial conduction and viscous dissipation (see Appendix 10-3). (*b*) Show how the heat exchanger can be designed if the solution to the partial differential equation for $T(r, z)$ can be obtained; i.e., show how the rate of heat transport $\dot{Q}_w$ in watts in the length L can be obtained analytically. (*c*) Show how the mean heat-transfer coefficient and mean Nusselt number for the exchanger can be obtained analytically. (*d*) Show how dimensional analysis can be used to correlate the heat-transfer coefficient experimentally if the analytical solution is not available. (*e*) Find asymptotic values of the Nusselt number.

SOLUTION (*a*) From Example 9-4 we obtain

$$-\frac{\mu}{r}\left[\frac{\partial}{\partial r}\left(r\frac{\partial v_z}{\partial r}\right)\right] = \text{const} = \frac{\mathcal{P}_0 - \mathcal{P}_L}{L} \equiv \Phi_M$$

Now $\partial T/\partial t$ is zero at steady state, and we neglect Φ_v. Then from Table I-7B we get

$$\rho\hat{C}_v\left(v_r\frac{\partial T}{\partial r} + v_z\frac{\partial T}{\partial z} + v_\theta\frac{\partial T}{r\,\partial\theta}\right) = k\left[\frac{\partial(r\,\partial T/\partial r)}{r\,\partial r} + \frac{\partial^2 T}{r^2\,\partial\theta^2} + \frac{\partial^2 T}{\partial z^2}\right]$$

We neglect $\partial^2 T/\partial\theta^2$ (due to symmetry) and $\partial^2 T/\partial z^2$ (axial conduction, for justification see Appendix 10-3). Letting $\hat{C}_v = \hat{C}_p$ and $v_r = 0 = v_\theta$ gives

$$\rho\hat{C}_p v_z\frac{\partial T}{\partial z} = k\frac{\partial(r\,\partial T/\partial r)}{r\,\partial r} \tag{10-37}$$

Table I-12 together with Tables I-2 and I-3 gives the same results with $\mathbf{v}\cdot\nabla T = v_z\partial T/\partial_z$.

(*b*) If we integrate the equation for v_z and use the boundary conditions, we get the parabolic velocity-profile equation (3-21). If we substitute this into the energy equation (10-37) and recall that the average velocity v_a is one-half the maximum [Eq. (3-28)], we get

$$2v_a\rho\hat{C}_p\left[1 - \left(\frac{r}{R}\right)^2\right]\frac{\partial T}{\partial z} = k\frac{\partial(r\,\partial T/\partial r)}{r\,\partial r} \tag{10-38}$$

which can be written in terms of the dimensionless variables

$$r^* \equiv \frac{r}{R} \qquad z^* \equiv \frac{z}{R} \qquad T^* \equiv \frac{T_R - T}{T_R - T_0}$$

as

$$\frac{\partial(r^*\,\partial T^*/\partial r^*)}{r^*\,\partial r^*} = (1 - r^{*2})\frac{\partial T^*}{\partial z^*}(\text{Pe})_H \tag{10-39}$$

where $(\text{Pe})_H \equiv 2R\rho v_a\hat{C}_p/k$ is called the *heat Peclet number*. Since z^* and $(\text{Pe})_H$ appear only together, and since $(\text{Pe})_H$ is constant, we can reduce the number of dimensionless groups by combining z^* and $(\text{Pe})_H$ into a single group $\zeta \equiv z^*/(\text{Pe})_H$.

Then Eq. (10-39) becomes

$$\frac{\partial(r^* \ \partial T^*/\partial r^*)}{r^* \ \partial r^*} = (1 - r^{*2}) \frac{\partial T^*}{\partial \zeta} \qquad \text{where}\dagger \ \zeta \equiv \frac{z^*}{(\text{Pe})_H}$$

The boundary conditions are

Condition	z	r	T	ζ or z^*	r^*	T^*
1	0	$0 < r < R$	T_0	0	$0 < r^* < 1$	1.0
2	≥ 0	0	$\frac{\partial T}{\partial r} \neq \infty$	≥ 0	0	$\frac{\partial T^*}{\partial r^*} \neq \infty$
3	≥ 0	R	T_R	≥ 0	1	0

Since Eq. (10-39) is first order in ζ, only one boundary condition on ζ is needed. Note that there is no boundary condition at $z = L$ in an exchanger of finite length since we have no information there; in fact our purpose is to find the temperature at $z = L$. Thus the three boundary conditions alone and the solution itself must apply to exchangers of any length from zero to infinity. This means that if we have the temperature as a function of z for an exchanger of infinite length, we merely need to find the temperature at $z = L$ for an exchanger of finite length L.

The solution to Eq. (10-39) has been obtained by separation of variables using the Frobenius method (see Appendix 10-4). This problem is called the *Graetz-Nusselt problem*. In terms of dimensionless variables, the solution[5] is

$$T^* = \sum_{i=1}^{\infty} B_i X_i(r^*, c_i) Z_i(\zeta, c_i) \tag{10-40}$$

in which

$$Z_i = \exp(-c_i^2 \zeta) \qquad r^* \equiv \frac{r}{R} \qquad \zeta \equiv \frac{z^*}{(\text{Pe})_H} \qquad z^* \equiv \frac{z}{R}$$

$$(\text{Pe})_H = \frac{2R\rho v_a \widehat{C}_p}{k} = \frac{2R\rho v_a}{\mu} \frac{\widehat{C}_p \mu}{k} = \text{Re Pr}$$

$$B_i = \text{const} = \frac{2}{c_i(\partial X/\partial c)|_{r^*=1}}$$

where $X_i(r^*, c_i)$ = eigenfunction

$(\text{Pe})_H$ = heat-transfer Peclet number

c_i = eigenvalue given as ith root of $X_i(1, c_i) = 0$

To find the bulk mean temperature at $z = z$, we use Eq. (10-36) and integrate.

†The group $Gz \equiv w\widehat{C}_p/zk \equiv \pi/2\zeta$ is often called the Graetz number.

In dimensionless variables we get

$$T_b^* = \frac{\int_0^1 T^*(r^*, \zeta) v^*(r^*) r^* \, dr^*}{\int_0^1 v^* r^* \, dr^*} \tag{10-41}$$

where $\quad T_b^* \equiv \dfrac{T_R - T_b}{T_R - T_0} \quad$ and $\quad v^* \equiv \dfrac{v_z}{v_a} = 2(1 - r^{*2})$

The result (see Appendix 10-4) is

$$T_b^* = \sum_{i=1}^{\infty} \frac{8 Z_i (dX_i/dr^*)|_{r^*=1}}{c_i^3 (\partial X_i/\partial c)|_{r^*=1}} \tag{10-42}$$

The values of $(dX_i/dr^*)|_{r^*=1}$ and $(\partial X_i/\partial c)|_{r^*=1}$ have been calculated and tabulated.[5] The first three are given in Table 10-2. For $i > 3$, $c_i = 4i + \frac{8}{3}$.

To find the total heat transport, we use Eq. (10-35)

$$\dot{Q}_w = wC_p(T_b - T_0)$$

If we define a dimensionless $\dot{Q}_w$ as

$$Q^* \equiv \frac{\dot{Q}_w}{wCp(T_R - T_0)} \tag{10-43}$$

this becomes

$$Q^* = \frac{T_b - T_0}{T_R - T_0} \equiv \frac{(T_R - T_0) - (T_R - T_b)}{T_R - T_0} = 1 - T_b^* \tag{10-44}$$

Hence if we have values of T_b^*, we can calculate Q^* and from Q^* we can calculate the total transport $\dot{Q}_w$.

As a result we can prepare a general design chart for this problem by calculating values of $\dot{Q}_w^*$ from the series solution, and plotting them versus ζ (see Fig. 10-3). Then in order to calculate the heat transport in a heat exchanger of given L, R, T_0, and T_R for an average velocity v_a and with fluid properties known, one merely calculates ζ_L, reads off the value of Q^* and then calculates $\dot{Q}_w$ from Q^*. Note that it is *not* necessary to calculate the heat-transfer coefficient first, but since

Table 10-2

i	c_i	$(dX_i/dr^*)\|_1$	$(\partial X_i/\partial c_i)\|_1$
1	2.704	−1.014	−0.501
2	6.68	1.349	0.371
3	10.67	−1.572	−0.318

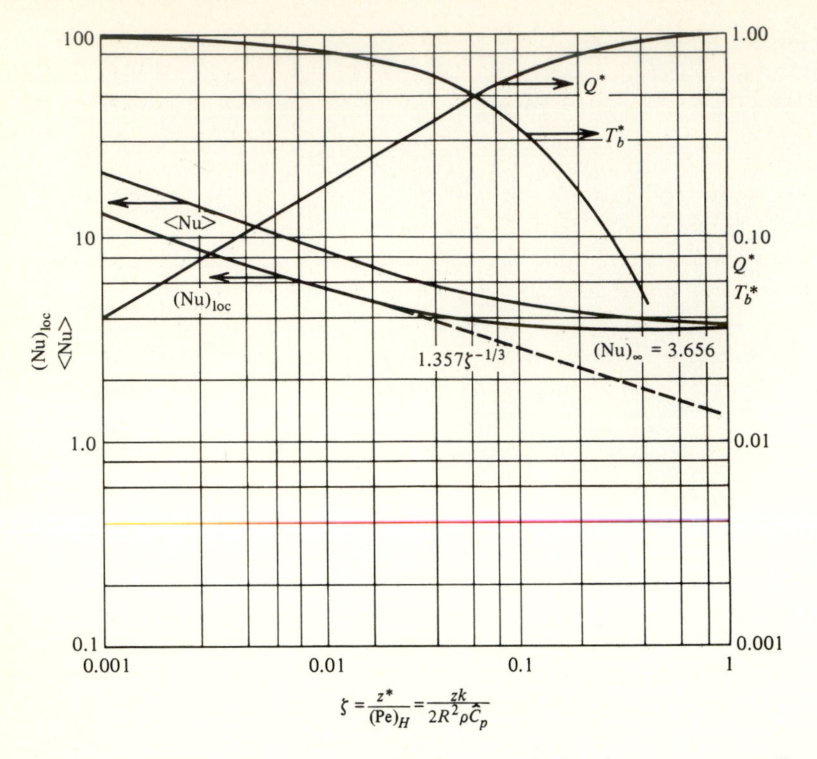

Figure 10-3 Design chart for laminar-flow heat transfer in tubes at constant wall temperature.

the use of heat-transfer coefficients is common engineering practice, we show how they can be calculated from the series solution.

(*c*) The reader should now read (or reread) Sec. 4-3 and go through the derivation of Eqs. (4-26*b*) to (4-26*h*). The *local* heat-transfer coefficient is given at a given z by equating Eq. (4-26) to Fourier's law at $r = R$.

$$h = \frac{(k\,\partial T/\partial r)|_{r=R}}{T_R - T_b} \tag{10-45}$$

The mean heat-transfer coefficient can be calculated most conveniently from Eq. (4-26*d*), which becomes, in terms of $\langle \mathrm{Nu} \rangle$,

$$\langle \mathrm{Nu} \rangle \equiv \frac{2R\langle h \rangle}{k} = \frac{w\widehat{C}_p}{z\pi k} \ln \frac{T_R - T_0}{T_R - T_b} \tag{10-46}$$

Since $w = \rho v_a A_c = \rho v_a(\pi R^2)$ and $\zeta \equiv (z/R)/(\mathrm{Pe})_H = z/(2R^2 v_a \rho \widehat{C}_p/k)$, in dimensionless form Eq. (4-26*d*) becomes

$$\langle \mathrm{Nu} \rangle = -\frac{\ln T_b^*}{2\zeta} \tag{10-47}$$

Since T_b^* depends only upon ζ, $\langle \mathrm{Nu} \rangle$ will depend only upon ζ. Hence we can use our series solution to prepare a general plot of T_b^* versus ζ. For a *given* exchanger of length L one could then calculate ζ_L, read off $\langle \text{Nu} \rangle(\zeta_L)$, calculate $\langle h \rangle_L$ and then calculate $\dot{Q}_w$ from $\langle h \rangle_L$ using Eqs. (4-26g) and (4-26h) at $\zeta = \zeta_L$ (see Fig. 10-3).

(d) In many cases it is not possible to solve the differential equations of change analytically to obtain $\langle \text{Nu} \rangle$. In such problems the equations of change can still be used to determine the dimensionless groups that should be involved in an experimental program and in the correlation of data.

Consider the dimensionless partial differential equation (10-39). By inspection we see that T^* will depend upon the Reynolds number, the Prandtl number, and the dimensionless length z^*. However, we note that z^*, Re, and Pr appear together as a single group $\zeta \equiv z^*/(\text{Re Pr})$. Hence T^* will depend upon only r^*, ζ, and the boundary conditions. Since no new dimensionless groups (such as ratios of dimensions) appear in the boundary conditions, T^* will depend only upon r^* and ζ.

Now T_b^*, as given by Eq. (10-41), will depend upon everything that T^* and v^* depend upon except that the variation with r^* will be eliminated by integration. (In the general case the integration limits may introduce geometric ratios, which usually appear in the boundary conditions as well.) Then T_b^* will depend only upon ζ, and since $\langle \text{Nu} \rangle$ will depend on T_b^* and ζ [by Eq. (10-47)], $\langle \text{Nu} \rangle$ will depend on ζ. Hence theoretically we need only take data on T_b versus z for a single fluid (say water) to obtain the *general* curve of T_b, $\dot{Q}_w$, or $\langle \text{Nu} \rangle$ versus ζ (see Fig. 10-4). Alternatively we could take data on T_b versus average velocity at a fixed value of z (as long as the flow remains laminar).

(e) The local Nusselt number can be shown (Appendix 10-4) to be given by [Eq. (A10-4-15)]

$$\text{Nu} = \frac{\displaystyle\sum_{i=1}^{\infty} B_i Z_i \left(\frac{dX_i}{dr^*}\right)\bigg|_{r^*=1}}{2\displaystyle\sum_{i=1}^{\infty} \frac{B_i Z_i}{c_i^2} \left(\frac{dX_i}{dr^*}\right)\bigg|_{r^*=1}} \tag{10-48}$$

At large values of ζ only one term in the infinite series is needed, and

$$\text{Nu} \to (\text{Nu})_\infty = \frac{c_1^2}{2} = 3.656 \tag{10-49}$$

Hence, the Nusselt number approaches a constant at large distances from the entrance. Note that as $\zeta \to \infty$, Eq. (10-42) becomes

$$T_b^* = \frac{8(-1.014)}{(2.704)^3(-0.501)} e^{-(2.704)^2 \zeta} = 0.819 e^{-7.312\zeta} \tag{10-50}$$

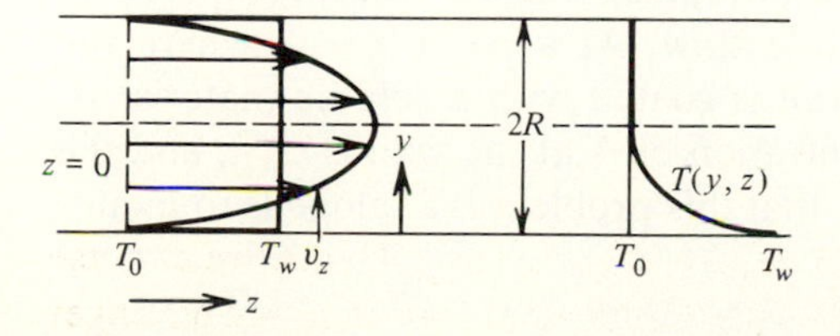

Figure 10-4 Heat transport in thermal entrance region.

Then
$$Q^* = 1 - 0.819e^{-7.312\zeta} \tag{10-51}$$

At the entrance ($x = 0$) the local Nusselt number is infinite since the gradient at the wall is infinite. An asymptotic solution has been obtained for $\zeta < 0.01$ and is given in Example 10-3. □

10-6 MASS TRANSFER

Using a molar balance over an element of volume, we obtain an eulerian equation of change for a species A [Eq. (8-119)]

$$\underbrace{\frac{\partial C_A}{\partial t}}_{\text{Accumulation}} + \underbrace{\nabla \cdot C_A \mathbf{v}}_{\text{Bulk flow}} + \underbrace{\nabla \cdot \mathbf{J}_A}_{\text{Diffusion}} \equiv \frac{\partial C_A}{\partial t} + \nabla \cdot N_A = \underbrace{R_A}_{\text{Generation}} \tag{10-52}$$

To convert this equation into lagrangian form we use the definition of D/Dt [Eq. (8-133)]

$$\frac{DC_A}{Dt} \equiv \frac{\partial C_A}{\partial t} + \mathbf{v} \cdot \nabla C_A \equiv \frac{\partial C_A}{\partial t} + \nabla \cdot C_A \mathbf{v} - C_A \nabla \cdot \mathbf{v}$$

Solving for $\partial C_A / \partial t$ and substituting into Eq. (10-52) gives

$$\frac{DC_A}{Dt} = -\nabla \cdot \mathbf{J}_A - C_A \nabla \cdot \mathbf{v} + R_A \tag{10-53}$$

For an incompressible fluid the term involving $\nabla \cdot \mathbf{v}$ is zero. Forms of the equation of continuity for species A in various coordinates are given in part A of Tables I-6 to I-8.

For a binary mixture of A and B with constant ρ and D_{AB} substitution of Fick's law (8-122) into Eq. (10-53) gives

$$\frac{DC_A}{Dt} = D_{AB} \nabla^2 C_A + R_A \tag{10-54}$$

This equation in various coordinate systems is given in part B of Tables I-6 to I-8. The equations of change are summarized as in Table I-12. Application of these equations to a situation analogous to Example 10-2 is given in Example 10-3.

Example 10-3: Forced-convection laminar-flow mass transfer An incompressible fluid is flowing through a pipe in laminar flow. At a point $z = 0$, where the velocity profile is fully developed, the wall is coated with a soluble material A, which dissolves in the liquid. The concentration of A at the wall is C_{Aw}, and the concentration of A at $z = 0$ is C_{A0}. Show that this problem is analogous to Example 10-2.

SOLUTION From Eqs. (10-53) or (10-54) and Table I-6 we obtain

$$\mathbf{v} \cdot \nabla C_{\text{A}} = D_{\text{AB}} \nabla^2 C_{\text{A}} + R_{\text{A}}$$

For flow in the z direction only, diffusion in the r direction, and no chemical reaction this becomes

$$v_z \frac{\partial C_{\text{A}}}{\partial z} = D_{\text{AB}} \frac{\partial (r\, \partial C_{\text{A}}/\partial r)}{r\, \partial r} \tag{10-55}$$

which can also be obtained by simplifying the mass A equation in part B of Table I-7. Together with its boundary conditions, Eq. (10-55) is directly analogous to Eq. (10-39) and its boundary conditions; therefore Eq. (10-40) will be the solution if we replace T^* with C^*_{A}. Also, Fig. 10-3 will apply to mass transport if we replace Nu with Sh and ζ with $z^*/(\text{Pe})_{\mathfrak{D}}$ where $\text{Pe}_{\mathfrak{D}} = \text{Re Sc}$. Likewise, we can replace Q^* with

$$\mathcal{W}^*_{\text{A}} \equiv \frac{\mathcal{W}_{\text{A}}}{(C_{\text{A}w} - C_{\text{A}0}) v_a A_{\text{cs}}} \tag{10-56}$$

and T^*_b with $C^*_{\text{A}b}$. □

10-7 HEAT AND MASS TRANSFER IN ENTRANCE REGION[4,15]†

Example 10-4: Heat transfer at entrance to a pipe In Example 10-2 we considered forced-convection heat transport for laminar flow in a pipe whose wall was at a constant temperature T_w. We now consider the *thermal* entrance region of the pipe and the application of boundary-layer theory to it. Our purpose is to find an expression for the Nusselt number that holds near the entrance, that is, for $z > 0$. In this region *the velocity profile is fully developed and parabolic*. The fluid enters at $z = 0$ with uniform temperature T_0 (see Fig. 10-4) and is immediately heated by the wall at T_w. Since this heating takes place near $y = 0$, the curvature of the wall can be neglected. Thus, if we let the distance from the wall be $y \equiv R - r$, Eq. (10-37) can be written

$$\rho \widehat{C}_p v_z \frac{\partial T}{\partial z} = k \frac{\partial (r\, \partial T/\partial r)}{r\, \partial r} = k \frac{\partial^2 T}{\partial y^2} \tag{10-57}$$

We now have an equation similar to that for flow past a flat plate [Eq. (9-155)] except that here $v_y = 0$ and T has replaced v_z. This means that Eqs. (9-155) and (10-57) are not analogous, nor are they so even with $v_y \neq 0$ since Eq. (9-155) is nonlinear [due to the $(v_z\, \partial v_z)/\partial z$ term]. However, the similarity between the equations suggests that a similarity transformation can be made to convert Eq. (10-57)

†This section may be omitted on first reading.

into an ordinary differential equation, as we did with Eq. (9-155). Thus, we can let

$$\eta \equiv \frac{y}{\delta(z)}$$

where $\delta(z)$ is a measure of the thermal-boundary-layer thickness.

Before attempting this transformation, we need to specify the velocity v_z to be used in Eq. (10-57). Since we are interested only in the region near the wall, it is common practice to assume that the profile near the wall is linear in y, or that

$$v_z = \dot{\gamma} y \qquad \text{where } \dot{\gamma} \equiv \left.\frac{dv_z}{dy}\right|_{y=0} = -\left.\frac{dv_z}{dr}\right|_{r=R}$$

Since $v_z = 2v_a[1 - (r/R)^2]$, we have $\dot{\gamma} = 4v_a/R$ and Eq. (10-57) becomes

$$\frac{\partial T}{\partial z} = \frac{\alpha R}{4v_a y}\frac{\partial^2 T}{\partial y^2} = \frac{\alpha\beta}{y}\frac{\partial^2 T}{\partial y^2} \qquad \text{where } \beta \equiv \frac{R}{4v_a} \tag{10-58}$$

Defining a dimensionless temperature as

$$T^* \equiv \frac{T_w - T}{T_w - T_0}$$

gives

$$\frac{\partial T^*}{\partial z} = \frac{\beta}{y}\frac{\partial^2 T^*}{\partial y^2} \tag{10-59}$$

The boundary conditions are:

Original:
$$T = \begin{cases} T_w & \text{at } y = 0 \text{ for } z > 0 \\ T_0 & \text{at } z = 0 \text{ for } y > 0 \\ & \text{and at } y = \infty \end{cases}$$

Transformed:
$$T^* = \begin{cases} 0 & \text{at } \eta = 0 \\ 1 & \text{at } \eta = \infty \end{cases}$$

Substituting $\eta \equiv y/\delta(z)$ into Eq. (10-57) and using the methods of Appendix 7-3 (see the Problems) gives

$$\delta = (3\beta C_1 z)^{1/3}$$

where C_1 is an arbitrary constant which is taken for convenience as 3. Then Eq. (10-57) becomes

$$\frac{d^2 T^*}{d\eta^2} + 3\eta^2 \frac{dT^*}{d\eta} = 0$$

whose solution (see the Problems) is

$$T^* = \frac{\int_0^\eta e^{-\eta^3}\, d\eta}{\Gamma(\frac{4}{3})} \tag{10-60}$$

where $\Gamma(\frac{4}{3})$ is the gamma function of $\frac{4}{3}$ (or 0.893) and where, by definition,

$$\Gamma(x) \equiv \int_0^\infty e^{-w} x^{w-1}\, dw$$

This leads to

$$\Gamma(p+1) = p\Gamma(p)$$

This result gives for the local heat-transfer coefficient

$$\text{Nu} \equiv \frac{2hR}{k} = -\frac{2R}{T_0 - T_w} \frac{\partial T}{\partial y}\bigg|_{y=0} = -2\frac{\partial T^*}{\partial y^*}\bigg|_{y^*=0}$$

which is a dimensionless gradient at the wall. Changing variables gives

$$\text{Nu} = 2R\left(\frac{dT^*}{d\eta}\frac{\partial \eta}{\partial y}\right)\bigg|_{y=0} = \frac{2R}{\Gamma(\frac{4}{3})\delta}$$

Substituting for δ and $\Gamma(\frac{4}{3})$ and rearranging gives

$$\text{Nu} = 1.357\zeta^{-1/3} \qquad \text{where } \zeta \equiv \frac{z^*}{(\text{Pe})_H} = \frac{z^*}{\text{Re Pr}}$$

The average Nusselt number is then

$$\langle \text{Nu} \rangle = \frac{2R\langle h \rangle}{k} = \frac{2R\int_0^z h\, dz}{kz} = \frac{\int_0^\zeta \text{Nu}\, d\zeta}{\zeta} = 2.036\zeta^{-1/3} = \tfrac{3}{2}\text{Nu}$$

These results can be expressed in terms of the j factor (as defined in Chap. 6)

$$\langle j_H \rangle \equiv \frac{\langle \text{Nu} \rangle}{\text{Re}(\text{Pr})^{1/3}} = \frac{2.036(\text{Re})^{1/3}}{\text{Re}(z^*)^{1/3}} = 2.036(\text{Re})^{-2/3}(z^*)^{-1/3} \tag{10-61}$$

This equation is plotted on j-factor charts like Fig. 6-15. □

Mass Transfer near Entrance

The analogous mass-transfer results can be obtained by the usual substitutions.

Noncircular Cross Section

Lightfoot[15] has generalized the preceding development to apply to noncircular cross sections by defining a dimensionless velocity gradient at the wall

$$\gamma^* \equiv \frac{\dot{\gamma} D_{\text{ch}}}{v_{\text{ch}}}$$

where D_{ch} and v_{ch} are characteristic lengths and velocities. (Thus for parallel plates $\gamma^* = 3$, $D_{\text{ch}} = Y$, and $\text{Re} \equiv Yv_a/\nu$.) This gives

$$\text{Nu} = \frac{\Gamma(\frac{4}{3})}{9^{1/3}}\left(\frac{\text{Re Pr}\,\gamma^* Y}{z}\right)^{1/3} \tag{10-62}$$

Example 10-5: Flat velocity profile: falling-film mass transfer[15,38]† Liquid films are sometimes used as mass-transfer devices. As an example consider a liquid B flowing under gravity on a vertical flat plate while a gas A is being absorbed into the film (see Fig. 10-5). If z is the distance along the film, y the distance from the gas-liquid interface, and δ the film thickness, the fully developed velocity profile can be found (Chap. 3) to be

$$v_z = v_M\left[1 - \left(\frac{y}{\delta}\right)^2\right]$$

where v_M is the velocity at the interface. At that location there is negligible transport (or drag), and the velocity gradient will be zero. This means that near the surfaces the velocity varies only slightly. Furthermore, since diffusion in liquids is slow, gas A does not penetrate very far. Consequently the velocity can be assumed constant over the region of interest.

The convective diffusion equation is then

$$v_z \frac{\partial C_A}{\partial z} = v_M \frac{\partial C_A}{\partial z} = D_{AB} \frac{\partial^2 C_A}{\partial y^2}$$

if axial diffusion is neglected in comparison to axial convection. The boundary conditions are found as follows:

At $y = 0$, we let $C_A = C_{As}$, the solubility of A in B. (1)
At $z = 0$, $C_A \approx 0$ (no A dissolved as yet). (2)
As $y \to \infty$, $C_A \to 0$ since gas A has not penetrated the entire film. (3)

†This example may be omitted on first reading.

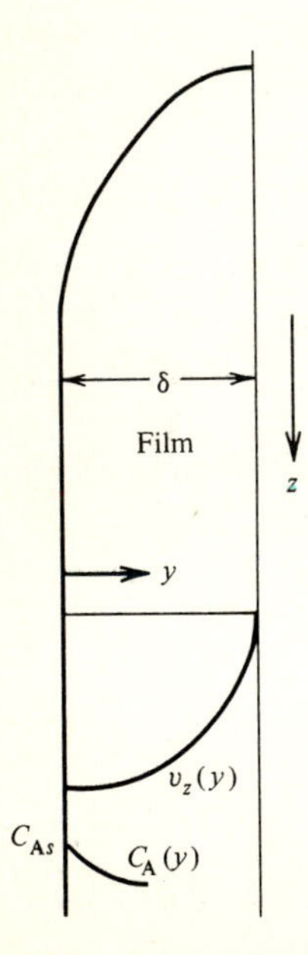

Figure 10-5 Mass transfer in falling film.

This equation and its boundary conditions will compare mathematically with the second of Eqs. (7-23), which we used to describe unsteady-state mass transport, if we replace $v_M(\partial C_A/\partial z)$ with $\partial C_A/\partial t$ by letting $z \equiv v_M t$. Previously we found a solution (see Appendix 7-3) by letting $\xi \equiv y/\sqrt{4D_{AB}t}$ [see the second of Eqs. (7-58)]. Therefore we here consider a solution in terms of

$$\xi \equiv \frac{y}{\sqrt{4D_{AB}z/v_M}}$$

If we let

$$C^* \equiv \frac{C_A}{C_{As}}$$

the boundary conditions are

$$C^* = \begin{cases} 1 & \text{at } \xi = 0 \qquad (4) \\ 0 & \text{at } \xi = \infty \qquad (5) \end{cases}$$

which agree with those used in Chap. 7. The differential equation then (Appendix 7-3) becomes

$$\frac{d^2C^*}{d\xi^2} + 2\xi\frac{dC^*}{d\xi} = 0$$

and the solution (Appendix 7-3) is

$$C^* = 1 - \operatorname{erf}\xi \tag{10-63}$$

The local Sherwood number is obtained by evaluating $(\partial C^*/\partial y)|_{y^*=0}$

$$\text{Sh} = \frac{1}{\sqrt{\pi}}\left(\text{Re Sc}\frac{D}{z}\right)^{1/2} \tag{10-64}$$

The corresponding equation for the heat-transfer analogy is

$$\text{Nu} = \frac{1}{\sqrt{\pi}}\left(\text{Re Pr}\frac{D}{z}\right)^{1/2} = \sqrt{\frac{2}{\pi}}\zeta^{-1/2} \qquad \text{where } \zeta \equiv \frac{z^*}{(\text{Pe})_H} \tag{10-65}$$

This result differs from the previous case (in which the velocity was assumed linear with distance from the wall) in that the power on ζ is $\frac{1}{2}$ rather than $\frac{1}{3}$; but is similar in that in each case the Nusselt number decreases with increasing ζ and increases with increasing Re and Pr. Also, in each case the boundary-layer thickness decreases as Re Sc z/D increases, showing that a thin boundary layer occurs for mass transfer in liquids, for which Sc may be on the order of 10^3. □

10-8 TRANSPORT IN ANISOTROPIC MEDIA†

For some materials the transport properties depend on the direction of transport. For example, the thermal conductivity of wood in the direction of the grain is

†This section may be omitted on first reading, but Eqs. (10-68) to (10-70) will be used later.

different from that perpendicular to the grain. Since the thermal conductivity has direction as well as magnitude, it can no longer be represented by a scalar. Instead we express it as a second-order tensor, so that its dot product with the gradient will correctly give a vector. Therefore we write

$$\mathbf{q} = -\underline{\mathbf{k}} \cdot \nabla T = -\underline{\mathbf{k}} \cdot \mathbf{G} \tag{10-66}$$

where $\mathbf{G} \equiv \nabla T$ is the gradient and $\underline{\mathbf{k}}$ is the second-order thermal-conductivity tensor. In index notation, if k_{ij} is a component of $\underline{\mathbf{k}}$,

$$\mathbf{q} = \sum_i q_i \boldsymbol{\delta}_i = -\sum_i \sum_j k_{ij} \boldsymbol{\delta}_i \boldsymbol{\delta}_j \cdot \sum_k G_k \boldsymbol{\delta}_k = \sum_i \sum_j k_{ij} G_j \boldsymbol{\delta}_i \tag{10-67}$$

or

$$q_i = -\sum_j k_{ij} G_j = -\sum_j k_{ij} \frac{\partial T}{\partial x_j} \tag{10-68}$$

which states that the heat flux in the i direction depends not only on the gradient in the i direction but also upon gradients in the other directions as well.

Expanding Eq. (10-57) gives

$$q_x = -k_{xx} \frac{\partial T}{\partial x} - k_{xy} \frac{\partial T}{\partial y} - k_{xz} \frac{\partial T}{\partial z} \tag{10-69}$$

$$q_y = -k_{yx} \frac{\partial T}{\partial x} - k_{yy} \frac{\partial T}{\partial y} - k_{yz} \frac{\partial T}{\partial z} \tag{10-70}$$

and similarly for q_{zz}. Note that in general, for anisotropic media, $\mathbf{q}$ and $\mathbf{G}$ are not in the same direction.

For heat transport in wood, we let k_p be the thermal conductivity of wood in a direction perpendicular to the grain and k_g be the thermal conductivity of wood in a direction along the grain. Then when the coordinate system is selected so that the X axis is perpendicular to the grain and the Y and Z axes are in the direction of the grain, we can write

$$\underline{\mathbf{K}} = \begin{bmatrix} k_p & 0 & 0 \\ 0 & k_g & 0 \\ 0 & 0 & k_g \end{bmatrix} \tag{10-71}$$

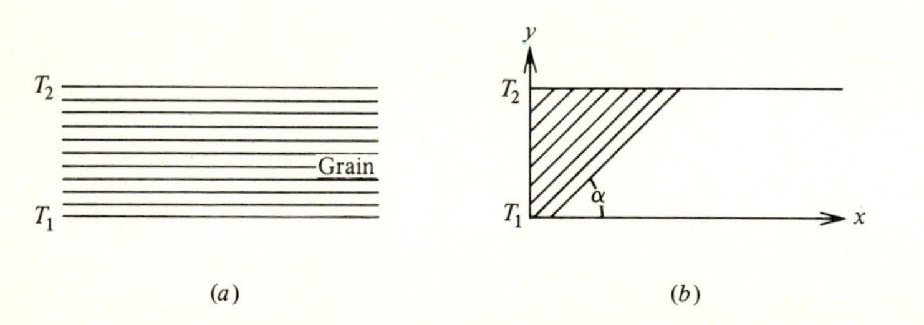

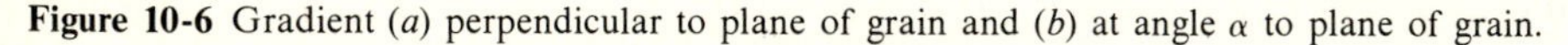

Figure 10-6 Gradient (a) perpendicular to plane of grain and (b) at angle α to plane of grain.

and $$q_X = -k_p \frac{\partial T}{\partial X} \qquad q_Y = -k_g \frac{\partial T}{\partial Y} \qquad q_Z = -k_g \frac{\partial T}{\partial Z} \tag{10-72}$$

These axes are called the *principal axes* of the tensor $\underline{\mathbf{K}}$. Figure 10-6*a* shows the coordinate system selected as the principal axes and Fig. 10-6*b* shows the principal axes at an angle α to the coordinate system. In Fig. 10-6*b* there will be off-diagonal components of $\underline{\mathbf{k}}$ that are nonzero, and hence the direction of $\mathbf{q}$ will be different from that of **G**.

10-9 EQUATIONS OF CHANGE IN TURBULENT FLOW

We indicated in Chap. 6 that a fluid in turbulent flow will exhibit fluctuations in the velocity, concentration, temperature, etc., measured at a given point. Thus the instantaneous velocity $\mathbf{v}$ can be represented as the sum of a time-averaged velocity $\overline{\mathbf{v}}$ and a fluctuating velocity $\mathbf{v}'$, or

$$\mathbf{v} = \overline{\mathbf{v}} + \mathbf{v}' \tag{10-73}$$

where application of the time-averaging process, defined in Appendix 6-1, gives

$$\overline{\mathbf{v}} = \frac{\int_t^{t+\Delta t} \mathbf{v}\, dt}{\Delta t} \tag{10-74}$$

Similar relationships can be written for concentration, temperature, and other properties.

Although the equations of change are derived for laminar flow, they can be applied to turbulent flow because of the experimental observation that the size of the smallest turbulent eddy is still quite large compared with the mean free path of a molecule. Therefore the fluid can still be considered a continuum, and derivatives with respect to the spatial variables are defined. This means that the equations of change, as derived, can be written for the laminar-flow conditions that exist at a point within the smallest eddy (or *turbule*).

However, the three-dimensional eddying nature of turbulence means that $\mathbf{v}'$ will always have nonzero components in all three directions, and therefore all three components of the EOM are, theoretically, required. Also since turbulent eddies range widely in size and frequency, and since the boundary conditions are unknown, these equations cannot be solved analytically. (Some attempt has been made to solve them numerically on a computer but because of the large capacity required, without much success.)

For this reason it is customary to time-average the equations of change to obtain equations in terms of the time-averaged potentials $\overline{\mathbf{v}}$, $\overline{C_A}$, and $\overline{T}$ since these are needed in practical problems. Such a procedure was followed for simple one-dimensional cases in Chap. 6 and is carried out for the more general three-dimensional case in Appendix 10-6. From this procedure one obtains equations that include the turbulent fluxes

$$\underline{\tau}^{(t)} \equiv \overline{\rho \mathbf{v}'\mathbf{v}'} \qquad \mathbf{q}^{(t)} \equiv \overline{\rho \widehat{C}_p T'\mathbf{v}'} \qquad \mathbf{J}_A^{(t)} \equiv \overline{C_A'\mathbf{v}'} \tag{10-75}$$

Although these turbulent fluxes appear in addition to the molecular or laminar fluxes, $\underline{\tau}^{(l)}$, $\mathbf{q}^{(l)}$, and $\mathbf{J}_A^{(l)}$, the laminar fluxes can be related to the gradients of the potentials by using the molecular-transport properties μ, k, and D_{AB}. Thus†

$$\overline{\underline{\tau}^{(l)}} = -\mu\overline{\underline{\Delta}} = -\mu[\nabla\overline{\mathbf{v}} + (\nabla\overline{\mathbf{v}})^T] \qquad \overline{\mathbf{q}^{(l)}} = -k\,\nabla\overline{T} \qquad \overline{\mathbf{J}_A^{(l)}} = -D_{AB}\,\nabla\overline{C}_A \tag{10-76}$$

The laminar-flow equations can be derived from the kinetic theory of gases (see Chap. 2). For turbulent flow an analogous model is often used to express the turbulent fluxes in terms of so-called *eddy* or *turbulent transport properties*. However, such a model has less validity since the mixing length (which is analogous to the mean free path) can approach the dimensions of the system and the flux is convective as well as diffusive; i.e., it should be proportional to the concentration as well as to the gradient. A model of this type has been discussed by Petty.[21]

Since at present there is no generally better model for turbulent transport, the use of the gradient model (called the *Bousinesq hypothesis*) is widespread. It can be summarized as follows.

In order to apply the equations of change to turbulent flow we

1. Replace $\mathbf{v}$, T, and C_A with the time-averaged values $\overline{\mathbf{v}}$, $\overline{T}$, and $\overline{C_A}$.
2. Replace the fluxes $\underline{\tau}$, $\mathbf{q}$, and $\mathbf{J}_A$ with the sum of the laminar and turbulent fluxes

$$\overline{\underline{\tau}} = \overline{\underline{\tau}^{(l)}} + \underline{\tau}^{(t)} \qquad \overline{\mathbf{q}} = \overline{\mathbf{q}^{(l)}} + \mathbf{q}^{(t)} \qquad \overline{\mathbf{J}_A} = \overline{\mathbf{J}_A^{(l)}} + \mathbf{J}_A^{(t)} \tag{10-77}$$

3. Express the molecular fluxes in terms of the molecular-transport properties μ, k, and D_{AB} and the turbulent fluxes in terms of the eddy transport properties. Since turbulence is generally not isotropic, the turbulent transport properties should be expressed as tensors to account for differences in the tendency toward transport in different directions. The flux laws then become (see Table I-17)

$$\underline{\tau}^{(t)} = -\underline{\underline{\mu}}^{(t)} : [\nabla\overline{\mathbf{v}} + (\nabla\overline{\mathbf{v}})^T] \qquad \mathbf{q}^{(t)} = -\underline{\mathbf{k}}^{(t)} \cdot \nabla\overline{T} \qquad \mathbf{J}_A^{(t)} = -\underline{\mathbf{D}}_{AB}^{(t)} \cdot \nabla\overline{C}_A \tag{10-78}$$

   Since little is known about the components of the above tensors, it is customary to use the scalars $\mu^{(t)}$, $k^{(t)}$, and $D_{AB}^{(t)}$
4. Use expressions for the variation of the eddy transport properties with position, e.g., the Prandtl mixing-length theory and a turbulence model for v' and l.
5. Solve the applicable equations of change, usually approximately or numerically.

† We use an overbar on the laminar flux and on the total flux to indicate that time-averaged potentials are involved.

Example 10-6: Turbulent-flow heat transfer in a pipe Consider the same physical situation as in Example 10-2 except that the flow is turbulent rather than laminar. (*a*) Simplify the energy equation so as to apply to this problem using a turbulent flux law. (*b*) Write the equation in dimensionless form and outline the solution. (*c*) Use dimensional analysis to determine what dimensionless variables the average Nusselt number would depend upon in turbulent flow.

SOLUTION (*a*) The energy equation (10-33) or Table I-12 can be applied to turbulent flow if we replace T and $\mathbf{v}$ with $\overline{T}$ and $\overline{\mathbf{v}}$ and $\mathbf{q}$ with $\overline{\mathbf{q}} = \mathbf{q}^{(l)} + \mathbf{q}^{(t)}$. Simplifying Eq. (10-32) by letting $\nabla \cdot \mathbf{v} = 0$ (incompressible fluid) and neglecting viscous dissipation gives (see Table I-12)

$$\rho \widehat{C}_v \frac{D\overline{T}}{Dt} \equiv \rho \widehat{C}_v \left(\frac{\partial \overline{T}}{\partial t} + \overline{\mathbf{v}} \cdot \nabla \overline{T} \right) = -\nabla \cdot \overline{\mathbf{q}} \tag{1}$$

Letting

$$\overline{v}_r = \overline{v}_\theta = 0 \qquad \overline{q}_\theta = 0 \qquad \overline{q}_z = 0 \qquad \widehat{C}_v = \widehat{C}_p$$

we obtain

$$\rho \widehat{C}_p \overline{v}_z \frac{\partial \overline{T}}{\partial z} = -\frac{\partial (r\overline{q}_r)}{r\,\partial r} \tag{2}$$

which can also be obtained from part A of Table I-7 by making the same assumptions. From Table I-12 or the second of Eqs. (10-77)

$$\overline{q}_r = \overline{q_r^{(l)}} + q_r^{(t)} \tag{3}$$

From the second of Eqs. (10-78)

$$\mathbf{q}^{(t)} = -\underline{\mathbf{k}}^{(t)} \cdot \nabla \overline{T}$$

Assuming isotropy leads to

$$q_r^{(t)} = -k^{(t)} \frac{\partial \overline{T}}{\partial r} \tag{4}$$

Substituting (3) and (4) into (2), and using Fourier's law for $q_r^{(l)}$ gives

$$\rho \widehat{C}_p \overline{v}_z \frac{\partial \overline{T}}{\partial z} = \frac{\partial [(k + k^{(t)}) r\, \partial \overline{T}/\partial r]}{r\,\partial r} \tag{10-79}$$

When we use the same dimensionless variables as in Example 10-2 and let

$$v^* \equiv \frac{\overline{v}_z}{v_a} \qquad k^* \equiv \frac{k + k^{(t)}}{k} \qquad \zeta \equiv \frac{z^*}{(\mathrm{Pe})_H}$$

where $(\mathrm{Pe})_H \equiv 2Rv_a/\alpha$, we can write Eq. (10-79) in dimensionless form as

$$\frac{\partial \left[\left(k^* r^* \frac{\partial T^*}{\partial r^*} \right) \right]}{r^*\,\partial r^*} = \frac{v^*}{2} \frac{\partial T^*}{\partial z^*} (\mathrm{Pe})_H = \frac{v^*}{2} \frac{\partial T^*}{\partial \zeta} \tag{5}$$

The boundary conditions are the same as those given for laminar flow in Example 10-2.

As in the Graetz solution for laminar flow, the dimensionless temperature is expressed as the product of a radial function $X(r^*)$ and an axial function $Z(\xi)$. The variables are then separated, yielding the eigenvalue c. The wall boundary condition is

$$X_i(1, c_i^2) = 0$$

whose roots are the c_i. The solution is then of the same form as Eq. (A10-4-1) or

$$T^*(r^*, \zeta) = \sum_{i=0}^{\infty} B_i X_i(r^*; c_i) Z_i(\zeta; c_i)$$

where $Z_i = e^{-c_i^2 \zeta}$.

The B_i can be determined by the orthogonality condition as in laminar flow. The eigenfunctions X_i will be the solutions of the equation

$$\frac{d(r^* k^*\, dX^*/dr^*)}{r^*\, dr^*} + c^2 x^* = 0 \qquad (6)$$

It is more difficult to solve the above equation than Eq. (A10-4-3), obtained in the case of laminar flow, because k^*, the dimensionless total thermal conductivity, varies widely with radial position, being 1 at the wall and very large near the center (see Chap. 6). Also, we need an equation for the velocity profile $\bar{v}_z(r)$ [or $v^*(r)$], which theoretically can be obtained from the momentum equation (as illustrated in Example 6-1) if the eddy viscosity $\mu^{(t)}(r)$ is known.

The solution to Eq. (10-79) was carried out[19,27] using the velocity and $k^{(t)}$ distributions recommended by Notter and Sleicher[27] [see Eqs. (6-61*c*) and (6-61*d*)]. The asymptotic Nusselt number for large z was then obtained from the first eigenvalue c_1 using the first of Eqs. (10-49)

$$(\mathrm{Nu})_\infty = \frac{c_1^2}{2} \qquad (7)$$

These calculated values of $(\mathrm{Nu})_\infty$ were represented by the curve-fit equations

$$(\mathrm{Nu})_\infty = \begin{cases} 4.8 + 0.0156(\mathrm{Pe})^{0.85}(\mathrm{Pr})^{0.08} & \text{for } 0.004 < \mathrm{Pr} < 0.1 \text{ and } 10^4 < \mathrm{Re} < 10^6 \quad (10\text{-}80) \\ 5 + 0.016(\mathrm{Re})^a(\mathrm{Pr})^b & \text{for } 0.1 < \mathrm{Pr} < 10^4 \text{ and } 10^4 < \mathrm{Re} < 10^6 \quad (10\text{-}81) \end{cases}$$

where $$a = \frac{0.88 - 0.20}{4 + \mathrm{Pr}} \qquad b = 0.33 + 0.5 \exp(-0.6\,\mathrm{Pr})$$

(*c*) In some turbulent-flow situations even an approximate solution is not possible and an empirical correlation for Nu must be obtained from experimental data. However, the experimental data required can be reduced by using dimensional analysis. For example, Eq. (5) indicates that the solution must be of the form

$$T^* = T^*(v^*, k^*, \zeta) \tag{8}$$

In order to integrate this equation we first need to find the variables upon which v^* and k^* depend. For the velocity, the EOM and the flux law yield

$$v^* = v^*(r^*, \mu^*) \tag{9}$$

where the eddy transport properties depend on Re and position, or

$$\frac{\mu + \mu^{(t)}}{\mu} \equiv \mu^* = \mu^*(\text{Re}, r^*) \tag{10}$$

For k^*, the molecular diffusivities α and ν are also involved, or

$$k^* = k^*(\text{Re}, r^*, \text{Pr}) \tag{11}$$

Combining Eqs. (8) to (11), we obtain

$$T^* = T^*(\text{Re}, \text{Pr}, r^*, \zeta) \tag{12}$$

The bulk mean temperature is given by Eq. (10-36) which, in dimensionless form, is

$$T_b^* = 2\int_0^1 T^* v^* r^* \, dr^* \tag{13}$$

From Eqs. (9) and (11) we obtain

$$T_b^* = T_b^*(\text{Re}, \text{Pr}, \zeta)$$

where the r^* dependence is eliminated in the integration.

Using Eq. (10-44), we find

$$Q^* = 1 - T_b^* = Q^*(\text{Re}, \text{Pr}, \zeta)$$

From Eq. (10-47)

$$\langle \text{Nu} \rangle = -\ln \frac{T_B^*}{2\zeta_L} = \langle \text{Nu} \rangle(\text{Re}, \text{Pr}, \zeta)$$

Now $$\zeta_L \equiv \frac{z_L^*}{(\text{Pe})_H} = \frac{z_L^*}{\text{Re Pr}} \qquad \text{where } z_L^* = \frac{L}{R}$$

and $$\langle \text{Nu} \rangle = f\left(\text{Re}, \text{Pr}, \frac{L/R}{\text{Re Pr}}\right) = f\left(\text{Re}, \frac{L}{D}, \text{Pr}\right)$$

Note that turbulent flow differs from laminar flow in that three dimensionless groups (Re, Pr, and L/D) are individually included whereas in laminar flow these groups can be combined into a single dimensionless variable $\zeta_L \equiv 2(L/D)/(\text{Re Pr})$. The difference occurs because in turbulent flow, the eddy transport properties depend upon Re and Pr but not ζ_L (or z^*).

As in laminar flow, at large L/D, the effect of length is not important and an asymptotic value is approached

$$\langle \text{Nu} \rangle \to \langle \text{Nu} \rangle_\infty(\text{Re}, \text{Pr})$$

This result agrees with the most widely used empirical equation, the Dittus-Boelter equation (6-90*d*). In turbulent flow the entrance length is quickly reached and usually neglected. □

10-10 TAYLOR'S THEORY OF TURBULENT DIFFUSION†

G. I. Taylor considered an idealized case in which the turbulence could be considered homogeneous in space and time, meaning that the average turbulence properties, e.g., the rms velocity fluctuations (or intensities), are independent of position and time. (Homogeneity in time is often called *stationarity*.) Consider a flow field in which there is flow only in the z direction and a concentration gradient only in the x direction. For simplicity assume that particles of fluid containing the diffusing material (energy or mass of A) originate at $x = 0$ and diffuse (see Fig. 10-7) laterally in time t a distance X, where

$$X = \int_0^t v\,dt \qquad \text{and} \qquad v = \frac{dX}{dt} \tag{10-82}$$

is the lateral fluctuating velocity (which may be positive or negative) (see Fig. 10-8). Thus, from the lagrangian point of view, as the particles are convected in the z direction with the velocity of the stream, they tend to diffuse varying distances X.

†This section can be omitted on first reading.

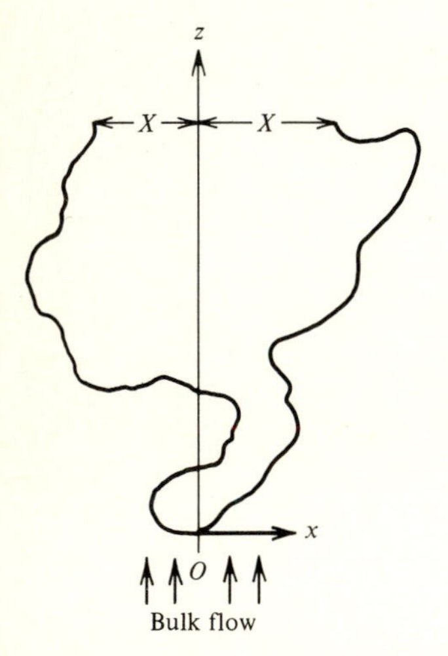

Figure 10-7 Paths of diffusing particles.

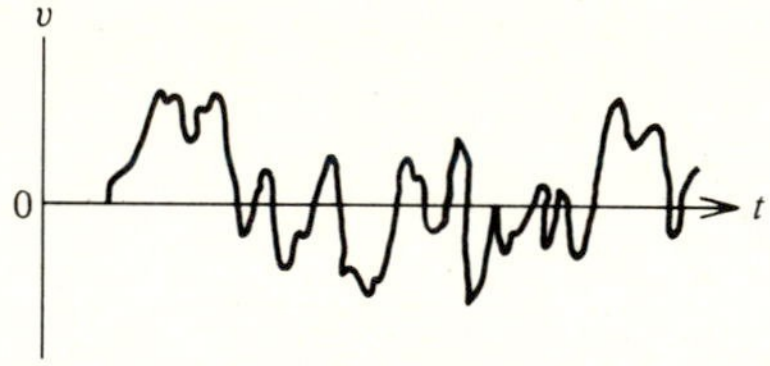

Figure 10-8 Fluctuating velocity in turbulent motion.

Since X can be either plus or minus, its mean value will be zero but $\overline{X^2}$, its mean-square value, will generally not be zero. We can see that $\overline{X^2}$ is a measure of the spread or variance σ_x^2 (Fig. 10-9) of the resulting concentration distribution. This spread will obviously be larger the larger the total time $\mathfrak{T}$ during which the diffusion occurs. The rate of change of the spread will be

$$\frac{\overline{dX^2}}{dt} = 2\overline{X\frac{dX}{dt}} = \overline{2Xv} \tag{10-83}$$

Taylor assumed that the mean concentration gradient in the x direction could be considered constant over a distance X and postulated that the average rate at which mass is being transferred by turbulent diffusion across an area perpendicular to the x axis is

$$J_{Ax}^{(t)} = -\overline{vX}\frac{d\bar{C}_A}{dx} = -\overline{vX}\frac{d\bar{C}_A}{dx} \tag{10-84}$$

where $\overline{vX}$ is the average† over all areas perpendicular to the x axis. Then the apparent eddy diffusivity is given by

$$D^{(t)} = \overline{vX} = \frac{1}{2}\frac{\overline{dX^2}}{dt} \tag{10-85}$$

Now let us consider the general case of lagrangian diffusion in which the observer is moving in the z direction with the velocity of the stream while the material, due to the lateral velocity fluctuations, diffuses laterally in the x direction (see Fig. 10-10). Using the general definition of a correlation coefficient [Eq. (A6-1-2)], we define a *lagrangian correlation coefficient* as

$$R_\tau \equiv \frac{\overline{v_t v_\tau}}{\overline{v^2}} \tag{10-86}$$

† It is customary in turbulence theory to assume that averages over space and over time are equal. This *ergodic hypothesis* is strictly true if the field is homogeneous and stationary. For lagrangian diffusion it is usually also assumed that one can then average over all particles that either start from the same origin at different starting times or start from different points at the same starting time.

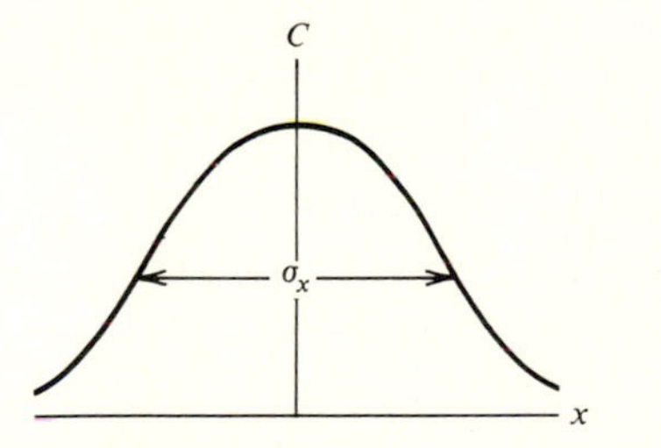

Figure 10-9 Standard deviation of concentration distribution.

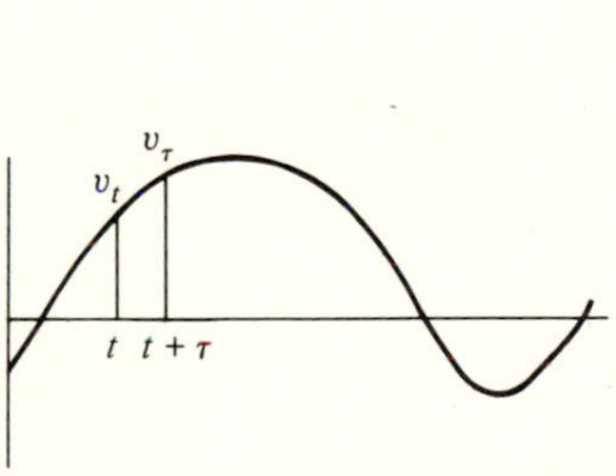

Figure 10-10 Velocities at t and $t + \tau$.

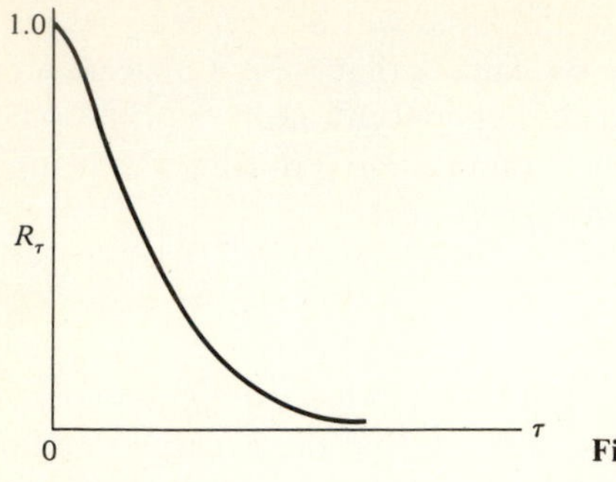

Figure 10-11 Correlation coefficient.

where v_t is the lateral velocity of an eddy at time t and v_τ is *the velocity* at time τ later. (Because of stationarity, $\overline{v_t^2} = \overline{v_\tau^2} = \overline{v^2}$.) Thus if time τ is small, we would expect that $R_\tau \to 1$, since $v_\tau \to v_t$. However as the time interval τ increases, it is likely that the eddy velocity will change sign and eventually there will be no memory of the velocity at time t; thus $R_\tau \to 0$ (see Fig. 10-11). Then from the second of Eqs. (10-82)

$$D^{(t)} = \overline{v_t X} = v_t \int_0^t v_\tau \, d\tau = \int_0^t \overline{v_t v_\tau} \, d\tau = \overline{v_\tau^2} \int_0^t R_\tau \, d\tau \tag{10-87}$$

and from Eq. (10-85)

$$\overline{X^2} = 2 \int_0^{\mathfrak{J}} D^{(t)} \, dt \tag{10-88}$$

or

$$\overline{X^2} = 2\overline{v^2} \int_0^{\mathfrak{J}} \int_0^t R_\tau \, d\tau \, dt \tag{10-89}$$

known as *Taylor's theorem of turbulent diffusion.*

At *small* values of time, the fluid motion will be correlated, or $R_\tau = 1$; then $\overline{X^2} \propto \mathfrak{J}^2$ and $D^{(t)} \propto t$ (see Fig. 10-12). At *large* values of time, there should be no correlation between v_t and v_τ and $R_\tau \to 0$; then $\overline{X^2} \propto \mathfrak{J}$, and $D^{(t)}$ is constant. This is consistent with observations on the spread of smoke from a fixed source, such as

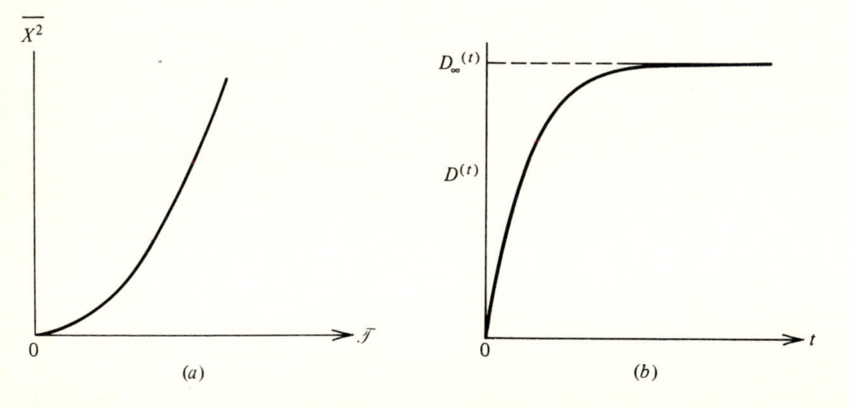

Figure 10-12 (*a*) Mean-square displacement; (*b*) turbulent diffusivity.

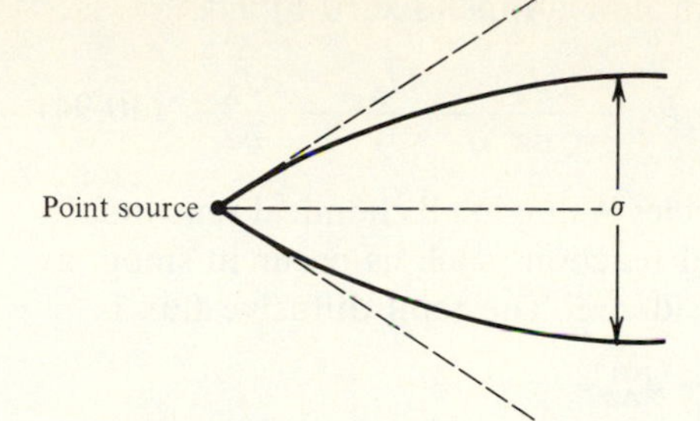

Figure 10-13 Dispersion from point source.

a stack; at small distances, the rms spread $(\overline{X^2})^{1/2}$ of the surface is a cone $\propto z$, but at large distances it is a paraboloid $\propto \sqrt{z}$ (see Fig. 10-13), where $z = v_z\mathfrak{J}$ is the distance downstream.

If we let $\mathfrak{J}_1$ be some large time after which $R_\tau = 0$, we can let the limiting value of $D^{(t)}$ be

$$D_\infty^{(t)} = \overline{v^2}\int_0^{\mathfrak{J}_1} R_\tau\, d\tau \equiv \overline{v^2}\Upsilon_L \qquad \text{where } \Upsilon_L \equiv \int_0^{\mathfrak{J}_1} R_L\, d\tau \tag{10-90}$$

is the area under the R_τ curve and Υ_L is called the *lagrangian time macroscale of turbulence*.

If we define a *lagrangian length macroscale* as

$$\sqrt{\overline{v^2}}\Upsilon_L \equiv \Lambda_L \tag{10-91}$$

we can write $D_\infty^{(t)}$ in terms of characteristic velocity and length scales, or

$$D_\infty^{(t)} = \sqrt{\overline{v^2}}\Lambda_L \tag{10-92}$$

Note that if we assume that

$$R_\tau = \exp\left(\frac{-|\tau|}{A}\right) \qquad A = \text{constant}$$

and insert it into Eq. (10-87), we get

$$D^{(t)} = \overline{v^2}(A - Ae^{-t/A}) \tag{10-93}$$

At *small* times the Taylor-series expansion gives

$$e^{-t/A} = 1 - \frac{t}{A} + \frac{1}{2}\frac{t^2}{A^2} \cdots$$

Then from Eq. (10-89), $\sqrt{\overline{X^2}}$ varies with $\mathfrak{J}$, and $D^{(t)}$ depends on t, as noted above. At large times, $\sqrt{\overline{X^2}}$ varies with $\sqrt{\mathfrak{J}}$, but $D^{(t)}$ is constant.

10-11 ATMOSPHERIC DISPERSION

An important application of turbulence theory is the prediction of pollutant concentrations in the atmosphere. A starting point is usually the diffusion equation (part A of Table I-6) modified for turbulent flow (Appendix 10-6) as†

$$\frac{\partial \bar{C}_A}{\partial t} + \bar{v}_x \frac{\partial \bar{C}_A}{\partial x} + \bar{v}_y \frac{\partial \bar{C}_A}{\partial y} + \bar{v}_z \frac{\partial \bar{C}_A}{\partial z} = \bar{R}_A - \frac{\partial \bar{J}_{Ax}}{\partial x} - \frac{\partial \bar{J}_{Ay}}{\partial y} - \frac{\partial \bar{J}_{Az}}{\partial z} \tag{10-94}$$

where $\bar{R}_A$ is the net rate of production of species A due to all chemical reactions in which A is involved. For the photochemical reactions such as occur in smog, as many as 20 to 30 reactions have been considered. The total diffusive flux is

$$\bar{J}_{Ax} = \bar{J}^{(l)}_{Ax} + J^{(t)}_{Ax}$$

Since turbulent transport is anisotropic [see Eq. (10-68)],

$$J^{(t)}_{Ax} = -D^{(t)}_{xx} \frac{\partial \bar{C}_A}{\partial x} - D^{(t)}_{xy} \frac{\partial \bar{C}_A}{\partial y} - D^{(t)}_{xz} \frac{\partial \bar{C}_A}{\partial z}$$

where $D^{(t)}_{xy}$ and $D^{(t)}_{xz}$ are the off-diagonal components of the turbulent diffusivity tensor and $D^{(t)}_{xx} \equiv D^{(t)}_x$ is a diagonal component.

The off-diagonal terms are usually neglected because little is known about them. Using similar expressions for $J^{(t)}_{Ay}$ and $J^{(t)}_{Az}$, and neglecting molecular diffusion compared with turbulent diffusion gives

$$\frac{\partial \bar{C}_A}{\partial t} + \bar{v}_x \frac{\partial \bar{C}_A}{\partial x} + \bar{v}_y \frac{\partial \bar{C}_A}{\partial y} + \bar{v}_z \frac{\partial \bar{C}_A}{\partial z} = \bar{R}_A + \frac{\partial}{\partial x}\left(D^{(t)}_x \frac{\partial \bar{C}_A}{\partial x}\right) + \frac{\partial (D^{(t)}_y \, \partial \bar{C}_A/\partial y)}{\partial y} + \frac{\partial (D^{(t)}_z \, \partial \bar{C}_A/\partial z)}{\partial z} \tag{10-94a}$$

This equation has been used[24,26] as the basis for an *urban diffusion model* of the Los Angeles area. To apply it one needs data on the wind velocity and on the turbulent diffusivities at *all* points in the atmosphere, on the strength of all sources, and on the rates of all reactions. It is usually assumed that $\bar{R}_A \equiv \bar{R}_A(\bar{C}_A, \bar{C}_p, \ldots, T)$ as discussed in Appendix 10-6 [see Eq. (A10-6-6)]. An equation can be written for each chemical species and can in principle be solved simultaneously for the concentration of each species as a function of x, y, z, and t. One solution method is to divide the urban airshed into a grid of horizontal incremental elements Δx and Δy and vertical elements Δz. Then one writes Eq. (10-94a) in terms of finite differences and calculates the change of C_A in the element $\Delta x\, \Delta y\, \Delta z$ over Δt.

The weak point of the above method is that one needs to know the turbulent diffusivities, which vary with wind velocity, distance from the ground, atmospheric conditions, roughness of the terrain, and travel time from a source. For this reason, another approach is often used in practical problems: The sources are classified as being point, line, or area sources. A typical point source might be a smoke stack, which is modeled as shown later; a typical line source might be a freeway, which can be modeled by integrating point sources over a path; and an

†An incompressible fluid is usually assumed.

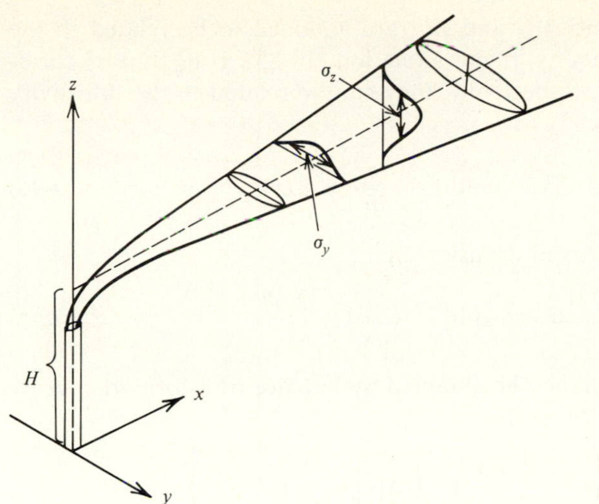

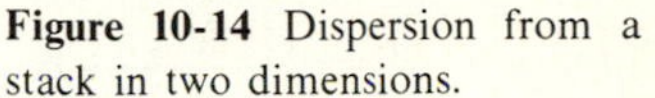
Figure 10-14 Dispersion from a stack in two dimensions.

area source might be a residential district, which can be treated as an area integral of point sources. Then at each reception point the contributions of each of the sources are added to give the total pollutant concentration.

The above method reduces the problem to that of modeling the point-source problem. The most commonly used model for such problems is the *Pasquill-Gifford model* (Fig. 10-14), based on the assumption that the plume spread in both y and z directions has a gaussian distribution. To obtain such a result analytically (in addition to the previously mentioned assumptions) one must assume:

1. z-Independent eddy diffusivities, where z = vertical distance
2. Steady state
3. Flat velocity profile, $\bar{v}_x$ = const
4. Negligible diffusion in the direction of the wind, which is the x direction
5. Total reflection of the plume at the earth's surface
6. Negligible chemical reaction

The diffusion equation under these conditions (see Example 10-7) is given by

$$\bar{v}_x \frac{\partial \bar{C}_A}{\partial x} = \frac{\partial}{\partial y}\left(D_y^{(t)} \frac{\partial \bar{C}_A}{\partial y}\right) + \frac{\partial}{\partial z}\left(D_z^{(t)} \frac{\partial \bar{C}_A}{\partial z}\right) \tag{10-95}$$

Example 10-7: Turbulent diffusion in the atmosphere Show how the equation of continuity for species A [Eq. (10-94a)] simplifies to Eq. (10-95) for the conditions of the Pasquill-Gifford model of transport from a stack.

Solution At steady state $\partial \bar{C}_A/\partial t = 0$. For negligible reaction, $R_A = 0$. For flow in the x direction only $\bar{v}_y = \bar{v}_z = 0$. Assuming that $\partial \bar{v}_x/\partial x = 0$ and neglecting transport in the direction of flow, we obtain Eq. (10-95). □

The diffusivity components $D_y^{(t)}$ and $D_z^{(t)}$ are assumed to be related to the so-called dispersion coefficients by Taylor's relation (10-85). If σ_y^2 and σ_z^2 correspond to Taylor's $\overline{Y^2}$ and $\overline{Z^2}$, respectively, they can be related to the diffusivity components $D_y^{(t)}$ and $D_z^{(t)}$ by

$$\frac{d\sigma_y^2}{dt} = 2D_y^{(t)} \qquad \text{and} \qquad \frac{d\sigma_z^2}{dt} = 2D_z^{(t)} \tag{10-96}$$

For constant velocity and no axial diffusion

$$\frac{d\sigma_y^2}{dx} = \frac{2D_y^{(t)}}{\bar{v}_x} \qquad \text{and} \qquad \frac{d\sigma_z^2}{dx} = \frac{2D_z^{(t)}}{\bar{v}_x} \tag{10-97}$$

The following solution can then be obtained by Laplace transformation or by combination of variables

$$C_A(x, y, z, H) = \frac{\dot{Q}_A}{2\sigma_y\sigma_z\bar{v}_x\pi} \exp\left[-\frac{1}{2}\left(\frac{y}{\sigma_y}\right)^2\right]\left\{\exp\left[-\frac{1}{2}\left(\frac{z-H}{\sigma_z}\right)^2\right] + \exp\left[-\frac{1}{2}\left(\frac{z+H}{\sigma_z}\right)^2\right]\right\} \tag{10-98}$$

where C_A = pollutant concentration, mol/vol
$\dot{Q}_A$ = source strength, mol/s
$\bar{v}_x$ = x-direction mean wind velocity
H = effective emission height
σ_y, σ_z = standard deviations of plume concentration in y and z directions, respectively

The dispersion coefficients and the diffusivities vary with distance downstream (according to Taylor's theory); but they also depend upon meteorological conditions, e.g., the "stability" of the atmosphere. If the temperature *decreases* with height by exactly the amount corresponding to the adiabatic expansion that would result from the decrease in pressure, there will be no tendency for either up or downward movement; this condition is called *neutral*. If the temperature decreases with z at a greater rate than neutral, there will be a tendency for the warm

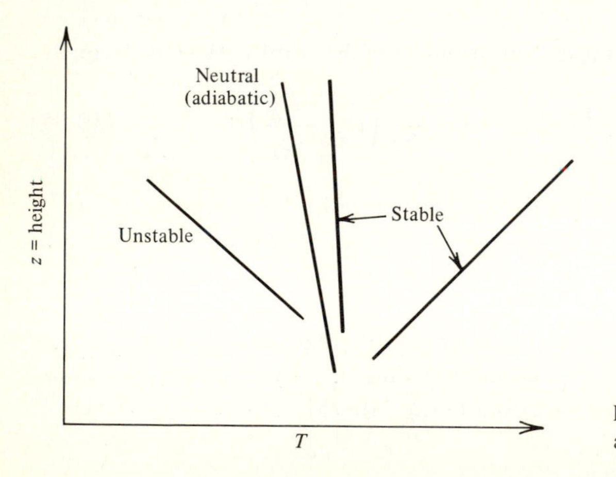

Figure 10-15 Temperature gradients in atmosphere.

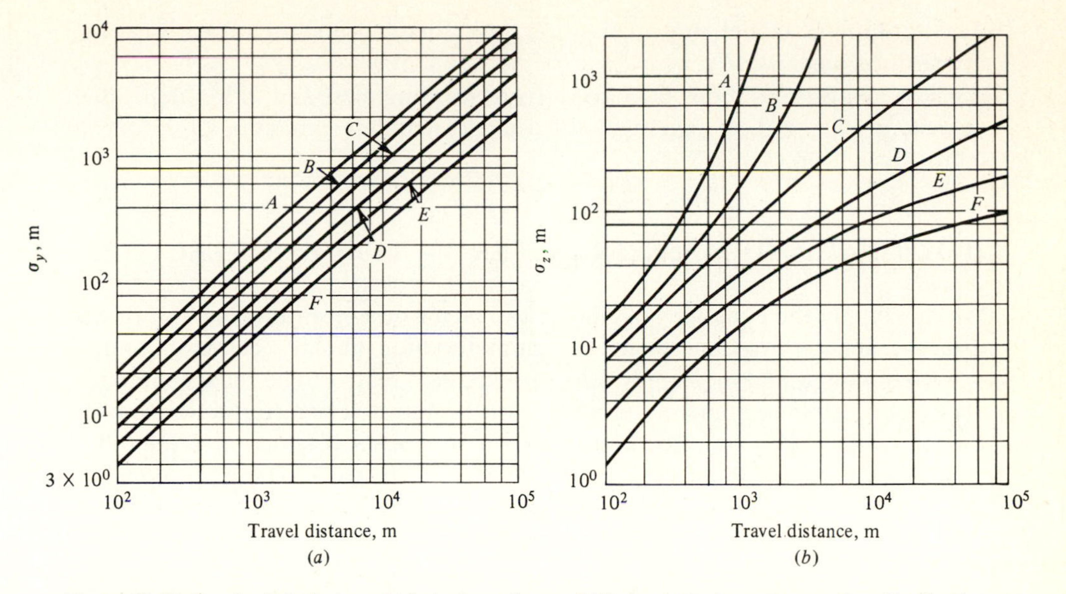

Figure 10-16 Standard deviation of (*a*) the lateral σ_y and (*b*) the vertical σ_z concentration distribution as a function of travel distance from a continuous point source. For Pasquill's diffusion categories *A* to *F* see Table 10-3. (*From D. B. Turner, "Workbook of Atmospheric Dispersion Estimates," USEPA Res., Triangle Park, N.C., 1970.*)

air to rise; this condition is called *unstable* (see Fig. 10-15). On the other hand, if the temperature gradient is less than the adiabatic, cold air will tend to sink toward the ground; this condition is called *stable*.

Relying on the work of Pasquill and Gifford, Turner[33] has prepared graphs of σ_y and σ_z versus the downwind distance (Fig. 10-16). The structure of turbulence in the atmosphere and the effect of other parameters is taken into account by defining six different stability classes, as shown in Table 10-3. At times the

Table 10-3 Relation of Pasquill turbulence types to weather conditions† for Fig. 10-16[33]

	Daytime insolation			Nighttime conditions	
Surface wind speed, m/s	Strong	Moderate	Slight	Thin overcast or $\geq \frac{4}{8}$ cloudiness‡	$\leq \frac{3}{8}$ cloudiness
2	*A*	*A–B*	*B*		
2	*A–B*	*B*	*C*	*E*	*F*
4	*B*	*B–C*	*C*	*D*	*E*
6	*C*	*C–D*	*D*	*D*	*D*
6	*C*	*D*	*D*	*D*	*D*

†Key to conditions: *A* = extremely unstable, *B* = moderately unstable, *C* = slightly unstable, *D* = neutral (applicable to heavy overcast, day or night), *E* = slightly stable, *F* = moderately stable.

‡The degree of cloudiness is defined as that fraction of the sky above the local apparent horizon which is covered with clouds.

Pasquill-Gifford model may give very large errors, but it can be used for rough approximations.

For neutral conditions the effect of surface roughness and of a z-dependent velocity profile has been investigated[10] using a computer simulation of the turbulent diffusion process.

10-12 HEAT AND MASS TRANSFER IN TUBULAR REACTORS

We now apply the equations of change to reactor design. In this section we first discuss a general two-dimensional design method for tubular reactors in laminar or turbulent flow and then apply it to a packed-bed reactor. The reader should be aware that there are numerous models in the literature for reactor design and the model we illustrate is selected not because it is necessarily superior to the others but because numerical results and experimental data are available for comparison. Our goal is to show how the equations of change can be used rather than to carry out an ideal design. Further details can be found in texts on reactor design.[6,28]

Tubular Reactor Design

Consider a tubular reactor which consists of a tube of circular cross section through which a reacting fluid is flowing (see Fig. 1-2). If the reaction is exothermal, the walls of the tube are surrounded by a cooling medium, which removes the heat of reaction. The fluid enters at some temperature T_0 and some concentration of limiting reactant C_{A0}. The velocity of the fluid in the direction of flow v_z is a function of the radial position r and is independent of axial distance z (if a sufficient entrance region has been provided). The presence of the cooling medium at the walls means that heat will be transported to the walls and a radial profile $T(r, z)$ will exist at each z. Since the temperature is higher at the center ($r = 0$), the rate of reaction, which increases rapidly with temperature, will be higher at the center. Therefore the principal product X will be formed in greater concentration at the center than near the walls. The resulting *concentration gradient* will give rise to *radial mass transport* and a concentration profile $C_X(r, z)$ for the product. The bulk mean temperature and concentration of a product will vary with z (as shown in Figs. 1-4 and 1-5). The latter is defined as

$$\langle C_X \rangle \equiv \frac{\mathcal{W}_X}{A_{cs} v_a}$$

or

$$\langle C_X \rangle \equiv C_{Xb} = \frac{\iint C_X(r, z) v_z(r)\, dA_{cs}}{\iint v_z\, dA_{cs}} \qquad \text{where } dA_{cs} = 2\pi r\, dr \qquad (10\text{-}99)$$

The *design problem* for a tubular reactor can be stated as follows:

1. What mean concentration of product $C_{Xb}(L)$ can be obtained in a reactor of length L for a given mass flow rate and given wall and entering temperature T_w and T_0?
2. What will be the total heat transferred at the wall $\dot{Q}_w$?

A *design method* might be outlined as follows:

1. Write a partial differential energy equation for the temperature $T(r, z)$, using the appropriate flux laws.
2. Write a partial differential mass equation for the concentration of a species i, $C_i(r, z)$, using the appropriate flux laws.
3. If the rate of each reaction is not known as a function of temperature and concentration, estimate it by methods given in kinetics texts.[6,28]
4. If the velocity profile is not known, solve the momentum equation for it if possible, using appropriate flux laws.
5. Solve these equations simultaneously to obtain the velocity, concentration, and temperature profiles at the exit $z = L$. Integrate over the cross section to find the average concentration and average temperature. From the average temperature the total heat transport $\dot{Q}_w$ can be obtained by an overall energy balance on the fluid.

Packed-Bed Reactor

Pseudo-homogeneous model We can treat a packed bed as being pseudo-homogeneous; i.e., we do not distinguish between solids and voids but look at it as a continuous region having transport properties averaged over both void and solid phases (see Fig. 10-17). We use the equations of change for mass and energy, as derived earlier, but interpret the concentration and temperature as being *local-volume-averaged*[32,37] values obtained by integrating over a small volume element containing both void and solid. Thus the local point concentration C_{Ap} will vary from zero (or nearly zero) in the solid phase to some peak value in a nearby void (or fluid) phase. Also, the point temperature T_p will vary with position in the void and solid phases. Then the C_A and T in the balance laws are interpreted as averages of C_{Ap} or T_p over an element of volume $\Delta\mathcal{V}$ large enough to represent both solid and fluid regions. Thus

$$C_A \equiv \frac{1}{\Delta\mathcal{V}} \iiint C_{Ap}\, d\mathcal{V} \qquad \text{and} \qquad T \equiv \frac{1}{\Delta\mathcal{V}} \iiint T_p\, d\mathcal{V} \qquad (10\text{-}100)$$

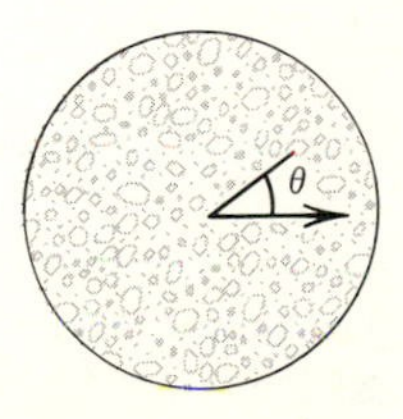

Figure 10-17 Cross section of packed column.

Instead of using some special symbol like angle brackets for a local volume average we will use a subscript p for the point value. We can see that we have a situation similar to the case of turbulent flow (where we examined fluctuations with respect to a *time*-averaged mean) except that here we have fluctuations with respect to a *volume*-averaged mean. Likewise, if we have a correlation between fluctuations of concentration or temperature and velocity, a flux (analogous to the turbulent flux) should occur. Other terms[37] may also appear but will be neglected.

For convenience we assume that in the packed bed the volume averaging can be done by integrating over angular position

$$C_{\mathrm{A}}(r, z) = \frac{1}{2\pi}\int_0^{2\pi} C_{\mathrm{A}p}(r, z, \theta)\, d\theta \qquad T(r, z) = \frac{1}{2\pi}\int_0^{2\pi} T_p(r, z, \theta)\, d\theta \qquad \text{(10-100}a\text{)}$$

where T_p and $C_{\mathrm{A}p}$ represent point values at any location in the cross section. The integration must be carried out over both void and solid regions since the temperature in the packing will generally differ from that in the void region.

We define an effective thermal conductivity tensor $\underline{\mathbf{k}}_e$ by the anisotropic flux law [Eqs. (10-66) or (10-68)]

$$\mathbf{q} = -\underline{\mathbf{k}}_e \cdot \nabla T \qquad \text{(10-101)}$$

We assume that the mixing resulting from the zigzag movement of elements of fluid through the void regions between the packing can be described by an effective eddy-diffusivity tensor $\underline{\mathbf{E}}$, defined by

$$\mathbf{J}_{\mathrm{A}} = -\underline{\mathbf{E}} \cdot \nabla C_{\mathrm{A}} \qquad \text{(10-102)}$$

Since both the mass and energy equations contain the reaction rate, and since the rate depends upon concentration and temperature, they must be solved simultaneously by a trial-and-error procedure.

For a simple tubular reactor (such as that shown in Fig. 1-2) assume that the following simplifications can be made:

1. Steady state $\rightarrow \partial/\partial t = 0$.
2. Fully developed flow $\rightarrow \partial v_z/\partial z = 0$.
3. Incompressible fluid $\rightarrow \rho = \text{const}, \widehat{C}_p = \widehat{C}_v = \text{const}$.
4. Cylindrical tube: no angular variation $\partial/\partial\theta = 0$.
5. One reaction $\rightarrow \Phi_R = -\Delta H_R R_{\mathrm{A}}$.

The equations of change† can then be written [Table I-7 or I-11 and Eq. (10-63)]

$$v_z \frac{\partial C_{\mathrm{A}}}{\partial z} = -\frac{\partial(J_{\mathrm{A}r} r)}{r\,\partial r} - \frac{\partial J_{\mathrm{A}z}}{\partial z} + R_{\mathrm{A}} \qquad \text{(10-103)}$$

$$\rho \widehat{C}_p v_z \frac{\partial T}{\partial z} = -\frac{\partial(r q_r)}{r\,\partial r} - \frac{\partial q_z}{\partial z} + (-\Delta H_R) R_{\mathrm{A}} \qquad \text{(10-104)}$$

† Although equations of change can be written for each species, they can be related through the use of the fractional conversion (or the extent of reaction) as the variable.

Using *flux laws* [Eqs. (10-101) and (10-102)], and neglecting off-diagonal components, we get

$$v_z(r)\frac{\partial C_A}{\partial z} = \frac{\partial[rE_r(r)\,\partial C_A/\partial r]}{r\,\partial r} + E_z(r)\frac{\partial^2 C_A}{\partial z^2} + R_A \tag{10-105}$$

$$\rho \widehat{C}_p \bar{v}_z \frac{\partial T}{\partial z} = \frac{\partial[rk_{er}(r)\,\partial T/\partial r]}{r\,\partial r} + k_{ez}(r)\frac{\partial^2 T}{\partial z^2} + (-\Delta H_R)R_A \tag{10-106}$$

Bulk flow | Radial transport | Axial transport | Reaction

Reaction rate and conversion For a packed bed, it is convenient to use a reaction rate r_c expressed in moles A per *unit mass* catalyst and unit time. Then if A is a reactant,

$$R_A = -r_c\rho_B \tag{10-107}$$

where ρ_B is the bulk density of the catalyst (mass of catalyst per volume of reactor). If we use the fractional conversion

$$X = \frac{C_{A0} - C_A}{C_{A0}} = \frac{C_X}{C_{A0}} \tag{10-108}$$

where C_{A0} is the initial reactant concentration, and neglect axial diffusion, the mass equation becomes

$$v_z(r)\frac{\partial X}{\partial z} = \frac{\partial[rE_r(r)\,\partial X/\partial r]}{r\,\partial r} + \frac{r_c\rho_B}{C_{A0}} \tag{10-109}$$

Similarly, for temperature we obtain

$$\widehat{C}_p v_z(r)\frac{\partial T}{\partial z} = \frac{\partial[rk_{er}(r)\,\partial T/\partial r]}{r\,\partial r} + (-\Delta H)r_c\rho_B \tag{10-110}$$

The boundary conditions are

Wall ($r = r_w$): $$\frac{\partial X}{\partial r} = 0 \qquad T = T_w \tag{10-111}$$

Entrance ($z = 0$): $$T = T_0 \tag{10-112}$$

Center ($r = 0$): $$\frac{\partial X}{\partial r} = 0 \qquad \frac{\partial T}{\partial r} = 0 \tag{10-113}$$

Dimensionless variables It is convenient to use dimensionless variables

$$r^* \equiv \frac{r}{r_w} \qquad k^* \equiv \frac{k_{er}}{\langle k_{er}\rangle} \qquad \gamma \equiv \frac{r_w}{d_p} \qquad z^* \equiv \frac{z}{r_w} \qquad E^* \equiv \frac{E}{\langle E_r\rangle}$$

$$v^* = \frac{v_z}{\langle v_z\rangle} \qquad (\mathrm{Pe}')_M \equiv \frac{d_p\langle v\rangle}{\langle E_r\rangle} \qquad (\mathrm{Pe}')_H \equiv \frac{d_p\langle v\rangle\rho\widehat{C}_p}{\langle k_{er}\rangle}$$

where $\qquad \langle k_{er} \rangle \equiv \frac{1}{A} \iint k_{er}\, dA \qquad \langle E_r \rangle \equiv \frac{1}{A} \iint E_r\, dA$

and where d_p is the pellet size (diameter), r_w is the tube radius, and $(\mathrm{Pe}')_M$ and $(\mathrm{Pe}')_H$ are the Peclet numbers for radial mass and heat transfer based respectively on $\langle E_r \rangle$ and $\langle k_{er} \rangle$ and on d_p. The differential equations then become

$$\frac{\partial X}{\partial z^*} = \frac{1}{v^*(\mathrm{Pe}')_M \gamma} \frac{\partial(r^* E^* \, \partial X/\partial r^*)}{r^* \, \partial r^*} + \frac{r_w r_c \rho_\mathrm{B} M_{0a}}{G_z y_{\mathrm{A}0} v^*} \tag{10-114}$$

$$\frac{\partial T}{\partial z^*} = \frac{1}{v^*(\mathrm{Pe}')_H \gamma} \frac{\partial(r^* k^* \, \partial T/\partial r^*)}{r^* \partial r^*} - \frac{r_w r_c \rho_\mathrm{B} \, \Delta H}{G_a \widehat{C}_p v^*} \tag{10-115}$$

where $G_a = \rho v_a$ = average mass velocity of feed
$y_{\mathrm{A}0}$ = mole fraction A in feed
M_{0a} = average molecular weight of feed

To solve these differential equations it is necessary to have expressions for $v^*(r^*)$, $E^*(r^*)$, $k^*(r)$, $(\mathrm{Pe}')_H$, and $(\mathrm{Pe}')_M$.

Transport properties and velocity profile in packed beds The mass Peclet number $(\mathrm{Pe}')_M$ has been correlated[11] by the equation

$$(\mathrm{Pe}')_M = 9.0 \left[1 + 19.4 \left(\frac{d_p}{D_t} \right)^2 \right] \tag{10-116}$$

If radiation is neglected, the heat Peclet number is given by[2]

$$\frac{\langle k_{er} \rangle}{\rho \widehat{C}_p d_p v_a} \equiv \frac{1}{(\mathrm{Pe}')_H} = \frac{\epsilon}{\mathrm{Pr}\ \mathrm{Re}'} + \frac{1}{(\mathrm{Pe}')_M} + \frac{1-\epsilon}{\mathrm{Re}} \frac{k_s^*}{\mathrm{Pr}} \frac{1 + 2k_p^*/\mathrm{Nu}'}{1 + 2k_s^*/\mathrm{Nu}'} \tag{10-117}$$

$$\mathrm{Re}' = \frac{d_p v_a}{\nu} \qquad k_s^* \equiv \frac{k_s}{k} \qquad k_p^* \equiv \frac{k_p'}{k}$$

where ϵ = void fraction
Pr = Prandtl number of fluid
k = thermal conductivity of fluid
k_s = thermal conductivity of pellet
k_p' = point-to-point contact thermal conductivity of solid
Nu' = modified Nusselt number = $h_c d_p/k$
h_c = heat-transfer coefficient between pellet and fluid surrounding it

The first term represents molecular conduction in the fluid phase, the second term is turbulent thermal diffusion (assumed equal to mass) through the void region, and the third term represents a series mechanism which consists of convection through the fluid film to the solid, conduction through the solid, conduction to another pellet by point contact, and convection from the solid through the film to the bulk fluid. The pellet Nusselt number is given by[14]

$$\mathrm{Nu}' = \begin{cases} 1.95(\mathrm{Re}')^{0.49}(\mathrm{Pr})^{2/3} & \mathrm{Re}' < 350 \qquad (10\text{-}118) \\ 1.06(\mathrm{Re}')^{0.59}(\mathrm{Pr})^{2/3} & \mathrm{Re}' > 350 \qquad (10\text{-}119) \end{cases}$$

The velocity profile $v^*(r^*)$ in a packed bed is influenced by the local void fraction $\epsilon(r)$, which varies with r because the packing does not orient itself near the wall the same as it does near the center. Experiments have indicated that the void fraction, which must approach unity at the wall, decreases rapidly to a minimum near the wall and then oscillates with a period of one pellet diameter as one moves toward the center. An equation for this variation has been developed by Cohen and Metzner,[8] who used it to obtain a local hydraulic radius and an average velocity for the wall, transition, and core regions. Their results generally agreed with the velocity data of Schwartz and Smith,[25] who measured velocities above the bed with circular anemometers. Although their experimental method tended to mask any oscillations, they found a maximum velocity at a distance of one particle diameter from the wall.

The velocity profile data of Schwartz and Smith have been correlated[12] in terms of γ as follows:

$$v^* = \frac{A_1 + A_2 r^{*(B+1)} - A_3 r^{*(B+2)}}{A_1 + [2A_2/(B+3)] - 2A_3/(B+4)} \tag{10-120}$$

where $B = 0.45\gamma^{1.5} \qquad A_1 = (B+2)^{-1} - \dfrac{\gamma - 1}{\gamma(B+1)}$

$$A_2 = \frac{\gamma - 1}{(B+1)\gamma} \qquad A_3 = (B+2)^{-1}$$

The data of Fahien and Smith have been correlated by Ahmed and Fahien[1] by the equations

$$E^* = \begin{cases} E_0^* + 3(E_M^* - E_0^*)\dfrac{r^{*2}}{r_M^{*2}} + 2(E_0^* - E_M^*)\dfrac{r^{*3}}{r_M^{*3}} & 0 < r^* < r_M^* \quad (10\text{-}121) \\ E_M^* \dfrac{1 - r^*}{1 - r_M^*} & r_M^* < r^* < 1 \quad (10\text{-}122) \end{cases}$$

where

$$r_M^* = 1 - \frac{2}{\gamma} \tag{10-123}$$

$$E_M^* = \frac{6(0.5 - 0.15E_0^* r_M^{*2})}{1 + r_M^* + 0.1r_M^{*2}} \tag{10-124}$$

$$E_0^* = \tfrac{9}{8} v_0^* (1 + 4.85\gamma^{-2}) \tag{10-125}$$

where v_0^* is obtained from Eq. (10-120). Thus $E^*(r^*)$ can be obtained from γ only.

The effective thermal conductivity was correlated by Ahmed and Fahien by the equations

$$k^* = \begin{cases} k_0^* + 3(k_M^* - k_0^*)\dfrac{r^{*2}}{r_M^{*2}} + 2(k_0^* - k_M^*)\dfrac{r^{*3}}{r_M^{*3}} & 0 < r^* < r_M^* \quad (10\text{-}126) \\ k_M^* - \dfrac{(k_M^* - k_w^*)(r^* - r_M^*)}{1 - r_M^*} & r^* > r_M^* \quad (10\text{-}127) \end{cases}$$

where $$k_M^* = \frac{3 - 0.9k_0^* r_M^{*2} - k_w^*(r_M^* - 3r_M^{*2} + 2)/(1 - r_M^*)}{1 + r_M^* + 0.1r_M^{*2}}$$

These equations show that the required parameters are k_w^*, k_0^*, and $\langle k_{er} \rangle$.

Numerical Integration

Equations (10-114) and (10-115) must be solved simultaneously along with a reaction-rate expression $r_c(X, T)$ (which couples them). We illustrate how this can be done by numerical integration.[28]

Example 10-8: Turbulent-reactor design: two-dimensional model A reactor for the conversion of SO_2 to SO_3 consists of a 2-in-nominal-diameter tube packed to a depth of 5.7 in with $\frac{1}{8}$-in cylindrical pellets containing 0.2% platinum. The tube is jacketed with boiling glycol so that the inside wall temperature is constant throughout its length at 193 °C. The entering gas mixture contains 6.60 mol % SO_2 and 93.4 mol % dry air. The measured mass velocity of gas through the bed is 350 lb/h · (ft² of total tube area). The temperature of the air entering the reactor is as follows:[28]

Radial position r^*	0.023	0.233	0.474	0.534	0.797	0.819	1.000
Entering gas temperature, °C	400.1	399.5	400.1	400.4	376.5	376.1	197.0

and the reaction rates are given in Table 10-4.

The bulk density of the catalyst as packed in the bed is 64 lb/ft³. The heat of reaction may be assumed constant and equal to −22,700 cal/g mol. Write equations for the temperature and conversion profiles and describe solution method.

SOLUTION ***Numerical integration*** In this method we divide the reactor into L axial increments of size Δz and N radial increments of size Δr (see Fig. 10-18). We use subscript n to indicate a radial increment and subscript l to indicate an axial

Table 10-4 Rate of reaction, g mol/h · (g catalyst)

	Conversion						
T, °C	0	10%	20%	30%	40%	50%	60%
360	0.0175	0.0120	0.00788	0.00471	0.00276	0.00181	
380	0.0325	0.0214	0.01433	0.00942	0.00607	0.00410	
400	0.0570	0.0355	0.02397	0.01631	0.0110	0.00749	0.00488
420	0.0830	0.0518	0.0344	0.02368	0.0163	0.0110	0.00745
440	0.1080	0.0752	0.0514	0.03516	0.0236	0.0159	0.0102
460	0.146	0.1000	0.0674	0.04667	0.0319	0.0215	0.0138
480		0.1278	0.0898	0.0642	0.0440	0.0279	0.0189

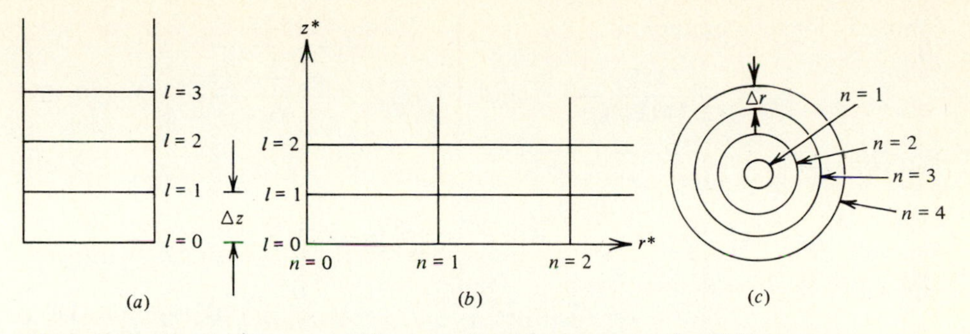

Figure 10-18 (*a*) Axial increments; (*b*) axial and radial increments; (*c*) radial increments.

increment. (Note that the volume increments are annular increments of size $2\pi r\,\Delta r\,\Delta z$.) This permits us to write a set of equations for the conversion and temperature for each increment, which are then solved simultaneously subject to the boundary conditions at the wall, center, and entrance.

Consider the conversion at a given value of r and z. We can approximate

$$\left(\frac{\partial X}{\partial z^*}\right)_{n,l+(1/2)} \qquad \text{with} \qquad \frac{X_{n,l+1} - X_{n,l}}{\Delta z^*}$$

This gives the average slope over Δz^*. In the radial direction if we let

$$\phi \equiv r^* E^* \frac{\partial X}{\partial r^*}$$

we can write

$$\left\{\frac{\partial[r^* E^*(r^*)\,\partial X/\partial r^*]}{\partial r^*}\right\}_{n,l} \equiv \left(\frac{\partial \phi}{\partial r^*}\right)_{n,l} \approx \frac{\phi_{n+(1/2),l} - \phi_{n-(1/2),l}}{\Delta r^*}$$

where
$$\phi_{n+(1/2),l} = r^*_{n+(1/2)} E^*_{n+(1/2)} \frac{X_{n+1,l} - X_{n,l}}{\Delta r^*}$$

and
$$\phi_{n-(1/2),l} = r^*_{n-(1/2)} E^*_{n-(1/2)} \frac{X_{n,l} - X_{n-1,l}}{\Delta r^*}$$

If we let

$$A_M \equiv \frac{\Delta z^*}{(\mathrm{Pe}')_M (\Delta r^*)^2 \gamma} \qquad \text{and} \qquad B_M \equiv \frac{r_w \rho_B\,\Delta z^*}{C_{A0}\langle v_z\rangle} \equiv \frac{r_w \rho_B \overline{M}_0\,\Delta z^*}{G_a\, y_{A0}}$$

and substitute the various differences into the conversion equation (10-114) and let $\overline{r}_c$ represent an average rate between l and $l + 1$, we get

$$X_{n,l+1} - X_{n,l} = \frac{A_M}{v^*_n r^*_n}[r^*_{n+(1/2)} E^*_{n+(1/2)}(X_{n+1,l} - X_{n,l}) - r^*_{n-(1/2)} E^*_{n-(1/2)}(X_{n,l} - X_{n-1,l})] + \frac{B_M \overline{r}_c}{v^*_n} \qquad (10\text{-}128)$$

Similarly for temperature, if

$$A_H \equiv \frac{\Delta z^*}{(\Delta r^*)^2(\mathrm{Pe}')_H \gamma} \qquad \text{and} \qquad B_H \equiv \frac{r_w \rho_B(-\Delta H)\,\Delta z^*}{G_a C_p} \tag{10-129}$$

Eq. (10-115) becomes

$$T_{n,l+1} - T_{n,l} = \frac{A_H}{v_n^* r_n^*}[r^*_{n+(1/2)}k^*_{n+(1/2)}(T_{n+1,l} - T_{n,l})$$

$$- r^*_{n-(1/2)}k^*_{n-(1/2)}(T_{n,l} - T_{n-1,l})] + B_H \frac{\overline{r}_c}{v_n^*} \tag{10-115a}$$

Boundary condition at center At $r = 0$, $\partial X/\partial r^* = 0$, and $\partial T^*/\partial r^* = 0$

$$\frac{\partial(r^* E^*\,\partial X/\partial r^*)}{r^*\,\partial r^*} = E^* \frac{\partial^2 X}{\partial r^{*2}} + \frac{E^*}{r^*}\frac{\partial X}{\partial r^*} + \frac{\partial X}{\partial r^*}\frac{\partial E^*}{\partial r^*}$$

As $r^* \to 0$,

$$\frac{\partial X/\partial r^*}{r^*} \to \frac{\partial^2 X/\partial r^{*2}}{1}$$

by L'Hôpital's rule. Using a half increment

$$\left(\frac{\partial^2 X}{\partial r^{*2}}\right)_{r=0} \approx \frac{(\partial X/\partial r^*)_{1/2} - (\partial X/\partial r^*)_0}{\Delta r^*/2}$$

leads to

$$\left(\frac{\partial X}{\partial r^*}\right)_{1/2} \approx \frac{X_1 - X_0}{\Delta r^*}$$

whence

$$\frac{\partial(r^* E^*\,\partial X/\partial r^*)}{r^*\,\partial r^*} = 2E_0^* \left(\frac{\partial^2 X}{\partial r^{*2}}\right)_0 = \frac{4E_0^*(X_1 - X_0)}{(\Delta r^*)^2}$$

Then

$$X_{0,l+1} - X_{0,l} = \frac{4A_M E_0^*(X_{1,l} - X_{0,l})}{v_0^*} + \frac{B_M \overline{r}_c}{v_0^*} \tag{10-130}$$

Similarly for temperature

$$T_{0,l+1} - T_{0,l} = \frac{4A_H k_0^*(T_{1,l} - T_{0,l})}{v_0^*} + \frac{B_H \overline{r}_c}{v_0^*} \tag{10-131}$$

Boundary condition at wall At $r^* = 1$, $\partial X/\partial r^* = 0$ (no flux). If $\phi \equiv r^* E^*\,\partial X/\partial r^*$,

$$\left(\frac{\partial \phi}{\partial r^*}\right)_N \approx \frac{\phi_N - \phi_{N-(1/2)}}{\frac{1}{2}\Delta r^*}$$

Because of the boundary condition $\phi_N = 0$,

$$\phi_{N-(1/2)} = r^*_{N-(1/2)}E^*_{N-(1/2)}\frac{X_N - X_{N-1}}{\Delta r^*}$$

$$\left(\frac{\partial\phi}{\partial r^*}\right)_N = -\frac{2}{(\Delta r^*)^2}(r^*_{N-(1/2)}E^*_{N-(1/2)})(X_N - X_{N-1})$$

Then

$$X_{N,l+1} - X_{N,l} = \frac{2A_M r^*_{N-(1/2)}E^*_{N-(1/2)}}{v^*_N r^*_N}(X_{N-1} - X_N)_l + \frac{B_M \bar{r}_c}{v^*_N} \qquad (10\text{-}132)$$

Actually $v^* = 0$ at $r = r_w$, but we want the *average* velocity between $N - \frac{1}{2}$ and N (since the region between $N - \frac{3}{2}$ and $N - \frac{1}{2}$ is accounted for by the equation for $N - 1$); for example, by interpolation, if $N = 5$,

$$v^*_N = \tfrac{1}{2}v^*_{N-(1/2)} = \tfrac{1}{2}v^*\Big|_{r^*=0.9}$$

Similarly
$$E^*_{N-(1/2)} = E^*\Big|_{r^*=0.9}$$

and
$$T_{N,l+1} = T_{w,l+1} \qquad \text{at} \begin{cases} r^* = 1 \\ T_N = T_w \end{cases} \qquad (10\text{-}133)$$

Alternate boundary condition at wall Another boundary condition at the wall is often used. If it is assumed that there is a very thin stationary fluid film at the wall, then T_w is the temperature on the inner surface of this film and we can write

$$k^*_N \frac{\partial T}{\partial r_*}\Big|_1 = -\text{Bi}\,(T^*_N - T^*_w) \qquad \text{at } r^* = 1 \qquad (10\text{-}134)$$

where $\text{Bi} = h_w r_w / k_e$ is the *Biot number*. The difficulty with this boundary condition is that it introduces an additional parameter, Bi, which is not very accurately known and which depends on z for short beds.

Equations (10-128) to (10-133) or (10-134) must be solved simultaneously along with an equation (or data) on the rate $r_c(X, T)$. Smith[28] has carried out such calculations using a double stepwise numerical integration. He assumed that the velocity and the diffusivities do not vary with r^*; thus $1 = E^* = k^* = v^*$. He obtained $(\text{Pe}')_M = 9.6$ from the correlation of Fahien and $(\text{Pe}')_H = 4.5$ from the equation of Argo and Smith.[2]

Smith used $N = 5$ radial increments and let Δz be 0.6 in. These gave $A_H = 0.404$, $A_M = 0.185$, $B_H = -798$, and $B_M = 4.38$. Equations (10-128) and (10-115*a*) then are

$$X_{n,l+2} = X_{n,l} + \frac{0.185}{r^*_n}[r^*_{n+(1/2)}(X_{n+1,l} - X_{n,l}) - r^*_{n-(1/2)}(X_{n,l} - X_{n-1,l})] + 4.38\bar{r}_c \qquad (10\text{-}35)$$

$$T_{n,l+1} = T_{n,l} + \frac{0.404}{r_n^*}[r_{n+(1/2)}^*(T_{n+1,l} - T_{n,l}) - r_{n-(1/2)}^*(T_{n,l} - T_{n-1,l})] + 798\bar{r}_c \quad (10\text{-}136)$$

Examination of them indicates that if conversion and temperature at the inlet $z = 0$ ($l = 0$) at radial positions $n + 1$ and $n - 1$ are known, the conversion X_n and temperature T_n at $l = 1$ can be calculated when $\bar{r}_c$ is known or assumed. As a basis, the rate at $z = 0$ can be used. This gives an estimate of the temperature and conversion leaving the increment. Thus the rate leaving the increment can be obtained and averaged with the rate entering the increment. If this average agrees with the assumed average rate to the desired accuracy, one goes to the next radial increment. If it does not, the process is repeated until it does. After calculations have been made for all radial increments at $l = 0$, one goes to the next axial increment $l = 1$.

To illustrate, we write Eqs. (10-135) and (10-136) at the inlet $z = 0$ ($l = 0$) for radial position $n = 3$ ($r^* = 0.6$). Then

$$X_{3,1} - X_{3,0} = \frac{0.185}{0.6}[0.7(1)(0 - 0) - 0.5(1)(0 - 0)] + 4.38\bar{r}_c = 4.38\bar{r}_c$$

$$T_{3,1} - 400 = \frac{0.404}{0.6}[0.7(376 - 400) - 0.5(400 - 400)] + 789\bar{r}_c$$

$$= \frac{0.404}{0.6}(-16.8) + 798\bar{r}_c = -11.2 + 798\bar{r}_c$$

From the data, r_c (at 0 percent conversion and $T = 400$) is 0.055. We may either use this as a first assumption of $\bar{r}_c$ or assume that the average rate will be something like 0.046. Then

$$X_{3,1} = 4.38(0.046) = 0.201$$

$$T_{3,1} = 400 - 11 + 798(0.046) = 426°\text{C}$$

From the data at 426°C, 20.1 percent conversion, $r_c = 0.04$, and

$$\bar{r}_c = \tfrac{1}{2}(0.055 + 0.040) = 0.0475$$

which is close enough.

The results obtained by Smith are compared in Fig. 10-19 with the experimental data and the results for a one-dimensional model that neglects radial transport. The discrepancy is probably due to the failure to use a lower effective thermal conductivity at the wall. □

Collocation

The above equations have also been solved with improved results using the method of orthogonal collocation by Ahmed and Fahien[1] and by Young and Finlayson,[39] who included axial transport as did Wang and Stewart.[34]

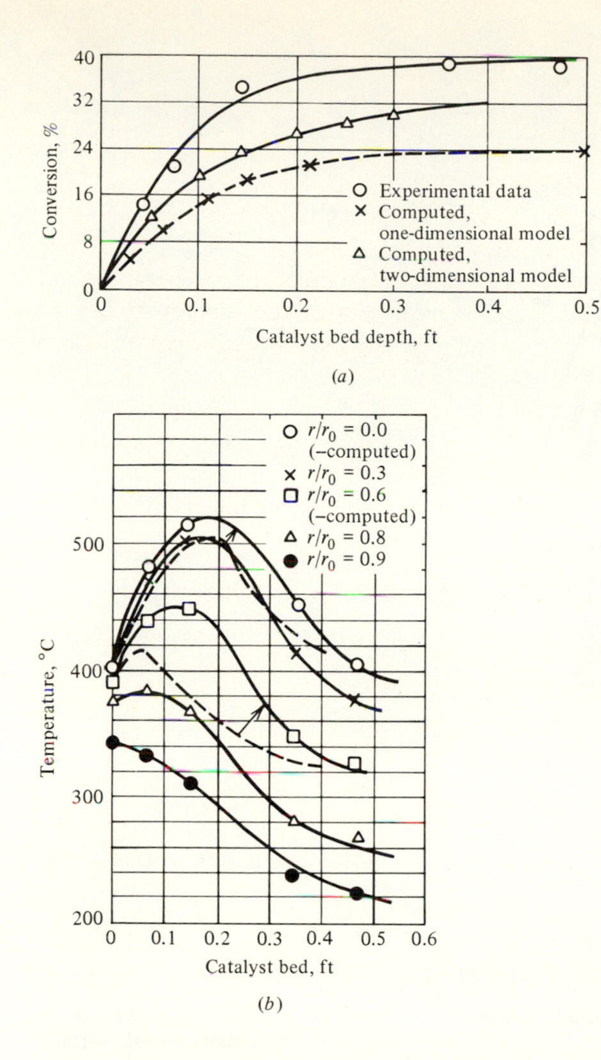

Figure 10-19 (*a*) Comparison of calculated experimental values and (*b*) longitudinal temperature profiles in SO_2 reactor. $G = 350\ \mathrm{lb/h \cdot ft^2}$, $\frac{1}{8}$-in catalyst particles in 2-in reactor. (*From J. M. Smith, "Chemical Engineering Kinetics," McGraw-Hill, New York, 1981.*)

10-13 TRANSPORT TO OR FROM SUBMERGED OBJECTS

In Sec. 9-20 we discussed flow around a submerged object, and for creeping flow past a sphere we gave the solution to the Navier-Stokes equations. From this result we were able to derive Stokes' law for the drag force acting on the sphere. Now let us consider a submerged object, such as a sphere, which is exchanging heat and/or mass with the fluid flowing past it (see Fig. 10-20*a*). In addition to the momentum boundary layer that forms around the object, there may in general be a mass and/or thermal boundary layer, in which the concentration and/or temperature may vary rapidly with distance from the interface. Our goal is to predict these profiles and to use them to find (1) the total rate of transport of mass and/or heat and (2) the Nusselt numbers for mass and/or heat. In order to do this it is first necessary to know the velocity distribution, which in principle can be found

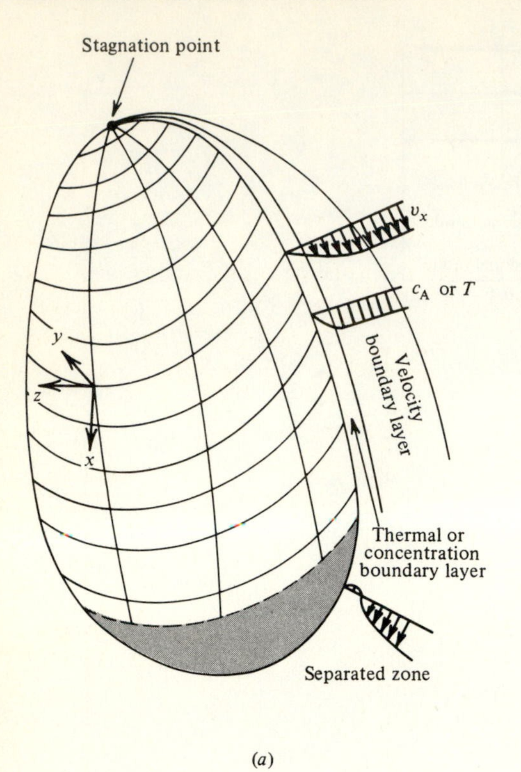

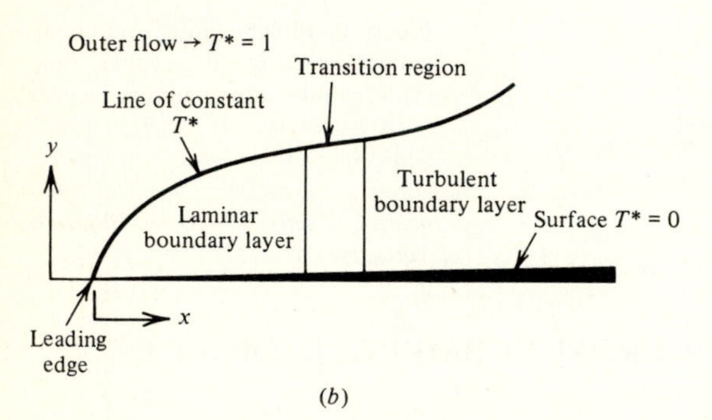

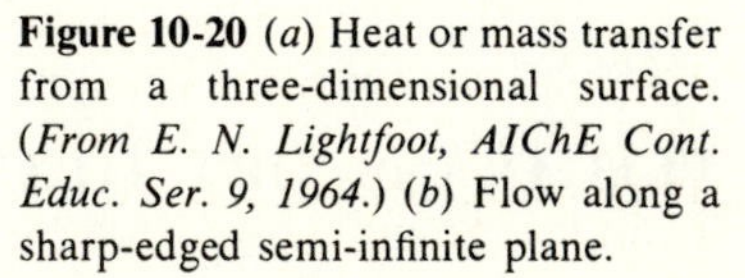

Figure 10-20 (*a*) Heat or mass transfer from a three-dimensional surface. (*From E. N. Lightfoot, AIChE Cont. Educ. Ser. 9, 1964.*) (*b*) Flow along a sharp-edged semi-infinite plane.

from the Navier-Stokes equation (as we did in Example 9-11 for creeping flow). If these equations cannot be solved, the velocity is assumed to vary linearly with distance from the interface or to be constant.

In reality the problem is complicated by the possible existence of four zones: a forward stagnation point, a laminar boundary layer, a turbulent boundary layer, and a separated zone. We will confine our discussion to the laminar-boundary-layer zone (except for creeping flow past a sphere, where the velocity for the entire region is known).

In general the equations of change simplify to

$$\nabla \cdot \mathbf{v} = 0 \qquad \mathbf{v} \cdot \nabla \mathbf{v} = \nu \nabla^2 \mathbf{v} - \frac{\nabla p}{\rho} \qquad \mathbf{v} \cdot \nabla T = \alpha \nabla^2 T \qquad \mathbf{v} \cdot \nabla C_\text{A} = D_\text{AB} \nabla^2 C_\text{A}$$

Continuity Momentum Energy Moles A

The general assumption is that the flow is two-dimensional, there being a velocity component v_x in the direction of flow along the surface and another, v_y, in a direction normal to the surface. For flow around a sphere the flow is assumed to be axisymmetrical; the v_r component is replaced by v_x and the v_θ component by v_y. For a curved surface x and y represent a curvilinear orthogonal coordinate system.

Although convection in the directions normal and parallel to the interface is considered, diffusion in directions other than normal to the surface is neglected. This differs from our previous example, where convection was considered only in the direction parallel to the surface. The addition of a convection term for flow normal to the surface means that, as in momentum transfer, the equation of continuity is needed to relate the velocity components to each other.

Since the energy and mass equations will have the same form, it is necessary to solve the equations for only one case and we will use Nu and Pe without subscripts to refer to either heat or mass transfer.

The energy equation can be made dimensionless by letting

$$T^* \equiv \frac{T - T_s}{T_\infty - T_s} \qquad \nabla^* = D\nabla \qquad \mathbf{v}^* \equiv \frac{\mathbf{v}}{v_\infty}$$

where T_∞ = approach temperature
v_∞ = approach velocity
T_s = surface temperature
D = characteristic length

Then

$$v_\infty \mathbf{v}^* \cdot \frac{\nabla^* T^* (T_\infty - T_s)}{D} = \frac{\alpha \nabla^{*2} T^* (T_\infty - T_s)}{D}$$

or

$$\frac{D v_\infty}{\alpha} \mathbf{v}^* \cdot \nabla^* T^* = \nabla^{*2} T^* \tag{10-137}$$

where

$$\frac{D v_\infty}{\alpha} \equiv (\text{Pe})_H = \text{Peclet number} = \text{Re Pr}$$

Thus if a solution for $\mathbf{v}^*$ were available, this could theoretically be used in the energy equation to find T^*, from which a local Nusselt number could be obtained from

$$(\text{Nu})_\text{loc} = \frac{hD}{k} = \frac{q_w D}{k(T_\infty - T_s)} = -\nabla^* T^* \bigg|_w = -\frac{\partial T^*}{\partial y^*}\bigg|_{y^*=0} \tag{10-138}$$

This equation could then be integrated over the surface of the object to find the average Nusselt (or Sherwood) number

$$\langle \text{Nu} \rangle = \frac{1}{A} \int (\text{Nu})_{\text{loc}} \, dA \tag{10-139}$$

If the diffusivity is small (that is, Pe is large, as is especially true in mass transfer), the diffusion-boundary-layer thickness, which varies as $(\text{Pe})^{-1}$, is very thin and only γ, the velocity gradient at the surface, is needed. As in our two previous examples, the cases usually considered are (1) the linear velocity profile, which results in a solution

$$\text{Nu} = (\text{Pe})^{1/3} f(\text{Re, position})$$

and (2) the flat velocity profile which results in

$$\text{Nu} = (\text{Pe})^{1/2} f(\text{Re, position})$$

For high Re, it can be shown[15] that the EOM can be put into a dimensionless form independent of Re, in which case solutions of the form

$$\text{Nu} = \text{Nu}((\text{Re})^{1/2}, (\text{Pr})^{1/3}, \text{position, geometry}) \tag{10-140}$$

are obtained.

For the special case of flow in the x direction past a flat plate at zero incidence (Fig. 10-20b), the equations are

$$\frac{\partial v_x^*}{\partial x^*} + \frac{\partial v_y^*}{\partial y^*} = 0 \tag{10-141}$$

$$\text{Pe} \left(v_x^* \frac{\partial T^*}{\partial x^*} + v_y^* \frac{\partial T^*}{\partial y^*} \right) = \frac{\partial^2 T^*}{\partial y^{*2}} \tag{10-142}$$

$$\text{Re} \left(v_x^* \frac{\partial v_x^*}{\partial x^*} + v_y^* \frac{\partial v_x^*}{\partial y^{*2}} \right) = \frac{\partial^2 v_x^*}{\partial y^{*2}} \tag{10-143}$$

Using a similarity transformation, we obtain[4]

and $$(\text{Nu})_{\text{loc}} = 0.332(\text{Re})^{1/2}(\text{Pr})^{1/3} \qquad \langle \text{Nu} \rangle = 2(\text{Nu})_{\text{loc}} \tag{10-144}$$

For creeping flow around a sphere Friedlander[13] has obtained

$$\langle \text{Nu} \rangle = \frac{(3\pi)^{2/3}(\text{Pe})^{1/3}}{4(2^{1/3})\Gamma(\frac{4}{3})} = 0.991(\text{Pe})^{1/3} \tag{10-145}$$

assuming a linear velocity profile and obtaining γ by differentiating the creeping-flow solution [Eq. (9-230)] for v_x. In Chap. 4 we found that for a stagnant fluid ($\text{Re} \to 0$) $\langle \text{Nu} \rangle \to 2.0$. Therefore it is customary to combine this result with the previous one to obtain

$$\langle \text{Nu} \rangle = 2.0 + 0.991(\text{Pe})^{1/3} \tag{10-146}$$

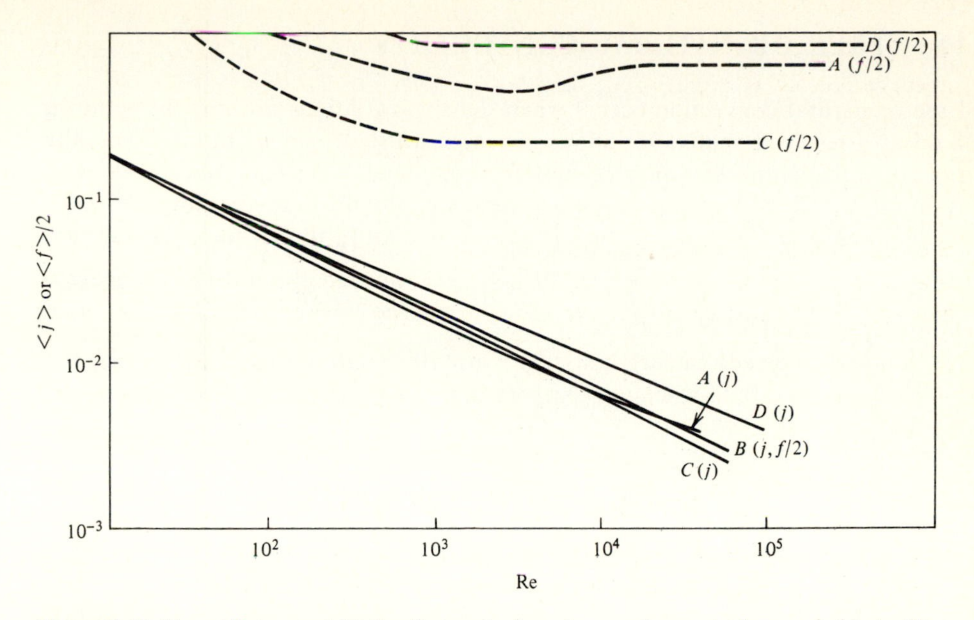

Figure 10-21 Mean j factors and friction factors for forced convection past submerged objects. (*From E. N. Lightfoot, AIChE Cont. Educ. Ser. 9, 1964.*)

Table 10-5 Explanation of Fig. 10-21

Shape	Re	$f/2$	j	Eq.	Remarks
Cylinder	$\frac{Dv_\infty}{\nu}$	$\frac{F_D}{DL(\frac{1}{2}\rho v_\infty^2)}$	$\text{Nu}(\text{Re})^{-1}(\text{Pr})^{-1/3}$	A	No creeping-flow limit
Flat plate	$\frac{xv_\infty}{\nu}$	$\frac{\tau_0}{2(\frac{1}{2}\rho v_\infty^2)}$	$\text{Nu}(\text{Re})^{-1}(\text{Pr})^{-1/3}$	B	$j = f/2$
Sphere	$\frac{Dv_\infty}{\nu}$	$\frac{F_D}{2(\pi R^2)(\frac{1}{2}\rho v_\infty^2)}$	$(\text{Nu} - 2)(\text{Re})^{-1}(\text{Pr})^{-1/3}$	C	$\langle\text{Nu}\rangle = 2 + (\text{Pe})^{1/3}$ (creeping flow; large Pe)
Packed beds	$\frac{D_p v_0}{6\nu(1-\varepsilon)}$	$\frac{-\Delta\mathcal{P}\, D_p \varepsilon^3}{2\rho v_0^2 L(1-\varepsilon)}$	$\text{Nu}(\text{Re})^{-1}(\text{Pr})^{-1/3}$	D	$\langle\text{Nu}\rangle \approx A\,\text{Pe}^{1/3}$ (creeping flow; large Pe)

Other cases are discussed by Lightfoot[15] and by Stewart[29] in an excellent review paper. Figure 10-21 and Table 10-5 summarize results for common shapes.

For simultaneous mass, heat, and momentum transport[4] the results depend on whether mass is being transferred into or out of the phase by condensation or evaporation, thereby changing the boundary-layer thickness.

10-14 FREE OR NATURAL CONVECTION

Free or natural convection occurs when density variations are present within a fluid as a result of spatial variations in temperature or concentration. Thus, the gravity term in the equation of motion is no longer constant but varies with position. Let T_R be a reference temperature, e.g., the mean temperature or ambient temperature. Then if the fluid is at rest and at T_R throughout, the EOM is

$$\nabla p = \rho_R g \tag{10-147}$$

where ρ_R is the density at T_R.

Since the velocity in such a system is usually small, we can assume that Eq. (10-147) will also hold when the temperature is not constant. Substituting it into the EOM gives

$$\rho \frac{D\mathbf{v}}{Dt} = -\nabla \cdot \underline{\tau} - \rho_R \mathbf{g} + \rho \mathbf{g} \tag{10-148}$$

Also it is assumed that, as a first approximation,

$$\rho = \rho_R + \left(\frac{\partial \rho}{\partial T}\right)_p \bigg|_{T_R} (T - T_R) \tag{10-149}$$

The coefficient of volume expansion is defined as

$$\beta \equiv \frac{1}{\hat{\mathcal{V}}}\left(\frac{\partial \hat{\mathcal{V}}}{\partial T}\right)_p = \frac{1}{1/\rho}\left[\frac{\partial(1/\rho)}{\partial T}\right]_p = -\frac{1}{\rho}\left(\frac{\partial \rho}{\partial T}\right)_p \tag{10-150}$$

For an ideal gas, $\beta = T^{-1}$. Substituting Eqs. (10-149) and (10-150) into Eq. (10-148) gives

$$\rho_R \frac{D\mathbf{v}}{Dt} = -\nabla \cdot \underline{\tau} - \rho_R \beta_R g (T - T_R) \tag{10-151}$$

where it is assumed that only the gravity force is effected by the density change. (This is called the *Boussinesq approximation.*)

Dimensional Analysis

At steady state, assuming constant μ, the EOM becomes

$$\rho_R \mathbf{v} \cdot \nabla \mathbf{v} = \mu \nabla^2 \mathbf{v} - \rho_R \beta_R g (T - T_R) \tag{10-152}$$

The energy equation is

$$\rho \hat{C}_p \mathbf{v} \cdot \nabla T = k \nabla^2 T \tag{10-153}$$

if viscous dissipation is neglected. We can write these equations in dimensionless form if we define a characteristic velocity $\mathbf{v}_c$ and characteristic distance or length L_c. Letting

$$\mathbf{v}^* \equiv \frac{\mathbf{v}}{v_c} \qquad \nabla^* \equiv L_c \nabla \qquad T^* = \frac{T - T_R}{T_1 - T_R} \tag{10-154}$$

leads to

$$\rho_R v_c^2 \mathbf{v}^* \cdot \frac{\nabla^* \mathbf{v}^*}{L_c} = \frac{\mu}{L_c^2} \nabla^{*2} \mathbf{v}^* v_c - \rho_R \beta_R g(T_1 - T_R) T^* \tag{10-155}$$

or

$$\mathbf{v}^* \cdot \nabla^* \mathbf{v}^* = \frac{\mu}{L_c v_c \rho_R} \nabla^{*2} \mathbf{v}^* - \frac{L_c \beta_R g(T_1 - T_R)}{v_c^2} T^* \tag{10-156}$$

Since there is no physical velocity (such as v_a) to serve, we let $v_c = \nu_R / L_c$ and $L_c = L$, so that

$$\mathbf{v}^* \cdot \nabla \mathbf{v}^* = \nabla^{*2} \mathbf{v}^* - \frac{L^3 \beta_R g(T_1 - T_R)}{\nu_R} T^* = \nabla^{*2} \mathbf{v}^* - \text{Gr}\, T^* \tag{10-157}$$

where

$$\text{Gr} \equiv \frac{L^3 \beta g(T_1 - T_R)}{\nu^2} \tag{10-158}$$

is the Grashof number. (The subscript R has been dropped for convenience.) Thus

$$v^* = v^*(x^*, y^*; \text{Gr}, T^*) \tag{10-159}$$

Similarly the energy equation gives

$$\rho \widehat{C}_p v_c \mathbf{v}^* \cdot \frac{\nabla^*}{L} T^*(T_1 - T_R) = \frac{k}{L^2} \nabla^{*2} T^* (T_1 - T_R) \tag{10-160}$$

or

$$\mathbf{v}^* \cdot \nabla^* T^* = \frac{\alpha}{v_c L} \nabla^* T^* = \frac{1}{\text{Pr}} \nabla^{*2} T^* \tag{10-161}$$

Then

$$T^* = T^*(x^*, y^*, v^*; \text{Pr})$$

Using Eq. (10-159) gives

$$v^* = v^*(x^*, y^*; \text{Gr}, \text{Pr}) \qquad \text{and} \qquad T^* = T^*(x^*, y^*; \text{Gr}, \text{Pr}) \tag{10-162}$$

Then the Nusselt number

$$\langle (\text{Nu})_L \rangle \equiv \frac{\langle h \rangle L}{k} = \frac{q_w L}{(T_1 - T_R) k} = -\frac{\left.\dfrac{\partial T}{\partial y}\right|_0 L}{T_1 - T_R} = -\left.\frac{\partial T^*}{\partial y^*}\right|_0 \tag{10-163}$$

or

$$\langle (\text{Nu})_L \rangle = \langle (\text{Nu})_L \rangle (\text{Gr}, \text{Pr}) \tag{10-164}$$

i.e., the Nusselt number depends on Gr and Pr.

Example 10-9: Free convection from a vertical flat plate A flat vertical plate of height L and width B is maintained at a temperature T_0 while the surrounding air is at temperature T_1. Since $T_0 > T_1$, the fluid near the plate will be rising due to the bouyancy force (see Fig. 10-22). Write the equations of change and find the rate of heat transfer from the plate and the Nusselt numbers.

SOLUTION Let y be the distance from the plate and x the distance up along the plate. Then the equations of change are

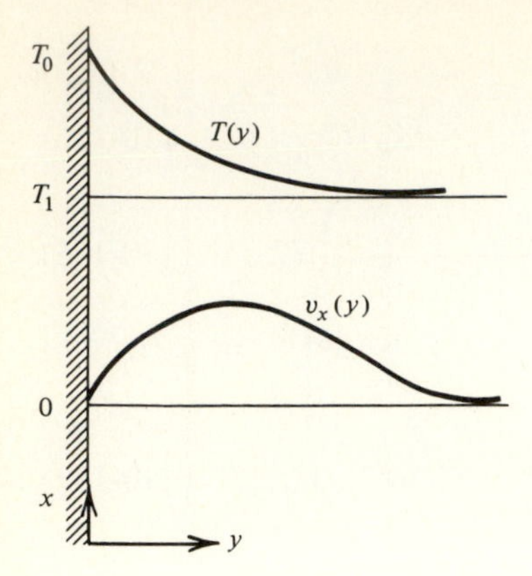

Figure 10-22 Velocity and temperature profiles for free convection to a vertical plate. T and v_x also vary with x.

$$\frac{\partial v_x}{\partial x} + \frac{\partial v_y}{\partial y} = 0 \tag{10-165}$$

$$\rho\left(v_x \frac{\partial v_x}{\partial x} + v_y \frac{\partial v_x}{\partial y}\right) = \mu \frac{\partial^2 v_x}{\partial y^2} + \rho_R g \frac{T - T_1}{T_1} \tag{10-166}$$

$$\rho \widehat{C}_p \left(v_x \frac{\partial T}{\partial x} + v_y \frac{\partial T}{\partial y}\right) = k \frac{\partial^2 T}{\partial x^2} \tag{10-167}$$

where viscous dissipation, axial momentum, and heat diffusion are neglected and $\beta_R = T_1^{-1}$ and $T_R \equiv T_1$. Pohlhausen[22] solved the above equations using a similarity transformation and the stream function. He let

$$\eta \equiv \frac{Cy}{x^{1/4}} \qquad \psi \equiv 4\nu C x^{3/4} f(\eta) \qquad C \equiv \left[\frac{g(T_0 - T_1)}{4\nu^2 T_1}\right]^{1/4} \qquad T^* \equiv \frac{T - T_1}{T_0 - T_1} \tag{10-168}$$

This gave ordinary differential equations for $T^*(\eta;\ \mathrm{Pr})$ and $f(\eta)$ which were solved numerically with the boundary conditions

$$f(0) = 0 = f'(0) \qquad f'(\infty) = 0 \qquad T^*(\infty) = 0 \tag{10-169}$$

The results agreed well with experimental data and indicated that both the velocity and thermal-boundary-layer thickness vary with $x^{1/4}$.

The heat-flow rate at the wall for the entire plate of width B is

$$\dot{Q}_{wt} = B \int_0^L q_0(x)\, dx = -kBC(T_0 - T_1) \frac{\partial T^*}{\partial \eta}\bigg|_0 \int_0^L x^{-1/4}\, dx$$
$$= \tfrac{4}{3} \Gamma B L^{3/4} C k (T_0 - T_1)$$

where $\Gamma \equiv (\partial T^*/\partial \eta)|_0 = \Gamma(\mathrm{Pr})$. For air ($\mathrm{Pr} = 0.733$), $\Gamma = -0.508$.

The mean Nusselt number at $x = L$ is then

$$\langle(\mathrm{Nu})_L\rangle = \frac{\dot{Q}_{wt}}{Bk(T_0 - T_1)} = \tfrac{4}{3}\Gamma L^{3/4} C k = \tfrac{4}{3}\Gamma \left[\frac{L^3 g(T_0 - T_1)}{\nu^2 T_1}\right]^{1/4} = \tfrac{4}{3}\Gamma (\mathrm{Gr})^{1/4} \tag{10-170}$$

where $$\mathrm{Gr} \equiv \frac{L^3 g(T_0 - T_1)}{\nu^2 T_1} \equiv 4\, CL^3 \tag{10-171}$$

is the Grashof number, usually written

$$\mathrm{Gr} \equiv \frac{gL^3(T_0 - T_1)\beta}{\nu^2} \tag{10-172}$$

For air,

$$\langle(\mathrm{Nu})_L\rangle = 0.478(\mathrm{Gr})^{1/4} \tag{10-173}$$

The above holds for laminar flow for which $\mathrm{Gr\,Pr} < 10^9$. □

Vertical Cylinders

When the data for several fluids and for vertical cylinders as well as plates are plotted, the following correlations are obtained[9]

$$\langle(\mathrm{Nu})_L\rangle = \begin{cases} 0.555(\mathrm{Ra})^{1/4} & \mathrm{Ra} < 10^9 \quad (10\text{-}174) \\ 0.0210(\mathrm{Ra})^{2/5} & \mathrm{Ra} > 10^9 \quad (10\text{-}175) \end{cases}$$

where $\mathrm{Ra} \equiv \mathrm{Gr\,Pr}$ is the Rayleigh number. When $\mathrm{Ra} \approx 10^9$, there is a transition from laminar to turbulent flow.

Combined Forced and Free Convection

We have given equations for pure free convection in this section and for pure forced convection in earlier sections; however, in many practical situations, both forced and free convection occur. Figure 10-23 shows the various regimes.

Mass Transfer

For free-convection mass transfer, analogous equations can be written. The mass-transfer Grashof number is

where $$(\mathrm{Gr})_{\mathrm{AB}} \equiv \frac{L_c^3 \zeta \rho_g^2 (x_{\mathrm{A}1} - x_{\mathrm{A}R})}{\mu^2} \qquad \zeta_{\mathrm{A}} \equiv -\frac{1}{\rho}\left(\frac{\partial \rho}{\partial x_{\mathrm{A}}}\right)_T \tag{10-176}$$

is the coefficient of volume expansion due to mass-fraction gradients.

Cellular Convection†

Cellular convection occurs when the Rayleigh number is large. Consider a fluid confined between two large parallel horizontal plates.[20,23] We let $z = 0$ be the midplane. The upper plate at $z = -L/2$ is maintained at T_1 while the lower plate is maintained at a temperature T_2 where $T_2 > T_1$. As a result of the temperature difference, the fluid will tend to circulate, forming "roll" cells that have a certain

† R. Narayanan collaborated on this subsection.

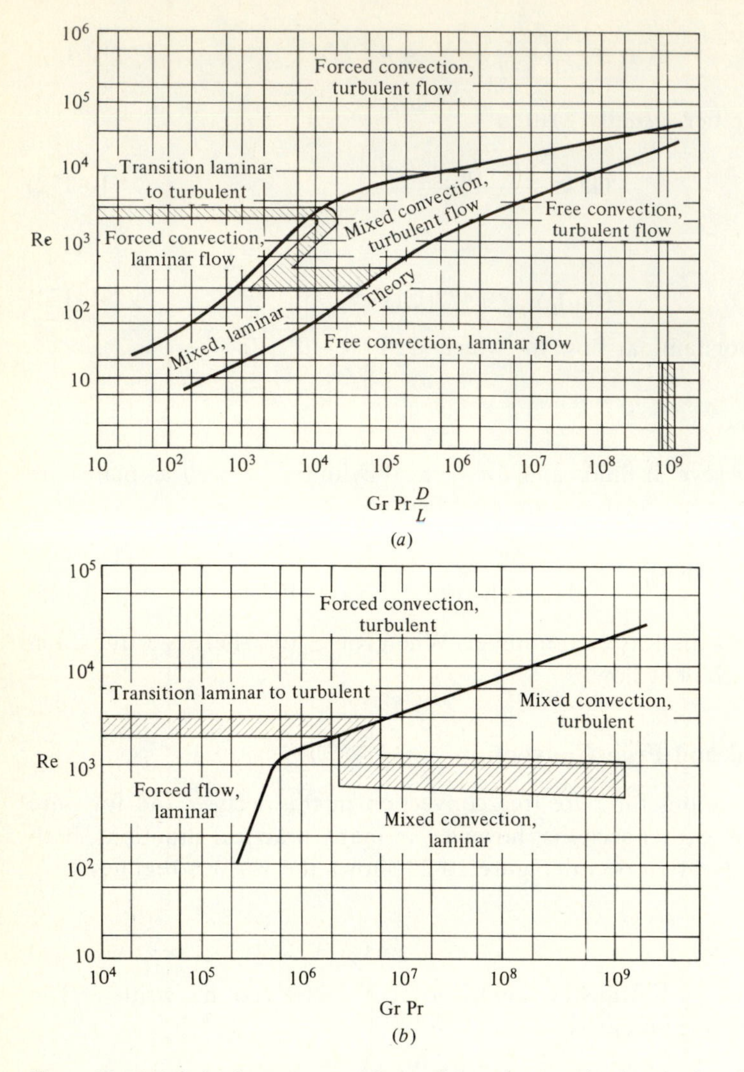

Figure 10-23 Relative importance of free and forced convection in tube flow for $10^{-2} < \mathrm{Pr}\, D/L < 1$. [*From B. Metais and E. R. G. Eckert, J. Heat Trans.*, **86:**295 (1964).]

spacing between them. Since the plates can be assumed to be infinite in the x and y directions, it is convenient to use cylindrical coordinates and to consider two-dimensional axisymmetric motion. We wish to find the cell spacing and the equation $r(z)$ for a streamline as well as the Ra for which convection begins.

Let us also assume that

1. The fluid is newtonian, with a high Prandtl number.
2. The density change with respect to temperature affects only the body-force term.
3. All other fluid properties are constant.

We will use the dimensionless variables

$$T^* \equiv \frac{T - T_2}{T_1 - T_2} \qquad \nabla^* \equiv L\nabla \qquad \mathcal{P}^* = \frac{(\mathcal{P} - \mathcal{P}_1)L^2}{\rho_2 \alpha \nu} \qquad \mathbf{v}^* = \frac{\mathbf{v}}{\alpha/L} \qquad t^* = \frac{t}{t_c}$$

$$\text{Ra} = \frac{L^3(-\beta_2)(T_2 - T_1)g}{\nu\alpha} = \text{Rayleigh number}$$

$$\text{Pr} = \frac{\widehat{C}_p \mu}{k} = \text{Prandtl number}$$

where β_2 = coefficient of volume expansion evaluated at lower plate
ρ_2 = density evaluated at lower plate temperature
$\mathcal{P}$ = dynamic pressure
t_c = characteristic time $= L^2/\alpha$

The equations of change are

Continuity:
$$\nabla^* \cdot \mathbf{v}^* = 0 \tag{10-177}$$

Motion:
$$\frac{1}{\text{Pr}}\frac{\partial v_r^*}{\partial t^*} + \frac{1}{\text{Pr}}\left(v_r^* \frac{\partial v_r^*}{\partial z^*} + v_z^* \frac{\partial v_r^*}{\partial z^*}\right) = \frac{-\partial \mathcal{P}^*}{\partial r^*} + \nabla^{*2} v_r^* \tag{10-178}$$

$$\frac{1}{\text{Pr}}\frac{\partial v_r^*}{\partial t^*} + \frac{1}{\text{Pr}}\left(v_r^* \frac{\partial v_r^*}{\partial r^*} + v_z^* \frac{\partial v_z^*}{\partial z^*}\right) = \frac{-\partial \mathcal{P}^*}{\partial z^*} + \nabla^{*2} v_z^* + \text{Ra}\, T^* \tag{10-179}$$

Energy:
$$\frac{\partial T^*}{\partial t^*} + \mathbf{v}^* \cdot \nabla^* T^* = \nabla^{*2} T^* \tag{10-180}$$

The assumption of large Pr will cause the left-hand sides of Eqs. (10-178) and (10-179) to be dropped. Then by cross differentiation and using the equation of continuity we can show that Eq. (10-179) reduces to

$$0 = \text{Ra}\left(\frac{\partial^2 T^*}{\partial r^{*2}} + \frac{1}{r^*}\frac{\partial T^*}{\partial r^*}\right) + \nabla^{*4} v_z^* \tag{10-181}$$

To linearize we let

$$T^* = T_0^* + \theta \qquad \text{and} \qquad v_z^* = v_0^* + u = u \tag{10-182}$$

where T_0^* and v_0^* represent the profile in the quiescent state (no flow) and θ and u are small deviations from it. We substitute Eq. (10-182) into Eqs. (10-181) and (10-180) and neglect the product of all deviation terms. Then

$$0 = \text{Ra}\left(\frac{\partial^2 \theta}{\partial r^{*2}} + \frac{1}{r^*}\frac{\partial \theta}{\partial r^*}\right) + \nabla^{*4} u \tag{10-183}$$

$$\frac{\partial \theta}{\partial t^*} - u = \frac{\partial^2 \theta}{\partial r^{*2}} + \frac{1}{r^*}\frac{\partial \theta}{\partial r^*} + \frac{\partial^2 \theta}{\partial z^{*2}} \tag{10-184}$$

where we have used the fact that $\partial T_0^*/\partial z^* = -1$. The boundary conditions at $z^* \pm \frac{1}{2}$ are

$$u = 0 \qquad \text{no flow} \tag{10-185}$$

$$\frac{\partial u}{\partial z^*} = 0 \qquad \text{no slip and use of continuity} \tag{10-186}$$

$$\theta = 0 \qquad \text{temperature deviations vanish} \tag{10-187}$$

Now let us use a separation-of-variables method. We assume that

$$\theta = \hat{\theta}(z^*)f(r^*)e^{\lambda t^*} \qquad \text{and} \qquad u = \hat{u}(z^*)f(r^*)e^{\lambda t^*} \tag{10-188}$$

θ and u must have the same functionality in r^*. If we apply Eqs. (10-188) to Eq. (10-184), we obtain

$$\left[\frac{d^2}{dz^{*2}} + \left(\frac{d^2}{dr^{*2}} + \frac{1}{r^*}\frac{d}{dr^*}\right) - \lambda\right]f\hat{\theta} = -f\hat{u} \tag{10-189}$$

By further decomposition we can show that

$$\left(\frac{d^2}{dz^{*2}} - a^2 - \lambda\right)\hat{\theta} = -\hat{u} \tag{10-190}$$

$$\left(\frac{d^2}{dr^{*2}} + \frac{1}{r^*}\frac{d}{dr^*}\right)f + a^2 f = 0 \tag{10-191}$$

It can be shown that λ is real. We would like to know the values of Ra for which we get a cellular or secondary flow; i.e., for which value of Ra does $\lambda = 0$? This represents no growth and no decay of the deviation variables and gives the critical $\Delta T = T_2 - T_1$ for which convection begins.

Therefore set $\lambda = 0$; after some algebra Eqs. (10-183) and (10-190) become

$$\left(\frac{d^2}{dz^{*2}} - a^2\right)\hat{\theta} = -\hat{u} \qquad \left(\frac{d^2}{dz^{*2}} - a^2\right)^2\hat{u} = a^2\,\text{Ra}\,\hat{\theta} \tag{10-192}$$

which can be combined to give

$$\left(\frac{d^2}{dz^{*2}} - a^2\right)^3\hat{u} = -a^2\,\text{Ra}\,\hat{u} \tag{10-193}$$

and the boundary conditions become

$$\hat{u} = 0 = \frac{d\hat{u}}{dz^*} \qquad \text{at } z^* = \pm\tfrac{1}{2} \tag{10-194}$$

Use of $\hat{\theta} = 0$ and the second of Eqs. (10-192) gives

$$\left(\frac{d^2}{dz^{*2}} - a^2\right)^2\hat{u} = 0 \qquad \text{at } z^* = \pm\tfrac{1}{2} \tag{10-195}$$

Equation (10-193) is a homogeneous equation subject to homogeneous-condition equations (10-194) and (10-195). Clearly, $\widehat{u} = 0$ is a solution, but other solutions may exist for particular values of Ra. There are, in fact, infinitely many values of Ra which yield nonzero $\widehat{u}$. Further consideration of Eq. (10-193) and a numerical computation[20] yield the least value of Ra for which $\widehat{u}$ is nonzero. This is equal to 1708 and corresponds to $a = 3.12$.

The corresponding solution to u is

$$u(r^*, z^*) = (\cos g_0 z^* - 0.06152 \cosh g_1 z^* \cos g_2 z^* + 0.104 \sinh g_1 z^* \sin g_2 z^*) f(r^*) \qquad (10\text{-}196)$$

where $\quad g_0 = 3.9 \quad g_1 = 5.2 \quad g_2 = 2.13 \quad f(r^*) = J_0(ar^*)$

Using the continuity equation, we can obtain v_r^* (and also ψ) and can show that

$$\psi = -u(r^*, z^*) \frac{r^* J_1(ar^*)}{af(r^*)} \qquad \text{where} \qquad v_r^* = \frac{1}{r^*}\frac{\partial \psi}{\partial z^*} \qquad v_z^* = \frac{-1}{r^*}\frac{\partial \psi}{\partial r^*} \qquad (10\text{-}197)$$

The streamlines for the roll cells are obtained by setting ψ equal to a constant and obtaining $r^*(z^*)$ such that Eq. (10-197) is satisfied.

The roll-cell boundaries occur when $\psi = 0$ or, equivalently for a fixed z^*, when $r^* = \alpha_n / a$, where α_n are roots of $J_1(\alpha_n) = 0$.

APPENDIX 10-1 DIFFERENTIATING AN INTEGRAL (LEIBNITZ EQUATION)

One Dimension

Consider an integral

$$\int_{a(t)}^{b(t)} f(x, t)\, dx \equiv I(t) \qquad (\text{A}10\text{-}1\text{-}1)$$

whose limits are a function of an independent variable such as time t. Suppose we wish to find the derivative dI/dt of the integral I. An equation derived by Leibnitz is

$$\frac{dI}{dt} \equiv \frac{d}{dt}\int_{a(t)}^{b(t)} f(x, t)\, dx = \int_{a(t)}^{b(t)} \frac{\partial f}{\partial t}\, dx + f(b)\frac{db}{dt} - f(a)\frac{da}{dt} \qquad (\text{A}10\text{-}1\text{-}2)$$

It can be derived by geometric means as follows.

Consider first the case in which a and b are constant (see Fig. A10-1-1). The integral I at a given time t is the area under the curve $f(x, t)$ between the limits a and b. But at time $t + \Delta t$ the function $f(x, t)$ has become the new curve $f(t + \Delta t)$. The change in the integral ΔI_1 is the hatched area between the curves $f(x, t + \Delta t)$

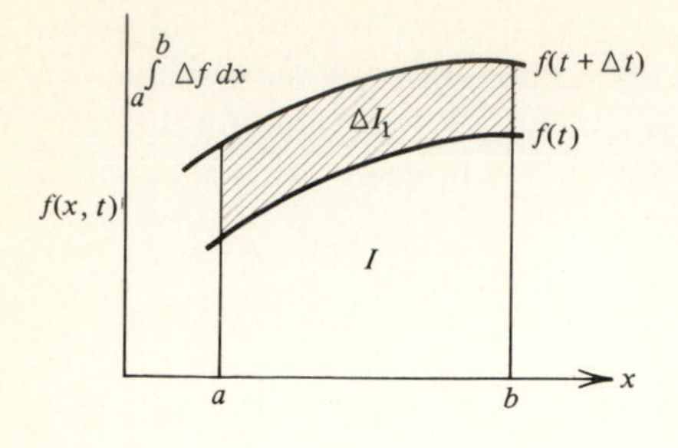

Figure A10-1-1 Change in interval for constant limits.

and $f(x, t)$, or

$$\Delta I_1 = \int_a^b [f(x, t + \Delta t) - f(x, t)]\,dx = \int_a^b \Delta f\,dx \qquad \text{(A10-1-3)}$$

Dividing by Δt and taking the limit as $\Delta t \to 0$ gives

$$\frac{dI_1}{dt} = \int_a^b \frac{\partial f}{\partial t}\,dx \qquad \text{(A10-1-4)}$$

Now consider the case where, during the time Δt, the limit $a(t)$ changes by an amount $\Delta a = a(t + \Delta t)$ and the limit $b(t)$ changes by an amount $\Delta b \equiv b(t + \Delta t) - b(t)$ (Fig. A10-1-2). The change in b results in an increase ΔI_2 in the area under the curve, and this increase is approximated by

$$\Delta I_2 = f(b)\,\Delta b + \Delta f(b)\,\Delta b \qquad \text{(A10-1-5)}$$

where the first term is the hatched area in Fig. A10-1-2 and the second term is the crosshatched area. Since the latter is a difference of higher order, it can be neglected and

$$\Delta I_2 = f(b)\,\Delta b$$

The change in a results in a decrease ΔI_3 in the area under the curve approximated by

$$\Delta I_3 = f(a)\,\Delta a + \Delta f(a)\,\Delta a \qquad \text{(A10-1-6)}$$

Again the latter term (crosshatched) is of higher order and can be neglected. Then the total change in the integral I is

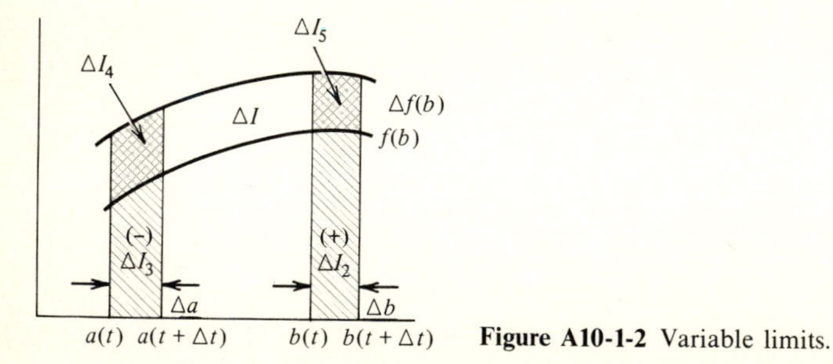

Figure A10-1-2 Variable limits.

$$\Delta I = \Delta I_1 + \Delta I_2 - \Delta I_3 = \int_a^b \Delta f \, dx + f(b)\,\Delta b - f(a)\,\Delta a$$

Dividing by Δt and letting it approach zero gives Eq. (A10-1-2). The derivative is then given by

$$\frac{dI}{dt} = \lim_{\Delta t \to 0} \frac{\Delta I}{\Delta t}$$

or
$$\frac{dI}{dt} = \int_{a(t)}^{b(t)} \frac{\partial f}{\partial t} dx + f(b)\frac{db}{dt} - f(a)\frac{da}{dt} \tag{A10-1-7}$$

Three Dimensions

We now consider an integral over a volume $\mathcal{V}$ that is changing with time, or

$$\iiint_{\mathcal{V}(t)} f(t)\, d\mathcal{V} \equiv I(t) \tag{A10-1-8}$$

We wish to evaluate dI/dt. For purposes of illustration we assume that the volume $\mathcal{V}(t)$ is a parallelepiped. Consider the faces perpendicular to the x axis (Fig. A10-1-3): One face is at $x = a$, and a parallel face is at $x = b$. The area of each face is A_x, which, for the moment, we will take as constant; i.e., the volume is changing with time (as in the one-dimensional case above). The rate at which the face at $x = b$ is moving is db/dt, and the rate at which the face at $x = a$ is moving is da/dt. We let w_x represent the x component of the velocity of the surface A_x

$$w_x\Big|_{x=b} = \frac{db}{dt} \quad \text{and} \quad w_x\Big|_{x=a} = \frac{da}{dt} \tag{A10-1-9}$$

Since the differential volume is $d\mathcal{V} = A_x\, dx$, we have

$$\frac{dI_x}{dt} = \frac{d}{dt}\int_{A_x}\int\int_a^b f(x, y, z, t)\, dx\, dA_x = \left[\frac{d}{dt}\int_a^b f(x, t)\, dx\right] A_x$$

$$= \left(\int_a^b \frac{\partial f}{\partial t} dx\right) A_x + f(b)\frac{db}{dt} A_x - f(a)\frac{da}{dt} A_x \tag{A10-1-10}$$

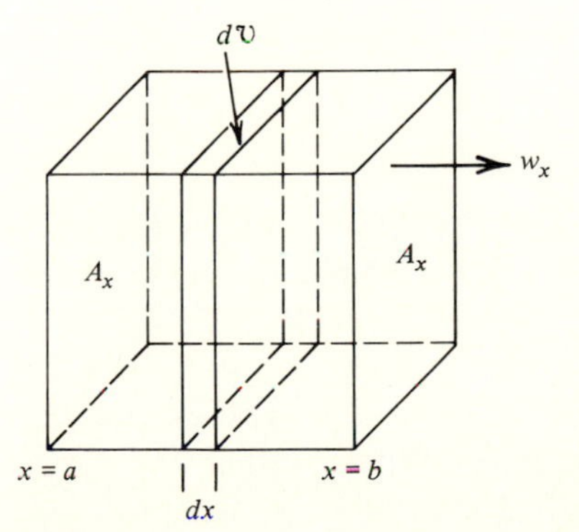

Figure A10-1-3 Change of volume with time.

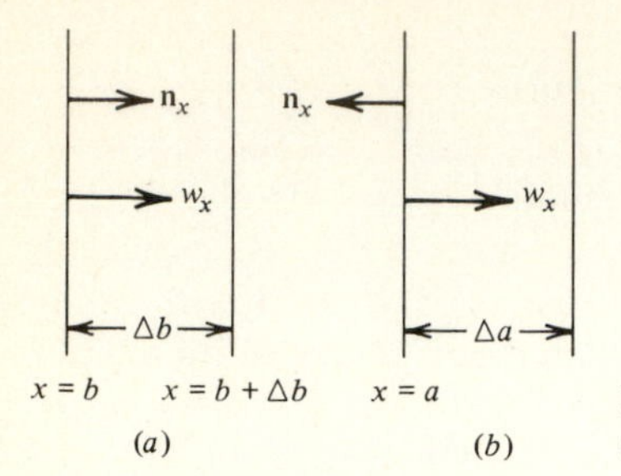

Figure A10-1-4 Change of volume (*a*) due to change in *b* and (*b*) due to change in *a*.

Now (Fig. A10-1-4*a*)

$$\frac{db}{dt}A_x = w_x A_x\bigg|_{x=b} = (\mathbf{w}_x \cdot \mathbf{S}_x)\bigg|_{x=b}$$

where

$$\mathbf{S}_x = \mathbf{n}_x A_x \qquad \mathbf{w}_x = w_x \boldsymbol{\delta}_x$$

Also (Fig. A10-1-4*b*)

$$\frac{da}{dt}A_x = w_x A_x\bigg|_{x=a} = -\mathbf{w}_x \cdot \mathbf{A}_x\bigg|_{x=a}$$

since $\boldsymbol{\delta}_x \cdot \mathbf{n}_x = -1$ at $x = a$. Then we can write Eq. (A10-1-10) more generally

$$\frac{dI_x}{dt} + \left(\int \frac{\partial f}{\partial t} dx\right) A_x + \int_{A_x}\int f\mathbf{w}_x \cdot d\mathbf{S}_x \qquad \text{(A10-1-11}a\text{)}$$

If we extend this procedure to the y and z directions, we obtain

$$\frac{d}{dt}\iiint_{\mathcal{V}(t)} f\, d\mathcal{V} = \iiint \frac{\partial f}{\partial t} d\mathcal{V} + \iint_{A(t)} f\mathbf{w} \cdot \mathbf{n}\, dA \qquad \text{(A10-1-11}b\text{)}$$

where $\mathbf{n}$ is the outwardly directed normal to surface $d\mathbf{S}$ moving with velocity $\mathbf{w}$. We note that $\mathbf{w} \cdot \mathbf{n}\, dA$ represents the rate at which a differential volume is swept out by the movement of an area dA with velocity $\mathbf{w}$.

If the surface moves with the velocity of a stream, $\mathbf{w} = \mathbf{v}$ and

$$\frac{D}{Dt}\iiint f\, d\mathcal{V} = \iiint \frac{\partial f}{\partial t} d\mathcal{V} + \iint f\mathbf{v} \cdot \mathbf{n}\, dA \qquad \text{(A10-1-12)}$$

which is known as the *Reynolds transport theorem* or the Leibnitz equation in three dimensions. It also applies if f is a vector or a tensor.

APPENDIX 10-2 DERIVATION OF EQUATIONS OF CHANGE USING VECTOR-TENSOR NOTATION

We have derived the equation of motion (9-32) and the total-energy equation (10-7) by using an element of volume in a cartesian coordinate system. The partial differential equations in terms of cartesian coordinate components were then written in vector-tensor form by using methods for expanding vector expressions (Appendix 8-2). It is important to note, however, that a vector-tensor equation must apply to *any* coordinate system since it describes the physical processes occurring within the fluid in terms of certain laws of nature, which are independent of the coordinate system selected. Therefore, it should be, and is, possible to derive these equations for an arbitrary element of volume by using vector expressions for the various physical quantities. In order to do this it is convenient to adopt the lagrangian point of view and to use two mathematical relationships: (1) the Gauss divergence theorem (Chap. 8) and (2) the Leibnitz equation (or Reynolds transport theorem) for differentiating an integral. The latter expression, derived in Appendix 10-1, is

$$\frac{D}{Dt}\iiint f\,d\mathcal{V} = \iiint \frac{\partial f}{\partial t}\,d\mathcal{V} + \iint f\mathbf{v}\cdot\mathbf{n}\,dA \qquad \text{(A10-2-1)}$$

where f may be a vector, tensor, or scalar.

Application of Transport Theorem to Momentum Transport

Consider an element of volume of constant mass moving with the velocity of the stream. A momentum-force balance gives [see Eq. (9-1)] the equation

$$\frac{1}{g_c}\frac{D}{Dt}\iiint \rho\mathbf{v}\,d\mathcal{V} = \iint \mathbf{t}_{(n)}\,dA + \frac{1}{g_c}\iiint \rho\mathbf{g}\,d\mathcal{V} \qquad \text{(A10-2-2)}$$

called *Cauchy's momentum equation*. Physically it says that the linear momentum of a system moving with the fluid velocity will increase with time as a result of surface forces and body forces acting on the system.

Using the Reynolds transport theorem, the term on the left can be written

$$\frac{1}{g_c}\iiint \frac{\partial(\rho\mathbf{v})}{\partial t}\,d\mathcal{V} + \frac{1}{g_c}\iint \rho\mathbf{v}\mathbf{v}\cdot\mathbf{n}\,dA \qquad \text{(A10-2-3)}$$

The first term on the right is

$$\iint \mathbf{t}_{(n)}\,dA = \iint \mathbf{n}\cdot\underline{\mathbf{T}}\,dA \qquad \text{(A10-2-4)}$$

Using the divergence theorem, both area integrals can be written in terms of volume integrals. Then Eq. (A10-2-2) becomes

$$\frac{1}{g_c}\iiint \frac{\partial(\rho\mathbf{v})}{\partial t}\,d\mathcal{V} + \frac{1}{g_c}\iiint \nabla\cdot\rho\mathbf{v}\mathbf{v}\,d\mathcal{V} = \iiint \nabla\cdot\underline{\mathbf{T}}\,d\mathcal{V} + \frac{1}{g_c}\iiint \rho\mathbf{g}\,d\mathcal{V} \qquad \text{(A10-2-5)}$$

Since the volume element is arbitrary,

$$\frac{1}{g_c}\frac{\partial(\rho\mathbf{v})}{\partial t} + \frac{1}{g_c}\nabla\cdot\rho\mathbf{v}\mathbf{v} = \nabla\cdot\underline{\mathbf{T}} + \frac{\rho\mathbf{g}}{g_c} \qquad \text{(A10-2-6)}$$

Using $\underline{\mathbf{T}} = -\underline{\boldsymbol{\delta}}p - \underline{\boldsymbol{\sigma}}$ gives

$$\frac{1}{g_c}\frac{\partial(\rho\mathbf{v})}{\partial t} + \frac{1}{g_c}\nabla\cdot\rho\mathbf{vv} = -\nabla p - \nabla\cdot\underline{\boldsymbol{\sigma}} + \frac{\rho\mathbf{g}}{g_c} \qquad \text{(A10-2-7)}$$

Furthermore, since $\underline{\boldsymbol{\sigma}} = \underline{\boldsymbol{\tau}}/g_c$,

$$\boxed{\frac{\partial(\rho\mathbf{v})}{\partial t} + \nabla\cdot\rho\mathbf{vv} = -g_c\nabla p - \nabla\cdot\underline{\boldsymbol{\tau}} + \rho\mathbf{g}} \qquad \text{(A10-2-8)}$$

We can recognize the term $\nabla\cdot\rho\mathbf{vv}$ as representing the net bulk transport of momentum.

We generally use an absolute system of units (for example, SI) in which $g_c = 1.0$ and $\underline{\boldsymbol{\tau}} = \underline{\boldsymbol{\sigma}}$.

Total-Energy Equation

Again consider an element of fluid of constant mass moving with the velocity of the stream. From the balance law for energy

$$\underbrace{\frac{D}{Dt}\iiint \rho\hat{E}'\,d\mathcal{V}}_{\substack{\text{Rate of change}\\ \text{of internal and}\\ \text{kinetic energy}}} = \underbrace{-\iint \mathbf{q}\cdot d\mathbf{S}}_{\substack{\text{Rate of}\\ \text{absorption}\\ \text{of heat}}} + \underbrace{\iint \mathbf{t}_{(\mathbf{n})}\cdot\mathbf{v}\,dA}_{\substack{\text{Rate work}\\ \text{is done by}\\ \text{surface forces}}} + \underbrace{\iiint \rho\mathbf{v}\cdot\mathbf{g}\,d\mathcal{V}}_{\substack{\text{Rate work}\\ \text{is done by}\\ \text{body forces}}} + \underbrace{\iiint \phi_H\,d\mathcal{V}}_{\substack{\text{Production}\\ \text{of}\\ \text{heat energy}}} \qquad \text{(A10-2-9)}$$

where $\hat{E}' \equiv \hat{U} + \hat{K}$
$\hat{U}$ = internal energy per unit mass
$\hat{K} = \frac{1}{2}v^2$ = kinetic energy per unit mass

We do not include potential energy in $\hat{E}'$ because it is accounted for in the term $\mathbf{v}\cdot\mathbf{g}$, which is the rate work is done by the body forces such as gravity.

The rate at which work is done by the viscous and pressure forces is given by the second term on the right, which can be written

$$\iint \mathbf{t}_{(n)}\cdot\mathbf{v}\,dA = \iint \mathbf{n}\cdot\underline{\mathbf{T}}\cdot\mathbf{v}\,dA = -\iint p\mathbf{n}\cdot\mathbf{v}\,dA - \iint \mathbf{n}\cdot\underline{\boldsymbol{\sigma}}\cdot\mathbf{v}\,dA \qquad \text{(A10-2-10)}$$

Application of the Reynolds transport theorem to the left-hand side of Eq. (A10-2-9) gives

$$\frac{D}{Dt}\iiint \rho\hat{E}'\,d\mathcal{V} = \iiint \frac{\partial(\rho\hat{E}')}{\partial t}d\mathcal{V} + \iint \rho\hat{E}'\mathbf{v}\cdot\mathbf{n}\,dA \qquad \text{(A10-2-11)}$$

If we convert all area integrals into volume integrals, we get

$$\iiint\left[\frac{\partial(\rho\hat{E}')}{\partial t} + \nabla\cdot\rho\hat{E}'\mathbf{v} - \nabla\cdot\mathbf{q} - \nabla\cdot p\mathbf{v} - \nabla\cdot\underline{\boldsymbol{\sigma}}\cdot\mathbf{v} + \rho\mathbf{v}\cdot\mathbf{g} + \Phi_H\right]d\mathcal{V} = 0$$

Since the volume is arbitrary, and since $\widehat{E}' \equiv \widehat{U} + v^2/2$, we have

$$\frac{\partial}{\partial t}\left[\rho\left(\widehat{U} + \frac{v^2}{2}\right)\right] + \nabla \cdot \left[\rho\left(\widehat{U} + \frac{v^2}{2}\right)\mathbf{v}\right] = -\nabla \cdot \mathbf{q} - \nabla \cdot p\mathbf{v} - \nabla \cdot \underline{\tau} \cdot \mathbf{v} + \rho\mathbf{v} \cdot \mathbf{g} + \Phi_H$$

Fixed-point accumulation of internal and kinetic energy	Bulk flow of internal and kinetic energy	Conduction	Pressure work	Viscous work	Gravity work	Generation

(A10-2-12)

APPENDIX 10-3 SCALING AND ORDERING THE ENERGY EQUATION

In Appendix 9-7 we discussed scaling and ordering the equation of motion in order to find the relative importance of each term. Our general procedure was to use characteristic values, or *scales,* that made the dimensionless variables obtained from them of the order of unity. When the equations were made dimensionless, dimensionless groups appeared as coefficient terms; small coefficients indicated that the term can be neglected since a small number times unity is a small number.

Let us now apply this procedure to the simple pipe heat exchange in Example 10-2. The energy equation is

$$\rho\widehat{C}_p\mathbf{v} \cdot \nabla T = -\nabla \cdot \mathbf{q} \qquad \text{or} \qquad \rho\widehat{C}_p v_z \frac{\partial T}{\partial z} = k\left[\frac{\partial(r\,\partial T/\partial r)}{r\,\partial r} + \frac{\partial^2 T}{\partial z^2}\right]$$

To determine when the axial-dispersion term can be neglected we let L be an axial length scale since z goes from 0 to L. Similarly we let $z^{**} \equiv z/L$ and $v^{**} \equiv v_z/v_M$, where v_M is the maximum velocity $2v_a$. Then

$$\frac{\rho\widehat{C}_p}{k} v_M v^{**} \frac{\partial T^*}{\partial z^{**}} \frac{T_0 - T_R}{L} = \frac{T_0 - T_R}{R^2}\left[\frac{\partial(r^*\,\partial T^*/\partial r^*)}{r^*\,\partial r^*} + \frac{R^2}{L^2}\frac{\partial^2 T^*}{\partial z^{**2}}\right]$$

or

$$\frac{2Rv_a}{2\alpha}\frac{R}{L}v^{**}\frac{\partial T^*}{\partial z^{**}} = \frac{\partial(r^*\,\partial T^*/\partial r^*)}{r^*\,\partial r^*} + \left(\frac{R}{L}\right)^2\frac{\partial^2 T^*}{\partial z^{**2}}$$

or

$$\underset{(1)}{\frac{1}{2}\text{Pe}\frac{R}{L}v^{**}\frac{\partial T^*}{\partial z^{**}}} = \underset{(2)}{\frac{\partial(r^*\,\partial T^*/\partial r^*)}{r^*\,\partial r^*}} + \underset{(3)}{\left(\frac{R}{L}\right)^2\frac{\partial^2 T^*}{\partial z^{**2}}}$$

or

$$\text{Pe}\frac{R}{L}v^*\frac{\partial T^*}{\partial z^{**}} = \frac{\partial(r^*\,\partial T^*/\partial r^*)}{r^*\,\partial r^*} + \left(\frac{R}{L}\right)^2\frac{\partial^2 T^*}{\partial z^{**2}}$$

Thus, term (3) will be small compared with term (2) when R/L is small. Multiplying through by $L/\text{Pe}\,R$ indicates that axial transport will be small compared with convective when the coefficient of $\partial^2 T^*/\partial z^{**2}$

$$\frac{R/L}{\text{Pe}} = \frac{(R/L)}{2Rv_a/\alpha} = \frac{\alpha}{2Lv_a} = \frac{1}{2(\text{Pe})_L}$$

is small, i.e., when v_a and L are large and α is small. Thus, the neglect of axial diffusion is less accurate near the entrance or for fluids of large α (liquid metals). A similar procedure can be used to determine when the viscous-dissipation term is negligible (see the Problems).

For turbulent flow a similar approach can be used if we let

$$\overline{q}_r^* \equiv \frac{R\overline{q}_r}{k(T_0 - T_R)} \equiv -\frac{R\overline{q}_y}{k(T_0 - T_R)}$$

$$\frac{\text{Pe}\, R\overline{v}^{**}}{2L}\frac{\partial \overline{T}^*}{\partial z^{**}} = \frac{\partial(r^*\overline{q}_r^*)}{r^*\,\partial r^*} \approx -\frac{\partial \overline{q}_y^*}{\partial y^*}$$

where curvature is neglected since we are primarily interested in the region near the wall. Then when Pe R/L is small, both sides of the equation are approximately zero and $\overline{q}_y^* = \overline{q}_0^*$ or $\overline{q}_y = \overline{q}_0 = \text{const}$, as assumed in Chap. 6.

APPENDIX 10-4 LAMINAR-FLOW HEAT TRANSFER WITH CONSTANT WALL TEMPERATURE

Solution to Graetz-Nusselt Problem

Since, for laminar flow, $v^* = 2(1 - r^{*2})$, the differential equation (10-38) can be written in dimensionless form (Appendix 10-3) as

$$\frac{\partial(r^*\,\partial T^*/\partial r^*)}{r^*\,\partial r^*} = (1 - r^{*2})\frac{\partial T^*}{\partial z^*}(\text{Pe})_H \equiv (1 - r^{*2})\frac{\partial T^*}{\partial \zeta} \tag{A10-4-1}$$

where $\zeta \equiv z^*/(\text{Pe})_H$ and $(\text{Pe})_H \equiv 2R\rho v_a \widehat{C}_p/k$ is the heat Peclet number. Using the separation-of-variables method, we let

$$T^*(r^*, \zeta) = X(r^*)Z(\zeta)$$

Then if we let primes stand for differentiation by the independent variable, Eq. (A10-4-1) gives

$$\frac{(r^*X'Z)'}{r^*} = (1 - r^{*2})XZ' \tag{A10-4-2}$$

If we divide by $(1 - r^{*2})XZ$, we find that the left-hand side will depend only on r^* and the right-hand side only on ζ; hence we can equate each to a constant, which we take as $-c^2$ (an eigenvalue)

$$\frac{(r^*X')'}{r^*(1 - r^{*2})X} = \frac{Z'}{Z} = -c^2$$

This gives the ordinary differential equations

$$(r^*X')' + c^2r^*(1 - r^{*2})X = 0 \tag{A10-4-3}$$

and

$$Z' + c^2Z = 0 \tag{A10-4-4}$$

whose solution is

$$Z = e^{-c^2\zeta} \tag{A10-4-5}$$

The solution to Eq. (A10-4-3) can be obtained in series by the Frobenius method. Moon and Spencer[18] let

$$X = \sum_0^\infty A_n r^{*n} \tag{A10-4-6}$$

The A_n can be related to A_0 by substitution of Eq. (A10-4-6) into Eq. (A10-4-3), or

$$X = \sum_0^\infty A_n r^{*n} = A_0 \left\{1 - \frac{c^2}{4}r^{*2} + \frac{1}{16}\left(c^2 + \frac{c^4}{4}\right)r^{*4} - \frac{1}{36(16)}\left(5c^2 + \frac{c^6}{4}\right)r^{*6} + \left[\frac{1}{64(16)}\left(c^4 + \frac{c^6}{4}\right) + \frac{5c^6 + c^8/4}{64(36)(16)}\right]r^{*8} + \cdots\right\} \tag{A10-4-7}$$

In general

$$A_n = \frac{c^2}{n^2}\left(A_{n-4} - A_{n-2}\right) \qquad \text{for } n \geq 4$$

Note that if $v^* = 1$ (flat velocity profile), we get a zero-order Bessel function whose series is

$$X = J_0(cr^*) = 1 - \frac{c^2}{4}r^{*2} + \frac{c^4}{32}r^{*4} - \frac{c^6}{36(64)}r^{*6} \cdots \tag{A10-4-8}$$

Using the boundary condition $T^* = 0$ at $r^* = 1$, we find the c_i as roots of the equation

$$X_i(c_i) = 0$$

where $c_1 = 2.704$. Then the solution is

$$T^* = \sum_{i=1}^\infty B_i X_i(r^*, c_i)Z_i(\zeta, c_i) \tag{A10-4-9}$$

where $Z_i = \exp(-c_i^2\zeta)$ and the B_i must be determined from the first boundary condition.

Equation (A10-4-3) can be shown to be a form of the Sturm-Liouville equation with weighting function $r^*(1 - r^{*2})$. Hence the eigenfunctions X_i form an orthonormal set. The orthogonality condition permits the evaluation of B_i from the boundary condition $T^* = 1.0$ at $\zeta = 0$. This gives

$$B_i = \frac{\int_0^1 X_i(r^*)(1 - r^{*2})r^*\,dr^*}{\int_0^1 [X_i(r^*)]^2(1 - r^{*2})r^*\,dr^*} = \frac{-2}{c_i(\partial x/\partial c)\Big|_{r^*=1}} \tag{A10-4-10}$$

Bulk Mean Temperature

From Appendix 4-3

$$T_b^* = \frac{\int_0^1 T^*(r^*, \zeta)v^*(r^*)r^*\,dr^*}{\int_0^1 v^* r^*\,dr^*} \tag{A10-4-11}$$

Using $v^* = 2(1 - r^{*2})$ and Eq. (A10-4-9) gives

$$T_b^* = 2\int_0^1 \left(\sum_{i=1}^{\infty} B_i X_i Z_i\right) v^* r^*\,dr^* = 4\sum_{i=1}^{\infty} B_i Z_i \int_0^1 (1 - r^{*2}) X_i r^*\,dr^*$$

Integrating Eq. (A10-4-3) gives

$$\int_0^1 (1 - r^{*2}) X r^*\,dr^* = c^{-2}\int_0^1 \frac{d(r^*\,dX/dr^*)}{dr^*}\,dr^* = c^{-2}\frac{dX}{dr^*}\bigg|_{r^*=1}$$

Then
$$T_b^* = -4\sum_{i=1}^{\infty} B_i Z_i c_i^{-2}\frac{dX_i}{dr^*}\bigg|_{r^*=1} \tag{A10-4-12}$$

The values of $(dX_i/dr^*)|_{r^*=1}$ have been calculated and tabulated[5] (Table 10-2).

Nusselt Number

Equating flow rate expressions at the wall we get

$$-k\frac{\partial T}{\partial r}\bigg|_W dA_W = h(T_R - T_b)\,dA_W \tag{A10-4-13}$$

Then
$$\text{Nu} \equiv \frac{2Rh}{k} = -\frac{2\partial T^*/\partial r^*|_{r^*=1}}{T_b^*} \tag{A10-4-14}$$

Now

$$T_B^* = -4\sum B_i \frac{Z_i}{c_i^2}\left(\frac{dX_i}{dr^*}\right)_1 \quad \text{and} \quad \frac{\partial T^*}{\partial r^*}\bigg| = 2\sum B_i Z_i \frac{dX_i}{dr^*}\bigg|_1$$

so that
$$\text{Nu} = \frac{\sum B_i Z_i (dX_i/dr^*)_1}{2\sum B_i (Z_i/c_i^2)(dX_i/dr^*)_1} \tag{A10-4-15}$$

Limiting Conditions

As $z \to \infty$ (for practical purposes when $\zeta > 0.25$), only one term ($i = 1$) in the series is important and

$$\text{Nu} \to (\text{Nu})_\infty = \frac{c_1^2}{2}$$

Then, from Table 10-2,

$$(\mathrm{Nu})_\infty = \frac{(2.704)^2}{2} = \frac{7.312}{2} = 3.656 \qquad \text{(A10-4-16)}$$

As $z \to 0$, $\partial T^*/\partial r^* \to \infty$, $T_B^* = 1$, Nu $\to \infty$ and $\langle \mathrm{Nu} \rangle \to \infty$.

An extension of this problem to a power-law fluid has been carried out.[17]

APPENDIX 10-5 TRANSPORT PHENOMENA IN MULTICOMPONENT SYSTEMS[3]

Equation of Continuity

The equation of continuity for a species i in a reacting multicomponent mixture of n species is obtained by a mass balance over an arbitrary element of volume. In terms of a coordinate system fixed in space (eulerian form) it is

$$\frac{\partial \rho_i}{\partial t} + \nabla \cdot \rho_i \mathbf{v} = -\nabla \cdot \mathbf{j}_i + r_i \qquad i = 1, 2, \ldots, n \qquad \text{(A10-5-1)}$$

in which

$$r_i = \sum_j \nu_{ij} R_j M_i \qquad \text{(A10-5-2)}$$

and

$$\mathbf{v} \equiv \frac{\sum \rho_i \mathbf{v}_i}{\sum \rho_i} = \text{mass average velocity} \qquad \text{(A10-5-3)}$$

where ρ_i = mass i per unit volume
r_i = production of species i, mass i per unit time and unit volume
$\mathbf{v}_i$ = velocity of species i
$\widetilde{R}_j$ = molar rate of jth reaction
ν_{ij} = stoichiometric coefficient or moles i appearing according to stoichiometry in jth reaction; the ν_{ij} are negative for reactants
M_i = molecular weight
$\mathbf{j}_i \equiv \rho_i(\mathbf{v}_i - \mathbf{v})$ = mass diffusive flux of i with respect to mass-average velocity

If we consider the lagrangian point of view, and use the definition of the substantial derivative [Eq. (8-133)],

$$\frac{D\rho_i}{Dt} \equiv \frac{\partial \rho_i}{\partial t} + \mathbf{v} \cdot \nabla \boldsymbol{\rho}_i = \frac{\partial \rho_i}{\partial t} + \nabla \cdot \boldsymbol{\rho}_i \mathbf{v} - \rho_i \nabla \cdot \mathbf{v} \qquad \text{(A10-5-4)}$$

Then the equation of continuity for species i in lagrangian form is

$$\frac{D\rho_i}{Dt} + \rho_i \nabla \cdot \mathbf{v} = -\nabla \cdot \mathbf{j}_i + r_i \qquad \text{(A10-5-5)}$$

Letting $\rho_i = \omega_i \rho$ and using the equation of continuity (8-136) gives

$$\rho \frac{D\omega_i}{Dt} = -\nabla \cdot \mathbf{j}_i + r_i \tag{A10-5-6}$$

If we sum Eq. (A10-5-6) over all species,

$$\sum r_i = 0 \qquad \sum \rho_i = \rho \qquad \text{and} \qquad \sum \mathbf{j}_i = \sum \rho_i \mathbf{v}_i - \sum \rho_i \mathbf{v} = 0$$

by the definition of $\mathbf{v}$ [Eq. (A10-5-3)]. Thus we get the equation of continuity of the fluid mixture (8-136). For an *incompressible* fluid, ρ = const and hence $\nabla \cdot \mathbf{v} = 0$. Then Eq. (A10-5-5) becomes

$$\frac{D\rho_i}{Dt} = -\nabla \cdot \mathbf{j}_i + r_i \tag{A10-5-7}$$

It is frequently more convenient to use molar units. If M_i is the molecular weight of species i,

$$C_i \equiv \frac{\rho_i}{M_i} \qquad R_i \equiv \frac{r_i}{M_i} \qquad \text{and} \qquad \mathbf{J}_i \equiv \frac{j_i}{M_i} \tag{A10-5-8}$$

then we can write Eq. (A10-5-1) as

$$\frac{\partial C_i}{\partial t} = -\nabla \cdot (\mathbf{J}_i + C_i \mathbf{v}) + R_i \tag{A10-5-9}$$

where $\mathbf{J}_i$ is the molar diffusive flux of i with respect to the mass-average velocity.

We can also write Eqs. (A10-5-1) and (A10-5-9) in terms of the following fluxes with respect to a fixed axis:

Mass:
$$\mathbf{n}_i \equiv \rho_i \mathbf{v}_i = \mathbf{j}_i + \rho_i \mathbf{v} \tag{A10-5-10}$$

Molar:
$$\mathbf{N}_i \equiv \frac{\mathbf{n}_i}{M_i} \equiv C_i \mathbf{v}_i = \mathbf{J}_i + C_i \mathbf{v} \tag{A10-5-11}$$

If we define a molar diffusive flux with respect to the molar-average velocity $\mathbf{v}^*$ as

$$\mathbf{J}_i^* \equiv C_i(\mathbf{v}_i - \mathbf{v}^*) \tag{A10-5-12}$$

where
$$\mathbf{v}^* \equiv \frac{\sum C_i \mathbf{v}_i}{\sum C_i} \tag{A10-5-13}$$

then, since
$$\mathbf{N}_i = \mathbf{J}_i + C_i \mathbf{v} = \mathbf{J}_i^* + C_i \mathbf{v}^* \tag{A10-5-14}$$

we have from Eq. (A10-5-9)

$$\frac{\partial C_i}{\partial t} = -\nabla \cdot \mathbf{N}_i + R_i \tag{A10-5-15}$$

and
$$\frac{\partial C_i}{\partial t} = -\nabla \cdot (\mathbf{J}_i^* + C_i \mathbf{v}^*) + R_i \tag{A10-5-16}$$

Equation of Motion

In the equation of motion $\widehat{\mathbf{F}}_i$, the body force per unit mass acting on species i, may not be the same for each species. For example, for an electrochemical system

$$\mathbf{F}_i = \mathbf{g} + \widetilde{N}\epsilon_i \nabla \frac{\Phi}{M_i \alpha} \tag{A10-5-17}$$

where $\widetilde{N}$ = Avogadro's number
ϵ_i = charge of i
Φ = electric potential
α = Faraday's constant,[7] (C/g equiv)

A momentum balance over a differential element of volume then gives

$$\frac{D\mathbf{v}}{Dt} \equiv \frac{\partial(\rho\mathbf{v})}{\partial t} + \nabla \cdot \rho\mathbf{v}\mathbf{v} = -\nabla \cdot \underline{\boldsymbol{\tau}} - \nabla p + \sum \rho_i \widehat{\mathbf{F}}_i \tag{A10-5-18}$$

Total-Energy Equation

If we let $\widehat{E} \equiv \widehat{U} + \widehat{K}$, where $\widehat{U}$ is the internal energy per unit mass and $\widehat{K} = v^2/2$ is the kinetic energy per unit mass, an energy balance over an element of volume neglecting electrical or nuclear heating gives

$$\frac{\partial}{\partial t}\left[\rho\left(\widehat{U} + \frac{v^2}{2}\right)\right] + \nabla \cdot \left[\rho\left(\widehat{U} + \frac{v^2}{2}\right)\mathbf{v}\right] = -\nabla \cdot \mathbf{q} - \nabla \cdot p\mathbf{v} - \nabla \cdot \underline{\boldsymbol{\tau}} \cdot \mathbf{v} + \sum_i \rho_i \mathbf{v}_i \cdot \widehat{\mathbf{F}}_i \tag{A10-5-19}$$

It is sometimes convenient to replace $\rho_i \mathbf{v}_i$ with $\mathbf{n}_i$.

Mechanical-Energy Equation

The total-energy equation includes the kinetic energy per unit mass $v^2/2$ and various work terms. As before, we can eliminate the kinetic energy by taking the dot product of $\mathbf{v}$ and the equation of motion in terms of $\rho(D\mathbf{v}/Dt)$. This gives the mechanical-energy equation

$$\rho \frac{D(v^2/2)}{Dt} = -\mathbf{v} \cdot \nabla \cdot \underline{\boldsymbol{\tau}} - \mathbf{v} \cdot \nabla p + \sum \rho_i \mathbf{v} \cdot \widehat{\mathbf{F}}_i \tag{A10-5-20}$$

Internal-Energy Equation

If we subtract the mechanical-energy equation from the total-energy equation, we get the *internal-energy equation*

$$\frac{D\widehat{U}}{Dt} = -\nabla \cdot \mathbf{q} - \underline{\boldsymbol{\tau}}\colon \nabla\mathbf{v} - p\,\nabla \cdot \mathbf{v} + \sum \mathbf{j}_i \cdot \widehat{\mathbf{F}}_i \tag{A10-5-21}$$

where we have used the identities in Eqs. (10-17) and (10-18).

Energy Equation in Terms of Enthalpy

To express the energy equation in terms of enthalpy we make use of the thermodynamic definition $\widehat{H} \equiv \widehat{U} + p/\rho$ and proceed as in Sec. 10-2, getting

$$\rho \frac{D\widehat{H}}{Dt} = -\nabla \cdot \mathbf{q} - \underline{\tau}:\nabla \mathbf{v} + \frac{Dp}{Dt} + \sum \mathbf{j}_i \cdot \widehat{\mathbf{F}}_i \tag{A10-5-22}$$

As indicated before, the term $-\underline{\tau}:\nabla \mathbf{v}$ represents the work lost through fluid friction, called the viscous dissipation.

Multicomponent Mixture

For a multicomponent mixture the total enthalpy is

$$H = H(T, p, m_1, \ldots, m_i, \ldots, m_n)$$

where m_i is the mass of species i. Then

$$dH = \left(\frac{\partial H}{\partial T}\right)_{p,m_i} dT + \left(\frac{\partial H}{\partial P}\right)_{T,m_i} dp + \sum_{i=1}^{n} \left(\frac{\partial H}{\partial m_i}\right)_{\substack{T,p,m_j \\ i \neq j}} dm_i \tag{A10-5-23}$$

Dividing by the total mass of the system gives

$$d\widehat{H} = \left(\frac{\partial \widehat{H}}{\partial T}\right)_{p,m_i} dT + \left(\frac{\partial \widehat{H}}{\partial p}\right)_{T,m_i} dp + \sum_{i} \left(\frac{\partial H}{\partial m_i}\right)_{\substack{T,p,m_j \\ i \neq j}} d\omega_i \tag{A10-5-24}$$

Substituting for the coefficients of dT, dp, and $d\omega_i$ gives

$$d\widehat{H} = \widehat{C}_p \, dT + \left\{\frac{1}{\rho} - T\left[\frac{\partial(1/\rho)}{\partial T}\right]_p\right\} dp + \sum \frac{\overline{H}_i}{M_i} d\omega_i \tag{A10-5-25}$$

where

$$\overline{H}_i \equiv \left(\frac{\partial H}{\partial m_i}\right)_{T,p,m_j} M_i$$

is the partial molal enthalpy of species i.

If we write the above equation in terms of the substantial derivative and in terms of the coefficient of volume expansion,

$$\beta \equiv \rho \left[\frac{\partial(1/\rho)}{\partial T}\right]_p \tag{A10-5-26}$$

we get

$$\rho \frac{D\widehat{H}}{Dt} = \rho \widehat{C}_p \frac{DT}{Dt} + (1 - \beta T)\frac{Dp}{Dt} + \sum \left(\frac{\overline{H}_i}{M_i}\right)_\rho \frac{D\omega_i}{Dt} \tag{A10-5-27}$$

and, since $\rho_i = \rho\omega_i$,

$$\rho \frac{D\omega_i}{Dt} \equiv \frac{D\rho_i}{Dt} - \omega_i \frac{D\rho}{Dt}$$

Substituting Eq. (A10-5-6) and the EOC gives

$$\rho \frac{D\omega_i}{Dt} = (-\rho_i \nabla \cdot \mathbf{v} - \nabla \cdot \mathbf{j}_i + r_i) + \omega_i \rho \nabla \cdot \mathbf{v} = -\nabla \cdot \mathbf{j}_i + r_i \tag{A10-5-28}$$

Substituting into Eq. (A10-5-27), we get

$$\rho \frac{D\hat{H}}{Dt} = \rho \hat{C}_p \frac{DT}{Dt} + (1 - \beta T)\frac{Dp}{Dt} + \sum \bar{H}_i(-\nabla \cdot \mathbf{J}_i + R_i) \tag{A10-5-29}$$

Substituting Eq. (A10-5-29) into the enthalpy equation (A10-5-22) gives

$$\rho \hat{C}_p \frac{DT}{Dt} = -\nabla \cdot \mathbf{q} - \boldsymbol{\tau}:\nabla \mathbf{v} + \beta T \frac{Dp}{Dt} - \sum \bar{H}_i(-\nabla \cdot \mathbf{J}_i + R_i) + \sum \mathbf{j}_i \cdot \hat{\mathbf{F}}_i \tag{A10-5-30}$$

We let

$$\mathbf{q} = \mathbf{q}^{(h)} + \sum \mathbf{j}_i \bar{H}_i \tag{A10-5-31}$$

where the second term on the right is called the *interdiffusion flux*. Then

$$\rho \hat{C}_p \frac{DT}{Dt} = -\nabla \cdot \mathbf{q}^{(h)} - \sum \mathbf{J}_i \cdot \nabla \bar{H}_i - \sum \bar{H}_i \tilde{R}_i - \boldsymbol{\tau}:\nabla \mathbf{v} + \beta T \frac{Dp}{Dt} + \sum \mathbf{j}_i \cdot \hat{\mathbf{F}}_i \tag{A10-5-32}$$

The term $\Sigma \bar{H}_i \tilde{R}_i$ can be recognized as the enthalpy change due to reaction per unit time and unit volume. The terms $\Sigma \mathbf{J}_i \cdot \nabla \bar{H}_i$, $\beta T(Dp/Dt)$, and $\boldsymbol{\tau}:\nabla \mathbf{v}$ are usually neglected. For an incompressible fluid $\beta = 0$. For an ideal gas $\beta = T^{-1}$.

We can write Eq. (A10-5-32) in eulerian form by applying Eq. (8-144). Then

$$\rho \hat{C}_p \frac{DT}{Dt} = \frac{\partial(\hat{C}_p \rho T)}{\partial t} + \nabla \cdot \rho \hat{C}_p T \mathbf{v} - \rho T \frac{D\hat{C}_p}{Dt} \tag{A10-5-33}$$

Frequently $\hat{C}_p$ can be considered constant. Also, we can let

$$\sum_i \bar{H}_i R_i = \sum_i \bar{H}_i \sum_j \nu_{ij} \tilde{R}_j = \sum_j \sum_i \bar{H}_i \nu_{ij} \tilde{R}_j \tag{A10-5-34}$$

or

$$\sum_i \bar{H}_i R_i = \sum_j (\Delta H_R)_j \tilde{R}_j = -\hat{Q}_R \rho \equiv \Phi_R \tag{A10-5-35}$$

where $-\Delta H_{Rj}$ = heat absorbed by jth reaction
$\hat{Q}_R$ = total heat of reaction absorbed per unit mass
Φ_R = total heat of reaction absorbed per unit volume

Then the energy equation can be written in the simplified form

$$\frac{\partial(\rho \hat{C}_p T)}{\partial t} + \nabla \cdot \rho \hat{C}_p T \mathbf{v} = -\nabla \cdot \mathbf{q}^{(h)} + \Phi_R \tag{A10-5-36}$$

Thermodynamics of Irreversible Processes

Equation of change for entropy Consider an element of volume $d\mathcal{V}$ in which the rate of production of entropy per unit volume is g. Let the entropy per unit mass be $\hat{\mathcal{S}}$ and the flux of entropy from an element of surface dA be $\boldsymbol{\sigma}$. Then our entropy balance gives†

$$\frac{D}{Dt}\iiint \rho\hat{\mathcal{S}}\,d\mathcal{V} = -\iint \boldsymbol{\sigma}\cdot\mathbf{n}\,dA + \iint g\,d\mathcal{V} \qquad \text{(A10-5-37)}$$

Using the Reynolds transport theorem and the Gauss divergence theorem, we obtain

$$\iiint\left[\frac{\partial}{\partial t}(\rho\hat{\mathcal{S}}) + \nabla\cdot\rho\hat{\mathcal{S}}\mathbf{v} + \nabla\cdot\boldsymbol{\sigma} - g\right]d\mathcal{V} = 0 \qquad \text{(A10-5-38)}$$

Since the volume is arbitrary, the integrand is zero. In lagrangian form it is

$$\rho\frac{D\hat{\mathcal{S}}}{Dt} = -\nabla\cdot\boldsymbol{\sigma} + g \qquad \text{(A10-5-39)}$$

Now let us try to determine the nature of the terms $\boldsymbol{\sigma}$ and g. To do so, we recall from thermodynamics that

$$dU = T\,d\mathcal{S} - p\,d\mathcal{V} + \sum\mu_i\frac{dm_i}{M_i} \qquad \text{(A10-5-40)}$$

where m_i is the mass of i. If we divide by the total mass of the mixture, solve for $d\hat{\mathcal{S}} = d\mathcal{S}/m_t$, and write the resulting equation as a substantial derivative, we get

$$\frac{D\hat{\mathcal{S}}}{Dt} = \frac{1}{T}\frac{D\hat{U}}{Dt} + \frac{p}{T}\frac{D(1/\rho)}{Dt} - \sum\frac{\mu_i}{TM_i}\frac{D\omega_i}{Dt} \qquad \text{(A10-5-41)}$$

If we use the equations of change for internal energy (A10-5-21), total mass (continuity), and mass of species i (A10-5-28), we get

$$\rho\frac{D\hat{\mathcal{S}}}{Dt} = -\frac{1}{T}\nabla\cdot\mathbf{q} - \frac{\underline{\boldsymbol{\tau}}:\nabla\mathbf{v}}{T} - \sum\frac{\mu_i}{TM_i}(-\nabla\cdot\mathbf{j}_i + r_i) + \frac{1}{T}\sum\mathbf{j}_i\cdot\hat{\mathbf{F}}_i \qquad \text{(A10-5-42)}$$

In order to compare with the entropy equation (A10-5-39) we first write

$$\nabla\cdot\frac{\mathbf{q}}{T} = \frac{\nabla\cdot\mathbf{q}}{T} - \frac{\mathbf{q}\cdot\nabla T}{T^2} \qquad \text{(A10-5-43)}$$

If we let $\mathbf{q} = \mathbf{q}^{(h)} + \Sigma\bar{H}_i\mathbf{J}_i$ and rearrange, we get

$$\frac{\nabla\cdot\mathbf{q}}{T} = \nabla\cdot\frac{\mathbf{q}^{(h)}}{T} + \frac{\mathbf{q}^{(h)}\cdot\nabla T}{T^2} + \sum\nabla\cdot\frac{\bar{H}_i\mathbf{J}_i}{T} + \sum\frac{\bar{H}_i\mathbf{J}_i\cdot\nabla T}{T^2} \qquad \text{(A10-5-44)}$$

†In this equation, $\boldsymbol{\sigma}$ is *not* the stress vector and g is *not* the gravity.

Also
$$\nabla \cdot \frac{\mu_i \mathbf{j}_i}{T} = \mathbf{j}_i \cdot \nabla \frac{\mu_i}{T} + \frac{\mu_i}{T} \nabla \cdot \mathbf{j}_i \tag{A10-5-45}$$

or
$$\sum \frac{\mu_i}{T} \nabla \cdot \frac{\mathbf{j}_i}{M_i} = \sum \nabla \cdot \frac{\mu_i}{T} \mathbf{J}_i - \sum \mathbf{J}_i \cdot \nabla \frac{\mu_i}{T} \tag{A10-5-46}$$

Substituting Eqs. (A10-5-45) and (A10-5-46) into (A10-5-42), we get

$$\rho \frac{D\hat{\mathcal{S}}}{Dt} = -\nabla \cdot \frac{\mathbf{q}^{(h)}}{T} - \frac{\mathbf{q}^{(h)} \cdot \nabla T}{T^2} - \sum \nabla \cdot \frac{\bar{H}_i \mathbf{J}_i}{T} - \sum \frac{\bar{H}_i \mathbf{J}_i \cdot \nabla T}{T^2} \tag{A10-5-47}$$

$$- \frac{\boldsymbol{\tau} : \nabla \mathbf{v}}{T} + \sum \nabla \cdot \frac{\mu_i \mathbf{J}_i}{T} - \sum \mathbf{J}_i \cdot \nabla \frac{\mu_i}{T} + \sum \frac{\mu_i R_i}{T} + \frac{1}{T} \sum \mathbf{j}_i \cdot \hat{\mathbf{F}}_i \tag{A10-5-48}$$

If we associate with $\nabla \cdot \boldsymbol{\sigma}$ those terms whose divergence is being taken, we find

$$\boldsymbol{\sigma} = \frac{\mathbf{q}^{(h)}}{T} + \sum \left(\frac{\bar{H}_i \mathbf{J}_i}{T} - \frac{\mu_i \mathbf{J}_i}{T} \right) = \frac{\mathbf{q}^{(h)}}{T} + \sum \mathbf{J}_i \bar{\mathcal{S}}_i \tag{A10-5-49}$$

where $\bar{\mathcal{S}}_i \equiv (1/T)(\bar{H}_i - \mu_i)$ is the partial molal entropy. The above equation indicates that the entropy flux is composed of (1) the reversible flux due to heat flux and (2) the flux due to mass diffusional processes. If the remaining terms in the equation of change for entropy are associated with g, we get

$$g = - \frac{\mathbf{q}^{(h)} \cdot \nabla T}{T^2} - \frac{\boldsymbol{\tau} : \nabla \mathbf{v}}{T} + \sum \frac{\mu_i R_i}{T} - \frac{1}{T} \sum \mathbf{J}_i \cdot \left(\bar{H}_i \frac{\nabla T}{T} + T \nabla \frac{\mu_i}{T} + \sum \mathbf{F}_i M_i \right) \tag{A10-5-50}$$

It is convenient to designate the term in parentheses as

$$\begin{aligned} \boldsymbol{\Lambda}_i \equiv \bar{H}_i \frac{\nabla T}{T} + T \nabla \frac{\mu_i}{T} + \sum \mathbf{F}_i M_i &= \bar{H}_i \frac{\nabla T}{T} - \frac{\mu_i \nabla T}{T} + \nabla \mu_i + \sum \hat{\mathbf{F}}_i M_i \\ &= \bar{\mathcal{S}}_i \nabla T + \nabla \mu_i + \sum \hat{\mathbf{F}}_i M_i \end{aligned} \tag{A10-5-51}$$

Further, if $R_i = \Sigma_j \nu_{ij} \tilde{R}_j$, where $\tilde{R}_j$ is the reaction rate for the jth reaction and ν_{ij} is the stoichiometric coefficient, the reaction-rate term can be written

$$\sum_i R_i \mu_i = \sum_i \left(\sum_j \nu_{ij} \tilde{R}_j \right) \mu_i = \sum_j \left(\sum_i \nu_{ij} \mu_i \right) \tilde{R}_j = \sum_j Y_j \tilde{R}_j$$

where
$$Y_j = \sum_i \nu_{ij} \mu_i$$

is called the *chemical affinity*.

The equation for entropy production can then be written

$$g = - \frac{\mathbf{q}^{(h)} \cdot \nabla \ln T}{T} - \frac{\boldsymbol{\tau} : \nabla \mathbf{v}}{T} - \frac{\sum \tilde{R}_j Y_j}{T} + \frac{\sum \mathbf{J}_i \cdot \boldsymbol{\Lambda}_i}{T} \tag{A10-5-52}$$

This equation indicates that the irreversible entropy production g can be expressed as the sum of the products of the fluxes and the driving forces, e.g., the flux $\mathbf{q}^{(h)}$ and the gradient $\nabla T/T$ or the flux $\underline{\tau}$ and the gradient $\nabla \mathbf{v}$. Then if we let $\mathcal{J}_m$ be the flux and X_m be a generalized gradient,

$$g = \frac{1}{T} \sum \mathcal{J}_m X_m \tag{A10-5-53}$$

The generalized fluxes and gradients can be listed as follows:

Transport process	$\mathcal{J}_m$	X_m
Heat	$\mathbf{q}^{(h)}$	$-\nabla \ln T$
Momentum	$\underline{\tau}$	$-\nabla \mathbf{v}$
Mass	$\mathbf{J}_i$	$-\mathbf{\Lambda}_i$
Reaction	$\tilde{R}_j$	Y_j

Since there are 9 components of $\underline{\tau}$, 3 components of $\mathbf{q}^{(h)}$, $3n_c$ components of $\mathbf{J}_i$, and n_s reactions occurring, there are $3n_c + n_s + 12$ fluxes and an equal number of driving forces. According to the thermodynamics of irreversible processes, the processes above can be coupled; e.g., a temperature gradient can produce a mass flux, or a concentration gradient can produce a heat flux.

Coupling between fluxes It is assumed as a postulate of the thermodynamics of irreversible processes that for conditions near equilibrium the fluxes are related linearly to the driving forces by

$$\mathcal{J}_m = \sum_n L_{mn} X_n \tag{A10-5-54}$$

where the L_{mn} are called the *phenomenological coefficients*. The diagonal terms ($m = n$) represent the direct effects, e.g., heat transport resulting from a temperature gradient. The off-diagonal components ($m \neq n$) apply to the coupled processes. According to *Onsager's reciprocal relations*, the L_{mn} are symmetrical, or

$$L_{mn} = L_{nm} \tag{A10-5-55}$$

Also, for isotropic systems, Curie's law states that coupling can occur only between tensors whose orders differ by an even number. Hence no coupling can occur between momentum flux (order 2) and either heat or mass (order 1), but coupling can occur between momentum flux and chemical reaction (order 0).

It should be noted that coupling can occur between the diffusive fluxes of the various species. Hence the flux of species i depends upon all the Λ_j for each of the j species. Note that the diffusion driving force contained in the Λ_j is not rigorously the concentration gradient but is the chemical-potential gradient. This is usually written in terms of the mole-fraction gradient ∇x_k as

$$\nabla \mu_j = \sum_{\substack{k=i \\ k \neq j}} \left(\frac{\partial \mu_j}{\partial x_k} \right)_{\substack{T,p,x_s \\ s \neq k,j}} \nabla x_k \tag{A10-5-56}$$

As a result of the coupling between mass fluxes, the following effects can be obtained: (1) a positive flux of species i with zero concentration gradient of species i and (2) a positive flux of species i with a positive concentration gradient of species i.[30,31,35]

Flux Laws for Multicomponent Systems

Energy flux For *pure* substances the energy flux in laminar flow is given by the conductive flux $\mathbf{q}^{(c)}$ plus the flux due to radiation $\mathbf{q}^{(r)}$, or

$$\mathbf{q} = \mathbf{q}^{(c)} + \mathbf{q}^{(r)} \tag{A10-5-57}$$

where

$$\mathbf{q}^{(c)} = -k\,\nabla T \tag{A10-5-58}$$

For *multicomponent* systems, rigorously one must also add terms for two additional effects: The first effect, called the *Dufour* or *diffusion-thermo effect,* is a heat flux $\mathbf{q}^{(x)}$ that occurs as a result of the *concentration* gradients of the various species, i.e., a coupling between the heat and mass fluxes. Since it is extremely small, it will not be discussed further. The second effect is the flux $\mathbf{q}^{(d)}$ that occurs due to the interdiffusion of species, or

$$\mathbf{q}^{(d)} = \sum \bar{H}_i \mathbf{J}_i \tag{A10-5-59}$$

Then

$$\mathbf{q} = \mathbf{q}^{(c)} + \mathbf{q}^{(r)} + \mathbf{q}^{(x)} + \mathbf{q}^{(d)} \tag{A10-5-60}$$

It is convenient to define a total-energy flux $\mathbf{e}$ by including the other terms that enter into the total-energy equation. Thus

$$\mathbf{e} = \mathbf{q} + \underline{\boldsymbol{\tau}} \cdot \mathbf{v} + p\mathbf{v} + \rho(\hat{U} + \hat{K})\mathbf{v} \tag{A10-5-61}$$

Usually $\underline{\boldsymbol{\tau}} \cdot \mathbf{v}$ and $\rho \hat{K}$ are negligible. If we also neglect $\mathbf{q}^{(r)}$ and $\mathbf{q}^{(x)}$, we get

$$\mathbf{e} = -k\,\nabla T + \sum \bar{H}_i \mathbf{J}_i + p\mathbf{v} + \rho \hat{U}\mathbf{v} \tag{A10-5-62}$$

But because

$$p + \rho \hat{U} = \rho \hat{H} = \sum C_i \bar{H}_i \tag{A10-5-63}$$

we have

$$\mathbf{e} = -k\,\nabla T + \sum \bar{H}_i \mathbf{J}_i + \sum C_i \bar{H}_i \mathbf{v} = -k\,\nabla T + \sum \mathbf{N}_i \bar{H}_i \tag{A10-5-64}$$

since $\mathbf{N}_i = \mathbf{J}_i + C_i \mathbf{v}$.

Mass fluxes Just as there is a contribution to the energy flux that arises as a result of concentration gradients, there is also a contribution to the mass flux $\mathbf{J}_i^{(T)}$ that results from a temperature gradient. This is called the *Soret effect* or the *thermal diffusion effect*. In addition, there are contributions to the diffusive flux due to mechanical driving forces. These are the pressure gradient, which causes a *pressure diffusion flux* $\mathbf{J}_i^{(p)}$, and the unequal external or body forces, which cause a *forced diffusion flux* $\mathbf{J}_i^{(F)}$.

The total molar flux is then

$$\mathbf{J}_i = \mathbf{J}_i^{(x)} + \mathbf{J}_i^{(T)} + \mathbf{J}_i^{(p)} + \mathbf{J}_i^{(F)} \tag{A10-5-65}$$

where $\mathbf{J}_i^{(x)}$ is the flux due to ordinary diffusion. However, in multicomponent mixtures the flux of species i depends on the concentration gradients of *all* the species present or, more rigorously, upon the mole-fraction gradients ∇x_k

$$\mathbf{J}_i^{(x)} = \frac{C^2}{\rho R_G T} \sum_{j=1}^{n} M_j D_{ij} \sum_{\substack{k=1 \\ k \neq j}} x_j \left(\frac{\partial \mu_j}{\partial x_k}\right)_{\substack{T,p,x_s \\ s \neq j,k}} \nabla x_k \tag{A10-5-66}$$

where μ_j = chemical potential of species j
D_{ij} = diffusivity of i through j with $D_{ii} = 0$
C = moles of mixture per volume
R_G = gas constant

For an ideal-gas mixture, $\partial \mu_j / \partial x_k = R_G T / x_j$ and this becomes

$$\mathbf{J}_i^{(x)} = \frac{C^2}{\rho} \sum_{j=1}^{n} M_j D_{ij} \nabla x_j \tag{A10-5-67}$$

The other terms are

$$\mathbf{J}_i^{(p)} = \frac{C^2}{\rho R_G T} \sum_{j=1}^{n} M_j D_{ij} \left[x_j M_j \left(\frac{\bar{\mathcal{V}}_j}{M_j} - \frac{1}{\rho} \right) \nabla p \right] \tag{A10-5-68}$$

$$\mathbf{J}_i^{(F)} = -\frac{C^2}{\rho R_G T} \sum_{j=1}^{n} M_j D_{ij} \left[x_j M_j \left(\hat{F}_j - \sum_{k=1}^{n} \frac{\rho_k}{\rho} \hat{\mathbf{F}}_k \right) \right] \tag{A10-5-69}$$

and

$$\mathbf{J}_i^{(T)} = -D_i^T \nabla \ln T \tag{A10-5-70}$$

where D_i^T is the thermal diffusion coefficient.

The forced-diffusion term would be of importance in ionic systems in which each ionic species might be acting under a different force $\hat{\mathbf{F}}_k$. For a binary consisting of species A and B, the flux of A is

$$\mathbf{J}_A = -\mathbf{J}_B = -\frac{C^2}{\rho R T} M_B D_{AB} x_A \left[\left(\frac{\partial \mu_A}{\partial x_A} \right)_{T,p} \nabla x_A - \frac{\rho_B}{\rho} (\hat{\mathbf{F}}_A - \hat{\mathbf{F}}_B) + \left(\frac{\bar{\mathcal{V}}_A}{M_A} - \frac{1}{\rho} \right) \nabla p \right] - D_A^T \nabla \ln T \tag{A10-5-71}$$

If we consider ordinary diffusion only,

$$\mathbf{J}_\mathrm{A} = -\frac{C^2}{\rho} M_\mathrm{B} D_\mathrm{AB} \left(\frac{\partial \mu_\mathrm{A}}{\partial x_\mathrm{A}}\right)_{T,p} \nabla x_\mathrm{A} \qquad \text{(A10-5-72)}$$

or

$$\mathbf{J}_\mathrm{A} = -\frac{C^2}{\rho} M_\mathrm{B} \mathfrak{D}_\mathrm{AB} \nabla x_\mathrm{A} \qquad \text{(A10-5-73)}$$

$\mathfrak{D}_\mathrm{AB}$ is the diffusion coefficient usually determined experimentally. For ideal solutions, these two diffusivities are equal.

If we convert ∇x_A in Eq. (A10-5-73) to mass-fraction driving force, we get

$$\mathbf{J}_\mathrm{A} = -\rho \mathfrak{D}_\mathrm{AB} \nabla \omega_\mathrm{A} \qquad \text{(A10-5-74)}$$

For constant ρ this becomes

$$\mathbf{J}_\mathrm{A} = -\mathfrak{D}_\mathrm{AB} \nabla C_\mathrm{A} \qquad \text{(A10-5-75)}$$

In engineering calculations of multicomponent mixtures it is common to use an effective diffusivity by letting B represent all components except species A.

An alternative method of defining multicomponent diffusivities is given by Lightfoot and Cussler.[16] In this so-called practical method[35]

$$J_i^{(x)} = \sum_{j=1}^{n=1} D_{ij} \nabla C_j$$

and $D_{ii} \neq 0$, whereas in Eq. (A10-5-66) $D_{ii} = 0$. As a result of the conservation of mass, only $n - 1$ of the concentrations or flux laws are independent and $\Sigma J_i^{(x)} = 0, \Sigma \omega_i = 1$. Thus for a ternary, if species 3 is taken as the solvent, one can write

$$J_1^{(x)} = -D_{11} \nabla C_1 - D_{12} \nabla C_2$$

It should be pointed out that it is also useful in some cases to define volume diffusive fluxes with respect to a *volume average velocity*. The principles involved are the same as those we have used for mass and molar averages; in general a diffusive flux is a flow rate per unit area relative to a specified average velocity which is obtained by weighting the species velocities in a specified manner.

APPENDIX 10-6 TIME AVERAGING EQUATIONS OF CHANGE

Mass Transport

The equation of continuity for a diffusing species A can be written for laminar flow [Eq. (10-52)] as

$$\frac{\partial C_\mathrm{A}}{\partial t} + \nabla \cdot C_\mathrm{A}\mathbf{v} = -\nabla \cdot \mathbf{J}_\mathrm{A}^{(l)} + R_\mathrm{A} \qquad \text{(A10-6-1)}$$

where C_A = concentration, moles per unit volume
$\mathbf{v}$ = mass-average velocity
$\mathbf{J}_A^{(l)}$ = molar diffusive flux with respect to mass-average velocity
R_A = net molar rate of production of species A per unit time and volume

We now postulate that the above equation applies at a point within the smallest eddy of a fluid in turbulent flow, but since conditions at this fixed point in space will be fluctuating with time, in order to use Eq. (A10-6-1) we will need to time-average the equation term by term as follows

$$\overline{\frac{\partial C_A}{\partial t}} + \overline{\nabla \cdot C_A \mathbf{v}} = -\overline{\nabla \cdot \mathbf{J}_A^{(l)}} + \overline{R_A} \qquad \text{(A10-6-2)}$$

Since the time average of the derivative is equal to the derivative of the time average,[36] the first term becomes $\partial \overline{C_A}/\partial t$. The second term becomes $\nabla \cdot \overline{C_A \mathbf{v}}$ if we interchange the time-averaging and differentiation processes. Then in $\overline{C_A \mathbf{v}}$ if we let

$$C_A = \overline{C_A} + C_A' \qquad \text{and} \qquad \mathbf{v} = \overline{\mathbf{v}} + \mathbf{v}'$$

we obtain

$$\overline{C_A \mathbf{v}} = \overline{(\overline{C_A} + C_A')(\overline{\mathbf{v}} + \mathbf{v}')} = \overline{\overline{C_A}\,\overline{\mathbf{v}}} + \overline{C_A' \overline{\mathbf{v}}} + \overline{\overline{C_A} \mathbf{v}'} + \overline{C_A' \mathbf{v}'} = \overline{C_A}\,\overline{\mathbf{v}} + \overline{C_A' \mathbf{v}'} \qquad \text{(A10-6-3)}$$

The second term above represents an additional contribution that may arise in turbulence due to correlation between concentration and velocity fluctuations. Since it is a vector and corresponds to a flux, we call it the *turbulent molar flux* (with respect to the mass average velocity) or

$$\mathbf{J}_A^{(t)} \equiv \overline{C_A' \mathbf{v}'} \qquad \text{(A10-6-4)}$$

The third term in Eq. (A10-6-2) is†

$$-\overline{\nabla \cdot \mathbf{J}_A^{(l)}} = -\nabla \cdot \overline{\mathbf{J}_A^{(l)}}$$

$$\overline{\mathbf{J}_A^{(l)}} = -D_{AB} \nabla \overline{C_A} \qquad \text{(A10-6-5)}$$

where for constant ρ for a first-order reaction the fourth term in Eq. (A10-6-2) can be written

$$\overline{R_A} = \overline{k_R(\overline{C_A} + C_A')} = k_R \overline{C_A}$$

but for a reaction between species A and B it becomes

$$\overline{R_A} = \overline{k_R(\overline{C_A} + C_A')(\overline{C_B} + C_B')} = k_R(\overline{C_A}\,\overline{C_B} + \overline{C_A' C_B'}) \qquad \text{(A10-6-6)}$$

Thus an additional term is obtained for turbulent flow if the concentration fluctuations are correlated.

†We use an overbar on $\overline{J_A^{(l)}}$ to emphasize that $\overline{J_A^{(l)}}$ depends on $\nabla \overline{C_A}$; that is, we are using time-averaged potentials. Since we *define* $\mathbf{J}_A^{(t)}$ as $\overline{C_A' v'}$, no overbar is needed.

On combining terms, Eq. (A10-6-2) becomes

$$\frac{\partial \overline{C_A}}{\partial t} + \nabla \cdot \overline{C}_A \overline{\mathbf{v}} = -\nabla \cdot \overline{\mathbf{J}_A^{(l)}} - \nabla \cdot \mathbf{J}_A^{(t)} + \overline{R_A} \qquad \text{(A10-6-7)}$$

It is convenient to define a total flux as

$$\overline{\mathbf{J}}_A \equiv \overline{\mathbf{J}_A^{(l)}} + \mathbf{J}_A^{(t)} \qquad \text{(A10-6-8)}$$

Then Eq. (A10-6-7) becomes

$$\frac{\partial \overline{C_A}}{\partial t} + \nabla \cdot \overline{C}_A \overline{\mathbf{v}} = -\nabla \cdot \overline{\mathbf{J}}_A + \overline{R_A} \qquad \text{(A10-6-9)}$$

If we compare the above equation with Eq. (A10-6-1), we see that we have merely replaced the concentration and velocity with the time-averaged values and have replaced the laminar flux with the time-averaged total flux.

Flux Laws

For laminar flow, Fick's law (for constant ρ) is [Eq. (A10-5-75)]

$$\mathbf{J}_A^{(l)} = -D_{AB} \nabla C_A \qquad \text{(A10-6-10)}$$

where D_{AB} is the molecular diffusivity. Analogously, for *isotropic* turbulent flow, one can define an eddy diffusivity by

$$\mathbf{J}_A^{(t)} = -D_{AB}^{(t)} \nabla \overline{C_A} \qquad \text{(A10-6-11)}$$

The total flux is then given by

$$\overline{\mathbf{J}}_A = -(D_{AB} + D_{AB}^{(t)}) \nabla \overline{C_A} \qquad \text{(A10-6-12)}$$

The use of a scalar eddy diffusivity implies that the turbulence is isotropic; i.e., its properties are independent of the orientation of the coordinate axes. In fact, however, turbulent flow is generally not isotropic, and it is more appropriate to use an eddy-diffusivity tensor $\underline{\mathbf{D}}_{AB}^{(t)}$, defined by

$$\mathbf{J}_A^{(t)} = -\underline{\mathbf{D}}_{AB}^{(t)} \cdot \nabla \overline{C_A} \qquad \text{(A10-6-13)}$$

The total flux is then

$$\overline{\mathbf{J}_A} = -(D_{AB}\underline{\boldsymbol{\delta}} + \underline{\mathbf{D}}_{AB}^{(t)}) \cdot \nabla \overline{C_A} \qquad \text{(A10-6-14)}$$

or

$$\overline{\mathbf{J}}_A = \underline{\mathbf{E}} \cdot \nabla \overline{C}_A \qquad \text{(A10-6-15)}$$

where $\underline{\boldsymbol{\delta}}$ is the unit tensor.

Equation of Motion

The equation of motion in eulerian form (9-32) can be time-averaged term by term as follows

$$\overline{\frac{\partial(\rho v_i)}{\partial t}} = -\overline{\frac{\partial}{\partial x_j}(\rho v_j v_i)} - \overline{\frac{\partial \tau_{ji}^{(l)}}{\partial x_j}} + \overline{\rho g_i} - \overline{\frac{\partial p}{\partial x_i}} \qquad \text{(A10-6-16)}$$

For momentum transport it is more convenient to use cartesian tensor notation and the summation convention described in Chap. 9; thus a repeated index means summation over that index [see Eq. (9-35)].

For simplicity we will assume that fluctuations in the pressure and density can be neglected; that is, $p = \bar{p}$ and $\rho = \bar{\rho}$. Using Eqs. (6-1) to (6-3*a*) and (A10-6-16) and the property that the average of a derivative equals the derivative of the average, we obtain

$$\rho \frac{\partial \bar{v}_i}{\partial t} = -\frac{\partial}{\partial x_j}(\rho \bar{v}_j \bar{v}_i) - \frac{\partial \overline{\tau_{ji}^{(l)}}}{\partial x_j} - \frac{\partial}{\partial x_j}(\rho \overline{v_i' v_j'}) + \rho g_i - \frac{\partial p}{\partial x_i} \tag{A10-6-17}$$

We note that the term $\rho \overline{v_i' v_j'}$ appears and that it is a turbulent momentum flux or stress that arises due to the correlation between the velocity fluctuations in the i direction and those in the j direction. This term is called the *Reynolds stress* and can be given the symbol

$$\rho \overline{v_i' v_j'} \equiv \tau_{ij}^{(t)} \tag{A10-6-18}$$

We observe that there are nine Reynolds stresses that form a tensor and that the diagonal components are $\rho \overline{v_i'^2}$. The quantity $\overline{v_i'^2}$ is also called the *turbulence intensity*. For *isotropic* flow the statistical properties do not depend on direction and are invariant to rotation or reflection of the axes; thus

$$\overline{v_1'^2} = \overline{v_2'^2} = \overline{v_3'^2} \tag{A10-6-19}$$

and $\tau_{ij}^{(t)} = 0$ when $i \neq j$. For convenience we define the total momentum flux as

$$\bar{\tau}_{ij} = \overline{\tau_{ij}^{(l)}} + \tau_{ij}^{(t)} \tag{A10-6-20}$$

The equation of motion can then be written

$$\frac{\partial (\rho \bar{v}_i)}{\partial t} + \frac{\partial}{\partial x_j}(\rho \bar{v}_j \bar{v}_i) = -\frac{\partial \bar{\tau}_{ij}}{\partial x_j} - \frac{\partial p}{\partial x_i} + \rho g_i \tag{A10-6-21}$$

or

$$\frac{\partial (\rho \bar{\mathbf{v}})}{\partial t} + \nabla \cdot \rho \bar{\mathbf{v}} \bar{\mathbf{v}} = -\nabla \cdot \bar{\underline{\tau}} - \nabla p + \rho \mathbf{g} \tag{A10-6-22}$$

If we compare the above equation with the corresponding equation (9-32) for laminar flow, we see that we have merely defined $\bar{\tau}_{ij}$ as the total flux and that the time-averaged velocities have been used instead of instantaneous velocities.

Flux Laws

For laminar flow of an incompressible newtonian fluid we have shown (Chap. 9) that

$$\tau_{ij}^{(l)} = -\mu \Delta_{ij} \tag{A10-6-23}$$

where Δ_{ij} is the rate-of-deformation tensor

$$\Delta_{ij} = \frac{\partial v_i}{\partial x_j} + \frac{\partial v_j}{\partial x_i} \tag{A10-6-24}$$

For isotropic turbulent flow, we could similarly define a turbulent or eddy viscosity $\mu^{(t)}$ by

$$\tau_{ij}^{(t)} = -\mu^{(t)}\overline{\Delta}_{ij} \tag{A10-6-25}$$

where $\mu^{(t)}$ is a scalar. However, since turbulence is not isotropic, it is more appropriate to use a fourth-order viscosity tensor

$$\tau_{ij}^{(t)} = -\mu_{ijkl}^{(t)}\overline{\Delta}_{kl} \qquad \text{nonisotropic turbulent} \tag{A10-6-26}$$

The total fluxes are then given by

$$\overline{\tau}_{ij} = \begin{cases} -(\mu + \mu^{(t)})\overline{\Delta}_{ij} & \text{isotropic} \quad \text{(A10-6-27)} \\ -(\mu\delta_{ijkl} + \mu_{ijkl}^{(t)})\overline{\Delta}_{kl} & \text{nonisotropic} \quad \text{(A10-6-28)} \end{cases}$$

where $\mu_{ijkl}^{(t)}$ is the eddy-viscosity tensor, δ_{ijkl} is a unit tensor, and

$$\overline{\Delta}_{kl} = \frac{\partial \overline{v}_k}{\partial x_l} + \frac{\partial \overline{v}_l}{\partial x_k} \tag{A10-6-29}$$

Energy Equation

The energy equation (10-33) for an incompressible fluid with no reaction can be written in eulerian form as

$$\frac{\partial(\rho\widehat{C}_p T)}{\partial t} + \nabla \cdot \rho\widehat{C}_p \mathbf{v}T = -\nabla \cdot \mathbf{q} - \underline{\tau}:\nabla\mathbf{v}$$

If we time-average this equation, we obtain a turbulent energy flux

$$\mathbf{q}^{(t)} \equiv \rho\widehat{C}_p\overline{T'\mathbf{v}'} \tag{A10-6-30}$$

For isotropic systems

$$\mathbf{q}^{(t)} = -k^{(t)}\,\nabla\overline{T} \tag{A10-6-31}$$

while for anisotropic systems

$$\mathbf{q}^{(t)} = -\underline{\mathbf{k}}^{(t)} \cdot \nabla\overline{T} \tag{A10-6-32}$$

The total heat flux is

$$\overline{\mathbf{q}} = \overline{\mathbf{q}^{(l)}} + \mathbf{q}^{(t)} \tag{A10-6-33}$$

or

$$\overline{\mathbf{q}} = \begin{cases} -(k + k^{(t)})\,\nabla\overline{T} & \text{isotropic} \quad \text{(A10-6-34)} \\ -(k\underline{\boldsymbol{\delta}} + \underline{\mathbf{k}}^{(t)}) \cdot \nabla\overline{T} & \text{anisotropic} \quad \text{(A10-6-35)} \end{cases}$$

Again we can modify the equation for laminar flow by replacing T and $\mathbf{v}$ with $\overline{T}$ and $\overline{\mathbf{v}}$ and using the total flux in place of the laminar flux.

The energy equation will also include the time average of the viscous-dissipation term, which for a newtonian incompressible fluid is

$$\overline{\underline{\tau}:\nabla\mathbf{v}} = \overline{(\overline{\underline{\tau}} + \underline{\tau}'):\nabla(\overline{\mathbf{v}} + \mathbf{v}')} = \overline{\underline{\tau}}:\nabla\overline{\mathbf{v}} + \overline{\underline{\tau}':\nabla\mathbf{v}'} + 0 + 0$$

For a newtonian fluid

$$\overline{\Phi_v^{(l)}} = -\underline{\tau}:\nabla\overline{\mathbf{v}} = \mu[(\nabla\overline{\mathbf{v}}) + (\nabla\mathbf{v})^T]:\nabla\overline{\mathbf{v}} \equiv \mu\overline{\Phi}_v^{(l)} = \mu[(\nabla\overline{\mathbf{v}}):(\nabla\overline{\mathbf{v}}) + (\nabla\overline{\mathbf{v}})^T:\nabla\overline{\mathbf{v}}]$$

$$= \mu\left(\frac{\partial\overline{v}_i}{\partial x_j}\frac{\partial\overline{v}_j}{\partial x_i} + \frac{\partial\overline{v}_i}{\partial x_j}\frac{\partial\overline{v}_i}{\partial x_j}\right)$$

Similarly

$$\Phi_v^{(t)} = \overline{\underline{\tau}':\nabla\mathbf{v}'} = \overline{\mu[\nabla\mathbf{v}' + (\nabla\mathbf{v}')^T]:\nabla\mathbf{v}'} \equiv \mu\phi_v^{(t)} = \mu\left(\overline{\frac{\partial v_i'}{\partial x_j}\frac{\partial v_j'}{\partial x_i}} + \overline{\frac{\partial v_i'}{\partial x_j}\frac{\partial v_i'}{\partial x_j}}\right)$$

Then

$$-\overline{\tau_{ij}\frac{\partial v_i}{\partial x_j}} = \mu\overline{\phi_v^{(l)}} + \mu\phi_v^{(t)} \tag{A10-6-36}$$

where

$$\overline{\phi_v^{(l)}} \equiv \frac{\partial\overline{v}_i}{\partial x_j}\frac{\partial\overline{v}_j}{\partial x_i} + \frac{\partial\overline{v}_i}{\partial x_j}\frac{\partial\overline{v}_i}{\partial x_j} \tag{A10-6-37}$$

and

$$\phi_v^{(t)} \equiv \overline{\frac{\partial v_i'}{\partial x_j}\frac{\partial v_j'}{\partial x_i}} + \overline{\frac{\partial v_i'}{\partial x_j}\frac{\partial v_i'}{\partial x_j}} \tag{A10-6-38}$$

are the laminar and turbulent viscous-dissipation functions. The time-averaged energy equation is then

$$\frac{\partial(\rho\widehat{C}_p\overline{T})}{\partial t} + \nabla\cdot\rho\widehat{C}_p\overline{\mathbf{v}}\overline{T} = -\nabla\cdot\mathbf{q} + \mu\overline{\phi_v^{(l)}} + \mu\phi_v^{(t)} \tag{A10-6-39}$$

PROBLEMS

10-1 Work out details of Example 10-1.

10-2 (*a*) Water is flowing through a 9.144-m-long heat exchanger of 0.0254 m diameter at Re = 1000. If Pr = 3.00, T_0 = 50°F, T_R = 100°F, k = 0.363 Btu/h·ft·°F, find T_{bL} in degrees Fahrenheit and $\dot{Q}$ in Btu per hour. Use Fig. 10-4*b*. Do *not* use a Nusselt number.

(*b*) Compare your result with that of Prob. 4-13.

10-3 Carry out the details of the solution in Example 10-4.

10-4 Devise a mass-transport analog to Example 10-4. State the physical process occurring, the differential equation, the dimensionless groups, and the expression for the average mass-transfer Nusselt number.

10-5 Use scaling and ordering to determine when the viscous-dissipation term in the energy equation can be neglected.

10-6 In certain applications, e.g., the design of monolith converters, a conduit of noncircular cross section is desirable because of the greater surface area available for transport. Cross sections considered are elliptical, rectangular, triangular, and hexagonal. An incompressible newtonian fluid of uniform temperature T_0 enters the heated section of a duct of length L whose wall is maintained at constant temperature T_w. At all locations the flow in the z direction is fully developed, laminar, and at steady state. The heat-transfer coefficient is not known, but the properties of the fluid (ρ, $\widehat{C}_p$, μ, k) can be predicted.

It is desired to develop charts that can be used to predict the total rate of heat transport $\dot{Q}_t$ over a length L in ducts of various cross section. In some cases, such as the circular cylinder, an analytical equation can be derived for fully developed laminar flow; in other cases an experimental program, based on dimensional analysis, may be needed. In order to demonstrate each approach, carry out the derivations below. Assume that the flow is fully developed, laminar, and steady state, that the fluid is newtonian with constant physical properties, and that there is flow $v_z(u_1, u_2)$ in the axial direction only; e.g., in a rectangular duct $u_1 \equiv x$, $u_2 \equiv y$.

(*a*) Demonstrate how to approach such problems *analytically*. Assume that the cross section is a rectangle of sides $2A$ and $2B$. Simplify the equations of change for continuity, momentum, and energy (in terms of velocity and temperature), neglecting axial conduction and viscous dissipation. If any terms are eliminated, state why.

(*b*) Write the equations needed to obtain $\dot{Q}_t$ if the energy equation can be solved for T.

(*c*) Define the average heat-transfer coefficient and *show* how it can be obtained if the equations of change can be solved analytically.

(*d*) Demonstrate your ability to deal with such problems *experimentally* for the case of a rectangular duct of sides A and B by writing the energy equation and boundary conditions in dimensionless form, letting the characteristic velocity be the average velocity, the characteristic distance be R_H (the hydraulic radius), and the dimensionless temperature be

$$T^* \equiv \frac{T - T_w}{T_0 - T_w}$$

(*e*) Define a dimensionless $\dot{Q}^*$ and show how it can be related to T^*.

(*f*) Find the dimensionless groups upon which $\langle \mathrm{Nu} \rangle \equiv \langle h \rangle 4\mathrm{R}_H/k$ will depend. Do not include any unnecessary groups.

(*g*) There is an old rule of thumb that if one has an expression for the Nusselt number for a circular cross section, this result can be used for a noncircular cross section by replacing the tube radius R with twice the hydraulic radius. Analyze this assumption on the basis of dimensional analysis of the equations of change. State whether the rule of thumb is rigorous; if not, describe the minimum experimental program needed to obtain plots that would predict the average heat-transfer coefficient for all rectangular cross sections and for all fluids and flow rates in the laminar region.

(*h*) State whether the above result could be used for mass transfer and, if so, how.

10-7 Work part (*a*) of Prob. 10-6 for a slit, outlining the analytical solution.

10-8 Work Prob. 10-6 for an ellipse with semimajor axis A and semiminor axis B.

10-9 Work Prob. 10-6 for an annulus of radii R_1 and R_2.

10-10 Work Prob. 10-6 for an arbitrary cross section of coordinates u_1 and u_2 with scale factors h_1 and h_2.

10-11 Work Prob. 10-6 for a segment of an annulus made by angles $\theta = 0$ and $\theta = \theta_1$.

10-12 Assume the same conditions as in Prob. 10-6 except that the flow is isotropic turbulent.

(*a*) Simplify the z component of the EOM and the energy equation (in terms of *fluxes*) and apply to this problem. (Be sure to modify *as needed* for *turbulent* flow.) Neglect axial conduction and viscous dissipation.

(*b*) Substitute the appropriate flux laws into the equations above.

(*c*) Describe how to obtain f and $\langle \mathrm{Nu} \rangle$ (in principle).

(*d*) State upon what variables the mean Nusselt number would now depend and give your reason for any differences over laminar flow.

10-13 An inexpensive method of cooling a hot surface is to contact the surface with a film of liquid, which flows down the surface under gravity. This problem is not directly analogous to the mass-transfer (gas-absorption) problem (Example 10-5) since the direction of transport is reversed. Show that this does not affect the equations involved or the solution if we let

$$T^* \equiv \frac{T - T_0}{T_s - T_0}$$

where T_0 is the temperature of the stream entering at $z = 0$ and T_s is the temperature of the surface.

10-14 For fully developed steady-state turbulent flow of an incompressible newtonian fluid in a tube it is often assumed that a scalar eddy viscosity can be used, that there is no radial pressure gradient, and that $\partial p/\partial z$ is constant. Determine whether the models

$$\tau_{ij}^{(t)} = -\sum_k \sum_l \mu_{ijkl}^{(t)} \bar{\Delta}_{lk} \qquad \text{and} \qquad \tau_{ij}^{(t)} = -\mu^{(t)}\Delta_{ij}$$

correctly predict (*a*) the existence of all components of $\tau_{ij}^{(t)}$ and (*b*) the pressure variation. Suggested approach:

1. Write $\tau_{ij}^{(t)}$ in terms of the velocity fluctuations and compare the results with predictions of each model.
2. Write the time-averaged r and z components of the equations of motion for this flow in terms of flux components.
3. Determine whether $\partial p/\partial z$ varies with r and z and compare with the predictions of each model.

10-15 A room is heated by electric cables located in a thin plate beneath the floor. The plate is a much better conductor than the wood. It is found that to maintain the temperature of the floor's upper surface at 22.2°C it is necessary to supply electricity at a rate of 0.13 kW per square meter of floor. If the wood is fir with the grain at an angle $\alpha = 60°$ with the floor, calculate the temperature of the lower floor surface ($y = 0$) (see the figure). *Note:* Thermal conductivities of fir wood parallel to and against the grain are,[1] respectively, 3.0×10^{-4} and 0.9×10^{-4} cal/s · cm · K.

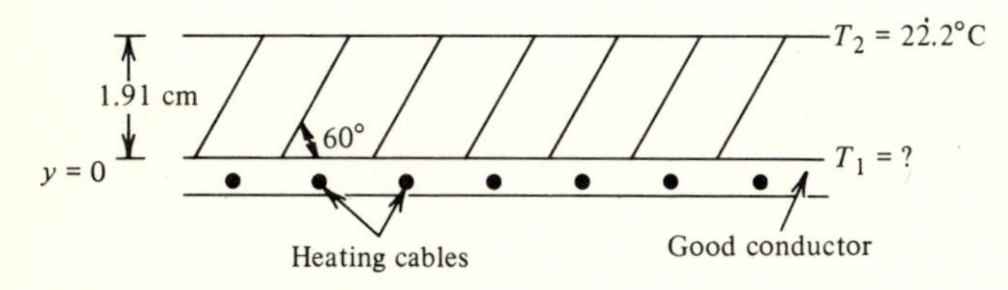

Figure P10-15 Problem 10-15.

10-16 A cylinder of wood is to be dried by high-frequency heating. The grain of the wood runs parallel to the axis of the cylinder, which is taken as the z direction. The x axis is also taken along the grain while the y axis is taken perpendicular to the grain. It is known that the thermal conductivity in the direction of the grain ($k_G = k_{xx} = k_{zz}$) is much greater than that in the direction perpendicular to the grain ($k_p = k_{yy}$). It will be assumed that Φ_e, the rate of internal heat generation per unit volume due to the electrical heating, is constant and uniform. Since the cylinder is long, conduction along its axis will be neglected. The temperature at the surface will be assumed constant and equal to T_s. The radius of the circular cross section of the cylinder is R.

(*a*) Write the steady-state energy balance in terms of components of **q**.

(*b*) Express **q** in terms of the temperature gradients.

(*c*) Using the results of Prob. 9-6 (on laminar flow in an ellipse), find the temperature distribution.

(*d*) Find the average temperature as a function of the heat generation Φ_e (using result of Prob. 9-7).

10-17 In a gas-absorption experiment, a gas A is being absorbed in a liquid B, which is flowing downward on the outside wall of a circular tube. The gas A is only slightly soluble in B, so that the viscosity of the liquid is not changed appreciably.

(*a*) Neglecting end effects and assuming newtonian behavior, show that the velocity distribution in the falling liquid film (Prob. 3-6) is

$$v_z = \frac{\rho g R^2}{4\mu} 1 - \left(\frac{r}{R}\right)^2 + 2\kappa^2 \ln \frac{r}{R}$$

(*b*) Obtain the differential equation describing the mass flux of A in the r and z directions in the liquid film.

(*c*) Assuming that A moves in the z direction primarily as a result of the downward flow of the film and that A moves in the r direction primarily by diffusion, substitute expressions for N_{Az} and N_{Ar} into the flux equation and obtain the differential equation describing $C_A(r, z)$.

(*d*) Since A is only slightly soluble in B, assume that A penetrates only a short distance into the liquid film. Under these conditions, A is present essentially only in the part of the film moving at the constant velocity $v_z|_{r=\kappa R} \equiv v_\kappa$. Substituting v_κ for v_z in the above equation, use separation of variables to show that the partial differential equation above is equivalent to the two ordinary differential equations

$$\frac{1}{G}\frac{dG}{dz} = -\beta^2 \quad \text{and} \quad v_\kappa D_{AB}\frac{1}{r}\frac{d}{dr}\left(r\frac{dF}{dr}\right) = -\beta^2 T$$

where $C_A = F(r)G(z)$ and β^2 is a separation constant.

(*e*) By making the substitution

$$w = \frac{\beta}{(D_{AB}v_\kappa)^{1/2}}$$

show that the equation involving $F(r)$ can be written

$$\frac{1}{r}\frac{d}{dr}\left(r\frac{dF}{dr}\right) + w^2 F = 0$$

which is a Bessel equation of zero order.

(*f*) Solving the two ordinary differential equations, show that the concentration distribution in the film is of the form

$$C_A(r, z) = [C_1 J_0(wr) + C_2 Y_0(wr)]e^{-\beta^2 z}$$

where C_1 and C_2 are constants and J_0 and Y_0 are zero order Bessel functions.

10-18 It is estimated that sulfur dioxide is being emitted at a rate of 80 g/s from a petroleum refinery from an average effective height of 60 m at 8:00 A.M. on an overcast winter morning with the wind at 6 m/s at $z = 1$ m. What is the ground-level concentration directly downwind from the refinery at a distance of 500 m?

10-19 If $R_L = \exp(-|\tau|/A)$, find expressions for $\overline{X^2}$ and $D^{(t)}$ assuming that A is known from experimental data.

10-20 Modify the dimensional analysis in Sec. 10-13 for mass transfer.

10-21 Modify the development in Sec. 10-14 for mass transfer.

10-22 Make the transformations indicated by Eqs. (10-168) to obtain the differential equations for $f(\eta)$ and $T^*(\text{Re}, \text{Pr})$.

10-23 (*a*) Complete the calculations in Example 10-7 for the first increment.

(*b*) Calculate the mean conversion.

(*c*) Repeat the calculations in parts (*a*) and (*b*), letting the velocity and diffusivities vary with radial position. Assume $k_w^* = 0.3$, $k_0^* = 0.8$.

10-24 Repeat Prob. 10-23 using a constant k_e but with a Biot number Bi = 3.

10-25 Using the large Prandtl-number assumption [Eqs. (10-177) to (10-179)], obtain Eq. (10-181).

10-26 Using Eqs. (10-184) and (10-188), obtain Eqs. (10-189) to (10-191).

10-27 Using Eq. (10-177) and the no-slip condition, obtain Eq. (10-186).

10-28 Show that a necessary condition for the quiescent state (no velocity gradients) to exist for a Boussinesq fluid is $Ra(\mathbf{F}\mathbf{x}\nabla T^*) = 0$, where **F** is a unit dimensionless vector acting in the direction of gravity.

REFERENCES

1. Ahmed, M., and R. W. Fahien: *Chem. Eng. Sci.,* **35**:889 (1980).
2. Argo, W. B., and J. M. Smith: *Chem. Eng. Prog.,* **49**:443 (1953).
3. Bird, R. B., C. F. Curtiss, and J. V. Hirschfelder: *Chem. Eng. Prog. Symp. Ser.,* **51**(16):70 (1955).
4. Bird, R. B., W. E. Stewart, and E. N. Lightfoot: "Transport Phenomena," Wiley, New York, 1960.
5. Brown, G. M.: *AIChE J.,* **6**:179 (1960).
6. Carberry, J. J.: "Chemical and Catalytic Reaction Engineering," McGraw-Hill, New York, 1976.
7. Chapman, T. W.: *AIChE J. Cont. Educ. Ser.* 4, p. 91, 1969.
8. Cohen, Y., and A. B. Metzner: *AIChE J.,* **27**:70 (1981).
9. Eckert, E. R. G., and T. W. Jackson: *NACA Report,* p. 1015, 1951.
10. Fahien, R. W., D. W. Kirmse, and S. B. Pahwa: *AIChE Symp. Ser.,* **72**(156):389 (1976).
11. Fahien, R. W., and J. M. Smith: *Chem. Eng. Prog.,* **49**:443 (1953).
12. Fahien, R. W., and I. Stankovich: *Chem. Eng. Sci.,* **34**:1350 (1979).
13. Friedlander, S. K.: *AIChE J.,* **7**:347 (1961).
14. Hougen, O., B. Gamson, and G. Thodos: *Trans. AIChE J.,* **31**:1 (1943).
15. Lightfoot, E. N.: *AIChE Cont. Educ. Ser.* 4, 1969.
16. Lightfoot, E. N., and E. L. Cussler: *Chem. Eng. Prog. Symp. Ser.,* **61**(58):66 (1965).
17. Middleman, S.: *Chem. Eng. Educ.,* **12**:166 (1978).
18. Moon, P., and D. E. Spencer: "Field Theory for Engineers," p. 146, Van Nostrand, Princeton, N.J., 1961.
19. Notter, R. H., and C. A. Sleicher: *Chem. Eng. Sci.,* **27**:2073 (1972).
20. Pellew, A., and R. V. Southwell: *Proc. R. Soc. Lond,* **A176**:312 (1940).
21. Petty, C. A.: Paper presented at AIChE Meeting, New Orleans, 1981.
22. Pohlhausen, E.: *Z. Angew. Math Mech.,* **1**:115 (1921).
23. Reid, W. H., and D. L. Harris: *Phys. Fluids,* **1**(2):102 (1953).
24. Roth, P. M., et al.: *Atmos. Environ.,* 8:97 (1974).
25. Schwartz, W. W., and J. M. Smith: *Ind. Eng. Chem.,* **45**:1209 (1953).
26. Seinfeld, J. H.: "Air Pollution: Physical and Chemical Fundamentals," McGraw-Hill, New York, 1975.
27. Sleicher, C. A., R. H. Notter, and M. D. Crippen: *Chem. Eng. Sci.,* **25**:845 (1970).
28. Smith, J. M.: "Chemical Engineering Kinetics," McGraw-Hill, New York, 1981.
29. Stewart, W. E.: Paper presented at Levich Conference, London, 1977.
30. Toor, H. L.: *AIChE J.,* **3**:197 (1957).
31. Toor, H. L.: *AIChE J.,* **10**:448, 460 (1964).
32. Tosum, I., and M. S. Willis: *Chem. Eng. Sci.,* **36**:781(1981).
33. Turner, H.: "Workbook of Atmospheric Dispersion Estimates," USEPA Res., Triangle Park, N.C., 1970.
34. Wang, J. C., and W. E. Stewart: Papers presented at AIChE Meeting, Chicago, Nov. 1980, and at AIChE Meeting, Los Angelos, Nov. 1982.
35. Weiland, Ralph H., and Ross Taylor: *Chem. Eng. Ed.,* **16**:158 (1982).
36. Whitaker, S.: "Introduction to Fluid Mechanics," chap. 6, Prentice-Hall, Englewood Cliffs, N.J., 1968.
37. Whitaker, S.: *Chem. Eng. Sci.,* **28**:139 (1973).
38. Yih, S. M., and R. C. Seagrave: *Int. J. Heat Mass Trans.,* **23**:749 (1980).
39. Young, L. C., and B. A. Finlayson: *Ind. Eng. Chem.,* **12**:412 (1973).

CHAPTER

ELEVEN

MACROSCOPIC AND POPULATION BALANCES

In engineering problems it is convenient to have available generalized *macroscopic* or overall balances of momentum, mass, and energy. These balances can be simplified to apply to individual practical problems, in a manner similar to that in which the equations of change were simplified in Chap. 10. As with the equations of change, the solution of a given engineering problem may require the use of more than one of the generalized balances.

There are two different approaches to deriving the macroscopic balances. The first is to apply the conservation law (say, for mass of i) on an overall basis, identifying all input, output, generation, and accumulation terms involved. For the mass balance this method of obtaining the macroscopic balances can be used without too much difficulty, but it requires considerably more physical intuition to obtain some of the other balances. In these cases, we use the second approach to derive the macroscopic balances from the equations of change. Recall that these equations were derived on a rate per unit volume basis and applied at any arbitrary point in the fluid. Consequently, to obtain the macroscopic balance, we need merely multiply by the differential volume and integrate over the total volume of the system.[2] To do so requires the use of the Gauss divergence theorem, the mathematical expression for differentiating an integral, and in some cases the use of thermodynamic functions.

11-1 DERIVATION OF THE MACROSCOPIC BALANCES†

Consider as our prototype system a fluid flowing through a plant (see Fig. 11-1) consisting of various items of equipment in series, e.g., a heat exchanger transfer-

† This section may be omitted on first reading.

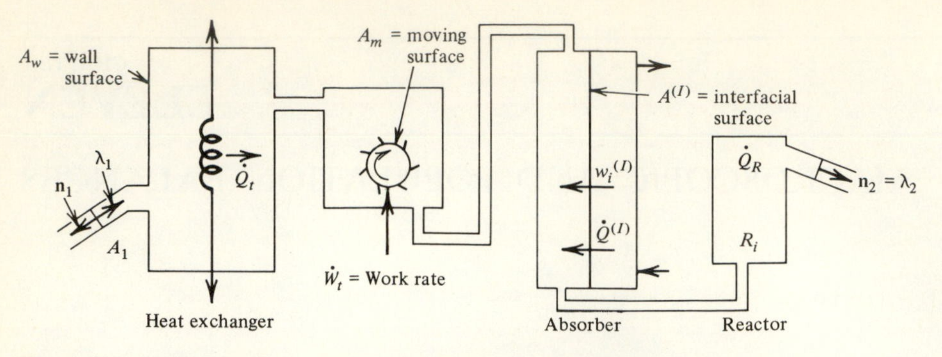

Figure 11-1 Generalized flow system. [*Adapted from R. B. Bird, Chem. Eng. Sci.*, **6**:*123* (*1957*).]

ring heat at rate $\dot{Q}_t$ from the surroundings to the system; a pump with moving surfaces A_m doing shaft work at rate $\dot{W}_t$ on the system; an absorber transferring to the system mass of species i at rate $w_i^{(I)}$ and heat at rate $\dot{Q}^{(I)}$; a reactor receiving heat $\dot{Q}_R$ from a reaction occurring at rate r_i; a typical inlet pipe or port of area A_1 and typical exit pipe of area A_2. The total wall area of the system is taken as A_w, and the total area is

$$A = A_e + A_w + A_m + A^{(I)} \tag{11-1}$$

where $A_e \equiv A_1 + A_2$ is the area of all entrances and exits. We are interested in deriving macroscopic or overall balances on the plant (or parts of it) for mass, energy, and momentum. To do so we start with the equations of change and integrate over the volume.

We will first derive general equations for unsteady-state conditions. The steady-state equations can then be obtained directly by setting the time derivatives equal to zero.

Macroscopic Mass Balance

Integrating the equation of continuity over the volume gives

$$\iiint \left(\frac{\partial \rho}{\partial t} + \nabla \cdot \rho \mathbf{v}\right) d\mathcal{V} = 0$$

Let us use the transport theorem (Appendix 10-1) on the first term and the Gauss divergence theorem (Chap. 8) on the second

$$\frac{d}{dt}\iiint \rho \, d\mathcal{V} - \iint_{A_m} \rho \mathbf{u} \cdot \mathbf{n} \, dA + \iint_{A} \rho \mathbf{v} \cdot \mathbf{n} \, dA = 0$$

where $\mathbf{u}$ = velocity of moving surface A_m. Let the total mass of the system be

$$m \equiv \iiint \rho \, d\mathcal{V}$$

Then
$$\frac{dm}{dt} + \iint_{A_m} \rho(\mathbf{v} - \mathbf{u}) \cdot \mathbf{n} \, dA + \iint_{A - A_m} \rho \mathbf{v} \cdot \mathbf{n} \, dA = 0$$

The first integral is zero if $\mathbf{u} = \mathbf{v}$; that is, the velocity of the moving surface is equal to the fluid velocity $\mathbf{v}$ (no slip). The last integral is over $A - A_m \equiv A_e + A_w + A^{(I)}$. Since the velocity is zero at the wall surfaces, the integral over A_w is zero. The integral over $A^{(I)}$ represents $w_t^{(I)}$, the net rate of mass transfer across the interface into the system, or

$$w_t^{(I)} \equiv -\iint_{A^{(I)}} \rho \mathbf{v} \cdot \mathbf{n}\, dA$$

The remaining integral over A_e represents the net rate of mass flow into the system, or

$$w_t \equiv -\iint_{A_e} \rho \mathbf{v} \cdot \mathbf{n}\, dA \tag{11-2}$$

Then
$$\frac{dm}{dt} = w_t^{(I)} + w_t \tag{11-3}$$

In general
$$w_t = \sum_p w_{Gp} \tag{11-3a}$$

where p is the pth port. Hence

$$w_t = w_{G1} + w_{G2} = -\iint_{A_1} \rho \mathbf{v} \cdot \mathbf{n}\, dA_1 - \iint_{A_2} \rho \mathbf{v} \cdot \mathbf{n}\, dA_2$$

Consider first the integral over A_1. If $\mathbf{v} = v_1 \boldsymbol{\lambda}_1$, where $\boldsymbol{\lambda}_1$ is a unit vector in the direction of flow, we have

$$w_{G1} = -\iint_{A_1} \rho_1 \mathbf{v} \cdot \mathbf{n}\, dA_1 = -\iint_{A_1} \rho_1 v_1 \boldsymbol{\lambda}_1 \cdot \mathbf{n}_1\, dA_1 = +\iint_{A_1} \rho_1 v_1\, dA_1$$

since $\boldsymbol{\lambda}_1 \cdot \mathbf{n}_1 = -1$; also since $\boldsymbol{\lambda}_2 \cdot \mathbf{n}_2 = +1$ at A_2, we have

$$w_{G2} \equiv -\iint \rho_2 \mathbf{v} \cdot \mathbf{n}\, dA_2 = -\rho_2 v_2 A_2$$

If we let w_1 and w_2 represent the magnitudes of w_{G1} and w_{G2}, respectively, we get

$$\frac{dm}{dt} = w_1 - w_2 + w_t^{(I)} \tag{11-4}$$

If we let $\Delta w \equiv w_2 - w_1$, we get

$$\boxed{\frac{dm}{dt} = -\Delta w + w_t^{(I)}} \tag{11-5}$$

and if we let the average value of the velocity over a cross section be $\langle v \rangle$ and assume that ρ is constant over the cross section, we can let

$$w_1 = \rho \langle v_1 \rangle A_1 \tag{11-6}$$

and similarly for w_2.

Mass of Species *i*

For a species i the equation of continuity is

$$\frac{\partial \rho_i}{\partial t} + \nabla \cdot \rho_i \mathbf{v} = r_i - \nabla \cdot \mathbf{j}_i$$

The average diffusive flux at entrances and exits $\langle j_i \rangle$ is usually negligible compared with the bulk flow. Also $\mathbf{j}_i = 0$ at solid surfaces (no diffusion through walls). With these assumptions a similar integration over the volume gives

$$\frac{dm_i}{dt} = -\Delta w_i + w_{it}^{(I)} + \langle r_i \rangle \mathcal{V} \tag{11-7}$$

where $w_{it}^{(I)}$ is the total mass transfer rate at the interface, or

$$w_{it}^{(I)} \equiv -\iint_{A^{(I)}} (\rho \mathbf{v}_i + \mathbf{j}_i) \cdot \mathbf{n}\, dA^{(I)} \tag{11-8}$$

and

$$\langle r_i \rangle \equiv \frac{\iiint r_i \, d\mathcal{V}}{\mathcal{V}} \tag{11-9}$$

is the total production rate of i. As before, m_i is the total mass of species i, and

$$w_i \equiv \rho_i \langle v \rangle A$$

Macroscopic Momentum Balance

Let us now integrate the momentum equation over the volume

$$\iiint \left(\underset{(1)}{\frac{\partial(\rho \mathbf{v})}{\partial t}} + \underset{(2)}{\nabla \cdot \rho \mathbf{v}\mathbf{v}} + \underset{(3)}{\nabla \cdot \underline{\tau}} + \underset{(4)}{\nabla p} + \underset{(5)}{\rho \mathbf{g}} \right) d\mathcal{V} = 0$$

Using the transport theorem and the Gauss theorem on terms (1) and (2) gives

$$\frac{d}{dt}\iiint \rho \mathbf{v}\, d\mathcal{V} - \iint_{A_m} \rho \mathbf{v}\mathbf{u} \cdot \mathbf{n}\, dA + \iint_{A} \rho \mathbf{v}\mathbf{v} \cdot \mathbf{n}\, dA$$

$$= \frac{d\mathbf{P}}{dt} + \iint_{A_m} \rho \mathbf{v}(\mathbf{v} - \mathbf{u}) \cdot \mathbf{n}\, dA + \iint_{A_e} \rho \mathbf{v}\mathbf{v} \cdot \mathbf{n}\, dA + \iint_{A_w} \rho \mathbf{v}\mathbf{v} \cdot \mathbf{n}\, dA$$

The integral over A_m is zero since $\mathbf{v} = \mathbf{u}$, and the integral over A_w is zero since $\mathbf{v} = 0$ at the wall. We neglect $\mathbf{F}^{(I)}$, the integral over $A^{(I)}$. If we let $\mathbf{v} = \boldsymbol{\lambda} v$, where $\boldsymbol{\lambda}$ is a unit vector in the direction of flow, we get

$$\iint_{A_e} \rho \mathbf{v}\mathbf{v} \cdot \mathbf{n}\, dA = \rho\langle v_1^2\rangle \boldsymbol{\lambda}_1 \boldsymbol{\lambda}_1 \cdot \mathbf{n}_1 A_1 + \rho\langle v_2^2\rangle \boldsymbol{\lambda}_2 \boldsymbol{\lambda}_2 \cdot \mathbf{n}_2 A_2$$

$$= \Delta(\rho\langle v^2\rangle \mathbf{A}) = \Delta\left(\mathbf{w}\frac{\langle v^2\rangle}{\langle v\rangle}\right)$$

where $\qquad \mathbf{A} \equiv \boldsymbol{\lambda} A \qquad \mathbf{w} = \rho\langle v\rangle \mathbf{A}$

Note that the vector $\mathbf{A}$ is equal to the vector $\mathbf{S} = n\mathbf{A}$ only if $\mathbf{n} = \boldsymbol{\lambda}$.

Using the Gauss theorem on term (3), letting $\mathbf{v} = \mathbf{u}$, and neglecting the integral over A_e gives

$$\iiint \nabla \cdot \underline{\tau}\, d\mathcal{V} = \iint_{A_m} \mathbf{n} \cdot \underline{\tau}\, dA + \iint_{A_w} \mathbf{n} \cdot \underline{\tau}\, dA \equiv \mathbf{F}_{vf}$$

where $\mathbf{F}_{vf}$ is the viscous force of the fluid on the wall. Term (4) gives

$$\iint_{A_w + A_m} p\mathbf{n}\, dA + \iint_{A_e} p\mathbf{n}\, dA$$

The first term is a normal force of the fluid on the solid $\mathbf{F}_{vf}$. The second term is

$$\iint_{A_e} \mathbf{n} p\, dA = \mathbf{n}_1 p_1 A_1 + \mathbf{n}_2 p_2 A_2 = -\boldsymbol{\lambda}_1 p_1 A_1 + \boldsymbol{\lambda}_2 p_2 A_2 \equiv \Delta(p\mathbf{A})$$

Let

$$\iint \mathbf{n} \cdot \underline{\tau}\, dA + \iint \mathbf{n} p\, dA \equiv \mathbf{F}_{vf} + \mathbf{F}_{pf} \equiv \mathbf{F}_f \equiv -\mathbf{F} \tag{11-10}$$

where $\mathbf{F}_f$ is the total force of the fluid on the solid.† If $\mathbf{F}_f$, $\mathbf{A}$, and $\mathbf{w}$ are vectors in the $\boldsymbol{\lambda}$ direction (as given by $\mathbf{v}$), we obtain

$$\boxed{\frac{d\mathbf{P}}{dt} = -\Delta\left(\mathbf{w}\frac{\langle v^2\rangle}{\langle v\rangle} + p\mathbf{A}\right) - \mathbf{F}_f + m\mathbf{g}} \tag{11-11}$$

where $\qquad \mathbf{P} \equiv \iiint_{\mathcal{V}} \rho \mathbf{v}\, d\mathcal{V} = \text{total momentum of the system} \qquad (11\text{-}12)$

Macroscopic Total-Energy Balance

The equation of change for total energy $\hat{E} = \hat{K} + \hat{U} + \hat{\psi}$ is

$$\rho \frac{D\hat{E}}{Dt} \equiv \frac{\partial(\rho \hat{E})}{\partial t} + \nabla \cdot \rho \hat{E}\mathbf{v} = -\nabla \cdot \mathbf{q} - \nabla \cdot \underline{\tau} \cdot \mathbf{v} - \nabla \cdot p\mathbf{v} + \Phi_H \tag{11-13}$$

When we multiply by $d\mathcal{V}$ and integrate over the volume, the left-hand side be-

†We use $\mathbf{F}$ to denote the force acting *on* the fluid; thus $\mathbf{F}g_c = \dot{\mathbf{P}}$, the rate of momentum gain, is analogous to $\dot{Q}_t$, the rate of heat gain.

comes

$$\frac{d}{dt}\iiint_{\mathcal{V}} \rho \hat{E}\, d\mathcal{V} - \iint_{A_e} \rho(\hat{\Psi} + \hat{U} + \hat{K})\mathbf{v}\cdot\mathbf{n}\, dA = \frac{dE}{dt} + \Delta(\rho\langle\hat{E}\rangle\langle v\rangle A)$$

Now

$$\langle\hat{E}\rangle \equiv \frac{\iint_A (\hat{\Psi} + \hat{U} + \hat{K})v\, dA}{\langle v\rangle A} = \langle\hat{\Psi}\rangle + \langle\hat{U}\rangle + \frac{1}{2}\frac{\langle v^3\rangle}{\langle v\rangle} \tag{11-14}$$

and so

$$\Delta(\rho\langle\hat{E}\rangle\langle v\rangle A) = \Delta\left(\langle\hat{\Psi}\rangle w + \langle\hat{U}\rangle w + \frac{1}{2}\frac{\langle v^3\rangle}{\langle v\rangle} w\right)$$

Then

$$\iiint_{A_w} - \nabla\cdot\mathbf{q}\, d\mathcal{V} = -\iint_{A_w} \mathbf{q}\cdot\mathbf{n}\, dA = \dot{Q}_t$$

We let the total work rate due to viscous stress be

$$\dot{W}_{vt} \equiv -\iiint \nabla\cdot\underline{\tau}\cdot\mathbf{v}\, d\mathcal{V} = -\iint_{A_m} \mathbf{n}\cdot\underline{\tau}\cdot\mathbf{v}\, dA \tag{11-15}$$

Also

$$-\iiint \nabla\cdot p\mathbf{v}\, d\mathcal{V} = -\iint_{A_e+A_m} p\mathbf{v}\cdot\mathbf{n}\, dA = -\Delta(p\langle v\rangle \mathbf{A}) + \dot{W}_{pt}$$

where $\dot{W}_{pt}$ is the total work rate due to pressure force, or

$$\dot{W}_{pt} \equiv -\iint_{A_m} p\mathbf{v}\cdot\mathbf{n}\, dA \tag{11-16}$$

Then since $p\langle v\rangle A \equiv wp/\rho$, the macroscopic balance is [from Eqs. (11-13), (11-14), (11-15), (11-16)]

$$\boxed{\frac{dE}{dt} = -\Delta\left[w\left(\hat{\Psi} + \hat{U} + \frac{p}{\rho} + \frac{1}{2}\frac{\langle v^3\rangle}{\langle v\rangle}\right)\right] + \dot{Q}_t - \dot{W}_t + \dot{Q}_H} \tag{11-17}$$

where $\dot{W}_t = \dot{W}_{pt} + \dot{W}_{Vt}$ is the total work rate and $\dot{Q}_H = \langle\Phi_H\rangle\mathcal{V}$ is the total heat-generation rate, in which

$$\langle\Phi_H\rangle \equiv \frac{\iiint \Phi_H\, d\mathcal{V}}{\mathcal{V}} \tag{11-18}$$

Macroscopic Mechanical-Energy Balance

We start with the equation of motion and take the dot product of it with $\mathbf{v}$ to get the kinetic-energy equation [see Eqs. (10-16), (10-17), (10-18)]

$$\rho\frac{DK}{Dt} \equiv \underset{(1)}{\frac{\partial(\rho\hat{K})}{\partial t}} + \underset{(2)}{\nabla\cdot\rho\hat{K}\mathbf{v}} = \underset{(3)}{-\nabla\cdot\underline{\tau}\cdot\mathbf{v}} + \underset{(4)}{\underline{\tau}:\nabla\mathbf{v}} + \underset{(5)}{\rho\mathbf{v}\cdot\mathbf{g}} - \underset{(6)}{\mathbf{v}\cdot\nabla p}$$

For terms (1) and (2) we use the transport and divergence theorems to obtain

$$\frac{d}{dt}\iiint \rho\hat{K}\,d\mathcal{V} - \iint_{A_m} \rho\hat{K}(\mathbf{v}-\mathbf{u})\cdot\mathbf{n}\,dA + \iint_{A_e} \rho\hat{K}\mathbf{v}\cdot\mathbf{n}\,dA \qquad (11\text{-}19)$$

If we let

$$\iiint \rho\hat{K}\,d\mathcal{V} \equiv K \qquad \mathbf{v}=\mathbf{u} \qquad \iint \tfrac{1}{2}\rho_1 v_1^3\,dA_1 \equiv \tfrac{1}{2}\rho_1\langle v_1^3\rangle A_1$$

then

$$\frac{dK}{dt} + \boldsymbol{\delta}_1\cdot\mathbf{n}_1\frac{\rho_1\langle v_1^3\rangle\langle v_1\rangle A_1}{2\langle v_1\rangle} + \boldsymbol{\delta}_2\cdot\mathbf{n}_2\frac{\rho_2\langle v_2^3\rangle\langle v_2\rangle A_2}{2\langle v_2\rangle} = \frac{dK}{dt} + \Delta\left(w\frac{\langle v^3\rangle}{2\langle v\rangle}\right)$$

Term (3) gives

$$-\iiint \nabla\cdot\underline{\boldsymbol{\tau}}\cdot\mathbf{v}\,d\mathcal{V} = -\iint_{A} \mathbf{n}\cdot\underline{\boldsymbol{\tau}}\cdot\mathbf{v}\,dA = -\iint_{A_m} \mathbf{n}\cdot\underline{\boldsymbol{\tau}}\cdot\mathbf{v}\,dA \equiv \dot{W}_V \qquad (11\text{-}20)$$

term (4) gives

$$-\iiint \underline{\boldsymbol{\tau}}:\nabla\mathbf{v}\,d\mathcal{V} \equiv E_V = \text{total viscous dissipation rate} \qquad (11\text{-}21)$$

and term (5) gives

$$\iiint \rho\mathbf{v}\cdot\mathbf{g}\,d\mathcal{V} = -\iiint \rho\mathbf{v}\cdot\nabla\hat{\psi}\,d\mathcal{V} = -\iiint \nabla\cdot\rho\mathbf{v}\hat{\psi}\,d\mathcal{V} + \iiint \hat{\psi}\,\nabla\cdot\rho\mathbf{v}\,d\mathcal{V}$$

$$= -\iint_{A} \rho\hat{\psi}\mathbf{v}\cdot\mathbf{n}\,dA - \iiint \hat{\psi}\frac{\partial\rho}{\partial t}\,d\mathcal{V} = \Delta(\rho\langle\hat{\psi}\rangle\langle v\rangle A) - \frac{d}{dt}\iiint \rho\hat{\psi}\,d\mathcal{V}$$

Term (6) requires integration of $\mathbf{v}\cdot\nabla p$ over a specified path. Three cases are of special interest: constant temperature, constant entropy, and constant density (incompressible fluid).

Constant-temperature case From thermodynamics for free energy, $G \equiv H - T\mathcal{S}$ and

$$d\hat{G} = -\hat{\mathcal{S}}\,dT + \frac{1}{\rho}\,dp$$

where $\hat{G}$ is the free energy per unit mass and $\hat{\mathcal{S}}$ the entropy per unit mass. Then at constant T

$$(d\hat{G})_T = \frac{1}{\rho}\,dp$$

and term (6) becomes

$$-\iiint \mathbf{v}\cdot\nabla p\,d\mathcal{V} = -\iiint \rho\mathbf{v}\cdot\frac{\nabla p}{\rho}\,d\mathcal{V}$$

$$= -\iiint \rho\mathbf{v}\cdot\nabla\widehat{G}\,d\mathcal{V} = -\iiint \nabla\cdot\rho\widehat{G}\mathbf{v}\,d\mathcal{V} + \iiint \widehat{G}\nabla\cdot\rho\mathbf{v}\,d\mathcal{V}$$

$$= -\iint \rho\widehat{G}\mathbf{v}\cdot\mathbf{n}\,dA + \iiint \widehat{G}\left(-\frac{\partial\rho}{\partial t}\right)d\mathcal{V}$$

$$= -\iint \rho\widehat{G}\mathbf{v}\cdot\mathbf{n}\,dA - \frac{d}{dt}\iiint \rho\widehat{G}\,d\mathcal{V} + \iint_{A_m} \rho\widehat{G}\mathbf{w}\cdot\mathbf{n}\,dA - \iiint \rho\frac{\partial\widehat{G}}{\partial t}\,d\mathcal{V}$$

$$= -\Delta(\rho\widehat{G}\langle v\rangle A) - \frac{d}{dt}\iiint \rho\widehat{G}\,d\mathcal{V} + \iiint \frac{\partial p}{\partial t}\,d\mathcal{V}$$

$$= -\Delta(w\widehat{G}) - \frac{d}{dt}\left[\iiint (\rho\widehat{G} - p)\,d\mathcal{V}\right] - \iint_{A_m} p\mathbf{v}\cdot\mathbf{n}\,dA$$

$$= -\Delta(w\widehat{G}) - \frac{d\mathcal{A}}{dt} + \dot{W}_{pt}$$

where $\mathcal{A}$ is the thermodynamic work function ($\mathcal{A} \equiv H - T\mathcal{S}$). Combining terms gives

$$\boxed{\frac{d}{dt}(K + \psi + \mathcal{A}) = -\Delta\left[\left(\frac{1}{2}\frac{\langle v^3\rangle}{\langle v\rangle} + \widehat{\psi} + \widehat{G}\right)w\right] + \dot{W}_t - E_v} \qquad (11\text{-}22)$$

where $\dot{W}_t = \dot{W}_{pt} + \dot{W}_{vt}$.

Constant-entropy case From thermodynamics

$$d\widehat{H} = \frac{dp}{\rho} - T\,d\widehat{\mathcal{S}} \qquad \text{or} \qquad (d\widehat{H})_{\mathcal{S}} = \frac{1}{\rho}dp$$

This results in

$$\boxed{\frac{d}{dt}(K + \psi + U) = -\Delta\left[\left(\frac{1}{2}\frac{\langle v^3\rangle}{\langle v\rangle} + \widehat{\psi} + \widehat{H}\right)w\right] + \dot{W}_t - E_v} \qquad (11\text{-}23)$$

Constant-density case (incompressible fluid) Term (6) becomes

$$-\iiint \mathbf{v}\cdot\nabla p\,d\mathcal{V} = -\iiint \nabla\cdot p\mathbf{v}\,d\mathcal{V} + \iiint p\,\nabla\cdot\mathbf{v}\,d\mathcal{V}$$

$$= -\iint p\mathbf{v}\cdot\mathbf{n}\,dA = -\Delta(p\langle v\rangle A) - \dot{W}_{pt} = +\frac{\Delta(wp)}{\rho} - \dot{W}_{pt}$$

Then

$$\boxed{\frac{d}{dt}(K + \psi) = -\Delta\left[\left(\frac{1}{2}\frac{\langle v^3\rangle}{\langle v\rangle} + \widehat{\psi} + \frac{p}{\rho}\right)w\right] - \dot{W}_t + E_v} \qquad (11\text{-}24)$$

The unsteady-state mechanical-energy balances at constant H and $\mathcal{S}$ have been generalized by Stewart (see Ref. 3), who let $dY = dp/\rho$, where Y is a thermodynamic property such as H or $\mathcal{S}$. This gives

$$\frac{d}{dt}(K + \psi + X) = -\Delta\left[\left(\frac{1}{2}\frac{\langle v^3\rangle}{\langle v\rangle} + \hat{\psi} + \hat{Y}\right)w\right] + \dot{W}_t - E_v \qquad (11\text{-}24a)$$

where $\qquad X \equiv \int \hat{X}\rho\, d\mathcal{V} \qquad$ and $\qquad \hat{X} = \hat{Y} - \dfrac{p}{\rho}$

Pings[6] has derived the alternate form

$$\frac{d(K + \psi)}{dt} = \sum_p \langle \hat{K} + \hat{\psi}\rangle w_{Gp} + \dot{W}_t - E_v + \int_{\mathcal{V}} \mathbf{v}\cdot\nabla p\, d\mathcal{V} + \sum_s p_s \frac{d\mathcal{V}_s}{dt} \qquad (11\text{-}24b)$$

where p_s is the pressure at the surface of the system and $d\mathcal{V}_s/dt$ is the rate of change of the volume of the system but at point S of boundary movement.

11-2 STEADY-STATE BALANCES

The steady-state balances can be obtained from the unsteady-state balances by letting $d/dt = 0$. If the energy balances are divided by w, a more conventional form is obtained in which the units of energy per unit mass appear instead of energy per unit time. In the SI this is equivalent to m^2/s^2, which is velocity squared; e.g., kinetic energy per unit mass. In the English system, typical units are ft · lbf/lb, which corresponds to work per unit mass, and force units must be multiplied by g_c. The following definitions are used in the steady-state balances shown in Table 11-1, which also summarizes the unsteady-state balances:

$$\hat{W}_t \equiv \frac{\dot{W}_t}{w} \qquad \hat{E}_v \equiv \frac{E_v}{w} \qquad \hat{Q} \equiv \frac{\dot{Q}}{w}$$

$$\Delta\hat{G} = \int_{p_1}^{p_2} \frac{dp}{\rho} \text{ at const } T \qquad (11\text{-}25)$$

$$\Delta\hat{H} = \int_{p_1}^{p_2} \frac{dp}{\rho} \text{ at const } \mathcal{S}$$

where p_1, p_2 are pressures at inlet and outlet.

Bernoulli's Equation

Originally derived for flow along a streamline assuming an inviscid fluid, this famous equation can more easily be obtained by letting $\hat{E}_v = 0$ in the mechanical-energy balance. Since there is no shaft work, this gives

$$\Delta\left(\tfrac{1}{2}v^2 + \frac{p}{\rho} + gz\right) = 0 \qquad (11\text{-}26)$$

Table 11-1 Macroscopic balances for a single species

	Unsteady state		Steady state	
Mass	$\frac{dm}{dt} = -\Delta w + w_t^{(I)}$	(11-4)	$\Delta w = w_t^{(I)}$	(1)
Momentum	$\frac{d\mathbf{P}}{dt} = -\Delta\left(\frac{\langle v^2\rangle}{\langle v\rangle}\mathbf{w} + p\mathbf{A}\right) - \mathbf{F}_f + m\mathbf{g}$	(11-11)	$\mathbf{F}_f = -\Delta\left(\frac{\langle v^2\rangle}{\langle v\rangle}\mathbf{w} + pA\right) + m\mathbf{g}$	(2)
Total energy	$\frac{dE}{dt} = -\Delta\left[\left(\widehat{U} + p\widehat{V} + \frac{1}{2}\frac{\langle v^3\rangle}{\langle v\rangle} + \widehat{\psi}\right)w\right] + \dot{Q}_t + \dot{W}_t$	(11-17)	$\Delta\left(\widehat{U} + p\widehat{V} + \frac{1}{2}\frac{\langle v^3\rangle}{\langle v\rangle} + \widehat{\psi}\right) = \widehat{Q}_t + \widehat{W}_t$	(3)
Mechanical energy balances				
Isothermal	$\frac{d}{dt}(K + \psi + A) = -\Delta\left[\left(\frac{1}{2}\frac{\langle v^3\rangle}{\langle v\rangle} + \widehat{\psi} + \widehat{G}\right)w\right] + \dot{W}_t - E_v$	(11-22)	$\Delta\left(\frac{1}{2}\frac{\langle v^3\rangle}{\langle v\rangle} + \widehat{\psi} + \widehat{G}\right) - \widehat{W}_t + \widehat{E}_v = 0$	(4)
Isentropic	$\frac{d}{dt}(K + \psi + U) = -\Delta\left[\left(\frac{1}{2}\frac{\langle v^3\rangle}{\langle v\rangle} + \widehat{\psi} + \widehat{H}\right)w\right] + \dot{W}_t - E_v$	(11-23)	$\Delta\left(\frac{1}{2}\frac{\langle v^3\rangle}{\langle v\rangle} + \widehat{\psi} + \widehat{H}\right) - \widehat{W}_t + \widehat{E}_v = 0$	(5)
Incompressible	$\frac{d}{dt}(K + \psi) = -\Delta\left[\left(\frac{1}{2}\frac{\langle v^3\rangle}{\langle v\rangle} + \widehat{\psi} + \frac{p}{\rho}\right)w\right] + \dot{W}_t - E_v$	(11-24)	$\Delta\left(\frac{1}{2}\frac{\langle v^3\rangle}{\langle v\rangle} + \widehat{\psi} + \frac{p}{\rho}\right) - \widehat{W}_t + \widehat{E}_v = 0$	(6)

Note: $\Delta\widehat{\psi} = g\,\Delta h$; $(\Delta\widehat{G})_T = \int_{p_1}^{p_2}\frac{dp}{\rho}$; $(\Delta\widehat{H})_{\mathcal{S}} = \int_{p_1}^{p_2}\frac{dp}{\rho}$.

Table 11-2 Multicomponent macroscopic balances

Mass	$\frac{dm}{dt} = -\Delta w + w_t^{(I)} \qquad \frac{dm_i}{dt} = -\Delta w_i + w_{it}^{(I)} + \langle r_i \rangle \mathcal{V}$
Momentum	$\frac{d\mathbf{P}}{dt} = -\Delta\left(\frac{\langle v^2\rangle}{\langle v\rangle}\mathbf{w} + p\mathbf{A}\right) - \mathbf{F}_f + m_t\mathbf{g} + \dot{\mathbf{P}}^{(I)}$
Total energy	$\frac{dE}{dt} = -\Delta\left[\left(\hat{U} + p\hat{V} + \frac{1}{2}\frac{\langle v^3\rangle}{v} + \hat{\psi}\right)w\right] + \dot{Q}_t + \dot{W}_t + \dot{Q}^{(I)} + \dot{Q}_R$
Mechanical energy balances	
Isothermal	$\frac{d}{dt}(K + \psi + \mathcal{A}) = -\Delta\left[\left(\frac{1}{2}\frac{\langle v^3\rangle}{\langle v\rangle} + \hat{\psi} + \hat{G}\right)w\right] + \dot{W}_t - E_v + B^{(I)}$
Isentropic	$\frac{d}{dt}(K + \psi + U) = -\Delta\left[\left(\frac{1}{2}\frac{\langle v^3\rangle}{\langle v\rangle} + \hat{\psi} + \hat{H}\right)w\right] + \dot{W}_t - E_v + B^{(I)}$
Incompressible	$\frac{d}{dt}(K + \psi) = -\Delta\left[\left(\frac{1}{2}\frac{\langle v^3\rangle}{\langle v\rangle} + \hat{\psi} + \frac{p}{\rho}\right)w\right] + \dot{W}_t - E_v + B^{(I)}$

It can also be written in differential form by replacing the Δ operator with differentials

$$\tfrac{1}{2}d(v^2) + \frac{dp}{\rho} + g\,dz = 0 \tag{11-27}$$

11-3 MULTICOMPONENT SYSTEMS

For multicomponent systems the macroscopic balances are modified to include transfer through interfaces, as indicated in Table 11-2.

These interfacial transfer quantities are defined as follows:

$$w_i^{(I)} \equiv -\iint_{A^{(I)}} (\rho\mathbf{v}_i + \mathbf{j}_i)\cdot\mathbf{n}\,dA^{(I)} \qquad \dot{Q}^{(I)} \equiv -\iint_{A^{(I)}} \mathbf{n}\cdot\sum_{i=1}^{s} C_i\mathbf{v}_i\hat{H}_i\,dA^{(I)}$$

$$\dot{\mathbf{P}}^{(I)} \equiv \iint_{A^{(I)}} \mathbf{n}\cdot\rho\mathbf{v}\mathbf{v}\,dA^{(I)} \qquad B^{(I)} \equiv \iint_{A^{(I)}} \mathbf{n}\cdot\rho\hat{E}\mathbf{v}\,dA^{(I)} \tag{11-28}$$

11-4 APPLICATIONS OF MOMENTUM AND MECHANICAL-ENERGY BALANCES

Macroscopic balances are commonly used in engineering applications because of their usefulness and simplicity. They are useful because they can give us what we want in most applications, namely, macroscopic quantities such as (1) the total transport to or from the system of heat, mass, and/or momentum or (2) the average temperature, concentration, or velocity of the system. They are simple in that, unlike the macroscopic equations of change, they are not partial differential equations. But that simplicity is somewhat misleading, because in systems where gradients of temperature, concentration, or velocity occur the distribution of these potentials must be known for rigorous calculation of the average potentials or of the transport rates at the boundaries of the system. If the potential distribution cannot be calculated, it is often estimated or assumed; or a transfer coefficient is assumed or estimated. The value of using the macroscopic balance then is that it can take the place of assuming transfer coefficients or potential distributions.

The macroscopic mechanical-energy balance can be applied either to an entire plant (Fig. 11-1), to obtain, for example, the total energy loss and hence the pumping requirements, or to individual subsystems of the plant, e.g., pipelines, fittings, flowmeters, expansions, contractions, pumps, compressors, turbines, and packed beds. We first illustrate its use in finding the energy loss in straight pipes and other subsystems and then show how these are added to obtain the energy loss and power required for an entire piping system.

Example 11-1: Application to straight conduits or pipes An incompressible fluid is flowing downward at steady state through a conduit of length L. The flow is fully developed. Let the inlet and outlet cross-sectional areas be A_1 and A_2, respectively. Apply the mass, momentum, and mechanical-energy balances to the conduit.

SOLUTION Since there is no interfacial transport, the *mass* equation for steady state gives [Eq. (11-4) or Eq. (1) of Table 11-1]

$$\frac{dm}{dt} = 0 = -\Delta w = w_1 - w_2 \qquad \text{or} \qquad w = \text{const} = \rho\langle v\rangle A$$

Since $A_1 = A_2$ and $\rho_1 = \rho_2$, $\langle v_1\rangle = \langle v_2\rangle$, which agrees with the third of Eqs. (8-112). The macroscopic *momentum* balance in the direction of flow ($+z$) at steady state is [Eq. (11-11) or Eq. (2) of Table 11-1]

$$F_{fz} = -\Delta\left(\frac{\langle v^2\rangle}{\langle v\rangle} w_z + pA_z\right) + mg_z \qquad \text{where } \langle v^2\rangle \equiv \frac{\iint v_z^2\, dA_z}{\iint dA_z} \tag{11-29}$$

If the flow is *laminar* and the pipe is circular (Chap. 3),

$$v_z(r) = 2\langle v\rangle\left[1 - \left(\frac{r}{R}\right)^2\right]$$

Then it can be shown by integration (see the Problems) that

$$\langle v^2\rangle = \tfrac{4}{3}\langle v\rangle^2 \tag{11-30}$$

If the flow is *turbulent,* the velocity profile is usually sufficiently flat to assume that $\overline{v}_z = \langle \overline{v}_z\rangle$ and consequently that

$$\langle \overline{v}^2\rangle = \langle \overline{v}\rangle^2 \tag{11-31}$$

Then $$\Delta\left(\frac{\langle \bar{v}^2 \rangle}{\langle \bar{v} \rangle} w\right) = \frac{\langle \bar{v}_2^2 \rangle}{\langle \bar{v}_2 \rangle} w_2 - \frac{\langle \bar{v}_1^2 \rangle}{\langle \bar{v}_1 \rangle} w_1$$

But since the flow is *fully developed,* the entrance and outlet velocity profiles will be the same and there will be no *net* bulk transport of momentum; hence this term will be zero. The second term is

$$-\Delta(pA) = -p_2A_2 + p_1A_1 = -(p_2 - p_1)A$$

The F_{fz} term is the fluid force [Eq. (11-10)]. In the last term, the total mass m of the system is ρAL, and $g_z = g$ if the flow is downward. Then the momentum balance in the z direction is

$$F_{zf} = F_f = (p_1 - p_2)A + \rho ALg \tag{11-32}$$

This equation is equivalent to the one we previously derived for drag force [Eq. (3-23)]. Note that the actual force of the *solid* on the fluid, $F_{\hat{f}z} = -F_{fz}$, is negative, which means that it acts opposite to the direction of flow. However, we usually deal with the drag force $F_D = |F_f|$.

The mechanical-energy balance at steady state (constant ρ) [Eq. (6) of Table 11-1] can be written

$$0 = \Delta\left(\frac{1}{2}\frac{\langle v^3 \rangle}{\langle v \rangle}\right) + \Delta\hat{\psi} + \frac{\Delta p}{\rho} - \widehat{W}_t + \widehat{E}_v = 0 \tag{11-33}$$

where $\widehat{W}_t \equiv \dot{W}_t/w$, $\widehat{E}_v \equiv E_v/w$, $-\dot{W}_t$ = shaft work, i.e., work done by moving equipment such as pumps, compressors, turbines, mixers, etc. The net viscous dissipation or lost work due to friction is

$$E_v = -\iiint_{\mathcal{V}} \underline{\tau} : \nabla \mathbf{v} \, d\mathcal{V}$$

and $$\langle v^3 \rangle \equiv \frac{\iint v_z^3 \, dA_z}{\iint dA_z}$$

For *laminar* flow in a circular pipe it can be shown that

$$\langle v^3 \rangle = 2\langle v \rangle^3 \tag{11-34}$$

For *turbulent* flow the velocity profile is sufficiently flat to permit assuming that $\bar{v}_z = \langle \bar{v}_z \rangle$, and

$$\langle \bar{v}^3 \rangle = \langle \bar{v} \rangle^3 \tag{11-35}$$

If the flow is *fully developed,* the entrance and outlet velocity profiles will be identical and hence

$$\Delta\frac{\langle v^3 \rangle}{\langle v \rangle} = \frac{\langle v_2^3 \rangle}{\langle v_2 \rangle} - \frac{\langle v_1^3 \rangle}{\langle v_1 \rangle} = 0$$

The second term represents the change in potential energy per unit mass between the exit and entrance, or

$$\Delta\hat{\psi} \equiv \hat{\psi}_2 - \hat{\psi}_1 = 0 - gL = -gL$$

Since there are no moving parts in the system, the shaft work $\hat{W}_t = 0$. Hence the mechanical-energy balance [Eq. (11-33)] becomes

$$0 = 0 - gL + \frac{p_2 - p_1}{\rho} + 0 + \hat{E}_v = 0 \tag{11-36}$$

The momentum-balance equation (11-29) can be solved for $p_1 - p_2$, giving

$$p_1 - p_2 = \frac{F_f}{A_z} - \rho L g$$

Substituting into the mechanical-energy balance (11-33) gives

$$\hat{E}_v = \frac{F_f}{\rho A_z} = \frac{-F}{\rho A_z} \tag{11-37}$$

The friction factor is defined in Eqs. (4-5) and (4-8) as

$$f \equiv \frac{F_D/A_w}{\frac{1}{2}\rho\langle v\rangle^2}$$

We can solve this equation for $F_D = |F_f|$ and substitute into the previous one to obtain

$$\hat{E}_v = \tfrac{1}{2}\langle v\rangle^2 \frac{A_w}{A_z} f \tag{11-38}$$

Letting $A_w/A_z = ZL/A_z = L/R_H$, where Z is the perimeter and $R_H \equiv A_z/Z$ is the hydraulic radius, gives

$$\hat{E}_v = \tfrac{1}{2}\langle v\rangle^2 \frac{L}{R_H} f \tag{11-39}$$

Since, in general, $\hat{E}_v$ will be proportional to $\frac{1}{2}\langle v\rangle^2$, it is convenient to define a term e_v, called the *friction-loss factor,* by

$$\hat{E}_v \equiv \tfrac{1}{2}\langle v\rangle^2 e_v \tag{11-40}$$

For *conduits* of hydraulic radius R_H

$$e_v = \frac{L}{R_H} f \tag{11-41}$$

where the friction factor f is a function of Reynolds number and of any geometric ratios required.

Equation (11-37) tells us how to relate F_f, the friction force in the macroscopic momentum balance to E_v, the energy loss in the macroscopic mechanical-energy balance. This energy loss can then be added together with those of other pipes and items of equipment to obtain a total energy loss for a piping system. However, in order to obtain F_f or F_D one still needs to know the friction factor; it can be obtained by the use of a microscopic balance to get the velocity profile (as was done in Chaps. 3, 4, and 6 for laminar and turbulent flow) or by using an empirical correlation, which requires extensive data taking. □

Application to Change in Cross Section, Fittings, etc.

For flow through fittings, e.g., valves and meters, and for expansion or contraction losses, tables or charts of $\hat{e}_v$ are available from manufacturers. In some cases they can be obtained theoretically by application of the macroscopic balances, e.g., for a sudden expansion, as described in the next example.

Example 11-2: Friction loss in sudden expansion Consider the sudden expansion (Fig. 11-2) of an incompressible fluid due to a sudden enlargement from A_1 to A_2 of the cross section of the pipe in which it is flowing. It is desired to relate $\hat{E}_v$, the friction loss per unit mass, to the ratio of the cross sections $\beta \equiv A_1/A_2$ and to the average velocity $V_2 \equiv \langle v_2 \rangle$ in the downstream pipe. Assume steady-state turbulent flow.

SOLUTION We first apply the *mass balance* [Eq. (1), Table 11-1] to obtain

$$w_1 = \rho V_1 A_1 = \rho V_2 A_2 = w_2$$

We let

$$\frac{V_2}{V_1} = \frac{A_1}{A_2} \equiv \beta = \frac{D_1^2}{D_2^2} \tag{11-42}$$

When we apply the *momentum* balance [Eq. (2), Table 11-1], the gravity term drops out since the balance is made in the direction of flow. Also, for turbulent flow we can use Eq. (11-31). Then the fluid force is

$$F_f = -\Delta(Vw) - \Delta(pA) = (V_1 - V_2)w + p_1 A_1 - p_2 A_2 \tag{11-43}$$

But the fluid force is defined by Eq. (11-10) as

$$\mathbf{F}_f = \iint p\mathbf{n}\, dA + \iint \mathbf{n} \cdot \underline{\tau}\, dA$$

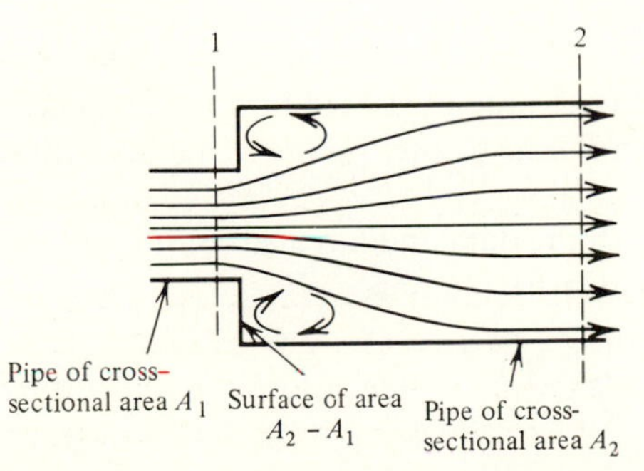

Figure 11-2 Flow in a sudden enlargement.

The last term is the viscous force acting on the inside wall surface; it is assumed to be negligible since the area it acts on is small. The first term is the force acting on the inside area $A_2 - A_1$. It is usually assumed that the average pressure on this area is equal to the pressure entering at A_1, or

$$F_f = -p_1(A_1 - A_2) \tag{11-44}$$

Equating (11-44) and (11-38) and letting $w = \rho_2 V_2 A_2$, we get

$$F_f = -p_1 A_2 + p_1 A_1 = (V_1 - V_2)\rho V_2 A_2 + p_1 A_1 - p_2 A_2$$

Then

$$(p_2 - p_1)A_2 = \rho V_2 A_2 (V_1 - V_2)$$

or

$$p_2 - p_1 = \rho V_2^2\left(\frac{1}{\beta} - 1\right) \tag{11-45}$$

In order to find $\widehat{E}_v$ we need to use the mechanical-energy balance [Table 11-1 or Eq. (11-33)]. We use Eq. (11-31) (for turbulent flow) and let the potential-energy change be zero. Since there is no pump, turbine, or other device doing or receiving work, $\widehat{W}_t = 0$. Then the mechanical-energy balance gives

$$\tfrac{1}{2}(V_2^2 - V_1^2) + \frac{p_2 - p_1}{\rho} + \widehat{E}_v = 0 \tag{11-46}$$

Substituting Eq. (11-45) for $p_1 - p_2$ gives

$$\tfrac{1}{2}V_2^2\left(1 - \frac{1}{\beta^2}\right) + V_2^2\left(\frac{1}{\beta} - 1\right) = -\widehat{E}_v$$

or

$$\widehat{E}_v = \tfrac{1}{2}V_2^2\left(\frac{1 - \beta}{\beta}\right)^2 \tag{11-47}$$

Since $\frac{1}{2}V^2$, the kinetic energy per unit mass of fluid, usually appears in such equations, it is common practice to factor it out by letting

$$e_v \equiv \frac{\widehat{E}_v}{\frac{1}{2}V_2^2} = \left(\frac{1 - \beta}{\beta}\right)^2 = \left(\frac{A_2}{A_1} - 1\right)^2 \tag{11-48}$$

Note that as $A_2 \to \infty, \beta \to 0$ and $e_v \to \infty$ but also $V_2 \to 0$. Therefore it is recommended that one use $e_v = 1$ for $\beta < \frac{1}{2}$. Values of e_v for other situations are given in Table 11-3 (for *turbulent* flow). The group $\frac{1}{2}V_2^2$ is called a *velocity head.* Thus e_v represents the friction loss $\widehat{E}_v$ in terms of velocity head. One velocity head is roughly the loss experienced in 50 diameters of pipe by a fluid in fully developed turbulent flow. Equation (11-48), called the *Borda-Carnot equation,* agrees within 20 percent with experimental data.

Martin[4] has compared this equation with the data and has also considered the problem of the sudden *contraction,* for which he obtains

$$e_v = \left(\frac{2}{m} - \frac{1}{\beta} - 1\right)^2 \tag{11-49a}$$

Table 11-3 Values[5] of friction-loss factor e_v

Sudden changes in cross-sectional area:†	
Rounded entrance to pipe	0.04
Sudden contraction	See Eq. (11-49*a*)
Sudden expansion	$\left(\frac{1}{\beta}-1\right)^2$
Orifice (sharp-edged)	$\frac{2.7(1-\beta)(1-\beta^2)}{\beta^2}$
Fittings and valves:	
90° elbows (standard)	0.75
90° elbows (square)	1.3
45° elbow (standard)	0.35
Globe valve (open)	6.0
Gate valve (open)	0.17

† $\beta = A_1/A_2$, and $\langle \bar{v} \rangle$ is the velocity upstream from the enlargement. Note that as $A_2 \to \infty$, $\beta \to 0$ and $e_v \to \infty$ but also $V_2 \to 0$. Therefore, use $e_v = 1$ for $\beta < \frac{1}{2}$.

where m is given by

$$\frac{1 - m/\beta}{1 - \beta^{-2}} = \left(\frac{m}{1.2}\right)^2 \tag{11-49b}$$

□

Example 11-3: Application to pump and compressor design If we let A_1 be the cross-sectional area before the entrance to a pump (plane 1) and let A_2 be the cross-sectional area at a plane outside the equipment (plane 2), the mechanical-energy equation (11-33) becomes (see Fig. 11-3)

$$\int_{p_1}^{p_2} \frac{dp}{\rho} + g\,\Delta h + \frac{\Delta}{2}\frac{\langle v^3 \rangle}{\langle v \rangle} + \hat{E}_v - \hat{W}_t = 0$$

where Δh, the difference in elevation between the two points, is usually small and can be neglected. Also, the change in kinetic energy is usually negligible if the fluid is incompressible. Then the shaft work per unit mass is

$$\hat{W}_t = \int_{p_1}^{p_2} \frac{dp}{\rho} - \hat{E}_v \tag{11-50}$$

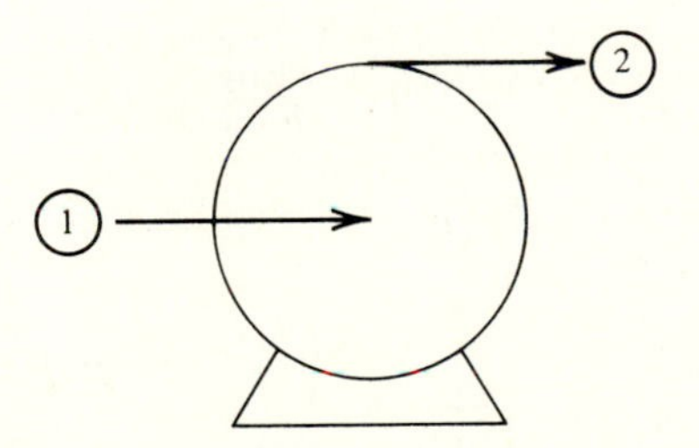

Figure 11-3 Pump stations 1 and 2.

The friction loss is *not* negligible but is difficult to calculate theoretically. Hence an *ideal* work is obtained by assuming $\hat{E}_v = 0$. The ideal work is then divided by an efficiency to get the actual work. The work is usually obtained in units such as ft · lbf/lb or feet of head. This is usually multiplied by w to obtain work rate (in, say, ft · lbf/h) and then converted to horsepower. The integral must be evaluated over the actual path. For an incompressible fluid, it becomes $\Delta p/\rho$.

For a compressor operating adiabatically, an isentropic path can be assumed. If the kinetic and potential energies are negligible and the fluid is an ideal gas with $\hat{C}_p/\hat{C}_v \equiv \gamma$, the ideal work is

$$\dot{W}_I = -\int_{p_1}^{p_2} \frac{dp}{\rho} = \frac{p_2}{\rho_1} \frac{\gamma}{\gamma - 1}\left[\left(\frac{p_2}{p_1}\right)^{(\gamma-1)/\gamma} - 1\right] \tag{11-51}$$

The derivation and integration steps are carried out in most thermodynamics texts. □

11-5 APPLICATION OF MECHANICAL ENERGY BALANCE TO PIPING SYSTEMS

For a *piping system* consisting of conduits of various lengths and hydraulic radii and of various fittings, valves, etc., the total friction losses can be added using Eqs. (11-39) and (11-40)

$$\hat{E}_v = \underbrace{\sum_i \left(\tfrac{1}{2}\langle v^2\rangle \frac{L}{R_H} f\right)_i}_{\text{All conduits}} + \underbrace{\sum_i \left(\tfrac{1}{2}\langle v^2\rangle e_v\right)_i}_{\text{All fittings}} \tag{11-52}$$

If the expressions for $\hat{E}_v$ for all pipes or conduits and for all fittings, expansions, contractions, etc., are substituted into the energy equation (11-33), we obtain

$$\Delta(\tfrac{1}{2}V^2) + g\,\Delta h + \int_{p_1}^{p_2} \frac{dp}{\rho} - \hat{W}_t + \underbrace{\sum_i \tfrac{1}{2}V_i^2 \frac{L_i}{R_{Hi}} f_i}_{\text{All conduits}} + \underbrace{\sum_i \tfrac{1}{2}V_i^2 e_{vi}}_{\text{All fittings, valves, etc.}} = 0 \tag{11-53}$$

In the first summation V_i is the average velocity in the ith pipe segment. In the second summation V_i is the average velocity *downstream* from the fitting.

Frequently Eq. (11-53) is modified by defining an *equivalent length* L_{eq} of straight pipe that will provide the same resistance as the fitting. Thus

$$\hat{E}_v = \tfrac{1}{2}V^2 e_v = \tfrac{1}{2}V^2 f \frac{L_{eq}}{R_H} \qquad \text{or} \qquad e_v = \frac{f L_{eq}}{R_H}$$

and

$$\frac{L_{eq}}{R_H} = \frac{e_v}{f} \tag{11-54}$$

This indicates that actually the friction-loss factor e_v depends upon the friction factor. However, the effect on the final result is small since we are dealing with a relatively small term in the equation. Also, the friction factor varies little with Re in highly turbulent flow, making it possible to use the average value. Tables of L_{eq}/R_H are available from pipe manufacturers.

Example 11-4: Application to piping system Consider the piping system illustrated in Fig. 11-4 containing a gate valve, a pump, and two 90° elbows. Assume the pressures at points 1 and 2 are both atmospheric. Calculate the horsepower of the pump if water is to be pumped through the system at 500 gal/min and the pipe is 5 in Schedule 40 steel.

SOLUTION From tables we find the actual diameter of the pipe to be 5.047 in. The cross-sectional area is then

$$A_z = \pi \frac{D^2}{4} = \frac{3.14(5.047)^2}{4(144)} = 0.1390 \text{ ft}^2$$

The volumetric rate of flow is

$$\dot{\mathcal{V}} = \frac{500 \text{ gal/min}}{(7.48 \text{ ft}^3/\text{gal})(60 \text{ s/min})} = 1.11 \text{ ft}^3/\text{s}$$

The average velocity is

$$V = \frac{\dot{\mathcal{V}}}{A} = \frac{1.11}{0.1390} = 8.02 \text{ ft/s}$$

From the mass balance $V_1 = V_2$. The Reynolds number then is

$$\text{Re} = \frac{(5.047/12)(8.02)(62.4)}{1.0(6.72)(10^{-4})} = 312{,}000$$

where the viscosity of the water is taken as 1 CP or $6.72(10^{-4})$ lb/ft · s (at 20°C).

The relative roughness is $k/D = [1.5(10^{-4})]/(5.047/12) = 3.6(10^{-4})$, and from Fig. 6-12, the friction factor is 0.0045.

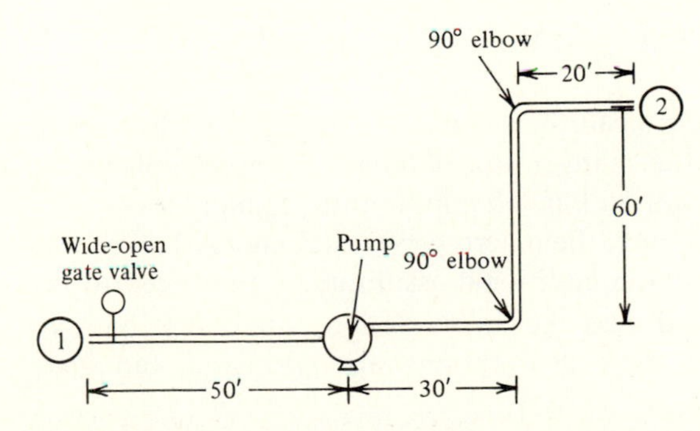

Figure 11-4 Piping system for Example 11-4.

The friction loss in the 160 ft of pipe is then given by the first of Eqs. (11-54)

$$\widehat{E}_v = \frac{1}{2}\frac{V^2 L}{g_c R_H} f = \frac{1}{2}\frac{(8.02)^2(160)(0.0045)}{(32.2)[5.047/(12)(4)]} = 6.84 \text{ ft} \cdot \text{lbf/lb}$$

For the fittings, $e_v = 0.9$ for each 90° elbow and $e_v = 0.2$ for the open gate valve (Table 11-2). Then,

$$(\widehat{E}_v)_{\text{fittings}} = \frac{1}{2}\frac{V^2}{g_c}[2(0.9) + 0.2]$$

$$= \frac{1}{2}\frac{(8.02)^2}{32.2}(2.0) = 2.0 \text{ ft} \cdot \text{lbf/lb}$$

Since the net change in elevation is 60 ft, the potential-energy term is:

$$\frac{g\,\Delta h}{g_c} = 60 \text{ ft} \cdot \text{lbf/lbm}$$

Both the pressure drop and kinetic-energy terms are zero. The energy balance Eq. (11-53) is therefore,

$$0 + 60 + 0 - \widehat{W}_t + 6.84 + 2.0 = 0$$

or

$$\widehat{W}_t = 68.8 \text{ ft} \cdot \text{lbf/lb}$$

The sign is positive since work is done on the system.

To express this in terms of power, we need to multiply by the mass rate of flow or $w = \dot{\mathcal{V}}\rho = 1.11(62.4) = 69.6$ lb/s.

The horsepower required is

$$\frac{\widehat{W}_t}{550} = \frac{68.8(69.6)}{550} = 8.7 \text{ hp}$$

If the pump has an efficiency of 40 percent, the actual horsepower required would be

$$\frac{8.7}{0.4} = 21.8 \text{ hp}$$

□

11-6 APPLICATION OF TOTAL-ENERGY EQUATION

Example 11-5: Application to heat transfer in a pipe Consider a fluid flowing at steady state upward through a heat-exchanger pipe of length L whose walls are at temperature T_w. The fluid enters with bulk mean temperature T_0 and leaves with bulk mean temperature T_{bL}. (*a*) Simplify the macroscopic total-energy balance as it applies to this problem. (*b*) State the additional assumptions that need to be made to obtain the commonly used heat-exchanger design equation (Chap. 4) $\dot{Q}_w = w\widehat{C}_p(T_{bL} - T_0)$. State when these assumptions are *not* valid and give examples.

SOLUTION (*a*) At steady state $d/dt = 0$, and the left-hand side is zero. Also, $w_1 = w_2 = w = \text{const}$. Since there is no shaft work, $\widehat{W}_t = 0$; and since flow is upward, $\Delta\psi = gL$. Then, from Table 11-1

$$\dot{Q}_w = \left(\Delta\widehat{H} + \tfrac{1}{2}\Delta\frac{\langle v^3\rangle}{\langle v\rangle} + gL\right)w$$

The change in enthalpy of the fluid is given by thermodynamics

$$\Delta\widehat{H} = \int_{T_{b0}}^{T_{bL}} \widehat{C}_p\, dT_b + \int_{p_0}^{p_L} \left(\frac{\partial H}{\partial p}\right)_T dp$$

The last term is zero for ideal gases and very small for liquids. For nonideal gases it can be evaluated from thermodynamic charts or from an equation of state.

(*b*) To obtain the design equations, we need to make the following assumptions:

1. The change in potential energy $w\,\Delta\widehat{\psi}$ is negligible compared with the change in enthalpy. In ordinary heat exchangers this assumption is valid, but it may not be valid, for example, in a natural-gas well 2 km deep.
2. The kinetic-energy term is negligible; i.e., the inlet profile is the same as the exit velocity profile. For fully developed flow of an incompressible fluid this assumption is rigorously correct. For flow in the entrance region of a pipe, however, the velocity profile entering would not be the same as that leaving. Also, for the flow of a compressible gas, e.g., natural gas, in a cross-country pipeline several miles long, the velocity profile would change because the drop in pressure would cause a decrease in density.
3. The fluid is either incompressible or an ideal gas; that is, $d\widehat{H} = \widehat{C}_p\, dT_b$.
4. The temperature difference is so small that $\widehat{C}_p$ is either not a function of temperature or a mean value can be used such that

$$\Delta\widehat{H} = \widehat{C}_p(T_{bL} - T_{b0})$$

The first three assumptions give

$$\dot{Q} = (\Delta\widehat{H})w \tag{11-55}$$

and the fourth gives the desired result. □

11-7 MACROSCOPIC ANGULAR-MOMENTUM BALANCE

The equation of change for angular momentum, as derived from the law of conservation for angular momentum (assuming symmetry of the stress tensor), is

$$\frac{\partial}{\partial t}(\mathbf{x}\times\mathbf{v}\rho) = -\nabla\cdot(\mathbf{x}\times\rho\mathbf{v}\mathbf{v})^T - \nabla\cdot(\mathbf{x}\times\underline{\tau})^T - \nabla\cdot\mathbf{x}\times p\underline{\boldsymbol{\delta}} - \mathbf{x}\times\rho\mathbf{g} \tag{11-58}$$

Rate of increase in angular momentum	Net addition of angular momentum	Net viscous torque	Net pressure torque	Net gravity torque

If we integrate over the volume, we get[3]

$$\frac{d}{dt}(\Gamma) = -\Delta\left[\left(\frac{\langle v^2\rangle}{\langle v\rangle}\mathbf{w} + p\mathbf{A}\right)(\mathbf{x}\times\mathbf{n})\right] + \dot{\Gamma}^{(I)} - \Upsilon_f + \Upsilon_g \tag{11-56}$$

for which $\Gamma = \frac{1}{\mathcal{V}}\iiint \rho\hat{\Gamma}\,d\mathcal{V}$ and $\hat{\Gamma} = \mathbf{x}\times\mathbf{v}$

where Γ = angular momentum per unit mass
Υ_c = net torque of fluid on solid
Υ_g = net torque due to gravity
$\dot{\Gamma}^{(I)}$ = angular momentum added across interface (usually negligible)

11-8 APPLICATION OF MACROSCOPIC ANGULAR-MOMENTUM BALANCE

The macroscopic angular-momentum balance is useful in the design of rotating machinery such as centrifugal pumps, turbines, etc.

Example 11-5: Design of centrifugal pumps A centrifugal pump basically consists of rotating blades or impellers (Fig. 11-5*a*, *b*) that impart an angular velocity to a flowing fluid. The resulting kinetic energy is converted into an increase in pressure in accordance with Bernoulli's equation. The head developed by the pump is equal to the change in pressure divided by the density of the fluid plus the change in kinetic energy. The units are ft · lbf/lbm, commonly expressed in practice as feet of head. It is equivalent to the potential energy of a column of the fluid of that height in feet. Note that the head must be expressed in terms of a particular fluid, such as water.

At point 1 the fluid enters the pump at the axis and is turned 90° so as to enter

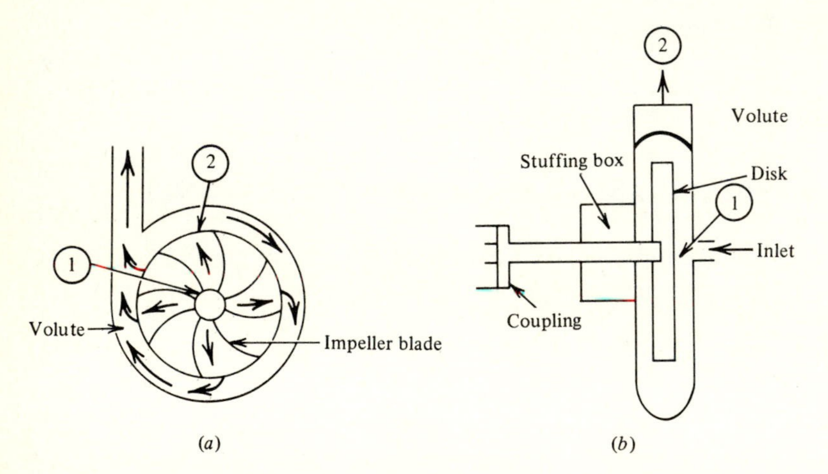

Figure 11-5 (*a*) Centrifugal pump; (*b*) side view.

the disk at velocity V_1. As it rotates with the disk, it flows rapidly toward the outer region, called the *volute* (Fig. 11-5). The macroscopic balance will be applied over the disk between the entrance or eye of the pump (point 1) and the exit at the volute (point 2). We will use cylindrical coordinates and let the z axis be the axis of rotation of the pump. The r direction is measured radially from the center, and θ is the angular direction measured clockwise about the axis. As the fluid enters at $r = R_1$, its largest velocity component is in the r direction but it will have a small angular component $v_{\theta 1}$. It then flows generally parallel to solid vanes or impeller blades that split the stream into segments. These blades are usually curved and are rotating with the disk at the operative speed of the pump in revolutions per minute. Since this speed is constant, the overall process is at steady state and $d/dt = 0$. Also, $w = w_2 = w_1 = \rho_1 \langle v_1 \rangle A_1$, where $A_1 = 2\pi R_1 b$ and b is the depth of the fluid on the disk. The z component of the macroscopic balance is appropriate since the axis of rotation is the z axis; then $\hat{\mathbf{\Gamma}}_z = (\mathbf{x} \times \mathbf{v})_z = r v_\theta \boldsymbol{\delta}_z$.

In pump design an ideal work or head is first calculated by considering only the change in angular momentum of the fluid as it passes through the pump. If $\langle v^2 \rangle / \langle v \rangle \approx \langle v \rangle$ and $(\mathbf{x} \times \mathbf{w})_z = (r \rho v_\theta A) \boldsymbol{\delta}_z$, the macroscopic angular-momentum balance can be simplified to

$$\boldsymbol{\Upsilon}_{fz} = \Delta(w r v_\theta) \boldsymbol{\delta}_z = w \, \Delta \hat{\Gamma} \boldsymbol{\delta}_z = w(R_2 v_{\theta 2} - R_1 v_{\theta 1}) \boldsymbol{\delta}_z \tag{11-57}$$

In general, work rate in a rotating system is given by

$$\dot{W} = \boldsymbol{\Upsilon} \cdot \boldsymbol{\omega} \tag{11-58}$$

where $\boldsymbol{\Upsilon}$ is the torque and $\boldsymbol{\omega}$ is the angular velocity of rotation. Since we are dealing only with one component, we can use the scalar magnitude Υ and ω. Then the ideal work is

$$\dot{W}_I = \omega w (R_2 v_{\theta 2} - R_1 v_{\theta 1}) \tag{11-59}$$

The ideal head developed by the pump is the ideal work divided by the mass rate of flow, or $\dot{W}_I / w = \hat{W}_I$. The ideal head is then

$$\hat{W}_I = \omega (R_2 v_{\theta 2} - R_1 v_{\theta 1}) \tag{11-60}$$

Usually $R_1 v_{\theta 1}$ is neglected, and the angular velocity of the pump is related to its linear velocity u_2 at the outlet by $\omega = u_2 / R_2$. Then

$$\hat{W}_I = u_2 v_{\theta 2} \tag{11-61}$$

By geometry it can be shown (see Fig. 11-6 and the nomenclature below) that

$$v_{\theta 2} = u_2 - v_{r2} \cot \beta_2 \tag{11-62}$$

where β_2 is an angle that measures the curvature of the blade; i.e., the tangent to the impeller at its tip makes an angle β_2 with the tangent to a circle traced out by the tip as it rotates. The volume rate of flow is given by

$$\dot{\mathcal{V}} = - \iint_A \mathbf{n} \cdot \mathbf{v} \, dA \tag{11-63}$$

where A is A_1 or A_2.

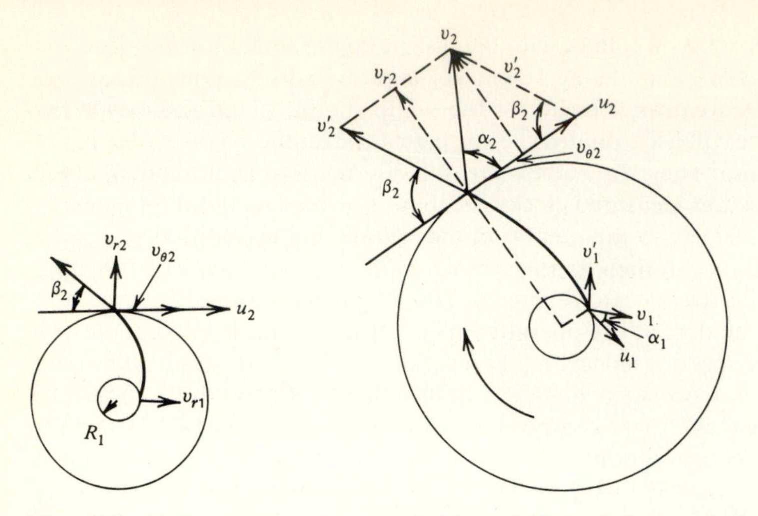

Figure 11-6 (*a*) Components of vector velocities and (*b*) detailed view of velocity components.

Pump nomenclature

$\mathbf{v}_1'$ = velocity of entering stream relative to impeller
$\mathbf{v}_2'$ = velocity of leaving stream relative to impeller
$\mathbf{u}_1$ = velocity of impeller vane at entrance relative to ground, $|\mathbf{u}_1| = \omega R_1$
$\mathbf{u}_2$ = velocity of impeller vane at exit relative to ground, $|\mathbf{u}_2| = \omega R_2$
$\pi - \beta_2$ = angle between v_2' and u_2
$\mathbf{v}_1 = \mathbf{v}_1' + \mathbf{u}_1$ = velocity of entering fluid relative to ground
$\mathbf{v}_2 = \mathbf{v}_2' + \mathbf{u}_2$ = velocity of leaving fluid relative to ground
α_1 = angle between $\mathbf{v}_1$ and $\mathbf{u}_1$
α_2 = angle between $\mathbf{v}_2$ and $\mathbf{u}_2$
$\mathbf{v}_2$ = velocity of flowing fluid relative to ground
$\mathbf{v}_{r2}$ = *flow velocity,* radial component of velocity of fluid relative to ground
$\mathbf{v}_{\theta 2}$ = *whirl velocity,* tangential component of $\mathbf{v}_2$, or $v_2 \cos \alpha_2 = u_2 - v_2' \cos \beta_2$
$v_{r2} = v_2 \sin \alpha_2 = v_2' \sin \beta_2$
Δh = head developed
b = channel width (height of vane)
N = revolutions per minute □

11-9 DIFFERENTIAL MACROSCOPIC BALANCES

In equipment where the flow is basically in one direction (absorbers, extractors, tubular reactors, pipelines, etc.) it is frequently more convenient to integrate the equations of change over the cross-sectional area than over the volume, in order to obtain an ordinary differential equation that can be solved more easily by numerical methods of integration than the partial differential equations of change. These equations are in terms of average concentration, temperature, and velocity as a function of the flow-direction coordinate z and apply over a differential length dz. They may therefore be called *differential macroscopic balances*. It should be noted that they can also be obtained by writing the macroscopic balances in differential form, but this method makes it unclear how the averages should be obtained if there are large radial gradients. We therefore illustrate the former method for a tubular reactor.

Differential Energy and Mass Balances in a Plug-Flow Reactor

A *plug-flow reactor* is a tubular reactor in which it is assumed that the gradients over the cross section are negligible and that a bulk cross-sectional average velocity v_b, concentration C_{Ab}, and temperature T_b can be used instead of point values. One can then obtain ordinary differential equations for the latter two average values as a function of axial position by integrating the equation of change over the cross section $A_z \equiv A$.

Mass At steady state the equation for moles of species A becomes (for constant ρ)

$$v_z \frac{\partial C_A}{\partial z} = \frac{\partial (C_A v_z)}{\partial z} = \widetilde{R}_A$$

since $J_{Ar} = 0 = J_{Az}$ and the only component of $\mathbf{v}$ is v_z. Multiplying by dA and integrating gives

$$\iint \frac{d}{dz}(C_A v_z)\, dA = \iint \widetilde{R}_A\, dA$$

or

$$\frac{d}{dz}(\langle C_A \rangle \langle v_z \rangle) A = \langle \widetilde{R}_A \rangle A \qquad \text{where } \langle \widetilde{R}_A \rangle \equiv \frac{\iint \widetilde{R}_A\, dA}{A}$$

If we multiply by dz and let $A\, dz = d\mathcal{V}$, we get

$$A\, d(\langle C_A \rangle \langle v_z \rangle) = \langle \widetilde{R}_A \rangle\, d\mathcal{V} \tag{11-64}$$

We note that $\langle C_A \rangle \langle v_z \rangle$ represents the number of moles of A reacted or produced in the differential volume $d\mathcal{V}$. Therefore in Smith's reactor design text (Ref. 11, p. 112) the left-hand side is usually written in terms of X, the fractional conversion of A (the moles of A per mole of A in feed), and F, the moles of A in feed per hour, as

$$F\, dX = \langle \widetilde{R}_A \rangle\, d\mathcal{V} \tag{11-65}$$

The design equation is then

$$\frac{\mathcal{V}}{F} = \int \frac{dX}{\langle \widetilde{R}_A \rangle} \tag{11-66}$$

To integrate we need to know how the rate varies with X and (if nonisothermal) with T. The integration is normally carried out numerically. If only one reaction is occurring, $\langle \widetilde{R}_A \rangle = r_X$, where r_X is the rate based on fractional conversion.

Energy To obtain the differential energy balance we simplify the equation of change (part A of Table I-7) to obtain

$$\rho \widehat{C}_p v_z \frac{\partial T}{\partial z} = -\frac{\partial (r q_r)}{r\,\partial r} + \Phi_R$$

Assuming that $\rho \widehat{C}_p v_z$ is constant, multiplying by dA, and integrating the left-hand side gives

$$\iint \frac{\partial(\rho \widehat{C}_p v_z T)}{\partial z}\,dA = \frac{d}{dz}\iint \rho \widehat{C}_p v_z T\,dA = \langle \rho \rangle \langle \widehat{C}_p \rangle \langle v_z \rangle \frac{d\langle T \rangle}{dz} A = w \langle \widehat{C}_p \rangle \frac{d\langle T \rangle}{dz} \tag{11-67}$$

where $w \equiv \langle \rho \rangle \langle v_z \rangle A$ is the total mass rate of flow. At steady state w will be constant. Integrating the left-hand side over dA gives the heat transport at wall per unit length

$$\int_0^{r_w} \frac{\partial (r q_r)}{r\,\partial r} 2\pi r\,dr = 2\pi r q_r \Big|_0^{r_w} = 2\pi r_w q_w = Z q_w = \frac{d\dot{Q}_w}{dz} \tag{11-68}$$

where $d\dot{Q}_w = U\,dA_w (\langle T \rangle - T_s)$
U = overall heat-transfer coefficient
Z = perimeter
dA_w = heat-transfer area at wall $= 2\pi r_w\,dz = A\,dz$
T_s = temperature of surroundings

Also
$$\iint \Phi_R\,dA = -\iint \Delta \widetilde{H}_R \widetilde{R}_A\,dA = -\Delta \widetilde{H}_R \langle \widetilde{R}_A \rangle A \tag{11-68a}$$

If we combine Eqs. (11-67), multiply by dz, and let $A\,dz = d\mathcal{V}$, we get

$$w \langle \widehat{C}_p \rangle\, d\langle T \rangle = U\,dA_w (T_s - \langle T \rangle) + (-\Delta \widetilde{H}_R) \widetilde{R}_A\,d\mathcal{V} \tag{11-69}$$

If we use the mass equation to express $\langle \widetilde{R}_A \rangle$ in terms of conversion, we get

$$w \langle \widehat{C}_p \rangle\, d\langle T \rangle = U(T_s - \langle T \rangle)\,dA_w + (-\Delta \widetilde{H}_R)\,F\,dX \tag{11-70}$$

This equation is equivalent to that given in Ref. 11, p. 228.

The forms of the design equations, as given by Aris,[1] can be obtained similarly.

Tubular Reactor with Axial Dispersion

In the ideal plug-flow reactor, we assumed that concentration, temperature, and velocity variations across the cross section could be neglected. In practice, however, there will be a radial diffusion of mass, energy, and momentum due to the gradients that exist. There are two ways this lateral (or radial) transport can be handled: (1) Write the equations of change to include a *radial-diffusion* flux and then express this flux in terms of a radial diffusivity (or thermal conductivity). This is called the *general diffusion model* and is the most rigorous, but it requires more computer time for solution. It is discussed in Sec. 10-9. (2) Treat the radial diffusion as an effective axial dispersion. This model, called the *axial-dispersion, dispersed-plug-flow,* or *Taylor diffusion model* will be discussed later.

Mass First it is useful to understand how the effective axial dispersion occurs. To illustrate the effect, consider laminar flow in a pipe, which, as we have seen, results in a parabolic velocity profile. Consequently the fluid at the centerline will have twice the average velocity while the fluid at the wall will have zero velocity. Hence, the material near the center will move through the tube much faster than material at the walls. If the concentration of a species is relatively higher at the center than at the walls, this species would on the average tend to move longitudinally at a faster rate than the average velocity. In accordance with our definition of diffusion as being a flux with respect to an average velocity, this relative movement can be considered an effective diffusion or a dispersion. Taylor[12] quantitatively described it by

$$\mathcal{W}_A \equiv \iint C_A(v_z - \langle v_z \rangle)\, dA \tag{11-71}$$

where dA is the cross-sectional area ($= 2\pi r\, dr$ for a pipe) and $\mathcal{W}_A$ is the moles of A dispersed axially per unit time.

In his papers Taylor was originally interested in the axial mixing like that occurring in oil pipelines when one petroleum product is followed by another. The mixing that takes place at the interface means that a certain amount of reprocessing of the two original products is required.

In his original work, Taylor neglected axial *diffusion* and chemical reaction and used a radial diffusivity E_r, which is constant in laminar flow but depends upon radial position in turbulent flow. The diffusing material is presumed to have some arbitrary concentration at time $t = 0$ at location $z = 0$ and to be convected axially by a velocity $v_z(r)$. Taylor reasoned that the axial dispersion flux should be proportional to the axial gradient of the average concentration C_{Aa} over the cross section, where

$$C_{Aa} \equiv \frac{\int_0^{r_w} C_A r\, dr}{\int_0^{r_w} r\, dr} \tag{11-72}$$

Then

$$\mathcal{W}_A \equiv -E_{ax} A \frac{\partial C_{Aa}}{\partial z} \tag{11-73}$$

where E_{ax} is called the *axial dispersivity*. For long tubes and for unsteady-state flow without reaction Taylor assumed that E_{ax} is constant. For short steady-state reactors E_{ax} is not constant, however, and an axial-dispersion model is not recommended.

Unsteady state For a long tubular reactor in which the reactants are introduced at $z = 0$ at $t = 0$, the equation of change of species A is

$$\frac{\partial C_A}{\partial t} + \frac{\partial (C_A v_z)}{\partial z} = -\frac{\partial (r J_{Ar})}{r\,\partial r} + \frac{\partial J_{Az}}{\partial z} + R_A$$

Integrating the first term over the cross section gives

$$\iiint \frac{\partial C_A}{\partial t}\, dA = \frac{\partial}{\partial t} \iint C_A\, dA = \frac{\partial C_{Aa}}{\partial t} A$$

We let

$$\iint \frac{\partial}{\partial z}(C_A v_z)\, dA = \frac{\partial}{\partial z} \iint C_A v_z\, dA$$

$$= \frac{\partial}{\partial z}\left[\iint C_A(v_z - \langle v_z \rangle)\, dA + \iint C_A \langle v_z \rangle\, dA\right]$$

$$= \frac{\partial \mathcal{W}_A}{\partial z} + \langle v_z \rangle A \frac{\partial C_{Aa}}{\partial z}$$

where $\mathcal{W}_A$ = effective axial dispersion of A. Then from Eqs. (11-71) and (11-73) we write

$$\mathcal{W}_A \equiv \iint C_A(v_z - \langle v_z \rangle)\, dA \equiv E_{ax} A \frac{\partial C_{Aa}}{\partial z} \qquad \text{(11-74)}$$

The diffusion term is

$$\int_0^{r_w} \left[\frac{\partial (r J_{Ar})}{r\,\partial r} + \frac{\partial J_{Az}}{\partial z}\right] 2\pi r\, dr = 0 + \frac{\partial}{\partial z} \langle J_{Az} \rangle A$$

The first term is zero because there is no mass transfer through the wall. $\langle J_{Az} \rangle$ is the axial flux due to non-Taylor dispersion, averaged over the cross section. For a homogeneous reactor, this flux would include the molecular diffusion plus the turbulent diffusion.

Axial diffusion also may occur in a packed bed due to the wandering or zigzag random-walk movement of the fluid through the packing interstices. For simplicity we will use an effective axial diffusivity E_z to describe the non-Taylor axial diffusion. Then, after averaging, we get

$$\langle J_{Az} \rangle = -E_z \frac{\partial C_{Aa}}{\partial z}$$

If we let the total axial diffusive flux be $E_{tz} = E_{ax} + E_z$, the design equation is

$$\frac{\partial C_{Aa}}{\partial t} - E_{tz} \frac{\partial^2 C_{Aa}}{\partial z^2} + \langle v_z \rangle \frac{\partial C_{Aa}}{\partial z} = \langle \widetilde{R}_A \rangle \qquad \text{(11-75)}$$

By a similar process we can derive the following differential macroscopic energy balance

$$\rho \hat{C}_p \frac{\partial T_a}{\partial t} - k_{tz} \frac{\partial^2 C_{Aa}}{\partial z^2} + \rho \hat{C}_p v_a \frac{\partial T_a}{\partial z} = \langle \Phi_R \rangle - \frac{q_w Z}{A} \tag{11-76}$$

where $\langle \Phi_R \rangle = \langle (-\Delta H)_R R_A \rangle$ and $k_{tz} = k_{\text{ax}} + k_z$

Compressible-Gas Flow

If the fluid flowing through a pipe is compressible, the equation of continuity can be integrated over the cross section at steady state as follows

$$0 = \iint \left(\frac{\partial(\rho v_z)}{\partial z} + \frac{\partial(r \rho v_r)}{r\,\partial r} \right) dA_z$$

or, since $dA_z = 2\pi r\,dr$,

$$0 = \frac{d}{dz} \int_0^R \rho v_z (2\pi r\,dr) + 2\pi \int_0^0 d(r v_r) = \frac{dw}{dz} + 0$$

Thus if ρ changes with z, v_z will also change with z, but $w = \text{const}$ at steady state. This result can be generalized to any cross section by the use of the Gauss divergence theorem.

Likewise the momentum balance in the z direction can be integrated over the cross section with the help of the equation of continuity to give, at steady state

$$\frac{d\langle \rho v_z^2 \rangle}{dz} = -\frac{d\langle \tau_{zz} \rangle}{dz} - \frac{\tau_w Z}{A_z} - \frac{dp}{dz} + \langle \rho \rangle g_z \tag{11-77}$$

It is usually assumed that $d\langle \tau_{zz} \rangle / dz$ is negligible and that for turbulent flow

$$\frac{d\langle \rho v \rangle^2}{dz} = \langle \rho \rangle v_a \frac{dv_a}{dz}$$

Then

$$\langle \rho \rangle v_a \frac{dv_a}{dz} = -\frac{d\mathcal{P}}{dz} - \frac{\tau_w Z}{A} \tag{11-78}$$

Since the fluid is compressible, this equation must be integrated over the length of the tube using the mass-balance equation $dw = 0$ (for constant temperature) and an equation of state, such as the ideal-gas law, $p = \rho R_G T/M$. Usually τ_w is related to the friction factor by Eq. (4-9). Alternatively a differential mechanical-energy balance may be written by adding the term $d\hat{E}_v$ to Eq. (11-27) and using Eq. (11-38). For turbulent flow, the result is the same.

11-10 POPULATION BALANCE

In many kinds of engineering equipment, e.g., crystallizers and reactors, in addition to the spatial coordinates and time it is necessary to consider the size distribution, age, or other property of a species in the development of a conservation or continuity equation. The resulting equation is termed a *general population balance* or a *population balance for countable entities*.[8] We first illustrate the development of the equation for the simple case of a crystallizer, where only one additional property, size, is needed. We then generalize the equation to apply to any number of properties.

If we let the characteristic length L of a particle be the additional property, we can define a *distribution function* $\Psi(x, y, z, t, L)$ as the fraction of particles of length L per unit volume and unit length. Then the *fraction of particles* in an element of volume $dx\,dy\,dz$ and having sizes between L and $L + dL$ is

$$\Psi\,dx\,dy\,dz\,dL$$

Since all particles must be in the entire volume $\mathcal{V}$ and must have a particle size between 0 and ∞,

$$\int_0^\infty \Psi\,dx\,dy\,dz\,dL \equiv \int_R \Psi\,dR = 1$$

where dR represents an element of volume-space and "size-space."

If we assume that there is no net generation or attrition of particles, the net accumulation of particles of size L will be zero, or

$$\frac{d}{dt}\int_R \Psi\,dR = 0 \qquad \text{particles/unit time} \tag{11-79}$$

We now write the Leibnitz equation for one dimension (Appendix 10-1), first in the customary form and then in an equivalent form

$$\begin{aligned} \frac{d}{dt}\int_{a(t)}^{b(t)} \Psi(x, t)\,dt &= \int_{a(t)}^{b(t)} \frac{\partial \Psi}{\partial t}\,dx + \Psi(b)\frac{db}{dt} - \Psi(a)\frac{da}{dt} \\ &= \int_{a(t)}^{b(t)} \left[\frac{\partial \Psi}{\partial t} + \frac{d}{dx}\left(\frac{dx}{dt}\Psi\right)\right]dx \end{aligned} \tag{11-80}$$

For R-space we can write

$$\frac{d}{dt}\int_{R(t)} \Psi\,dR = \int_R \left[\frac{\partial \Psi}{\partial t} + \sum_i \frac{\partial}{\partial \xi_i}\left(\frac{d\xi_i}{dt}\Psi\right)\right]dR \tag{11-81}$$

where ξ_i represents x, y, z, or L. Now, as usual, $dx/dt = v_x$, etc., or

$$\frac{d\xi_i}{dt} = v_i \tag{11-82}$$

For $\xi = L$ we can interpret $v_L = dL/dt$ as the time rate of change (growth) of a crystal of size L. Then, if Eq. (11-79) applies, the left-hand side of Eq. (11-81) is zero

$$\int_R \left[\frac{\partial \Psi}{\partial t} + \sum_i \frac{\partial}{\partial \xi_i}(v_i \Psi)\right]dR = 0$$

Since dR is arbitrary,

$$\frac{\partial \Psi}{\partial t} + \sum_{i=1}^{3} \frac{\partial}{\partial x_i}(v_i \Psi) + \frac{\partial}{\partial L}(v_L \Psi) = 0$$

or

$$\boxed{\frac{\partial \Psi}{\partial t} + \nabla \cdot \Psi \mathbf{v} + \frac{\partial}{\partial L}(v_L \Psi) = 0} \tag{11-83}$$

We can generalize the above equation by considering that several properties ξ_i are involved other than the space coordinates and time. For example, the description of a microbial population in a city sewage system would require taking into account various ages, masses, etc., of the microorganisms. Furthermore, one would have to include, for the region R, a birth function B (and a death function D), defined as the fraction of entities created (or destroyed) per unit time and unit volume per unit of each property.

If there were m coordinates and $m - 3$ property coordinates, region dR would be defined as

$$dR = dx\, dy\, dz\, d\xi_4 \cdots d\xi_m \tag{11-84}$$

If Ψ is the fraction of particles per unit volume and unit property space

$$\int_R \Psi(x, y, z, \xi_4, \cdots, \xi_m, t)\, dR = 1 \tag{11-85}$$

then

$$\frac{d}{dt} \int_R \Psi\, dR = \int_R (B - D)\, dR \tag{11-86}$$

Proceeding as before, we see that if $v_j = d\xi_j/dt$, the application of the multidimensional Leibnitz rule gives

$$\int_R \left[\frac{\partial \Psi}{\partial t} + \sum_{i=1}^{3} \frac{\partial}{\partial x_i}(v_i \Psi) + \sum_{j=4}^{m} \frac{\partial}{\partial \xi_j}(v_j \Psi) - B + D \right] dR = 0 \tag{11-87}$$

or

$$\boxed{\frac{\partial \Psi}{\partial t} + \nabla \cdot \mathbf{v}\Psi + \sum_{j=4}^{m} \frac{\partial}{\partial \xi_j}(v_j \Psi) - B + D = 0} \tag{11-88}$$

Note that the units on the above are particles per unit time and volume and property space.

Macroscopic Balance

Frequently we are interested only in values averaged over the volume $\mathcal{V}$ of a system. We define the average value of Ψ as

$$\bar{\Psi} \equiv \frac{1}{\mathcal{V}} \iiint_{\mathcal{V}} \Psi\, d\mathcal{V} \tag{11-89}$$

We now integrate the microscopic balance over the volume (but *not* over dR). The first term is

$$\iiint \frac{\partial \Psi}{\partial t} d\mathcal{V} = \frac{d}{dt}\iiint_{\mathcal{V}} \Psi \, d\mathcal{V} - \iint_A \mathbf{n} \cdot (\mathbf{w}\Psi) \, dA = \frac{d}{dt}(\bar{\Psi}\mathcal{V})$$

since $\mathbf{w} = 0$ if there are no moving surfaces. The second term is transformed into an area integral by using the divergence theorem

$$\iiint_{\mathcal{V}} \nabla \cdot \Psi \mathbf{v} \, d\mathcal{V} = \iint_A \Psi \mathbf{n} \cdot \mathbf{v} \, dA$$

Now the velocity of the fluid $\mathbf{v}$ is zero at all surfaces except at the entrance pipe, of cross-sectional area A_i, and the exit pipe, of area A_{ex}. We assume that the property Ψ is constant over A_i and A_{ex} and we let v_i be the average velocity over A_i and v_{ex} be the average velocity at A_{ex}. Then, since $\mathbf{n} \cdot \mathbf{A} = -A_i$ at A_i and $\mathbf{n} \cdot \mathbf{A} = +A_{ex}$ at A_{ex}, we get

$$\int \nabla \cdot \Psi \mathbf{v} \, d\mathcal{V} = -\int_{A_i} v\Psi \, dA_i + \int_{A_{ex}} v\Psi \, dA_{ex} = v_{ex}\Psi_{ex}A_{ex} - v_i\bar{\Psi}_i A_i$$

Then we can write a *macroscopic balance*

$$\boxed{\begin{aligned}\frac{1}{\mathcal{V}}\frac{d}{dt}(\mathcal{V}\bar{\Psi}) + \sum_{j=4}^{m}\frac{\partial}{\partial \xi_j}(v_j\bar{\Psi}) + \bar{D} - \bar{B} &= \frac{1}{\mathcal{V}}(v_i\bar{\Psi}_i A_i - v_{ex}\bar{\Psi}_{ex}A_{ex}) \\ &= \frac{Q_i\bar{\Psi}_i - Q_{ex}\bar{\Psi}_{ex}}{\mathcal{V}}\end{aligned}}$$

(11-90)

$$\text{where} \qquad Q_{ex} \equiv v_{ex}A_{ex} \qquad Q_i \equiv v_i A_i \qquad \bar{D} = \frac{\iiint D \, d\mathcal{V}}{\mathcal{V}}$$

Note that the units are the same as in Eq. (11-88)

The previous equation was derived on the assumption that there are no changes in the volume of the system $\mathcal{V}$. However, such changes can occur due to (1) accumulation at a free interface or (2) changes in the volume occupied by the entities. If the net result of these changes is $\Psi(d\mathcal{V}/dt)$, the macroscopic balance for a mixed suspension becomes

$$\frac{\partial \bar{\Psi}}{\partial t} + \sum_{j=4}^{m}\frac{\partial}{\partial \xi_j}(v_j\bar{\Psi}) + \bar{\Psi}\frac{d \ln \mathcal{V}}{dt} = \bar{B} - \bar{D} + \frac{\sum Q_j\bar{\Psi}_j}{\mathcal{V}} \qquad (11\text{-}91)$$

where $Q_j = v_j A_j$ is the volumetric input (+) or output (−) of entity-free fluid.

Application to Crystallizers

Consider the problem[9] of k agitator-crystallizers operating in series. Nucleation occurs in each tank, operation is steady state, and there is perfect mixing in each tank. The only property coordinate of interest is the particle size L. On the assumption of perfect mixing, we can use the macroscopic population balance [Eq. (11-91)] and simplify it in the following manner:

$$\frac{\partial \bar{\Psi}}{\partial t} = 0 \text{ due to steady state} \qquad \frac{d \ln \mathcal{V}}{dt} = 0$$

$$\bar{B} = \bar{D} = 0 \text{ due to problem conditions} \qquad \xi \equiv L \qquad v \equiv \frac{dL}{dt}$$

Thus Eq. (11-91) reduces (for tank j) to

$$\underset{\text{Accumulation}}{\frac{d}{dL}\left(\frac{\partial L}{\partial t}\right)\bar{\Psi}_j} = \underset{\text{Input}}{\frac{Q_{j-1}\bar{\Psi}_{j-1}}{\mathcal{V}_{j-1}}} - \underset{\text{Output}}{\frac{Q_j \Psi_j}{\mathcal{V}_j}} \tag{11-92}$$

In terms of mass flow rates $w_j = Q_j \rho$, this becomes

$$\frac{d}{dL}\left(\frac{\partial L}{\partial t}\bar{\Psi}_j\right) = \frac{w_{j-1}\bar{\Psi}_{j-1}}{H_{j-1}} - \frac{w_j \bar{\Psi}_j}{H_j}$$

where $H_j \equiv \mathcal{V}_j \rho$ is the mass holdup in tank j. If we assume that *McCabe's ΔL law* is obeyed, the growth rate $r = \partial L/\partial t$ is independent of L and can be factored out. Also, for simplicity, we shall assume that the residence time $H/w = \mathfrak{I}$ is the same for each tank. Then

$$r\frac{d}{dL}(\bar{\Psi}_j) = \frac{1}{\mathfrak{I}}(\bar{\Psi}_{j-1} - \Psi_j) \tag{11-93}$$

and

$$\frac{d\bar{\Psi}_j}{dL} = \frac{\bar{\Psi}_{j-1} - \bar{\Psi}_j}{r\mathfrak{I}}$$

If we define a dimensionless size $x = L/r\mathfrak{I}$, we have

$$\frac{d\bar{\Psi}_1}{dx} = -\bar{\Psi}_1 \qquad \text{and} \qquad \frac{d\bar{\Psi}_k}{dx} = \bar{\Psi}_{k-1} - \bar{\Psi}_k \tag{11-94}$$

the solution to which is

$$\bar{\Phi}_k = c_1 e^{-x} + c_2 x e^{-x} + \cdots + c_k x^{k-1} e^{-x}$$

where c_j can be expressed in terms of nuclei density $\bar{\Psi}^0$ in each tank $= \Psi$ at $x = 0$. Thus in general

$$\bar{\Psi}_k = e^{-x}\left(\bar{\Psi}_k^0 + \frac{\bar{\Psi}_{k-1}^0 x}{1!} + \frac{\bar{\Psi}_{k-2}^0 x^2}{2!} + \cdots + \frac{\bar{\Psi}_1^0 x^{k-1}}{(k-1)!}\right) \tag{11-95}$$

We can write $\bar{\Psi}_k$ in a more useful form by multiplying it by x^3 and normalizing to

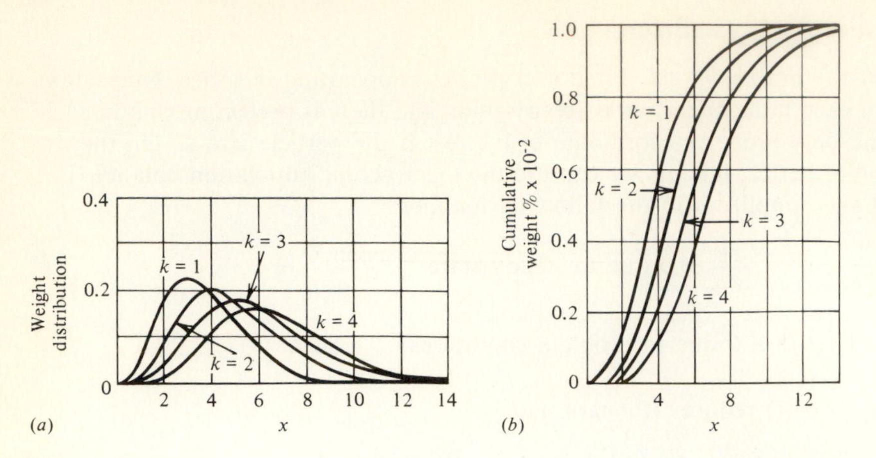

Figure 11-7 Multitank operation and nucleation only in the first tank: (*a*) weight distribution versus dimensionless size and (*b*) cumulative weight percentage versus dimensionless size. [*From A. D. Randolph and M. A. Larson, AIChE J.,* **8**:*639* (*1962*).]

get a weight percentage distribution

$$100\omega = \frac{x^3\bar{\Psi}_k}{\int_0^\infty x^3\bar{\Psi}_k\,dx} \tag{11-96}$$

If nucleation occurs only in the *first* tank, only the last term in Eq. (11-95) appears and

$$\bar{\Psi}_k = \frac{\bar{\Psi}_1^0 x^{k-1}e^{-x}}{(k-1)!} \tag{11-97}$$

and the weight percent distribution is given in Fig. 11-7*a*. The maximum dimensionless crystal size can be shown to be $x_{\max} = k + 2$ (see the Problems). This suggests the possibility of producing an arbitrary distribution, e.g., bimodal, by means of controlled nucleation in each tank.

The cumulative weight percentage distribution for nucleation only in the first tank is shown in Fig. 11-7*b*. It is found by integrating from zero size $x = 0$ up to arbitrary x. For a further discussion of crystallizers and population balances see Refs. 7 and 10.

PROBLEMS

11-1 It is desirable to maintain a reacting fluid at a moderate temperature while flowing through a 6-in pipeline over a 762-m distance to a reactor (see Fig. P11-1). For this purpose a simple double-pipe heat exchanger is made by enclosing the 6-in pipe with a 10-in pipe and passing cooling water through the annulus. The cooling water is pumped from a reservoir (as shown) at a rate of 12.0 ft^3/min at 68°F. The water is pumped through a 2-in-ID Schedule 40 pipeline containing three 90° elbows and a fully open gate valve. If the pressure *leaving* the exchanger is to be 16.0 lb/in^2 abs, determine:

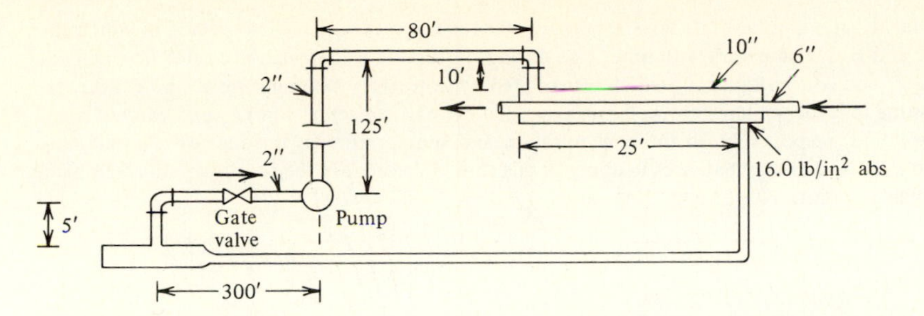

Figure P11-1 Problem 11-1.

(*a*) Velocity in feet per second in the pipelines
(*b*) Reynolds number in the pipeline and in the exchanger
(*c*) Whether the flow is turbulent
(*d*) Equivalent length of fittings and valves
(*e*) Total friction loss in the pipeline in ft·lbf/lb
(*f*) Horsepower required for the pump if its efficiency is 40 percent

11-2 Repeat Prob. 11-1 if the flow rate is 1.2 ft^3/min.

11-3 Simplify the unsteady-state macroscopic balances to obtain balances for the unsteady-state situations in Chap. 7:
(*a*) Slab with each side at a different temperature (Sec. 7-1)
(*b*) Couette-flow startup (Sec. 7-3)
(*c*) Laminar flow in an annulus with fluid originally at rest (Example 7-4)

11-4 Repeat Prob. 11-3 for (*a*) unsteady-state heat transport in a fluid (Chap. 8) and (*b*) unsteady-state drying of a porous solid (Chap. 8).

11-5 Derive Eq. (11-24*a*) for the generalized mechanical-energy balance.

11-6 For laminar flow find the ratio $\langle v^2 \rangle / \langle v \rangle$.

11-7 Repeat Prob. 11-6 for $\langle v^3 \rangle / \langle v \rangle$.

11-8 Repeat Prob. 11-6 for slit flow.

11-9 Repeat Prob. 11-7 for slit flow.

11-10 Repeat Prob. 11-6 for turbulent flow using the Blasius equation.

11-11 Consider flow in the entrance region, as in Example 9-8. The velocity profile entering at $z = 0$ is flat ($v_z = \langle v_z \rangle$), and the velocity profile at $z = L_e$ is parabolic. Write simplified macroscopic mass, momentum, and mechanical-energy balances for this case.

11-12 A centrifugal pump is pumping 2000 gal/min of a fluid of specific gravity 0.90 at 1300 r/min. The pump has an impeller of 0.3048 m outside diameter, and the effective width of the channels is 2.54 cm. The impeller vanes are swept back to an outlet vane angle $\beta_2 = 22.5°$. The liquid enters the impeller along the shaft and changes direction through a right angle in a plane perpendicular to the impeller disk just before entering the vanes. Calculate (*a*) the ideal head developed by the impeller and (*b*) the actual head developed by the impeller if the whirl velocity is seven-tenths of the ideal value as a result of circulation losses.

11-13 Starting with the species A continuity equation, derive a differential macroscopic balance for each phase in a gas absorber (Chap. 4).

11-14 For a simple pipe heat exchanger, derive a differential macroscopic balance by simplifying the general equation.

11-15 If there are k crystallizer tanks operating in series, show that the maximum weight percentage is given by $x_{max} = k + 2$ *if* nucleation occurs only in the first tank.

11-16 A continuous-flow reactor is a stirred tank in which the reacting stream enters at a constant mass rate w_f and leaves at a constant rate w. A semibatch reactor is a continuous-flow reactor in which the temperature and conversion vary with time, e.g., during startup. Consider such a reactor originally at temperature T_0 and concentration C_0 which is being fed by a stream at temperature T_f and concentration C_f. Assume that the contents are well mixed, so that the exit concentration C_{ex} and exit temperature T_{ex} are equal, respectively, to the concentration and temperature in the reactor. Simplify the macroscopic mass and energy balances to apply to this case and compare the result with the following equation given by Smith (Ref. 11, p. 251)

$$w_f \widehat{C}_p (T_f - T_{ex}) + w \, \Delta H_R (X_f - X_{ex}) = m \widehat{C}_p \frac{dT_{ex}}{dt} + \Delta H_R R_A \mathcal{V}$$

where $X \equiv (C_f - C_{ex})/C_f$ is the conversion.

11-17 Derive Eq. (11-77).

REFERENCES

1. Aris, R. A.: "Elementary Reactor Analysis," Prentice-Hall, Englewood Cliffs, N.J., 1969.
2. Bird, R. B.: *Chem. Eng. Sci.,* **6**:123 (1957).
3. Bird, R. B.: *Chem. Eng. Prog. Symp. Ser,* **61**(58):1 (1965).
4. Martin, J. J.: *Chem. Eng. Educ.,* **8**:138 (1974).
5. Perry, R. H., and C. H. Chilton: "Chemical Engineers' Handbook," McGraw-Hill, New York, 1973.
6. Pings, C. J.: *Chem. Eng. Sci.,* **17**:949 (1962).
7. Ramkrishna, D.: *Chem. Eng. Educ.,* **12:**14 (1978).
8. Randolph, A. D.: *Can. J. Chem. Eng.,* December 1964, p. 280.
9. Randolph, A. D., and M. A. Larson: *AIChE J.,* **8**:644 (1962).
10. Randolph, A. D., and M. A. Larson: "Theory of Particulate Processes," Academic, New York, 1971.
11. Smith, J. M.: "Chemical Engineering Kinetics," McGraw-Hill, New York, 1981.
12. Taylor, G. I.: *Proc. R. Soc.,* **A219**:186 (1953).

SOME SI CONVERSION FACTORS

Mass	1 lb = 453.6 g = 0.4536 kg
Length	1 ft = 0.3048 m
	1 in = 2.540×10^{-2} m
Area	1 ft^2 = 0.0929 m^2
	1 in^2 = 6.45×10^{-4} m^2
Force	1 lbf = 4.448 N = 4.448×10^5 dyn = 4.448 kg · m/s^2
Pressure	1 Pa = 1 N/m^2 = 10^{-5} bar
	1 lbf/in^2 = 6.895×10^3 N/m^2 = 6.895×10^{-2} bar
	1 atm = 1.0133×10^5 N/m^2 = 1.0133×10^5 Pa = 1.0133 bars
Energy	1 ft · lbf = 1.356 J
	1 erg = 10^{-7} J
	1 cal = 4.184 J = 4.184×10^7 ergs
	1 Btu = 1.054×10^3 J = 1.054×10^{10} ergs
Heat capacity	1 Btu/lbm · °F = 4.187 kJ/kg · K = 1 cal/g · °C
Thermal conductivity	1 Btu/h · ft · °F = 1.730 W/m · K
	1 Btu · in/h · ft^2 · °F = 0.1441 W/m · K
Viscosity	1 poise (P) = 1 g/m · s = 10^{-1} Pa · s = 10^{-1} N · s/m^2
	1 centipoise (cP) = 10^{-3} Pa · s = 10^{-3} N · s/m^2
	1 lbm · s/ft^2 = 47.88 N · s/m^2
	1 lbf/ft · s = 1.488 N · s/m^2
	1 N · s/m^2 = 1 (kg · m/s^2)(s/m^2) = 1 kg/m · s
Diffusivity	1 ft^2/h = 2.5806×10^{-5} m^2/s
Power	1 hp = 745.7 W = 745.7 J/s
	1 Btu/h = 0.2931 W = 0.2931 J/s
Heat flux	1 Btu/h · ft^2 = 3.155 W/m^2

INDEX

When authors' names are not given in the text, page numbers are in *italic*.